T0140240

Handbook of Human Centric Visualization

Weidong Huang

Editor

Handbook of Human Centric Visualization

Foreword by Peter Eades

 Springer

Editor
Weidong Huang
CSIRO ICT Centre
Marsfield, NSW, Australia

ISBN 978-1-4899-9920-7   ISBN 978-1-4614-7485-2 (eBook)
DOI 10.1007/978-1-4614-7485-2
Springer New York Heidelberg Dordrecht London

Printed on acid-free paper

Springer is part of Springer Science+Business Media (www.springer.com)

Foreword

Visualization research promises to help humans to explore and comprehend information, thus making vast stores of data useful to humankind.

Much of this visualization research follows a simple pattern: (1) "here is an important data set," (2) "here is a picture of this data set," (3) "wow, the picture looks cool!"

Huang's book does contain some of the world's coolest pictures of data. But the book goes further: it considers the much deeper questions of *how* humans read visualizations, and *why* we use visualization. The very nature of visualization is examined. Visualization is not just a technology; it is a human communication mechanism.

In particular, the question of what kind of scientific methodology should be used to evaluate visualizations is considered in detail.

Such topics are beyond the gamut of Computer Science, and can only be answered by a multidisciplinary approach. The book has chapters written by researchers in a variety of disciplines, from psychology to business, from philosophy to engineering.

The book is revolutionary in its scale and breadth.

Peter Eades
September 3, 2012.

Preface

We visualize data for human appreciation and understanding. In other words, all visualizations are meant to be human centric. However, human centric visualizations do not come automatically.

To ensure that a visualization is human centric, we need proper theories and principles to guide the process of the visualization design. Once the visualization is produced, we need methods and measures to verify whether the design objectives are indeed achieved. Rapid advances in display technology and computer power have enabled researchers and practitioners to produce visually appealing pictures. However, the effectiveness of those pictures in conveying the embedded information to end users has been relatively less explored.

Handbook of Human Centric Visualization aims to contribute to the human side of the visualization research. It addresses issues related to design, evaluation, and application of visualizations. Topics include visualization theories, design principles, evaluation methods and metrics, human factors, interaction methods, and case studies. This cutting-edge book is an edited volume whose contributors include well-established researchers worldwide, from diverse disciplines including psychology, education, visualization, and human-computer interaction.

This book consists of twenty-nine chapters, which are grouped into the following seven parts:

    I. Visual Communication
   II. Theory and Science
 III. Principles, Guidelines, and Recommendations
 IV. Methods
  V. Perception and Cognition
 VI. Dynamic Visualization
VII. Interaction

The main features of this book can be summarized as follows:

1. Provides a comprehensive overview of human centric visualization
2. Represents latest developments and current trends in the field

3. Presents visualization theories
4. Covers design principles and guidelines
5. Presents evaluation methodologies and case studies
6. Includes contributions from leading experts and active researchers from a range of disciplines

This book is designed for a professional audience composed of practitioners, lecturers, and researchers working in the field of computer graphics, visualization, human-computer interaction and psychology. Undergraduate and postgraduate students in science and engineering focused on this topic will also find this book useful as a comprehensive textbook or reference.

Weidong Huang

Editorial Board

Acknowledgement

I wish to thank all authors who submitted their valuable work for consideration. Without their effort and contribution, this book would not have been possible. I received 84 exciting chapter proposals in total covering almost every aspect of human centric visualization. Unfortunately, due to limitations on the size of the book and the range of topics to be addressed, many good proposals were not considered further.

I thank our valued members of the International Advisory Editorial Board. Their world class reputation and generous support for this project have helped to attract the high-quality submissions and receive the overwhelming responses from the wide scientific community.

Editing a book like this is a time-consuming and sometimes lonely process. However, throughout this process, I have consistently received positive feedback and support from many individuals. Some recommended potential authors. Some helped or offered to help in shaping the book. All this has helped to keep my spirits high throughout the project. I thank them all. I would also like to extend my special thanks to Chaomei Chen, Jian Chen, Mary Czerwinski, Andreas Kerren, Stephen Kosslyn, Fred Paas, Helen Purchase, Barbara Tversky, Jack van Wijk, and Kang Zhang for their help in one way or another.

I thank Professor Peter Eades for writing the foreword.

My sincere thanks and appreciation are also extended to Springer editors Susan Lagerstrom-Fife and Courtney Clark for their professional help and support.

Weidong Huang

Contents

Contributors

Shirley Agostinho Research interests: Learning Design (a formalism for documenting teaching and learning practice to facilitate sharing, adaptation, and reuse by teachers); how learners can apply cognitive load theory design principles in online learning environments to manage their cognitive load. University of Wollongong, Wollongong, Australia

Wolfgang Aigner Research interests: information visualization; visual analytics; human-computer interaction; user-centred design; interaction design. Vienna University of Technology, Vienna, Austria

Richard Arias-Hernandez Simon Fraser University, Surrey, Canada

Simon Attfield Research interests: sensemaking; interacting with information; visual analytics. Middlesex University, London, UK

Alexander P. Auchus Research interests: memory disorders and dementia. University of Mississippi Medical Center, Jackson, USA

Paul Ayres Research interests: cognitive load theory, multimedia learning, educational psychology. University of New South Wales, Sydney, Australia

Juhee Bae North Carolina State University, Raleigh, USA

Ann Blandford Research interests: situated use of technology; sensemaking; interacting with information; visual analytics. University College London, London, UK

Tanja Blascheck University of Stuttgart, Stuttgart, Germany

Jim Blythe University of Southern California, Los Angeles, USA

Sahar Bokosmaty Research interests: application of cognitive load theory design principles in the teaching and learning of mathematics. University of Wollongong, Wollongong, Australia

Sebastian Bremm Research interests: visual analytics, visual feature descriptor analysis, visual tree comparison, multivariate data. Technische Universität Darmstadt, Darmstadt, Germany

Michael Burch Research interests: information visualization, software visualization, dynamic graph visualization, evaluation, visual analytics. University of Stuttgart, Stuttgart, Germany

Haipeng Cai Research interests: scientific visualization and programming language design. University of Southern Mississippi, Hattiesburg, USA

Juan Cristobal Castro-Alonso Research interests: cognitive load theory, multimedia learning, instructional design, animations, motor skills. University of New South Wales, Sydney, Australia

Jian Chen Research interests: human-centred computing issues in scientific visualization, information visualization, and three-dimensional user-interaction. University of Maryland Baltimore County, Baltimore, USA

Matthew Cooper Research interests: interactive visualization and virtual and augmented reality. Linköping University, Linköping, Sweden

Matthew Coxon Research interests: human memory and learning: visuospatial working memory; learning with augmented reality; and learning in virtual environments. York St. John University, UK

Stephen Boyd Davis Research interests: theory and practice of representation; uses of spatiality; chronographics; design research. Royal College of Art, London, UK

Mark Dittmer Research interests: human-centred informatics; interactive visualizations; human-information interaction. Western University, London, Canada

Sarah Faisal Research interests: visualization; visual analytics. University College London, London, UK

Paolo Federico Research interests: information visualization; visual analytics; human-computer interaction; dynamic networks. Vienna University of Technology, Vienna, Austria

Dieter W. Fellner Research interests: visual computing, digital libraries, semantics in modelling, visual analytics. Technische Universität Darmstadt, Fraunhofer IGD, Darmstadt, Germany

Sebastian Fiebig Research interests: visual analytics, visual tracking of user actions, data mining, statistical inference, insight provenance. Technische Universität Darmstadt, Darmstadt, Germany

Brian Fisher Simon Fraser University, Surrey, Canada

Nigel Foreman Research interests: spatial cognition; neuropsychology; virtual reality applications; research paradigms; ethics and international psychology. Middlesex University, London, UK

Camilla Forsell Research interests: perceptually motivated constraints on visualization and development of new methods and metrics for evaluation. Linköping University, Linköping, Sweden

Carla M.D. S. Freitas Research interests: graph visualization, social networks analysis, volumetric and illustrative visualization, and visualization evaluation. Federal University of Rio Grande do Sul, Porto Alegre, Brazil

Emden R. Gansner AT&T Labs – Research, New Jersey, USA

Joseph H. Goldberg Research interests: eye tracking for usability evaluation; visualization design and evaluation; visual and cognitive complexity. Oracle America, Applications UX, Redwood Shores, USA

Tera Marie Green Simon Fraser University, Surrey, Canada

Robert Haworth Research interests: Human-computer interaction; interactive visualizations; digital cognitive games. Western University, London, Canada

Jonathan I. Helfman Agilent Technologies, Santa Clara, USA

Yifan Hu AT&T Labs – Research, New Jersey, USA

Weidong Huang Research interests: visual perception and cognition; HCI; Visualization and human factors. CSIRO ICT Center, Marsfield, NSW, Australia

Christopher D. Hundhausen Washington State University, Pullman, USA

Stephen G. Kobourov University of Arizona, Arizona, USA

Liliya Korallo Research interests: uses of virtual environments (VEs) in Education; spatial cognition and applications to teaching and learning of chronology; VEs in assessment and treatment of mental health conditions. Middlesex University, London, UK

David Krackhardt Carnegie Mellon University, Pittsburgh, USA

Simone Kriglstein Vienna University of Technology, Vienna, Austria

Arjan Kuijper Research interests: visual computing, mathematical basis of computer vision and image analysis, scale-space. Technische Universität Darmstadt, Fraunhofer IGD, Darmstadt, Germany

David H. Laidlaw Research interests: applications of visualization, modelling, computer graphics, and computer science to other scientific disciplines. Brown University, Providence, USA

Richard K. Lowe Research interests: animations and multimedia learning, visual literacy and learning, learning from pictures and diagrams in education. Curtin University, Sydney, Australia

Katerina Mania Research interests: Perceptual and selective rendering algorithms, human-centred simulation engineering, fidelity metrics of synthetic scenes, neuro-correlates of fidelity, eye tracking in synthetic worlds, 3D interface design, serious games, visualization, digital cultural heritage, latency psychophysics, synthetic characters. Technical University of Crete, Chania, Greece

Cathleen McGrath Loyola Marymount University, Los Angeles, USA

Silvia Miksch Research interests: visual analytics; information visualization; interaction methods; time-oriented data; plan management. Vienna University of Technology, Vienna, Austria

Magnus Moar Research interests: locative media; novel interfaces; children's use of new technologies. Middlesex University, London, UK

Michelle L. Nugteren Erasmus University Rotterdam, Rotterdam, The Netherlands

Fred Paas Research interests: cognitive load theory, instructional design, training of complex cognitive and motor skills, multimedia learning, embodied cognition. Erasmus University Rotterdam, Rotterdam, The Netherlands

University of Wollongong, Sydney, Australia

Paul Parsons Research interests: human-centred informatics; interactive visualizations; visual analytics; human-information interaction; design of cognitive activity support tools. Western University, London, Canada

Marcelo S. Pimenta Research interests: user-centred design, usability evaluation, software visualization. Federal University of Rio Grande do Sul, Porto Alegre, Brazil

Margit Pohl Vienna University of Technology, Vienna, Austria

Michael Raschke University of Stuttgart, Stuttgart, Germany

Ronald A. Rensink Research interests: visual perception, information visualization, visual attention, visual memory, machine vision, philosophy of science. University of British Columbia, Vancouver, Canada

Dominique L. Scapin Research interests: user-centred design, human-computer interaction. INRIA Rocquencourt, Le Chesnay Cedex, France

Katharina Scheiter Knowledge Media Research Center, Tübingen, Germany

Kamran Sedig Research interests: human-centred informatics; interactive visualizations; visual analytics; human-information interaction; design of cognitive activity support tools; digital cognitive games; information-rich artefacts. Western University, London, Canada

Michael Smuc Research interests: information design, human-information interaction, visualization evaluation, graph comprehension, dynamic network analysis. Danube University Krems, Krems, Austria

Huib K. Tabbers Erasmus University Rotterdam, Rotterdam, The Netherlands

Sharon Tindall-Ford Research interests: cognition and instruction; how human movement can facilitate learning (from a cognitive load theory perspective). University of Wollongong, Wollongong, Australia

Melanie Tory University of Victoria, Victoria, Canada

Barbara Tversky Research interests: spatial cognition and language, event perception and cognition, diagrammatic reasoning, visual communication, sketching, gesture. Columbia Teachers College, New York, USA

Stanford University, Stanford, USA

J. Ángel Velázquez-Iturbide Research interests: educational software for programming education; innovative instruction for programming education; software visualization. Universidad Rey Juan Carlos, Madrid, Spain

Tatiana von Landesberger Research interests: visual analytics, visual graph analysis, multivariate data. Technische Universität Darmstadt, Darmstadt, Germany

Benjamin Watson North Carolina State University, Raleigh, USA

Daniel Weiskopf Research interests: visualization, visual analytics, GPU methods, computer graphics, and special and general relativity. University of Stuttgart, Stuttgart, Germany

Florian Windhager Research interests: knowledge visualization, dynamic network analysis, information design, time topography. Danube University Krems, Krems, Austria

Lukas Zenk Research interests: dynamic network analysis, organizational networks, network visualization, innovation management. Danube University Krems, Krems, Austria

Part I
Visual Communication

Visualizing Thought

Barbara Tversky

Abstract Depictive expressions of thought predate written language by thousands of years. They have evolved in communities through a kind of informal user testing that has refined them. Analyzing common visual communications reveals consistencies that illuminate how people think as well as guide design; the process can be brought into the laboratory and accelerated. Like language, visual communications abstract and schematize; unlike language, they use properties of the page (e.g., proximity and place: center, horizontal/up-down, vertical/left-right) and the marks on it (e.g., dots, lines, arrows, boxes, blobs, likenesses, symbols) to convey meanings. The visual expressions of these meanings (e.g., individual, category, order, relation, correspondence, continuum, hierarchy) have analogs in language, gesture, and especially in the patterns that are created when people design the world around them, arranging things into piles and rows and hierarchies and arrays, spatial-abstraction-action interconnections termed *spractions*. The designed world is a diagram.

1 Introduction

Communication in the wild is a sound and light show combining words, prosody, facial expressions, gestures, and actions. Although it is often presumed—think of the "letter of the law" and transcripts of trials–that meanings are neatly packaged

This paper is reprinted with permission from *Topics in Cognitive* Science, 2011,3, 499–535.

B. Tversky (✉)
Columbia Teachers College, 525 W. 120th, New York, NY 10025, USA

Department of Psychology Bldg 420, 450 Serra Mall, Stanford University, Stanford, CA 94305-2130, USA
e-mail: btversky@stanford.edu

W. Huang (ed.), *Handbook of Human Centric Visualization*,
DOI 10.1007/978-1-4614-7485-2_1, © Springer Science+Business Media New York 2014

into words joined by rules into utterances, in fact, other channels of communication carry significant aspects of meaning, despite or perhaps because of the fact that they cannot be neatly packaged into units strung together by rules (e.g., [25, 46, 68, 92, 93]). Prosody, as in irony or sarcasm, can overrule and even reverse meanings of words, as can facial expressions. Pointing can replace words, for things, for directions, and more, so that natural descriptions, narratives, or explanations cannot be fully understood from the words alone (e.g., [34]). Gestures go beyond pointing, they can show size, shape, pattern, manner, position, direction, order, quantity, both literally and metaphorically. They can express abstract meanings, mood, affect, evaluation, attitude, and more. Gestures and actions convey this rich set of meanings by using position, form, and movement in space. Communication can happen wordlessly, as in avoiding collisions on busy sidewalks or placing items on the counter next to the cash register to indicate an intention to buy. In fact, the shelf next to the cash register is designed to play a communicative role. Standing next to a circle of chatting acquaintances can be a request to join the conversation. Opening the circle is the group's wordless response. Rolling one's eyes can signify, well, rolling one's eyes. Communication in the wild combines and integrates these modes, usually seamlessly, with each contributing to the overall meaning (e.g., [25, 35, 46, 68, 92, 140, 142]).

Gestures and actions are especially convenient because their tools, like the tools for speech, are free, and they are always with us. But gestures, like speech, are fleeting; they quickly disappear. They are limited by what can be produced and comprehended in real time. These limitations render gestures abstract and schematic. Visualizations, on paper, silk, parchment, wood, stone, or screen, are more permanent, they can be inspected and reinspected. Because they persist, they can be subjected to myriad perceptual processes: compare, contrast, assess similarity, distance, direction, shape, and size, reverse figure and ground, rotate, group and regroup; that is, they can be mentally assessed and rearranged in multiple ways that contribute to understanding, inference, and insight. Visualizations can be viewed as the permanent traces of gestures; both embody and are embodied. Like gesture, visualizations use position, form, and actions in space to convey meanings (e.g., [155]). For visualizations, fleeting positions become places and fleeting actions become marks and forms. Here, we analyze the ways that place and form constrain and convey meaning, meanings that are based in part in actions.

Traces of visual communication go far back into prehistory. Indeed, they are one of the earliest signs of culture. They not only precede written language, but also served as the basis for it (e.g, [42, 114]). Visual communications come in myriad forms: animals in cave paintings, maps in petroglyphs, tallies on bones, histories on columns, battles in tapestries, messages on birch bark, journeys in scrolls, stories in stained glass windows, dramas in comics, diagrams in manuals, charts in magazines, graphs in journals. All forms of communication entail design, as the intent of communication is to be understood by others or by one's self at another time. Communication design, then, is inherently social, because to be understood by another or by self at another time entails fashioning communications to fit the presumed mental states of others or of one's self at another time.

Diagrams, along with pictures, film, paintings in caves, notches in wood, incisions in stone, cuttings in bone, impressions in clay, illustrations in books, paintings on walls, and of course words and gestures, externalize thought. They do this for many reasons, often several simultaneously. Some are aesthetic: to arouse emotions or evoke pleasure. Some are behavioral: to affect action or promote collaboration. Some are cognitive: to serve as reminders, to focus thoughts, to reorganize thoughts, to explore thoughts. Many are communicative: to inform both self and others.

Because depictions, like other cultural artifacts (e.g., [31, 101]), have evolved over time, they have undergone an informal but powerful kind of natural user testing, produced by some, comprehended by others, and refined and revised to improve communication by a community of users. Similar processes have served and continue to serve to design and re-design language (cf. [25]). Features and forms that have been invented and reinvented across cultures and time are likely to be effective. Analyzing these depictive communications, then, can provide valuable clues to designing new ones. It can save and inspire laboratory work, as well as the tasks of designers. What's more, the natural evolution of communication design can be brought into the laboratory and accelerated for specific ends (see [152]).

Oddly, this rich set of visual forms has traditionally been discussed in the domain of art, along with painting, drawing, and photography. Increasingly, that discussion has expanded to include diagrams, charts, film, graphs, notational systems, visual instructions, computer interfaces, comics, movies, and more, to take into account the mind that perceives, conceives, and understands them, and to ripple across domains (e.g., [6, 12, 33, 47, 76, 90, 95, 124, 126, 128, 143–145, 167, 169, 173]). Similarly, discussions of human communication have historically focused on language, typically narrowly conceived as words and sentences, and have only recently broadened to include prosody, gesture, and action (e.g., [5, 16, 25, 46, 68, 92]).

Unlike symbolic words, forms of visual communication, notably diagrams and gestures, often work by a kind of resemblance, that is, sharing features or associations, typically visuo-spatial features, with the meanings they are intended to convey (a claim of some philosophic controversy, e.g., [49, 58, 168]). The proverbial "big fish" is indicated in gesture by expanding the fingers or hands horizontally, thus capturing the approximate relative horizontal extent of the fish, but ignoring its other properties. How the fish swam to try to get away is abstracted and conveyed differently, perhaps by embodying the fish and its movements. Similarly, the shape, dimensions, and even actions of the fish can be abstracted in a variety of ways to the page. The fish example illustrates another property of visual communication. In capturing features of the world, visual communications are highly selective; they omit information, normally information that is regarded as less essential for the purposes at hand. They abstract and schematize not only by omission but also by exaggeration and even by additions. Maps, for example, are not simply shrunken aerial photographs. Maps selectively omit most information, houses, trees, fields, mountains, and the like, but also many of the twists and turns of roads or coastlines; they disproportionately enlarge roads and rivers to make them visible;

they turn entire metropolises into dots. Maps may also add features like government boundaries and topological levels that are not visible.

In other words, maps, like many other kinds of visualizations, distort the "truth" to tell a larger truth. The processes that abstract, schematize, supplement, and distort the world outside onto the world of a page, filtering, leveling, sharpening, categorizing, and otherwise transforming, are the same processes the nervous system and the brain apply to make sense of the barrage of stimuli the world provides. Attention is selective, ignoring much incoming information. The perceptual systems level and sharpen the information that does come in; for example, the visual system searches for the boundaries that define figures by sharpening edges and corners, by filling in gaps, by normalizing shapes. Cognition filters, abstracts, and categorizes, continuing this process, and symbol systems carry these processes further. Long things don't necessarily get long names, though children often expect them to (e.g., [141]). Tallies eliminate the identity of objects, recording them just as instances, though tallies preserve a one-to-one correspondence that Arabic numerals, more convenient for calculations, do not.

The virtues of visual communications have been extolled by many (e.g., [72, 81, 101, 113, 146–148, 151]). As noted, they are cultural artifacts created in a community [31, 87, 101, 162], fine-tuned by their users (e.g., [154]). They can provide a permanent, public record that can be pointed at or referred to. They externalize and clarify common ground. They can be understood, revised, and manipulated by a community. They relieve limited capacity short-term memory, they facilitate information processing, they expand long-term memory, they organize thought, they promote inference and discovery. Because they are visual and spatial, they allow human agility in visual-spatial processing and inference to be applied to visual-spatial information and to metaphorically spatially abstract information.

In contrast to purely symbolic words, visual communications can convey some content and structure directly. They do this in part by using elements, marks on a page, virtual or actual, and spatial relations, proximity and place on a page, to convey literal and metaphoric elements and relations. These ways of communicating meanings may not provide definitions with the rigor of words, but rather provide suggestions for meanings and constraints on them, giving them greater flexibility than words. That flexibility means that many of the meanings thus conveyed need context and experience to fully grasp. A line in a route map has a different meaning from a line in a network and from a line in a graph, though, significantly, all connect. Nor is the expressive power of visual communication as great as that of language (e.g., [130]); abstract or invisible concepts like forces, traits, counterfactuals, and negations are not easily conveyed unambiguously in depictions. Even so, conventions for conveying these kinds of concepts have evolved as needed, in road signs, mathematics, science, architecture, engineering, and other domains, a gradual process of symbolization akin to language.

What are the tools of depictions, especially diagrams? How do they communicate? The components of visual communication are simple: typically, a flat surface, prototypically, a page (or something analogous to a page like a computer screen) and marks or forms placed on it (e.g., [64, 148, 149, 164]). Each of these, place and form, will be analyzed to show how they can represent meanings that are literal and metaphoric, concrete and abstract. The interpretations will be shown to depend on content and context, on Gestalt or mathematical properties of the marks in space, on the place of the marks on the page, as well as the information processing capacities and proclivities of the mind. The foundations and processes of assigning meaning can be revealed, then, by recurring inventions and by errors and biases in interpretation, that is, by uses and misuses, by successes and failures. The analysis of inventions of visual communication can provide directions for the design of visual communications.

Because assigning meaning, whether from description or depiction, is in part a reductive process–the space of possible meanings is greater than the space of ways to express meanings—misuses, misinterpretations, and misunderstandings are as inevitable as successes, and both are instructive. Expressing meanings, then, entails categorization. Categories create boundaries where none exist, some instances are included and others, even close ones, are not. The consequence of categorization is to increase the perceived similarity of members included in the category and to exaggerate the perceived distance between members and non-members. Although the focus here is on meanings conveyed through place and forms, the meanings are deeper, they are conceptually spatial, some more literal, some more metaphorical, so that they have parallels in other ways of using space as well, in words, in actions, and in gesture, in the virtual space created by gesture and the mental space created by words (e.g., [40, 77, 155]). First we will discuss place in space, and then forms in space.

2 Place in Space

2.1 Organizing Space in the World

2.1.1 Spatial Actions Create Meaningful Patterns

Three quarters of a million years ago, a group of hominims living in the northern Jordan River valley separated the activities of communal life into different spatial areas, cooking activities in one area and tool-building activities in another [4]. Each of these spaces was subdivided, again by function. This primate society created unintended visual communications about their lives to archeologists living generations afterwards. Before the page, there was space itself. Perhaps the simplest way to use space to communicate is to arrange or rearrange things in it. An early process is grouping things in space using proximity, putting similar things in close proximity and farther from dissimilar things, actions that reflect the Gestalt laws of perception. These separated spatial groupings signal separate associated things. "Close" family members and friends sit nearer to one another than strangers. The flatware tray in a drawer of most kitchens allows arranging the knives together in one pile and separating them from the pile of forks and the pile of spoons. Drawers in the bedroom allow arranging the socks together and separating them from other articles of clothing that are also grouped and piled by kind. Shelves and drawers allow hierarchical organization, one shelf for canned goods, another for baking supplies, further organized inside by kind and recency of purchase, in two dimensions. Table settings distribute various items in one-to-one correspondences, each setting gets a plate, a glass, a knife, a fork, a spoon, and a napkin. Themes as well as categories are spatially organized, things for cooking spatially separated from things for sleeping. Larger spaces, homes and villages, are arranged in two and three dimensions, turning inhabited spaces into diagrams, vertical patterns of windows on buildings and horizontal patterns of streets on the ground. We rearrange things in space to capture attention and to affect behavior in the present as well as the future, for example, putting the letters to be mailed by the door or the bills to be paid on the top of the desk (or desktop) or lining up the ingredients for a recipe in order of use (e.g., [72]), ordinal mappings of time and actions in time onto space. Written text is spatially arranged to reflect the organization of thought, spaces between words and sentences, larger spaces between paragraphs. Greek text describing mathematics was written formulaically, fixed orders of semantic forms, often in rows, that formed tables for reasoning [98]. Even babies do it; many discover "in your face" early on. When they want attention, they center their faces in the face of the person whose attention they seek, directly in the line of vision. A fundamental service of space, hence meaning of space, is proximity to me. I can perceive and act on the things and beings that are close to me, in reach of the body, primarily eyes, hands, feet, and, for beings, voice. For my actions (and my perception), the best position is centered in front of me. These many deliberate organizations of space serve to direct attention, to augment memory, to facilitate and organize actions, and to communicate to ourselves or to others.

One implication of this analysis is that action underlies perception. The actions of organizing space for many ends into groups, hierarchies, orders, correspondences, continua, and the like create spatial patterns that are far more regular than those created by nature, thus a signal that they are created by sentient minds. These regular spatial patterns conform to the Gestalt laws of perception, augmenting their perceptibility. Things that entail similar actions, whether socks or knives, are grouped together, creating stacks or piles, and things that entail ordered actions are lined up in that order, along a line, creating a temporal continuum on a horizontal surface. Perception of stacks and lines is enhanced by the Gestalt Laws of good continuation or common fate.

Not much farther afield, architecture can be viewed as an advanced form of arranging things in space (in three dimensions) for a number of reasons, among them, to inform and to facilitate or constrain action. Department stores put like things close to each other, separating them from different things. The grouping is hierarchical, men's clothes together, women's clothes together, and within each, shirts in one place and outerwear in another. Architectural spaces are also designed to affect behavior. Elevators are placed in eyesight of entrances, desired corridors are broad and well-lit. Departments in department stores were once geometrically organized along parallel and perpendicular paths, presumably because such an organization facilitates way-finding (e.g., [146]). Increasingly, they seem to be organized like Chinese gardens, in zigzagging meandering paths. In Chinese gardens, a meandering organization of space creates surprises and the impression of a larger space to be contemplated and enjoyed. In department stores, a meandering organization undoubtedly interferes with way-finding and provides more temptations to purchase. In architectural designs, the plan, a horizontal slice, serves action and the elevation, a vertical slice, serves aesthetics (e.g., [7]).

Spaces are also arranged and designed for symbolic and aesthetic reasons. The square patterns that cultures as distant as China and Rome used to build their cities, with roads aligned north–south and east–west, seem to serve several ends at once, cognitive, aesthetic, and symbolic. Other patterns that are consequences of organizing space appear and reappear across cultures in ceramics, weaving, basket-making, and architecture (as well as poetry and music), especially patterns that have geometric repetitions and symmetries (e.g., [8, 48]).

Spaces are also created on the fly, to serve behavior (the reminders on the desk) or communication. Arrernte speakers in Australia routinely draw the locations and movements of their conversation topics in the sand [172]. When people describe locations of places or events involving actions, like football plays or accidents, they often use whatever small objects are at hand, coins, salt shakers, Lego blocks, or fingers to represent the locations and movements of whatever they are describing, creating a map on a surface. If pencil and paper are handy, they often sketch instead (e.g., [69]).

2.1.2   Conception, Action, Perception, Communication, and Meaning

People, then, design and redesign the spaces they inhabit, arranging them and rearranging them to serve a variety of ends. The spaces they create are a visible embodiment of the abstract concepts underlying the organizations. These spaces form regular patterns that resonate with principles of perceptual organization. The close couplings of action, conception, and perception support meanings and afford communication. The examples above are few from many, but they illustrate some of the core phenomena. People put like things together, often into piles, rows, or bins, and separate them from different things. They cluster by kind, often hierarchically. They order things in rows or piles in a variety of ways, depending on their purpose, ingredients in order of use, photograph albums in order of time, bills to be paid in order of importance. These acts select single features and create single dimensions or continua out of disparate things. People also arrange things by themes, and distribute sets of items in one-to-one correspondences. They choose distances and sizes in three dimensions. Whether informally in conversation or more formally on maps and architectural plans, people also map locations in the world onto a representing world, models or diagrams. These same kinds of organization in space, clusters, orders, maps, and more, are used to locate things on a page to represent and communicate ways things are organized and related in the mind as well as in the world.

2.2   Organizing the Space of a Page, More Literally

As we have seen, people arrange and rearrange the things in the spaces around them into clusters, orders, and more complex organizations for cognitive, social, aesthetic, and symbolic ends. People do the same with the space of a page, for things that are literally spatial as well as things that are metaphorically spatial. In contrast to the space of the world, the space of a page is two-dimensional, though it allows conveying three and more dimensions. Conceptually, the two dimensions of a page are defined with respect to a viewer's frame of reference and a page oriented horizontally, left-right and top-bottom (or up-down) (cf. [6, 8]) Conceptually, there is also a page-centric frame of reference: center, periphery.

We begin with an early, basic organization of the space of a page or virtual page, what can be called pictorial space, used to map and represent the visible world. Think first of ancient paintings of animals in the rugged ceilings of caves or the tadpole figures of people drawn by children all over the world (e.g., [67]). Several aspects of place on the page will be analyzed through prevalent examples: up/down, left/right, center, and proximity among them. Some of those uses benefit thought, some uses conflict, and some even hinder, but all are a testament to the cognitive power of place and marks on the page.

2.2.1   Pictorial Space

Perhaps the earliest and simplest and still the most common way to use space in depictions is to map the space of the viewed world to a surface, what is traditionally called a picture. This mapping takes the three-dimensional world into a two-dimensional one, the page, a transformation that is undoubtedly facilitated by the fact that the world captured by the retina and the rest of the visual system is a two-dimensional mapping of the three-dimensional world from a particular perspective. Mapping pictorial space to the page puts things on the ground at the bottom of the page and things in the sky at the top, just as at an easel that holds the page in the plane of the world. Put horizontally on a table, the space of the page is mapped so the ground is close to the viewer, "at the bottom," and the sky is far from the viewer, "at the top" (cf. [122]). This correspondence applies the notion of "upright" to the page. It is such a compelling organization of space that upside down pictures are harder to recognize and remember, and especially faces of individuals, stimuli of special significance in our lives (e.g., [17, 59, 109]).

When placed horizontally, as on a table or desk, the actual space of the page conflicts with the actual surrounding space as the ground-to-sky bottom-to-top dimension of the page is no longer literally vertical as it is in the world. Nevertheless, the mapping of vertical to horizontal where ground is close to the viewer's perspective is conceptually powerful, so that the opposite mapping is regarded as upside-down. The pull of the picture plane is so strong that students in a course in information design use it implicitly in diagramming information systems.

In diagrams of information systems, what must be shown are the topological relations among the system components [100]. The actual locations of components are irrelevant; all that matters is the connections among them, indicated by lines. Nevertheless, designers' sketches frequently map physical locations, for example, placing a truck that transmits information at the bottom of the page, as if on the ground, and a satellite at the top, as if in the sky. Although a organizing a sketch using pictorial space may aid comprehension of the components of the system, it could prevent designers from "seeing" and using other organizations of components that might make better sense for the design.

2.2.2 Maps

Like the making of pictures, the making of maps entails shrinking a viewed environment as well as selecting and perhaps distorting important features and omitting others [148]. However, the making of maps requires more, beginning with taking a perspective not often seen in real life, a perspective from above, looking down. Maps, even ancient ones, typically include far more than can be seen from a single viewpoint, so that the making of maps also entails integrating many different views to convey a more comprehensive one. Despite these challenges, evidence of maps, typically petroglyphs as they survive the ravages of time, goes back at least 6,000 years (e.g., [15]) and of architectural plans nearly that far. Although maps often represent a horizontally extended world on a horizontal surface, they are frequently placed vertically ("upright"), requiring the same transformation that pictorial representations do (but without gravity and a conceptual up and down). Even though arbitrary, the conventional north-up orientation of maps has both cognitive and practical consequences; north-up maps are easier for many judgments (e.g., [124]).

Maps are one of the most ancient, modern, and widespread means of visual communication, and serve as an illustrative paradigm for many aspects of visual communication. Ancient as they are, maps represent remarkable feats of the human mind, the products of powerful mental transformations. Although human experience is primarily from within environments, a perspective that has been called egocentric, route, or embedded, maps take a viewpoint from outside environments, above them, a perspective that has been called extrinsic, allocentric, or survey. Thus, the making of maps and the understanding of maps entail a dramatic switch of perspective, one that takes remarkably little effort for well-learned environments (e.g., [83, 140]). What's more, just as spontaneous descriptions of space mix perspectives, using route and survey expressions in the same clause [139], maps (as well as pictorial and other external representations) often show mixed perspectives; for example, many ancient and modern maps of towns and cities show the network of roads from an overhead view and key buildings from a frontal view (e.g., [148]). Like Cubist and post-Cubist art, maps can show different views simultaneously in ways that violate the rules of perspective, but that may promote understanding of what is portrayed.

More commonly, maps show a single perspective, a two-dimensional overview of a three-dimensional world. Designers of spaces, architects, seem to work and think in two dimensions at a time, plans or elevations [7]. Architectural plans map an overview of a design; they show the relations among entrances, walls, furniture, and the like, and are used for designing behavior, for the functional aspects of buildings and complexes. Elevations show how structures will be viewed from the outside, and are important for designing aesthetic aspects of buildings [7].

Producing and comprehending maps require other major mental transformations, integrating and shrinking a large environment, one that typically can't be seen at a glance, to a small one that can fit onto a piece of paper. Even preschoolers are able to perform some of these mental feats, for example, using a schematic map to find a hidden toy (e.g., [29]). The creation of maps requires yet another mental feat, abstracting the features that are important, that need to be included in the external representation, and eliminating those that do not. The uses of maps range widely, road maps, weather maps, maps of spread of populations of people, of plants, of diseases, maps for hiking, for surveillance of water, of earthquakes, of soil quality, and more. The features that are essential to include vary with the use; for some uses, mountains can be omitted but roads must be included, and for others, mountains need to be preserved but roads can be eliminated. Similarly, some kinds of maps add information not directly visible in the environment, contour lines for topography of the ground or for weather fronts. Many of the same mental processes used in creating and using external representations parallel those used in creating and using mental ones (e.g., [123]), though there are naturally differences as well. And, like mental representations, external representations constrain as well as enable understandings and interpretations. The very same processes that facilitate comprehension and communication, of inclusion and elimination, of leveling and sharpening, of addition and subtraction, also focus and constrain the meanings, with inevitable consequences of misunderstandings, misinterpretations, and error.

2.3   Organizing the Space of a Page, More Metaphorically

Traditional pictures, architectural plans, and maps are literally spatial in the sense that they represent things that are visible in the world, typically preserving shapes and spatial relations among and within the forms. Such mappings are derived from the spatial world through the mind, by schematizing or abstracting information from the spatial world. At another extreme are mappings that are regarded as abstract or metaphorically spatial. Such mappings are constructions, derived from mental representations in the mind through similar schematizing processes to forms and places on the page. For concepts that are not literally spatial, form and place are freed of any need to resemble "reality." Nevertheless, the uses of form and place in conveying meanings are constrained by certain psychological correspondences, perceptual, cognitive, and social. Many of these metaphorically spatial concepts are evident in spatial language: someone is at the top of the class, another has fallen into a depression, friends grow close or apart; a field is wide open, a topic is central to a debate (e.g., [77]). Those constraints and some of their consequences will be discussed in the subsections below on organization of space as well as in the subsequent section on Forms in Space. We continue now with a discussion of certain properties of the page, and how they are used to convey meanings.

2.3.1 Proximity: Category and Continuum

Perhaps the most fundamental way that space is used to create abstract meaning is proximity; things that are closer conceptually are placed closer on a page. As in organizing real space, proximity can be used hierarchically to organize metaphoric spaces, first to create clusters, groups, or categories of similar things (like the stack of shirts on a shelf), and then to create clusters, groups, or categories of categories (like the men's department). Grouping by proximity is commonly used on the space of the page. The letters of one word are separated from the letters of another word by a space, making reading easier. Ideas are further separated on the page by paragraph indentation. Similar uses of space occur in writing and comprehending math equations, where spacing affects the order of carrying out mathematical operations (e.g., [79, 80]).

Often the things to be represented are ordered, thus represented on a continuum: countries by size, events by dates. When things are ordered conceptually, they can be arranged in an order on a page, forming a continuum. If some pairs of the ordered things are conceptually closer and others conceptually farther, proximity can be used to represent the closeness of the pairs on the conceptual relationship. This spatial progression forms the conceptual basis for simple mathematics as well as for graphing, conveying mathematical concepts on a page (e.g., [30])

How should orders be arrayed? The very shape of a page suggests three kinds of arrays: horizontal, vertical, and central–peripheral. The salient dimensions of the world reinforce the horizontal and vertical, and certain properties of vision reinforce center–periphery. Rep- resenting orders entails selection of spatial dimension as well as selection of a direction within a dimension, issues to be discussed in the following sections.

2.3.2 Central-Peripheral

A center-outward organization reflects the organization of the retina, with the fovea, the point of greatest acuity, at the center. Acuity, hence attention, is at the center of the visual field, with acuity and attention declining in all directions from the center. That people organize space center-outwards seems inevitable. Just like the toddler placing her face smack in the middle of someone's field of view, putting something in the center of a page puts it liter- ally and figuratively in the center of focus of the eye and of attention. Symbolic centers are ubiquitous, from the angels around God to the etiquette of seating arrangements at a formal dinner for a visiting dignitary [8]. Early in the twentieth century, an African king wished to prove the modernity of his country by having it surveyed to make a map. On learning that the capital of the country was not in the county's geographic center, he ordered that its location on the map be moved to be more central [174]. Mandalas, common in Hindu and Buddhist traditions, represent the cosmos or the spiritual world, with spiritual symbols at the center. They not only symbolize the cosmos but also serve as meditation aids, centering meditation on the center of the mandala [36]. Greek and Roman vases place important figures in the center and less important to the sides [125], as do advertisements and paintings from all over the globe. Language does this too, of course; we have been talking about the center and the periphery, both literally and figuratively. These spatial features of vision become conceptual features of thought, central or peripheral, a kind of embodiment.

A central–peripheral organization may coordinate well with a single focus of attention and the organization of the eye, but it is not well suited for ordering, either of attention or of things. The periphery extends in all directions from the center without an explicit direction or ordering. At the extreme, a center–periphery organization is dichotomous: central and important versus less central and important. Some mandalas have concentric rings that are ordered outwards, but there is no clear ordering within each ring. Vases, advertisements, and the like are organized by pictorial space as well as by center–periphery, so that the periphery extends leftwards and rightwards (andor downwards and upwards) from the center rather than in all directions as in a mandala. A horizontal (or vertical) organization simplifies, but since the start point is the center, there is no explicit way to integrate the orderings of things to either side. In addition, the human visual system is especially sensitive to horizontal and vertical, less so to oblique lines (e.g., [61]). Perhaps for these reasons, complete orderings tend to use a straight line, horizontal or vertical, one of the edges of the page as a guide, and to begin at one end or the other. It is worth noting that written languages, which typically require serial order, use vertical columns or horizontal rows.

2.3.3 Page Parallels

The central–peripheralmore important–less important arrangement of space has the advantage of centering the most important, the highest on some attribute, but the disadvantage of making it difficult to compare the orders of those in the periphery, as the order descends in more than one direction. Using one of the dimensions of the page for ordering makes the start point and direction explicit and easy to follow, but it raises the dual questions of which dimension, vertical or horizontal, and where to start. Those decisions are influenced by a number of factors. Some seem to be general across cultures, for example, primacy to up, the location of gods in most cultures. Others seem to be more influenced by culture, for example, horizontal direction, right to left or left to right.

To investigate the spontaneous use of spatial dimensions to convey abstract ones, children from 4 years old to college age from three language cultures, English-speaking Americans, Hebrew-speaking Israelis, and Arabic-speaking Israelis, were asked to place stickers on a square page to indicate the relations of three instances on each of four dimensions: spatial, temporal, quantitative, and preference ([157]; for similar work on generating mathematics, see [62]) Because English is written from left to right but Hebrew and Arabic are written from right to left, the study also examined the effects of writing order on inventions of graphs. For the spatial task, the experimenter first positioned three small dolls in a row in front of the child and asked the child to place stickers on the page to represent the locations of each of the dolls. All the children per- formed the spatial mapping task with no difficulty. Then the children were asked to represent the more abstract concepts spatially. For representing time, the experimenter sat next the child and asked the child to think about the times of the day for breakfast, for lunch, and for dinner. For representing quantity, the experimenter asked the child to think about the amount of candy in a handful, in a bagful, or on the shelf in the supermarket. For representing preference, the experimenter asked the child to think about a television show he or she really liked, did not like at all, or sort of liked. Then the experimenter put a sticker in the middle of the page for the middle value, lunch or the amount of candy in a bagful, or the so-so TV show and asked the child to put a sticker on the page for the other two extreme values, one at a time, in counter-balanced order.

A few of the youngest children did not put the stickers representing three examples on a line; instead they scattered the stickers over the page or put one on top of the other, indicating that they did not see the instances as ordered on a continuum. Scattering the stickers across the page suggests that the children saw the instances as three different categories and piling them on top of each other suggests that the children saw the instances as a single category, say meals or candy or TV shows. Either arrangement indicates that the children used space categorically but not ordinally. Most of the preschool children and all of the older children and adults did place the stickers (or dots, for the adults) on a virtual line, thereby using one of the dimensions of the space of the page to represent the underlying dimension. Children represented the more concrete dimension, time, as a line earlier than the more abstract dimensions, quantity, and preference in that order.

A second experiment assessed whether children could map interval as well as order [157]. They were first asked to place stickers to indicate the locations of the three small dolls, when two were placed quite close to each other, but relatively far from the third. Even the youngest children used spatial proximity to repre- sent interval in the placement of stickers. Then the children were asked to represent instances of temporal, quantitative, and preference concepts that were unequally spaced. For example, they were asked to place stickers to represent the time for breakfast, morning snack, and dinner. Despite heavy-handed prompting, only at 11–12 years of age did children reliably place stickers closer for instances closer on the dimension and place stickers farther for instances farther on the dimension.

Together, the results indicated that children spontaneously use spatial proximity and linear arrays to represent categorical, ordinal, and interval properties of abstract dimensions. With increasing age, children's representations progress from categori- cal to ordinal to interval. Their graphic productions are true inventions; that is, they do not correspond to the graphing conventions that older children are exposed to in school. For example, the directions of increases in their graphic inventions, to which we turn now, were not consistent across dimensions within or across children nor did they universally proceed from left to right.

2.3.4 Direction in Space: Horizontal

Center–periphery uses direction, from the center outwards to the periphery to indicate importance or closeness to God. Center–periphery mappings work well for vague cases, where the center is the highlight and the exact ordering of the cases in the periphery is not of concern. But if it is, a spatial order that is easy to discern is preferable. We have seen that children and adults mapped orders of spatial, temporal, quantitative, and preference concepts onto lines. For the case of time, the preferred orientation was horizontal across cultures and ages. Mapping time to horizontal, evident even in Chinese, a language written in columns (e.g., [14]), is likely to have a basis in motion, which for humans and most creatures and natural phenomenon, is primarily horizontal. Motion is in space, on the plane, and takes time. In many senses, space, time, and motion are intertwined and sometimes interchangeable. Knowledge of space frequently comes from motion in time, from exploring environments and piecing together the parts. Spatial distance is often expressed as time, a 20 min walk or an hour's drive. That said, concepts of space appear to be primary, and concepts of time derived from concepts of space [13], perhaps because space can be viewed and time cannot. Time is a neutral dimension, and, as shall be seen, the vertical dimension appears to be preferred for evaluative concepts and the horizontal dimension for neutral concepts. Nevertheless, although time is primarily represented horizontally, as shall be seen, there are cases where time is represented vertically; for each dimension, there is a preferred directionality.

In the studies of Tversky et al. (1991) [157], children and adults from all three language cultures preferred to map time horizontally. However, the direction of temporal increases reflected cultural habits, specifically, the order of reading and writing. English speakers typically arrayed temporal events from left to right and Arabic speakers from right to left, corresponding to the direction of writing in those languages. Hebrew speakers were split. Although writing proceeds right to left in Hebrew, numbers proceed left to right, as in Western languages. For the Arabic populations in this study, arithmetic is taught right to left until 5th grade, when it is reversed to conform to Western conventions. In addition, Hebrew characters are formed left to right, whereas Arabic characters are formed right to left, and Hebrew-speaking Israelis are more likely to be exposed to Western left to right languages.

The influence of reading order appears for a wide range of concepts, especially those related in some way to time. Counting, like writing, is serial, and takes place in time. The mental number line has an implicit spatial ordering evident in speed of calculations, left-to-right in readers of languages that go from left to right, and the opposite for languages that go from right to left and absent in illiterates (e.g., [30, 179]). Temporal order of events is gestured left to right in native Spanish speakers but right to left in native Arabic speakers, even when speaking Spanish (e.g., [111, 112]). Writing order affects perception of motion (e.g., [85, 94]), perceptual exploration and drawing (e.g., [22, 96, 166]), aesthetic judgments (e.g., [22, 23, 97]), emotion judgments [110, 165], judgments of agency, power, and speed [20, 21, 55, 86, 131], and art [20, 91]. A variety of factors correlated with reading order seem to underlie these effects. The effects of reading order on perception of apparent motion and of speed and on perceptual organization seem to derive from long-term reading habits. The effects of reading order on judgments of agency, where figures on the left are seen as more powerful, seem to derive from language syntax, where the actor is typically earlier in the sentence than the recipient of action.

The respondents in study by Tversky et al. (1991) [157], children and adults, did not use a graphic template to map abstract relations to the page. Mappings of quantity and preference, in contrast to mappings of time, did not reflect reading order. Speakers of all three languages were equally likely to map quantity and preference from right to left, left to right, and down to up. That is, their horizontal mappings corresponded to writing order only half the time, for both language orders. And vertical mappings were also used frequently, with large quantities and preferred alternatives at the top. Mapping increases in quantity or preference from up to down was avoided by all cultures especially for quantity and preference, for reasons elaborated below.

Writing order is one mapping of order to the page, a weaker one that depends on culture. In the large cross-cultural study of spontaneous mappings, it appeared only for temporal concepts [157]. Even there, although English speakers tended to map order left to right and Arabic speakers right to left in correspondence with writing and with numerals, Hebrew speakers did not show a strong preference, most likely because they were familiar with cultural artifacts ordered both ways. In contrast to the vertical dimension with its strong asymmetry defined by gravity and corresponding to people's upright posture, the horizontal dimension has only weak asymmetries. Although the horizontal surface of the world is very salient, it has no privileged direction, unlike the vertical direction defined by gravity. The front–back axis of the body has strong asymmetries, but the left–right axis is more or less symmetric. Handedness is a notable exception; however, it is primarily behavioral rather than visible, and it varies across people with biases that depend on handedness (e.g., [19]. The plasticity across cultures of left–right horizontal mappings sup- ports the claim that for the page, directional bias along the horizontal axis is weaker than directional bias along the vertical axis, hence influenced by cultural factors such as writing reading direction.

The plasticity of the horizontal left–right (or right–left) dimension, suggested by its influence from cultural factors, is no doubt partly due to the absence of salient left–right asymmetries in the body or the world (e.g., [24, 38]). It seems to be reinforced by a salient fact about human communication, either with other humans or with graphics. Communication normally happens face to face, where my left and right are the reverse of yours or the reverse of that depicted. So although godly figures are depicted or described with angels on his right and the devil on his left, his right is the viewers' left. Some languages do not even distinguish right from left, leading to different organizations of space (e.g., [84]). A number of factors, then, converge to render mappings to the horizontal dimension to be more flexible than those to the vertical dimension.

2.3.5 Direction in Space: Vertical

By contrast, the use of the vertical to express asymmetric evaluative concepts like power, strength, and quality is evident in a broad range of gestures and linguistic expressions across cultures and has a basis in the nature of the world and the things in it, including ourselves (e.g., [24, 27, 38, 77, 136, 148, 149]). Gravity makes it more difficult to go up than to go down, so that it takes power, strength, health, and energy to go upwards. People, along with many other animals and plants, grow taller and stronger as they reach adulthood, and taller people tend to be stronger. People who are healthy and happy have the energy to stand tall and people who are weak or ill or depressed slump. Piles of money or other things grow higher as their numbers increase. Remember that children and adults alike used the vertical to represent increases in quantity and preference, with large quantities and preferences at the top, never the reverse. In a more complex graphing task, children and adults preferred steeper lines, those that incline more upwards, to represent greater rates [39, 41]. On the whole, more power, better health, greater strength, and more money are good, and less of all that is bad. This maps lower numbers to lower values and to lower spatial positions and higher numbers to higher values and higher spatial positions. The starting point is the ground. Low numbers are bad and high numbers are good. Gestures such as high five and thumbs up reflect the correspondence of upwards with positive value.

But mappings to vertical can conflict, with consequent confusions. The world and our experiences in it provide reasons for beginning at the top. People's major perceptual and conceptual machinery, our eyes, our ears, and our brains, are at the top of our bodies. Reading order enters here as well; most written languages begin at the top, whether they go left to right or right to left, whether they are written in rows or columns. Numbering, then, can begin at the top or begin at the bottom. So familiar are the two mappings to numbers that we hardly notice the contradiction: the number one player, the one at the top, is the one with the highest number of points. Rises in unemployment or inflation are bad, but are mapped upwards because of rising numbers. These alternative mappings to vertical were seen in a survey of common diagrams in college textbooks for biology, earth sciences, and linguistics [149]. Almost all the diagrams of evolution had man (yes, man) at the top and almost all of the geological eras had the present era at the top; that is, each kind of diagram began at the bottom with prehistory, and depicted the culmination of "progress" at the top [138]. Earlier time was at the bottom, later time at the top. Although evolutionary trees have man at the top, we speak of the "descent of man" not the ascent of man. In contrast, linguistic trees, like family trees, typically had the progenitor language at the top and the language derived from it descending downwards. For linguistic and family trees, time begins at the top. In memory, the concept "depth of processing," which suggests that lower is more abstract and meaningful, is synonymous with the concept of "levels of processing," which specifies that higher levels of processing are deeper, more abstract, and meaningful. Deep thought occurs at high levels of thinking.

Although there are multiple mappings of abstract dimensions and relations to direction in space, there are also some consistencies. Notably, horizontal and vertical are chosen for ordering, not diagonal or circular or some other path through space, undoubtedly related to the privileged status of horizontal and vertical in vision (e.g., [61]). The horizontal direction, the primary plane for motion, human and other, is readily mapped to time and more frequently used for other neutral concepts. By contrast, the vertical dimension formed by gravity is readily mapped to quantity and force, and more frequently used for evaluative concepts like quantity and preference. The vertical direction has salient and far-reaching asymmetries in the world and in human perception and behavior, with multiple correspondences from evaluative concepts like strength, power, health, and wealth to the upwards direction. The horizontal dimension has fewer asymmetries in the world and in human perception and behavior so weaker, cultural variables affect direction, notably, the direction of reading, writing, and arithmetic, and to some extent, handedness. These spatial meanings are reflected in language and in gesture as well. While both those on the politically left and those on the politically right will agree that it is better to be on top, they will disagree on whether left or right is better.

2.4   Mapping Meaning to Space

A variety of examples have shown that people readily map meaning to space, and to the space of a page. They use spatial properties of the page to relay a range of ideas, abstract, and concrete: proximity, place, linear arrays, horizontal, vertical, and direction to group categories, show relationships, illustrate orders, convey conceptual distance, express value, and more. We have already accumulated a small catalog of meaningful mappings to space: depictive or geographic, clumps for categories, center to catch attention or convey importance, lines for orders, distance/proximity in space to reflect distance/proximity on an abstract dimension, horizontal for time and concepts related to time, vertical for strength, quantity, force, power, and concepts related to them. Direction matters, too: Concordant with the vertical asymmetry of the world created by gravity and the human experience of living in the world, up is readily associated with increases in amount, strength, goodness, and power. The horizontal dimension of the world is more neutral, so less strongly tied to abstract concepts and more susceptible to cultural influences such as reading order and handedness. But there are caveats on these mappings. For one thing, they are incomplete and variable; different features may be mapped on different occasions. Hence, these map- pings can conflict, especially when associated with number; a high score can determine who is first. These correspondences are natural in the sense that they have been invented and reinvented across cultures and contexts, they have origins in the body and the world, and they are expressed in spatial arrangements, spatial language, and spatial gestures.

3 Forms in Space

Now we turn from the space of the page to marks on the page, to examine how marks convey a range of meanings, like space, by using natural correspondences. Although the simplest marks are dots or lines, the most common now and throughout history are undoubtedly what have been referred to as pictograms, icons, depictions, or likenesses, from animals on the ceilings of caves to deer on road signs. Marks on a page have been termed signs, which refer to objects for minds that interpret them, by Peirce, who distinguished three kinds of them (e.g., [51]). An icon denotes an object by resemblance, an index, such as a clock or thermometer, denotes an object by directly presenting a quality of an object, and a symbol, a category that includes certificates as well as words, denotes an object by convention.

Here, we first discuss some properties and uses of likenesses or icons, and then turn at greater length to a specific kind of symbol, which we have called a glyph (e.g., [151, 160]). Glyphs are simple figures like points, lines, blobs, and arrows, which derive their meanings from their geometric or gestalt properties in context. Glyphs are especially important in diagrams because they allow visual means of expressing common concepts that are not easily conveyed by likenesses. Glyphs have parallels to certain kinds of gestures, for example, points that suggest things that can be conceived of as points or linear gestures that suggest relationships between things. They also bear similarities to words like point and relationship whose meanings vary with context.

Marks, whether likenesses or glyphs, like lines and circles, have visual characteristics other than shape that increase their effectiveness in conveying meaning. An important feature is size. The greater the size, the greater the chance of attracting attention. The toddler knows not only that centrality captures attention, but size as well. The toddler wanting attention puts her face close, blocking other things in the visual field. Size, like centrality, can also indicate importance. Greek vases use both centrality and size; the major figure is larger and in the center, with the others arrayed to either side in decreasing order of importance. Larger bar graphs represent greater quantity or higher ratings. Additional salient visual features, like color, boldness of line, highlighting, and animation, also serve to attract attention and convey importance.

3.1 Likenesses

Even sketchy likenesses can be readily recognized by the uninitiated. A toddler who had never seen pictures but could label real objects recognized simple line drawings of common objects [58]. Depictions have other impressive advantages over words in addition to being readily recognized: They access meaning faster [127] and enjoy greater distinctiveness and memorability (e.g., [107]). Perhaps because of their advantages for establishing meaning and memory, likenesses are so compelling that they are produced even when not needed and even when drawing them increases time and effort: in diagrams of linear and cyclical processes produced by undergraduates [71], in diagrams of information systems by graduate students in design [100].

Likenesses have been creatively integrated into more abstract representations of quantitative data by Neurath and his Vienna Circle and later colleagues in the form of isotypes [99]. Isotypes turn bars into depictions, for example, the number of airplanes in an army or yearly production of corn by a country is represented by a proportional column (or row) of schematic airplanes or corn plants.

Just as likenesses can facilitate comprehension and memory, they can also interfere. Because depictions are specific and concrete, including them when they are not essential to the meaning of a diagram can inhibit generalization, to sets of cases not depicted. By contrast, glyphs, because they are abstractions, can encourage generalization. Capturing the objects in the world and their spatial arrays in diagrams is compelling and has some communicative value, but it can interfere or even conflict with the generalizations or abstractions diagrams are meant to convey. An intriguing example comes from diagrams of the water cycle in junior high science textbooks collected from around the world (Chou, Vikaros, & Tversky, unpublished data). The typical water cycle diagram includes mountains, snow, lakes, sky, and clouds. On the one hand, these diagrams intend to teach the cycle of evaporation of surface water, formation of clouds, and precipitation. They use arrows to indicate the directions of evaporation and precipitation. On the other hand, they also want to show the water cycle on the geography of the world. As a consequence, the arrows ascend and descend everywhere, so that the cyclicity is obscured. In studies investigating interpretations of slope in diagrams of the atmosphere, students' inferences were more influenced by the conceptual mapping of rate to slope than by the geographic mapping [41]. In producing diagrams, for example, of a pond ecology, when groups work in pairs, the compelling iconicity evident in individual productions often disappears [117]. Diagrams produced by dyads become more abstract, most likely because the irrelevant or distracting iconicity is idiosyncratic and the abstractions shared. The conflict between visualizing the world and visualizing the general phenomena that occur in the world is especially evident when diagrams are used to convey the invisible such as evaporation and gravity. With all the challenges of conveying the visible, conveying the invisible, time, forces, values, and the like presents even more challenges. Glyphs are ideal for visually conveying the invisible. They are not iconic, they do not depict the visible world, so they do not confuse or distract, yet they share many of the advantages of visual communication over purely symbolic communication, notably rapid access to meaning. We turn now to many examples of using glyphs to visually convey invisible and abstract concepts.

3.2   Meaningful Glyphs

We shift now from the complex and representative to the simple and abstract. Probably the simplest mark that can be made on the page is a dot, a mark of zero dimensions. Slightly more complicated, a line, a single dimension, followed by various two-dimensional or three-dimensional forms. These simple marks and others like them that we have termed glyphs have context-dependent meanings suggested by their Gestalt or mathematical properties [149, 151]. On a map of the United States, New York City can be represented as a point, or the route from New York to Chicago as a line, or the entire city can be represented as a region, containing points and lines indicating, for example, roads, subway stops, and subway lines. Continuing, New York City can also be diagrammed as a three- dimensional space in which people move. Like many other spatial distinctions, this set of distinctions has parallels in language and gesture, parallels that suggest the distinctions are conceptual and widely applicable. Regarding an entity in zero, one, two, or three dimensions has implications for thought. In a paper titled, "How language structures space," Talmy (2000) [137] pointed out that we can conceptualize objects in space, events in time, mental states, and more as zero-, one-, or two-dimensional entities. In English, prepositions are clues to zero-, one-, two- (and three-) dimensional thinking, notably at, on, and in. She waited at the station, rode on the train, rose in the elevator. She arrived at 2, on time, and was in the meeting until dinner. She was at ease, on best behavior, in a receptive mood. Visual expressions of dimensionality are common in diagrams, as they abstract and express key conceptual components.

3.2.1   A Visual Toolkit for Routes: Dots and Lines

Dots, lines, and regions abound in diagrams. Dots and lines, nodes and links or edges are the building blocks of route maps. They also form a toolkit for a related set of abstractions, networks of all kinds. To uncover the basic visual and verbal vocabularies of route maps, students outside a dormitory were asked if they knew how to get to a nearby fast food restaurant. If they did, they were asked either to draw a map or to write directions to get there. A pair of studies confirmed that dots and lines, nodes and links, are the basic visual vocabulary of route maps, and that each element in the visual vocabulary for route directions corresponds to an element in the basic verbal vocabulary for route directions [158, 159]. Notably, although the sketch maps could have been analog, they were not; turns were simplified to right angles and roads were either straight or curved. Land- marks were represented as dot-like intersections identified by street names or as nonspecific shapes. Short distances with many turns were lengthened to show the turns, and long distances with no actions were shortened. Thus, the route maps not only categorized continuous aspects of the world, they also distorted them. Interestingly, the verbal directions were similarly schematized. Distances were specified only by the bounding landmarks; turns were specified only by the direction of the turn, not the degree. The consensus visual vocabulary consisted of lines or curves, L, T, or + intersections, and dots or blobs as landmarks. The corresponding verbal vocabulary consisted of terms like "go straight" or "follow around" for straight and curved paths, "take a," "make a," or "turn" for the intersections, and named or implicit landmarks at turning points. The vocabulary of gestures used to describe routes paralleled the visual and verbal vocabularies [155]. These close parallels between disparate modes of communication suggest that the same conceptual structure for routes underlies all of them.

A second study provided students with either the visual or the verbal toolkit, and asked them to use the toolkit to create instructions for several dozen destinations, near and far [83]. They were asked to supplement the toolkits if needed. In spite of that suggestion, very few students added elements; they succeeded in using the toolkits to create a variety of new directions. Although the semantics (vocabularies) and syntax (rules of combing semantic elements) of route maps and route directions were similar, their pragmatics differs. Route maps cannot omit connections; they must be complete. Route directions can elide; for example, in a string of turns, one end-point is the next start-point, so it is not necessary to mention both.

Why do directions that are so simplified and distorted work so well? Because they are used in a context, and the context disambiguates [150]. This is another general characteristic of diagrams; they are designed to be used by a specific set of users in a specific context. Indeed, part of the success of route maps and route directions is that they have been developed in communities of users who collaborate, collectively and interactively producing and comprehending, thereby fine-tuning the maps and directions, a natural kind of user-testing that can be brought into the laboratory and accelerated [153].

The success of the visual and verbal toolkits for creating route maps and route directions has a number of implications. It has already provided cognitive design principles—paths and turns are important; exact angles and distances are not— for creating a highly successful algorithm for on-line on-demand route directions [1]. It suggests that maps and verbal directions could be automatically translated from one to the other. It is encouraging for finding similar visual and verbal intertranslatable vocabularies for other domains, such as circuit diagrams or musical notation or even domains that are not as well structured domains such as assembly instructions, chemistry, and design. It suggests empirical methods for uncovering domain-specific visual and verbal semantics, syntax, and pragmatics. Finally, it shows that certain simple visual elements have meanings that are spontaneously produced and interpreted in a context. Some of these visual elements have greater generality. Lines are naturally produced and interpreted as paths connecting entities or landmarks that are represented as dots. Hence their widespread use, from social networks, connections among people, to computer networks, connections among computers or components of computers, and more.

3.2.2 Lines Connect, Bars Contain

As Klee put it, "A line is a dot that went for a walk." Lines are also common in graphs, again, as paths, connections, or relations. So are dots and bars. Graph lines connect dots rep- resenting entities with particular values on dimensions represented by the lines. The line indicates that the entities are related, that they share a common dimension, but have different values on that dimension. Bars, in contrast to lines, are two-dimensional; they are containers that separate their contents from those of others. In graphs, bars indicate that all the instances inside are the same and different from instances contained in other bars. To ascertain whether people attribute those meanings to bars and lines, in a series of experiments, students were shown a single graph, either a line graph or a bar graph, and asked to interpret it [176]. Some of the graphs had no content, just A's and B's. Other graphs displayed either a discrete variable, height of men and women, or a continuous variable, height of 10- and 12-year-olds. Because lines connect and bars contain and separate, students were expected to favor trend descriptions for data presented as lines and favor discrete comparisons for data presented as bars, especially for the graphs without content. For the content-free graphs, the visual forms, bars or lines, had major effects on interpretations, with far more trends for lines and discrete comparisons for bars. More surprisingly, the visual forms had large effects on interpretations of graphs with content, in spite of contrary content. For example, using a line to connect the height of women and men biased trend interpretations, even, "as you get more male, you get taller." These were comprehension tasks. Mirror results were obtained in production tasks, where students were provided with a description, trend, or discrete comparison, and asked to produce an appropriate graph. More students produced line graphs when given trend descriptions and bar graphs when given discrete comparisons, as before, in spite of contrary content. The meanings of the visual vocabulary, lines or bars, then, had a stronger effect on interpretations and productions than the conceptual character of the data. When the glyph, line or bar, matched the content, there were more appropriate interpretations and when the glyph did not match the content, there were more inappropriate interpretations (for other issues with bars and lines, see [118, 119]).

3.2.3 Lines can Mislead

Because glyphs such as lines, dots, boxes, and arrows, induce their own meanings, they are likely to enhance diagrammatic communication when their natural mean- ings are consistent with the intended meaning and to interfere with diagrammatic communication when the natural meanings conflict with the intended meanings. This interaction was evident in the case of bar and line graphs for discrete and continuous variables, where the interpretations of the visual glyphs trumped the underlying structure of the data when they conflicted. Mismatches between the natural interpretations of lines as paths or connections and the intended interpre- tations in diagrams turn out to underlie difficulties understanding and producing certain information systems designs. A central component of information system design is a LAN or local area network, common in computer systems in every institution. All of the components in a LAN are interconnected so that each can directly transmit and receive information from each other. A natural way to represent that interconnectivity would be lines between all pairs of components. For large systems, this would quickly lead to a cluttered, indecipherable diagram. To insure legibility, a LAN is diagrammed as if a clothesline, a horizontal line, with all the interconnected components hanging from it. However, when students in information design are asked to generate all the shortest paths between components from diagrams containing a LAN, many make errors. A common error demonstrates a strong bias from the line glyph. The shortest paths many students generate show that they think that to get from one component on a LAN to another, they must pass through all the spatially intermediate components, much like traveling a route, to go from 10th St to 30th St one must pass 11th, 12th, 13th, and so on [28, 100]. Here, again, the visual trumps the conceptual and misleads.

Lines have mixed benefits in other cases, for example, in interpreting evolutionary diagrams where they can lead to false inferences [102, 103, 106]. Yet another example comes from visualizations of space, time, and agents, diagrams that are useful for keeping track of schedules, suspects, pollen, disease, migrations, and more [70]. In one experiment, information about the locations of people over time was presented either as tables with place and time as columns or rows and dots representing people as entries or as tables with lines connecting individuals from place to place over time. Because lines connect, one might expect that the lines would help to keep track of movements of each individual. In one task, participants were asked to draw as many inferences as they could from the diagrams; in another they were asked to verify whether a wide range of inferences was true of the diagrams. At the end of the experiment, they were asked which interface they preferred for particular inferences. Overall, participants performed better with dots than with lines both in quantity of inferences drawn and in speed and accuracy of verification. However, and consonant with expectations, there was one exception, one kind of inference where dots lost their advantage, inferences about the sequence of locations of individuals. For temporal sequence, lines were as effective and as preferred as dots. Nevertheless, the lines interfered with generating and verifying other inferences. In another experiment, participants were asked to generate diagrams that would represent the locations of individuals over time. Most spontaneously produced table-like visualizations, notably without lines. As for preferences, participants preferred the visualizations with dots over those with lines except for temporal sequences. These findings suggest that popular visualizations that rely heavily on lines, such as parallel coordinates (e.g., [63]) and especially parallel sets (e.g., [11]), should be used with caution, and only when the lines are meaningful as connectors.

Arrows are asymmetric lines. As a consequence, arrows suggest asymmetric relation- ships. Arrows enjoy several natural correspondences that provide a basis for extracting meaning. Arrows in the world fly in the direction of the arrowhead. The residue of water erosion is a network of arrow-like lines pointing in the direction of erosion. The diagonals at the head of an arrow converge to a point. Studies of both comprehension and production of arrows show that arrows are naturally interpreted as asymmetric relationships. In a study of comprehension, students were asked to interpret a diagram of one of three mechanical systems, a car brake, a pulley system, or a bicycle pump [57]. Half of each kind of the diagram included arrows, half did not. For the diagrams without arrows, students gave structural descriptions, that is, they provided the spatial relations of the parts of the systems. For the diagrams with arrows, students gave functional descriptions that provided the step-by-step causal operations of the systems. The second study provided a description, either structural or functional, of one of the systems and asked students to produce a diagram. Students produced diagrams with labelled parts from the structural descriptions but produced diagrams with arrows from the functional descriptions. Both interpretation and production, then, showed that arrows suggest asymmetric temporal or causal relations.

One of the benefits of arrows can also cause difficulties; they have many possible meanings. Arrows suggest many possible asymmetric relations [57]. Their ambiguity can cause misconceptions and confusion. Arrows are used to label or focus attention; to convey sequence; to indicate temporal or causal relations; to show motion or forces; and more. How many meanings? Some have proposed around seven (e.g., [170]), others, dozens (e.g., [60]). A survey of diagrams in introductory science and engineering texts revealed that many diagrams had different meanings of arrows in the same diagram, with no visual way to disambiguate them [156].

Circles, with or without arrows, can be viewed as another variant on a line, one that repeats with no beginning and no end. As such, circles have been used to visualize cycles, processes that repeat with no beginning and no end. The common etymology of the two words, circle and cycle, is one sign of the close relationship between the visual and the conceptual. However, the analogies, like many analogies, are only partial. Circles are the same at every point, with no natural divisions and no natural direction. Yet when we talk about cycles, we talk about them as discrete sequences of steps, sometimes with a natural beginning. Hence, cycles are often visualized as circles with boxes, text, or pictograms conveying each stage of the process.

A series of studies on production and comprehension of visualizations of cyclical and linear processes asked participants to produce or interpret appropriate marks on paper [71]. In a set of studies, participants were asked to fill in circular diagrams with four boxes at 12 o'clock, 3 o'clock, 6 o'clock, and 9 o'clock with the four steps of various cyclical processes, everyday (e.g., washing clothes, seasons) and scientific (e.g., the rock cycle, the water cycle). They did this easily. Although circles have no beginning, many cycles there have a conceptual beginning, and students tended to place that at 12 o'clock, and then proceed clockwise. Conversely, when asked to interpret labeled circular diagrams, they began at 12 o'clock and proceded clockwise, except when the "natural" starting point of a cycle, for example, the one-cell stage of mitosis, was at another position. In a second set of studies, students were given blank pages and asked to produce diagrams to portray cyclical processes, like the seasons or the seed-to-plant-to-seed cycle, as well as linear processes, like making scrambled eggs or the formation of fossil fuel. Both cycles and linear processes had four stages. Unsurprisingly, most students portrayed the linear processes in lines, but, more surprisingly, most portrayed the cyclical processes as lines as well, without any return to the beginning. Heavy-handed procedures, presenting only cyclical processes, calling them such, and listing the stages vertically, brought the frequency of circular diagrams to 40 %. Changing the list of stages so that the first stage was also the last, as in "the seed germinates, the flower grows, the flower is pollinated, a seed is formed, the seed germinates," induced slightly more than 50 % of participants to draw the stages in a circle, but still, more than 40 % drew lines. There is strong resistance to producing circular diagrams for cycles, even among college students. In the final study, participants were provided with a linear or circular diagram of four stages of a cycle, and asked which they thought was better. Over 80 % of participants chose the circular display. This is the first case we have found where production and preference do not match, though production lags comprehension in other domains, notably, language acquisition.

Why do people prefer circular diagrams of cycles but produce linear ones? We speculate that linear thinking is easier than circular; that is, it is easier to think of events as having a beginning, a middle, and an end, a forward progression in time, than it is to think of events as returning to where they started and beginning all over again, without end. Events occur in time, time marches relentlessly forward, and does not bend back on itself. Each day is a new day, each seed a new seed; it is not that a specific flower emanates from a seed and then transforms back into one. Thinking in circles requires abstraction, it is not thinking about the individual case, but rather thinking about the processes underlying all the cases. What is more, the sense in which things return to where they started is different in different cases. Every day has a morning, noon, and night, but each morning, noon, and night is unique. A cell divides into two, and then each of those cells undergoes cell division. For clothing and dishes, however, the very same articles of clothing and the very same dishes undergo washing, drying, get put away each time. Viewing a circular diagram enables that abstraction, and once people "see" it (the diagram and the underlying ideas), they prefer the abstract depiction of the general processes to the more concrete depiction of the individual case.

3.2.4 Boxes and Frames

Earlier, we saw that people interpret bars as containers, separating their contents from everything else. Boxes are an ancient noniconic depictive device, evident explicitly in stained glass windows, but even prior to that, in Roman wall frescoes. Frames accentuate a more elementary way of visually indicating conceptual relatedness, grouping by proximity, for example, the spaces between words. Framing a picture is a way of saying that what is inside the picture has a different status from what is outside the picture. Comics, of course, use frames liberally, to divide events in time or views in space. Comics artists sometimes violate that for effect, deliberately making their characters pop out of the frame or break the fourth wall, sometimes talking directly to the reader. The visual trope of popping out of the frame makes the dual levels clear, probably even to children: The story is in the frames, the commentary outside [171]. Speech balloons and thought bubbles are a special kind of frame, reserved for speech or thought; as for other frames, they serve to separate what is inside from what is outside. Frames, like parentheses, can embed other frames, hierarchically, indicating levels of conceptual spaces, allowing meta-levels and commentaries. Boxes and frames serving these ends abound in diagrams, in flow charts, decision trees, networks, and more.

3.2.5 Complex Combinations of Glyphs

As was evident from the visual toolkit for routes, glyphs can be combined to create complex diagrams that express complex thoughts and systems. Like combining words into sentences, combining glyphs into systems follows domain-specific syntactic rules (e.g., [159]). Networks of lines and nodes, more abstractly, concepts and connections between concepts, are so complete and frequent that they constitute a major type of diagram. Others types of diagrams include the following: hierarchies, a kind of network with a unique beginning and layers of asymmetric relations, such as taxonomies and organization charts; flow charts consisting of nodes and links representing temporal organizations of processes and outcomes; decision trees, also composed of nodes and links, where each node is a choice. A slightly different type of diagram is a matrix, a set of boxes organized to represent the cross-categorization of sets of dimensions or attributes. These organized sets of glyphs and space constituting diagrammatic types appear to match, to naturally map, conceptual organizations of concepts and relations. That is, for networks, hierarchies, and matrices, students were able to correctly match a variety of conceptual patterns onto the proper visualization [104, 105].

Note that many of these visual complex combinations of glyphs, for example, bar and line graphs, social and computer networks, decision and evolutionary trees, have no pictorial information whatsoever, yet they inherit all the advantages of being visual. They enable human application of visuospatial memory and reasoning skills to abstract domains.

3.2.6 Sketches

The aim of most of the diagrams discussed thus far is to convey certain information clearly in ways that are easily apprehended, from route directions to data presentations to scientific explanations. Another important role for visualizations of thought is to clarify and develop thought. This kind of visualization is called a sketch because it is usually more tentative and vague than a diagram. Sketches in early phases of design even of physical objects, like products and buildings, are frequently just glyphs, lines and blobs, with no specific shapes, sizes, or distances (e.g., [45, 116]). Designers use their sketches in a kind of conversation: They sketch, reexamine the sketch, and revise [116]. They are intentionally ambiguous. Ambiguity in sketches, just like ambiguity in poetry, encourages a multitude of interpretations and reinterpretations. Experienced designers may get new insights, see new relationships, make new inferences from reexamining their sketches, a positive cycle that leads to new design ideas, followed by new sketches and new ideas [132–135]. Ambiguity can help designers innovate and escape fixation by allowing perceptual reorganization and consequent new insights, a pair of processes, one perceptual, finding new figures and relations, and one conceptual, finding new interpretations, termed "constructive perception" [133, 134, 161].

3.2.7 Glyphs: Simple geometric forms with related meanings

Diagrams and other forms of visual narratives are enhanced by the inclusion of a rich assortment of schematic visual forms such as dots, lines, arrows, circles, and boxes, whose meanings derive from and are constrained by their Gestalt or mathematical properties within the confines of a context. The meanings they support, entities, relations, asymmetric relations, processes, and collections, are abstract, so apply to many domains. They encourage the kind of abstractions needed for inference, analogy, generalization, transfer, and insight. They have analogs in other means of recording and communicating ideas, in language and in gesture, suggesting that they are elements of thought.

There are other abstract visual devices, infrequent in diagrams, but common in graphic novels and comics, lines suggesting motion, sound, fear, sweat, emotions, and more (e.g., [90]). Some of these, like the lines, boxes, and arrows discussed above, have meanings suggested by their forms. Motion lines, for example, seem to have developed as a short-hand or schematization of the perceptual blurring of viewed fast motion. Others, like hearts for love, are more symbolic. The concepts conveyed by the diagrammatic schematic forms are not as readily depictable as objects or even actions.

Those glyphs, such as dots, lines, arrows, frames, and circles, that enjoy a consensus of context-dependent meanings evident in production and comprehension seem to derive their meanings in ways similar to the ways pictograms establish meanings, overlapping features. Among the properties of lines is that they connect, just as relationships, abstract or concrete, connect. Among the properties of boxes is that they contain one set of things and separate those from other things. What is in the box creates a category, leaving open the basis for categorization to the creator or interpreter. The box implies that the things in the box are more related or similar to each other than to things out of the box. The box might contain a spatial region, a temporal slice, a set of objects. These mappings of meaning, the transfer of a few of the possible features from the object represented to the representing glyph, are partial and variable. The consequence is variability of meaning, allowing ambiguity and misconception. A case in point is uses of arrows, which map asymmetric relations. But there are a multitude of asymmetric relations, temporal order, causal order, movement path, and more. In well- designed diagrams, context can clarify, but there are all too many diagrams that are not well designed.

The concepts suggested by glyphs have parallels in language and gesture with the same trade-offs between abstraction and ambiguity. Think of words, notably spatial ones that parallel glyphs, like relationship or region or point. A romantic relationship? A mathematical relationship? Here, context will likely disambiguate, but not on all occasions. There is good reason why spatial concepts, whether diagrammatic or linguistic or gestural, have multiple meanings; they allow expression of kinds of meanings that apply to many domains.

Much has been said on what depictions do well: make elements, relations, and transformations of thought visible, apply human skills in visuospatial reasoning to abstract domains, encourage abstraction, enable inference, transfer, and insight, promote collaboration. But many concepts essential to thought and innovation are not visible. A key significance of glyphs is that they can visualize the invisible, entities, relations, forces, networks, trees, and more.

4 Processing and Designing Diagrams

4.1 Processing Diagrams

Good design must take into account the information-processing habits and limitations of human users (e.g., [18, 73, 74, 108, 120, 160]). The page is flat, as is the visual information captured by the retina. Reasoning from 3D diagrams is far more difficult than reasoning from 2D diagrams whether depictive (e.g., [44]) or conceptual (e.g., [118]). Language, visual search, and reasoning are sequential and limited, so that continuous animations of explanatory information can cause difficulties (e.g., [2, 3, 52, 54, 115, 149]).

Ability matters. Spatial ability is not a unitary factor, and some aspects of spatial thinking, especially performing mental transformations and integrating figures, matter for some situations and others for others (e.g., [53, 56, 75, 134]). Different spatial, and undoubtedly conceptual, abilities are needed for different kinds of tasks and inferences that involve diagrams.

Expertise matters. It can trade off with ability. As noted, diagrams, like language, are incomplete and can be abstract, requiring filling in, bridging inferences. Domains include implicit or explicit knowledge that allows bridging, encouraging correct interpretations and discouraging incorrect ones. The significance of domain knowledge was illustrated in route maps and holds a fortiori in more technical domains (e.g., [26]).

Working memory matters. Although, as advertised, external representations relieve working memory, they do not eliminate it. Typically, diagrams are used for comprehension, inference, and insight. All involve integrating or transforming the information in diagrams, processes that take place in the mind, in working memory. Imagine multiplying two three- digit numbers, even when the numbers are before your eyes, without being able to write down the product of each step (see [121]).

Structure matters. When diagrams are cluttered with information, finding and integrating the relevant information takes working memory capacity. Schematization, that is, removing irrelevant details, exaggerating, perhaps distorting, relevant ones, even adding relevant but invisible information, can facilitate information processing in a variety of ways. Aerial photographs make poor driving maps. Schematization can reduce irrelevancies that can clutter, thereby allowing attention to focus on important features, increasing both speed and accuracy of information processing (e.g., [32, 126, 149]).

Sequencing matters. Conveying sequential information, important in history, science, engineering, and everyday life, poses special challenges. Sometimes a sequence of steps can be shown in a single diagram; Minard's famous diagram of Napoleon's unsuccessful campaign on Russia is a stellar example. Time lines of historical events are another common successful example. Depicting each step separately and connecting them, often using frames and arrows, is another popular solution, from Egyptian tomb paintings showing the making of bread to Lego instructions. Both separating and connecting require careful design. People segment continuous organized action sequences into meaningful units that connect perception and action, by changes in scene, actor, action, and object (e.g., [9, 10, 153, 163, 177, 178]). A well-loved solution to showing processes that occur over time is to use animations. Animations are attractive because they appear to conform to the Congruity Principle: They use change in time to show change in time, a mentally congruent relation [149]. However, as we have just seen, the mind often segments continuous processes into steps (e.g., [153, 176]), suggesting that step-by-step presentation is more congruent to the way the mind understands and represents continuous organized action than continuous presentation. The segmentation of routes by turns and object assembly by actions provide illustrative examples. Animations can suffer two other shortcomings: They are often too fast and too complex to take in, violating the Apprehension Principle, and they show, but do not explain [159]. Even more than in static diagrams, visualizing the invisible, causes, forces, and the like, is difficult in animations. And, indeed, a broad range of kinds of animations for a broad range of content have not proved to be superior to static graphics (e.g., [89, 129, 149, 156]).

Multi-media matters. Depictions and language differ in many ways, some discussed earlier, among them, expressiveness, abstraction, constraints, accessibility to meaning (e.g., [130]). As we have seen, many meanings may be easier to convey through diagrams, but diagrams can also mislead. Diagrams usually contain words or other symbolic information; the visuals, even augmented with glyphs, may not be sufficient. Maps need names of countries, towns, or streets. Network diagrams need names of the nodes and sometimes the edges. Economic graphs need labels and numeric scales to denote years or countries or financial indices. Anatomical diagrams need names of muscles and bones. But diagrams often need more than labels and scales. Although arrows can indicate causes and forces, the specific forces and causes may need language. In addition, redundancy often helps (e.g., [2, 3, 88]). Just as diagrams need to be carefully designed to be effective, so does language.

4.2   Designing Diagrams

The previous analyses of place and form in diagrams were based on historical and con- temporary examples that have been invented and reinvented across time and space. They have been refined by the generations through informal user testing in the wild. The analyses provide a general guideline for designing effective diagrams: Use place in space and forms of marks to convey the kinds of meanings that they more naturally convey. For example, use the vertical for evaluative dimensions, mapping increases upwards. Use the horizontal for neutral dimensions, especially time, mapping increases in reading order. Use dots for entities, lines for relations, arrows for asymmetric relations, boxes for collections.

Disambiguate when context is not sufficient. Although helpful, these are general guidelines often not sufficient for specific cases. The previous analyses of the evolution and refinement of diagrams also suggest methods to systematically develop more specific guidelines when needed, to formalize the natural user testing cycle—produce, use, refine—and bring it into the laboratory by turning users into designers. One project used this procedure for developing cognitive design principles for assembly instructions [156]. Students first assembled a TV cart using the photograph on the box. They then designed instructions to help others assemble the cart. Other groups of students used and rated the previous instructions. Analysis of the highly rated and effective instructions revealed the following cognitive principles: Use one diagram per step, segment one step per part, show action, show perspective of action, and use arrows and guidelines to show attachment and action. A computer algorithm was created to construct assembly diagrams using these guidelines, and the resulting visual instructions led to better performance than those that came with the TV cart. These cognitive principles apply not just to assembly diagrams but more broadly to visual explanations of how things behave or work. Moreover, the cycle of producing, using, and refining diagrams is productive in improving diagrams even with a single person [65, 66, 82, 161].

5   Diagrams as a Microcosm of Cognition

Diagrams and other depictions are expressions and communications of thought, a class that includes gesture, action, and language. In common with gesture and action, diagrams use place and form in space to convey meanings, concrete and abstract, quite directly. This paper has presented an analysis and examples of the ways that place and form create meanings, an analysis that included the horizontal, vertical, center–periphery, and pictorial organization of the page as well as the dots, lines, arrows, circles, boxes, and likenesses depicted on a page. In combination, they enable creating the vast variety of visual expressions of meaning, pictures, maps, mandalas, assembly instructions, highway signs, architectural plans, science and engineering diagrams, charts, graphs, and more. Gestures also use many of these features of meaning, but they are more schematic and fleeting; diagrams can be regarded as the visible traces of gestures just as gesturing can be regarded as drawing pictures in the air.

The foundations of diagrams lie in actions in space. People have always organized things and spaces to serve their ends: securing, storing, and preparing food, making and using artifacts, designing shelter, navigating space. The consequences of these actions are the creation of simple geometric patterns in space, patterns that are good gestalts, and that are readily recognized. The patterns invite abstract interpretations: Groups signal similar features or related themes, orders signal dimensions or continua, distributions signal one-to-one or one- to-many correspondences. The creation and interpretation of these patterns form the rudiments of abstract thought: categories, relationships, orderings, hierarchies, dimensions, and counting (e.g., [30, 37, 43, 50, 62, 78]). The spatial patterns can be manipulated by the hands or by the mind (e.g., [123, 152]) to create further abstractions; they form spatial-action representations for the abstractions that underlie the feats of the human mind, a three-way interaction that can be termed *spraction*. Spractions, then, are actions in space, whether on objects or as gestures, that create abstractions in the mind and patterns in the world, intertwined so that one primes the others. Like language, spractions support and augment cognition and action; unlike language, they do so silently and directly. The arrangements and organizations used to design the world create diagrams in the world: The designed world is a diagram.

Acknowledgements The author is indebted to many colleagues, collaborators, and commentators, including Maneesh Agrawala, Jon Bresman, Herb Clark, Danny Cohen, Jim Corter, Stu Card, Felice Frankel, Nancy Franklin, Pat Hanrahan, Mary Hegarty, Julie Heiser, Angela Kessell, Paul Lee, Julie Morrison, Jeff Nickerson, Jane Nisselson, Laura Novick, Ben Shneiderman, Penny Small, Masaki Suwa, Holly Taylor, Jeff Zacks, and Doris Zahner. The author is also indebted to the following grants for facilitating the research andor preparing the manu- script: National Science Foundation HHC 0905417, IIS-0725223, IIS-0855995, and REC 0440103, the Stanford Regional Visualization and Analysis Center, and Office of Naval Research NOOO14-PP-1-O649, N000140110717, and N000140210534.

References

1. Agrawala, M., & Stolte, C. (2001). Rendering effective route maps: Improving usability through generalization. Proceedings of SIGGRAPH 2001 (pp. 241–250). New York: ACM.
2. Ainsworth, S. (2008a). How do animations influence learning? In D. H. Robinson & G. Schraw (Eds.), Recent innovations in educational technology that facilitate student learning (pp. 37–67). Charlotte, NC: Information Age Publishing.
3. Ainsworth, S. (2008b). How should we evaluate multimedia learning environments?. In J.-F. Rouet, R. Lowe, & W. Schnotz (Eds.), Understanding multimedia comprehension. New York: Springer.

4. Alperson-Afil, N., Sharon, G., Kislev, M., Melamed, Y., Zohar, I., Ashkenazi, S., Rabinovich, R., Biton, R., Werker, E., Hartman, G., Feibel, C., & Goren-Inbar, N. (2009). Spatial organization of hominin activities at Gesher Benot Ya'aqov, Israel. Science, 326, 1677–1680.
5. Argyle, M. (1988). Bodily communication. London: Routledge.
6. Arnheim, R. (1974). Art and visual perception: A psychology of the creative eye. Berkeley: University of California Press.
7. Arnheim, R. (1977). The dynamics of architectural form. Berkeley: University of California Press.
8. Arnheim, R. (1988). The power of the center: A study of composition in the visual arts. Berkeley: University of California Press.
9. Barker, R. G. (1963). The stream of behavior as an empirical problem. In R. G. Barker (Ed.), The stream of behavior (pp. 1–22). New York: Appleton-Century-Crofts.
10. Barker, R. G., & Wright, H. F. (1954). Midwest and its children: The psychological ecology of an American town. Evanston, IL: Row, Peterson and Company.
11. Bendix, F., Kosara, R., & Hauser, H. (2006). Parallel sets: Interactive exploration and visual analysis of categori- cal data. IEEE Transactions on Visualization and Computer Graphics, 12, 558–568.
12. Bertin, J. (1981). Graphics and graphic-information-processing. New York: Walter de Gruyter.
13. Boroditsky, L. (2000). Metaphoric structuring: Understanding time through spatial metaphors. Cognition, 75, 1–28.
14. Boroditsky, L. (2001). Does language shape thought?: Mandarin and English speakers' conceptions of time. Cognitive Psychology, 43, 1–22.
15. Brown, L. (1979). The story of maps. New York: Dover.
16. Card, S. K., Mackinlay, J. D., & Shneiderman, B. (1999). Readings in information visualization: Using vision to think. San Francisco: Morgan Kaufman.
17. Carey, S., Diamond, R., & Woods, B. (1980). Development of face recognition—A maturational component? Developmental Psychology, 16, 257–269.
18. Carpenter, P. A., & Shah, P. (1998). A model of the perceptual and conceptual processes in graph comprehension. Journal of Experimental Psychology: Applied, 4, 75–100.
19. Casasanto, D. (2009). Embodiment of abstract concepts: Good and bad in right- and left-handers. Journal of Experimental Psychology: General, 138, 351–367.
20. Chatterjee, A. (2001). Language and space: Some interactions. Trends in Cognitive Science, 5, 55–61.
21. Chatterjee, A. (2002). Portrait profiles and the notion of agency. Empirical Studies of the Arts, 20, 33–41.
22. Chokron, S., & De Agostini, M. (2000). Reading habits influence aesthetic preference. Cognitive Brain Research, 10, 45–49.
23. Chokron, S., & De Agostini, M. (2002). The influence of handedness on profile and line drawing directionality in children, young, and older normal adults. Brain, Cognition, and Emotion, 48, 333–336.
24. Clark, H. H. (1973). Space, time, semantics, and the child. In T. E. Moore (Ed.), Cognitive development and the acquisition of language (pp. 27–63). New York: Academic Press.
25. Clark, H. H. (1996). Using language. Cambridge, England: Cambridge University Press.
26. Committee on Support for Thinking Spatially. (2006). Learning to think spatially. Washington, DC: The National Academies Press.
27. Cooper, W. E., & Ross, J. R. (1975). World order. In R. E. Grossman, L. J. San, & T. J. Vance (Eds.), Papers from the parasession on functionalism (pp. 63–111). Chicago: Chicago Linguistic Society.
28. Corter, J. E., Rho, Y.-J., Zahner, D., Nickerson, J. V., & Tversky, B. (2009). Bugs and biases: Diagnosing misconceptions in the understanding of diagrams. Proceedings of the Cognitive Science Society. Mahwah, NJ.
29. De Loache, J. S. (2004). Becoming symbol-minded. Trends in Cognitive Science, 8, 66–70.

30. Dehaene, S. (1997). The number sense: How the mind creates mathematics. New York: Oxford University Press.
31. Donald, M. (1991). Origins of the modern mind. Cambridge, MA: Harvard University Press.
32. Dwyer, F. M. (1978). Strategies for improving visual learning. State College, PA: Learning Services.
33. Elkin, J. (1999). The domain of images. Ithaca, NY: Cornell University Press.
34. Emmorey, K., Tversky, B., & Taylor, H. A. (2000). Using space to describe space: Perspective in speech, sign, and gesture. Journal of Spatial Cognition and Computation, 2, 157–180.
35. Engle, R. A. (1998). Not channels but composite signals: Speech, gesture, diagrams and object demonstrations are integrated in multimodal explanations. In M. A. Gernsbacher & S. J. Derry (Eds.), Proceedings of the twentieth annual conference of the cognitive science society. Mahwah, NJ: Erlbaum.
36. Fontana, D. (2005). Meditating with mandalas. London: Duncan Baird Publishers.
37. Frank, M. C., Everett, D. L., Fedorenko, E., & Gibson, E. (2008). Number as a cognitive technology: Evidence rom Piraha language and cognition. Cognition, 108, 819–824.
38. Franklin, N., & Tversky, B. (1990). Searching imagined environments. Journal of Experimental Psychology: General, 119, 63–76.
39. Gattis, M. (2002). Structure mapping in spatial reasoning. Cognitive Development, 17, 1157–1183.
40. Gattis, M. (2004). Mapping relational structure in spatial reasoning. Cognitive Science, 28 (4), 589–610.
41. Gattis, M., & Holyoak, K. J. (1996). Mapping conceptual to spatial relations in visual reasoning. Journal of Experimental Psychology: Learning, Memory, and Cognition, 22, 1–9.
42. Gelb, I. (1963). A study of writing, 2nd ed. Chicago: University of Chicago Press.
43. Gelman, R., & Gallistel, R. (1986). The child's understanding of number. Cambridge, MA: Harvard.
44. Gobert, J. D. (1999). Expertise in the comprehension of architectural plans. In J. Gero & B. Tversky (Eds.),Visual and spatial reasoning in design (pp. 185–205). Sydney, Australia: Key Centre of Design Computing and Cognition.
45. Goel, V. (1995). Sketches of thought. Cambridge, MA: MIT Press.
46. Goldin-Meadow, S. (2003). Hearing gesture: How our hands help us think. Cambridge, MA: Harvard University Press.
47. Gombrich, E. (1961). Art and illusion. Princeton, NJ: Princeton University Press.
48. Gombrich, E. (1979). The sense of order: A study in the psychology of decorative art. Oxford, England: Phaidon.
49. Goodman, N. (1978). Languages of art: An approach to a theory of symbols. New York: Bobbs-Merrill
50. Gordon, P. (2004). Numerical cognition without words. Evidence from Amazonia. Science, 306, 496–499.
51. Hartshorne, C., & Weiss, P. (Eds.) (1960). Collected papers of Charles Sanders Peirce. Cambridge, MA: Harvard University Press.
52. Hegarty, M. (1992). Mental animation: Inferring motion from static displays of mechanical systems. Journal of Experimental Psychology: Learning, Memory, and Cognition, 18, 1084–1102.
53. Hegarty, M. (2010). Components of spatial intelligence. In B. H. Ross (Ed.) The psychology of learning and motivation. Vol. 52. Pp. 265–297. San Diego: Academic Press.
54. Hegarty, M., Kriz, S., & Cate, C. (2003). The roles of mental animations and external animations in understanding mechanical systems. Cognition and Instruction, 2, 325–360.
55. Hegarty, P., Lemieux, A. F., & McQueen, G. (2010). Graphing the order of the sexes: Constructing, recalling, interpreting, and putting the self in gender difference graphs. Journal of Personality and Social Psychology, 93, 375–391.
56. Hegarty, M., & Waller, D. (2006). Individual differences in spatial abilities. In P. Shah & A. Miyake (Eds.),Handbook of visuospatial thinking. Cambridge, England: Cambridge University Press.

57. Heiser, J., & Tversky, B. (2006). Arrows in comprehending and producing mechanical diagrams. Cognitive Science, 30, 581–592.
58. Hochberg, J., & Brooks, V. (1962). Pictorial recognition as an unlearned ability: A study of one child's performance. American Journal of Psychology, 75, 624–628.
59. Hochberg, J., & Galper, R. E. (1967). Recognition of faces: I. An exploratory study. Psychonomic Science, 6, 156–163.
60. Horn, R. E. (1998). Visual language. Bainbridge Island, WA: MacroVu, Inc.
61. Howard, I. P. (1982). Human visual orientation. New York: Wiley.
62. Hughes, M. (1986). Children and number: Difficulties in learning mathematics. Oxford, England: Blackwell.
63. Inselberg, A., & Dimsdale, B. (1990). Parallel coordinates: A tool for visualizing multi-dimensional geometry,Visualization '90. Proceedings of the First IEEE Conference on Visualizations, 23–26, 361–378.
64. Ittelson, W. H. (1996). Visual perception of markings. Psychonomic Bulletin & Review, 3, 171–187.
65. Karmiloff-Smith, A. (1979). Micro-and macro-developmental changes in language acquisition and other representational systems. Cognitive Science, 3, 91–118.
66. Karmiloff-Smith, A. (1990). Constraints on representational change: Evidence from children's drawing. Cognition, 34, 1–27.
67. Kellogg, R. (1969). Analyzing children's art. Palo Alto, CA: National Press.
68. Kendon, A. (2004). Gesture: Visible action as utterance. Cambridge, England: Cambridge University Press.
69. Kessell, A. M., & Tversky, B. (2006). Using gestures and diagrams to think and talk about insight problems. Proceedings of the Meetings of the Cognitive Science Society.
70. Kessell, A. M., & Tversky, B. (2008). Cognitive methods for visualizing space, time, and agents. In G. Stapleton, J. Howse, & J. Lee (Eds.), Theory and application of diagrams. Dordrecht, The Netherlands: Springer.
71. Kessell, A. M., & Tversky, B. (2009). Thinking about cycles: Producing sequences but preferring circles.
72. Kirsh, D. (1995). The intelligent use of space. Artificial Intelligence, 73, 31–68.
73. Kosslyn, S. M. (1989). Understanding charts and graphs. Applied Cognitive Psychology, 3, 185–223.
74. Kosslyn, S. M. (2006). Graph design for the eye and the mind. Oxford, England: Oxford University Press.
75. Kozhevnikov, M., Kosslyn, S., & Shephard, J. (2005). Spatial versus object visualizers: A new characterization of cognitive style. Memory and Cognition, 33, 710–726.
76. Kulvicki, J. (2006). Pictorial representation. Philosophy Compass, 10, 1–12.
77. Lakoff, G., & Johnson, M. (1980). Metaphors we live by. Chicago: University of Chicago Press.
78. Lakoff, G., & Nunez, R. (2000). Where mathematics comes from. New York: Basic Books.
79. Landy, D., & Goldstone, R. L. (2007a). Formal notations are diagrams: Evidence from a production task. Memory & Cognition, 35, 2033–2040.
80. Landy, D., & Goldstone, R. L. (2007b). How abstract is symbolic thought? Journal of Experimental Psychology: Learning, Memory, & Cognition, 33, 720–733.
81. Larkin, J. H., & Simon, H. A. (1987). Why a diagram is (sometimes) worth ten thousand words. Cognitive Science, 11, 65–99.
82. Lee, K., & Karmiloff-Smith, A. (1996). The development of external symbol systems: The child as a notator. In R. Gelman & T. Kit Fong Au (Eds.), Perceptual and cognitive develoment (pp. 185–211). San Diego: Academic Press.
83. Lee, P. U., & Tversky, B. (2005). Interplay between visual and spatial: The effects of landmark descriptions on comprehension of routesurvey descriptions. Spatial Cognition and Computation, 5 (2 & 3), 163–185.
84. Levinson, S. C. (2003). Space in language and cognition: Explorations in cognitive diversity. Cambridge, England: Cambridge University Press.

85. Maass, A., Pagani, D., & Berta, E. (2007). How beautiful is the goal and how violent is the fistfight? Spatial bias in the interpretation of human behavior Social Cognition, 25, 833–852.
86. Maass, A., & Russo, A. (2003). Directional bias in the mental representation of spatial events: Nature or culture? Psychological Science, 14, 296–301.
87. Mallery, G. (18931972). Picture writing of the American Indians. (Originally published by Government Printing Office). New York: Dover.
88. Mayer, R.E. (2001). Multimedia learning. Cambridge, England: Cambridge University Press.
89. Mayer, R. E., Hegarty, M., Mayer, S. Y., & Campbell, J. (2005). When passive media promote active learning: Static diagrams versus animation in multimedia instruction. Journal of Experimental Psychology: Applied, 11, 256–265.
90. McCloud, S. (1994). Understanding comics. New York: Harper Collins.
91. McManus, I. C., & Humphrey, N. (1973). Turning the left cheek. Nature, 243, 271–272.
92. McNeill, D. (1992). Hand and mind: what gestures reveal about thought. Chicago: University of Chicago Press.
93. McNeill, D. (2005). Gesture and thought. Chicago: University of Chicago Press.
94. Morikawa, K., & McBeath, M. (1992). Lateral motion bias associated with reading direction. Vision Research, 32, 1137–1141.
95. Murch, W. (2001). In the blink of an eye. Beverly Hills, CA: Silman-James Press.
96. Nachshon, I. (1985). Directional preferences in perception of visual stimuli. International Journal of Neuroscience, 25, 161–174.
97. Nachshon, I., Argaman, E., & Luria, A. (1999). Effects of directional habits and handedness on aesthetic preference for left and right profiles. Journal of Cross Cultural Psychology, 30, 106–114.
98. Netz, R. (1999). Linguistic formulae as cognitive tools. Pragmatics and Cognition, 7, 147–176.
99. Neurath, O. (1936). International picture language: The first rules of isotype. London: Kegan Paul, Trench,Trubner & Co., Ltd.
100. Nickerson, J. V., Corter, J. E., Tversky, B., Zahner, D., & Rho, Y.-J. (2008). The spatial nature of thought: Understanding information systems design through diagrams. In R. Boland, M. Limayem, & B. Pentland (Eds.), Proceedings of the 29th International Conference on Information Systems.
101. Norman, D. A. (1993). Things that make us smart. Reading, MA: Addison-Wesley.
102. Novick, L. R. (2001) Spatial diagrams: Key instruments in the toolbox for thought. In D. L. Medin (ed.), The psychology of learning and motivation, Vol. 40. (pp. 279–325). New York: Academic Press.
103. Novick, L. R., & Catley, K. M. (2007). Understanding phylogenies in biology: The influence of a Gestalt percep- tual principle. Journal of Experimental Psychology: Applied, 13, 197–223.
104. Novick, L. R., & Hurley, S. M. (2001). To matrix, network, or hierarchy, that is the question. Cognitive Psychology, 42, 158–216.
105. Novick, L. R., Hurley, S. M., & Francis, M. (1999). Evidence for abstract, schematic knowledge of three spatial diagram representations. Memory and Cognition, 27, 288–308.
106. Novick, L. R., Shade, C. K., & Catley, K. M. (2011). Linear versus branching depictions of evolutionary history: Implications for design. Topics in Cognitive Science, 3, 536–539.
107. Paivio, A. (1986). Mental representations. New York: Oxford.
108. Pinker, S. (1990). A theory of graph comprehension. In R. Freedle (Ed.), Artificial intelligence and the future of testing (pp. 73–126). Hillsdale, NJ: Erlbaum.
109. Rock, I. (1973). Orientation and form. New York: Academic Press.
110. Sakhuja, T., Gupta, G. C., Singh, M., & Vaid, J. (1996). Reading habits affect asymmetries in facial affect judgements: A replication. Brain and Cognition, 32, 162–165.
111. Santiago, J., Lupianez, J., Perez, E., & Funes, M. J. (2007). Time (also) flies from left to right. Psychonomic Bulletin & Review, 14, 512–516.
112. Santiago, J., Roman, A., Ouellet, M., Rodríguez, N., & Perez-Azor, P. (2008). In hindsight, life flows from left to right. Psychological Research. DOI: 10.1007/s00426-008-0220-0

113. Scaife, M., & Rogers, Y. (1996). External cognition: How do graphical representations work? International Journal of Human-Computer Studies, 45, 185–213.
114. Schmandt-Besserat, D. (1992). Before writing, Volume 1: From counting to cuneiform. Austin: University of Texas Press.
115. Schnotz, W., & Lowe, R. K. (2007). A unified view of learning from animated and static graphics. In R. K. Lowe & W. Schnotz (Eds.), Learning with animation: Research and design implications. New York: Cambridge University Press.
116. Schon, D. A. (1983). The reflective practitioner. New York: Harper Collins.
117. Schwartz, D. (1995). The emergence of abstract representations in dyad problem solving. The Journal of the Learning Sciences, 4, 321–354.
118. Shah, P., & Carpenter, P. A. (1995). Conceptual limitations in comprehending line graphs. Journal of Experimental Psychology: General, 124, 43–61.
119. Shah, P., & Freedman, E. (2010). Bar and line graph comprehension: An interaction of top-down and bottom-up processes. Topics in Cognitive Science.
120. Shah, P., Freedman, E., & Vekiri, I. (2005). The comprehension of quantitative information in graphical displays. In P. Shah & A. Miyake (Eds.), The Cambridge handbook of visuospatial thinking (pp. 426–476). New York: Cambridge University Press.
121. Shah, P., Mayer, R. E., & Hegarty, M. (1999). Graphs as aids to knowledge construction: Signaling techniques for guiding the process of graph comprehension. Journal of Educational Psychology, 91, 690–702.
122. Shepard, R. N., & Hurwitz, S. (1984). Upward direction, mental rotation, and discrimination of left and right urns in maps. Cognition, 18, 161–193.
123. Shepard, R. N., & Podgorny, P. (1978). Cognitive processes that resemble perceptual processes. In W. Estes (Ed.), Handbook of learning and cognitive processes, Vol. 5. (pp. 189–237). Hillsdale, NJ: Erlbaum.
124. Sholl, M. J. (1987). Cognitive maps as orienting schema. Journal of Experimental Psychology: Learning, Memory and Cognition, 13, 615–628.
125. Small, J. P. (1997). Wax tablets of the mind. New York: Routledge, Paul.
126. Smallman, H. S., St. John, M., Onck, H. M., & Cowen, M. (2001). "Symbicons": A hybrid symbology that combines the best elements of symbols and icons. Proceedings of the 54th annual meeting of the human factors and ergonomics society (pp. 110–114).
127. Smith, M. C., & Magee, L. E. (1980). Tracing the time course of picture-word processing. Journal of Experimental Psychology: General, 109, 373–392.
128. Stafford, B. (2007). Echo objects: The cognitive work of images. Chicago: University of Chicago Press.
129. Stasko, J., & Lawrence, A. (1998). Empirically assessing algorithm animations as learning aids. In J. Stasko, J. Domingue, M. H. Brown, & B. A. Price (Eds.), Software visualization (pp. 419–438). Cambridge, MA: MIT Press.
130. Stenning, K., & Oberlander, J. (1995). A cognitive theory of graphical and linguistic reasoning: Logic and implementation. Cognitive Science, 19, 97–140.
131. Suitner, C., & Maass, A. (2007). Positioning bias in portraits and self-portraits: Do female artists make different choices? Empirical Studies of the Arts, 25, 71–95.
132. Suwa, M., & Tversky, B. (1996). What architects see in their sketches: Implications for design tools. In Human factors in computing systems: Conference companion (pp. 191–192). New York: ACM.
133. Suwa, M., & Tversky, B. (2001). Constructive perception in design. In J. S. Gero & M. L. Maher (Eds.), Computational and cognitive models of creative design V (pp. 227–239). Sydney, Australia: University of Sydney.
134. Suwa, M., & Tversky, B. (2003). Constructive perception: A skill for coordinating perception and conception. In Proceedings of the Cognitive Science Society Meetings.
135. Suwa, M., Tversky, B., Gero, J., & Purcell, T. (2001). Seeing into sketches: Regrouping parts encourages new interpretations. In J. S. Gero, B. Tversky, & T. Purcell (Eds.), Visual and spatial reasoning in design(pp. 207–219). Sydney, Australia: Key Centre of Design Computing and Cognition.

136. Talmy, L. (1983). How language structures space. In H. L. Pick Jr & L. P. Acredolo (Eds.), Spatial orientation: Theory, research and application (pp. 225–282). New York: Plenum.
137. Talmy, L. (2000). Toward a cognitive semantics, Vols 1 & 2. Cambridge, MA: MIT Press.
138. Taylor, H. A., & Tversky, B. (1992a). Descriptions and depictions of environments. Memory and Cognition, 20, 483–496.
139. Taylor, H. A., & Tversky, B. (1992b). Spatial mental models derived from survey and route descriptions. Journal of Memory and Language, 31, 261–282.
140. Taylor, H. A., & Tversky, B. (1997). Indexing events in memory: Evidence for index preferences. Memory, 5, 509–542.
141. Tolchinsky-Landsmann, L., & Levin, I. (1987). Writing in four- to six-year-olds: Representation of semantic and phonetic similarities and differences. Journal of Child Language, 14, 127–144.
142. Tomasello, M. (2008). Origins of human communication. Cambridge, MA: MIT Press.
143. Tufte, E. R. (1983). The visual display of quantitative information. Cheshire, CT: Graphics Press.
144. Tufte, E. R. (1990). Envisioning information. Cheshire, CT: Graphics Press.
145. Tufte, E. R. (1997). Visual explanations. Cheshire, CT: Graphics Press.
146. Tversky, B. (1981). Distortions in memory for maps. Cognitive Psychology, 13, 407–433.
147. Tversky, B. (1995). Cognitive origins of graphic conventions. In F. T. Marchese (Ed.), Understanding images (pp. 29–53). New York: Springer-Verlag.
148. Tversky, B. (2000). Some ways that maps and graphs communicate. In C. Freksa, W. Brauer, C. Habel, & K. F. Wender (Eds.), Spatial cognition II: Integrating abstract theories, empirical studies, formal methods, and practical applications (pp. 72–79). New York: Springer.
149. Tversky, B. (2001). Spatial schemas in depictions. In M. Gattis (Ed.), Spatial schemas and abstract thought (pp. 79–111). Cambridge, MA: MIT Press.
150. Tversky, B. (2003). Navigating by mind and by body. In C. Freksa, W. Brauer, C. Habel, & K. F. Wender (Eds.), Spatial Cognition III: Routes and navigation, human memory and learning, spatial representation and spatial reasoning (pp. 1–10). Berlin: Springer Verlag.
151. Tversky, B. (2004). Semantics, syntax, and pragmatics of graphics. In K. Holmqvist & Y. Ericsson (Eds.), Language and visualisation (pp. 141–158). Lund, Sweden: Lund University Press.
152. Tversky, B. (2005). Functional significance of visuospatial representations. In P. Shah & A. Miyake (Eds.), Handbook of higher-level visuospatial thinking (pp. 1–34). Cambridge, England: Cambridge University Press.
153. Tversky, B., Agrawala, M., Heiser, J., Lee, P. U., Hanrahan, P., Phan, D., Stolte, C., & Daniele, M.-P. (2007). Cognitive design principles for generating visualizations. In G. Allen (Ed.), Applied spatial cognition: From research to cognitive technology (pp. 53–73). Mahwah, NJ: Erlbaum.
154. Tversky, B., Corter, J. E., Nickerson, J. V., Zahner, D., & Rho, Y. J. (2008a). Transforming descriptions and diagrams to sketches in information system design. In G. Stapleton, J. Howse, & J. Lee (Eds.), Theory and application of diagrams. Dordrecht, The Netherlands: Springer.
155. Tversky, B., Heiser, J., Lee, P., & Daniel, M.-P. (2009). Explanations in gesture, diagram, and word. In K. R. Coventry, T. Tenbrink, & J. A. Bateman (Eds.), Spatial language and dialogue (pp. 119–131). Oxford, England: Oxford University Press.
156. Tversky, B., Heiser, J., Lozano, S., MacKenzie, R., & Morrison, J. (2007). Enriching animations. In R. Lowe & W. Schnotz (Eds.), Learning with animation. Cambridge, England: Cambridge University Press.
157. Tversky, B., Kugelmass, S., & Winter, A. (1991). Cross-cultural and developmental trends in graphic produc- tions. Cognitive Psychology, 23, 515–557.
158. Tversky, B., & Lee, P. U. (1998). How space structures language. In C. Freksa, C. Habel, & K. F. Wender (Eds.), Spatial cognition: An interdisciplinary approach to representation and processing of spatial knowledge (pp. 157–175). Berlin: Springer-Verlag.
159. Tversky, B., & Lee, P. U. (1999). Pictorial and verbal tools for conveying routes. In C. Freksa & D. M. Mark (Eds.), Spatial information theory: Cognitive and computational foundations of geographic information science (pp. 51–64). Berlin: Springer.

160. Tversky, B., Morrison, J. B., & Betrancourt, M. (2002). Animation: Can it facilitate? International Journal of Human Computer Studies, 57, 247–262.
161. Tversky, B., & Suwa, M. (2009). Thinking with sketches. In A. Markman (Ed.), Tools for innovation. Oxford, England: Oxford University Press.
162. Tversky, B., Suwa, M., Agrawala, M., Heiser, J., Stolte, C., Hanrahan, P., Phan, D., Klingner, J., Daniel, M.-P., Lee, P., & Haymaker, J. (2003). Sketches for design and design of sketches. In U. Lindemann (Ed.), Human behavior in design: Individuals, teams, tools (pp. 79–86). Berlin: Springer.
163. Tversky, B., Zacks, J. M., & Hard, B. M. (2008). The structure of experience. In T. Shipley & J. M. Zacks (Eds.), Understanding events. (pp. 436–464). Oxford, England: Oxford University.
164. Tversky, B., Zacks, J., Lee, P. U., & Heiser, J. (2000). Lines, blobs, crosses, and arrows: Diagrammatic commu- nication with schematic figures. In M. Anderson, P. Cheng, & V. Haarslev (Eds.), Theory and application of diagrams (pp. 221–230). Berlin: Springer.
165. Vaid, J., & Singh, M. (1989). Asymmetries in the perception of facial effects: Is there an influence of reading habits? Neuropsychologia, 27, 1277–1286.
166. Vaid, J., Singh, M., Sakhuja, T., & Gupta, G. C. (2002). Stroke direction asymmetry in figure drawing: Influence of handedness and reading writing habits. Brain and Cognition, 48, 597–602.
167. Wainer, H. (1992). Understanding graphs and tables. Educational Researcher, 21, 14
168. Walton, K. (1990). Mimesis as make-believe. Cambridge, MA: Harvard University Press.
169. Ware, C. (2008). Visual thinking for design. Burlington, MA: Morgan Kaufman.
170. Westendorp, P., & van der Waarde, K. (2000 2001). Icons: Support or substitute? Information Design Journal, 10, 91–94.
171. Wiesner, D. (2001). The three pigs. New York: Clarion.
172. Wilkins, D. P. (1997). Alternative representations of space: Arrernte narratives in sand and sign. In M. Biermans & J.v.d. Weijer (Eds.), Proceedings of the CLS opening academic year 1997–1998 (pp. 133–162). Nijmegen, The Netherlands: Nijmegen Tilburg Center for Language Studies.
173. Winn, W. D. (1987). Charts, graphs and diagrams in educational materials. In D. M. Willows & H. A. Haughton (Eds.), The psychology of illustration. New York: Springer-Verlag.
174. Woodward, D., & Lewis, G. M. (1998). History of cartography. Vol. 2. Book 3: Cartography in the traditional Africa, America, Arctic, Australian, and Pacific societies. Chicago: University of Chicago Press.
175. Zacks, J., Levy, E., Tversky, B., & Schiano, D. J. (1998). Reading bar graphs: Effects of depth cues and graphical context. Journal of Experimental Psychology: Applied, 4, 119–138.
176. Zacks, J., & Tversky, B. (1999). Bars and lines: A study of graphic communication. Memory and Cognition, 27, 1073–1079
177. Zacks, J., & Tversky, B. (2001). Event structure in perception and conception. Psychological Bulletin, 127, 3–21.
178. Zacks, J., Tversky, B., & Iyer, G. (2001). Perceiving, remembering and communicating structure in events. Journal of Experimental Psychology: General, 136, 29–58.
179. Zebian, S. (2005). Linkages between number, concepts, spatial thinking, and directionality of writing: The SNARC effect and the REVERSE SNARC effect in English and monoliterates, biliterates, and illiterate Arabic speakers. Journal of Cognition and Culture, 5, 165–190.

Gryphon: A 'Little' Domain-Specific Programming Language for Diffusion MRI Visualizations

Jian Chen, Haipeng Cai, Alexander P. Auchus, and David H. Laidlaw

Abstract We present Gryphon, a 'little' domain-specific programming language (DSL) for visualizing diffusion magnetic resonance imaging (DMRI). A key contribution is its compositional approach to customizing visualizations for evolving analytical tasks. The language is designed for non-programmer, here brain scientists for exploratory studies. The semantics of Gryphon includes a simple set of keywords derived from brain scientists vocabulary while performing imaging tasks of mapping data to graphic marks such as color, shape, value, and size. A pilot study with two neuroscientists suggested that Gryphon was easy to learn, though some additional functions and interface components are needed to empower brain scientists.

J. Chen (✉)
Assistant professor, Computer Science and Electrical Engineering, University of Maryland
Baltimore County, Baltimore, MD, USA
e-mail: jichen@umbc.edu

H. Cai
Graduate student, School of Computing, University of Southern Mississippi,
Hattiesburg, MS, USA
e-mail: haipeng.cai@eagles.usm.edu

A.P. Auchus
Professor, Department of Neurology, University of Mississippi Medical Center,
Jackson, MS, USA
e-mail: aauchus@umbc.edu

D.H. Laidlaw
Professor, Computer Science Department, Brown University, Providence, RI, USA
e-mail: dhl@cs.brown.edu

W. Huang (ed.), *Handbook of Human Centric Visualization*,
DOI 10.1007/978-1-4614-7485-2_2, © Springer Science+Business Media New York 2014

1 Introduction

Brain researchers have long used diffusion magnetic resonance imaging (DMRI) techniques to study connectivity of brain structures. DMRI is a MRI technique that measures motion of water molecules in tissue [21]. Experimental evidence has shown that water diffusion is anisotropic in organized tissues, such as white matter and muscle, and that reconstructing the orientation and curvature of white matter can provide detailed information about neural pathways. The curves (or fibers) are portrayed graphically using streamline tracing algorithms or glyphs such as hyperstreamlines initialized at seed points to reveal fiber tracts. Tracts following similar directions form fiber bundles [22]. Numerous studies have shown that tract-based analysis of the brain white matter can offer a deeper insight into brain structures.

Understanding the structures of fiber tracts, however, has proven difficult for several reasons that lead to the needs for effective visualization and interaction techniques. First, the advances in image capturing and processing techniques permit the display of human brain features at millimeter scales, generating highly dense visualizations. A whole-brain tractography can have about 10,000 tubes within the volume of a human head. Without effective visualizations, brain scientists are left with occluded images that impede their view of the data. Second, grouping, trimming, and labeling of the numerically computed fibers are needed for the tractography to provide information about the connections between brain regions. Fiber tracking reliability can further vary with imaging resolution, noise, and patient orientation as well as decreased anisotropy that occur with disease. Therefore, interaction techniques are also crucial for performing some operations to reach certain regions of interest (ROI) in the brain for the tasks at hand [1].

One approach to addressing the sheer data complexity in DMRI is to provide a large number of geometric processing, registration, segmentation, and clustering algorithms to improve data analysis [6]. Further coupling these algorithms with graphical user interfaces (GUIs) made possible more efficient data queries and analysis workflow [32]. This paper takes a complementary approach to focus on a programming language where the brain scientists can customize visualization that has good visual encodings to convey more precise information. In general, relying the same visualization for all needed tasks is impossible. For example, streamtubes are not always good visual representations to reduce clutter, though they improve orientation perception [12]. Tensor shapes, sizes, and their spatial locations also influence legibility of the visual displays. If visual encoding is not carefully constructed and applied to data, the reader may become misled by the visualization [28].

We describe Gryphon, a domain-specific programming language that lets brain scientists quickly compose a visualization tailored to their brain DMRI data. We report a prototypical implementation in this work exemplified in Fig. 1. A Gryphon program has a sequence of steps, each of which carries out a single high-level visualization operation, e.g., grouping, aggregation, and other modification on

```
LOAD "/home/lucy/clusteredBundles.data"
Select "CC,CST,IFO"
Update size BY FA IN "CST"
Update depth BY color IN "CC"
Update shape BY ribbon IN "IFO"
```

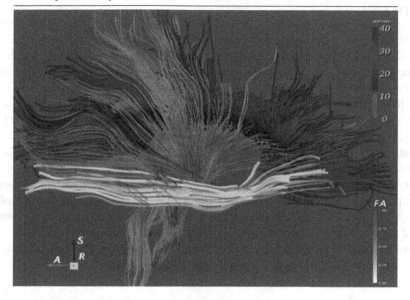

Fig. 1　Mixed color encoding of various fiber bundles

graphical elements such as color, size, value, shape, etc. Our language design makes three contributions: (1) a compositional approach to understanding visual encoding in scientific visualizations; (2) the design process to capture the domain-specific semantics to design this (we hope) easy-to-use "little" language; and (3) the language, Gryphon, itself to exemplify our design methods.

2　Background and Related Work

2.1　DMRI Visualization Toolkits

The first approach to addressing the sheer data complexity in DMRI is to design data processing, registration, segmentation, and clustering algorithms to improve data analysis. Some algorithms and toolkits are generic. SCIRun, MeVisLab, and Amira are general-purpose tools that provide comprehensive workflows in which users can directly manipulate the data [16, 17, 30]. Others are domain specific for DMRI data analysis. OpenWalnut supports diverse data processing and analyses of DMRI, computed tomography (CT), MRI, and functional MRI (fMRI), that on the one hand emphasizes speed and extensibility, similar to Amira and, on the other hand, handles fluid data processing workflow, similar to MeVisLab and SCIRun [13]. Diderot is perhaps the first language tool that provides a rather complete set of algorithms for tensor field processing. It also introduces parallel algorithm for more efficient computing [6].

MedINRIA and DTIStudio also include data-processing pipelines for DMRI visualization for tract generation, processing of tracking data, and visualizations [15, 32]. A set of tools on the neuroimaging tools and resources (NITRC) website are uploaded regularly to help problem solving in specific neurological diseases (http://www.nitrc.org). TractVis, 3D Slicer, and ExplorDMRI have distinct functionality for measuring uncertainty [18,25,33]. vIST/e (http://bmia.bmt.tue.nl/software/viste/) is useful for high-angular-resolution diffusion imaging analysis (HARDI) for glyphs, tube visualizations and seeding controls and fusing DMRI/HARDI visualizations. Users can interactively explore and observe both local tensor field information and global details about brain structures.

Our goal for Gryphon is to focus on visualization design driven by human visual perceptions, to provide a DMRI visualization tool similar to ColorBrewer for color design [14]. We wish to either make use of known good design practice or run a set of experiments to collect design principles that will guide our visualization tool design for brain scientists detecting information from imaging. The data processing methods can be preprocessing steps before data are visualized.

2.2 Visualization Languages

There are two ways to study visualization language: using language as a framework to map data to visual features to perception [3], and using language as a programming tool for computers to draw visualizations on the screen [19].

These two language aspects were formally linked since APT, which is the first automated graphic presentation tool to characterize graphic presentations, provide a set of criteria for deciding the role of each visible sign or symbol placed in a graphic by studying effectiveness and efficacy, and finally implement the programming language to 'recommend' good visualizations [19]. Wilkinson also describes a set of grammatical rules for defining graphics that mainly focus on the data analysis process [34], similar to Bertins semiology for graphics mark drawings [3].

Many graphics presentation tools have used APTs style of analysis for formal graphical languages. For example, SAGE expanded APT's data-centric visualization to a task-centric presentation system that responds to user queries for information and generates explanations of the changes occurring in quantitative molding systems such as project modeling and financial spreadsheets [29]. The aim of Processing, D^3, and ProtoVis is information visualization to design multivariate data visualizations or information graphs [4,5,26].

Our Gryphon programming language is similar to APT [19], D^3 [5], or Processing [26] in being built from the semiotic perspective: color, shape, value, texture work as sign vehicles for users to represent some parameter measurements. We expand the concept to scientific uses in DMRI visualizations to display spatial, physically-based, and spatial data versus their abstract and chosen representations [31]. Additionally, depth and occlusions in 3D that make it difficult to see all the data necessitate certain graphical tricks [35]. Another difficulty with adding 3D is that orientation becomes an issues: it makes easy to get lost providing the ability to move around in space. Thus, validating, measuring, and constructing effective 3D visualizations in general are still unsolved problems for dense dataset visualizations.

From the language design perspective, our Gryphon is also a 'little' language [2] that is simple for brain scientists to build, document, learn, and use. All instructions are captured following a human-centered computing design process to capture domain semantics. All programming instructions operate on small chunks of data of DMRI geometries.

Recently, Metoyer et al. report from an exploratory study a set of design implications for creating visualization languages and toolkits [20]. Their findings inform visualization language design through the way participants describe visualizations and their inclination to use ambiguous and relative, instead of definite and absolute, terms that can be refined later via a feedback loop. We have obtained similar findings and have designed Gryphon to reflect high-level semantics and spatial data expression.

3 Gryphon: A DMRI Visualization Language

3.1 Language Design

3.1.1 Linguistic Analysis

The first contribution of Gryphon lies in its simple syntax and semantics, carefully selected by two methods: experts's reviews while performing some tasks and experts' comments on the use of shapes, color, size, value, and texture. These methods assume that the brain scientists programming effort would be reduced if the computer language enables them to express semantics in natural expressions explicitly similar to the way they think about them, and share some design goals with the natural programming environment [23].

We recorded about 5 h of audio with four brain scientists who were asked to query brain DMRI tractography visualizations. Think-aloud protocol was enforced when they spoke about their intent. Their speech semantics were coded, tagged, and ranked using a speech tagger and a statistical parser. The output of this process was a parser tree that represent the structure of the sentence. Note that since the parser was statistical, it attempted to resolve ambiguities, such as prepositional phrase attachments. The parse tree was then converted into a representation that was simply

Table 1 Gryphon language symbols and keywords: these keywords are not case-sensitive

Verbs	load, selection locate, update, calculate
Prepositions	in, out
Conjunctives	by, with
Selection rules	[], <, <=, >, >=, ==, =, +, −
Build-in routines	AvgFA, AvgLA, NumFiber
Constants	shape, color, size, depth, FA, LA, sagittal, axial, coronal
	CC, CST, CG, IFO, ILF, DEFAULT, RESET

a split of the words in the sentence, showing the words that they depended on and the words that were dependent on them. For example, say that we wish to select the structure corpus callosum (CC) which was inferior to the cortical spinal tracts. A neurologist would say, "*Select the inferior structure of CC next to the CST*". Such spatial reference was frequently used when the brain scientists talked about embedded structures.

The next phase of the analysis involved converting the dependency structure into a semantic representation. The semantic representation was a description of the top-ranked verbs and nouns, and relative spatial relationships that converged to a relatively small semantic space. We further queried the neuroscientists subjective preferences for colors, shape, and texture values for them to understand DMRI streamtube visualizations, keeping in mind the visual analysis of how structure affected human understanding.

As an example, we asked them about spatial relationships when using multiple visual mappings of depth values to size (the further away, the smaller) and color (the deeper, the redder in a blue-to-red color scale). Their comments included "*it is misleading to have the different sizes while colors are present to discern depth*", and "*I would rather have it (the size) stay the same as I spin it (the model) around*", "*Visual mapping of depth to color is preferable since I like it (the model) with colors. That is what I need to look at, and I think that color is a good idea but prefer color by orientation (of the fiber segments)*".

We also hoped that each sentence in the language could record a segment of meaningful actions in the neuroscientists' workflow notation, that neuroscientists could revisit and reuse for new analyses. From our interview, brain scientists felt that current measurement matrices were sufficient, such as those reported in Correia et al. [7], yet the interaction and visualization were somewhat limited for allowing them to query effectively.

The analyses above produce the core content of Gryphon, the simple set of language symbols and keywords shown in Table 1. Five key verbs describe key actions brain scientists would like for DMRI visualizations. Prepositions are used to define the scope of the actions and conjunctives to connect statement terms. Selection rules in Gryphon are exactly the same as those in elementary math. Specifically, "[ ]" is the range operator to give numerical bounds in conditional expressions, and "+" and "−" are relative increment and decrement for brain scientists to shift their previous operations spatially rather than giving absolute coordinates, e.g., moving a cutting plane by 5 mm. Some built-in routines or keywords are chosen following the tensor field measurement matrix in Correia et al. [7], e.g., AvgGA and AvgLA calculate the average FA (fractional anisotropy) and average LA (linear anisotropy) of fibers accordingly and NumFiber counts the number of fibers in a fiber bundle. Gryphon also recognizes some clustered fiber bundles, such as corpus callosum (CC), corticospinal tracts (CST), cingulum (CG), inferior longitudinal fasciculus (ILF), and inferior frontal occipital fasciculus (IFO).

3.1.2 Data Model and Input

Gryphon focuses on visual transformations in 3D visualizations. In our current implementation, fibers are clustered in terms of brain anatomy; each fiber is manually tagged with an anatomical cluster identity as either one of the five major bundles of CC, CST, CG, IFO, or ILF or not. In practice, Gryphons ability to recognize the constants for the major anatomical bundles depends on these cluster tags in the structure of the data model input. However, our language design is not restricted to visualizing clustered data: Gryphon can be expanded to be adaptable to an unclustered data model.

In a Gryphon program, the first step is to indicate the source of data model by giving the name of a data file. As an example, a Gryphon data input statement is written as:

```
normalBrain = LOAD "/home/lucy/normalS1.dat",
```

where the LOAD command parses the input file and creates data structures that fully describe the data model, including identifying the clustering tags. This input specification statement can also update the current data model at the beginning of the visualization pipeline if it is not the first step in a Gryphon script. The evaluation is optional and, when provided, saves the result to a variable, here normalBrain, for later reference.

3.1.3 Encoding Composition

According to Bertin's semiotic theory [3], graphically encoding data with key retinal elements of color, size, orientation, texture, value, and shape is critical in the legibility of two-dimensional (2D) graphical representations. In 3D visualizations, occlusion, an important factor in depth perception, has a detrimental impact on overall legibility, and depth cues are an ordinal dimension in the design space of 3D occlusion management for visualization. Numerous good visualization design adopted certain occlusion management [10] mechanisms, such as halo effects [11].

We have contributed to symbolic mapping of color, size, and shape for 2D graphical legibility enhancement and depth encoding, also via common visual

```
SELECT "ALL"
UPDATE shape BY LINE IN "CST"
UPDATE size BY FA IN "CG"
UPDATE color BY FA IN "IFO"
UPDATE depth BY transparency IN "CG"
UPDATE depth BY value IN "CC" WITH 0.2,0.8
UPDATE depth BY color IN "ILF"
```

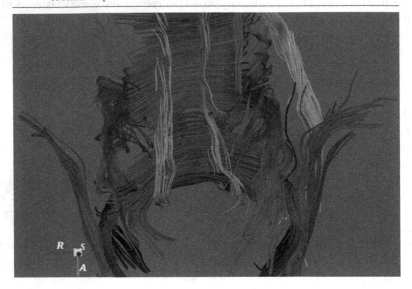

Fig. 2 Another example showing the mixed encoding by using the separation principles to depict fiber bundles

elements such as color, size, value (amount of ink), and transparency, as depth cues for occlusion reduction in the 3D environment. As already shown in the previous example scripts, Gryphon allows brain scientistis to freely customize DMRI visualizations using either a single data-encoding scheme alone or compound encoding scheme by flexibly combining multiple encoding methods.

In composing or exploratory process with DMRI visualizations, brain scientists often attempt to compare a ROI across datasets captured at a different time or from another patient and would like to differentiate them from each other, thus a depth-enhanced visualization could help generate legible displays for the multiple ROIs. In addition, on other occasions users have difficulty in perceiving information along the depth direction.

Gryphon's data-encoding flexibility is also driven by the perceptual principle of shape choices, which states that it is easy for human eyes to perceive categorical data in different shapes. Figure 2 demonstrated some encoding approaches supported by the language, though the picture itself might not represent a good visualization. Suppose a brain scientist has composed the streamtube visualization of a brain DMRI data set with default data encoding (uniform size, color, and shape without depth cues) and now wants the overall encoding scheme to differ across fiber

Table 2 Combinational rules of constants in the UPDATE statement

var1	var2	var3
shape	line, tube, ribbon	–
color	FA, LA	–
size	FA, LA	minimal, scale
depth	size, color, value, transparency	lower, upper
DEFAULT	–	–
RESET	–	–

bundles. In order to achieve this effect, an example Gryphon snippet can be written as follows:

In the resulting visualization in Fig. 2, each of the five major bundles will differ visually from others since all these bundles are encoded differently (these mixed encoding approach is subject to experimental verification). Often times, once a ROI is filtered out, it is also necessary to examine the selected fibers more carefully. For this purpose, Gryphon allows brain scientists to impose various data-encoding schemes upon data of interests. Such visualization customization is performed by the UPDATE command, which works in an immediate mode and updates the view after execution. The general UPDATE syntax pattern is:

```
UPDATE var1 BY var2 WITH para1,...,paraN IN|OUT target,
```

where *var1* is an attribute, such as shape, color, size, depth, of the current visualization to be modified, and *var2* gives how the actual updating operation is to be performed relative to *var1*. The parameter list, *para1...N* statement allows specific *UPDATE* to a particular data encoding operation. Like the target specification (optional with all commands as stated before), the *BY* clause and *WITH* clause are both optional. Table 2 lists all possible combinations of *var1*, *var2* and associated parameter list developed currently. In the table, *lower, upper* gives the bound of depth mapping and minimal, scale indicates the minimum and the scale of variation in size encoding. *DEFAULT* and *RESET*, when accompanying the verb *UPDATE*, act as a command to revoke all data-filtering and data-encoding operations respectively. The following script shows how to inspect the change in FA along fibers in a ROI by mapping FA value to tube size; this yields an alternative representation of the FA variation in that ROI, compared to some conventional coloring approach [9].

```
UPDATE RESET
partialILF = LOCATE "FA in [0.5,0.55]" OUT "ILF"
UPDATE size BY FA IN "partialILF"
```

3.1.4 Considering Interactivity

The third design contribution lies in the support of some typical interaction tasks: (1) checking integrity of neural structures of a brain as a whole; (2) examining fiber orientation in a ROI or fiber connectivity across ROIs; (3) comparing fiber bundle sizes between brain regions; (4) tracing the variation of DMRI quantities such as FA along a group of fibers; and (5) picking particular fibers according to a quantitative threshold.

When using DMRI visualizations, neurologists look at not only the whole data model, but also some regional details, for example the location of pathological conditions. To reach ROIs in brain DMRI visualizations, brain scientists often narrow down the view scope to a relatively large anatomical area in the first place and then dive into a specific ROI. In visualizations in which neural pathways are depicted as streamtubes, the ROIs are usually clusters of fiber bundles.

For instance, at the beginning of a visualization exploration, one of our brain scientist collaborators wants to look into frontal lobe fibers within the intersection of two fiber bundles, CST and CC, and will ignore all other regions of the model. Further, suspicious of fibers with average FA under 0.5 for a cerebral disease with which the brain is probably afflicted, the brain scientist continues examining the suspect fibers. Later on, the scientist focuses on the small fiber region to see how it differs from typical ones, for instance in terms of orientation and DMRI parameter measures in the evaluation metrics. Gryphon supports this process through high-level primitives, such as SELECT and common arithmetical conditional operators, including the range operator. Gryphon mainly contains facilities for step-by-step data filtering with these primitives. For example, supposed the user above wants to explore the fibers of interest, the Gryphon program will look like:

```
SELECT "FA < 0.5" IN "CST"
SELECT "FA < 0.4" IN "CC"
```

As a result, fibers in both specific bundles with average FA under 0.5 will be highlighted to help brain scientists focus on the local data being explored. In addition to this, the user can customize the visualization of the filtered fibers through various visual encoding methods using the *UPDATE* syntax. This is particularly useful when the brain scientist wants to keep the data already found in focus before moving to explore other relevant local data so as to add more fibers into the focus region, or when the scientist simply seeks a more legible visualization of the data. The instance below, following the same example, illustrates how better depth perception achieved by a type of depth encoding, together with a differentiating shape encoding, are added up to the two selected fiber bundles:

```
SELECT "FA < 0.5" IN "CST"
SELECT "FA < 0.4" IN "CC"
UPDATE depth BY color IN "CST"
UPDATE shape BY ribbon IN "CC"
```

This brief sequence of commands can help the brain scientists locate desirable fiber tracts with high accuracy while allowing flexible customization upon current visualizations, similar to some slider dragging selection in interfaces [15]. In this case, tracts of interest (TOIs) are first focused and then further differentiated for more effective exploration. In general, Gryphons design emphasizes this task-driven process of visualization exploration, which fits the thinking process of end users of the present visualizations. Our neurologist collaborators frequently filter data in order to reach an ROI in their DMRI visualizations.. Gryphon offers two commands for data filtering: *SELECT* and *LOCATE*. The data-filtering syntax pattern in Gryphon is as follows:

```
SELECT condition|spatialOperation IN|OUT target
result = LOCATE condition IN|OUT target
```

These two commands have similar functionality but different semantics: *SELECT* executes filtering in an immediate mode by highlighting target fibers, while *LOCATE* performs an offline filtering operation, retrieving target fibers and sending the result to a variable without causing any change in the present visualization. Also, *SELECT* provides relative spatial operations through moving anatomical cutting planes. In fact, it combines these two commands into one while differentiating the two semantics (by recognizing the presence of variable evaluation and taking spatial operations as an alternative to the condition term). However, we have kept these two commands separate based on brain scientists' input asking for a more straightforward understanding of the semantics and more easily learned language usage, for example.

```
SELECT "LA <= 0.72" IN "ALL"
partialILF = LOCATE "FA in [0.5,0.55]" OUT "ILF"
```

The *SELECT* statement filters fibers in the entire DMRI model with average anisotropy greater than 0.72 (by putting them in the contextual background) and highlights all other fibers. In comparison, the *LOCATE* statement does not update the visualization but picks up fibers outside the ILF bundle having average FA value in the specified range. Note that when no specific data encoding is applied, different colors are assigned to ROI fibers in different major bundles in Gryphon so that one ROI can be distinguished from another when more than one is highlighted. Also, filtered fibers are still semitransparent as the contextual background rather than being removed from the visualization.

3.1.5 Spatial References

The fourth design choice of Gryphon is that it is a language in which brain scientists are able to operate with relative spatial terms. Brain scientists frequently use spatial terms such as parasagittal, in, out, mid-axial and near coronal, etc. in their descriptions of DMRI visualizations in the 3D space. They also use a set of relative positioning terms, such as *above, under, on top of, across*, and *between*, etc., and more domain-specific ones such as *frontal, posterior, dorsal*, etc. At present, Gryphon contains a subset of these spatial terms.

In 3D data models such as those from DMRI, spatial relationships between data components are one of the essential characteristics (as is typical of 3D scientific data in genera). Accordingly, in composing a DMRI visualization one must be able to use spatial operators with domain-conventional terms in order to describe the process of visualization authoring. In response, Gryphon supports spatial operations by combining two approaches. First, three visible cutting planes that provide guidance in the three conventional anatomical views, the axial, coronal, and sagittal views, are integrated in the visualization view. Then, flexible manipulating operations upon the three planes are built into Gryphons spatial syntax definitions. This enables brain scientists to navigate in the dense 3D data model with a highly precise filtering fashion through numerical input. For instance, suppose the streamtube representation of a DMRI model being programmed is derived using unit seeding resolution from DMRI volumes of size $256 \times 256 \times 31$ captured at voxel resolution $0.9375 \times 0.9375 \times 4.52$ mm, and suppose both the axial and coronal planes are located at their initial position so that nothing is cut along these two views. In order to examine a suspect anomaly in the brain region of the occipital lobe, a brain scientists can filter the data model so that approximately only this region is kept. Relative movements can be imposed similarly on the sagittal plane as well using the code below.

```
SELECT "coronal +159.25"
SELECT "axial -27.5"
SELECT "sagittal +183.2"
```

3.1.6 Flat Control Structure

Gryphon is designed to provide a declarative language environment for brain scientists who may not have programming skills. Therefore, we purposely eliminate the conditional and iterative structures from the language design of Gryphon and keep only the most intuitive sequential structure. This gives Gryphon a flat control structure. Meanwhile, Gryphon uses high-level semantics to overcome its weakness in expressing user task requirements. First, the need for an iterative structure usually stems from the need to operate on multiple targets. In Gryphon, the operation target is a common term in all syntax patterns to indicate the scope of the data. These scopes are defined as enumeration and default terms in Gryphon syntax patterns. On the one hand, with enumeration, brain scientists simply list all targets in the target term, thus avoiding iteration. For example, suppose a user intends to select three bundles and then to change the size encoding for two of them; a Gryphon script can include:

```
suspfibers = LOCATE FA in 0.2 0.25 in "CST, ILF"
UPDATE size BY FA in "suspfibers"
```

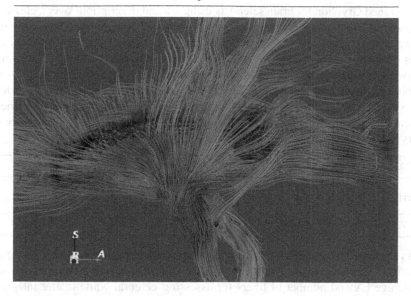

Fig. 3 LOCATE function only stores the data without necessarily making a selection

```
SELECT "CST,CC,CG"
UPDATE size BY FA IN "CST,CG"
```

On the other hand, with term default, when a target term is lacking in a single statement, *ALL* is taken as the default scope, meaning the entire data model will be the regions to be manipulated. This rule is applicable to all types of Gryphon statements, that target term is optional in all Gryphon syntax patterns.

Second, requirement for a conditional structure comes from brain scientists' requests for a way to express conditional processing. For example, they often filter fibers according to *FA* thresholds. In Gryphon, a conditional expression can be flexibly embedded in a statement. We have shown in previous examples how to embed conditional expressions in *SELECT* statements. For syntactic simplicity, condition is expressed in *UPDATE* statements indirectly through variable reference, as the following example snippet shows. Here *LOCATE* is an alternative to *SELECT* but it results in storage of the fibers filtered into a variable for later reference instead of highlighting those fibers immediately, as *SELECT* does (Fig. 3).

3.1.7 Fully Declarative Language

We designed Gryphon for brain scientists using natural descriptions over a classical programming language: elements close to those in a computer programming language have been changed to be as declarative as possible. In Gryphon, all types of statements are designed to follow a consistent pattern: begun by a verb, followed by operations and, optionally, ended by data target specification, with optional evaluation of statement result to a variable for later reference if provided. This syntax consistency has been applied to the data measurement statement where invocation of built-in numerical routines is involved. To measure the number of fibers in a selected bundle, for instance, instead of writing

```
CALCULATE NumFibers("CST"),
```

write

```
CALCULATE NumFibers IN "CST"
```

As shown above, besides visually examining the graphical representations, brain scientists often need to investigate the DMRI data in a quantitative manner, such as average FA and number of fibers for assessing cerebral white-matter integrity. Accordingly, Gryphon provides built-in numerical routines to calculate some of the DMRI metrics in Correia et al. [7]. The following pattern shows the Gryphon data analysis syntax:

```
val = CALCULATE ${metricRoutine}$ IN|OUT target
```

MetricRoutine can be one of *AvgFA*, *AvgLA* and *NumFibers* to represent fiber integrity. In this syntax pattern, keeping the resulting value by evaluation is optional and sometimes useful when referred to afterwards. For example, in order to sum the fibers with average *FA* falling within a particular range and then figure out the average LA of these target fibers, a brain scientist can write in Gryphon to show its visualization in Fig. 4.

3.2 Language Implementation

Gryphon is declarative in general form, with support of some programming language features, such as variable referencing and arithmetical and logical operations. Current implementation does not support a fully featured interpreter or compiler but by a string-parsing-based translator of descriptive text to visualization pipeline components and manipulations. The core of Gryphon is implemented on top of the Visualization Toolkit (VTK) using C++. The rendering engine is driven by the visualization pipeline and legacy VTK components ranging from various geometry filters to data mappers. Moreover, the support of language features, such as compositional data encoding, a group of new pipeline components like those for

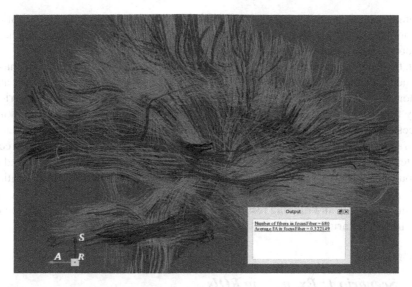

Fig. 4 Interface embedded to show query results

view-dependent per-vertex depth-value ordering has been added on top of related VTK classes, and some legacy VTK components have been tailored for specific needs of Gryphon visualizations.

In particular, the Gryphon script interpreter is implemented as data filters using the VTK visualization pipeline. As such, interpreting a Gryphon script involves translating the text, according to defined syntax and semantics, to data transformations in the VTK pipeline. For data encoding flexibility, multiple VTK data transformation pipelines have been employed.

Finally, the programming interface is implemented using Qt for C++. Interactions like triggering the execution of a Gryphon program, serializing and deserializing the text script, etc. are all developed with Qt widgets, although interactions with the visualization itself are handled using legacy VTK facilities with necessary extensions. Since our language targets non-programmer debugging skills are not expected of users. Consequently, instead of building a full-blown debugging environment as seen in almost all integrated development environments (IDEs), we employee an output window to prompt users with all error messages caused by invalid syntax or unrecognized language symbols. We have used GUI utilities of Qt for C++ to dump, after running a script, to process the error messages.

4 Scenarios of Use

We describe several sample task scenarios that can be performed by brain scientists on a brain DMRI model using the Gryphon language. The usage scenarios associated with the sample tasks are representative of typical real-world visualization tasks of brain scientists with expertise in DMRI. The usages range from visualization customization and exploration to DMRI data analysis and cover the main language features and functionalities of the current Gryphon implementation.

In the following scenarios, a vascular neurologist (and a Gryphon end user), has a geometrical model derived from a brain DT-MRI data set and wants to com-pose and explore visualizations of the data for diagnostic purposes. In each of the scenarios, Lucy fulfills his task by programming a Gryphon script that describes his thinking process for that task and then clicks the Run button to execute the script. Lucy programs with Gryphon syntax references showing in a help window and corrects any term that is typed incorrectly with the assistance of error messages displayed in the output window. Once the script is interpreted correctly, either the visualization is changed or numerical values appear in the output window as the results of script execution. Scripts and running results are presented at the end of the description of each usage scenario.

4.1 Scenario 1: Examining ROIs

One common task brain scientists perform is to examine particular regions of interest (ROIs) rather than the whole brain. In this task, Lucy is interested only in all fibers within the temporal lobe area that belong to the CG bundle and CST fiber bundles in the parietal lobe area that have average LA value no greater than a threshold to be determined. The SELECT command with relative spatial operations using the anatomical planes enables Lucy to reach precisely the ROIs she desires. She first aims to filter fiber tracts outside the temporal and parietal area by adjusting the three cutting planes with relative movements and then starts trying to reach the exact target fiber tracts using both fiber bundle filters and conditional expression related to LA. Lucy begins with an estimate for the undecided LA threshold and then keeps refining until she gets the accurate selection of target fibers. In the end, she has a runnable script written in Gryphon (Fig. 5).

4.2 Calculating Metrics

In addition to visual examinations, neurologists often request quantitative investigations of their DMRI models. In this scenario, Lucy attempts to check white-matter integrity in a brain model to improve the limited reliability of DMRI tractography. For a rough estimation of the integrity, she uses the CALCULATE command to retrieve the size, in terms of the number of fibers, and average FA of both the whole brain and representative bundles. With the average FA she has requested before, Lucy goes further to use it to kick out CST fibers with average FA below the bundle-wise average. Lucy writes the script in Fig. 6 to get the result.

```
LOAD "/home/lucy/braindti.data"
SELECT "axial +63.35"
SELECT "sagittal +70"
SELECT "coronal -48.5"
SELECT "sagittal -0.25"
SELECT "axial +7.2"
SELECT "CG"
SELECT "LA <= 0.275" in "CST"
```

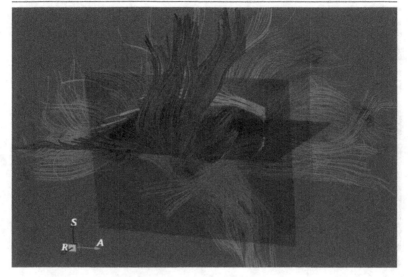

Fig. 5 Select regions of interest by relative unit specification

5 Discussion

5.1 Experimental Studies

The design was performed collaboratively with brain scientists. We performed a 2-h study with brain scientists. Generally, they liked its flexibility and found it easy to learn. They would like to encode more parameters calculated from their new studies. They would also prefer to have a graphical interface where these parameters can be put in directly.

5.2 Integrating Perceptual Principles

Current visualization methods in Gryphon are limited and are not guided fully by perceptual principles. In the long term, the Gryphon language should follow the rules that visualization should require minimal intervention on the part of the

```
LOAD "/home/lucy/braindti.data"
SELECT "all"
Calculate NumFibers
Calculate AvgFA
cstFAavg = Calculate AvgFA in "CC"
Calculate NumFibers in "CST"
Update reset in "all"
Select "FA>= cstFAavg in "CC"
```

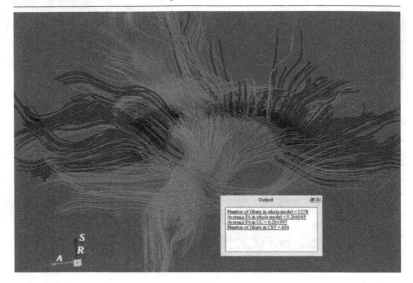

Fig. 6 Brain DMRI matrix calculation

designers, similar to that of APT. APT and its Tableau environment embodies a genuinely prescriptive theory of designing graphical encoding based on results from perceptual studies by analysis of data types. Similarly, ColorBrewer prescribes colors by data types, while providing a flexible user interface to adjust the values [14]. Our current implementation has limited availability of integrated principles, yet provides a language at a higher level so that brain scientists do not have to control low-level encoding details.

Several improvements include better color design, for example using new color-embedding methods to show fiber orientations [8]. The lighting needs to be improved to improve spatial structure presentation. Some perceptual-related rendering can be represented, e.g., ambient lighting to control dense line rendering for showing spatial relationships [24] or volume visualizations where structures are more pronounced [27]. It should be possible to generate ideas as such by visual composition in our Gryphon language.

5.3 Bidirectional Text (Programming Syntax) and Visualization Environment

Gryphon can be extended to rely on the coexistence of text display and visualization to form a bidirectional notation environment. By this we mean that the text and the visualization can augment each other, with the text becoming what is called "secondary notations" from software engineering and the visualization showing the results. The text scripts can be used to exhibit workflows and steps in arriving at the visualization that might otherwise be less accessible. On the other hand, the programmable notation is also our programming language that can carry a simple syntax based on brain scientists' tasks.

6 Conclusion

We have presented Gryphon, a simple domain-specific programming language for exploring 3D DMRI visualizations. We described the design process and results. Empirical studies suggested that user interfaces and more intuitive interaction techniques are desirable to have a more useful language.

Acknowledgements The authors thank the participants for their time and effort, Drs. Juebin Huang, Stephen Correia, and Judy James for their help on task analyses. We also thank Katrina Avery for her editorial support. This work was supported in part by NSF IIS-1018769, IIS-1016623, IIS-1017921, OCI-0923393, EPS-0903234, DBI-1062057, and CCF-1785542, and NIH (RO1-EB004155-01A1).

References

1. Akers, D.: CINCH: A cooperatively designed marking interface for 3d pathway selection. In: Proceedings of the 19th annual ACM symposium on User interface software and technology, pp. 33–42. ACM (2006)
2. Bentley, J.: Programming pearls: little languages. pp. 711–721. ACM (1986)
3. Bertin, J.: Semiology of graphics: diagrams, networks, maps (1983)
4. Bostock, M., Heer, J.: Protovis: A graphical toolkit for visualization. IEEE Transactions on Visualization and Computer Graphics 15(6), 1121–1128 (2009)
5. Bostock, M., Ogievetsky, V., Heer, J.: D^3 data-driven documents. IEEE Transactions on Visualization and Computer Graphics 17(12), 2301–2309 (2011)
6. Chiw, C., Kindlmann, G., Reppy, J., Samuels, L., Seltzer, N.: Diderot: a parallel dsl for image analysis and visualization. In: Proceedings of the 33rd ACM SIGPLAN Conference on Programming Language Design and Implementation, pp. 111–120. ACM (2012)
7. Correia, S., Lee, S., Voorn, T., Tate, D., Paul, R., Zhang, S., Salloway, S., Malloy, P., Laidlaw, D.: Quantitative tractography metrics of white matter integrity in diffusion-tensor mri. Neuroimage 42(2), 568 (2008)

8. Demiralp, C., Hughes, J.F., Laidlaw, D.H.: Coloring 3D line fields using Boy's real projective plane immersion. IEEE Trans. on Visualization and Computer Graphics (Proc. Visualization '09) **15**(6), 1457–1463 (2009)
9. Demiralp, Ç., Zhang, S., Tate, D., Correia, S., Laidlaw, D.: Connectivity-aware sectional visualization of 3d dti volumes using perceptual flat-torus coloring and edge rendering. Eurographics (2006)
10. Elmqvist, N., Tsigas, P.: A taxonomy of 3d occlusion management for visualization. IEEE Transactions on Visualization and Computer Graphics **14**(5), 1095–1109 (2008)
11. Everts, M., Bekker, H., Roerdink, J., Isenberg, T.: Depth-dependent halos: Illustrative rendering of dense line data. IEEE Transactions on Visualization and Computer Graphics **15**(6), 1299–1306 (2009)
12. Forsberg, A., Chen, J., Laidlaw, D.: Comparing 3d vector field visualization methods: A user study. IEEE Transactions on Visualization and Computer Graphics **15**(6), 1219–1226 (2009)
13. Goldau, M., Wiebel, A., Hlawitschka, M., Scheuermann, G., Tittgemeyer, M.: Visualizing DTI parameters on boundary surfaces of white matter fiber bundles. In: Signal Processing, Pattern Recognition, and Applications/722: Computer Graphics and Imaging. ACTA Press (2011)
14. Harrower, M., Brewer, C.: Colorbrewer.org: an online tool for selecting colour schemes for maps. The Cartographic Journal **40**(1), 27–37 (2003)
15. Jiang, H., Van Zijl, P., Kim, J., Pearlson, G., Mori, S.: DtiStudio: resource program for diffusion tensor computation and fiber bundle tracking. Computer Methods and Programs in Biomedicine **81**(2), 106–116 (2006)
16. Johnson, C., Parker, S., Weinstein, D.: Large-scale computational science applications using the scirun problem solving environment. Citeseer (2000)
17. Koenig, M., Spindler, W., Rexilius, J., Jomier, J., Link, F., Peitgen, H.: Embedding vtk and itk into a visual programming and rapid prototyping platform. In: Proceedings of SPIE, vol. 6141, p. 61412O (2006)
18. Leemans, A., Jeurissen, B., Sijbers, J., Jones, D.: ExploreDTI: a graphical toolbox for processing, analyzing, and visualizing diffusion MR data. In: Proceedings 17th Scientific Meeting, International Society for Magnetic Resonance in Medicine, vol. 17 (2009)
19. Mackinlay, J.: Automating the design of graphical presentations of relational information. ACM Transactions on Graphics (TOG) **5**(2), 110–141 (1986)
20. Metoyer, R., Lee, B., Riche, N., Czerwinski, M.: Understanding the verbal language and structure of end-user descriptions of data visualizations. ACM CHI (2012)
21. Mori, S.: Introduction to diffusion tensor imaging. Elsevier Science (2007)
22. Mori, S., van Zijl, P.: Fiber tracking: principles and strategies–a technical review. pp. 468–480. Wiley Online Library (2002)
23. Myers, B., Pane, J., Ko, A.: Natural programming languages and environments. pp. 47–52. ACM (2004)
24. Peeters, T., Vilanova, A., Strijkers, G., ter Haar Romeny, B.: Visualization of the fibrous structure of the heart. In: Vision, Modeling and Visualization, pp. 309–316 (2006)
25. Pieper, S., Halle, M., Kikinis, R.: 3d slicer. In: IEEE International Symposium on Biomedical Imaging: Nano to Macro, pp. 632–635. IEEE (2004)
26. Reas, C., Fry, B.: Processing: a programming handbook for visual designers and artists. The MIT Press (2007)
27. Rheingans, P., Ebert, D.: Volume illustration: Nonphotorealistic rendering of volume models. IEEE Transactions on Visualization and Computer Graphics **7**(3), 253–264 (2001)
28. Rogowitz, B., Kalvin, A.: The which blair project: A quick visual method for evaluating perceptual color maps. In: Proceedings of the conference on Visualization, pp. 183–190. IEEE Computer Society (2001)
29. Roth, S., Kolojejchick, J., Mattis, J., Goldstein, J.: Interactive graphic design using automatic presentation knowledge. In: Intelligent User Interfaces, p. 237 (1998)
30. Stalling, D., Westerhoff, M., Hege, H., et al.: Amira: A highly interactive system for visual data analysis. pp. 749–67 (2005)

31. Tory, M., Moller, T.: Rethinking visualization: A high-level taxonomy. In: IEEE Symposium on Information Visualization, pp. 151–158. IEEE (2004)
32. Toussaint, N., Souplet, J., Fillard, P., et al.: Medinria: Medical image navigation and research tool by inria. In: Proceedings of MICCAI, vol. 7, pp. 1–8 (2007)
33. Wang, R., Wedeen, V.: Diffusion toolkit and trackvis. Proceedings of the International Society for Magnetic Resonance in Medicine **3720** (2007)
34. Wilkinson, L., Wills, G.: The grammar of graphics. Springer Verlag (2005)
35. Zheng, L., Wu, Y., Ma, K.: Perceptually based depth-ordering enhancement for direct volume rendering. IEEE Transactions on Visualization and Computer Graphics (to appear) (2012)

Viewing Abstract Data as Maps

Emden R. Gansner, Yifan Hu, and Stephen G. Kobourov

Abstract From telecommunications and abstractions of the Internet to interconnections of medical papers to on-line social networks, technology has spawned an explosion of data in the form of large attributed graphs and networks. Visualization often serves as an essential first step in understanding such data, when little is known. Unfortunately, visualizing large graphs presents its own set of problems, both technically in terms of clutter and cognitively in terms of unfamiliarity with the graph idiom. In this chapter, we consider viewing such data in the form of geographic maps. This provides a view of the data that naturally allows for reduction of clutter and for presentation in a familiar idiom. We describe some techniques for creating such maps, and consider some of the related technical problems. We also present and discuss various applications of this method to real data.

1 Introduction

In an increasingly technological world, we find ourselves dealing with large, multivariate data sets in order to make informed decisions. For many, tables of numbers are a cue for the eyes to glaze over. Even experts can have difficulties determining patterns in "raw" data. For this reason, the statistics community has, over several centuries, developed a variety of visualizations for statistical data [5] which can expose correlations and structures that otherwise might be missed. Recently, a complementary effort in the information visualization community has

E.R. Gansner (✉) • Y. Hu
AT&T Labs – Research, Florham Park, NJ, USA
e-mail: erg@research.att.com; yifanhu@research.att.com

S.G. Kobourov
University of Arizona, Tucson, AZ, USA
e-mail: kobourov@cs.arizona.edu

W. Huang (ed.), *Handbook of Human Centric Visualization*,
DOI 10.1007/978-1-4614-7485-2_3, © Springer Science+Business Media New York 2014

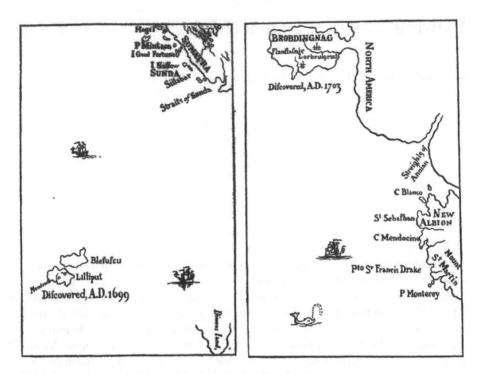

Fig. 1 Lilliput and Brobdingnag (Thanks to Project Gutenberg)

expanded the number of visual metaphors [30, 43]. The key to all of this work is to take the many dimensions inherent in the data and reduce it down to the two or three that are accessible to the human eye, and doing it in a fashion that reveals or at least maintains the data's most salient features.

One approach to making this information more accessible to the human is to rely on more familiar, less technical visual metaphors, in essence, to map abstract data into a more concrete or physical space, ones tangible to and experienced by the viewer. Not surprisingly, various researchers in the geographical information science community (GIS) were among the pioneers in promoting this style of visualization [27, 40, 41], which they termed *spatialization*.

Constraining the metaphor even more, one can consider how to present abstract data in the context of a geographic map. At the simplest level, this might involve merging geographic and abstract data [12]. Cartograms [24, 25] provide another level of abstraction, in which quantitative information about a geographic region is represented by area or distance, usually requiring a distortion of the geography. These appear particularly attractive to the popular press.

More interesting is to derive a pseudo-geographic map from the abstract data. People seem generally intrigued by maps, and seeing fictional or abstract information portrayed on a map makes it more real. Authors, from Swift (Fig. 1) to Hardy to Tolkien, have long provided maps to guide the reader through their world.

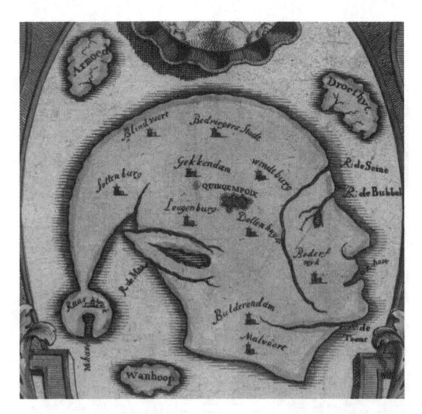

Fig. 2 Aspects of eighteenth century financial bubbles as a map (http://bigthink.com/strange-maps/554-the-fools-head-map-a-fossil-of-the-financial-bubbles-of-1720)

More relevantly, there is also a very long tradition of displaying abstract data in the form of a map. One early example is the Fool's Head Map (Fig. 2) diagramming the financial bubbles of the early 1700s. This blends real geography (the Thames and Seine Rivers) with allegorical aspects such as "Crazy Town" and islands of "Poverty" and "Despair." More recent examples include Cardelli's map of programming language concepts used as the cover for Ullman's ML book [44] and Randall Munroe's take on the evolution of online communities (Fig. 3). Clearly, creating these maps relied on the talent, knowledge and wit of someone with artistic talents.

For use in information visualization, it is necessary that we be able to automate the production of such maps, with an eye for aesthetics but also with the acceptance that some genius of the hand-made map will be missing. Various techniques have been devised for producing such maps. For example, from the GIS community, there is the notable work of Skupin, Fabrikant and others [10, 35, 37–39]. Adding to intuition, there is some evidence that map-based displays of abstract data provide an aid to comprehension [9].

Fig. 3 Munroe's map of online communities (http://xkcd.com/802/)

Until the recent past, much data visualization dealt with simple attributed data. That is, the data was viewed as a collection of records, each record being one of a small number of types, which determined the fields it had. A further complication has been added that now many data sets have a graph or explicit relational structure as well. (Theoretically, a graph can be represented using a simple attributed data model, and in practice, the data often contained a graph implicitly. Making the graph explicit also implies the desire to visualize the graph explicitly.) Canonical examples of graph-based data are the various graphs induced by the Internet, the friendship graph induced from Facebook, or the "following" relation from Twitter. Driven by the increasing presence and importance of graphs, in software engineering, biology, telecommunications, social networks, etc., there has been a great deal of work in the theory and practice of drawing graphs [1], including graphs with thousands of vertices.

a
b

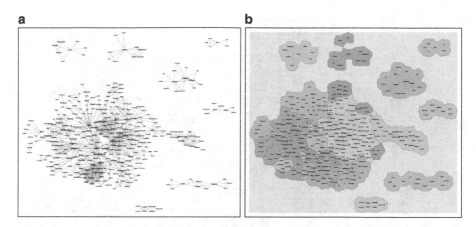

Fig. 4 A node-link drawing compared to a map representation of the same graph

As with other forms of data, representing graphs as maps can make the data more accessible to the reader, replacing a typical node-link diagram or point-cloud visualization with a more compelling drawing. A further impetus for considering graphs as maps arises naturally from standard graph drawing. Figure 4a displays a typical graph layout using a node-link diagram. The graph represents the author collaborations between 1994 and 2004 at the Symposium on Graph Drawing. The drawing exhibits the connected components, and closely related nodes are indicated by proximity, but cluster structure is only hinted at. It is not difficult to already see a map there, so why not go the next few steps and arrive at the rendering in Fig. 4b? In this version, the cluster structure is obvious. Coloring the nodes in the node-link drawing would still only imply the clusters. The map representation makes the clusters explicit as well as indicating strong cluster relations where two clusters share a border.

In the remainder of this chapter, we will explore a technique for displaying graphs as geometric maps. The creation of the basic geometry is described in the next section. Section 3 addresses some of the auxiliary problems that arise in making a good map, such as how to best color the regions or provide additional features to help the reader. Unlike in the real world, the geography of much abstract data is in constant motion, changing with each packet or phone call. We look at the issue of dynamic maps in Sect. 4. Section 5 applies the methods described to real-world data and shows how such maps can provide insights into the data. We conclude with a few thoughts in Sect. 6.

2   Making the Map

The technique we describe here, which we refer to as GMap, allows us to generate map-like representations from an abstract graph. In particular, given a graph with weighted edges, such as how similar or dissimilar two books are based on customer

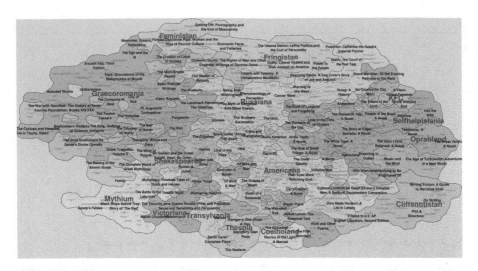

Fig. 5 A map of books related to "1984" from Amazon.com

purchases, we produce a drawing with a map-like look, with countries that enclose similar objects, outer boundaries that follow the outline of the vertex set, and inner boundaries that have the twists and turns found in real maps. A typical example is given in Fig. 5, showing just under 1,000 books, with edges determined by Amazon.com's record of related purchases. Our maps also can have lakes, islands, and peninsulas, similar to those found in real geographic maps. The technique is applicable outside of the domain of graphs; it can typically be used on most high-dimensional data sets.

This technique is a framework in the true sense of the word, rather than a specific algorithm. It consists of three main steps. The first two steps are fairly generic and can be achieved by a variety of existing algorithms. The last step is tuned to creating a map and involves a special-purpose algorithm. When finished, we have a drawing that has the basic appearance of a geographic map. With the addition of colors for the countries, perhaps coastal shading, mountains or some other effects, we have an acceptable imitation of a map. As noted above, these final features will be covered in Sect. 3.

Much of our presentation is narrative and informal. We refer the interested reader to the articles [14, 15, 19] for technical details and more references and examples. A prototype implementation is available as part of the open source Graphviz software package [16, 18].

2.1 Laying Out the Boundaries

For the first step, we take as input a graph or high-dimensional data set, and embed it into the plane. The statistics and scientific modeling communities have extensively explored this problem and provide many ways of doing this. Possible embedding algorithms include principal component analysis, multidimensional scaling (MDS), force-directed algorithms, or non-linear dimensionality reductions such as Locally Linear Embedding and Isomap.

The second step takes this collection of points in the plane and aggregates them into clusters. Here, it is important to match the clustering algorithm to the embedding algorithm. For example, a geometric clustering algorithm such as k-means [31] may be suitable for an embedding derived from MDS, as the latter tends to place similar points in the same geometric region with good separation between clusters. On the other hand, with an embedding derived from a force-directed layout, a modularity based clustering [32] could be a better fit. The two algorithms are strongly related, and therefore we can expect vertices that are in the same cluster to also be physically close to each other in the embedding.

In the third step, we use the two-dimensional embedding together with the clustering to create the actual map by delineating country boundaries, carving continental outlines, and separating islands from continents. This can be accomplished with the help of plane partitioning techniques such as Voronoi diagrams, along with the addition of new algorithmic techniques to ensure realistic looking outer and inner boundaries. We want to create a map, with inner boundaries separating points not in the same cluster and outer boundaries preferably following the general outline of the point set. A naive approach for creating the map is to form the Voronoi diagram of the vertices based on the embedding information, together with four points on the corners of the bounding box. This is illustrated in Fig. 6a. Such maps often have sharp corners, and angular outer boundaries. We can generate more natural outer boundaries by adding random points to the current embedding. A random point is only accepted if its distance from any of the real points is more than some preset threshold. Note that this step can be implemented efficiently using a suitable space decomposing data structure, such as a quadtree. This leads to boundaries that follow the shape of the point set. In addition, the randomness of the points on the outskirts gives rise to some randomness of the outer boundaries, thus making them more map-like, as seen in Fig. 6b. Furthermore, depending on the value of the threshold, this step can also result in the creation of lakes and fjords in areas where vertices are far apart from each other. Nevertheless, some inner boundaries remain artificially straight.

At this point, we still note the undesirable feature that the "countries" all have roughly the same area (Fig. 6b), whereas we might prefer some areas to be larger than others (e.g., due to the importance of the entities they represent). As an illustration, in Fig. 6, we assume that "node 1" is more important than the other two nodes, and use a larger label for that area.[1] To make areas follow the shape of the labels, we first generate artificial points along the bounding boxes of the labels as shown in Fig. 6c. To make the inner boundaries less uniform and more map-like,

---

[1] A weighted Voronoi diagrams can be used to make the area of each Voronoi cell proportional to its weight. We do not use this approach, however, because we want the Voronoi cell to also contain a specific shape, e.g., the bounding box of a label.

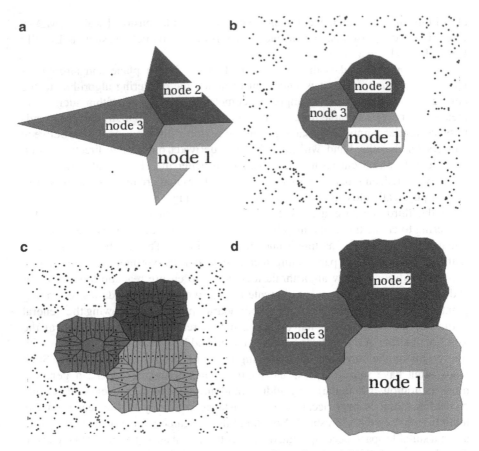

Fig. 6 (**a**) Voronoi diagram of vertices and corners of bounding box; (**b**) better construction of outer boundaries through placement of random points; (**c**) Voronoi diagram of vertices and points inserted around the bounding boxes of the labels; (**d**) the final map

we perturb these points randomly instead of running strictly along the boxes. Here Voronoi cells that belong to the same vertex are colored in the same color, and cells that correspond to the random points on the outskirt are not shown. Cells of the same color are then merged to give the final map in Fig. 6d. Note that instead of the bounding boxes of labels, we could use any 2D shapes, e.g., the outlines of real countries, in order to obtain a desired look and proportion of area, as long as these shapes do not overlap.

We note that not all real maps have complicated boundaries. For example, boundaries of the western states in the United States often have long straight sections. We believe that irregular boundaries are more typical of historical and geographic boundaries, and lead to more map-like results. But this is a matter of personal taste and our technique can generate maps of both styles.

When mapping vertices that contain cluster information, in addition to merging cells that belong to the same vertex, we also merge cells that belong to the same cluster, thus forming regions of complicated shapes, with multiple vertices and labels in each region. At this point we can add more geographic components to strengthen the map metaphor. For instance, in places where there is significant space between vertices in neighboring clusters, we can add lakes, rivers, or mountain ranges to the map to indicate the distance.

With the regions determined, we have a representation of the data in which closely related objects, as determined by the graph topology and possibly edge weights, are drawn closely together. This geometric information is then used to discover clusters among the objects. To emphasize the clusters, each is represented as a collection of geometric regions.

When projecting high dimensional data into low dimensional space, distance distortion is inevitable, and the resulting figure will often have some anomalies and distortions. Thus, some strongly related objects may be separated by seemingly unrelated objects. For example, in Fig. 13, we see several Richard Pryor shows connected to the purple country by color but lying outside of the country's main region. These shows are also closely connected to shows in other countries further down the map and are therefore pulled away. Such fragmentation is inherent in the embedding and clustering algorithms used in the first two steps. However we have proposed ways [14] to use the clustering information to adjust the layout, so that the regions of countries are more contiguous, at the expense of some loss of relational information captured in the original embedding.

3 Map Features

Once the map geometry is in place, we can add additional graphical attributes to the drawing in order to enhance its clarity, to serve as keys to the abstract data, or to simply make it more aesthetically appealing. To this end, one natural approach is to employ additional cartographic or topographic conventions, such as overlaying mountains, rivers or roads, applying coastal shading, or generating a relief map such as the one shown in Fig. 15.

One feature common to almost all maps is a coloring of the regions to emphasize commonality or separation. Thus, in past centuries, one could rely on all the states of the British Empire being colored pink. Achieving a good coloring for our artificial maps brings its own set of problems, which we now address.

3.1 Map Coloring

In this subsection we consider the problem of assigning good colors to the countries in our maps. The Four Color Theorem states that only four colors are needed to color any map so that no neighboring countries share the same color. It is implicitly assumed that each country forms a contiguous region. However, this result is of limited use to us because countries in our maps are often not contiguous. For instance, in Fig. 13 as we previously noted, we have several Richard Pryor shows that belong to the *Saturday Night Live* cluster, but are separated due to his connections with films in other regions. In cases where one cluster is represented by several disjoint regions we must use the same color for all regions to avoid ambiguity. Thus, four colors (or even five or six) are not enough. Instead, we will have to use one unique color for each cluster to avoid ambiguity.

Estimation of the number of colors an "average human" can discriminate, when color pairs are presented side by side, ranges from tens of thousands to a million. However, the number of colors a person can differentiate, when similar colors are not immediately next to each other, is far smaller. A further limiting factor is that 5 % of males are color blind, which rules out certain coloring schemes. Finally, some coloring schemes are used more often than others in maps, reducing the number of colors even more.

In coloring our maps, we start with a coloring scheme from ColorBrewer [3], and generate as many colors as the number of countries by blending the base colors. As a result our color space is piecewise linear and discrete. It remain to be decided which color should be assigned to which country. Because the number of countries can be as many as 30 in many examples, and because we blend a few distinctive colors to form a discrete 1D array of colors, two consecutive colors in the linear array of colors are similar to each other. When applying these colors to the map, we want to avoid coloring neighboring countries with such adjacent pairs of colors. Although two non-neighboring countries with similar colors can lead the viewer to believe that they are disjoint regions of the same country, this problem diminishes when the two countries are sufficiently far apart, as it is unlikely that distant regions that are far away belong to the same cluster. With this in mind, we define the *country graph*, $G_c = \{V_c, E_c\}$, to be the undirected graph where countries are vertices, and two countries are connected by an edge if they share a non-trivial boundary. We then consider the problem of assigning colors to nodes of G_c so that the color distance between nodes that share an edge is maximized.

More formally, let C be the color space, i.e., a set of colors; let $c : V_c \to C$ be a function that assigns a color to every vertex; and let $w_{ij} \geq 0$ be weights associated with edges $\{i, j\} \in E_c$, indicating how important it is to color node i and j with distinctive colors. Let $d : C \times C \to R$ be a color distance function. Define the vector of color distances along edges to be

$$v(c) = \{w_{i,j} \, d(c(i), c(j)) \mid \{i, j\} \in E_c\}.$$

Then we are looking for a color function that maximizes this vector with respect to some cost function. Two natural cost functions are:

$$\max_{c \in C} \sum_{\{i,j\} \in E_c} w_{i,j} \, d(c(i), c(j))^2 \quad \text{(2-norm)}$$

and

$$\max_{c \in C} \min_{\{i,j\} \in E_c} w_{i,j}\, d(c(i), c(j)) \quad \text{(MaxMin)}$$

The weights along the edges can be used to model the undesirable effect of two nearby but not connected countries having very similar colors by making the country graph a complete graph, and assigning edge weights to be the inverse of the distance between two countries.

Dillencourt et al. [6] investigated the case where all colors in the color spectrum are available. They proposed a force-directed model aimed at selecting $|V_c|$ colors as far apart as possible in the color space. In our map coloring problem, however, we are limited to "map-like" colors for aesthetic reasons, and our color space is discrete. Therefore, for simplicity, we model our coloring problem as one of vertex labeling, where our color space is $C = \{1, 2, \ldots, |V_c|\}$, and the color function we are looking for is a permutation that maximizes the labeling differences along the edges. The cost functions we consider are

$$\max_{\pi} \sum_{\{i,j\} \in E_c} w_{i,j} (\pi_i - \pi_j)^2, \quad \text{(2-norm)} \tag{1}$$

and

$$\max_{\pi} \min_{\{i,j\} \in E_c} w_{i,j} |\pi_i - \pi_j|, \quad \text{(MaxMin)} \tag{2}$$

where π_i is the i-th element of the permutation π of $\{1, 2, \ldots, |V_c|\}$.

It turns out that the MaxMin problem (2) is known as the antibandwidth problem, and arises in a number of practical applications. For example, it belongs to the family of obnoxious facility location problems. Here the "enemy" graph is one for which nodes are people and there is an edge between two people if and only if they are enemies. The problem is to build each person a house along a road so that the minimal distance between enemies is maximized [4]. Another example is the radio frequency assignment problem in which the nodes correspond to transmitters and the edges are between interfering transmitters; the objective is to assign the frequencies so that those for the interfering transmitters are as different as possible.

This antibandwidth maximization problem is NP-Complete [29]. In the literature, theoretical results have been presented for some special graphs, including paths, cycles, rectangular grids, special trees and complete bipartite graphs (see, for example, [34] and the references therein).

For more general graphs, heuristics algorithms are being developed. Hu et al. [20] have developed an algorithm GSpectral (Greedy Spectral) that is based on computing the eigenvector corresponding to the largest eigenvalue of the Laplacian associated with the graph and then using a greedy refinement algorithm. Duarte, Martí, Resende and Silva [7] have proposed a linear integer programming formulation and several heuristics based on GRASP (Greedy Randomized Adaptive Search

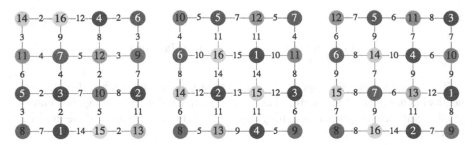

Fig. 7 Coloring schemes RANDOM, SPECTRAL, and SPECTRAL+GREEDY. Each node is colored by the color index shown as the node label. Edge labels are the absolute difference of the endpoint labels

Procedure) with path relinking. They present some high-quality computational results for general graphs, although the run-times for their relatively modest-sized test problems (graphs with fewer than 9,000 nodes) are quite high (typically several minutes for their fastest approach applied to their largest problems). Scott and Hu [36] presented a faster heuristic with the rough idea of finding a pseudo diameter of the graph first, then ordering the corresponding level sets in an alternating fashion, followed by a greedy refinement. The algorithm was found to give comparable ordering to GRASP, but works for much larger graphs.

Here we describe the GSpectral algorithm. The algorithm is motivated by the fact that the complementary problem of finding a permutation that *minimizes* the labeling differences along the edges is well-studied. For example, in the context of minimum bandwidth or wavefront reduction ordering for sparse matrices, it is known that the problem is NP-hard, and a number of heuristics [23, 28, 42] were proposed. One such heuristic is to order vertices using the Fiedler vector. This is found to be very effective when combined with a refinement strategy. Motivated by this approach, we approximate (1) by

$$\max_c \sum_{\{i,j\}\in E_c} w_{i,j}(c_i - c_j)^2, \text{ subject to } \sum_{k\in V_c} c_k^2 = 1 \qquad (3)$$

where $c \in R^{|V_c|}$. This continuous problem is solved when c is the eigenvector corresponding to the largest eigenvalue of the weighted Laplacian of the country graph, while the Fiedler vector (the eigenvector corresponding to the second smallest eigenvalue) minimizes the objective function above. Once (3) is solved, we use the ordering of the eigenvector as an approximate solution for (1). We call this algorithm SPECTRAL.

Figure 7 illustrates three coloring schemes on a 4×4 unweighted grid graph given 16 colors in some discrete spectrum. A random assignment of colors, RANDOM, does reasonably well, but has one edge with a color difference of 2. SPECTRAL performs better, with the minimum color difference of 4. However there are still 2 edges with a color difference of only 4. It is easy to see that SPECTRAL can

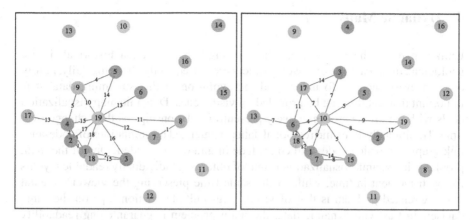

Fig. 8 Applying coloring schemes for the country graph corresponding to the map in Fig. 4b. *Left*: SPECTRAL. There are two edges of color difference 1. *Right*: SPECTRAL+GREEDY, the smallest color difference along any edges is now 4. Node labels are the color index given to the node, and edge label are the absolute difference of the node color index. Nodes are positioned at the center of the polygons in Fig. 4b

be improved (e.g., swapping colors 6 and 2 would improve the measurements according to both cost functions). With this in mind we propose GREEDY, a greedy refinement algorithm based on repeatedly swapping pairs of vertices, provided that the swap improves the coloring scheme according to one of the two cost functions. Starting from a coloring scheme obtained by SPECTRAL and applying GREEDY often leads to significant improvements.

So far we have been using a simple grid graph to illustrate the algorithms. The actual country graphs are usually more complex. Figure 8 (left) gives the country graph corresponding to the map in Fig. 4b, with color assignment given by SPECTRAL. There are two edges of color difference 1. Applying the GREEDY algorithm guided by the MaxMin cost function to the result of SPECTRAL gives Fig. 8 (right). Now the minimum color difference along any edge is 4, a large improvement. This is indeed the coloring scheme used to create Fig. 4b.

The GREEDY algorithm has a high computational complexity as we consider all possible $O(|V_c|^2)$ pairs of vertices for potential swapping. Since recomputing the cost functions can be done in time proportional to the sum of degrees of the pair on nodes considered for swapping, the overall complexity of GREEDY is $O(|V_c|^2 + |E_c|^2)$. Because the country graph G_C is typically much smaller than the underlying graph G, GREEDY is still quite fast and all maps in this chapter were colored using SPECTRAL+GREEDY, the GSpectral algorithm.

We note in passing that GREEDY is flexible enough to be used with any other cost functions. For example, the MaxMin cost function could be modified to measure the distance between two colors in terms of their Euclidean distance in the RGB or Lab color space, instead of the index difference.

4 Dynamic Maps

Unlike maps of the real world, where changes happen on a historical, if not geological, time scale, the data we consider here is frequently changing daily, hourly or even every second. To understand the evolution of this streaming data, it is important that stability can be provided by visual cues. Dynamic map visualization deals with the problem of effectively presenting relationships as they change over time. Traditionally, dynamic relational data is visualized by animations of node-and-link graphs, in which nodes and edges fade in and out as needed. One of the main problems in dynamic visualization is that of obtaining individually readable layouts for each moment in time, while at the same time preserving the viewer's mental map. A related problem is that of visualizing multiple relationships on the same dataset. Just as with dynamic data, the main problem is guaranteeing readability while preserving the viewer's mental map. Representations based on the geographic map metaphor could provide intuitive and appealing visualizations for dynamic data and for multiple relationships on the same dataset.

We give some motivation for dynamic map layout first, then describe a heuristic to promote dynamic *cluster stability*, an optimal color assignment algorithm to maximize *color stability* between maps, and heuristics to improve *layout stability*. Additional details can be found in [21].

4.1 Dynamic Maps: A Motivation

Consider the problem of computing a "good" distance measure between a set of known DNA samples that is based on multiple similarity measures (e.g., NRY and mtDNA), with the goal of creating a "canonical map" of the DNA space spanned by these samples. In this map, DNA samples are nodes, two nodes are close to each other if they have a high similarity, and groups of similar nodes are clustered into "countries."

Next an unknown DNA sample can be compared to the known ones and then placed on the map, in a way that minimizes its distance to the most similar known samples. In order to do this, we must compute such a "good" distance measure from multiple similarity metrics, for example, by assigning weights to each metric and taking a weighted sum, or some non-linear combination thereof. Once appropriate weights have been assigned we can create the canonical map where we will place unknown DNA samples.

Thus the main problem here is figuring out how to appropriately combine a set of different similarity metrics. Given two different similarity metrics on the same set of DNA samples, a simple way to visualize them is to create two static maps. This, however, is not very helpful to the scientists who would like to understand the correspondences and differences between these two metrics, as node positions on the static maps are likely to be unrelated. In addition, color assignment for the countries are random, making it even harder to understand the relationships.

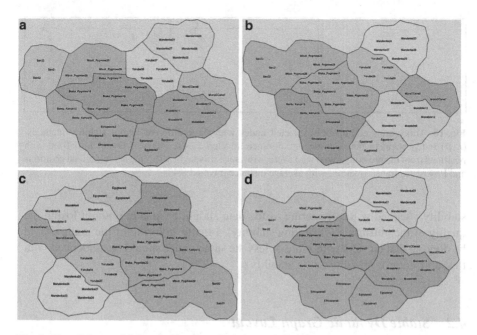

Fig. 9 Visualizing multiple maps requires both layout stability and color stability. (**a**): original map. (**b**): a new map based on different similarity data, and with node layout computed independently. It is difficult to see the corresponding nodes in the two maps. (**c**): the new map computed to optimize node layout stability with regard to (**a**), which makes it possible to compare nodes, while clusters are still hard to compare. (**d**): the new map with optimal node layout and color assignments, which makes it easy to compare with (**a**); e.g., it is clear that two clusters in the top left of (**a**) are now merged

The maps in Fig. 9 show the nature of the problem. In Fig. 9a, 39 subjects are embedded in 2D space based on mtDNA similarity, using multi-dimensional scaling, and clustered and mapped. In Fig. 9b we have re-embedded the subjects using a different similarity metric (NRY DNA similarity), independent of Fig. 9a. Compared with Fig. 9a, the layout changed significantly. Furthermore, although the same color palette is used, colors are assigned independently, making it even harder to figure out the relationship between Fig. 9b,a. In Fig. 9c, the embedding of NRY DNA similarity is done to minimize the difference to that of the embedding based on the mtDNA similarity measure, making it possible to see that node positions are largely unchanged. However due to the color assignment, it is still difficult to compare it with the map in Fig. 9a. Finally, in Fig. 9d, colors are properly matched such that clusters with mostly the same nodes are colored using the same colors. This makes it easy to compare Figs. 9a,d. For example, we can clearly see that two countries in the top left are now merged in a single country.

From this example, we can see that to give the viewer a stable mental map when viewing dynamic maps, we first have to ensure that the layout is done such that the same node should appear at the same or a nearby position if possible (layout

Fig. 10 Trajectories of randomly selected nodes with three different layout stability methods. (*Left*) independent layout with average distance traveled 21.41; (*Middle*) layout initialized with positions from the previous frame with average distance traveled 13.19; (*Right*) initialized positions and Procrustes transformation, with average distance traveled 8.43

stability); secondly, the clustering of the data should be stable, without losing the quality of the clustering (clustering stability). Finally, coloring of the maps should be done so that clusters with more or less the same nodes should be associated with the same color if possible (color stability).

4.2 Stable Dynamic Graph Layout

Abstractly, the problem of dynamic map layout is that of computing node positions, which is related to the well-known readability versus mental map preservation problem for dynamic graph drawing. Traditionally, given a sequence of graphs, one can compute node positions for the current graph in the sequence by starting with the node positions from the previous graph in the sequence and followed by local node position refinement. One shortcoming of such an approach is that even with the node-position initialization, two consecutive graphs in the sequence with very similar topologies can have very different drawings, causing node-jumping between frames, and failing to preserve the mental map. One technique that can help to moderate the change in position is to apply a Procrustes transformation of the coordinates of the nodes, so that the new layout matches the old layout as much as possible by using scaling, translation and rotation.

Another dynamic layout approach is to "anchor" some, or all, of the nodes, or to limit node movement by adding artificial edges linking graphs in different time frames [8]. However, such approaches can introduce biases that were not in the data itself, which is undesirable when analyzing highly sensitive real-world data, such as DNA similarity.

To evaluate the different layout stability approaches, we compare the trajectories of a set of randomly selected nodes from the dataset in Fig. 9. Figure 10 (left) shows such node trajectories, where the position of a node in the new graph is obtained by an independent MDS computation of the two layouts. Figure 10 (middle) shows the node trajectories, when using an MDS layout of the current frame, where the position of each node is initialized with the position obtained from the previous frame. Finally, Fig. 10 (right) shows node trajectories, where the position of each node is initialized with the position obtained from the previous iteration and combined with a Procrustes transformation to fit the previous frame.

In all cases we experimented with, the last strategy was the best one, its trajectories the least jittery. We quantify these strategies by computing the average node-travel distance per frame (over all nodes in the graph, not just the random sample shown in the figure). In this example the distances traveled are 21.41, 13.19 and 8.43 pixels, respectively. This confirms that there are non-trivial improvements when we use the initial nodes position together with a Procrustes transformation.

4.3 Stable Clustering

We now consider the clustering problem on dynamic graphs, where the changes are adding/removing nodes, adding/removing edges, and modifications in node weights and edge weights. For the purpose of mental map preservation, we seek to preserve the clustering structure between the iterations as much as possible, provided that doing so does not result in suboptimal clustering.

One commonly used clustering for graph data is based on minimizing the modularity of a partition of the nodes. Here the modularity of a partition is defined as

$$Q = \frac{1}{2m} \sum_{i,j \in V} [w_{i,j} - \frac{k_i k_j}{2m}] \delta(C(i), C(j)) \tag{4}$$

where $w_{i,j}$ is the weight of the edge between i and j. The scalar $k_i = \sum_j A_{i,j}$ is the sum of the weights of the edges attached to node i, and $C(i)$ is the cluster node i is assigned to. The $\delta-$ function $\delta(C(i), C(j))$ is 1 if $C(i) = C(j)$ and 0 otherwise, and $m = \frac{1}{2} \sum_{i,j} A_{i,j}$ is the sum of all edge weights.

We describe a simple heuristic to combine the two objectives of modularity and cluster stability for dynamic clustering. This heuristic is a dynamic variation of the agglomerative clustering algorithm of Blondel et al. [2]. Heuristics are a reasonable approach, as the dynamic modularity clustering problem is also NP-Hard [17].

We begin with each node as a singleton. During the first level of clustering, we consider merging only node pairs which belong to the same cluster in the clustering of the previous iteration. When no more node pairs are left for merging, the current clustering is used to construct a "contracted graph" with each cluster as a super node and appropriately adjusted adjacencies and edge weights. We proceed iteratively with the contracted graph as input. The clustering of the previous iteration is explicitly used in the first level and afterwards we apply the algorithm of Blondel et al. [2].

We evaluate the effectiveness of our heuristic with a measure of cluster similarity given by Rand [33]. This measure is based on node-pair clustering as follows. Let C and C' denote two clusterings of a graph G, and let S_{11} denote the set of pairs that are clustered together in both clusterings, and S_{00} denote the set of pairs that are in different clusters. Then the Rand distance between the two clusterings is given by

$$rand(C, C') = 1 - \frac{2(|S_{11}| + |S_{00}|)}{n(n-1)} \tag{5}$$

The value will be 0 if the two clusterings are identical, and 1 if one clustering is a singleton clustering and the other one is that of all nodes in the same cluster.

With the data from Fig. 9, we evaluated the quality between each pair of successive iterations and averaged these values over all successive pairs. Without our heuristic, the average Rand measure was 0.0631, and with the heuristic, it was 0.0252. This shows an improvement of a little more than 60 % with the heuristic.

4.4 Stable Map Coloring

Color stability, that is, using the same color for countries on the two maps that share most of their nodes, is an essential ingredient in visualizing dynamic maps. In order to maintain color stability, we need to match the best pairs of clusters in different maps.

Given two maps, let C_{old} and C_{new} be vectors representing clustering information of these two maps. We have to minimize the number of nodes whose cluster is different in C_{old} and C_{new}. Let $s(C_{old}, C_{new})$ be the number of nodes that do not undergo clustering change.

$$s(C_{old}, C_{new}) = \sum_{u \in V} \delta(C_{old}(u), C_{new}(u)); \tag{6}$$

$\delta(u, v) = 1$ if $u = v$, and 0 otherwise.

The cluster matching problem is to find a permutation Π of the clustering C_{new}, such that $\Pi(C_{new})$ maximizes $s(C_{old}, \Pi(C_{new}))$. For example, let $C_{old} = \{1, 1, 2, 2, 3\}$ be the clusters assigned to the five nodes v_1, v_2, v_3, v_4, v_5; let $C_{new} = \{2, 2, 1, 3, 4\}$ be the new clustering in which v_3 and v_4 split into two clusters. Clearly $s(C_{old}, C_{new}) = 0$. The optimum matching is the permutation $\Pi : \{1, 2, 3, 4\} \rightarrow \{2, 1, 4, 3\}$. The resulting clustering, $\Pi(C_{new}) = \{1, 1, 2, 4, 3\}$ gives $s(C_{old}, \Pi(C_{new})) = 4$.

The problem can be modeled with a maximum weighted matching (MWM) of a bipartite graph. The corresponding bipartite graph G_C has node set $\{1, 2, \ldots, |C_{old}|\} \times \{1, 2, \ldots, |C_{new}|\}$. The edge weight, $w(i, j)$, corresponds to the number of nodes that are common between cluster i of C_{old} and cluster j of C_{new}.

$$w(i, j) = \sum_{u \in V} \phi(i, j, u) \tag{7}$$

$\phi(i, j, u) = 1$ if $C_{old}(u) = i$ and $C_{new}(u) = j$.

The maximum weighted bipartite matching of G_C gives a matching Π between the clusters C_{old} and C_{new} that will maximize $s(C_{old}, \Pi(C_{new}))$. The MWM for bipartite graphs can be found using the Hungarian algorithm [26]. For bipartite graphs, an efficient implementation of the Hungarian algorithm using Fibonacci heaps [11] runs in $O(mn + n^2 \log n)$, where m and n are the number of edges and nodes in G_C, respectively. If we assume that a cluster in the old clustering does not split into more than a constant number of clusters in the new clustering, then $m = O(n)$. This yields a $O(n^2 \log n)$ algorithm for MWM. Since $w(i, j)$ are all integers in the range 0 to $|V|$, the algorithm by Gabov and Tarjan algorithm [13] for MWM can be implemented with $O(n^{\frac{3}{2}} \log(n|V|))$ complexity. In practice, the number of clusters is typically small and the Hungarian algorithm is fast enough.

5 Case Studies

In this section, we walk through some sample views of graphs as maps derived from real-world data and note aspects found in the drawings that might be perceived by a typical user. As noted above, the technique is intended for fairly large graphs.

Gleaning information from the maps, as with any large data set, typically involves an interactive, multi-scale process, similar to that used for exploring geographic maps. One can view the map at small scale to sense the overall layout, the major regions, and how they relate to each other. One then zooms in to see local detail, and to traverse the map along small features. At some point, one may zoom out again to put the local details into a global context.

Based on this style of use, our figures are most effectively displayed as a large image, often a meter or more in width, or via an interactive viewer. In the former case, the user can physically move to change the scale. In the latter case, the viewer provides the scale change and, at the same time, can provide some version of semantic zoom, so that more detail is added the more the user zooms in. In addition, an interactive viewer can provide such additional features as textual search or links connecting a feature on the map to some external information. For example, clicking one of the books shown in Fig. 5 might take the user to the books entry at Amazon.com or to a Wikipedia article on the author.

It is important to note that in all cases the countries and their geography in the resulting maps are not part of the input data, but emerge from the graph layout and clustering algorithms. This gives the user a potential tool to discover structure based solely on local data. On the other hand, if a desired clustering is known, this can be used as the basis for "country" construction.

5.1 Maps and Recommendations

Recommender systems provide a motivating reason for displaying graphs as maps. Many content providers, both to assist their customers in making choices and to motivate them to make more selections, have systems to suggest additional picks based on various individual and group statistics, processed and refined with various algorithms. Typically, the user is provided with a small list or table of options, perhaps with some associated numbers giving some clue as to how the selections were made.

We feel that the map metaphor can give the user the relevant information in a more familiar way, with country placement and the underlying edges suggesting the connections. In addition, with the appropriate GUI, the user has access to a large volume of data, rather than a pruned list. With the full map, the user is not limited to a small region but can explore the map, following connections far from initial centers of interest. Guided to the movie *The Exorcist*, the user may wander down to *Babylon 5*.

The examples in the next two sections can be viewed as visual bases for recommender systems for movies and television shows, respectively. Indeed, the provenance of the movie data is directly tied to recommender systems.

5.2 A Map of Movies

Figure 11 shows a map derived from the data used for the Netflix competition to invent a better recommender algorithm. The underlying graph uses movies (and television shows) as nodes. Closely related shows are connected with an edge. In addition, the edges are weighted based on how strong the connection is. The base graph contains 11,283 nodes and 71,449 edges. Using a minimum edge weight as a threshold, we obtain a graph with 11,831 edges. Most of this graph resides in a single connected component with 3,407 nodes and 11,116 edges. It is this final graph that is used in the figure.

Zooming in, one can readily identify various countries. In the north, one finds a country of teen/adult animation (Fig. 12) containing the likes of *The Simpsons*, *Futurama* and *South Park*, with some Dave Chappelle shows pulled into the mix. These last form a segue to the purple country to the right (Fig. 13) containing shows with a certain style of adult humor, such as *Saturday Night Live*, *George Carlin* and *Chris Rock*.

At the bottom of the map, we find two adjacent countries (Fig. 14). The more southerly consists of classic, space-based science fiction shows such as *Star Wars*, *Star Trek*, *Stargate* and *The X-Files*. To the north we discover the eerier science fiction of *The Twilight Zone* and *The Outer Limits*. Moving clockwise around the periphery of the map, we encounter clearly defined regions of Japanese films; *Mystery Science Theater 3000* shows; Michael Moore documentaries; cerebral

Fig. 11 A map of movies and TV shows (3,407 nodes, 11,116 edges)

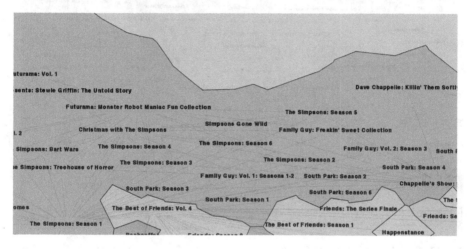

Fig. 12 Mostly teen/adult animation

British detectives such as Campion and Lord Wimsey abutting a whole separate country of British mysteries; Ken Burns documentaries; horror films such as *Nightmare on Elm Street*; and several contiguous countries of juvenile fare.

5.3 Personalized Recommendations

Most of the maps we have seen so far have been constructed independent of any particular person, with data based on the aggregate behavior of many people. This information can then be tailored to an individual. For example, starting from a map of related TV shows, we could generate a personalized heat map version where regions of low interest are colored with cool colors, and regions containing highly

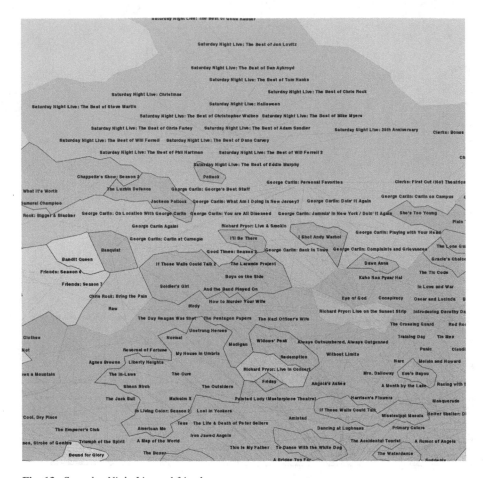

Fig. 13 *Saturday Night Live* and friends

recommended shows are colored with a hot color. Figure 15 shows such a heat map, where shows are scored using a factorization based recommender [22], with dark colors for shows that score low, and light for shows that score high. Such maps would be generated dynamically based on the viewing preference of an individual, and based on what TV shows are available at this moment in time, much like a personalized weather forecast, but for TV shows. These maps uniquely capture the viewing preferences of the user or household, and evolve as the availability of TV shows, and the user's taste, change with time. We can also generate a heat map profile, determined by how often the user watches certain shows over a fixed time period, say, a week or a month. Handling such fluid scenarios well requires the dynamic techniques discussed in Sect. 4 to be well honed.

Note that this approach could be extended to social networks: What recommendations could be made that might appeal to my friends on Facebook as a group?

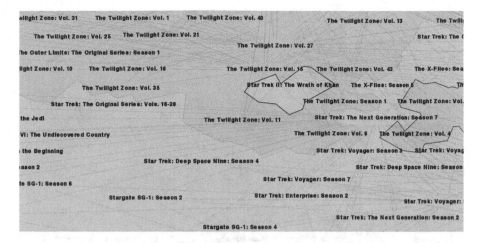

Fig. 14 Parts of two science fiction countries

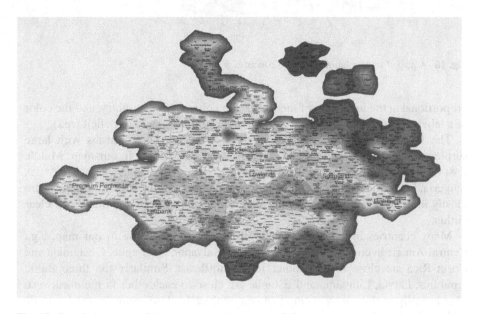

Fig. 15 Sample heat map giving personalized recommendations

5.4 Trade

Figure 16 is a map visualizing the trade relations between all countries. Bilateral trade data between each of the 209 countries and its top trading partners were acquired from Mathematica's `CountryData` package. The font size of a label is

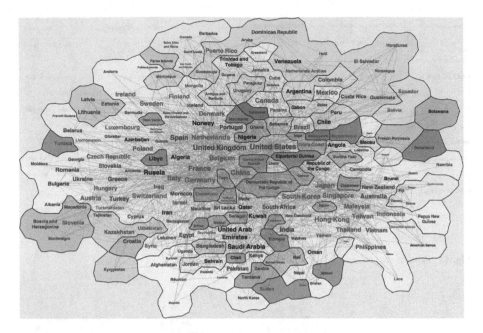

Fig. 16 A map of trade relations between countries

proportional to the logarithm of the total trade volume of the country, and the color of a label reflects whether a country has a trade surplus (black) or deficit (red).

The label color gives an easy way to spot the oil-rich countries with large surpluses, which are distributed all over the world as well as in our map: Middle East (Saudi Arabia, Kuwait), Europe (Russia), South America (Venezuela), Africa (Nigeria, Equatorial Guinea). On the other hand, the countries with huge deficits are mostly in Africa (Sierra Leone, Senegal, Ethiopia) with the United States, the clear outlier.

Many countries in close geographic proximity end up close in our map, e.g., Central American countries like Honduras, El Salvador, Nicaragua, Guatemala and Costa Rica are close to each other in the northeast. Similarly the three Baltic republics, Latvia, Lithuania and Estonia, are close to each other in the northwest. This is easily explained by noting that geographically close countries tend to trade with each other. There are easy-to-spot exceptions: North Korea is not near South Korea, Israel is not particularly close to Jordan or Syria.

The G8 countries (Canada, France, Germany, Italy, Japan, Russia, United Kingdom, and the United States) are all in close proximity to each other in the center of the map. Two of the largest and closest countries in our map are China and the United States. Clearly, the proximity is due to the very large trade volume rather than geographic closeness. All these countries are in the largest cluster which is dominated by European countries in the west, Asian countries in the east, and Middle Eastern countries in the south.

Interestingly, we see from the map that African countries are distributed in several clusters in close proximity to China (a major trading partner to many African countries), the United States (trading less with Africa these days), and around former colonizers (e.g., Togo, Cameroon and Senegal, which are all close to France). On the other hand, Caribbean and South and Central American countries form several clusters in the north of the map. In addition, these clusters are mostly contiguous, essentially forming a supercluster. This differentiation between Latin America and Africa is clearly brought out by the GMap figure.

Finally, we note that the periphery of the map contains small countries from around the world, and countries with few trading partners.

6 Final Thoughts

One rule that most researchers in information visualization discover is that people tend to be most comfortable and adept when they see data the way they have always seen it. Ignoring this can lead to a "demonstrably" better but alien visualization being rejected by its intended users. For us, this provides a strong argument for the effectiveness of displaying graphs and clusters as geographic maps to promote a more human centric visualization. At present, we have mostly anecdotal and non-scientific evidence that this is a preferable way for viewing relational data. But we have seen people spend long periods of time poring over these maps. It would be desirable to perform more extensive user studies to explore how well maps compare to other metaphors, and to explore more ways to use maps. At the same time, there is still much work to be done enhancing and tuning the underlying algorithms, especially in the context of dynamic graphs.

Acknowledgements We would like to thank Stephen North and Chris Volinsky for helpful discussions and encouragement.

References

1. Battista, G.D., Eades, P., Tamassia, R., Tollis, I.G.: Algorithms for the visualization of Graphs. Prentice-Hall (1999)
2. Blondel, V., Guillaume, J., Lambiotte, R., Lefebvre, E.: Fast unfolding of communities in large networks. Journal of Stat. Mechanics: Theory and Experiment **2008**, P10,008 (2008)
3. Brewer, C.: ColorBrewer - selecting good color schemes for maps. www.colorbrewer.org
4. Cappanera, P.: A survey of obnoxious facility location problems. Technical Report TR-99-11, Dipartimento di Informatica, Universit a di Pisa (1999)
5. Cleveland, W.S.: Visualizing Data. Hobart Press, Summit, New Jersey, U.S.A. (1993)
6. Dillencourt, M.B., Eppstein, D., Goodrich, M.T.: Choosing colors for geometric graphs via color space embeddings. In: 14th Symposium on Graph Drawing (GD), pp. 294–305 (2006)
7. Duarte, A., Martí, R., Resende, M., Silva, R.: GRASP with path relinking heuristics for the antibandwidth problem. Networks (2011). Doi: 10.1002/net.20418

8. Erten, C., Harding, P.J., Kobourov, S.G., Wampler, K., Yee, G.V.: Graphael: Graph animations with evolving layouts. In: G. Liotta (ed.) Graph Drawing, *Lecture Notes in Computer Science*, vol. 2912, pp. 98–110. Springer (2003)

9. Fabrikant, S.I., Montello, D.R., Mark, D.M.: The distance-similarity metaphor in region-display spatializations. IEEE Computer Graphics & Application **26**, 34–44 (2006)

10. Fabrikant, S.I., Montello, D.R., Mark, D.M.: The natural landscape metaphor in information visualization: The role of commonsense geomorphology. JASIST **61**(2), 253–270 (2010)

11. Fredman, M.L., Tarjan, R.E.: Fibonacci heaps and their uses in improved network optimization algorithms. J. ACM **34**, 596–615 (1987). DOI http://doi.acm.org/10.1145/28869.28874. URL http://doi.acm.org/10.1145/28869.28874

12. Fuchs, G., Schumann, H.: Visualizing abstract data on maps. In: Proceedings of the Information Visualisation, Eighth International Conference, IV '04, pp. 139–144. IEEE Computer Society, Washington, DC, USA (2004). DOI 10.1109/IV.2004.152. URL http://dx.doi.org/10.1109/IV.2004.152

13. Gabow, H.N., Tarjan, R.E.: Faster scaling algorithms for network problems. SIAM J. Comput. **18**, 1013–1036 (1989). DOI 10.1137/0218069. URL http://portal.acm.org/citation.cfm?id=75795.75806

14. Gansner, E.R., Hu, Y.F., Kobourov, S.G.: Gmap: Drawing graphs as maps. http://arxiv1.library.cornell.edu/abs/0907.2585v1 (2009)

15. Gansner, E.R., Hu, Y.F., Kobourov, S.G., Volinsky, C.: Putting recommendations on the map - visualizing clusters and relations. In: Proceedings of the 3rd ACM Conference on Recommender Systems. ACM (2009)

16. Gansner, E.R., North, S.C.: An open graph visualization system and its applications to software engineering. Softw., Pract. Exper. **30**(11), 1203–1233 (2000)

17. Görke, R., Maillard, P., Staudt, C., Wagner, D.: Modularity-driven clustering of dynamic graphs. In: 9th Symp. on Experimental Algorithms, pp. 436–448 (2010)

18. *Graphviz* graph visualization software. www.graphviz.org/

19. Hu, Y., Gansner, E.R., Kobourov, S.G.: Visualizing graphs and clusters as maps. IEEE Computer Graphics and Applications **30**(6), 54–66 (2010)

20. Hu, Y., Kobourov, S., Veeramoni, S.: On maximum differential graph coloring. In: Proceedings of the 18th international conference on graph drawing (GD'10), pp. 274–286. Springer-Verlag (2011)

21. Hu, Y., Kobourov, S., Veeramoni, S.: Embedding, clustering and coloring for dynamic maps. In: Proceedings of IEEE Pacific Visualization Symposium. IEEE Computer Society (2012)

22. Hu, Y.F., Koren, Y., Volinsky, C.: Collaborative filtering for implicit feedback datasets. In: 8th IEEE International Conference on Data Mining (ICDM), pp. 263–272 (2008)

23. Hu, Y.F., Scott, J.A.: A multilevel algorithm for wavefront reduction. SIAM Journal on Scientific Computing **23**, 1352–1375 (2001)

24. Keim, D.A., Panse, C., North, S.C.: Medial-axis-based cartograms. IEEE Computer Graphics and Applications **25**(3), 60–68 (2005)

25. van Kreveld, M.J., Speckmann, B.: On rectangular cartograms. Comput. Geom. **37**(3), 175–187 (2007)

26. Kuhn, H.W.: The hungarian method for the assignment problem. Naval Research Logistics Quarterly **2**(1–2), 83–97 (1955). DOI 10.1002/nav.3800020109. URL http://dx.doi.org/10.1002/nav.3800020109

27. Kuhn, W., Blumenthal, B.: Spatialization: spatial metaphors for user interfaces. In: Conference companion on Human factors in computing systems: common ground, CHI '96, pp. 346–347. ACM, New York, NY, USA (1996). DOI 10.1145/257089.257361. URL http://doi.acm.org/10.1145/257089.257361

28. Kumfert, G., Pothen, A.: Two improved algorithms for envelope and wavefront reduction. BIT **35**, 1–32 (1997)

29. Leung, J.Y.T., Vornberger, O., Witthoff, J.: On some variants of the bandwidth minimization problem. SIAM J. Comput. **13**, 650–667 (1984)

30. Lima, M.: Visual Complexity: Mapping Patterns of Information. Princeton Architectural Press (2011)
31. Lloyd, S.: Last square quantization in pcm. IEEE Transactions on Information Theory **28**, 129–137 (1982)
32. Newman, M.E.J.: Modularity and community structure in networks. Proc. Natl. Acad. Sci. USA **103**, 8577–8582 (2006)
33. Rand, W.M.: Objective criteria for the evaluation of clustering methods. J. of the American Statistical Association pp. 846–850 (1971)
34. Raspaud, A., Schröder, H., Sýkora, O., Török, L., Vrt'o, I.: Antibandwidth and cyclic antibandwidth of meshes and hypercubes. Discrete Mathematics **309**, 3541–2552 (2009)
35. Salvini, M.M., Gnos, A.U., Fabrikant, S.I.: Cognitively plausible spatialization of network data. In: Proceedings of the 20th International Cartographic Conference (2011)
36. Scott, J., Hu, Y.: Level-based heuristics and hill climbing for the antibandwidth maximization problem. Technical Report RAL-TR-2011-019, Ritherford Appleton Laboratory, UK (2011)
37. Skupin, A.: A cartographic approach to visualizing conference abstracts. IEEE Computer Graphics & Application **22**(1), 50–58 (2002)
38. Skupin, A.: The world of geography: Visualizing a knowledge domain with cartographic means. Proc. National Academy of Sciences **101**(Suppl. 1), 5274–5278 (2004)
39. Skupin, A.: Discrete and continuous conceptualizations of science: Implications for knowledge domain visualization. Journal of Informetrics **3**(3), 233–245 (2009)
40. Skupin, A., Buttenfield, B.P.: Spatial metaphors for visualizing information spaces. In: Proc. AUTO-CARTO 13, pp. 116–125 (1997)
41. Skupin, A., Fabrikant, S.I.: Spatialization. In: Handbook of Geographic Information Science, pp. 61–80. Blackwell Publishers (2008)
42. Sloan, S.W.: An algorithm for profile and wavefront reduction of sparse matrices. International Journal for Numerical Methods in Engineering **23**, 239–251 (1986)
43. Steele, J., Iliinsky, N.: Beautiful Visualization: Looking at Data through the Eyes of Experts, 1st edn. O'Reilly Media, Inc. (2010)
44. Ullman, J.D.: Elements of ML programming - ML 97 edition. Prentice Hall (1998)

Part II
Theory and Science

Individual Differences and Translational Science in the Design of Human-Centered Visualizations

Tera Marie Green, Richard Arias-Hernandez, and Brian Fisher

Abstract In this chapter, we discuss a research framework borrowed from medicine, called translational science, and how it may be used to develop more useful research protocols for the study of user interface cognition and visual analytical reasoning. Translational science incorporates laboratory research, field studies, and other empirical protocols into a holistic research program which ambitiously incorporates the study of individual and collaborative cognition in a longitudinal and/or ethnographic approach to interactive visualization research and design. To introduce how translational science fits into human centric visualization design and evaluation, we discuss research methods it would employ. We also explore the unique variabilities that affect both the human-visualization interaction and visualization-mediated human to human collaboration through our reported research. These variabilities—or individual differences—complicate the study of user interface cognition and make a more holistic approach like translational science necessary. A current and on-going translational science program is described, and we discuss its unique challenges and contributions.

1 Introduction

Visualizing information or shared knowledge can be an impactful and valuable method in the very human cognitive processes of making sense of new data, looking for patterns, developing hypotheses around these patterns, and generalizing conclusions which become the basis for action. Each of these cognitive tasks is complex and influenced by a myriad of factors. And given that users often gather around a visualization to reason together, differences within and between the users

T.M. Green (✉) • R. Arias-Hernandez • B. Fisher
School of Interactive Arts + Technology, Simon Fraser University, 250-13450 102nd Ave,
Surrey, BC V3T 0A3, Canada
e-mail: terag@sfu.ca; ariash@sfu.ca; bfisher@sfu.ca

W. Huang (ed.), *Handbook of Human Centric Visualization*,
DOI 10.1007/978-1-4614-7485-2_4, © Springer Science+Business Media New York 2014

also become influential in how the visualization is used and what value is attributed to its use. In this chapter, we explore current research in the interaction between the information visualization and the user as well as groups of users gathered around visualizations. In addition, we introduce a framework for a more holistic evaluation of interactive visualizations called Translational Science, borrowed from medicine, which not only evaluates current visualizations, but also adds to an understanding of how humans use visualizations to analyze and inform future interactive visualization design.

When considering the how best to design and evaluate visualization interfaces so that they may best communicate and facilitate interaction with data requires an understanding of the analytic process, the data being analyzed and used, and the human analyst(s) that will seeking to derive conclusions through interaction with the visualization. This understanding is not a straightforward endeavor.

There is no standard unit for human cognition; human analysts vary in the way they approach, organize, and utilize information. Further, there is the analytical process—the iterative, recursive narrative of analysis that starts with the first attempt at sensemaking and ends with hypotheses, conclusions, or other analyst-defined end-goals. In visualization, this analytical process is complicated by the use of visualization of large (often noisy) data and collaboration with other humans, who each have their own individual differences. In visualization, the analytical process is not likely to be a cursory affair, and so neither should the design and evaluation of visualization interfaces be superficial or one-dimensional. To build this understanding, exploratory and evaluative study should be holistic, in situ, and in vivo. And these studies, in part or in whole, should tackle the entire visualization process, from the analytical process, use of data, and the individual differences of the human analyst.

Note how much of this understanding involves a comprehension of human cognition. Despite the overwhelming research emphasis in the visualization and visualization on graphics and computational analysis, the visualization cannot do the analysis by itself or on its own. It is a tool, and potentially a very powerful one. But it is not the human analyst. What the human brings to the table is unique and cannot be supplanted by graphics or computation. Humans supply the motivation: goals, direction, and priorities. Motivation will vary from task to task and from day to day. Humans solve problems. Computers can tackle unambiguous problems that can be solved following rule sets, including fuzzy rules. But when the tasks get complex or conflicting, humans can reorganize information to solve problems in ways not even they can narrate or explain; this happens during 'a-ha!' moments of spontaneous insight [56] and other types of information reorganization. Further, a computer can categorize with uncertain rules, but it is humans that can create mental models and adapt them on the fly, inferring rules where necessary [57], and then use those models to reason through complex problems. These cognitive processes may not be completely understood but they can be observed and mapped, which informs the design process. We will explore this core idea throughout what follows.

2 Research for Design

Currently, the understanding the analytical process in visualization tends to be limited to developer-centered decisions, domain-only literature reviews, or to discussions with users and experts. A brief survey of the literature easily provides examples. Andrienko, Andrienko, and Wrobel developed a toolkit for movement data analysis after a literature review of related work [1]. This is rather common; developers look at what's available and develop something that extends a concept or meets a specific need. Sometimes this need is computational [12], and sometimes the developers decide their approach to interaction or visual display is preferable [33]. Some tools are developed after a conversation with an expert that uses a process the developer would like to visualize. iPCA, a tool designed to support factor analysis, was created after conversations with experts in one domain [17]. WireVis, a tool that was developed for financial analysis was developed after conversations with a few members of one banking department [4]. Personal experience and anecdotal evidence suggests that this modus operandi is similar to many visualization design cycles.

This approach has been at least somewhat successful in part because even if the visualization is difficult to work with, analytical reasoning is so adaptive that humans can learn to use inferior tools [29]. A skilled analyst can make some use of an interface if it meets even minimum requirements and the analyst has had some time to learn the visualization. So designing and evaluating interfaces that communicate effectively to and with the human analyst requires more than proof that the analyst can get through the interface or solve a laboratory task. For this reason, understanding the analytic process requires much more than a description of programming requirements. Visualization interface design is indeed at least part tool creation. But because visualization tasks are complex to the point of impossible, a thorough understanding of the problem or problem solutions that the interface is designed to support is important.

This understanding of analytical process can include a detailed description based off of interviews with would-be users, but in many cases it should include some method in which designers work hand-in-hand with the analysts they would work with. This is especially true of visualization that would support collaboration between multiple users, as the analytical process becomes more complex when cognition becomes social. What is garnered from these studies is a better understanding of the analytical process, as well as how the individual differences of multiple users contribute to or impede the current analysis.

For examples of field research protocols, we turn other disciplines such as Interaction Design. (See Fig. 1.) There are a variety of methods available, but for the scope of this short chapter we will look at only a handful of designs including Activity Theory and Participatory Design. The social group as the interface to human cognition is a phenomenon more studied in anthropology than in cognitive sciences, even though since the 1980s cognitive anthropologists and other cognitive scientists have emphasized the importance of incorporating this fact into mainstream

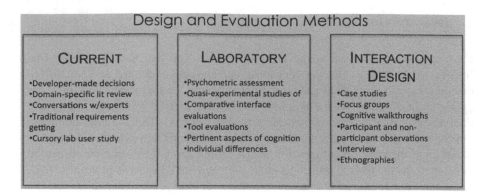

Fig. 1 Design and evaluation methods for visualization

cognitive science [28, 37]. Each of these systems of study have research tools or rotocols which assist in breaking down the analytical process into subprocesses.

One tradition in cognitive science that emphasizes the role of society and culture on the shaping of human cognition is Activity Theory [21], which is based on Leontiev's [23] psychological analyses of activity and Vygotsky's [40] studies on child development, from which he derived his insights on the 'zone of proximal development.' Activity Theory [21] breaks the analytic process down into subject, object, activity, and operations. Skilled analysts are *subjects* which work on artifacts such as computers. Every observable action the analyst makes either with or without the object is an *activity*. What motivates the subject to do an activity is their *object* (as in *objective*). When an activity is done so often that it becomes practiced and rote, it becomes an *operation*.

Activity Theory sees all activity as *mediated* through tools and interfaces and other artifacts. It is through activity with artifacts such as an interactive visualization that the analyst or subject creates a context to work within. The context is a "transformative" engagement in which the subject acts mediated through artifacts and with other subjects to create a synergy unique to that context [21].

Activity Theorists postulate that the human mind is shaped by the social and cultural context. For example, for Vygotsky, it is the immediate social group that provides the cognitive scaffolding required for children to develop more complex cognitive structures. Absent the necessary interactions with other individuals, these cognitive structures will also be absent [40]. Activity theorists also postulate that activity, the main driver of cognition, is motivated by the individual's social and cultural context. In other words, the objects or goals that orient human activities as well as the human activities by themselves are shaped by and originate from social and cultural structures. As members of a social and cultural collective we adopt goals, which make sense of what we do for us and for others, and we conform to socially and culturally accepted activities to accomplish these goals [27]. Therefore, human activity-mediated cognition is inseparable from social and cultural structures.

What makes Activity Theory intriguing for Visualization is that it provides a coding schema for the analytical process, which appeals to the computational and analytic bent of the domain. The coding schema breaks down into small understandable pieces that allow the process to be narrated, to be walked through a timeline, and to be reduced to a series of concrete parts. In that vein, Activity Theory does not purport to the analytical process intrinsic motivation; all motivation—or *object*—is assumed to be derivable from observation of activities and context.

Participatory Design is similar to Activity Theory in its attempt to understand the analytic process at depth, but it's approach is less observatory and is more hands on. It is longitudinal, as any intensive interaction with the users of complex visual interfaces is likely to be. Rather than examine the analytic process so that it might be broken down into component parts, Participatory Design sees understanding the analytic process as a goal to be achieved. And it achieves this understanding by engaging the process and the stakeholders directly.

There are a variety of user-centered methods that would consider themselves Participatory Design; the one under discussion was first defined by Ehn. "Users participating in design" and "designers participating in use" [9] depend on all stakeholders, including designers, interacting with each other and the analytic process. This interaction takes place through what Wittgenstein [43] called "language games." These language games are part of the fabric of our everyday lives as we interact with others in our world, and like most games, language games have rules. These rules depend on the participation of everyone, and accordingly, everyone has a say in what the rules should be. In this way, Participatory Design puts a heavy emphasis on the democratic process.

In the world of visualization these "design games" would often be embodied. Visualization is an interdisciplinary field. Its design games would be similarly interdisciplinary. Complex concepts discussed by one expert may not be immediately understood by the diverse stakeholder group. For example, a cognitive scientist describing how mental models are used to infer rules may find that her technical jargon much less useful than a diagram of the analytical process that she walks the group through. A designer may act out the functionality of the interface design, rather than just describe it. These embodiments are difficult to capture or understand if the research protocol is third person or behind glass.

Participatory design depends on cooperation as much as it does on democracy. Stakeholders are expected to learn from each other, to share knowledge and ideas, and to reach consensus. Over time, the group will likely develop its own language which reflects the nature of the design and which is understood by every member. Participatory Design may even extend the analytical process, as the members increase their joint understanding of the game. New artifacts may be created and the language may adapt to reflect the new understanding. Neither Activity Theory nor Participatory Design is a superficial method; both take time and effort, but the understanding gained through the collaboration with observation and stakeholders would inform visualization development in ways that the current methods cannot.

Whether Participatory Design or similar user-centered design approaches, Interaction Design offers exploratory tools that are not commonly used in visualization design. To date, design in Visualization has been focused on logistics such as how many attributes can be mapped onto a glyph or on scaling data so that more can be visualized on the screen. While much more attention is slowly being paid to users and human cognition, very little rigorous exploratory design research is being conducted. Thus, visualization interfaces continue to be developed that may or may not achieve their desired goal. The right questions are never asked; thus, the correct answers seem elusive.

Another tradition in cognitive science that emphasizes the social and cultural shaping of cognition is that of the ethnographic studies of distributed cognition, such as those of Lave's [22] and Hutchin's [20]. The main tenet of this tradition is that human cognition is a phenomenon that is not bounded to internal states located in the human brain. On the contrary, human cognitive processes emerge as a result of the situated coupling of internal (i.e. mental) representations-processes with external representations-processes [32]. The external representations are normally embodied in physical form in a material object, or artifact (e.g. maps, charts, compasses, and stars used in naval navigation) or they may come from another human actor (e.g. making use of somebody else's memory). The external processes can also be implemented either by artifacts (e.g. calculators, computers) or by other human agents. According to this tradition, an understanding of cognition cannot be limited to internal, mental representations but it should also include interactions among humans, and interactions between humans and artifacts. In other words, cognition is not the exclusive result of processes occurring inside the human brain but the result of situated interactions across individuals of a social group and between individuals and their physical environment.

Cultural anthropologists and cognitive anthropologists have also demonstrated the need to incorporate cultural differences as shapers of individual and social differences in human cognition. For example, Nisbett's cross-cultural work in visual perception in Asians and Westerners [28] casts doubt in previously conceived notions of a universal, hard-wired, perceptual apparatus in humans that is independent of cultural differences. Henrich, Heine and Norenzayan also conducted a wide literature review that reported on substantial variability in psychological and behavioral measures among different cultures even in basic or fundamental aspects of psychology such as visual perception [19]. Thus, it seems that even at the perceptual level, there is a variability that could be associated to cultural factors.

These insights from activity theory, distributed cognition, and cognitive anthropology have shown the pressing need to introduce the social and cultural character of cognition as an element of analysis in studies of visual analytic reasoning, not only in situations of collaborative visual analysis, where its application is more obvious, but also in studies of individual visual analysis, where the influence of social and cultural factors has traditionally been neglected.

3   Research for Evaluation

It's not surprising then, given that visual analysis interfaces are often created without a thorough comprehension of the analytical process, that the evaluation of these interfaces also tends to fall short. Current approaches are superficial. A short series of tasks completely unrelated to the analytic process the interface was designed to support is usually the laboratory task administered to the $\sim$10 participants (which are often students around the laboratory). Some examples of this style of evaluation would include the previously referenced iPCA, whose evaluation used 12 participants who were asked to point out at least one positive one positive correlation and an outlier [17] No evaluation of the tool with real-world tasks or with experts was conducted. Other tool evaluations (Data Meadows) used between 2 and 18 participants to verify the presumed usefulness of their tool or visualization system or tool [10, 17].

Evaluation of visualization interfaces tends to focus on aspects specific to the visualization being evaluated, or on comparing the visualization to some other tool or interface. Plaisant has divided the evaluation of information visualization into roughly four groups: controlled experiments, usability evaluation, controlled experiments comparing two of more tools, and case studies of tool [36]. She considers usability testing and controlled "experiments" to be the core of this type of evaluation [36].

What makes visualization evaluation less than sufficient is not necessarily the types of evaluation that are currently being employed. What makes current evaluation ineffective is that (a) we are not certain how to define "effective" quantitatively, (b) visualization systems are complex and must be studied in vivo at some point, (c) limited research understanding of analytic cognition, and (d) a poor understanding of the human analyst and the individual differences that make each "user" unique.

The field of visualization is not certain how to define an "effective" interface. What very often happens is superiority is asserted but never proven. "Effective" requires that we define "ineffective." And thanks to the flexibility and adaptability of human reasoning, most interfaces can be learned to some degree [29]. Thus many interfaces are pronounced effective even if only because the participants got through the tasks, no matter the participant's performance. This is a poor sort of construct validity. Null hypothesis testing is often used because statistical certainty in the murky waters of evaluative uncertainty is comforting. When the numbers, however, do not prove statistically significant, it is quite tempting to pronounce the interface effective anyway (see for example [17]). Until the domain of visualization defines "effective", it will never know how to evaluate its interfaces.

Evaluating interfaces is an involved proposition at any level. But with visualization interfaces it is especially so. Visualization involves analytical cognition and interactive visualization. It is impossible to evaluate for one of these systems in a single study, let alone both or the interaction of the two. The variables of cognition can quite involved, and begin with the individual differences that the human brings to the interface, then to perceptual logics and categorization, followed by mental model creation, rule inferencing, hypothesis generation and analysis, and finally the dissemination of conclusions, with recursion and iteration at any point (and perhaps every point) in the analytic process. When one considers the breadth of cognitive engagement in the endeavor of visualization, it's easy to see that evaluation should include much more than a cursory study reported at the end of a conference paper.

In addition to human cognition, there is the evaluation of the visualization itself, which requires, once again, that researchers define nebulous concepts like "good" and "better." What makes one interface superior to another? There are concrete factors like scalability and graphics. And then there are more ambiguous factors like scaffolding working memory and supporting insight generation. Without a thorough understanding of the analytic process, it would be difficult at best to build a protocol that would evaluate the usefulness of an interface in communicating data and analyses to the human analyst. There are many aspects of the visualization involved in this communication: what data are presented, how they are presented, and how the analyst interacts with what are presented.

Further, visual interfaces, as we have already seen, are tools, artifacts, or objects which are acted upon. Studying a tool in a laboratory is of limited use; it is important to see how those tools hold up in a real-world environment over time. And when the interactive complexity of human-computer and human-computer-human collaboration are considered, laboratory recreation is difficult. Further, a laboratory user study is not the best environment to recreate a scenario in which experts collaborate with each other using the visualization under real world stresses this is best done in situ, and often over time. Interaction design, Human Computer Interaction, and anthropological disciplines have developed several methodologies to evaluate holistically analytical processes with objects or artifacts. These include simple protocols such as cognitive walkthroughs [42]and focus groups, as well as participant and non-participant observation [6] to recording tasks in process in situ for post-hoc detailed coded analysis . Other ethnographic tools allow the entire culture of the analysis to be broken down into analyzable parts. Periodic follow-ups and interviews to test the impact of the tool over time [36]. There are no shortage of methods to study how visualization supports (or doesn't support) the analytical process and collaboration. Until the visualization object is tested in the field by the experts who use it, the interface or system cannot be said to have been truly evaluated.

4 Individual Differences

One very good reason for understanding human cognition and the analytical process before designing and evaluating visualization is that analysts and other users are not at all the same in how they perceive, organize and use information. Differences start with perception; humans perceive visual stimuli differently depending on how it is presented [31]. What this also means is that when data are presented visually, the representation becomes a framing that primes the human analyst see "see" data a particular way; this has long been common knowledge in visual and news media (e.g. [8, 34]), but to date, even this simple concept of framing has been largely ignored by the visualization community. Framing sets analysts up to see data a certain way, which impacts all reasoning and decisions made about data from that point forward. The ripple effect could be slight, or it could be profound. Nobody knows, because the differences are not acknowledged, let alone evaluated.

Framing and the cognitive response to it is one difference that the visualization can be designed to minimize or take advantage of. And the inherent differences between users go much deeper and can be more pervasive than that. From differences in personality to differences in learning style to differences in academic and cognitive ability, these inherent differences impact every aspect of cognition when human analysts interact with visual interfaces.

Personality factors have long been believed to effect cognitive performance demonstratively. Personality factors for the purposes of our discussion may be defined as "relatively enduring styles of thinking, feeling, and acting." [25]. What makes these factors interesting to analytical cognition is that, by adulthood, they are inherent and pervasive, and, according to some research, universal to human experience, regardless of culture or language [25, 26]. The most commonly recognized and widely accepted model of personality is referred to as the Five-Factor Model (FFM): extraversion (sociability), agreeableness (trust, affection, and pro-social behavior), conscientiousness (goal-orientation and impulse control), neuroticism (moodiness and anxiety), and openness (imagination and insight). These five factors have been studied, assessed, and critiqued in a wide variety of behavioral research.

Other personality factors do not attempt to systemically describe personality, but only some aspect of thinking, feeling, and acting. Rotter's Locus of Control evaluates the degree to which a person feels in control of or responsible for personal and life events [45]. The Beck Anxiety Inventory was designed to separate trait anxiety from clinical depression [46]. Trait anxiety is the tendency to be anxious all the time, as compared to state anxiety, which is triggered by specific life happenings. The Self-Regulation Scale assesses the degree to which a person can stay focused and on task even when distracted by uncertainty or emotional interruptions [49]. And the Scale of Tolerance-Intolerance of Ambiguity [48] is a similar but different assessment how a person self-perceives new, complex or impossible problems or events, and whether she views those events as threatening. These are just a few examples of personality constructs which have been studied and assessed by the behavioral sciences. Whether specific factors or systems of factors, one set of individual differences that impacts the analytical process. And these differences can predispose his use of information in knowable ways.

Learning and cognitive styles directly contribute to how humans manipulate information as well. Styles differ from personality traits in that they do not so much describe the person as they describe a tendency in how information is perceived, organized, and used. These styles refer to a specific aspect or aspects of cognition, not the entire description of the person. An awareness of learning and cognitive styles are vital to our understanding of the analytical process, as they tend to influence what information is deemed important and how that information is disseminated and integrated into hypotheses and conclusions. For example, one broad assessment is the Index of Learning Styles, which uses four continuums to measure how the participant preferred to see and use information and knowledge [50]. The first continuum is Sensing(practicality, facts and empiricism) versus Intuitive (theory and ideas). The next is the Visual versus Verbal, or a preference for graphics and pictures or words and text. The third is Active(learning by doing) versus Reflective (learning by thinking or reflecting). And the last is Sequential versus Global, which measure the degree to which the participant prefers to tackle problems in a bottom-up, step-by-step fashion to a top-down, holistic perspective. Each of these continuums describes a proclivity for learning and organizing. And especially in the case of the Visual versus Verbal continuum, the implications for visual interface interaction are obvious.

Inherent differences such as personality factors have been found to impact cognitive behavior and performance. Personality factors such as Big Five Neuroticism were found to impact how participants perceived visual scenes [7, 24]. Heaven & Quintin found that personality factors could predict a degree of racial prejudice, which could be described is a rule inferred from visual categorization [15]. All five factors of the FFM were found to be associated with information seeking behaviors in graduate students [16]. Palmer similarly found that information searchers could be grouped based on measured learning and search styles [30]. The five factors of the FFM were also reported as related to students' ability to apply, study, and analyze information [35]. The Thematic Apperception Test (a projective picture interpretation measure of general personality) was reported to differentiate between problem-solvers and their problem-solving styles [51].

And finally, academic or cognitive ability has been shown to influence how rational humans can be. "Rationality" in psychology is a term used to describe (a) how prescriptive a person's reasoning process is [52] and (b) whether a person can explain why the reached the answer they reached [11]. Stanovich [51] has argued, based on his research and others', that how a person scores on measures of intelligence or cognitive ability can predict what types of reasoning a person will likely use (pp. 149–151). Persons who measure higher in analytical ability tend also to be more "rational" or normative, that is to say, they tend to solve problems in what is believed to be the most efficient and accurate way ([52], pp. 150–151).

These inherent differences are brought to the interface. (See Fig. 2.) They cannot be controlled but they may be measured and understood. And these differences can be aggregated and their interactions evaluated. That is to say, from a battery of psychometric items, there are usually a subgroup that have a measureable impact on the analytical process. This will likely be a compilation of items from the psychological constructs we've already discussed (see for example [14]), tailored for the observed and known environmental and institutional differences. The matrix of these measured and known differences is a "personal equation of interaction" which can be used to inform a visualization design that communicates analysis is a way that the analyst reasons and learns best. The human is no longer conforming to the interface; the interface at long last is conforming to the human. And because

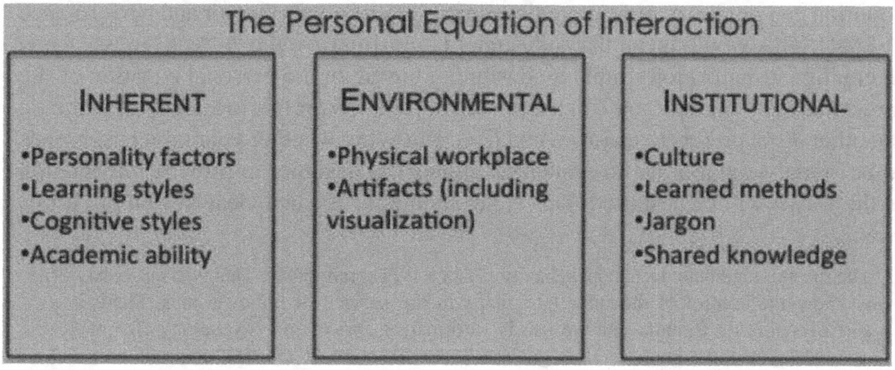

The Personal Equation of Interaction

INHERENT	ENVIRONMENTAL	INSTITUTIONAL
•Personality factors •Learning styles •Cognitive styles •Academic ability	•Physical workplace •Artifacts (including visualization)	•Culture •Learned methods •Jargon •Shared knowledge

Fig. 2 Aspects of the personal equation

experts in a given field can tend to share a similar personal equation (e.g. [13]), designing for the differences of expert domains becomes a possibility. And for other types of visualization (such as end-user commercial designs), the personal equation would allow for real-time individuation, adjusting the interaction or the visualization to the individual user's personal equation.

Inherent differences, or differences in ability, style and personality factors, are only one of three major types of individual difference that contribute to a personal equation of interaction and impact visualization. (See Fig. 2.) Environment and institutional differences also impact the analytical process. Environmental differences are any aspect of the context or physical environment that impact the analytic process. This would include interactive visualization, other visual interfaces, and the presence and input of other human analysts. This interacts with the 3rd type of individual difference. Institutional differences are what differentiates the expert domain from others: its culture, its proprietary language, and social interaction. Institutional differences tend to be learned differences that become rote and persistent over time: shared knowledge, methods, assumptions, and jargon frame the analytic process and encourage information to be used (or excluded) in knowable ways. Environmental and institutional differences interrelate with each other and with the inherent differences within each analysts to create a complete personal equation of interaction. While each type of difference is usually evaluated separately and often by different disciplines of human behavior, an awareness that they do interact is necessary to understanding the impact of each individual difference separately.

Thus, given these differences, assuming that all users approach and use an interface the same way is at best inhibitory and at worst catastrophic to the design and evaluation of interactive visualizations. Without understanding how users will use what is designed and developed, the entire process is in the dark. Taking the time to understand the personal equation of the analyst being designed for brings motivation, accountability, and inspiration to the development process. And the benefits do not stop there. Some evidence suggests that designing with a personal equation can predict or almost predict some types of user behavior and performance [14, 44]. This would mean that a designer could improve the "effectiveness" of an interactive visualization simply by designing based on the personal equation of the target user. For example, 60% of the human population prefers to learn with pictures; the other 40 % prefers to learn via text [38]. Knowing whether your target user tends to be visual-spatial or audio-sequential allows the designer to provide information in the way it will be best used. This is not such a far-fetched idea. Newell put it this way:

> The most fundamental fact about behavior is that it is programmable. That is to say, behavior is under the control of the subject to shape in the service of his own ends. There is a sort of symbolic formula that we use in information processing psychology. To predict a subject you must know: (1) his goals; (2) the structure of the task environment; and (3) the invariant structure of his processing mechanisms. From this you can pretty well predict what methods are available to the subject; and from the method you can predict what the subject will do. [53]

Thus, with a personal equation of interaction, we contribute to both a knowledge of the analytical process, a knowledge of the task and data involved, and a knowledge of the human analyst.

5 Variability in Collaborative Cognition

It's easy to see how the individual differences we have just described impact on collaborative analysis as well. Collaboration as a social process is an intrinsic component of visual analysis. Data creation, visual representations, the framing of problems, analysis, and solutions are all social constructions in which what is determined to be "good" or "best fit" are decided not by an individual but by the group or a subset of the group. Each individual with his or her individual differences influence the activity of the group, and this conflation of differences among members flavors the analytical process in ways that may be unique to the group. For example, mental models and schemata are created within social contexts and in interactions with other group members. Differences between individual perspectives must be negotiated. And the usefulness and value of artifacts, such as visualizations, is determined by the social group. Even, reasoning styles and cognitive biases can be found as being determined by the interaction between the individual and the social group in which analysts work. Rob Johnston [18], for example, in his study of intelligence analysis, found that the organizational structure of the CIA rewarded the support of hypothesis coming from figures of authority even in front of contradictory evidence, something that institutionalizes and encourages confirmation bias at the individual level. Thus, it is important not only to consider how the sole user interacts with a visualized representation, but also how the collaborative effort uses and judges visualizations and the institutionalized methods of reasoning with visualizations through the life process of their analyses. The importance and impact of the visualization may change multiple times as the needs of the collaboration change.

6 Translational Science

This use of visualization over time and throughout the analysis illustrates a key argument for the evaluation of visual analysis tools over time. Analytic cognition—decision-making about and with information—is not a discrete process with a distinct beginning and end. New information revises old hypotheses and improves on current generalities. Once the ideas are tested in real life scenarios or exposed to broader critique, the conclusions drawn may need refining. Generally accepted policies may change as the needs of the problem or the collaborative group change. Thus it is not only necessary to consider the use of visualizations in a single human-computer dyad, as an artifact and actor in collaboration, but also as part of a life cycle of information-sharing that leads to evolving solutions and ever-improving analytical processes. However, current research in visualization lacks a framework for evaluation that can inform visualization design throughout such a cycle (Fig. 3). How could human centered visualization design approach such a task?

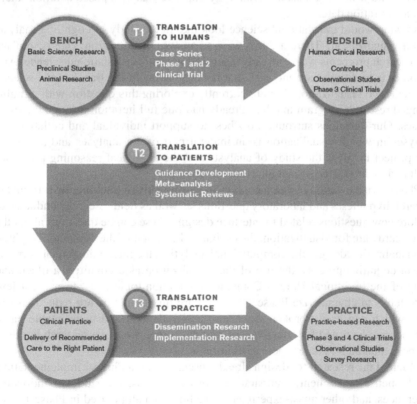

Fig. 3 The translational science model from medicine (From http://www.ahscstrategicplan.org/strategicgoal4.aspx)

Translational science comes from the field of medicine, where significant null-hypothesis testing in the lab or a satisfactory use case in the field could never suffice in determining the success of tools and methods in clinical practice. It is called "translational" because it "translates" the knowledge or technique under study to a bigger and bigger audience (user group, as it were) with each new stage or phase. Translation science divides its evaluation into three phases [54], which we will briefly summarize before going through each Phase individually. Phase 1 is laboratory research, in which research questions are asked and answered. These answers are applied in the field in Phase 2, during which solutions are tested in situ under the stress of time and consequences. Any solution that does not survive this pressure cooker is sent back to Phase 1. Phase 3 doesn't happen until Phases 1 and 2 are satisfactorily completed. In Phase 3, these implemented solutions become part of policy (e.g. become best practices and accepted protocol). For medical translational science, the process ends there. But for human centered visualization, we would propose one last step, making the science cyclic. Questions arising in Phase 3 need to be answered, informing new research in Phase 1. Visualization is an emerging science, and if nothing else, as technology improves, any proposed solution likely has an expiration date.

So what would translational science look like in the study of cognitive analysis with a visualization? There are two ways to answer this question. One is to place the research protocols we've already discussed inside of translational science. And the other is to discuss current work that is being doing in the translational science of interactive visualization. We are currently exploring this question with a multi-pronged research program that has already had one full iteration through the three Phases. Our questions surround how best to support individual and collaborative analyses in which visualization is an integral part of the analyses and synthesis. Our project involves the study of analysts and the analytical reasoning processes mediated by interactive visual interfaces at Boeing.

Phase 1 visualization research consists predominantly of traditionally formulated research hypotheses and laboratory questions, as well as field research conducted to explore new questions related to interface design. These can be questions about the use of hardware for visualization, the design and creation of the computer graphics, the visualization design, the computational analytics, the interaction metaphors, the human cognitive process, the use of the visualization as a cognitive artifact, and many of the traditional Human Computer Interaction topics, just to name a few. What tends to characterize Phase 1 research are questions that are reductionist or tasks which can easily broken down into easily defined and measureable parts. Methods can be tested for effectiveness, and the components of a research interface can be evaluated for some aspect of usefulness.

Much of the research for design already discussed has a Phase 1 implementation. Psychometric assessment, comparative tool evaluations, the study of individual differences, and other quasi-experimental methods can all be used in Phase 1. (See Fig. 4.) Case studies, observations, and ethnographies can also be Phase 1 research, if the objective is to explore and learn new information that can be applied in the field. In addition, in human research, the researcher seeks to create laboratory

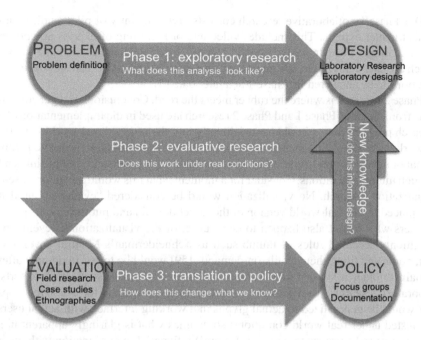

Fig. 4 Translational science for human centric visualization

questions or tasks which can be in some way generalized to a real world problem or question. Many of the usual critiques of laboratory research apply to Phase 1 research, including questions of content and construct validity as well as the generalization of study results to real world problems. However, large questions are difficult to study in their inherent complexity; some reduction is required to effectively evaluate or compare elements of the question. This is especially true of human null-hypothesis testing, which requires a substantial effort and comparatively large numbers of participants in order to generate results deemed significant.

In our on-going translational research, Phase 1 includes our study of individual differences explored how intrapersonal differences impact visualization interaction. Termed the Personal Equation of Interaction, iterative tasks such as interface learnability, target identification, and inference with categories are administered after each participant competes a large psychometric measure of common personality factors and cognitive styles [14]. This research is limited to laboratory tasks, which have some generalizability to common visualization tasks. This may include simple questions within active interfaces, such as trace finding or trial and error learning. Or it may involve asking users to compare one real-world exemplar to another in a variety ways, which involves induction and inference. The results of these studies inform more rigorous studies which break down previous tasks into smaller questions for closer inspection.

Our Phase 1 collaborative research consists predominantly of pilot studies with subject matter experts. This includes video and/or recording of experts at work in vivo and in situ. Further, we have conducted structured and unstructured interviews, which are later analyzed to inform Phase 2 research, as well as to create new research questions to be answered by Phase 1 at a later date [3].

Phase 2 research is where the rubber meets the road. Conclusions and generalizations from previous Phase 1 and Phase 2 research are used in the implementation of a research program that is implemented directly into the targeted real-world scenario of workplace. What this means is that assumptions and guidelines garnered from the laboratory get a real-time, rigorous test of usefulness over a period of time and through multiple iterations. Consider for a moment what this would mean for current visualization research. No visualization would be considered "evaluated" until it was placed in the real-world version of the target domain and put through its paces by users who do not also happen to be a subset of the visualization's developers. Commonly accepted rules of thumb such as Schneiderman's Mantra: "overview first, zoom and filter, then details-on-demand" [59] would be tested in domain after domain through a variety of interfaces and scores of non-visualization experts. Laboratory research in aspects of interface cognition, such as the color red pops first would be generalized to actual glyphs in a working interface with actual users and tested under real world conditions; sometimes what is glaringly apparent in a reductionist tasks proves to be less (or more!) influential in conjunction with other variables of interface design and cognitive process. How long Phase 2 continues depends on the type of question and the real-world scenarios; stakeholders make this decision concordance with research goals.

Each type of protocol previously discussed for interface evaluation are incorporated in a human centric visualization Phase 2. Because so much of the analysis conducted through and with visualization is collaborative, protocols from the anthropological sciences are particularly valuable when Phase 2 research begins in situ. Participant and non-participant evaluation, ethnographic studies of distributed cognition, and cultural studies can all be a part of evaluating the interactive visualization as an artifact in collaborative analysis.

In our research program, previous research in grounded theory, activity theory and collaborative learning has been utilized in the design of an ethnography of collaborative visual analysis in the wild. Working with safety analysts at Boeing, Inc. and using Tableau, we have developed a method pairing an information visualization (i.e. Tableau) expert with an aircraft safety analysis expert [2]. Working together in this Phase 2 research to find solutions to real-life tasks and problems around Tableau in the workplace at Boeing, the Pair Analytics team is evaluated through a longitudinal analysis of their work, exploring both the interaction within the human collaboration and the human-computer-human triad. The general process could be described in the following way. First, an initial ethnography of analytical work identifies who the analysts are, what kind of questions they determine to be critical, the kind of data used to answer those questions, the analytical methods and reasoning styles used during the analysis, the ways the environment is used to support the analysis (i.e. use of tools), and the social and cultural factors that facilitate or constrain the analysis (e.g. institutional methods for analysis or communication of analysis, impact of authority, influence of social factors such as gender, expertise or age on analysis, styles of collaboration, etc.). Second, a longitudinal series of paired analytical sessions are established between visual analytical experts and subject matter experts to conduct an exploratory and collaborative visual analysis. An analytical question identified in the fieldwork is chosen to drive the analysis as well as the necessary dataset and the most appropriate visual analytical tools available to answer the analytical question addressed to the specific dataset. Each session is recorded and captured in audio, video of the collaboration, and screen capture of all of the interactions with the visualization tool. Third, the data of each pair analytics session is coded using a coding scheme that we have developed to study the way visual analytic tools facilitate or impede processes in collaborative analyses, such as testing hypotheses, confronting anomalies, advancing the analysis, and finding insights [3]. The coding scheme is theoretically grounded in cognitive science theory for joint actions [5]. The results of this Phase 2 take the form of refinements of the models of collaborative analytical reasoning and insights to improve the design of interactive visual interfaces to increase the effectiveness of coordinated visual analysis [3].

Successful Phase 3 research includes Melville's elusive white whale. If a solution to a problem proves to be a real solution to the problem, or a novel concept proves superior to what came before it, it moves from research to policy. It becomes part of the accepted "ground truth" at this stage and not before. By tackling problems in the lab and then field before implementing them system wide, it is more difficult to make critical errors or stumble upon idiosyncrasies in the solution that require it to be reformulated. So often in visualization research, especially where human cognition is involved, we trust our personal intuition or lean on assumptions from colleagues about how humans reason. However, as we have discussed repeatedly in this chapter, it is dangerous to assume that what seems intuitive is scientifically correct, just as it fundamentally incorrect to assume that all humans think and reason the same way, or that they think the same way that we do. Only through testing our solutions outside of our own paradigms and mental sets and putting our solutions in the hands of users who will depend on them can we assert that we have accomplished anything of merit.

Phase 3 research might include any of the research protocols already discussed, if the goal is incorporating this new proven knowledge or technology into the policy and practices of the industry or academic domain. The best demonstration of Phase 3 validity is likely to be how this new "ground truth" is used in the next iteration of Phase 1 exploratory research. If the Phase 3 "truth" proves to hold up when tested in novel situations or environments in the Phase 1 lab, then its value is only buttressed. If it does not, then the Phase 3's failure to produce ground truth is also a subject for Phase 1 research. In this way, a translational science of human centric visualization it much more cyclic than linear.

For example, based on our Phase 2 research at Boeing, the safety team has taken the successful use of Phase 2 research and analyses and changed the way the organization works, implementing new best practices in a Phase 3 translation. For example, visual analytic tools such as Tableau have now been incorporated to the process of aircraft safety analysis as a result of the success of some of our past pair analyses on the impact of bird strikes on air incidents and accidents and the results of this analysis have changed the design of Boeing aircrafts as well as the pilot manuals [41]. While we have not pushed forward any large research initiatives into Phase 3, this is, to some degree, to be expected. Only what succeeds survives; very little of what fails ever sees the light of day. Further, translational science is not a one-off study or quick evaluation. It tends, rather, to be longitudinal, taking months and/or years to accomplish. Our translational research is very much a project in progress, but we have already seen how its use changes the way we view research and approach applied research questions and solutions.

As we discussed previously, Phase 3 is not the end. Technology changes so rapidly, and the needs of both the individual and the corporation tend to shift as the world, culture and technology shifts. It is to be expected, then, that Phase 3 research would spawn new questions or raise new possibilities that need to be further tested and evaluated. Thus we assert that Phase 3 may indeed inform the next round of Phase 1 research questions in the translational science cycle.

7 Conclusion

As we have seen, human centric visualization is designed to work with adaptive, flexible human cognition in a collaboration that often includes other artifacts and multiple users or analysts. What research is conducted to ensure that this collaboration works to it greatest potential is usually focused on narrow questions or small tasks, which is not nearly sufficient to inform interface design or evaluate the finished visualization. Human analysts bring some variability to the keyboard in the way they approach information and use what they know, which impacts not only how the visualization design is used but how useful it is. Knowing whether a visualization is effective requires (a) defining what effectiveness is and (b) evaluating a visualization in a real-world situation long enough to know if the visualization or visualization technique (interaction, display, etc.) is effective. This will likely take time and more than one method of evaluation. To that end we have introduced translational science as a framework for designing and evaluating visualization. Translational science walks a visualization or visualization from inception to real-world use in stages which translate the visualization from a research tool to a tool in use with a small number of real users to a tool which changes the way we do visualization or visualization techniques.

References

1. Andrienko G, Andrienko N, Wrobel S 2007 Visual analytics tools for analysis of movement data. *ACM SIGKDD Explorations Newsletter – Special issue on visual analytics 9(2):* 38–46
2. R. Arias-Hernández, T.M. Green, B. Fisher, Pair Analytics: Capturing Reasoning Processes in Collaborative Visualization. *Proceedings of the 44th Annual Hawaii International Conference on System Sciences.* IEEE Digital Library 2011a.
3. R. Arias-Hernandez, L.T. Kaastra, and B. Fisher, Joint Action Theory and Pair Analytics: In-vivo Studies of Cognition and Social Interaction in Collaborative Visualization. In: L. Carlson, C.Hoelscher, and T. Shipley (Eds.), *Proceedings of the 33rd Annual Conference of the Cognitive Science Society* Austin TX: Cognitive Science Society. pp. 3244–3249. 2011b.
4. R. Chang, M. Ghoniem,, R. Kosara, W. Ribarsky, J. Yang, E. Suma, C. Ziemkiewicz, D. Kern, A. Sudjianto, WireVis: Visualization of categorical, time-varying data from financial transactions. In *Proceedings of the 2007 IEEE Symposium on Visual Analytics Science and Technology* Sacramento, CA, October 2007, 155–162. 2007.
5. H.H. Clark. *Using language.* Cambridge: Cambridge University Press. 1996.
6. J. Cooper, R. Lewis, and Urquhart. Using participant and non-participant observation to explain information behavior. *Information Research, 2004.* 1–18. 2004.
7. Corn AL 1983 Visual function: A theoretical model for individuals with Low Vision. Journal of Visual Impairment and Blindness 77(8): 373
8. Dunegan K 1993 Framing, cognitive modes, and image theory: Toward an understanding of a glass half full. *Journal of Applied Psychology 78(3):* 491–503
9. P. Ehn, Scandinavian Design: on participation and skill. *In P. Adler & T. Winograd (Eds.). Usability: turning technologies into tools.* Oxford: Oxford University Press. 1991.
10. Elmqvist N, Stasko J, Tsigas, P 2008 DataMeadow: A visual canvas for analysis of large-scale multivariate data. *Information Visualization 7:* 18–33
11. J. St B. T. Evans and P.C. Wason. Rationalization in a reasoning task. *British Journal of Psychology, 67(4),* 479–486. 1976.
12. Y.H. Fua, M.O. Ward, and E.A. Rundensteiner. Hierarchical parallel coordinates for exploration of large datasets. *Proceedings of the IEEE Conference on Visualization 1999,* October 24–27, San Francisco, CA. 43–50. 1999.
13. J.A. Guirao-Goris, and G. Duarte-Climents. The Expert Nurse Profile and Diagnostic Content Validity of Sedentary Lifestyle: The Spanish Validation, *International Journal of Nursing Terminologies and Classifications* 18(3), (July-September, 2007). 84–92. 2007.
14. T.M. Green, D.H. Jeong, and B. Fisher Using personality factors to predict interface learning performance, *Proceedings of Hawai'i International Conference on System Sciences 43,* January 2010, Koloa, Hawai'i,. 1–10. 2010.
15. Heaven P.C.L, Quintin D.S 2002 Personality factors predict racial prejudice. *Personality and Individual Differences* 34: 625–634
16. J. Heinstrom, Five personality dimensions and their influence on information behavior. *Information Research 9(1).* paper 165 [Available at http://InformationR.net/ir/9-1/paper165.html] 2000.
17. D.H. Jeong, W. Dou, F. Stukes, W. Ribarsky, H.R. Lipford, and R. Chang. Evaluation the relationship between user interaction and financial visual analytics. *Proceedings the of IEEE Visual Analytics Science and Technology,* Columbus, OH, October 24–27. 83–90. 2008.
18. R. Johnston. *Analytic Culture in the US Intelligence Community: an Ethnographic Study.* Washington, D.C.: The Center for the Study of Intelligence - CIA. 2005.
19. J. Henrich. S.J. Heine and A. Norenzayan. The Weirdest People in the World? *Behavioral and Brain Sciences* 33, Cambridge: Cambridge University Press. 61–135. 2010.
20. E. Hutchins. *Cognition in the Wild.* Cambridge: The MIT Press. 1995.
21. V. Kaptelinin, and B. A. Nardi. *Acting with Technology: Activity Theory and Interaction Design.* Cambridge: The MIT Press. 2009.

22. J. Lave, *Cognition in Practice: Mind, Mathematics and Culture in Everyday Life*. Cambridge: Cambridge University Press. 1988.
23. Leontiev AN 1978 *Activity, Consciousness, and Personality*. Englewood Cliff s N.J, Prentice Hall
24. Macia. Visual perception of landscape: sex and personality differences *In Elsner, G.H. and Smardon, R.C. (eds.), Proceedings of our national landscape: a conference on applied techniques for analysis and management of the visual resource*. Incline Village, NC. April 23–25. 279–285. 1979.
25. R.R. McCrae and P.T. Costa Jr. Personality trait structure as a human universal. *American Psychologist 52(5).* 509–516. 1997.
26. McCrae R.R, Costa P.T 1987 Validation of the five-factor model of personality across instruments and observers. *Journal of Personality and Social Psychology 52(1);* 81–90
27. G.H. Mead. *Mind, Self, and Society*. Chicago: University of Chicago. 1934.
28. R.E. Nisbett. *The Geography of Thought: How Asians and Westerners Think Differently … and Why*. New York: Free Press. 2003.
29. D. Norman. *The Design of Everyday Things*. London: MIT Press. 1998.
30. J. Palmer. Scientists and information: II. Personal factors in information behavior. *Journal of Documentation,* 3, 254–275. 1991.
31. Po B.A, Fisher B.D, Booth K.S 2005 A two visual systems approach to understanding voice and gestural interaction. *Virtual Reality* 8: 231–241
32. M. Scaife, and Y. Rogers. External Cognition: How Do Graphical Representations Work? *International Journal of Human-Computer Studies* 45(2): 185–213. 1996.
33. M.C. Schatz, A.M. Phillippy, B. Scneiderman, and S. L. Salzberg. Hawkeye: An interactive visual analysis tool for genome assemblies. *Genome Biology, 8(3). 2007.*
34. Scheufele DA 2000 Agenda-setting, priming, and framing revisited: Another looked at cognitive effects of political communication. *Mass Communication & Society 3(2–3)*
35. H.C. Schouwenburg. *Personality and academic competence*. Poster presented at the seventh meeting of the International Society for Study of Individual Differences, Warsaw, Poland. 1995.
36. B. Shneiderman and C. Plaisant. Strategies for evaluating information visualization tools: multi-dimensional in-depth long-term case studies. *Proceedings of the 2006 AVI workshop on Beyond time and errors: novel evaluation methods for information visualization*. Italy. May 23–26. 1–7. 2006.
37. Shore. *Culture in Mind: Cognition, Culture, and the Problem of Meaning*. Oxford: Oxford University Press. 1996.
38. L.K. Silverman. Identifying visual-spatial and auditory-sequential learners: A validation study. In N. Colangelo & S. G. Assouline (Eds.), *Talent development V: Proceedings from the 2000 Henry B. and Jocelyn Wallace National Research Symposium on Talent Development*. Scottsdale, AZ: Gifted Psychology Press 2000.
39. Suchman L 1987 *Plans and Situated Actions*. Cambridge University Press, Cambridge Mass
40. Vygotsky L 1978 *Mind in Society: The Development of Higher Psychological Processes*. Harvard University Press, Cambridge, Mass
41. A.T. Wade, and R. Nicholson. Improving Airplane Safety: Tableau and Bird Strikes. In: *IEEE Information Visualization 2010 Conference Compendium (Discovery Exhibition)*, October 24–29, Salt Lake City, UT, USA 2010.
42. Warton, J. Rieman C. Lewis, and P. Polson. The cognitive walkthrough method: A practitioner's guide. In: *Usability Inspection Methods*. John Wiley & Sons: New York. 105–140. 1994.
43. L. Wittgenstein. *Philosophical investigations*. Oxford. Basil Blackwell & Mott. 1953.
44. C. Ziemkiewicz, R.J. Crouser, A.R. Yauilla, S.L. Su, W. Ribarsky, W. and R. Chang. How locus of control influences compatibility with visualization style. *Proceeding of the IEEE Visual Analytics Science and Technology,* October 23–28. Providence, RI. 81–90. 2011.
45. J.B. Rotter, Generalized expectancies for internal versus external control of reinforcement. Psychological Monographs 80, 609. (1966).

46. A.T. Beck, N. Epstein, G. Brown, & R.A. Steer. An inventory for measuring clinical anxiety: psychometric properties. *Journal of Consulting and Clinical Psychology*, *56*(6), 893–897. 1988.
47. R. Schwarzer, M. Diehl, & G.S. Schmitz. The self-regulation scale. *Gesundheitspsychologie*, (Berlin:Freie Universitat). 1999.
48. S. Budner, Intolerance of ambiguity as a personality variable. *Journal of Personality*, *30*, 29–50. 1962.
49. T.A. Litzinger, S.H. Lee, J.C. Wise, & R.M. Felder. A psychometric study of the Index of Learning Styles. *Journal of Engineering Education*, *96*(4), 309–319. 2007.
50. K.E. Stanovich, *Who is rational? Studies of individual differences in reasoning*. Mahwah, NJ: Lawrence Erlbaum. 1999.
51. A. Newell. You can't play 20 questions with nature and win: Projective comments on the papers of this symposium. *Visual Information Processing* (W.G. Chase (Ed.).). New York: Academic Press. 1973.
52. Marincola F 2003 Translational Medicine: A two-way road. *Journal of Translational Medicine* 1(1): 1
53. Fisher B, Green T.M, Arias-Hernandez R 2011 Visual Analytics as a Translational Science *Topics in Cognitive Science 3(3):* 609–625
54. T.M. Green and B. Fisher, Towards the Personal Equation of Interaction: The impact of personality factors on visual analytics interface interaction, *IEEE Visual Analytics Science and Technology (VAST) 2010*.
55. B Shneiderman, The Eyes Have It: A Task by Data Type Taxonomy for Information Visualizations. In *Proceedings of the IEEE Symposium on Visual Languages*, pp 336–343, 1996.
56. R. Chang, C. Ziemkiewicz, T.M. Green, and W. Ribarsky. "Defining insight for visual analytics." *Visualization Viewpoint, Computer Graphics and Application*, 29(2): 14–17. 2009.
57. P. Cherubini and P.N. Johnson-Laird. "Does everyone love everyone? The psychology of iterative reasoning." Think Reasoning, 10: 31–53. 2004.

Evaluating Visualization Environments: Cognitive, Social, and Cultural Perspectives

Christopher D. Hundhausen

Abstract Computer-based visualization environments enable their users to create, manipulate, and explore visual representations of data, information, processes and phenomena. They play a prominent role in the practices and education of many science, technology, engineering, and mathematics (STEM) communities. There is a growing need to evaluate such environments empirically, in order not only to ensure that they are effective, but also to better understand how and why they are effective. How does one empirically evaluate the effectiveness of a visualization environment? I argue that choosing an approach is a matter of finding the right *perspective* for viewing human use of the visualization environment. This chapter presents three alternative perspectives—*Cognitive, Social, and Cultural*—each of which is distinguished by its own intellectual tradition and guiding theory. In so doing, the chapter has three broad goals: (a) to illustrate that different research traditions and perspectives lead to different definitions of effectiveness; (b) to show that, depending upon the research questions of interest and the situations in which a visualization environment is being used, each perspective can prove more or less useful in approaching the empirical evaluation of the environment; and (c) to provide visualization researchers with a repertoire of evaluation methods to draw from, and guidelines for matching research methods to research questions of interest.

1 Introduction

Computer-based *visualizations* of data, processes and phenomena play a prominent role in many science, technology, engineering, and mathematics (STEM) communities. For example, geologists work with visualizations of plate tectonics

C.D. Hundhausen (✉)
Human-centered Environments for Learning and Programming (HELP) Lab,
School of Electrical Engineering and Computer Science, Washington State
University, Pullman, WA 99164-2752, USA
e-mail: hundhaus@wsu.edu

W. Huang (ed.), *Handbook of Human Centric Visualization*,
DOI 10.1007/978-1-4614-7485-2_5, © Springer Science+Business Media New York 2014

in order to better understand and predict earthquakes (see, e.g., [1]). Meteorologists work with visualizations of clouds, wind, and pressure in order to better understand and predict the weather (see, e.g., [2]). Computer scientists work with visualizations of computer programs and their execution in order to develop better software (see, e.g., [3]).

It follows that, in order to become future practitioners, STEM learners need to acquire skills in creating, interpreting, and working with such visualizations. Textbooks in STEM disciplines are replete with visual representations, and the teaching practices of many STEM disciplines revolve around the creation, manipulation, and use of visualizations, often through the use of software environments.

With respect to the use of visualizations within the practices, decision-making, and education of scientific disciplines, a key question for visualization researchers arises:

RQ 1: How to develop effective visualization environments?

Of course, answering that question requires one to consider a related question:

RQ 2: How to evaluate the effectiveness of visualization environments?

Concurring with previous writings on this topic (see, e.g., [4, 5]), I argue that addressing both of these questions depends intimately on understanding the purposes for which a visualization environment[1] is being enlisted. This entails not only understanding the particular *tasks* to be supported, but also the *broader context* in which the visualization environment is being used. However, in this chapter, I maintain that simply understanding the tasks and context is not enough; one also needs to define the very notion of *effectiveness*. The central argument is this: How one defines effectiveness hinges on the perspective from which one chooses to view human interaction with a visualization environment. Different perspectives lead to different definitions, which in turn lead to different approaches to evaluation. The problem for the visualization researcher, then, is to choose an appropriate perspective in light of the particular research questions of interest.

This chapter presents a framework of three alternative perspectives—*Cognitive*, *Social*, and *Cultural*—each distinguished by its own intellectual tradition and guiding theory (see Table 1). From the Cognitive Perspective,[2] a visualization environment's effectiveness depends on the extent to which it promotes task efficiency. From the Social Perspective, a visualization environment's effectiveness is tied to its ability to mediate human communication that establishes mutual intelligibility and resolves communication breakdowns efficiently. Finally, from the Cultural Perspective, a visualization environment's effectiveness rests on its ability

---

[1]Throughout this chapter, I use the term *visualization* to refer to an external representation of a phenomenon, process, idea, or data set. I use the term *visualization environment* to refer to a computer-based software tool that enables one to view, interact with, and explore such an external representation.

[2]Throughout this chapter, I will capitalize the three perspectives in order to emphasize that I am not using them as general terms, but rather in the specific senses defined in this chapter.

Table 1 Three perspectives for evaluating visualization environments

Perspective	Intellectual tradition	Theory of effectiveness	Evaluation method
Cognitive	Cognitive science	External cognition (see, e.g., [6, 7]) —visualizations are effective because they augment human memory and transform cognitive operations into easier perceptual operations	Cognitive modelling of visualization tasks to make predictions regarding human performance
Social	Sociology (especially ethnomethodology)	Situated cognition (see, e.g., [8, 9]) —visualizations are effective because they serve as powerful *mediational resources* [10] in human conversations, helping humans to establish a shared understanding of complex phenomena	Conversation and interaction analysis [11] of human communication mediated by visualization in order to determine role and value of visualization in mediating communication
Cultural	Anthropology (especially ethnography and cognitive anthropology)	Communities of practice [12, 13] —visualizations are effective because they facilitate increasingly central forms of participation in a community of practice; they can be used as a basis for gauging one's level of membership in a particular community of practice [14]	Consensus analysis of visualization "reading" and "writing" tasks to determine level of community membership

to facilitate participation in a community of practice, and ultimately to enable community newcomers to become central community members, thereby allowing the community to sustain itself.

By synthesizing and juxtaposing a diverse body of research that has studied and theorized about the role and value of visualization environments in human tasks, communication, and communities, this chapter aims to address the increasing need, within the visualization research community, to approach rigorously the challenging problem of empirical evaluation. In so doing, the chapter has three broad goals:

- To illustrate that different theoretical perspectives lead to different views of effectiveness;
- To show that, depending upon the research questions of interest and the situations in which a visualization environment is being used, each perspective can prove more or less useful; and
- To provide visualization researchers with practical advice on how to approach the problem of empirical evaluation: a repertoire of evaluation methods to draw from, and guidelines for matching research methods to research questions of interest.

The remainder of the chapter is organized as follows. In the next section, I briefly present each perspective, including a review of its foundational literature and a description of the evaluation method it advocates. In order to illustrate how the three lenses lead to different kinds of research investigations of effectiveness, I then present a case study in which I use each of the perspectives to evaluate a common visualization environment. The chapter concludes by by proposing a set of guidelines for matching research questions to appropriate evaluation perspectives.

2 Three Perspectives

In order to evaluate the effectiveness of a visualization environment, one must first define what one means by effectiveness by *operationalizing* the goals of the visualization environment. For example, some visualization environments, such as those used to solve problems, aim to promote *task efficiency*. This is often operationalized in terms of task completion time or error rates. Other visualization environments, such as *visual analytics* environments designed to help scientists make sense of large and complex data sets [15], aim to assist in generating new understanding and insight. These broad notions prove trickier to operationalize, since they are difficult to describe precisely. Still other visualization environments, such as those designed to help STEM learners comprehend disciplinary phenomena (see, e.g., [16–18]), aim to promote *learning*, which is often operationalized in terms of knowledge acquisition and transfer. The point is that, whatever the desired goals of a visualization environment might be, the first step toward evaluating its effectiveness entails concretely specifying, in observable and measurable terms, the dependent variables in terms of which to measure the visualization environment's effectiveness.

While operationalizing key outcomes is essential to conducting successful empirical evaluations, I believe we ought to go further by building on *theories of effectiveness* that can account for those outcomes. Building on theories not only helps us to design better visualization environments; it also allows us to generalize our evaluation results, so that others can learn from them.

Within the context of evaluating human interaction with visualization environments, what theoretical perspectives are relevant and useful? Below, I briefly describe three—labeled broadly as *Cognitive*, *Social*, and *Cultural*—that have guided past investigations of visualization effectiveness in the literature. The aim of this discussion is to serve as a *primer* for visualization researchers who need to perform empirical evaluations, with references to key literature for those interested in delving deeper.

2.1 The Cognitive Perspective

The Cognitive Perspective has its roots in cognitive science, a multidisciplinary field concerned with better understanding human cognition. To this end, researchers who work from this perspective have traditionally performed empirical studies of humans performing tasks, and used those studies as foundation for constructing *cognitive models* of human performance (see, e.g., [19, 20]). In such models, humans are viewed as computational agents with associated memories and processing units whose characteristics are derived from basic psychological research [21]. By carefully analyzing human tasks, researchers are thus able to transform human tasks into computational procedures that can be used both to account for human cognition, and to predict human performance.

In the 1980s, a line of cognitive science research turned its attention to better understanding the value of external representations in facilitating human tasks (see, e.g., [6, 22]). While the original research focused on diagrams, subsequent work has also considered animations (see, e.g., [23]) and virtual reality [7]. Likewise, while the original work aimed to account for the ways in which external visual representations influence the cognitive processing of *internal* representations, the focus has since shifted to include the interplay of *internal* and *external* representations, the latter of which are routinely manipulated and updated during human tasks. This new line of inquiry has come to be known as *external cognition* [7, 24].

According to the Cognitive Perspective, a visualization environment is effective insofar as it promotes efficient cognitive processing, which, in turn, results in efficient human task performance. On this view, the advantages of two- and three-dimensional visual representations, as opposed to one-dimensional textual representations, lie in their ability to capitalize on the efficiency with which the human perceptual system can perform visual search and recognition [6]. If they are designed well, visual representations can reduce task time by enabling humans to replace costlier mental procedures with more efficient perceptual ones.

A simple example, adapted from [22], serves to illustrate the potential cognitive efficiency differences promoted by alternative representations. Suppose that you are charged with the task of finding flight connections between Pittsburgh and Mexico City, subject to the constraint that no layover can be longer than 4 h. Further, suppose that you are not allowed use modern web search for this task. Instead, you must use either (a) a text-based list of flights that includes their starting and ending place and time, or (b) a graphical representation in which time is encoded on the x-axis, and flights appear on the y-axis, with each flight represented as a bar whose length corresponds to its duration, and whose start and end points are labelled with their start and end locations. In this graphical representation, flights with similar starting and ending locations are grouped together spatially on the y-axis.

To perform this task using the textual list of flights, you must sequentially search the list for connecting flights, and you must perform mental arithmetic to compute layover times. In contrast, you can complete the task more efficiently with the graphical representation. Since the ending destination and starting destination of flights are spatially grouped, finding viable connections becomes a matter of perceptual recognition. Likewise, computing layover times is transformed from a mental calculation to a perceptual one: you need only observe the width of the whitespace between adjacent flight bars.

In sum, for the Cognitive Perspective, the effectiveness of a visualization environment must evaluated with respect to one or more specific tasks. For each task, one must specify a detailed human procedure for completing the task using both the visualization environment and an alternative representation that is used as a point of comparison. One can then create cognitive models of each alternative procedure. By assigning empirically-derived time estimates (as in, e.g., the Keystroke Level Model [21]) to each step of the human procedure, one can obtain concrete time predictions for task performance. Thus, the Cognitive Perspective furnishes a theoretical account of human task behaviour that can be used both to guide the design of effective (i.e., cognitive efficiency-promoting) visualization environments, and to derive theoretical predictions against which to compare actual human performance in empirical studies of effectiveness.

2.2 The Social Perspective

Whereas the Cognitive Perspective focuses on the impact of visualization environments on individual human task efficiency, the Social Perspective is concerned with role and value of visualization environments in facilitating human communication. The particular version of the Social Perspective considered in this chapter is based on *Situated Action Theory*. Rooted in a branch of sociology known as Ethnomethodology, the perspective was first formulated for the field of human-computer interaction through the studies of Winograd and Flores [9] and Suchman [8], and has since become highly influential in studies of computer-supported collaborative work and learning [25].

Situated Action Theory holds that "our everyday social practices render the world publicly available and mutually intelligible" ([8], p. 57) In other words, "objectivity is a product of systematic practices, or [people's] methods for rendering [their] unique experience and relative circumstances mutually intelligible" ([8], p. 57). Thus, instead of attempting to discover and enumerate preexisting norms that govern people's behavior (the approach of traditional social science), Situated Action Theory focuses squarely on understanding the "ethnomethods" by which people make sense of the world. With respect to human communication, the implication is clear: whereas social science has traditionally taken the objective reality of social facts as the *basis* for human communication, Situated Action Theory treats objective reality—or mutual intelligibility—as the fundamental *result* of such communication.

Given the theoretical perspective just described, the essence of Situated Action Theory as a theory of effectiveness for visualization environments can be summarized by three basic tenets:

- *Knowledge is socially constructed.* Communication is not merely the transmission of preexisting knowledge. Rather, communication fundamentally entails two or more parties' negotiating a shared understanding. There exists no preexisting social reality that is merely confirmed by such negotiation. Instead, the mutual intelligibility that arises out of such negotiation *is* the social reality.
- *Conversational participants draw on an array of communicative resources, including visualizations, to build shared understandings.* A key communicative resource is visualization environments, which serve to mediate conversations by providing a shared representation that conversational participants can modify, annotate, and refer to as they construct shared understandings [26]. Other communicative resources include the basic organizational regularities of conversation documented by conversation analysts, including turn-taking and adjacency pairs [27], as well as contextualization cues such as gesture, gaze, and speech prosidy [28].
- *Communication breakdown is fundamental to communication.* Communication breakdown occurs when conversational participants' understandings of what is being talked about diverge; it is a completely normal occurrence in human communication. As Suchman [8] points out, communication can succeed amidst such breakdowns because conversational participants "work, moment by moment, to identify and remedy the inevitable troubles that arise" [8, p. 83].

To evaluate the effectiveness of a visualization environment from this perspective requires one to study the detailed, moment-to-moment interactions between humans and the visualization environment. To that end, Jordan and Henderson [11] outline a rigorous methodology called *interaction analysis* for studying human-human and human-computer interaction from this perspective. In brief, the evaluation methodology involves (a) videotaping the interactions of interest, (b) carefully reviewing the videotape (preferably collaboratively in a group) to identify "hotspots" that might be indicative of effectiveness; (c) transcribing the hotspots; and finally (d) analyzing the hotspots for evidence of the visualization environment's ability to assist humans in negotiating mutual intelligibility. Using this methodology, one can construct an empirical case, firmly grounded in evidence, of the role and value of a visualization environment in facilitating human collaboration and communication, and ultimately in helping humans to construct understandings of what is being visually represented by the environment.

2.3 The Cultural Perspective

The two perspectives considered so far focus their attention on individuals or small groups performing tasks. The Cultural Perspective assumes an even broader lens by considering the role of visualizations at the scale of entire *communities of practice* [13]. In many communities, including those consisting of electric circuit designers, air traffic controllers, and algorithm designers, visual representations play a prominent role. The Cultural Perspective gauges the effectiveness of the visual representations in terms of their ability *to sustain and reproduce* such communities.

The Cultural Perspective described here builds on the *Situated Learning Theory* proposed by Lave and Wenger [12], which posits that a community of practice is sustained and reproduced by providing opportunities for newcomers to participate, in increasingly central ways, in the practices of the community. On this view, a visualization environment is seen as effective insofar as they facilitate opportunities for increasingly central participation in the community.

Constructing and meaningfully interpreting visualizations constitute two prominent forms of community participation facilitated by visualizations. Empirical studies of visualization provide evidence that, as newcomers become more expert members of a community, they become more conversant in the visual representations of their community, enabling them to construct and interpret those visualizations in similar ways [29–31]. According to the Cultural Perspective, a robust and vibrant community of practice—one that is able to sustain and reproduce itself over time—will provide sufficient opportunities for newcomers to learn how to construct and interpret such visualizations by participating, in increasingly central ways, in practices mediated by those visualizations.

It follows that one way to gauge the extent to which a community is succeeding in this regard is to study the ways in which a variety of individuals with varying levels of membership in the community construct and interpret the visual representations of the community. One would expect little variance in the ways in which the most central members of a given community of practice read and construct community representations. Conversely, one would expect considerable divergence in the ways in which newcomers in that community read and construct the community's representations. Over time, if the community is sustaining and reproducing itself, one would expect newcomers' practices to converge upon those of its central members [12].

In order to gauge the agreement of community members with respect to visualization reading and writing tasks, one can build upon Cultural Consensus

Theory [14], a formalized, consensus-based model of community that has evolved out of research in cognitive anthropology. According to Consensus Theory, each community of practice has a *semantic domain*—an organized set of symbols (e.g., notations, words, and graphical representations) for referring to a "common conceptual sphere" ([14], p. 315). On this view, the semantic domain associated with a given community of practice is a matter of *consensus*; the conceptual spheres characteristic of a community of practice, as well as the symbol systems appropriate for referring to them, are precisely those on which the members of the community agree.

Consensus Theory derives a formal statistical model for assessing the "cultural competence" of the individual members of a community of practice. This assessment is based on the answers that a sample of community informants furnishes to a set of questions. The questions must be carefully chosen so that they address a body of knowledge on which the community of practice is assumed to agree. However, unlike the statistical models traditionally applied to test-taking, Consensus Theory's statistical model does not assume an objective truth against which informants' answers are to be measured. Rather, the model uses informants' answers as a basis for constructing a *cultural truth*, according to which informants' cultural competence can then be assessed.[3]

Consensus Theory's statistical model constructs such a cultural truth based on patterns of agreement among informants. The assumption is that "the correspondence between the answers of any two informants is a function of the extent to which each is correlated with the [cultural] truth" ([14], p. 316). In other words, the most central members of the community, whose cultural knowledge is the most "complete," are highly likely to agree with each other by offering identical answers to the questions. In contrast, less central members of the community are less likely to agree both with each other, and with central members of the community.

Thus, a general procedure for performing a Consensus Analysis of a community of practice around the activities of visualization construction and interpretation involves three steps: first, developing visualization construction and interpretation tasks designed to elicit cultural knowledge that members of the community are assumed to share; second, administering that survey to a sample of community members; and third, analyzing informants' answers using Consensus Theory's statistical model. Such an analysis can yield two key results: (a) an empirical account of the meaning of a given communities visual representations—the mappings between visual elements and semantics; and (b) evidence of the extent to which a given community enables newcomers to become more central members by facilitating graphical readership and construction skills.

---

[3] Hence, the use of the term *informant* (as opposed to, say, *subject*) is deliberate; it underscores the fact that participants in a Consensus Study are *informing* the researcher of their culture, rather than the researcher *subjecting* them to a test.

3 Applying the Perspectives: A Case Study

In order to illustrate the ways in which these perspectives can be used to evaluate visualization effectiveness in practice, I now present a case study that draws on a line of my past research into the use of algorithm visualization in computer science education [30]. The case study focuses on the construction and interpretation of algorithm visualizations for the purpose of understanding the *bubblesort*, a simple sorting algorithm (see Fig. 1). Recall that the bubblesort algorithm places a list of items in ascending order by making successive passes through the list. In each pass, it compares adjacent items, swapping them if they are out of order. Through this series of exchanges, the largest item "bubbles up" to the end of the list. After the nth iteration of the algorithm, the largest n items will have found their rightful place in the list. After $n-1$ passes, the entire list is guaranteed to be sorted.

In order to help explain this algorithm, suppose that one constructs the classic algorithm visualization depicted in Fig. 1. In this visualization, the sticks represent the elements to be sorted; stick height represents element magnitude. As the visualization progresses, adjacent stick elements are flashed, and smoothly switch places if they are out of order. By the end of the visualization, stick elements form an ascending or descending ramp, indicating that they are in order. Let us now evaluate this visualization from each of the three perspectives described in the previous section.

3.1 Evaluation from Cognitive Perspective

With respect to this effectiveness of this visualization, the key question from the Cognitive Perspective is this: *Does the visualization promote cognitive task efficiency?* In order to get at that question, we first need to identify one or more tasks that the visualization is hypothesized to support. For the purpose of this example, let's assume that it will be used to aid a novice in performing the following *tracing task*:

```
1:   BUBBLE-SORT(A)
2:   for j ← n - 1 to 1 do
3:      for i ← 1 to j do
4:         if (A[i] > A[i+1]) then
5:            exchange A[i] ↔ A[i+1]
6:         end if;
7:      end for;
8:   end for;
9:   end BUBBLE-SORT;
```

Fig. 1 Pseudocode description of the bubblesort algorithm

Table 2 Correct result of tracing task for input data {7, 4, 10, 3, 1}

Pass #	Exchanges within pass	Array state at end of pass
1	{7,4} {7,10} {10,3} {10,1}	{4,7,3,1,10}
2	{7,3} {7,1}	{4,3,1,7,10}
3	{4,3} {4,1}	{3,1,4,7,10}
4	{3,1}	{1,3,4,7,10}

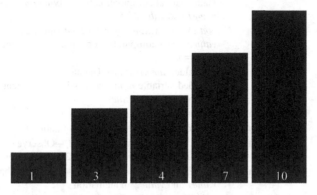

Fig. 2 "Sticks" visualization of the bubblesort algorithm

Specify the series of exchanges that takes place in each pass of the bubble-sort's inner loop (lines 3–7 in Fig. 1), along with the state of the array at the end of each pass, assuming the following set of input data: 7, 4, 10, 3, 1.

Table 2 presents the correct trace for this task.

Whether the algorithm visualization in Fig. 2. promotes cognitive task efficiency is a relative question; it can only be answered by comparing alternatives. For the purpose of this example, I will compare it against the pseudocode description of the algorithm shown in Fig. 1. In the remainder of this section, I construct cognitive models to predict task performance using each alternative representation, use those models as a basis for evaluating the effectiveness of the visualization, and finally highlight important limitations of the evaluation.

3.1.1 Human Procedures

The two alternative representations in Figs. 1 and 2 support completely different human procedures for performing the trace task. The pseudocode representation requires the user to manually trace through the code line-by-line, updating external representations of key variable values along the way. To perform such a trace, the user must have the ability to read and mentally simulate pseudocode. In contrast,

Table 3 Operators for human trace task procedure

Operator	Used with	Description
RV	Pseudocode	*RecordValue(name,value)*—Jot down a variable name and associated value
LV	Pseucocode	*LookUpValue (name)*—Look up the value of a recorded variable *name* by searching the list of variable values
MV	Pseucocode	*ComputeValue(value2, op, value1)*—Perform mathematical computation by applying *op* (+,−,/,*) to *value1* and *value2*
CV	Pseudocode	*CompareValues(value1,op,value2)*—Compare *value1* to *value2 using* comparative operator *op* (<, <=, >, >=, =)
UV	Pseucocode	UpdateValue(*name,value*)— Update the value of a recorded variable *name* by crossing out current value and jotting down *value*
PA	Visualization	Click "Play" button to start animation
SA	Visualization	Click "Stop" button to stop/pause animation
OC	Visualization	*ObserveComparison(value1, value2)*—Observe comparison of *value1* and *value2* in animation
OE	Visualization	*ObserveExchange(value1, value2)*—Observe exchange of *value1* and *value2* in animation
RT	Both	*StartNewTraceRow()*—Start a new row in the trace table
RE	Both	RecordExchange(value1, value2)—Update the trace table by recording an exchange between value1 and value2
RA	Both	RecordArray(array)Update the trace table by recording the state of array

the algorithm animation allows the user to glean the required tracing information from observing the animation unfold. However, in order to glean this information, the user must know how to interpret the visual language of the animation (sticks, comparisons and exchanges) within the context of the trace. In addition, the user must be able to recognize the point at which a pass of the algorithm's inner loop has completed, as this is where the array state must be recorded in the task.

Table 3 defines a set of operators that a human can use to perform the trace task using the two alternative representations. Notice that each representation affords a different set of operators. When performing the task with pseudocode, one hand-traces the pseudocode by recording, computing, looking up, and updating variable values (the first five operators in the table). In contrast, when performing the task with the visualization, one interacts with and observes the algorithm animation (the next four operators in the table). In both cases, one must record the trace on paper (the final three operators in the table).

Using the operators in Table 3 as a basis, Figs. 3 and 4 present general human procedures for performing the trace tasks with the two alternative representations. While these procedures are written in pseudocode, note that only those operators from Table 3 (written in **bold** in these figures) are actually performed by humans; the other pseudocode elements are is used to express the repetition and conditionality of the human procedure.

```
HumanTraceWithPseudocode(A, size)
01: for x ← 1 to size
02:   RV("A[x]", A[x]) //Record array values in trace notes
03: end for
04: RV("n",size) //Record value of n in trace notes
05: RV("j",MV(size,-,1)) Record value of j in trace notes
06: do
07:   RV(i,1) //Record value of i in trace notes
08:   do //Look up & compare adjacent values
09:     if (CV(LV("A[i]"),>,LV("A[MV("i",+,1)]")))
10:       RE("A[i]",A[i+1]) //Record exchange in trace table
11:       UV("A[i]",A[i+1]) //Update value in trace notes
12:       UV("A[i+1]",A[i]) //Update value in trace notes
13:     end if
14:     UV("i",MV(LV("i"),+,1)) //Update value in trace notes
15:   While (CV("I","<=",LV("j"))
16:   RA(A)//Record current state of A to trace table
17:   UV("j",MV(LV("j"),-,1))
18:   if (CV("j",>=,1)
19:     RT() //Start new row of trace table
20:   end if
21: while j >= 1 //Human operator not included here since it
                 //was computed on line 18
```

Fig. 3 General human procedure for performing trace task with pseudocode

```
HumanTraceWithVisualization(A, size)
01: j ← size - 1
02: PA(); //Start animation
03: do
04:   i ← 1
05:   do
06:     OC(A[i],A[i+1]) //Observe comparison
07:     if (j < size - 1) AND (i = 1)
08:       SA()//Pause animation to record array state
09:       RA(A)//Update trace table with current array state
10:       RT() //Create new row of trace table
11:       PA()
12:     end if
13:     if (a[i] > a[i+1])
14:       OE(a[i],a[i+1]) //Observe exchange
15:       SA() //Pause animation to record exchange
16:       RE(a[i],a[i+1]) //Record exchange
17:       PA() //Restart animation
18:     end if
19:     i ← i + 1
20:   while (i <= j)
21:   j ← j - 1
22: while (j >= 1)
```

Fig. 4 General human procedure for performing trace task with visualization

With respect to the pseudocode procedure (Fig. 3), note the following assumptions and conventions. First, it assumes that a human will externalize key elements of the tracing task by keeping trace notes, as indicated by the comments on the lines of the procedure that reference these trace notes. Second, variables used in the human trace notes are placed in double quotes in order to distinguish them from variables used in the pseudocode. Third, in a given step of the pseudocode procedure, it is assumed that values that were used in the preceding step remain in working memory, and hence do not need to be looked up or computed.

Also note two assumptions made by the visualization procedure (Fig. 4). First, the procedure assumes that the animation must be paused whenever the viewer needs to record trace information. In practice, it may be the case that the viewer could record this information without pausing the animation. Second, since the bubblesort animation provides no explicit visual indication that a given iteration has completed, the procedure assumes that the *first* comparison of a given iteration provides the clearest visual indication that the previous iteration has completed. At this point (lines 7–12), the procedure has the viewer record the current state of the array and start a new row of the trace.

3.1.2 Comparison of the Human Procedures

Recall that, from the Cognitive perspective, visual representations can promote an efficiency advantage by replacing mental operations with less costly perceptual operations [6]. As Table 4 makes explicit, the bubblesort animation promotes such an efficiency advantage in three ways. First, rather than needing to perform mental calculations (searches, comparisons, and additions) to compute comparisons, the visualization viewer merely has to *observe* them in the animation. Likewise, rather than having to explicitly update trace notes to indicate an exchange of two values, the visualization viewer merely has to *observe the* exchange. Finally, the visualization viewer is freed from the "bookkeeping" of creating and updating an external representation of the algorithm's execution; such bookkeeping is done "for free" by the animation.

As Table 4 highlights, the human procedure supported by the visualization is more efficient than the one supported by pseudocode. However, it would be nice to make the comparison more explicit by associating empirically-derived time estimates with each human operation, as is done, for example, by the GOMS and KLM [21] models. Unfortunately, while time estimates for some of these human operations appear in the literature (e.g., SA and PA amount to the combination of P_1 and P in the KLM [21]), such estimates are not available for many others, owing to the fact that they involve interactions with paper and not computers. Deriving time estimates for the human operations in the two procedures is beyond the scope of this chapter; I leave it as an exercise for the interested reader.

Table 4 Comparison of procedures supported by pseudoccode and visualizaton

Bubblesort operation	Pseudocode procedure (see Fig. 3)	Visualization procedure (see Fig. 4)
Comparison	`09:CV(LV("A[i]"),>,LV(A[MV(i,+,1)]))`	`06: OC`
Exchange	`10:UV("A[i]",A[i+1])`	`14: OE`
	`11:UV("A[i+1]",A[i])`	`15: SA`
		`17: PA`
Trace bookkeeping	`//Keep track of array in notes`	—
	`01: for x ← 1 to size`	
	`02:RV("A[x]", A[x])`	
	`03: end For`	
	`04:RV("n",size)`	
	`10:UV("A[i]",A[i+1])`	
	`11:UV("A[i+1]",A[i])`	
	`//Keep track of variable j in notes`	
	`05:RV("j",MV(size,-,1))`	
	`15:UV("j",MV(LV("j"),-,1))`	
	`16: if (CV(j,>=,1)`	
	`//Keep track of variable i in notes`	
	`07:RV(i,1)`	
	`13:UV("i",MV(LV("i"),+,1))`	
	`14: While (CV(i,"<=",LV(j))`	

3.1.3 Discussion

The evaluation method advocated by the Cognitive Perspective involves constructing and comparing detailed human procedures supported by alternative representations. A strength of this approach is that it makes explicit the potential advantages of visualization environments in terms of the human performance they promote within specific tasks. Visualization researchers who are considering the design of a new visualization can use this kind of evaluation to perform a reality check on its viability: Will the visualization really provide an advantage over the representation(s) presently being used? By creating a cognitive model early on in the design process, a visualization researcher has a golden opportunity to gain early insight into whether and how to proceed.

At the same time, the evaluation approach advocated by the Cognitive Perspective has several important limitations. First, because it necessarily assumes practiced, error-free performance, it cannot account for learning, errors, distractions, or fatigue—all factors that are present in the real world. Second, as underscored by Tversky and Morrison's analysis of the cognitive benefits of animation [32], one needs to be careful when comparing two alternative representations that are not informationally equivalent. This issue is highlighted in the example just presented. As a static representation, the pseudocode in Fig. 1 presents an abstraction of the general Bubblesort procedure—one that can be used to derive that procedure for any input data set. In contrast, the sticks animation in Fig. 2 presents a concretization of the Bubblesort procedure for a specific input data set. Clearly, the information made available by the sticks animation more closely fits the tracing task than the pseudocode. Conversely, if the analysis had considered the effectiveness of the two representations in supporting the task of writing of a Java implementation of the Bubblesort algorithm, the pseudocode representation would have been a closer fit.

A final limitation of the evaluation method advocated by the Cognitive perspective relates to ecological validity: In practice, tasks are not always as well-defined as they need to be in order for cognitive modeling to work. The preceding algorithm visualization example underscores this issue. In educational practice, learners generally use algorithm visualizations like the one in Fig. 2 to *interactively explore* an algorithm (as in, e.g., [33]), not to solve a specific tracing task. Constructivism, the educational theory on which such practice is based, holds that learners will challenge, rearrange, and construct their own mental models of the algorithm's procedural behaviour through such interactive exploration [34]. However, modelling this complex process of knowledge integration is beyond the scope of the Cognitive Perspective, which focuses on the use of specific representations to solve specific problems. Thus, the foregoing example necessarily focused on an artificially constrained usage scenario for algorithm visualization. At the same time, it is important to acknowledge that, in many *problem-solving* (as opposed to learning) scenarios, the use of visualizations to accomplish specific tasks is common.

3.2 Evaluation from Social Perspective

With respect to this effectiveness of the bubblesort visualization, the key question from the Social Perspective is this: *Does the visualization enable conversational participants to establish a shared understanding of the bubblesort algorithm?* For the purpose of this example, suppose that a pair of novice computer scientists, Jen and Mia, have been given a pseudocode description of the algorithm that (Fig. 1), along with the bubblesort animation software tool (see Fig. 2 for a snapshot). They have been asked to work as a team to complete the same tracing task used in the previous section:

Specify the series of exchanges that takes place in each pass of the bubble-sort's inner loop (lines 3–7 in Fig. 1), along with the state of the array at the end of each pass, assuming the following set of input data: 7, 4, 10, 3, 1.

Table 5 presents a transcription of part of Jen and Mia's interaction with the Bubblesort animation and the Bubblesort pseudocode as they try to complete the task. Because my aim here is to illustrate an evaluation method, and not to present

Table 5 Transcription of dramatized interaction with bubblesort animation

The users	Available to software	The animation software	
Not available to the software		Available to user	Design rationale
1: J: Let's see how this works.	Hits "Play" button	"7" and "4" flash	Show comparison
2: M: Okay.			
3:		"7" and "4" swap	Show exchange
4:		"7" and "10"flash	Show comparison
5:		"10" and "3" flash	Show comparison
6:		"10" and "3" swap	Show exchange
7: M: What is happening?		"10" and "1" flash	Show comparison
8: J: Looks like the big stick is moving right.		"10" and "1" swap	Show exchange
9:		"4" and "7" flash	Show comparison
10: J: This is moving kinda fast. Do you know what's going on?		"7" and "3" flash	Show comparison
11: J: [Looks at instruction sheet] So, we're supposed to record each exchange in the inner loop.	Hits "Pause" button	"7" and 3" swap / Animation pauses	Show exchange / User wants to pause animation

(continued)

Table 5 (continued)

| The users | The animation software | | |
Not available to the software	Available to software	Available to user	Design rationale
12: M: Mmm, hmm.			
13: J: I think that happens when the sticks change places.			
14: M: Yes, but what about the flashes?			
15: J: They seem to happen before the sticks swap.			
16: M: I did notice that. But I also saw sticks flashing at other times.			
17: J: Hmm, maybe let's start it over. (2 sec. pause)Okay, they flashed.	Hits "Start Over" button	"7" and "4" flash	User wants to start over; show comparison
18: M: And swapped.		"7 and "4" swap	Show exchange
19: J: Okay, let's record this. (Writes "{7,4}" in trace.)	Hits "Pause" button	Animation pauses	User wants to pause animation
20: M: (Looks at trace). I think that's right.	Hits "Play" button	"7" and "10" flash	Show comparison
		"10" and "3" flash	Show comparison
21: J: Hey, did you see that?	Hits "Pause" button	Animation pauses	Users wants to pause animation
22: M: Yep, there was a flash but no swap.			
23: J: (Looks at pseudocode.) (20 sec. pause.) (Pointing to line 4)I think this "if" statement is comparing each pair of items.			

24: M: (*Looking at pseudocode*) (3 sec. pause). So maybe the sticks flash when they are being compared?

25: J: (Pointing to line 5) But they only swap when the first one is bigger.

26: M: Makes sense to me. Let's watch and see.

27: J: Okay	Hits "Play" button	"10" and "3" swap	Show exchange
28: M: They swapped.			
29: J: I'll write this down. (*Writes "{10,3}" in trace.*)	Hits "Pause" button	Animation pauses	User wants to pause animation
30: J: Restart.	Hits "Play" button	"10" and "1" flash	Show comparison
		"10" and "1" swap	Show exchange
31: M: Another swap.	Hits "Pause" button	Animation pauses	User wants to pause animation
32: J: (*Writes "{10,1}" in trace.*)			
33: J: Okay.	Hits "Play" button	"4" and "7" flash	Show comparison
		"7" and "3" flash	Show comparison
		"7" and "3" swap	Show exchange
34: M: Looks like another swap	Hits "Pause" button	Animation pauses	User wants to pause animation

35: J: (*Writes "{7,3}" in trace.*) (*2 sec. pause*)

Table 5 (continued)

The users		The animation software	
Not available to the software	Available to software	Available to user	Design rationale
36: J: Okay	Hits "Play" button	"7" and "1" flash	Show comparison
		"7" and "1" swap	Show exchange
37: M: Another swap.	Hits "Pause" button	Animation pauses	User wants to pause animation
38: J: (*Writes "{7,1}" in trace.*) (*5 sec. pause*)I just noticed something. This table has a column for filling in the state of the array at the end of each pass.			
39: M: What's a pass? (*Both look at pseudocode.*) (*30 sec. pause*)			
40: J: I don't really understand these "for" loops, but I did notice that after the largest stick moved to the right-hand side, the flashes started again from the left.			
41: M: Yeah, it was like it started over.			
42: J: So maybe a "pass" ends when the largest stick bubbles to the right?			
43: M: Makes sense.			
44: J: So we gotta pay attention to when that happens, so I can pause the animation and write down the values.			
45: M: Okay.			

Table 6 Interaction log derived from analysis of transcript in Table 5

Line(s)	Event
7–10	They acknowledge that, at first glance, they don't know what's going on in the animation. They look at the task instructions and orient themselves to the task
13	They correctly map stick exchanges to exchanges in the trace
14–17	They acknowledge that they don't know what the flashes mean, and decide to explore further
19	They record first entry in trace
23–25	They posit the meaning of the flash events by connecting them to the pseudocode
29–38	They step through the animation, writing down swaps as they seem them, and (erroneously) recording all swaps in first trace row
38–42	They negotiate a shared understanding of "pass," a term used in the trace task instructions

empirical results, the transcription is a *dramatization*. While it is loosely based on actual empirical studies of the human use of algorithm visualization [30, 31, 35], it has been strategically developed to highlight the analysis method advocated by the Social Perspective.

Following the analysis framework of Suchman [8], the table contains four columns. The table deliberately separates the parts of the interaction that are mutually available to the users and the software (middle two columns) from the parts that are available only to one party (first and fourth columns). In so doing, it highlights asymmetries in the interaction: the software does not have access to questions and issues that arise in column 1, just as the users do not have direct access to the (implicit) design rationale underlying the software's output in column 4.

3.2.1 Interaction Analysis

To illustrate the evaluation method advocated by the Social Perspective, I now turn to an *interaction analysis* [11] of the episode depicted in the transcript. Recall that *interaction* is a multi-pass collaborative effort involving both a team of researchers and (often) the participants in the episode under study. Assume here that, through this process, the episode transcribed in Table 5 was singled out as significant because of the insight it provides into the role and value of the bubblesort animation in helping learners to perform the tracing task.

Table 6 presents a content log of significant events in the transcript. The first episode noted in the table (lines 7–10) illustrates that the significance of the animation events with respect to the trace task was not immediately clear. Jen and Mia are seeing the flashes and animated exchanges, but are unable to meaningfully interpret these events. Acknowledging this, they re-read the task instructions, and

quickly deduce that the animated exchanges they have just witnessed are to be
recorded in their trace (lines 11–13):

```
J: [Looks at instruction sheet] So, we're supposed to
   record each exchange in the inner loop.
M: Mmm, hmm.
J: I think that happens when the sticks change places.
```

However, they also identify a mystery to be solved (lines 14–17):

```
M: Yes, but what about the flashes?
J: They seem to happen before the sticks swap.
M: I did notice that. But I also saw sticks flashing at
   other times.
J: Hmm, maybe let's start it over.
```

After correctly recording the first exchange in their trace table (line 19), they
observe an anomaly on line 21 that prompts them to delve deeper into the mystery
of the flashing sticks:

```
J: Hey, did you see that?
M: Yep, there was a flash but no swap.
```

They consult the pseudocode for help. After a 20 s pause, Jen identifies the "if"
statement on line 4 as playing an important role in the algorithm's decision-making
(line 23). When Mia then connects the statement to the flash event (line 24), the pair
has constructed a working understanding of the significance of the flash events.

At this point, the pair shows evidence that they think they will be able to complete
the task. Indeed, they enter a start-pause cycle in which they record four successive
exchanges to their trace table. However, on line 38, Jen notices that their trace is
missing a key piece of information:

```
J: This table has a column for filling in the state of
   the array at the end of each pass.
```

This leads to Mia's confusion—presumably shared by Jen—over the meaning
of the term "pass." Again, the pair looks to the pseudocode for help. This time,
however, the pseudocode proves unhelpful. After a 30 s pause, Jen declares that
she doesn't understand "these for loops" (line 40), but that she does remember
something potentially significant about the animation:

```
J: I did notice that after the largest stick moved to
   the right-hand side, the flashes started again from
   the left.
```

This recollection leads to the following exchange, in which the pair negotiates
the meaning of "pass" within the context of the animation:

```
M: Yeah, it was like it started over.
```

```
J: So maybe a "pass" ends when the largest stick
   bubbles to the right?
M: Makes sense.
```

The episode concludes with Jen reducing this meaning into a practice that will allow the pair to complete the tracing task:

```
J: So we gotta pay attention to when that happens, so I
   can pause the animation and write down the values.
```

3.2.2 Discussion

The Social Perspective proceeds from the idea that the value of a visualization environment lies in its ability to serve as a *mediational resource* [26] in helping humans negotiate shared understandings of phenomena. As illustrated by the foregoing analysis, this perspective requires one to study the detailed, moment-to-moment interactions between humans, visualization environments, and other objects in the environment. While such an evaluation approach requires a significant investment of the researcher's time, there is a reward for researchers who are willing and able to put in the time: a sophisticated and nuanced understanding, firmly grounded in documentary evidence, of the role and value of visualization in specific scenarios of use. Such an understanding can ultimately help researchers to make progress in the field by serving as the basis for new theories of visualization use and effectiveness (see, e.g., [10, 36, 37]).

The foregoing analysis considered a fictitious episode adapted from prior studies of human visualization. As such, it highlights two key findings of that prior work that convey the flavor of the results one can expect to draw from this kind of analysis. First, in contrast to the implicit assumption made by the Cognitive Perspective—that a viewer will readily understand the significance of each animation event with respect to the trace task—the episode highlighted that the meaning of each visualization event must *constructed* through interaction with the environment. At the beginning of the episode, participants conceded that they could not make sense of the visualization relative to the tracing task at hand. The rest of the episode portrayed the intricate process by which they gradually ascribed significance to visualization events.

This leads to a second general result of the analysis: in contrast to the naïve assumption that a visualization environment can be used on its own, the episode illustrated that visualization environments are often used in concert with other resources in task scenarios. In the fictitious scenario considered here, it is fair to conclude that Jen and Mia would not have been able to do the tracing task with the bubblesort animation alone. Instead, the bubblesort animation raised many questions of interpretation: What does the flashing mean? Where does the end of a "pass" occur? Only through interleaved interactions with the animation, pseudocode, and task instructions were Jen and Mia ultimately able to resolve these questions and make progress on the task.

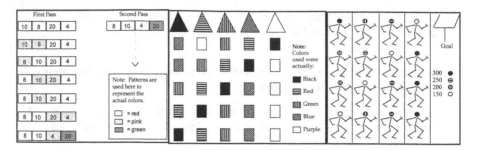

Fig. 5 "Number," "Color" and Football visualizations of the Bubblesort [30]

3.3 Evaluation from Cultural Perspective

With respect to this effectiveness of the bubblesort visualization, the key question from the Cultural Perspective is this: *In what ways does the visualization support increasingly central participation in a community of practice?* To illustrate how this question might be addressed from the Cultural Perspective, I will explore the use of a visualization "writing" task involving bubblesort visualizations as a means of gauging one's level of membership in the community of "schooled" computer scientists who share a common understanding of the standard algorithms and data structures taught in undergraduate computer science programs.

3.3.1 Visualization "Writing" Task

A central form of participation in the community of "schooled" computer scientists is the ability to *construct* algorithm visualizations that are consistent with those constructed by community experts. To gauge this form of community membership, one could design an empirical study in which participants are presented with the Bubblesort pseudocde and asked to do the reverse of the previous task: construct an algorithm visualization of the bubblesort.

In prior empirical studies [30, 31], my colleagues and I asked computer science graduate students to construct a bubblesort visualization for the purpose of explaining the algorithm to an introductory-level computer science student who is unfamiliar with the algorithm. Participants used simple art supplies (colored pens, colored construction paper, scissors) to construct their visualizations. They "executed" their homemade visualizations by doing such things as (a) moving construction paper "cutouts" across the table, (b) gesturing, (c) marking up their visualizations dynamically with a pen, and (d) providing a play-by-play narrative of the action as it unfolded. Figure 5 presents snapshots of the three visualizations that we observed in the study.

On the surface, the visualizations presented in Fig. 5 appear to vary widely. However, by conducting a semantic-level analysis, we can map these visualizations to their underlying semantics, as shown in Tables 7 and 8. Table 7 maps the objects of

Table 7 Mapping the lexical objects and attributes of the bubblesort visualizations to their semantics (Adapted from [30])

Semantics	"Number" Lexicon	"Color" Lexicon	"Football" Lexicon
Entry in array "a"	Square	Square	Stick figure
Value of entry in array "a"	Number symbol	Color	Color (as weight)
Array "a"	Contiguous row of squares	Non-contiguous row of squares	Contiguous row of figures
Inner loop pass history	Rows of sorting elements	—	—
Outer loop pass history	Columns of rows	Rows of sorting elements	Rows of sorting elements
Legend explicating ordering on sort elements	—	Triangles with color spectrum	Column of color/player weight pairs

the visualizations, along with the objects' attributes, to their underlying semantics. Table 8 maps significant visualization transformations to lines of pseudocode. As these tables indicate, even though the three visualizations differ widely at a lexical level, they all portray the bubblesort algorithm in terms of a similar semantics. The next section illustrates that it is possible to use this kind of semantic level analysis as input to a Consensus Analysis.

3.3.2 Quantifying Agreement

In order to use Consensus Theory's statistical model to assess an informant's "cultural competence," we need a means of quantifying the extent to which two individuals agree. To that end, we can make use of the semantic-level analysis technique just described. With respect to the "reading" task, two participants are said to agree to the extent that they can view the sticks animation and perform a similar semantic-level analysis on the representation. More formally, let s_1 be the set of semantic primitives (i.e., lexical-to-semantic mappings) that informant i_1 gleans from viewing a representation, and let s_2 be the set of semantic primitives that informant i_2 gleans from viewing the same representation. Then the *proportion of agreement* between informants p_1 and p_2 can be defined as the proportion of semantic primitives in s_1 and s_2 that are identical:

$$\text{Proportion of agreement}(p_1, p_2) = \frac{|s_1 \cap s_2|}{|s_1 \cup s_2|} \tag{1}$$

In the case of visualization "writing" task, the researcher, not the informant, performs the semantic level analysis. Other than this difference, the agreement between two informants can be calculated in an analogous fashion: Two participants are said to agree to the extent that the representations that they construct have semantic primitives in common. More formally, let s_1 and s_2 be the set of

Table 8 Mapping the lexical transformations of the bubblesort visualizations observed to their semantics (Adapted from [30])

Semantics	"Number" Lexicon	"Color" Lexicon	"Football" Lexicon
Do outer loop	Start new column of rows	Create new row of squares	Create new row of football players
Do inner loop	Create new row of squares	—	—
Reference elements	Color elements pink	—	Location of football
Compare elements	—	—	Intuitions about how player size relates to running, tackling, and fumbling
Exchange elements	Exchange numbers	Exchange colors	Ball carrier advances by tackling next football player in line (thereby exchanging positions with that player)
Don't exchange elements	—	—	Fumble football to next player in line
Terminate outer loop	Color square in correct order green	—	—
Terminate Sorting	Ordering of natural numbers, all squares green	Color squares match legend	Players ordered by weight

Table 9 Computing the pair-wise proportion of agreement among the visualizations presented in Fig. 5

Pair	% Agreement (Objects/Attributes)	% Agreement (Trans.)	% Agreement (Total)
Number-Color	4/6 (67 %)	3/6 (50 %)	7/12 (58 %)
Number-Football	4/6 (67 %)	4/8 (50 %)	8/14 (57 %)
Color-Football	5/6 (83 %)	3/6 (50 %)	8/12 (67 %)

semantic primitives determined to exist in the visualizations of participant p_1 and p_2, respectively. Then Eq. 1 expresses the proportion of agreement between the visualizations of p_1 and p_2.

To illustrate how one might apply this definition of agreement, Table 9 computes the pair-wise levels of agreement among the three visualizations presented in Fig. 5, based on the semantic-level analysis presented in Tables 7 and 8.

3.3.3 Sample Consensus Study

Based on the ideas just discussed, I now sketch out how one might design and analyze a Consensus Study of visualization "writing" in order to measure the effectiveness of alternative pedagogical treatments in facilitating students' convergence on expert competence in an undergraduate computer science course. In order to establish the "cultural truth" in these tasks, a group of experts (computing instructors and professionals) would first complete visualization writing tasks. The results of a semantic level analysis of their tasks would be fed to Consensus Theory's statistical model, in order to determine the "cultural truth" for these tasks—that is, the "culturally correct" set of semantic primitives that experts include in the visualizations that they construct.

One would now be in a position to conduct a study of visualization writing in the course. Using a between-subjects, repeated-measures design, one would divide students into two groups, each of which completes an alternative set of learning exercises. Before completing these learning exercises, both groups would complete visualization construction tasks for the first of the two algorithms. The results of a semantic-level analysis of these tasks would be fed into a Consensus Analysis, so that we can determine students' baseline cultural competence with respect to the expert informants in the study. Then, upon completing their respective pedagogical treatments, both student groups would complete construction tasks for the second of the two algorithms. Once again, the results of a semantic-level analysis of these tasks would be fed into a Consensus Analysis, in order to determine students' "final" cultural competence with respect to the expert informants in the study.

At this point, one would be in a position to perform statistical tests for significant differences between the two groups. If the average change in one group's level of cultural competence (that is, the difference between its final and baseline cultural competence) is significantly higher than the average change in the other group's level of cultural competence, then that would be grounds for concluding that one pedagogical treatment is more effective than the other.

3.3.4 Discussion

The foregoing proposal to use visualization "writing" tasks to gauge convergence upon a cultural consensus raises at least three general issues. First, the semantic level analysis technique relies on the *interpretation* of a researcher who is also an expert in the community in which the study is being performed. To ensure that the technique is uniformly applied, and to reduce research bias, multiple researchers trained in the technique would ideally perform independent semantic-level analyses on informants' representations; the *intraclass reliability coefficient* (see, e.g., [38]) of the analysts could then be incorporated into the Consensus Analysis.

A second issue is that the proposed approach may, in fact, oversimplify the notion of semantic agreement. Indeed, Tversky [39] has developed a formal model for assessing the agreement between two sets of semantic features that is far more sophisticated than Eq. 1 (set intersection divided by set union). In particular, given a domain of feature (i.e. semantic primitive) sets $\{s_1, s_2, \ldots, s_n\}$, Tversky defines a function $s(a,b)$ to be the relative similarity between feature sets a and b. The function $s(a,b)$ assigns a value to each pair of feature sets, such that $s(a,b) > s(c,d)$ means that a is more similar to b than c is to d. The function $s(a,b)$ is defined as a *linear contrast* between the semantic primitives in set a and b:

$$s(a, b) = \theta f(a \cap b) - \alpha f(a - b) - \beta f(b - a) \tag{2}$$

where $f(a \cap b)$ counts the number of primitives that a and b have in common; $f(a-b)$ counts the number of primitives belonging to a but not to b; $f(b-a)$ counts the number of primitives belonging to b but not to a; and θ, α, and β are weightings indicating the relative importance of these three entities. Notice that, in contrast to Eq. 1, Eq. 2 provides a framework for weighting the relative importance of various features of a semantic domain. In performing a Consensus Analysis, one would do well to experiment with the Tverskian model, in addition to using the less sophisticated formula presented in Eq. 1.

A third issue raised by the approach has to do with the kind of statement it is able to make about visualization effectiveness. The Cognitive and Social Perspectives lead to statements about the ability of given a given visualization to support task efficiency and human-human interaction. In contrast, the Cultural Perspective leads to statements not about the effectiveness of a given visualization, but rather about the effectiveness of the practices of a community that use that visualization. Thus, if the goal of the researcher is to scrutinize the design of a specific visualization environment, then the Cultural Perspective will be unable to help.

4 Implications: Matching Research Questions to Perspectives

This chapter has presented three alternative perspectives for evaluating the effectiveness of visualization environments. Each perspective arises out of a distinct intellectual tradition, gives rise to a distinct theory of visualization effectiveness, and suggests a distinct methodology for evaluating effectiveness. By applying each perspective to a case study of visualization use, the chapter has illustrated the different kinds of statements about effectiveness to which each perspective can lead.

As illustrated by the preceding case study, no perspective is superior; each furnishes a theoretical orientation and evaluation method that can prove useful for evaluating effectiveness. The task for the visualization researcher is to identify the perspective that is the best *match* for the research questions of interest. A synthesis of the chapter's presentation and application of the perspectives leads to the following guidelines for matching research questions to an appropriate perspective:

- If the research question focuses on the ability of a visualization environment to promote human efficiency in well-defined human tasks, then the Cognitive Perspective will likely provide the most useful lens for evaluating effectiveness. The researcher can break the tasks down into human procedures, develop predictive models of human performance, and ultimately make design decisions based on those models.
- If the research question focuses on the use of a visualization environment for collaborative tasks that use the environment to build understanding and insight, and/or to make decisions, then the Social Perspective will likely provide the most useful lens for evaluating effectiveness. The researcher can videotape the collaborative use of the environment, and carefully analyze the video record in order to build an empirical account of the role and value of the visualization environment in facilitating collaboration, communication, and mutual intelligibility.
- If the research question focuses on the use of a visualization environment in sustaining a community of practice, then the Cultural Perspective will likely provide the most useful lens for evaluating effectiveness. The researcher can have community participants engage in visualization "reading" and "writing" tasks both to document the semantic domain of the community, and to gauge the centripetal movement of newcomers toward the center of the community.
- Multiple perspectives can be enlisted to build a more nuanced account of effectiveness. For example, a researcher may be interested both in efficient task performance, and in the ability of visualizations to assist collaborating scientists in building insight. In such cases, it may make sense both to build cognitive models to predict task performance with a given visualization (Cognitive Perspective), and to study video accounts of the collaborative use of the visualization in such tasks (Social Perspective). Likewise, if a researcher is interested in both the collaborative use of a visualization environment, and its role in sustaining a community of practice, the researcher could choose both to study video recordings of the collaborative use of the environment, and to use those video recordings as a basis for measuring community consensus.

The success of any evaluation effort rests on the researcher's ability both to formulate the appropriate research questions for the given situation, and to collect the appropriate empirical data to address those questions. By presenting a framework of three perspectives, this chapter has aimed both to illustrate a range of possible empirical approaches, and to provide initial guidance on choosing the right one.

References

1. A. Chourasia, S. Cutchin, Y. Cui, R. W. Moore, K. Olsen, S. M. Day, J. B. Minster, P. Maechling, and T. H. Jordan, Visual Insights into High-Resolution Earthquake Simulations, *IEEE Computer Graphics and Applications*, vol. 27, no. 5, pp. 28 –34, 2007.
2. K. Riley, D. Ebert, C. Hansen, and J. Levit, Visually accurate multi-field weather visualization, In *Proceedingsof IEEE Visualization 2003*, Los Alamitos, CA: IEEE Computer Society Pess, pp. 279–286, 2003.
3. S. Eick, J. Steffen, and E. Sumner, Seesoft: A Tool for Visualizing Line-Oriented Software Statistics, *IEEE Transactions on Software Engineering*, vol. 18, no. 11, pp. 957–968, 1992.
4. H. Lam, E. Bertini, P. Isenberg, C. Plaisant, and S. Carpendale, Empirical Studies in Information Visualization: Seven Scenarios, *IEEE Transactions on Visualization and Computer Graphics*, vol. 30, no. November, pp. 2479–2488, 2011.
5. L. A. Treinish, Task-specific visualization design, *IEEE Computer Graphics and Applications*, vol. 19, no. 5, pp. 72–77, 1999.
6. J. H. Larkin and H. A. Simon, Why a diagram is (sometimes) worth ten thousand words, *Cognitive Science*, vol. 11, pp. 65–99, 1987.
7. M. Scaife and Y. Rogers, External cognition: how do graphical representations work?, *International Journal of Human-Computer Studies*, vol. 45, pp. 185–213, 1996.
8. L. A. Suchman, *Plans and Situated Actions: The Problem of Human-Machine Communication*. New York: Cambridge University Press, 1987.
9. T. Winograd and F. Flores, *Understanding Computers and Cognition*. New York: Addison-Wesley, 1987.
10. J. Roschelle, Learning by collaborating: Convergent conceptual change, *Journal of the Learning Sciences*, vol. 2, no. 3, pp. 235–276, 1992.
11. B. Jordan and A. Henderson, Interaction analysis: Foundations and practice, *Journal of the Learning Sciences*, vol. 4, no. 1, pp. 39–103, 1995.
12. J. Lave and E. Wenger, *Situated Learning: Legitimate Peripheral Participation*. New York: Cambridge University Press, 1991.
13. E. Wenger, *Communities of Practice: Learning, Meaning and Identity*. Cambridge: Cambridge University Press, 1998.
14. A. K. Romney, S. C. Weller, and W. H. Batchelder, Culture as consensus: A theory of culture and informant accuracy, *American Anthropologist*, vol. 88, no. 2, pp. 313–338, 1986.
15. P. C. Wong and J. Thomas, Visual analytics, *IEEE Computer Graphics and Applications*, vol. 24, no. 5, pp. 20–21, 2004.
16. R. Ben-Bassat Levy, M. Ben-Ari, and P. Uronen, The Jeliot 2000 program animation system, *Computers & Education*, vol. 40, no. 1, pp. 1–15, 2003.
17. C. D. Hundhausen, P. Agarwal, R. Zollars, and A. Carter, The design and experimental evaluation of a scaffolded software environment to improve engineering students' disciplinary problem-solving skills, *Journal of Engineering Education*, under review.
18. V. Michalchik, A. Rosenquist, R. Kozma, P. Kreikemeier, P. Schank, and B. Coppola, Representational resources for constructing shared understandings in the high school chemistry classroom, In *Visualization: Theory and practice in science education*, J. Gilbert, M. Nakhleh, and M. Reiner, Eds. New York: Springer, pp. 233–282, 2008.

19. M. C. Chuah, B. E. John, and J. Pane, Analyzing graphic and textual layouts with GOMS: Results of preliminary analysis, In *CHI'94 Conference Companion*, New York: ACM Press, pp. 323–324, 1994.

20. J. R. Anderson, M. Matessa, and C. Lebiere, ACT-R: A theory of higher level cognition and its relation to visual attention, *Human-Computer Interaction*, vol. 12, no. 4, pp. 439–462, 1997.

21. S. K. Card, T. P. Moran, and A. Newell, *The Psychology of Human-Computer Interaction*. Hillsdale, NJ: Lawrence Erlbaum Associates, 1983.

22. S. M. Casner and J. H. Larkin, Cognitive efficiency considerations for good graphic design, In *Cognitive Science Society Proceedings*, Hillsdale, NJ: Erlbaum, pp. 275–282, 1989.

23. M. Hegarty, Mental animation: inferring motion from static displays of mechanical systems, *Journal of Experimental Psychology: Language, Memory, and Cognition*, vol. 18, pp. 1084–1102, 1992.

24. J. Zhang and D. Norman, Representations in distributed cognitive tasks, *Cognitive Science*, vol. 18, pp. 87–122, 1994.

25. B. Nardi, Studying context: A comparison of activity theory, situated action models, and distributed cognition, In *Context and Consciousness: Activity Theory and Human-Computer Interaction*, Cambridge, MA: The MIT Press, pp. 69–102, 1996.

26. J. Roschelle, Designing for cognitive communication: Epistemic fidelity or mediating collaborative inquiry?, In *Computers, Communication and Mental Models*, D. Day and D. K. Kovacs, Eds. London: Taylor & Francis, pp. 13–25, 1996.

27. J. Heritage, Recent developments in conversation analysis, *Sociolinguistics*, vol. 15, pp. 1–16, 1985.

28. J. Gumperz, *Discourse Strategies*. Cambridge: Cambridge University Press, 1982.

29. T. R. G. Green, M. Petre, and R. K. E. Bellamy, Comprehensibility of Visual and Textual Programs: A Test of Superlativism Against the 'Match-Mismatch' Conjecture, In *Empirical Studies of Programmers: Fourth Workshop*, pp. 121–146, 1991.

30. S. A. Douglas, C. D. Hundhausen, and D. McKeown, Toward empirically-based software visualization languages, In *Proceedings of the 11th IEEE Symposium on Visual Languages*, Los Alamitos, CA: IEEE Computer Society Press, pp. 342–349, 1995.

31. Z. D. Chaabouni, A user-centered design of a visualization language for sorting algorithms, University of OregonEditor, 1996.

32. B. Tversky and B. Morrison, Can animations facilitate?, *Int. J. Hum.-Comput. Stud.*, vol. 57, no. 4, pp. 247–262, 2002.

33. A. W. Lawrence, A. N. Badre, and J. T. Stasko, Empirically evaluating the use of animations to teach algorithms, In *Proceedings of the 1994 IEEE Symposium on Visual Languages*, Los Alamitos, CA: IEEE Computer Society Press, pp. 48–54, 1994.

34. M. Ben-Ari, Constructivism in computer science education, *J. Comput. Math. Sci. Teach.*, vol. 20, no. 1, pp. 45–73, 2001.

35. C. D. Hundhausen, Integrating algorithm visualization technology into an undergraduate algorithms course: Ethnographic studies of a social constructivist approach, *Computers & Education*, vol. 39, no. 3, pp. 237–260, 2002.

36. D. Suthers and C. Hundhausen, An experimental study of the effects of representational guidance on collaborative learning processes, *Journal of the Learning Sciences*, vol. 12, no. 2, pp. 183–219, 2003.

37. C. D. Hundhausen, Using end user visualization environments to mediate conversations: A 'Communicative Dimensions' framework., *Journal of Visual Languages and Computing*, vol. 16, no. 3, pp. 153–185, 2005.

38. P. E. Shrout and J. L. Fleiss, Intraclass correlations: Uses in assessing rater reliability, *Psychological Bulletin*, vol. 86, no. 2, pp. 420–428, 1979.

39. A. Tversky, Features of similarity, *Psychological Review*, vol. 84, pp. 327–352, 1977.

On the Prospects for a Science of Visualization

Ronald A. Rensink

Abstract This paper explores the extent to which a scientific framework for visualization might be possible. It presents several potential parts of a framework, illustrated by application to the visualization of correlation in scatterplots. The first is an *extended-vision thesis*, which posits that a viewer and visualization system can be usefully considered as a single system that perceives structure in a dataset, much like "basic" vision perceives structure in the world. This characterization is then used to suggest approaches to evaluation that take advantage of techniques used in vision science. Next, an *optimal-reduction thesis* is presented, which posits that an optimal visualization enables the given task to be reduced to the most suitable operations in the extended system. A systematic comparison of alternative designs is then proposed, guided by what is known about perceptual mechanisms. It is shown that these elements can be extended in various ways—some even overlapping with parts of vision science. As such, a science of some kind appears possible for at least some parts of visualization. It would remain distinct from design practice, but could nevertheless assist with the design of visualizations that better engage human perception and cognition.

R.A. Rensink (✉)
Departments of Computer Science and Psychology, University of British Columbia,
2366 Main Mall, Vancouver, BC, V6T 1Z4 Canada
e-mail: rensink@cs.ubc.ca; rensink@psych.ubc.ca

W. Huang (ed.), *Handbook of Human Centric Visualization*,
DOI 10.1007/978-1-4614-7485-2_6, © Springer Science+Business Media New York 2014

1 Introduction

Is there a best way to visually display a given dataset for a given task, and if so, can we find it? Considerable effort has been expended on this issue over the years.[1] The result has been a set of specialized disciplines concerned with the design of displays in various domains, such as cartography, diagram design, statistical graphics, visual interface design, and information visualization (e.g., [1–4]). These disciplines have achieved considerable success, often resulting in designs that are highly *effective*— i.e., that enable performance that is rapid, accurate, and relatively effortless [5].

But many important questions remain unanswered. What is the best way to measure how a given visualization works? How could we find the perceptual and cognitive factors that limit its performance? Could we determine if its design is optimal? Answering such questions will require something more than intuition and *post hoc* measures. And more than design guidelines. It will require a framework that is systematic and rigorous—ideally, one that is *scientific* in the best sense of the word, sensitive to the nature of the visualization task, the computational issues involved, and the nature of the human viewer. The issue considered here is the extent to which this might be possible for visualization. Could it—or at least part of it— even be developed into a science?

To answer this, it may first be worth considering what a science is. Few disciplines are as systematic and rigorous as physics. What then allows a domain to be considered a science? One commonly-accepted criterion is for the domain to have a framework—a *paradigm*—that explicitly describes (1) a related set of entities in the world that are of interest, (2) the kinds of questions that can be asked about them, and (3) possible ways of answering these questions [6, 7]. In other words, a science is not so much a particular body of knowledge, but an *organized way of thinking* about a coherent set of issues. A framework of this kind would ideally consist of a set of characterizations, theories and practices that are consistent with each other and connected by their reference to a common set of issues.

In what follows, several potential elements of such a framework are proposed for visualization; as a test of their suitability and consistency with each other, each is applied to the case of correlation in scatterplots.[2] The first concerns the issue of how visualization can best be characterized. An *extended-vision thesis* is proposed,

---

[1]Historically, much of this work focused on *graphic displays, which* convey information using the geometric and radiometric properties of an image (as opposed to simple text alone). Meanwhile, more modern work focuses on *visual displays*, which rely on the extensive use of visual intelligence for their interpretation. For purposes here, graphic displays and visual displays are considered much the same, with the former term emphasizing the means, and the latter the ends.

[2]A single framework for all aspects of visualization (e.g., usability) is problematic, owing to the heterogeneous nature of the components (e.g., perceptual vs. motor mechanisms) and the possible lack of specificity in the tasks for which it might be used [32]. Discussion here focuses on a more restricted set of issues, viz., the extent to which visualization can enable a human to perceive some well-defined structure in a dataset. This abstracts away from details of particular tasks, and so increases the chances of a systematic framework for at least some parts of visualization. Vision

which posits that the viewer and visualization system can be considered a single system that enables structure in a dataset to be perceived in much way as "basic" vision enables perception of structure in the world. This characterization is then used as the basis of the next element: a more thoroughgoing approach to evaluation, informed by methodology drawn from vision science. Turning to issues of design, an *optimal-reduction thesis* is introduced, which states that an optimal visualization can be considered as one that reduces the given task to the most appropriate set of operations in the extended system. The next element shows how this view can motivate ways of assessing the effectiveness of various design parameters, with the results providing insight into the underlying perceptual mechanisms. Finally, it is shown that these elements can be extended to several other kinds of visualization, and that some can even begin to overlap parts of vision science. As such, it appears that at least some aspects of visualization can be handled in a more integrated and systematic way, one that can help make better use of the perceptual and cognitive abilities of the human viewer.

1.1   The Need for a Systematic Framework

Before considering particular elements, it may be helpful to say a few things about the need for a systematic framework in the first place. It might be thought, for example, that such a framework is unnecessary: designers have long explored the space of possible designs, and by now have reasonably good intuitions about what is optimal, or at least highly effective. Or it might be thought that we as observers have extensive (and perhaps privileged) experience with the operation of our visual systems, and so could authoritatively decide on the issues relevant for any particular design.

But although our intuitions about design—especially those derived from long experience—are important, they are incomplete. To begin with, many devices commonly used in static displays (e.g., box plots, small multiples) are relatively recent, so that intuition has had relatively little time to develop. And ways of visualizing complex structures such as networks are not only recent, but involve a considerable degree of complexity, making the possibility of effective intuitions even less likely. The same concerns apply equally well to dynamic displays and interactive systems (e.g., [8, 9]). All these suggest a design space of such high dimensionality that it has not been—and may never be—completely explored. Guidelines are emerging to help with this (e.g., [10]). But a systematic framework could help create such guidelines, and perhaps even go beyond them.

It might be thought that a framework of some kind might be based directly on our beliefs about how we see. But there are likewise limitations here. For example,

---

science uses a similar approach, focusing on well-defined functions rather than on ways that vision might help carry out some poorly-defined task [21].

although we have a strong impression that we build up a complete "picture" of our surroundings, the visual system does not operate this way: instead of a dense accumulation of data, we instead likely use a dynamic, just-in-time representation, where only a few coherent structures exist at any moment [11]. Moreover, evidence is increasing that conscious perception is only one aspect of how we see, with considerable visual intelligence in processes that operate without any conscious involvement [12]. As such, intuitions about vision—and more generally, about perception and cognition—are not enough to create a viable framework for this purpose. Something more systematic is needed, something that can enable the effective use of knowledge about the main psychological factors involved,

1.2   The Applicability of a Systematic Framework

It might be argued that even if a systematic framework of some kind were possible, it would not be of much use: how could it be general enough to apply to domains as diverse as cartography and statistical graphics, yet specific enough to guide particular designs? And what about new developments, such as a speedup in rendering, or a discovery about visual attention? Wouldn't the whole framework need to be rebuilt each time they occur?

In other design disciplines, the existence of a systematic framework for visualization is problematic neither in principle nor in practice. Architecture, for example, has long had such a framework. An architect can incorporate physical constraints into the design of a building, say, to guarantee that it will not fall over due to imbalances in weight distribution. Doing so does not interfere with design in any real way— it does not prohibit anything that is physically viable. Rather, constraints such as those based on physical forces or material properties can be applied to any design, determining whether it is viable, and sometimes even whether it is optimal. There is no a priori reason why a similar approach would not also work for visualization.

Indeed, some systematicity already exists in the design process for visualization. Guidelines exist for particular applications, such as designing a map or a graph (e.g., [1, 13, 14]); some even include explicit discussion of perceptual mechanisms, so as to enable adaptation to particular circumstances (e.g., [2, 10, 15]). A complementary approach starts with a particular perceptual mechanism and connects it to various tasks (e.g., [16, 17]). A third approach is based on general relations that exist between different kinds of tasks and different perceptual mechanisms (e.g., [5, 18]). However, none of these approaches is entirely quantitative, nor does it address all issues. What is needed is something that incorporates the best of all these, and enables visualization to be treated in a comprehensive, integrated way.

2 Extended Vision

A first step towards a more systematic framework would be to characterize—as far as possible—exactly what visualization is. Loosely speaking, visualization (in the sense considered here) can be described as *the transforming of a problem into graphical form, so as to engage the visual intelligence of a human viewer*. Said another way, the goal of visualization is to translate a given problem into the language of human vision and cognition. The effectiveness of a given design—a given *graphical representation*—is then determined by the extent to which it can be created (typically, in near real-time) while still enabling the most appropriate perceptual and cognitive mechanisms to be engaged on the task at hand (cf. [5, 19]).

Consider the ages and heights of a set of people. When these are represented via position (Fig. 1a), several trends—such as height increasing with age up to about age 20—are immediately apparent. In contrast, when these data are represented via length (Fig. 1b), these relations are virtually impossible to see. The effective design somehow engages perceptual mechanisms that are more suitable for the task. But what exactly does this mean?

Addressing such questions in a meaningful way requires a characterization of visualization that can enable sufficient articulation of the underlying issues. One such possibility is the *extended-vision thesis*: the viewer and the visualization system can be considered a single information-processing system that enables the viewer to perceive structure in the given dataset much as they would perceive structure in the world using "basic" vision. For such an "extended" system, the input is the dataset under consideration (the ages and heights of a group of people, say), the output some function of this input (correlation between age and height), and

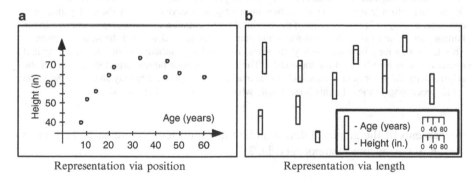

Representation via position   Representation via length

Fig. 1 Different graphical representations. (a) Age is represented by *horizontal position*, height by *vertical position*; size of the graphic items is irrelevant. Using this representation, the relation between age and height is immediately apparent. (b) Age is represented by length *above* the interior line, height by length *below*; position of the items is irrelevant. Here, the relation between the two quantities is far less obvious, even though the data represented are the same

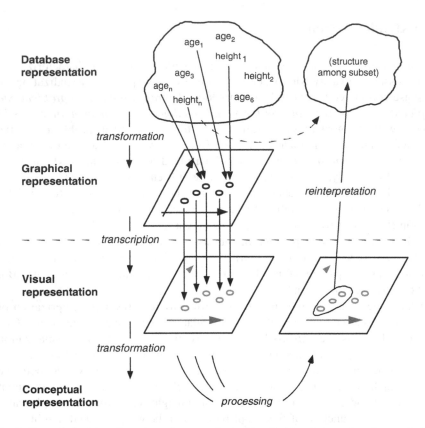

Fig. 2 Extended vision (Based on the example in Fig. 1). Translation of a given task into perceptual and cognitive operations involves three stages: (1) *transformation* of the relevant aspects of the data into a *graphical representation* on a graphical device (e.g., a sheet of paper, or a computer display), (2) *transcription* of the graphical representation to a *visual representation* in the human viewer, and (3) further *transformation* and *processing* to determine the structure present. This latter step may be done entirely via visual mechanisms, although higher-level, conceptual operations may also be used for some tasks. The result can then be reinterpreted in terms of the original data. (If processors are available in the machine component, they may also carry out some initial processing of the data; this is essentially what occurs in visual analytics)

the contents of the graphical device an intermediate step that connects the human viewer to the machine component (Fig. 2).[3]

---

[3]In some cases (e.g., cartography), the "machine" component may be distributed over a several different devices—and perhaps even the occasional human—with only the result contained in the display (e.g., a map). From a functional point of view, such a configuration is still considered a single visualization system. The performance of such "off-line" systems differs from "on-line" systems only in one respect: because the time scales of the two components are quite different, the effectiveness of an off-line display can usually be assessed in terms of the speed, accuracy, and

An obvious difference between a basic and an extended system is that the projection of information onto the eye of the viewer moves from the initial stage to an intermediate one. But the goal remains the same: extract useful structure in the underlying data. Interestingly, selection of which data to examine (and perhaps process), along with its transformation into graphical form typically occurs via a sequence of stages—the visualization pipeline [5]—in largely the same way that human vision transforms light into a particular perceptual structure (cf. [20, 21]). As such, the architecture of an extended system is an "extended pipeline" created by the concatenation of all these processing stages, with each focusing on a different aspect of the data (cf. [22]).

Note that the information flow between processing elements need not be unidirectional: feedback generally exists between elements in the machine component, as well as in human vision (cf. [5, 23]). Feedback likewise exists between the human and machine elements in any system that is interactive, resulting in a rough architectural consistency throughout the entire system. Indeed, the continual interactions between the components of an extended system that occur during sensemaking [5] are highly analogous to the interactions between the components of a basic system that occur during visual perception [24, 25]. As such, the interactive aspects of visualization (central to e.g., information visualization or visual analytics) are captured by this characterization in a reasonably natural way.

Because these interactive aspects involve a wide range of issues touching on much of cognitive psychology, they are not discussed in great detail here. To keep discussion focused on the main question here—viz., the possibility of a science of visualization—emphasis is placed on "static" noninteractive aspects. This recapitulates the development of vision science, where color and form perception were investigated before more interactive—and generally, more complex—processes such as scene perception [11].

3 Systematic Evaluation

Another important element of any framework for visualization involves the issues related to evaluation. Evaluation of visualization has taken on a variety of forms, ranging from careful quantitative measurement (e.g., [26]) to simple verification of basic functionality (see e.g., [27]). Such variety is understandable: tasks differ in the degree to which they can be described quantitatively, and thus, the degree to which quantitative measures can be applied [28, 29]. And sometimes the goal is merely to verify that a given visualization can be used to some extent. But if the goal is to carefully compare different designs to determine which is the best, or to understand why one works better than another, preference should be given to highly informative measurements. But if so, what exactly should be measured, and how?

---

effort exerted by the human component alone. (Depending on the situation, however, it may be necessary to take into account such things as the cost of producing the display.)

This is in general a complex issue, making it difficult to find definitive answers. Fortunately, however, decades of work in vision science have been spent on developing high-precision and robust techniques to measure how well graphical structures can be perceived (e.g., [8], Appendix C; [30], Appendix A). The deep similarity between vision and visualization posited by the extended-vision thesis suggests that many of these approaches (along with their foundations in measurement theory) could be applied almost directly to the evaluation of visualizations, resulting in the development of evaluation techniques with a high degree of utility. Said another way: vision science not only offers possible mechanisms to help with visualization design [8], but also possible methodologies.

In what follows, this point will be illustrated via its application to scatterplots. Although scatterplots appear to be simple (at least, once learned), and are widely used to enable correlation to be easily seen (e.g., [1, 31]), relatively little is known about how well they work, or why. As such, the evaluation of these provides an ideal example of what might be done.

3.1 Task Specification

The first step in a rigorous evaluation of a perceptual system—extended or not—is a clear specification of its *function*. In the case of scatterplots, something like "enable discovery of interesting structure in the data" might initially be thought sufficient. But what does "interesting" mean? Finding outliers? Finding trends? If trends, is it correlation or something else? And if correlation, what kind? Spearman? Pearson? Some tasks are inherently vague; requests to "just find something interesting" are not uncommon [29, 32]. But a precise specification should be attempted whenever possible. It can often be achieved.

The next step is to determine an appropriate set of *inputs* for testing. (In the jargon of vision science, these would be the *stimuli*.) Ideally, these are representative of the data encountered in everyday applications; failing that, they should at least have sufficient range to enable determination of how performance depends on various properties of interest. In the case of scatterplots, inputs might be specified as a set of ordered (scalar) pairs drawn from a bivariate gaussian distribution with particular means, variances, and correlation, along with a particular number of pairs in each set. The number of instances and viewers tested should be large enough to ensure sufficient statistical power in the results of the evaluation. In addition, the representation used to display these values would require specification of all properties pertaining to its graphical nature, such as dot size and color.

The final step would then be to specify an appropriate set of *measures* by which performance could be evaluated. In the case of scatterplots, evaluation would determine how well the given representation supports the perception of correlation in the test dataset, based on measures such as accuracy, variability, and timecourse.

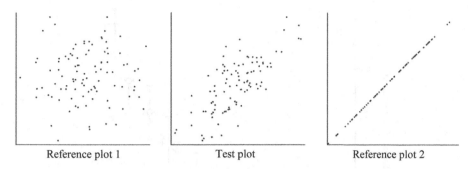

Fig. 3 Example of bisection task. Observers adjusted the correlation of the central test plot until its correlation was halfway between those of the reference plots. Plots were 5° × 5° in extent. Here, the value of the test plot is r = 0.74, corresponding to the subjective midpoint g = 0.5

3.2 Accuracy

One of the most basic performance measures (one often used in evaluating visualizations) is *accuracy*: how well on average the extended system extracts structure from a given dataset.[4] For the case of Pearson correlation r in scatterplots, various studies of accuracy have been carried out (e.g., [26, 33, 34]). Almost all tested observers by asking for direct numerical estimates, viz., a number between 0 and 1 corresponding to the correlation seen in the scatterplot (For a review, see [35]). Several important results have been discovered this way, including the finding that perceived correlation g tends to underestimate physical correlation r (especially for intermediate levels of correlation), and that essentially no correlation is perceived when |r| < 0.2.

However, the use of direct numerical estimates is usually not optimal. To begin with, its central assumption—that numbers can consistently be assigned to perceived magnitudes—may be incorrect [36]. Indeed, assigning a number to a perceived quantity is a somewhat unnatural task; human perception typically focuses on relations rather than absolutes [37]. Consequently, a better approach—one often used in vision science—might be *bisection*, a technique that takes advantage of the ability of humans to easily and accurately determine the midpoint of a structure. In [38], for example, observers were shown a display containing two *reference plots* (one with a high level of correlation, one with a low) along with a *test plot* between these. Observers adjusted the correlation of the test plot until it appeared to be halfway between the correlations of the reference plots (Fig. 3).

---

[4] Although inaccuracies are usually caused by mechanisms in the human viewer, they can also be due to the machine component—e.g., insufficient sampling, or a bias in an algorithm. According to the extended-vision thesis, the source is irrelevant: the measure of interest is based on the performance of the *entire* system. For the most part, discussion here focuses on the human viewer, since this is typically the largest source of inaccuracy. But if need be, the accuracy of the machine component could also be evaluated. Similar considerations apply to other measures of performance.

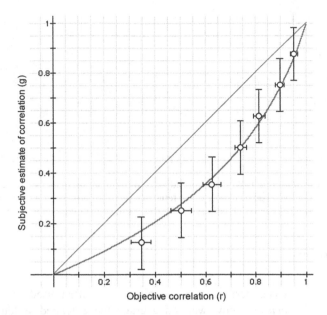

Fig. 4 Perceived correlation via bisection. The curve describing the perceived correlation is $g(r) = \ln(1-br)/\ln(1-b)$; best fit to data is for $b = 0.88$. The *straight line* $g(r) = r$ is for comparison; accuracy is based on the difference between this function and $g(r)$. Note the severe underestimation of correlation, especially around $r = 0.6$

Experiments were run with 20 observers, on scatterplots containing 100 dots. In a first round of testing, observers judged the halfway point between the extremes $r = 0$ and $r = 1$. (These corresponded to $g = 0$ and $g = 1$, respectively; the halfway point was $g = 1/2$). A second round applied this test recursively, with each observer now asked to find the value that appeared to be halfway between $g = 0$ and $1/2$, and the value between $g = 1/2$ and 1, resulting in estimates for $g = 1/4$ and $g = 3/4$ respectively. Finally, this method was applied again to determine the values for the subjective estimates $g = 1/8$, $3/8$, $5/8$, and $7/8$. All conditions were appropriately counterbalanced (For details, see [38]).

Results are shown in Fig. 4. Consistent with the results of other studies, an underestimation of correlation appears, especially around $r = 0.6$. The results are also broadly consistent with two previous proposals for perceived correlation: the square function $g(r) = r^2$ [33, 34], and a more complex double-power function $g(r) = 1 - (1-r)^a(1+r)^b$, where a and b are free parameters [39]. But the data were precise enough to show that a better fit was with the logarithmic function

$$g(r) = \ln(1 - br)/\ln(1 - b) \tag{1}$$

where b is a real number such that $0 < b < 1$; $g(r) = r$ when $b = 0$ [38].

The accuracy for a given design can be determined by the difference between this function and the physical correlation r. Note that it can be described by a single value (b), covering the entire range of correlations possible. A measure can even be determined based on the relative frequency at which various correlations might be encountered in a task [38].

Interestingly, Eq. 1 is a form of *Fechner's law*, which states that perceived magnitude is proportional to the logarithm of physical magnitude; it applies to the perception of several simple properties, such as brightness [40]. To make this more explicit, Eq. 1 can be rewritten as

$$g(u) = \ln(u)/\ln(1-b) \tag{2}$$

where $0 < b < 1$, and $u = 1-br$, the distance *away* from complete correlation. The dependence on u suggests that the relevant factor may be related to the *dispersion* of the dots in the scatterplot, rather than correlation per se. As such, the greater sensitivity of the bisection technique not only provides a better estimate of accuracy (and possibly, a better way to describe it), but also begins to cast some light on the underlying mechanisms.

3.3 Variability

Although accuracy is important for evaluation, other measures are also useful. One of these is *variability*, the extent to which the extended system gives the same answers when given the same data. (Recall that the mean values of these determine accuracy.) Equivalently, it describes the *discriminative power* of the system—i.e., its ability to distinguish between data values that are somewhat similar. Variability can also provide insight into the maximum and minimum values the system might provide, and how often these might occur.

For scatterplots, variability can be assessed by an approach commonly used in vision science: determining how much two properties must differ in order to be *discriminated*, i.e., to see that they are not the same. More precisely, for any objective correlation r, the goal is to find the *just noticeable difference* (*jnd*), the value of Δ for which correlations r and r $\pm$ Δ can be discriminated 75 % of the time ([30], Appendix A). The greater the jnd, the greater the separation needed to see that two scatterplots have different correlations, and thus, the greater the variability.

Variability was evaluated this way using 20 observers, on scatterplots containing 100 dots each [38]. A set of base correlations was examined, ranging from r = 0 to r = 1 in increments of 0.1. For each base correlation, two side-by-side scatterplots were shown—one with the base, the other with a variant correlation. Observers were asked to select the scatterplot appearing to be more highly correlated (Fig. 5). All conditions were appropriately counterbalanced (For details, see [38]).

The results of this test are shown in Fig. 6. Here, the absolute value of the jnd is plotted against the adjusted correlation r_A, the average correlation of the

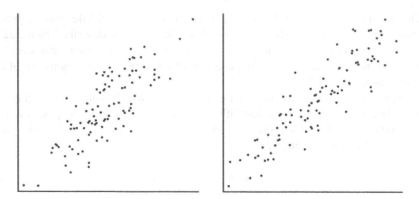

Fig. 5 Example of discrimination task. Observers are asked to choose which scatterplot is more highly correlated. Plots were 5° × 5° in extent. In this example, base correlation is 0.8; jnd is from above (i.e., the variant against which the base is tested has a higher correlation)

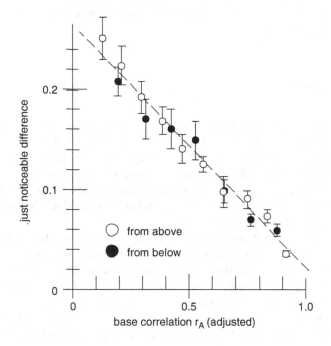

Fig. 6 Discriminability of correlation in scatterplots. *White dots* indicate that estimates are made by comparing a scatterplot with the base correlation against a higher correlation; *black dots* indicate that the comparison is made against a lower correlation. As is evident, these give the same estimates. Best fit of slope k is k = −0.22. Error bars denote standard error of the mean. Importantly, the value of b for the best-fitting line (b = 0.91) is similar to the value of b for accuracy (b = 0.88)

scatterplots being compared ($r_A = r + jnd(r)/2$). Portrayed this way, two patterns become evident. First, there is no dependence of the absolute value of the jnd on the direction of the variant—the same value is found regardless of whether the variant has a correlation higher or lower than the base. Second, a striking linearity exists in the way that jnd depends on adjusted correlation. This behavior can be described to a high degree of precision by

$$jnd\ (r) = k\ (1/b - r_A) \tag{3}$$

where k and b are real numbers such that $0 < k, b < 1$. Note that for both k and b, smaller values denote lesser variability, in that jnd is lower.

Interestingly, as in the case of accuracy, the relevant variable appears to be the distance *away* from complete correlation. Indeed, Eq. 3 can be rewritten

$$jnd\ (u) = ku \tag{4}$$

where $0 < k < 1$, and, as before, $u = 1 - br$, with $0 < b < 1$. This is a form of *Weber's Law*, which states that jnd is proportional to perceived magnitude; this appears to hold for the perception of several simple physical properties, such as vibration frequency and brightness [40]. In any event, the simplicity of this behavior allows variability to be described by just two scalar values: k and b.

Under some conditions, Weber's law for discrimination leads to Fechner's law for perceived quantity [41]. This appears to be the case for correlation, in that not only do both laws hold, but they are systematically related, with the value of b for perceived quantity having much the same value as its counterpart for discrimination [38]. Thus, only two values (k and b) are needed to describe accuracy and variability.

3.4 Timecourse

Although accuracy and variability are sometimes all that needs to be evaluated, other aspects of performance can also be important. One of these is *timecourse*, the minimum time needed to extract a structure from a dataset, such that more time will not lead to further improvement in performance. Although there may be delays in the machine component, speed of performance is usually limited by the human viewer. As such, measuring the minimum time needed to determine a structure from its graphical representation will often be an important part of evaluation. Not only can the result inform decisions about timing in dynamic displays, but it may also provide some information about the underlying mechanisms. Indeed, studies of timecourse are a common way to investigate various aspects of visual perception [30].

Returning to the case of scatterplots, the form of the laws describing performance suggests that correlation is—or at least, is associated with—a perceptually simple property. As such, it might be determined extremely quickly. To examine this possibility, discrimination was measured for scatterplots shown for controlled

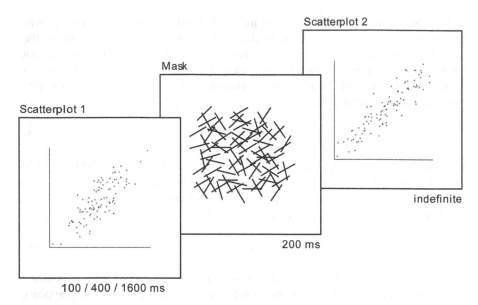

Fig. 7 Measurement of timecourse of correlation perception. Plots were $5° \times 5°$ in extent. Observers were asked to choose which scatterplot is more highly correlated (b = 0.88)

amounts of time [42]. Here, scatterplots were presented sequentially instead of side-by-side; the first was presented for 100, 400, or 1,600 milliseconds (ms) before being followed by a mask that effectively stopped it from being processed further. A second scatterplot was then presented and remained on until the observer responded (Fig. 7). Jnds for these were measured as before.

Results for 20 observers are given in Fig. 8. These show jnd to be linear for all timescales examined. Performance for the 400 and 1,600 ms conditions was almost identical. Performance for the 100 ms condition showed a slight deterioration, but was otherwise much the same, indicating that the process was largely complete by that time. As such, these results indicate that correlation in scatterplots can indeed be determined by the visual system quite rapidly—likely within the 150 ms typical for processes such as object recognition and estimation of averages (e.g., [43, 44]).

4   Optimal Reduction

Although the extended-vision thesis can help understand what visualization is, and more systematic techniques can help with evaluation, they cannot address all issues dealing with visualization, such as those concerned with optimal visual representation. For example, even when a clear specification of a task exists, they still cannot help specify what an optimal representation might be, or whether a particular representation is the best one possible. Of course, it may often be

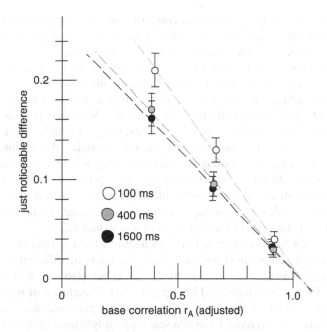

Fig. 8 Discrimination as a function of presentation time (Data from [42]). *Error bars* denote standard error of the mean. Note that performance is virtually identical for the 400 and 1,600 ms presentations. Although performance for 100 ms differs to some extent, it is still fairly similar, with a high degree of linearity

that an optimal representation simply does not exist. But it is nevertheless worth investigating the extent to which such representations might exist and might be found, or failing that, at least determine the extent to which representations would be able to avoid as many poor features as possible.

Traditionally, guidelines have been a major way to improve the likelihood of good design (e.g., [2, 8, 10]). And the relative suitability of a representation can be determined to some extent simply by evaluating a set of alternatives and selecting the one that yields the best performance. But such approaches cannot *guarantee* that a particular design is the best one, much less give a complete account of *why* it would (or would not) be optimal.

It is therefore worth returning to a more abstract perspective. As mentioned in Sect. 2, visualization can be characterized as the transformation of an original problem of finding abstract structure into one of finding geometrical structure. Once this geometrical structure has been found (by whatever means), it can be re-interpreted in terms of the original dataspace.[5] Considered in this way, visualization

---

[5]This assumes that the viewer can interpret the visual (or possibly other) representations used as a correlation. Developing such a "return route" to enable the inverse mapping may be an important part of training. Note that some of the performance limitations could occur at this stage, rather than the formation of the visual result proper.

is essentially a *similarity transform*—a commonly-used technique that transforms a given representation into one that enables solutions to be more easily found [45]. More precisely, it *reduces* the given problem to a simpler one.

This view then suggests that the issue of optimal representation might be approached in terms of an *optimal-reduction thesis*: the optimality of a particular design can be usefully viewed in terms of its ability to reduce a task to the most suitable set of operations in the extended system. (Note that this includes the processes used in the transformations themselves, many of which would be part of the machine component.) A given task can generally be reduced in various ways; in some sense this is similar to an exercise in programming, where different algorithms are used. However, different algorithms require different amounts of time or space (e.g., [46]); different kinds of processing elements create different levels of noise. Only a few possible designs—perhaps just one—will engage the most appropriate mechanisms [19]. The goal of the design process is to find this.

To make this a bit more concrete, return again to the example of correlation. Correlation can be determined not only via scatterplots of different designs (involving different kinds of axes or symbols, say), but also via devices such as parallel coordinates, bar charts, and line graphs [31]. The choice of representation determines the particular reduction—i.e., the particular processes used. Apart from the algorithms needed to create it, use of a scatterplot might involve operations in the human viewer such as grouping, determination of shape, and perhaps measurement of aspect ratio. Meanwhile, parallel coordinates might require isolating the lines between corresponding elements and determining the variance of their orientations. The goal for design is then to search through the space of possible processes to find those that lead to performance that is fastest, most accurate, and requires least effort from the viewer.

4.1 Operations Inventory

Perhaps the most direct way to approach design based on the optimal-reduction thesis is via an *operations inventory*. This would be a catalog of all possible operations (in both the human and the machine component) that could be applied in a given visualization. The description of each would include things such as its associated costs (in terms of e.g., time and space) and performance on various kinds of inputs. The goal would be to find the sequence of operations from this inventory that led to the fastest performance for the given task, consistent with constraints on overall performance (accuracy and variability, say). Finding this sequence would still be a problem, but it would at least be a more objective and well-defined one. In some ways, it would be similar to the problem faced by a compiler—viz., reducing a process to a given set of machine operations, such that it is carried out as quickly as possible.

Operations in the inventory could be organized into several sets. One might comprise basic elementary geometric and radiometric operations that could be directly applied to an image, e.g., grouping, determining the convex hull of a set of points, following a curve, finding the point of maximum intensity in a given region. More complex methods constructed from these could then form a second set—e.g., finding the set of similar shapes among items with various sizes and orientations. The inventory could also include constraints on interactions between operations, such as which ones cannot be applied concurrently. It might even contain a library of optimal—or at least relatively efficient—procedures that could be used for any visualization task (cf. [47]); another (not necessarily disjoint) set might contain tasks that can be carried out on the basis of visual operations alone [48]. Note that many operations could be carried out either by the machine or by the human (e.g., a figure could be rotated either graphically or mentally). In such cases, the particular choice would result from considering the costs and performance limitations with respect to the entire task.

Of course, the search for an optimal reduction will succeed only to the extent that such a reduction actually exists. It might be, for example, that instead of a single candidate, a *family* of candidates exists, corresponding to different trade-offs [49]. Or even if a unique candidate exists, it may not be possible to find it in reasonable time if the space of possible designs is too large. But as with intractable problems in general, however, there could exist a systematic approach to at least some aspects of the problem (e.g., [50]), or that would find candidates that are at least adequate [51].

Another limitation is that although there are many models of perceptual and cognitive processes in humans [52], much is still unknown [30]. As such, the evaluation of candidate designs will not be able to take everything into account. Nevertheless, existing knowledge might still enable candidates to be found that, even if not optimal, are still likely to be good.

4.2   General Representational Principles

A complementary approach to design—one that may overcome some of the limitations of an operations-based approach—is to focus not on operations, but on *form*, viz., the form of the graphical representation. Indeed, design has often relied on guidelines that essentially constrain the set of candidate representations considered (e.g., [10, 15, 53]). Many of these guidelines were developed for particular applications, such as tables and graphs. And although they are very important in these areas, they are often of limited generality. As such, they need to be complemented by a set of more general guidelines that would apply to any visualization.

A more general approach of this kind could be based entirely on (1) general information-processing principles, and (2) well-established knowledge about the perceptual or cognitive mechanisms of the human viewer [18]. The result would be a set of *general representational principles* (GRPs) that focuses not on a particular representation for a particular task, but on properties that *any* graphical representation should have for *any* task. Examples of these include:

- *Invertibility*. The graphical representation must support a 1:1 mapping between data values and the visual representations in the viewer. (For example, greater value along some data dimension would map to greater contrast, or to greater height.) If different values mapped to the same representation, information would be lost. Conversely, if different visual structures corresponded to the same data value in the context of a single task, different visual processes would be involved, causing interference.
- *Distinctness*. Values that need to be distinguished for the task must map onto visual representations that are distinct. (For example, if it is important to notice that two values are different, they must map to contrasts or positions that can easily be seen as different.) Otherwise, important information would be lost, or at the very least, performance would be slowed.
- *Uniformity*. Values along a single data dimension must map onto a single visual dimension, such as height, colour, or orientation. If different visual dimensions were used, different processes would become involved; both information and time would be lost trying to combine the results of these.
- *Ordering*. Data values that are ordered in some way must map to a visual property that is similarly ordered over the relevant range. (For example, greater value could map to greater height, if mapped to orientation or color, only a subset could be used, since these properties are cyclic.) If this is not done, the ability of the visual system to use perceptual order cannot be harnessed; indeed, if it operates against its natural ordering, performance could degrade substantially.
- *Separability*. Data involving separate dimensions must map to visual properties that are separable—i.e., can be attended separately (e.g., size and orientation). Otherwise, these values may become parts of a perceptually integrated structure, with the separate components then being difficult to access.

Such principles are not entirely new—several comprise the basis (often unstated) for much of good design (cf. [8, 18]). For example, the principle of separability is obeyed in the effective scatterplot of Fig. 1a, which uses the separable properties of horizontal and vertical for each data dimension. In contrast, the ineffective scatterplot of Fig. 1b violates this principle, using properties that become parts of a perceptually integrated structure—viz., the upper and lower parts of the structure corresponding to each glyph. Likewise, the principle of ordering has often been recommended for displays (e.g., [8, 18]), and the principle of distinctness is related to that of the smallest effective difference [54]. What is proposed here is that such principles be explicitly distinguished from task-specific constraints and identified as a distinct group. Moreover, given that they are based on universal considerations, it may also be possible to develop and organize them in a more systematic way (cf. [55]), or give them a more quantitative character.

GRPs constrain the kinds of transformations permitted between data values and graphical representations, eliminating many possible reductions right from the start. And because they are general, they will tend to be consistent with each other; the simultaneous use of several such principles will not lead to a clash, but to a more tightly-bounded space of possibilities, increasing the chances for an effective design. Such principles could also be easily combined with more task-specific ones, guiding design for particular applications.

5 Assessment of Alternative Designs

Given the difficulties faced in searching through all the alternatives possible for a design, and the fact that much is still unknown about the perceptual and cognitive mechanisms involved, the search for optimal—or even good—designs must be supplemented by empirical assessment. As in the case of evaluation, however, it may be possible to carry out such assessments in a relatively systematic way, informed in part by elements of vision science. These not only can suggest particular techniques, but also some of the design parameters to consider. Indeed, if alternatives are examined in an appropriate way, the results can also provide considerable insight into the mechanisms involved.

5.1 Different Parameter Values

To see how performance can be assessed for different values along a single design parameter, consider the issue of how many dots—or more precisely, symbols—a scatterplot should display. Too many would cause *overplotting*, where important information is crowded out. Displaying only a randomly-chosen subset of data points would solve this. But how many should be selected? Too few would cause correlation to be conveyed poorly. Consequently, it would be useful to know how performance depends on the number of dots displayed.

To investigate this, accuracy and variability was measured for a set of scatterplots containing various numbers of dots (12, 24, 48, 100, and 200), with the other parameters remaining fixed (Fig. 9) [56]. Evaluations were similar to those described in Sect. 3. The form of the laws governing accuracy and variability meant that only a few correlations needed to be tested: accuracy could be measured via the physical correlation corresponding to a perceived correlation of 1/2 (the first phase of the approach outlined in Sect. 3), while variability was measured using just three base correlations. This meant that all conditions could be tested on a single observer within a single experimental session. Such a *within-observer* design allows a far more sensitive measure of the effects of a parameter (in this case, number), in that it minimizes noise caused by the use of different observers [57].

Results for 12 observers are shown in Fig. 10. No differences were found in subjective perception (and thus, accuracy) for the various numbers of dots. (A slight difference appeared for n = 200, but was not statistically significant.) As might be expected on the basis of sampling, variability decreased as more dots were shown, but stabilized at 48 dots. Given that a within-observer design was used, this result

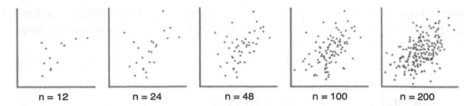

n = 12   n = 24   n = 48   n = 100   n = 200

Fig. 9 Examples of tests of number of dots. For each observer, accuracy and variability was measured for each condition, and then compared

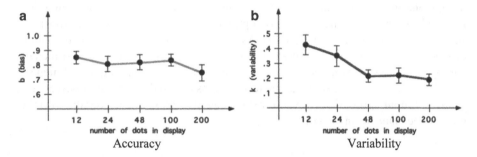

Accuracy   Variability

Fig. 10 Performance as a function of number of dots (Data from [56]). (a) Accuracy, as described by the measure b, derived from perceived correlation. (b) Variability, as described by the measure k. Each condition used the same set of observers

provides strong evidence that performance is not affected as long as the number of dots is 48 or more (up to 200), at least for the kind of dots tested here. A similar approach could also be used to investigate the upper limit, viz., how many dots can be displayed until overplotting begins to be a problem.

More generally, a systematic approach of this kind can be used for the assessment of any design parameter. For example, in a study examining the effect of dot (symbol) luminance, color, and size [56], observers were tested over a wide range of values for each parameter, along with a condition in which the values were mixed (Fig. 11).

Each parameter was tested separately, with 12 observers per parameter. Results showed an interesting amount of invariance: for all parameters tested, value had no measurable effect on either accuracy or variability [56]. Thus—at least for purposes of conveying correlation—designs varying along these dimensions appear to be equivalent.

Such invariance is also informative in terms of the possible mechanisms that underlie correlation perception in scatterplots. The indifference to size, for example, rules out the involvement of simple operations such as blurring, since the overall shape of the dot cloud does not seem to matter greatly. Instead, it suggests that the operations involved may rely primarily on the locations of the centers of the symbols.

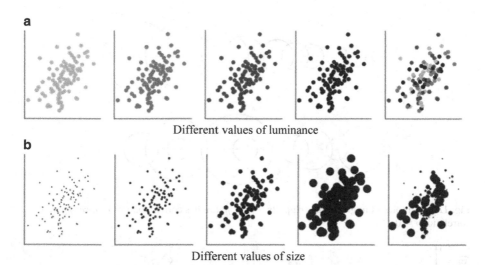

Fig. 11 **Examples of tests of parameter values**. (**a**) Luminance. Here, four different values were tested, along with a condition in which they were equally mixed. (**b**) Size. Again, four different values were tested, along with a condition in which they were equally mixed. For each of these, accuracy and variability was measured and compared across observers

5.2   Other Dimensions

This way of assessing alternative designs can in principle be applied more generally—for example, assessing designs that use different *dimensions*. To clarify what is meant here, note that a scatterplot represents the first data dimension by horizontal position and the second by vertical; for both data dimensions, space "carries" information. But it may be that carriers need not be spatial—several perceptual dimensions (or "visual variables") exist that are similar to position in many ways [8, 18, 30]. It might therefore be possible to use these dimensions to convey correlation, provided they obey the GRPs of Sect. 4.

As an example of this, consider "augmented stripplots" where the first data dimension is represented via horizontal position (as for scatterplots) and the second data dimension via size (diameter); correlation is then conveyed via the relation between these two properties (Fig. 12). Note the interaction here between visualization and vision science: the use of dimensions other than space is suggested by findings from vision science about perceptual dimensions; in return, the extent to which these dimensions can be used in a visualization design lets us learn more about their nature.

Owing to their isomorphism with scatterplots, augmented stripplots can be evaluated via the same kinds of techniques. Accuracy and variability were tested for a group of observers [56] based on correlations over the range 0–1, in increments of 0.1. Preliminary results from 18 observers are shown in Fig. 13.

Interestingly, both accuracy and variability appear to obey laws that are quite similar to those for scatterplots. Among other things, this means that the assessment

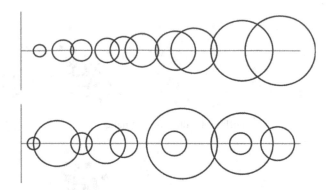

Fig. 12 Examples of augmented stripplots. *Upper figure* has correlation r = 1; *lower figure* has correlation r = 0

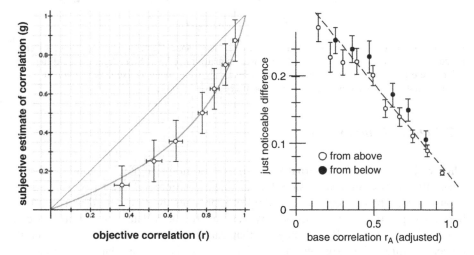

Fig. 13 Performance for augmented stripplots (Data from [56]). (**a**) Accuracy, as given by perceived correlation. Best-fit line to results is for b = 0.91 (cf b = 0.88 for scatterplots). (**b**) Variability, as given by jnd. Best-fine line is for k = 0.26 and b = 0.86 (cf. k = 0.22 and b = 0.91 for scatterplots). Both measures were obtained using the same set of observers

of different values of design parameters in these kinds of visualizations can be simplified: since only a few scalars (k and b) are sufficient to describe performance, only a few base correlations need to be tested. It also suggests that the perception of correlation in augmented stripplots is much the same as is it in scatterplots, perhaps reflecting an ability more general than previously believed. It remains an open question as to the mechanisms involved in these kinds of representations, and which other properties might also result in such behavior. But the approach sketched here is clearly capable of suggesting interesting new directions of research, and then examining them.

6   Limits of the Framework

The preceding sections have shown how several sets of issues in visualization—pertaining to both theory and practice—can be addressed by systematic approaches that can be considered parts of a common framework. These approaches can help us understand how visualizations work, how they might be evaluated, and even to some extent how they might be designed. But this has been shown in detail only for those aspects pertaining to correlation in scatterplots. Returning now to the question that motivated this paper: to what extent is a systematic framework—perhaps even a science—possible for visualization in general?

Clearly, the elements discussed here do not cover all aspects of visualization. For instance, issues concerning the nature and role of interaction have not been discussed in any great detail. More fundamentally, perhaps, little has been said about what would constitute an adequate explanation for *why* a particular visualization works (cf. vision science, where standards of explanation are much more developed [21]). The elements proposed here also say little about how an optimal design could actually be found, or even how to tell if one exists in the first place. And little consideration has been given to the nature of the design process itself (cf. [58–60]).

But the goal of this paper is simply to get a sense of whether at least some aspects of visualization could be handled via a scientific framework, and if so, what this might look like. The discussion here has shown that systematic and coherent approaches can be developed to address several important issues concerning the study and design of visualizations. The central issue now becomes the extent to which these elements could be developed into a framework of greater range and power.

6.1   Generality

The proposals here have largely used as their touchstone the determination of correlation in scatterplots. This was done primarily to show that these proposals are consistent with each other, and are useful for at least one real-world domain. In a way, this use of scatterplots is like the use of fruit flies by geneticists to speed up their investigation of complex systems: focus on a real system complex enough to involve the issues at hand, but simple enough to be studied in a relatively straightforward way. The issue now is the extent to which the approaches outlined above can be generalized to more than just these "fruit flies".

In regards to scatterplots themselves, these approaches could likely be used to investigate design parameters of various kinds—not only properties of individual symbols, but aspects such as the size and shape of the cloud formed by the set of datapoints. They could also be used to examine issues such as the effect of outliers on performance, or the presence of a second group of symbols representing an irrelevant population. Again, such investigations would likely yield not only important knowledge about how well scatterplots work, but also additional insights into the mechanisms involved.

Importantly, these approaches do not depend on anything particular to scatterplots or to correlation. They could therefore be applied to correlation as conveyed by other kinds of representation, such as bar charts, parallel coordinates, or line graphs [31]. Likewise, they could also be applied to other descriptive statistics, such as means and variances. They could even be applied to tasks such detection of outliers or the detection of two different populations in a given dataset. A considerable amount of work would be required to carry out such studies, but at least the general outlines of the approach are clear.

One possible limit, however, is the complexity of the structures (and related operations) to be visualized. For example, applying these approaches to the visualization of connectivity patterns in networks might lead to issues that are too complex to resolve, or at least, resolve in a reasonable amount of time. On the other hand, it appears that some kinds of complex structure can be effectively handled via metaphor, which engages the appropriate higher-level cognitive processes (e.g., [61]); if so, evaluation techniques for these might be modeled on those used in cognitive science. Settling this issue is an important goal for future work. It is likely connected the general issue of the extent to which different kinds of information can be represented visually [62].

Another possible limit is the extent to which approaches developed for "static" aspects of visualization can apply to dynamic aspects such as interaction. Interaction has its origins in the fact that there is often too much information to be displayed at any time, requiring an active process of exploration [5, 63]. A parallel situation exists in visual perception, where the inability to represent more than a fraction of the visual input in stable form means that vision must rely on a dynamic system in which objects are represented in a "just-in-time" manner [11]. This parallelism may explain why interaction can be such a natural activity if visualization is designed correctly; indeed, GRPs can be developed towards this end [64]. Going beyond this would not only require knowledge of how scenes are perceived, but also why a viewer might represent a particular object at a particular time. It is increasingly clear that these issues are complex. Whether they are insurmountable remains an open question.

6.2   Visualization, Vision, and Science

Given that at least some of what has been proposed here could likely be generalized further, to what extent could it also be developed into a truly scientific framework, or even an outright science of visualization?

As mentioned earlier, a science is essentially an organized way of thinking about a connected set of issues. Looking at visualization in terms of issues, at least three such sets can be identified.[6] Some appear able to support the development of a science; others do not.

The first set concerns visualization as *artifact*—i.e., as a system that already exists. Issues here include determining its characteristics, ascertaining how it works, and describing how it relates to others like it (i.e., taxonomy). Work has been done on some of these issues—e.g., taxonomies of data types [65], operation types [63], and algorithms [66]—and things could likely be developed further in these directions. Indeed, according to the extended-vision thesis, visualization systems can be treated much the same as basic vision systems, with a similar handling of basic entities, questions, and methodologies. In the case of vision (or biology more generally), coherence in subject matter is obtained in part by the need for systems to be reasonably effective at what they do, creating a tendency to converge on common architectures [21, 67, 68]. Similar pressures clearly exist on visualization systems. As such, the coherence needed for a science almost certainly exists. Moreover, it appears likely that systematic approaches could be developed for many tasks that are well-defined (such as correlation perception), resulting in an *artifact-level science* of visualization. Several of the approaches discussed here could be part of this. It is unclear how many such "islands" of well-defined tasks (or subtasks) could be handled this way, but if vision science is any guide, there would be quite a few.

A very different set of issues concerns the *practice* (or activity) of design— e.g., how to design a visualization that enables discovery of interesting structure in some set of high-dimensional data. Such activity is generally regarded as not a science [69]: the immense space of possibilities, the often ill-defined nature of what is requested of the system, and the need to include subjective factors such as aesthetic preferences make it difficult to carry out via a simple set of approaches. Design as applied to visualization inherits this [29, 32]. (For example, even defining a clear purpose for a visualization system can be difficult—if not impossible— in some circumstances [32].) But although this *design practice* of visualization cannot be a science, parts of it could nevertheless be made more systematic; effective design methods (cf. software programming) could make good use of what is known in related sciences—including the "islands" of the artifact-level science of visualization—as well as any other kinds of relevant knowledge. (Several of the approaches discussed in Sects. 4 and 5 could be part of this.) The result would be an open-ended activity similar in many ways to engineering, architecture, or any other organized design discipline [69].

A third set of issues concerns the *nature* of the design process (as opposed to issues that arise during a design activity). For example, how can design itself be characterized? Which aspects of the design process can be formalized, and which cannot? What kinds of design processes have been developed, and how do they relate to each other? It has been suggested that such issues could be the basis

---

[6]Other issues exist, such as those concerned with visualization as a technology [32]. However, these are not directly relevant to concerns about the nature of a scientific framework, and so are not discussed here.

for a science under some circumstances [70]; a few studies (e.g., [59, 60]) are possible beginnings of this in the context of visualization. The extent to which these developments will eventually result in a *design-level science* of visualization (or equivalently, a science of visualization design) is currently unclear. But there appear to be no a priori objections standing in the way.

Finally, it should be mentioned that the proposals discussed here also point towards the possibility of even closer connections between visualization and vision science. Knowledge about perceptual mechanisms can be usefully applied to visualization (e.g., [5, 9]), and Sects. 3 and 5 show this to be true for methodology as well. But there is still another connection that appears to be emerging. In Sect. 5, it was shown that systematic techniques could be used to determine how the number of dots in a scatterplot affects the accuracy and variability of correlation estimates. Although this was a study of how performance depended on different values of a design parameter (the number of dots), it could also be viewed as a controlled experiment in vision science, with the results shedding a bit of light on the mechanisms involved. Similar considerations apply to the other examples in that section, such as the study of how correlation can be conveyed by size as well as by position. More generally, then, if researchers in visualization expand their set of techniques to include those typically used by vision scientists (while keeping the same kind of stimuli), while vision scientists expand their set of stimuli to include those typically used by visualization researchers (while keeping the same kind of methodologies), the possibility arises of a class of studies that belong in both fields, with results of genuine interest to each.

Acknowledgements Many thanks to Stephen Few, Dave Kasik, Minjung Kim, Tamara Munzner, Vicki Lemieux, Michael Sedlmair, Ben Shneiderman, and Jack van Wijk for their comments on earlier versions of this paper. Thanks also to Tamara Munzner for feedback on the idea of the operations inventory, and to Jack van Wijk for interesting discussions about the limits of a possible science of visualization. Support for this work and several of the studies mentioned was provided by the Natural Sciences and Engineering Research Council of Canada (NSERC), and The Boeing Company.

References

1. W.S. Cleveland. *Visualizing Data*. Summit, NJ: Hobart Press, 1993.
2. A.M. MacEachren. *How Maps Work: Representation, Visualization, and Design*. New York: Guilford Press. pp. 51–149, 1995.
3. M. Massironi. *The Psychology of Graphic Images: Seeing Drawing, Communicating*. Matwah NJ: Erlbaum, 2002.
4. H. Wainer. *Graphic Discovery*. Princeton: University Press, 2005.
5. S.K. Card, J.D. Mackinlay, and B. Shneiderman. Information visualization. In S.K. Card, J.D. Mackinlay, and B. Shneiderman (Eds.) *Readings in Information Visualization: Using Vision to Think*. San Francisco: Morgan Kaufman. pp. 1–34, 1999.
6. T.S. Kuhn. *The Structure of Scientific Revolutions*, 2nd ed. Chicago: University of Chicago Press, 1970.

7. I. Lakatos. *The Methodology of Scientific Research Programmes: Philosophical Papers, Volume 1.* Cambridge: Cambridge University Press, 1978.
8. C. Ware. *Information Visualization: Perception for Design,* 2nd ed. San Francisco: Morgan Kaufman, 2004.
9. C. Ware. *Visual Thinking for Design.* San Francisco: Morgan Kaufman, 2008.
10. S.C. Few. *Information Dashboard Design: The Effective Visual Communication of Data.* Sebastopol, CA: O'Reilly Media, Inc., 2006.
11. R.A. Rensink. The dynamic representation of scenes. *Visual Cognition,* 7: 17–42, 2000.
12. A.D. Milner and M.A. Goodale. *The Visual Brain in Action.* Oxford: Oxford University Press, 1995.
13. C.A. Brewer. *Designing Better Maps: A Guide for GIS Users.* Redlands CA: ESRI Press, 2005.
14. N.B. Robbins. *Creating More Effective Graphs.* Hoboken, NJ: John Wiley & Sons, 2004.
15. S.M. Kosslyn. *Graph Design for the Eye and Mind.* Oxford: Oxford University Press, 2006.
16. R.A. Rensink. The management of visual attention in graphic displays. In C. Roda (Ed.), *Human Attention in Digital Environments.* Cambridge: University Press. pp. 63–92, 2010.
17. C.D. Wickens and J.S. McCarley. *Applied Attention Theory.* Boca Raton, FL: CRC Press, 2008.
18. J. Bertin. *Semiology of Graphics: Diagrams, Networks, Maps.* Madison WI: University of Wisconsin Press, 1983.
19. J.H. Larkin an H.A. Simon. Why a diagram is (sometimes) worth ten thousand words. *Cognitive Science,* 11: 65–99, 1987.
20. J.D. Foley, A. van Dam, S.K. Feiner, and J.F. Hughes. *Introduction to Computer Graphics.* Reading MA: Addison-Wesley, 1993.
21. D. Marr. *Vision: A Computational Investigation into the Human Representation and Processing of Visual Information.* San Francisco: W.H. Freeman, 1982.
22. J. Zhang and D. Norman. Representations in distributed cognitive tasks. *Cognitive Science,* 18: 87–122, 1994.
23. V. Di Lollo, J.T. Enns, and R.A. Rensink. Competition for consciousness among visual events: The psychophysics of reentrant visual processes. *Journal of Experimental Psychology: General,* 129: 481–507, 2000.
24. A.K. Mackworth. Vision research strategy: Black magic, metaphors, mechanisms, miniworlds, and maps. In A.R. Hanson, E.M. Riseman (Eds.), *Computer Vision Systems.* New York: Academic Press. pp. 53–60, 1978,
25. U. Neisser. *Cognition and Reality.* San Francisco: W.H. Freeman. pp. 20–24, 1976.
26. J. Li, J.-B. Martens, and J.J. van Wijk. Judging correlation from scatterplots and parallel coordinate plots. *Information Visualization,* 9: 13–30, 2010.
27. R.M. Baecker, J. Grudin, W. Buxton, and S. Greenberg. *Readings in Human-Computer Interaction: Toward the Year 2000.* San Francisco: Morgan Kaufman, 1995.
28. Carpendale. Evaluating information visualizations. In A. Kerren et al. (Eds.) *Information Visualization: Human-Centered Issues and Perspectives.* LNCS 4950. Berlin: Springer. pp. 19–45, 2008.
29. J.-D. Fekete, J.J. van Wijk, J.T. Stasko, and C. North. The value of information visualization. In A, Kerren et al. (Eds). *Information Visualization: Human-Centered Issues and Perspectives.* LNCS 4950. Berlin: Springer. pp. 1–18, 2008.
30. S.E. Palmer. *Vision Science: Photons to Phenomenology.* Cambridge MA: MIT Press, 1999.
31. R.L. Harris. *Information Graphics: A Comprehensive Illustrated Reference.* Oxford: Oxford University Press, 1999.
32. J.J. van Wijk. The value of visualization. *Proceedings IEEE Visualization 2005,* pp. 79–86, 2005.
33. P. Bobko, and R. Kerren. The perception of Pearson product moment correlations from bivariate scatterplots. *Personnel Psychology,* 32: 313–325, 1979.
34. I. Pollack. Identification of visual correlational scatterplots. *J. Experimental Psychology,* 59: 351–360, 1960.
35. M.E. Doherty, R.B. Anderson, A.M. Angott, and D.S. Klopfer. The perception of scatterplots. *Perception & Psychophysics,* 69: 1261–1272, 2007.

36. W. Ellermeier, and G. Faulhammer. Empirical evaluation of axioms fundamental to Stevens's ratio-scaling approach: I. Loudness production. *Perception & Psychophysics*, **62**: 1505–1511, 2000.
37. D. Laming. *The Measurement of Sensation*. Oxford: Oxford University Press, 1997.
38. R.A. Rensink, and G. Baldridge. The perception of correlation in scatterplots. *Computer Graphics Forum*, **29**: 1203–1210, 2010.
39. W.S. Cleveland, P. Diaconis, and R. McGill. Variables on scatterplots look more highly correlated when scales are increased. *Science*, **216**: 1138–1141, 1982.
40. S. Coren, L.M. Ward, and J.T. Enns. *Sensation and Perception*, 5th ed. New York: Harcourt Brace. pp. 15–49, 1999.
41. H.E. Ross. On the possible relations between discriminability and apparent magnitude. *British J. Mathematical and Statistical Psychology*, **50**: 187–203, 1997.
42. R.A. Rensink. Rapid Perception of Correlation in Scatterplots. *Journal of Vision,11*. [Vision Sciences Society, Naples, FL, USA. May 2011.]
43. S.C. Chong, and A. Treisman. Representation of statistical properties. *Vision Research*, **43**: 393–404, 2003.
44. S. Thorpe, D. Fize, and C. Marlot. Speed of processing in the human vision system. *Nature*, **381**: 520–522, 1996.
45. Z.A. Melzak. *Bypasses: A Simple Approach to Complexity*. NY: John Wiley & Sons, 1983.
46. S. Baase. *Computer Algorithms: Introduction to Design and Analysis*, 2nd ed. Reading MA: Addison-Wesley, 1988.
47. R. Amar, J. Eagan, and J. Stasko. Low-level components of analytic activity in information visualization. In *Proc IEEE Symposium on Information Visualization 2005*, pp. 111–117, 2005.
48. D.J. Kasik. Strategies for consistent image partitioning. *IEEE Multimedia*, **11**: 32–41, 2004.
49. R.A. Rensink and G. Provan. The analysis of resource-limited vision systems. *Proceedings of the Thirteenth Annual Conference of the Cognitive Science Society*, pp. 311–316. Chicago IL, USA, 1991.
50. M.R. Garey, and D.S. Johnson. *Computers and Intractability: A Guide to the Theory of NP-Completeness*. San Francisco: Freeman, 1979
51. H.A. Simon. *The Sciences of the Artifical*, 3rd ed. Cambridge MA: MIT Press. pp. 111–138, 1996.
52. W.D. Gray. *Integrated Models of Cognitive Systems*. New York: Oxford University Press, 2007.
53. S.C. Few. *Show Me the Numbers: Designing Tables and Graphs to Enlighten*. Oakland, CA: Analytics Press. pp. 92–130, 2004.
54. E.R. Tufte. *Visual Explanations*. Cheshire CT: Graphics Press, 1997.
55. J.D. Mackinlay. Automating the design of graphical presentations of relational information. *ACM Transactions on Graphics*, **5**: 110–141, 1986.
56. R.A. Rensink. Invariance Of Correlation Perception. *Journal of Vision,12*. [Vision Sciences Society, Naples, FL, USA. May 2012.].
57. P. Martin, and P. Bateson. *Measuring Behaviour: An Introductory Guide*, 2nd ed. Cambridge: University Press, 1993.
58. J.J. Garrett. *The Elements of User Experience: User-Centered Design for the Web*. Berkeley CA: Peachpit Press, 2002.
59. T. Munzner. A nested model for visualization design and validation. *IEEE Transactions on Visualization and Computer Graphics*, **15**: 921–924, 2009.
60. M. Sedlmair, M. Meyer, and T. Munzner. Design study methodology: Reflections from the trenches and the stacks. *IEEE Information Visualization Conference 2012*, Seattle, WA, USA.
61. W.B. Paley. Interface and mind. *it – Information Technology*, **51**: 131–141, 2009.
62. R. Arnheim. *Visual Thinking*. Berkeley: University of California Press, 1972.
63. J. Heer, and B. Shneiderman. Interactive dynamics for visual analysis: A taxonomy of tools that support the fluent and fleixble use of visualizations. *ACM Queue*, **10**: 30:30–30:55. 2012.
64. R.A. Rensink. Internal vs. external information in visual perception. *Proceedings of the Second International Symposium on Smart Graphics*, Hawthorne, NY, USA. pp. 63–70, 2002.

65. B. Shneiderman. The eyes have it: A task by data type taxonomy for information visualizations. *Proceedings of the 1996 IEEE Symposium on Visual Languages*. pp. 336–343, 1996.
66. M. Tory, and T. Möller. Rethinking visualization: A high-level taxonomy. *IEEE Symposium on Information Visualization*. pp. 151–158, 2004.
67. S. Conway Morris. *Life's Solution: Inevitable Humans in a Lonely Universe*. Cambridge: Cambridge University Press, 2004.
68. G. McGhee. *Convergent Evolution: Limited Forms Most Beautiful*. Cambridge MA: MIT Press, 2011.
69. N. Cross. Designerly ways of knowing: Design discipline versus design science. *Design Issues*, **17**: 49–55, 2001.
70. D. Grant. Design methodology and design methods, *Design Methods and Theories,***13**:1, 1979.

Part III
Principles, Guidelines and Recommendations

Toward a Better Understanding and Application of the Principles of Visual Communication

Juhee Bae and Benjamin Watson

Abstract Graphic designers have excelled at conveying visual messages for decades. We take their work for granted: the back of a Wheaties box, an informative poster, a newspaper front page (Fig. 1), slides and web pages. One of graphic design's primary goals is to create a "visual hierarchy" that gives structure to text and imagery, and improves viewer understanding. However, while graphic design methods have clearly been successful, there has been little work examining the reasons for that success, and few tools that help users apply those methods well. This is particularly surprising when we consider the ubiquity and indeed importance of this sort of visual communication in our lives. In this chapter, we lay the foundations for a research agenda that begins addressing these problems. We begin with a review of related research. In psychology, the Gestalt laws of visual grouping are clearly fundamental to the communication of knowledge structure, however little is known about how Gestalt laws work together to create complex hierarchies. Obviously design practice has a great deal to tell us, but the accepted theory of visual hierarchy is quite limited. We also examine pertinent research in human-computer interfaces and information visualization, especially of any related tools. We then propose a paradigm for studying the effective communication of complex visual structure, including possible measures of such effectiveness, and our own initial work along those lines. We conclude by discussing open questions in this line of research.

1 Introduction

Visual communication is happening all around us. Indeed, it is so ubiquitous, we are not usually conscious of it. It tells us how to use our skin care product (Fig. 2), encourages us to eat well, explains the latest political dispute, and tells us what is

J. Bae (✉) • B. Watson
North Carolina State University, Raleigh, NC, USA
e-mail: jbae3@ncsu.edu; bwatson@ncsu.edu

W. Huang (ed.), *Handbook of Human Centric Visualization*,
DOI 10.1007/978-1-4614-7485-2_7, © Springer Science+Business Media New York 2014

product label   poster

newspaper   warning sign

Fig. 1 Visual applications. (**a**) Wheaties box, (**b**) informative poster, (**c**) newspaper front page, and (**d**) warning sign

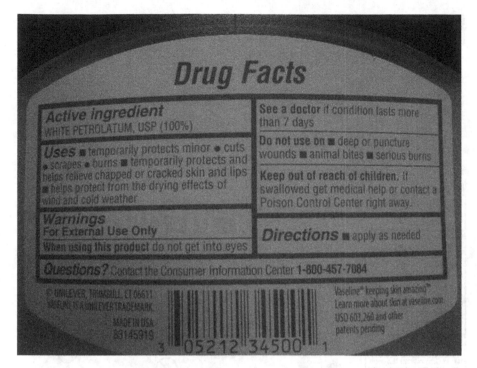

Fig. 2 Instructions on a skin care product

on offer (Fig. 3). These messages can appear on labels, posters, magazines, slides, and brochures. And thanks to design software and the web, they are easier to create and disseminate than ever. Yet despite its importance, visualization has not given much consideration to everyday visual communication, and its study in other fields is limited.

We believe the time is ripe for that to change. It is now widely recognized that society is experiencing a crisis of communication: technology is increasing the complexity of the information that we must communicate and understand, while failing to provide reliable tools to help us do so both widely and well. As the warning sign in Fig. 1d illustrates, ineffective visual communication has always been a problem. But the widespread discussion of "Death by PowerPoint" [30], culminating in a well known (and yet familiar) fiasco of a an astronaut assignment chart (Fig. 4), makes it that the problem has reached a new level.

Fortunately, several recent scientific and technical developments, across a number of fields, now make rigorous study of visual communication possible. In this chapter, we describe those developments, a number of research directions, and our own initial efforts in those directions.

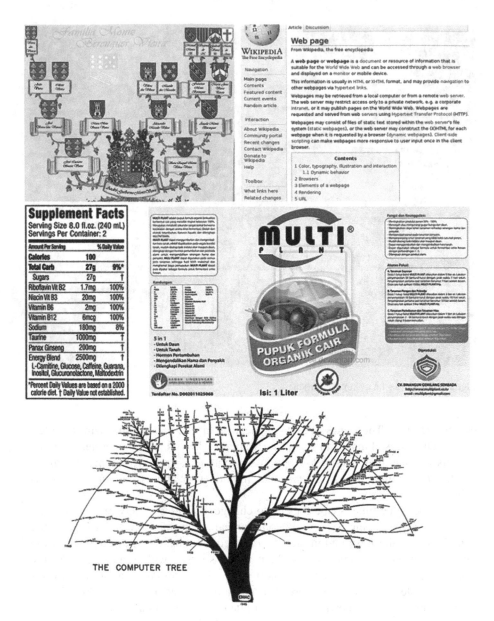

Fig. 3 Various applications of charts, web page, product labels and visualization

2 Related Work

In this review and in our own work, we treat the communication of knowledge as proceeding through two transforms: from informational to visual, and from visual to cognitive (Fig. 5). Informational knowledge is abstract, and contains the information

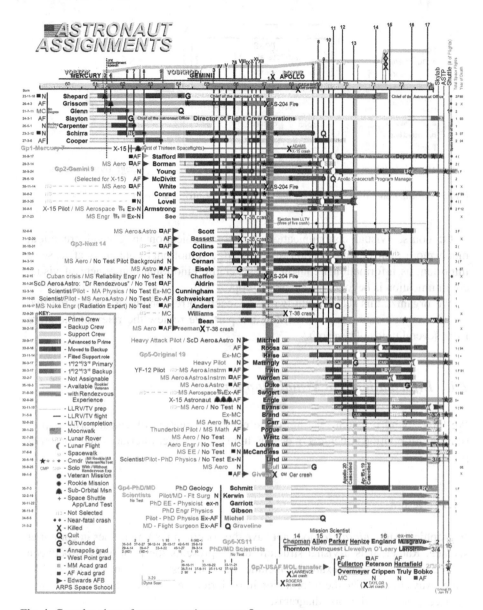

Fig. 4 Complex chart of astronaut assignment to fly

itself as well as its structure, order and emphasis; its visual is a representation of the abstract, making it possible to see the information and its structure, or and emphasis; and its cognitive representation the understanding which the viewer creates from the visual. Graphic designers make their living performing the informational to visual transform, while the viewer maps the visual to cognitive by perceiving and understanding the visual.

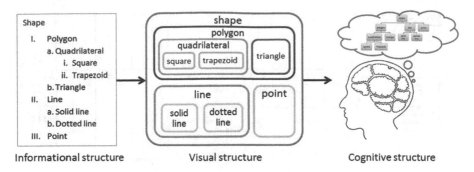

Fig. 5 Informational to visual transform and visual to cognitive transform

Of course, visual communication and communication in general are about much more than communication of knowledge: emotional content is at least as important as informational content. However the emotional component of visual communication is beyond this chapter's scope.

We organize this section into two parts, each corresponding to one of the two visual communication transforms. In each part, we review the relevant literature from several fields of research.

2.1 Informational to Visual Mapping

We have found research relating to the effective creation of visual forms representing information structures in design, human-computer interfaces and computer graphics.

2.1.1 Design

In graphic design, *hierarchy* does not have the specific meaning it does in computer science. Rather, it is a visual structure that represents the relationships between visible elements. Some examples include the tree, nest, and stair structures [14]. Other ways to organize the information into hierarchies are by node-link diagrams, matrices, and classification diagrams.

Williams introduces four fundamental elements of graphic design, several of which are crucial for creating visual hierarchy: contrast, repetition, alignment, and proximity [35]. Contrast varies visual components such as type, color, size, line thickness, shape, and space to create a pleasing or engaging variation. Repetition duplicates or echoes many of those same visual elements to unify regions and create aesthetic visual rhythms. Alignment arranges visual elements along linear or curvilinear paths to show visual connection. Finally, proximity adds organization to visuals by placing related elements closer together.

Everyday graphic design works under a common sense assumption: that visual communication will be most successful when the visual hierarchy is a clear and accurate representation of the information itself: particularly its structure, sequence and emphasis. Given the limited history of research in graphic design, particularly in the United States, the field has been content to accept decades of successful practice as validation of this assumption. Given today's communication crisis, we believe the time has come to examine this *graphic design hypothesis* in detail.

2.1.2 Human-Computer Interfaces

Several human-computer interface researchers have studied the use of design layouts to communicate application structure to users. In so doing, they began the examination of the graphic design hypothesis.

Grabinger's study [6] measured design effectiveness in terms of user learning, that is through knowledge of the acquisition, organization, and processing of information by users. Text was organized by line length, directive cues, paragraph indication, a status bar, line spacing, functional area, text columns, and illustrations. Learning improved as more organizational cues were used.

Ngo et al. [19] identified a set of aesthetic interface characteristics on which they built a model that predicts the acceptance of the user. The characteristics included balance, equilibrium, symmetry, sequence, cohesion, unity, proportion, simplicity, density, regularity, economy, homogeneity, rhythm, order and complexity. Ngo et al. defined a computational formula for each characteristic using objective measures such as the number and distance between objects, width and height of the frame and objects, as well as the area, location and number of objects. In testing on several interface layouts, their model accurately predicted aesthetic ratings by users.

Tullis [31, 32] developed a model to investigate the relationship between the user performance, subjective ratings and six measures of layout: overall density, local density, number of groups, size of groups, number of items, and layout complexity. Overall density is a ratio between the area covered with elements and the total screen area; the local density is a ratio between the area of a group of elements and the total screen area; and layout complexity represents the degree to which the screen elements are aligned (more alignment reduces complexity). When users were answering a simple question, Tullis found that users located interface elements more quickly when grouping was used, while subjective ratings were related to alignment and local density.

Although they did not specifically set out to study the graphic design hypothesis, the overall result of work by human-computer interface researchers provides some support for it. In particular, in Grabinger's work [6], learning improved with increased visual organization; while in Tullis' work [31, 32], additional grouping and alignment improved search performance and user preferences.

2.1.3 Graphics

Computer graphics has traditionally been the discipline of tool builders, producing 3D graphics APIs, 3D modeling packages, image processing programs, and 2D tools such as Powerpoint and Illustrator. Certainly these tools do make it easier to produce visual communication, but the question of whether or not the results are effective is debatable. In general, if one were to assert that visual communication is a language, we would maintain that most existing tools make it very simple to produce the words in that language, but provide little or no assistance with the language's grammar. When the user is a visual novice, the result is usually confusing, and often unintelligible.

Here we review the few tools we have found that do incorporate a deeper understanding of visual communication.

There is already a good history of research on layout tools for magazines and newspapers. According to Lok and Feiner [15], such tools have used abstract/spatial constraint solvers to set the position and size of the elements, often with grids as constraints, with machine learning or relational grammars as component technologies. Some research attempts to evaluate the effectiveness of layouts aesthetically [16, 19].

Jacobs et al. introduced an adaptive grid-based document layout tool [9] that builds a document page by selecting among adaptive templates that already incorporate style, layout, and constraints. Then, it inserts the actual page content into the template, and reformats it based on the page content, orientation and template constraints.

While these layout tools have been very successful, and do offer more assistance to users than better known tools by helping place the elements in a visual (often with the use of aesthetically-based constraints), they do not consider the semantic structure of the information being presented, or at best assume that the structure must be displayed in a standard format. Nevertheless future researchers may wish to use many of the same component technologies (e.g. solvers, machine learning) in building improved tools based on a deeper understanding of visual language.

2.2 Visual to Cognitive Mapping

We have found research relating to the way in which a perceived visual is transformed into a cognitive structure in the fields of psychology, education and visualization.

2.2.1 Overview

Researchers in neurophysiology and psychology have studied how our visual system organizes elements into groups [4,8,20–22,29]. Several researchers from other fields also recognize the importance of grouping.

Grouping techniques are important in helping to organize information [3].

A graphic should not show only the leaves; it should show the branches as well as the entire tree [5].

In psychology, the Gestalt laws [12, 13] established patterns of perception of organization that explain how people perceive objects as a group. The laws include similarity, proximity, common fate, symmetry, good continuation, and closure as well as following findings of connectedness and common region. In addition, researchers found that some of these grouping principles have precedence over others. For example, grouping by proximity precedes grouping by similarity, and grouping by connectedness precedes both grouping by proximity and similarity. To the best of our knowledge, most other combinations of grouping techniques have not been studied. The measurements used in grouping experiments were time performance, preferences, multi-dimensional scaling, and cluster analysis.

In education [17], spatial contiguity, one of the principles in multimedia learning, introduces a rule to locate related words and visuals closer together which corresponds with the grouping principle of proximity. The researcher found that placing relevant information together helps learners to remember and understand the concept and semantic meanings.

In visualization, researchers interested in determining the effectiveness of their work have begun creating guidelines and tools for evaluating their visuals.

2.2.2 Psychology

Gestalt Theory

In the early twentieth century, the Gestalt school of psychology developed Gestalt theory, which attempted to understand perception in terms of organizational laws or principles [12, 13]. The German word, *Gestalt* means a structure, configuration, or pattern of physical, biological, or psychological phenomena that is integrated so as to constitute a functional unit with properties not derivable by summation of its parts [2]. Some Gestalt laws are illustrated in Fig. 6.

They include law of proximity (Fig. 6b)—nearby elements are perceived together; law of similarity (Fig. 6c)—similar elements tend to be grouped together; law of closure (Fig. 6d)—closed contour to be grouped; law of continuity (Fig. 6g)—well-aligned contours are perceived to be grouped; law of common fate (Fig. 6h)—elements moving together tend to be grouped; and law of symmetry—symmetrically arranged pairs of elements are perceived as a group. Additional studies added connectedness (Fig. 6e) [22] and common region (Fig. 6f) [20] as grouping principles. Regarding to similarity, the attributes of the elements can be their lightness, hue, size, orientations or shapes. For instance, Fig. 6b shows an example of grouping by proximity which the closeness of black squares allows us to perceive four distinct groups. Figure 6c shows an example of grouping by similarity in hue which divides the figure into two groups of black squares and two groups of grayish squares.

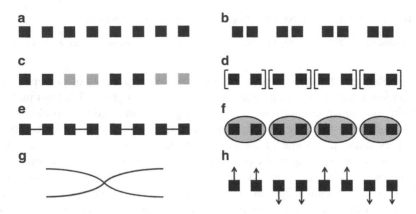

Fig. 6 Examples of gestalt principles. (**a**) No grouping. Grouping by (**b**) proximity, (**c**) color similarity, (**d**) closure, (**e**) connectedness, (**f**) common region, (**g**) good continuation, and (**h**) common fate

Figure 6d depicts grouping by closure which brackets surrounding two squares form four groups. Figure 6e shows four groups of connected squares. Figure 6f shows four groups of squares which each group has a common region by a boundary and grayish background. With continuity, we perceive two continual lines in Fig. 6g, not four discontinued line elements. With common fate, Fig. 6h shows two groups of upward squares and two groups of downward squares by perceiving elements with the same direction as a group.

Precedence of Gestalt Laws

Previous researchers performed numerous experiments that provided clues as to which Gestalt laws viewers apply before others. This essentially describes the central focus of our work, that is, how to combine grouping principles to create a visual structure. First, some researchers support the notion that grouping by proximity dominates similarity [4, 8]. Responses are faster to a task involving global structures formed by proximity rather than similarity of shape and luminance. Others found that proximity precedes similarity and global similarity operates before local similarity of geometric features (e.g., viewers perceive a global diamond shape, not squares, when small squares consist a diamond) [22]. Next, researchers conducted experimental studies to determine whether grouping by connectedness is perceived prior to grouping by proximity and similarity [7, 8, 22]. Palmer and Rock [22] proposed that connectedness operates prior to grouping principles such as proximity and similarity. For example, people tend to group two rectangles that are connected with a line than two rectangles closer to each other (Fig. 7).

However, Han [8] partially refuted this implication of connectedness always winning proximity. Grouping by connectedness performed better than weak proximity but when elements were grouped by strong proximity, it showed equally

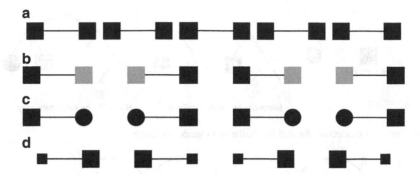

Fig. 7 Connectedness operates prior to (**a**) proximity, (**b**) similar color, (**c**) similar shape, and (**d**) similar size

fast reaction times as grouping by connectedness. Nevertheless, the advantage of grouping by connectedness revealed when the number of elements in the group increased. Han asked the participants to find a specific letter or to discriminate horizontal or vertical alignments, and found that response times were faster with grouping by connectedness than by weak proximity and similarity of shape, but not with strong proximity.

Kimchi studied how grouping works in local and global levels [11]. The researcher tested the implications of connectedness by matching tasks and visual search tasks with various numbers and sizes of elements. The study found that global configuration reveals when there are smaller and many local elements in the pattern. This is why we perceive a circle when we see a circular configuration consisting of square-shaped elements. While there are some principles that have obvious precedence over others, there are other grouping principles that compete against each other. Because of this ambiguity, the elements may be grouped differently depending on the viewer. For example, Fig. 8b, c show how people can draw different hierarchy from given elements (Fig. 8a) according to different grouping orders. The features of the elements that can be observed are shape and color differences. Figure 8b shows an hierarchy perceived by grouping with shape similarity. Since there are three rectangles and two circles (Fig. 8a), three rectangles are grouped at the left part and two circles are grouped at the right part of the tree. Then, among the three rectangles at left, two rectangles depicted in red can be grouped again. On the other hand, Fig. 8c shows a hierarchy perceived by grouping with color similarity. Since there are two rectangles and one circle in red and one rectangle and one circle in blue (Fig. 8a), elements in red are located at the left part and elements in blue are located at the right part of the tree. Then, among the red elements at left, two red elements shaped in rectangle are grouped again.

It is reasonable to create a grouping precedence by applying multiple grouping principles to the elements and see which grouping principle operates prior to others. Figure 9a shows an example that applies two grouping principles, common region and connectedness. Figure 9b, c additionally applies proximity with common region

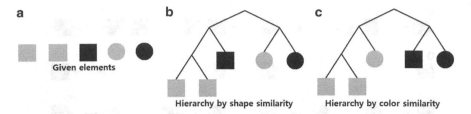

Fig. 8 Different hierarchies formed by similarity in shape and color

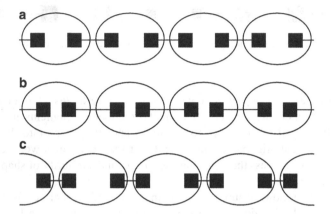

Fig. 9 Combination of grouping principles with proximity, common region, and connectedness. Notice the equivalent size of ovals for common region since the size affects grouping. (**a**) Grouping with common region and connectedness. (**b**) Grouping closer elements with common region and further elements with connectedness. (**c**) Grouping closer elements with connectedness and further elements with common region

and connectedness. It is necessary to conduct a measurable experiment and perform exact analysis but in general, common region seems to win other grouping methods (Fig. 9). How people resolve a visual cue competition situation is explained by its reliability [10]. He examined the visual depth cue and set two hypotheses that observers assess the cue more reliable when it is less ambiguous and when it is correlated with other cues in the environment. According to experiments in [11], global shape is more robust to noise or blur than the local shape, thus, people may consider it as more reliable. When one cue is weaker than the other, then the stronger cue wins when multiple cues exist simultaneously.

Recent research on perceptual grouping has introduced two processes: base grouping and incremental grouping [25]. Base grouping depends on neurons that are focused to feature conjunctions. It is fast and occurs in parallel. For example, responding to the orientation of a line element or shape feature does not require a lot of time. On the other hand, incremental grouping requires enhanced neuronal activity that labels the image elements with object-based attention. This process takes more time and the memory capacity is limited. The two groupings were experimented by letter identification and search for letters with same colored dots. In this study, a task for examining base grouping was to read the sentence although the letters were overlapped and displaced. According to the researchers' finding, the sentence was noticable despite the misplacement. However, giving an incremental grouping task to find letters with same colored dots requires more time since it is necessary to follow the letter strokes individually.

2.2.3 Education

A series of studies about better ways to understand presentation and improve learning has helped researchers to rethink the relationship between verbal and visual representation [17]. Mayer found limitations of verbal forms of presentation and attempted to study the best way to combine visual and verbal representations to enhance learning. He examined in two kinds of multimedia learning environments: a book-based environment with words and pictures and a computer-based environment with narration and animation. He introduces seven principles of design for multimedia presentation. Among these principles, the spatial contiguity principle is related to proximity in one of the gestalt laws of grouping. Spatial contiguity is to locate related words and pictures physically closer.

Spatial contiguity is verified by an experiment that resulted in a group of students who learned better and remembered more when related words and pictures are closer together on the page or the screen than further away. For the experiment, two groups of students were set. One group of students were exposed to a screen with text and pictures closer (integrated presentation group) while the other group of students were exposed to text and pictures further away (separated presentation group). The result encourages us to arrange the word and relative picture closer to each other as Mayer explains this as an aid to build cognitive connections between words and pictures. It supports the effectiveness of grouping by proximity from one of the Gestalt laws. Among the assumptions of a cognitive theory in multimedia learning, Mayer asserts that cognitive theory of multimedia learning depends on active processing. Active processing explains that human engages actively in learning by attending to relevant incoming information, organizing selected information into coherent mental representations, and integrating mental representations with other knowledge [17]. Moreover, explaining the cause-and-effect context along the text and illustration helps to provide a meaningful structure overall.

Mayer assessed these different presentations by the traditional measures of learning: retention and transfer tests. Retention test would ask the participants what they remember after viewing the presentation. For example, he used an explanation of the process of lightning as the lesson to be viewed by the students and asked them to write down how lightning develops. The retention score is examined by the meaning of what the learner wrote down but not the exact wording. The score is marked in percentage of correct answers divided from the total. Mayer was more interested in transfer test which verifies how much the learner understands the content and makes use of it to solve other problems. There can be a redesign question, troubleshooting question, prediction question, and conceptual questions. The students were given a paper to write down the answers to the questions. Again, the score is computed by percentage of the number of acceptable answers compared to the answer key divided to the total. In both tests, there were two people who scored the answer by consensus for objectivity.

2.2.4 Visualization

Researchers in visualization understand that they are visual communicators, and have recently begun asking if their imagery communicate well. With this motivation, most of their visual communication work focuses on analysis rather than synthesis.

Wattenberg and Fisher used a machine vision technique to extract perceptual organization from grayscale images [34]. They point out that the structure of visualization should match the structure of the data and convey its intention. Furthermore, Rosenholtz [28] introduced a quantitative model that applies perceptual grouping to an image and returns a visual structure of the grouping. The algorithm was applied to show the perceptual organization of text, graphs, and maps. With a paragraph of text, the predicted hierarchical segmentation revealed the hierarchical grouping of letters, words, lines, and header with the paragraph. They went on to ask the designers whether the existing models and tools of human perception helped the design process [27]. Not only did they provide the design guidelines, but also, unexpectedly, they support the way designers communicate each other by providing a common perceptual language. However, these models lack in experimental validation and need to compare the result with the viewers' perception and intention. In addition, Wattenberg and Fisher point out the need of the study on how to combine different grouping techniques. They also comment that computer vision work is inadequate for user interface designs and information graphics. Moreover, they mention the lack of work comparing models of perceptual grouping with human perception. The difference between their work and ours is that they perform visual analysis by building a tool to analyze existing visuals, while we perform visual synthesis to construct new visuals.

Ware discusses how to support efficient visual search with a multi-scale structure [33]. By featuring visual properties of a region, a positive response to pattern recognition and eye movement can be maintained. The researcher asserts that a multi-level structure helps individuals to remember the location of a particular object or region, and makes it easier to revisit places, even those seen for a short time. He also mentions semantic pattern mapping with spatial metaphors such as order, proximity, similarity, and size of the objects. The objects located closer to one another may mean that they have semantically similar concepts and related information. Objects with the same shape, color, or texture may have the same semantic meaning.

3   Open Questions and Challenges

As we hope this survey makes clear, if we are to address society's communication crisis in any real way, there are several important research goals and questions that remain. In this section, we list and briefly discuss them.

- *What is successful communication, and how can it be measured?* Even if we accept for the moment the definition of communication as the delivery of knowledge, the question of whether that delivery was successful is non-trivial. How much of that knowledge should the recipient retain? How long must they retain it? Should the knowledge be literal only, or should deeper connections be drawn between delivered facts? Education research (some of which we reviewed above) asks similar questions, but there is an entire field of research dedicated to communications itself that we have not considered here. However, neither communication, education, nor indeed psychology researchers have spent a great deal of time considering visual communication.
- *How should grouping techniques be combined to communicate complex structure?* As we have discussed above, psychology researchers have already examined some pairwise grouping combinations, such as proximity with similarity, connectedness with proximity, and connectedness with similarity. Some grouping techniques are closely related, for example it is difficult to align several objects without also placing them in some sort of proximity. However, the majority of combinations have not been studied, and design practice rarely uses two grouping techniques in isolation. Further study is needed to examine which grouping techniques work well together, and what implicit principles guide the combination of grouping techniques in design practice.
- *How does successful communication of order and importance in knowledge take place, and how does it interact with communication of structure?* In this chapter we have focused on visual communication of knowledge structure, but clearly sequence and importance are a regular and important part of visual communication. We are not aware of a great deal of research investigating these topics, though we are aware of some eye tracking work that is relevant.
- *Is graphic design's hypothesis true: is a faithful visual representation of knowledge semantics the best way to communicate it?* It may seem counterintuitive, but there is significant research supporting the notion that clarity and accuracy are not always the best ways to communicate: forcing viewers to work harder to understand may make that understanding last longer. On the other hand, perhaps one cannot assume that viewers will dedicate the cognitive energy to achieve that lasting understanding. Would it be possible to serve both dedicated/deeper viewers and passing/shallow viewers in one visual?
- *How should culture, audience, and medium affect visual communication?* Great communicators know how to tailor their appeal to different audiences. In particular, great visual communicators know how to specialize their messages to different mediums (e.g. poster, television, slide or web). How do they do this, and how should tools help less talented communicators do so?

- *How can this knowledge of visual communication be successfully embedded in design software?* As should be clear by now, even the subset of visual communication we focus on in this chapter is quite complex, with many apparently conflicting principles and constraints. Incorporating them into a successful tool will be challenging. The methods used for existing layout tools (briefly discussed above) may provide guidance.

We would be remiss if we did not again note that communication is about much more than communication of facts, their organization, order and relative importance. Communication is highly interpersonal, meaning that emotion may be the most important component of communication. We have not considered it at all here, meaning that there is an entire, closely related range of research questions ripe for the picking.

4 Our Own Initial Progress

4.1 Some Applied Experience

A graph is a huge grouping structure, which becomes challenging to view, navigate, and understand when the screen space is limited. We introduce an interface displaying a graph neighborhood in three columns (Fig. 10). A home node (usually arrived at via search) is located at the left column and nearby nodes are shown on middle (one link away) and right (two links away) columns. For example, with IMDb(Internet Movie Database), the graph is bipartite and consists of person and movie in which the task is to find which people worked on which movies. When a search key is an actor, the second column contains all the movies related with him while the third column contains all the actors/actresses he collaborated with based on the second column. When static, each column shows only three nodes (movies or actors/actresses), with the relatedness depicted only among the elements on the screen. A flicking interaction on the screen for each column enables viewing off-screen nodes (other movies or actors/actresses) while groups are indicated by grouping principles. We applied grouping methods such as color similarity, connectedness, position similarity, and texture similarity (Fig. 10b–e).

We learned that grouping techniques such as common region, proximity, closure, good continuation, and common fate are harder to depict many to many relationships. Nevertheless, we can make use of some grouping techniques based on visual similarity (e.g., color, texture, and position) and connectedness; however, it becomes challenging when it already contains complex visual content such as IMDb headshots. Moreover, the interactive settings containing a specific motion (e.g., vertical column scrolling) makes it challenging with some of the grouping techniques such as proximity, common fate, and good continuation. The data set from movies and music included many-to-many relationship and had limitations when it came to displaying the groups using the grouping principles mentioned

Fig. 10 Different explicit link representations. (**a**) Text: nodes with the same name are linked. (**b**) Color: nodes with the same nodes with the same color are linked. (**c**) Connectedness: nodes with lines between them are linked. (**d**) Proximity: nodes at/containing the same *vertical* position are linked. (**e**) Texture: nodes with/containing the same image are linked

above. For example, elements that are related need to move closer to each other when grouped by proximity. If there are three groups on the screen, the elements would need to clutter towards the middle or be replaced, which is even more confusing. With common region, it is possible to overlap colored regions as layers, but this would result in confusion of color mixing. Closure can be applied by surrounding the related elements by colored brackets, but it was assumed that this would not work well in distinguishing multiple groups. With common fate, a group can be indicated by moving related elements in the same direction, but has a limitation when an element is a member of two or more groups.

Results largely matched our expectations, with text-based and proximity-inspired links performing worst, texture- and color-inspired link depictions performing better and connectedness-inspired link lines performing best. However, users were only about 20% faster with link lines than with texture-inspired links containing thumbnails.

4.2   Basic Study of Communicating Knowledge Structure

Our approach is to systematically study the combinations of two or three Gestalt principles, measure the conveyed cognitive structure, and the resulting learning. We plan to use a fully crossed within subjects $5 \times 5 \times 5$ design. As a start, we will study the combination of two grouping principles. We chose five hierarchical examples including product labels and web pages that have two levels of grouping. First and second level grouping principles each has five levels: common region, connectivity, color similarity, proximity, and alignment.

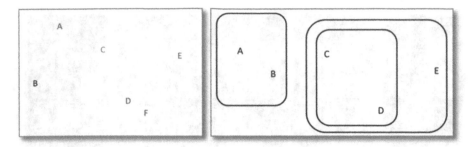

Fig. 11 Examples of grouping letters by color and common region

4.2.1 Test Bed

Researchers from various backgrounds have started to use Amazon's Mechanical Turk [1] to perform inexpensive experiments which can recruit thousands of people. The strength of Mechanical Turk is the number of participants that can perform the experiment which leads to the strength of statistical results. We plan to give the participants several examples so that they can arrange the hierarchical understanding to the drag-and-drop tree structure. Figure 11 illustrates example of grouping letters by color similarity and common region.

4.2.2 Measurement

Response Time or Search Time

Repetition discrimination is an indirect method that detects a repeating element in a row of elements [21]. It is indirect, since the participants are neither aware of the object of the study of grouping nor asked to report any groupings. In this study, the researchers used response time to measure the relative grouping effect when the same shapes are located in the same perceptual groups and in different groups. A larger difference in the reaction time between different conditions indicated a stronger grouping effect. However, the study focused on single level grouping, and therefore did not deal with multiple grouping principles. Another study on the visual searching of features and objects compared search times [29]. The study investigates the relationship between grouping and preattentive features such as color and shape. Preattentive features appear despite the grouping, however, if the task is to search conjunctive targets with more than two perceptual features, attention is required. Thus, searching conjunctive targets takes serial time as the number of group grows. The search becomes harder as each of the conjunctive feature is located between two groups. The study is related with grouping by similarity in color and shape. A study that developed a numerical model to evaluate graphical user interface screen measured participants' search time in looking for a given control (e.g., buttons, check boxes, radio buttons, etc.) to perform an instructed action [23]. The model combines screen factors such as element size, local density, alignment, and grouping, and produces a complexity score for the screen given. The researchers found that poor alignment and poor local density negatively affected time performance. They concluded that screens with grouping indications and alignment resulted in shorter search times, emphasizing the importance of grouping in visual layouts.

Ratings

Most of the studies that gathered data on response time also gathered data on preference ratings [21, 23]. Palmer and Beck [21] maintained to use ratings in most of their measures of grouping. The researchers asked the participants to rate the strength or degree of the grouping between the elements of the target pair. Parush et al. [23] not only measured the search time, but also let the participants rate their pairwise preferences on a seven-point scale given the design screens. Among the design factors, they found that alignment and grouping influenced subjective preferences, while density had a lower preference rate. Eventually, the weights obtained by the users' preferences were applied to recalculate the complexity score from their evaluation model.

Grouping Methods

Tree edit distance is a common metric to evaluate the similarity of two trees. Given two trees, it counts the number of operations such as insert, delete, and update the labels when one tree converts to another. While it is a NP-Hard problem for an unordered tree, it is generally accepted for a rooted ordered tree. A rooted ordered tree is when the order of the siblings from left to right is significant. A general used one is from Zhang and Shasha's algorithm [36]. Numerous grouping analyses have been conducted in anthropological studies. A study on folkbiology of fish allowed experts from different cultures to classify fish species [18]. Initially, the researchers prepared 44 species of fish on name cards. After sorting out the names that were not recognized by the participants, the researchers asked the participants to group the fish that live together and share a common habitat. They also asked for positive or negative relations (e.g., help or eat) between different kinds of fish using 19 categories of relationships. The participants were asked to form hierarchy of fish by repeatedly grouping and dividing existing groups. They used principal component analysis and multi-dimensional scaling(MDS) to assess consensus across the experts. This research on the classification of fish demonstrated a cultural difference, whereby Native American fish experts sort more ecologically by habitat, while majority-culture fish experts sort more by the characteristics of the fish.

Consensus analysis addresses anthropologists' investigation of an unknown culture and depends on informants' responses [26]. It derives a formal mathematical model to analyze the informants' consensus on questionnaire data, which provides individual competencies and an estimate of the correct answer to each question asked to the informants. Another study on texture classification has helped to identify how people perceive textural features in three dimensions [24]. The researchers used 12 characteristics (e.g., contrast, repetitive, random, fine, etc.) to classify the group of 56 textures on 9-point Likert scales. Then, the participants sorted the items into similar groups and described why the items in each group were similar. The researchers applied hierarchical cluster analysis and non-parametric MDS to the similarity matrix produced by the users' grouping selections of textures.

4.2.3   Combinations of Grouping Techniques

A simple illustration of the hierarchy of an object can be drawn as a tree. However, a tree-shape hierarchy does not fully utilize a page on a rectangular-shaped monitor screen, document, poster, and web pages. We propose to use Gestalt principles to group the objects in the hierarchy. We would like to determine whether the grouping precedence occurs when the observer performs a task on a combination of groupings. If different grouping principles are applied to different levels, we may find the grouping principles that are perceived before others. With a two level grouping, there would be $n \times n$ pairs of grouping techniques if there are n grouping techniques, including those of a grouping technique with itself. We plan to study which combinations are better in communicating the structures. Figure 12 shows plausible single-level groupings. The first row shows single coding, the second shows double coding, and the third and fourth show both double and triple coding.

4.2.4   How This Affects Learning

The purpose of measuring learnability is to determine the extent to which learners remember and understand the content of the presentation. This study can be used in addition to the measurement in section "Grouping Methods".

We follow the approach used in multimedia learning, wherein Mayer introduced a way to measure learning using retention and transfer tests [17]. The goal of the retention test is to see how well the observer remembers the content, while the transfer test checks how well the observer understands the content. Thus, we refer to the retention and transfer tests to measure how well the observer remembers and understands the content of the presentation. We plan to ask observers to write down the flow of the storyboard or the content of the presentation (i.e., retention test) among presentation slides, posters, and web pages. Moreover, we will ask them to answer several questions to provide solutions to problems related to the content (i.e., transfer test).

For a retention test, Mayer prepared blank sheets to write down the explanation, giving participants 6 min to do so. Likewise, we plan to give instructions such as: "Please give a written explanation of the cooling effects of dirt holes on gas turbine

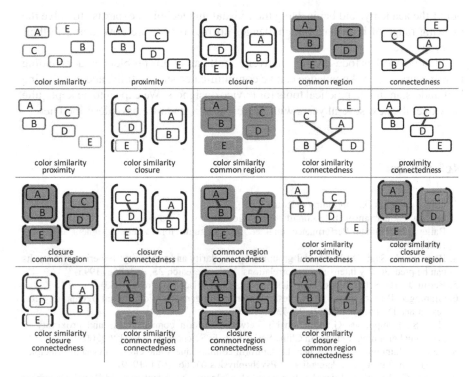

Fig. 12 Combination of five grouping principles with proximity, color similarity, closure, connectedness, and common region. Groups are *A* and *B*, *C* and *D*, and *E* itself

blades." Like Mayer, we will allow the participants 6 min to respond. For a transfer test, Mayer gives one question at a time on a sheet of paper and allows 2.5 min for participants to write down answers from what they understand. After the time limit, the question sheet with the answer is collected and the next question sheet is given. Some of the transfer questions may be: "What could you do to find the cooling effect?", "Suppose you see film cooling holes blocked. What is the cause?" and "What causes cooling effects of dirt holes on gas turbine blades?" We will instruct the participants to write down as many acceptable answers as possible.

We assume that similarity measurement from visual templates correspond with learnability. We expect that more similar informational and cognitive structure will produce higher retention and transfer scores.

5   Discussion and Future Work

This chapter has advocated for and described a program of research in everyday visual communication, one that will address the growing volume, complexity and ubiquity of information. We believe it is time for visualization researchers to move out of the academy, and begin using their visual and technical expertise to solve the growing range of informational challenges normal people are now facing in their everyday lives.

Our own work is focused on a relatively small part of this problem: understanding how the structure of knowledge can be communicated successfully, and embodying that understanding in practical tools for novice designers. We expect to be kept quite busy over the next several years with the research agenda we have described here.

References

1. https://www.mturk.com/mturk/welcome
2. http://www.merriam-webster.com/dictionary/gestalt
3. Bailey, R.: Human Performance engineering: A guide for system designers. Prentice-Hall (1982)
4. Ben-AV, M., Sagi, D.: Perceptual grouping by similarity and proximity: Experimental results can be predicted by intensity autocorrelations. Vision research **35**, 853–866 (1995)
5. Bertin, J.: The Semiology of Graphics. University of Wisconsin Press (1983)
6. Grabinger, R.: Computer screen designs: viewer judgements. Educational Technology Research and Development **41**(2), 35–73 (1993)
7. Han, S., Humphreys, G.: Relationship between uniform connectedness and proximity in perceptual grouping. Science in China Series C: Life Sciences **46**, 113–126 (2003)
8. Han, S., Humphreys, G.W., Chen, L.: Uniform connectedness and classical Gestalt principles of perceptual grouping. Perception & Psychophysics **61**, 661–674 (1999)
9. Jacobs, C., Li, W., Schrier, E., Bargeron, D., Salesin, D.: Adaptive Grid-Based Document Layout. In Proceedings of SIGGRAPH'03 pp. 838–847 (2003)
10. Jacobs, R.: What determines visual cue reliability? Trends in cognitive sciences **6**, 345–350 (2002)
11. Kimchi, R.: Uniform connectedness and grouping in the perceptual organization of hierarchical patterns. J. Exp. Psychol. Hum. Percept. Perform. **24**, 1105–1118 (1998)
12. Koffa, K.: Principle of Gestalt Psychology. London, Routledge & Kegan Paul Ltd. (1935)
13. Lidwell, W., Holden, K., Butler, J.: Gestalt Theory (Uber Gestalttheorie): 39–59, [Ellis, W.D.(Ed.) (1938): A sourcebook of Gestalt psychology. London: Routledge & Kegan Paul (1924)
14. Lidwell, W., Holden, K., Butler, J.: Universal Principles of Design. Rockport Publishers (2003)
15. Lok, S., Feiner, S.: A Survey of Automated Layout Techniques for Information Presentations. In Proceedings of SmartGraphics 2001 (2001)
16. Lok, S., Feiner, S., Ngai, G.: Evaluation of Visual Balance for Automated Layout. Proceedings of the 9th international conference on Intelligent user interface pp. 101–106 (2004)
17. Mayer, R.: Multimedia Learning. Cambridge University Press (2001)
18. Medin, D.L., Ross, N., Atran, S., et al.: Folkbiology of freshwater fish. Cognition **99**, 237–273 (2006)
19. Ngo, D., Teo, L., Byrne, J.: Evaluating Interface Esthetics. Knowledge and Information Systems **4**(1), 46–79 (2002)
20. Palmer, S.: Common region: A new principle of perceptual grouping. Cognitive Psychology **24**, 436–447 (1992)
21. Palmer, S., Beck, D.: The repetition discrimination task: An objective method for studying perceptual grouping. Perception and Psychophysics **69**(1), 68–78 (2007)
22. Palmer, S., Rock, I.: Rethinking perceptual organization: The role of uniform connectedness. Psychonomic bulletin & review **1**(1), 29–55 (1994)

23. Parush, A., Nadir, R., Shtub, A.: Evaluating the layout of graphical user interface screens: Validation of a numerical, computerized model. International Journal of Human Computer Interaction **10**(4), 343–360 (1998)
24. Rao, A., Lohse, G.: Towards a Texture Naming System: Identifying Relevant Dimensions of Texture. Vision Research **36**(11), 1649–1669 (1996)
25. Roelfsema, P., Houtkamp, R.: Incremental grouping of image elements in vision. Attention, Perception, & Psychophysics **73**(8), 2542–2572 (2011)
26. Romney, A., Weller, S., Batchelder, W.: Culture as consensus: a theory of culture and informant accuracy. Am. Anthropol. **88**, 313–338 (1986)
27. Rosenholtz, R., Dorai, A., Freeman, R.: Do Predictions of Visual Perception Aid Design? ACM Transactions on Applied Perception **8**(2) (2011)
28. Rosenholtz, R., Twarog, N.R., Schinkel-Bielefeld, N., Wattenberg, M.: An intuitive model of perceptual grouping for HCI design. In Proceedings of SIGCHI 2009 pp. 1331–1340 (2009)
29. Treisman, A.: Perceptual grouping and attention in visual search for features and for objects. Journal of Experimental Psychology: Human Perception and Performance **8**(2), 194–214 (1982)
30. Tufte, E.R.: The cognitive style of PowerPoint, vol. 2006. Graphics Press Cheshire, CT (2003)
31. Tullis, T.: The Formatting of Alphanumeric Displays: A Review and Analysis. Human Factors **25**(6), 657–682 (1983)
32. Tullis, T.: A system for evaluating screen formats: Research and application. Advances in human-computer interaction **2**, 214–286 (1988)
33. Ware, C.: Visual thinking for design. Morgan Kaufmann (2008)
34. Wattenberg, M., Fisher, D.: A model of multi-scale perceptual organization in information graphics. Infovis 2003 (2003)
35. Williams, R.: The Non-Designer's Design Book. Peachpit Press (2004)
36. Zhang, K., D., S.: Simple fast algorithms for the editing distance between trees and related problems. SIAM Journal of Computing **18**, 1245–1262 (1989)

Pep Up Your Time Machine: Recommendations for the Design of Information Visualizations of Time-Dependent Data

Simone Kriglstein, Margit Pohl, and Michael Smuc

Abstract Representing time-dependent data plays an important role in information visualization. Time presents specific challenges for the representation of data because time is a complex and highly abstract concept. Basically, there are two ways to support reasoning about time: time can be represented by space, and time can also be represented by time (animation). From the point of view of the users, both forms of representation have their strengths and weaknesses which we will illustrate in this chapter. In recent years, a large number of visualizations has been developed to solve the problem of representing time-dependent data. Nevertheless, it is still not clear which types of visualizations support the cognitive processes of the users. It is necessary to investigate the interactions of real users with visualizations to clarify this issue. The following chapter will give an overview of empirical evaluations and recommendations for the design of visualizations for time-dependent data.

1 Introduction

Representing time-dependent data plays an important role in information visualization. Time presents specific challenges because time is a complex and highly abstract concept. Reasoning about large numbers of co-occurring processes and their interrelationships, e.g., is difficult without an appropriate visualization tool. Several possibilities to solve this problem have been proposed in the literature. Basically, there are two ways to support reasoning about time:

S. Kriglstein (✉) • M. Pohl
Vienna University of Technology, Institute for Design and Assessment
of Technology, Vienna, Austria
e-mail: simone.kriglstein@tuwien.ac.at; margit@igw.tuwien.ac.at

M. Smuc
Danube University Krems, Krems an der Donau, Austria
e-mail: michael.smuc@donau-uni.ac.at

W. Huang (ed.), *Handbook of Human Centric Visualization*,
DOI 10.1007/978-1-4614-7485-2_8, © Springer Science+Business Media New York 2014

- **Space:** Time can be represented by space. In this case, space is used as a metaphor. For example, one of the most common solutions is the presentation of length of lines in space in order to show the length of time intervals.
- **Animation:** Time can also be represented by time. In this case, animation is used to represent developments over time.

Nevertheless, it is still not clear which types of visualizations support the cognitive processes of the users. Several studies evaluating users' interactions with visualizations of time-dependent data have been conducted, but so far, no systematic guidelines for the design of such systems exist. The following chapter will give an overview of empirical evaluations in this area and tries to come up with some tentative recommendations for the design of visualizations for time-dependent data. By empirical evaluations we mean studies conducted with real users. In this context, we concentrate on evaluation studies in the research areas: information visualization and visual analytics. We would like to point out that considerable research concerning the representation of temporal processes has also been conducted in areas like visuo-spatial cognition, graph comprehension and educational psychology. This research has influenced the discussion about the design of information visualizations. Nevertheless, we think that for the development of specific recommendations or guidelines it makes sense to restrict the scope to the more narrow area of information visualization and visual analytics.

We will first outline some of the specific characteristics of time-dependent data and different possibilities to represent such data. First, we will discuss evaluation studies describing interactions with visualizations using space as a metaphor, and animations. We will try to generalize the results gained from the literature research and derive some tentative recommendations. We prefer to talk about recommendations rather than guidelines. We think that there is still too little systematic research in this area to be able to come up with more explicit guidelines. It should be pointed out that some recommendations only apply to certain contexts. Animations, for example, are beneficial in some contexts and confusing in other contexts. Such constraints have to be identified and we think that recommendations based on empirical research with real users are necessary to ensure a high quality of the design of information visualizations.

2 Time-Dependent Data

Reasoning about time is a challenging activity because time is a highly abstract concept. Basically, there are two different ways to design information visualization systems representing time-dependent data in consideration of visual variables (cf. [5,9,45,46]). On the one hand, time can be represented using a spatial metaphor. On the other hand, time can be represented by time – that is, animation can be used to support the analysis of time-dependent data.

2.1   Space Metaphor

We often use space as a metaphor to reason about time [36]. The representation of time-dependent data has some specific characteristics, even when space is used to represent time and should also be considered in the design of such systems.

There are different possibilities how time can be represented in space. Timelines are, for example, a powerful metaphor to visualize events and their chronological order. In this context, line graphs play an especially important role. Shah et al. [67] describe that subjects have a tendency to interpret line graphs as trends even when the data do not support this (e.g., when the line graph depicts differences between females and males). In this way, well-designed visualizations can support finding insights in complex data and help users to draw inferences about data.

A basic distinction discussed extensively in the literature is between linear and cyclical time [37, 45]. The most common model which is being used when reasoning about time is the linear model. Nevertheless, time-dependent data very often represents cyclical processes. Sales figures in retail trade often follow cyclical patterns. Sales before Christmas often tend to be higher and during holidays lower. Usually, circular visualizations are used to support such processes in order to detect such patterns. Another distinction is between time points and time intervals [37, 45]. In information visualization, there is a large variety of visualizations related to time points, but there are still not many visualizations representing time intervals. Gantt charts are simple visualizations representing time intervals. Andrienko and Andrienko [5] also distinguish between ordered time versus branching time. Time is inherently ordered. This is reflected in most visualizations. There are applications, however, especially in simulations predicting future developments, which should take several different time paths into account.

There are many different possibilities to represent linear time, e.g. traces or small multiples [3]. Traces can use line width or brightness to represent the passing of time. The fading of older elements in general can also be used in other contexts to denote older items which are replaced by newer ones. Features like text labels can also be adopted to convey information of time. Figure 1 illustrates an example for the usage of traces. The example presents a spatiotemporal visualization approach for the analysis of the gameplay data for the educational game DOG*eometry* [75,76]. States of the game are presented as nodes, and arrows visualize the players' activity for the transition of one state to another state of the game. The width of arrows reflects the number of players. Colors of nodes represent specific states (e.g., orange for start node and blue for end node) and colors of the arrows present different types of players' activities (e.g., orange for reset).

Small multiples are a well known possibility to convey information of temporal developments. They can represent quite a large amount of complex information and have some advantages over animations [61], but again information cannot be seen at a glance. In addition, there are limits to the number of small multiples which can be represented on one screen.

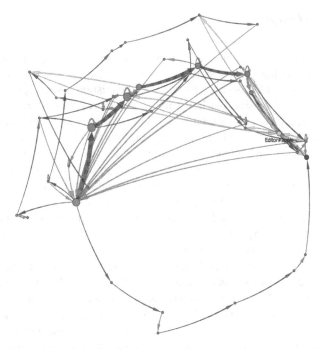

Fig. 1 Example for a node-link visualization approach to analyze spatiotemporal information of gameplay data [76] (Image courtesy of G. Wallner)

Time can also be added as a third dimension to a two-dimensional representation (space-time cube, [45]). The question whether 2D has more advantages than 3D is highly controversial [1, 3] and is discussed in more detail in Sect. 3.1.7. Nevertheless, there are certainly some contexts where a 3D application to represent time-dependent data is helpful.

2.2 Animation

Animation uses time to represent time. For example, animation support in the visualization system for gameplay data (cf. Fig. 1) allows users to analyze the distribution of players in regard to the different states during their playtime (see Fig. 2).

Animation is a sequence of static representations (frames) for the time steps in the data with the consideration to depict the information in a logic and predefined order [3, 60]. Further aspects are display date, duration, rate of change, frequency and synchronization [45]. For a fluid motion in animation 24–30 frames per second and for a slide show 2–4 frames per seconds are necessary [3, 31].

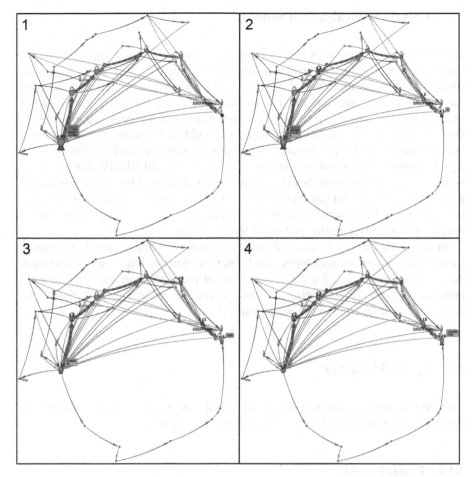

Fig. 2 Animation example to analyze players' movements in the game [76]. The distribution of the two colors of the player symbols reflects the number of female and male players. Screenshots were taken every 2 min (Image courtesy of G. Wallner)

The usage of animation has its advantages and disadvantages. On the one hand, it enables to represent more complex data on the screen because it is not necessary to include time as a dimension. On the other hand, users cannot see phenomena at a glance any more. The cognitive load on short term memory is highly increased. Users, therefore, often look at animations repeatedly to be able to make sense of the data [54].

3 Analysis of Evaluation Studies

Although the involvement of users already very early in the development process can increase the quality of the design and users' acceptance (cf. [2, 40, 42]), user studies are still difficult to find. A lot of research work has been done in measuring the system performance or user performance (e.g., task completion time, error rates, accuracy rates) or in presenting the design of the visualization approach with the help of scenarios/use cases. However, we could also observe that the interest and the steady call for publications in journals, conferences and workshops (see, e.g., Information Visualization Journal, IEEE VisWeek, and BELIV Workshop) to conduct such evaluation studies increased in the last years. The aim of conducted evaluation studies concentrated primarily on the comparison of different techniques (e.g. 2D versus 3D or static representation versus animation), to observe how users interact with the visualization and/or get feedback from them.

In this section we try to conclude and generalize our insights which we gained from evaluation studies with focus on information visualization for time-dependent data. The structure used for the presentation of the evaluation studies follows a bottom-up approach. Especially in the case of comparison studies, it is difficult to decide to which section a study belongs to.

3.1 Space Metaphor

This section will discuss findings of empirical studies evaluating visualizations especially for time-dependent data based on the space metaphor.

3.1.1 Linear Layouts

Linear charts for time-dependent data are nowadays probably the most popular graphs to display time series [17]. For linear layouts, a linear time axis and an axis to display quantitative information is used to represent data – allowing to solve typical tasks like the analysis of trends or shapes, finding minima and maxima but also comparison tasks [32, 35, 43] and prediction analysis [53].

As stated earlier, basic results from graph comprehension literature are not described in this section. This is especially the case for the analysis of single time series, since there is a vast amount of literature from graph comprehension on this topic (see Simkin and Hastie [68] for a good introduction) which would go beyond the scope of this paper and would provide only limited benefits for the design of visual analytics tools. Our focus lies more on the applicability and problem areas for temporal visual analytics methods and modern information visualization techniques. Therefore, we will concentrate mainly on empirical findings for layout techniques of multiple time series in this section, with emphasis on an efficient usage of space, the

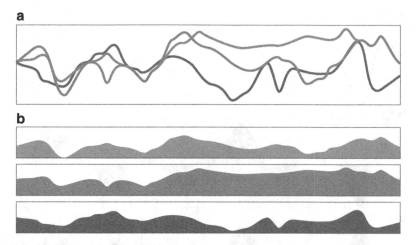

Fig. 3 Examples for (**a**) multiple line chart which share the same space and have one time axis in common (shared-space graphs) and (**b**) small multiples where the graphs are split and each graph represents a single time series with its own time axis (split-space graphs)

pros and cons of different layouts, the avoidance of clutter and overlaps, some non-standard techniques for coding or representing time dependent data and some notes about issues to deal with when interacting with line charts to represent time series data in compound with comprehensive visual analytics applications.

Following Javed et al. [35], there exist in general two ways for the layout of multiple line charts that represent time series (see Fig. 3): multiple line charts and small multiples. Both layouts have their strengths and weaknesses. For shared space graphs, one problem is the identification of a single time series without labels, since labeling but also interactions could be difficult when the lines overlap. Based on a comprehensive experimental series by Javed et al. [35], the authors reported empirical evidence for problems with clutter and overlap for shared-space layouts. But they also found that this layout performs well for comparison tasks with a local visual span, like comparison on one time point in multiple time series. On the other hand, the identification and labeling of single time series is easier in split-space graphs like small multiples, but low vertical resolution (in case of many time series the charts have to be squeezed to make more room) and the spacing between multiples could be problematic. Comparison turned out to be easier with larger visual spans, and – in contrast to shared space layouts – with this layout users performed well for high numbers of concurrent time series and for dispersed tasks.

The same is true for or horizon graphs, another split-space layout. With horizon graphs, based on the work of Saito et al. [63], a general weakness of split-space techniques can be compensated, at least to some extent. Split-space techniques generally suffer from their high demand for vertical space. With horizon graphs, it is possible to shorten the vertical extension of a small multiple by mirroring filled charts (negative values are flipped around zero, see Fig. 4). Then they could

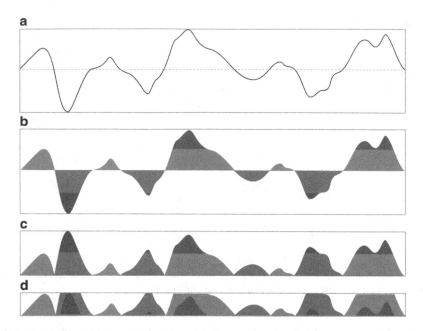

Fig. 4 Examples for (**a**) line chart, (**b**) line chart that is divided into two bands filled with different colors for positive and negative values, (**c**) line chart with mirrored negative values and (**d**) with compressed bands

be cut (often in half), with the upper band of the chart getting repainted and overlaid over the lower band (see Fig. 4). Heer et al. [32] found that mirroring doesn't hamper graphical perception, but it is doubling room. Heer and colleagues pointed out that some training is necessary. Although the usage of layered bands increased estimation time and error rate for larger chart sizes with more than 24 pixels (6.8 mm), it led to better estimation accuracies for smaller graphs. In general, layering should be limited to two bands for small chart sizes (six pixels), and maximally mirrored when used in large charts.

As a general result for small multiple layouts, Javed et al. [35] and also Heer et al. [32] reported a trade-off between time and accuracy when chart size is decreased. For smaller graphs, participants don't become slower (or even are getting faster), but correctness usually decreases. Another promising approach is the use of color-coding instead of lines (or areas) to represent quantitative data. When quantitative information is coded as color-fields, Lam et al. [43] concluded from their study results that color-coding turned out to be good for simple pattern finding tasks with a limited visual angle. For more complex patterns, users preferred and worked more accurately with small multiple time series. A mixture of color-coded and standard line diagrams – either embedded and only interactively explorable or side-by-side – did not facilitate visual search, supposedly due to higher transaction costs caused by the demands to interact.

When looking for the big picture in data, color-fields can increase accuracy and speed. Correll et al. [19] reported these results of their experiments and assumed that color-coding makes (perceptual) averaging over a time span easier, since it makes stronger use of pre-attentative processing, using the *low-level* perceptual system instead of mental computation.

Although there are some methods to mitigate the negative effects of small multiples, Javed et al. [35] recommended using other techniques or visual representations if too many time series should be displayed and visual clutter becomes too high, i.e. temporal queries [33] or aggregation [24].

3.1.2 Cyclical Layouts

Although cyclical layouts provide a compact representation of time-dependent data, the acceptance of such layouts depends on the working habits of the users. Zhao et al. [83] mentioned that breaking the habits is a big challenge because user feel confident with familiar representations that can also influence if the visualization is assessed as useful or not. Therefore, it can happen that users are more skeptical in regard to cyclical layouts over the well-known linear layouts. However, after users learn to read the cyclical layouts, the cyclical layouts got positive feedback in regard to periodic data (see, e.g., [14]).

Chin et al. [16] compared different layouts and they found out that a cyclical layout is helpful to detect temporal regions and periodic patterns. Furthermore, the participants solved the tasks more quickly, accurately, and with higher user satisfaction with the tested cyclical approach than with the tested linear timeline approach [16]. However, they also pointed out that cyclical layouts have a limited applicability for non-periodic data and problems with orientation can happen especially for spiral visualization approaches [16].

3.1.3 Time Intervals

Visualization of time intervals helps users to see the durations, the resultant interconnections, and also to identify unexpected sequences (see, e.g., [13,77,81]). Wang et al. [77] found out that the visualization of intervals influenced how the participants interpreted the data as if only time points were visualized. The usage of lines to visualize intervals was helpful to identify and to remember information about durations of events and to detect overlapping events easier [77]. This is also confirmed by Campbell et al. [13] who used rectangles for the presentation of time intervals.

Additional to line and rectangle representations, Gantt diagrams are noted as an effective way to visualize time intervals. Because of their popularity (e.g., for project management tasks) the learning effort is relatively low (see, e.g., [2,34]).

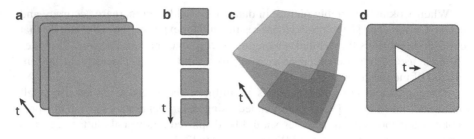

Fig. 5 Layout examples for (**a**) superimposition, (**b**) juxtaposition, (**c**) space-time cube and (**d**) animation (cf. Sect. 3.2)

3.1.4 Maps

Some popular ways to visualize time in space in combination with geo-spatial information is to use small multiples, visualizing time either as traces, trajectories or paths (superimposed on a 2D map) or by adding another dimension called the space-time cube [29] (cf. c in Fig. 5) or 2.5D visualization. Let us start with the latter two. Although there is a high demand for empirical results since 3D visualization is always a point of lively debates in the visualization community, only a rather small number of user studies have been carried out to determine the usefulness and weaknesses of 3D visualizations of temporal data (see also Sect. 3.1.7).

To fill this gap, Kristensson et al. [41] conducted a between-subjects experiment with 30 (novice) participants to shed more light on the strengths and weaknesses of 2D maps with traces and 2.5D visualizations. Kristensson et al. [41] concluded that the usage of 2D representations is less error-prone than space-time cubes for direct and simple tasks. As soon as the tasks get more complex and difficult in the sense that a deeper geo-spatial understanding of the data is needed to solve a task, the space-time cube outperforms the 2D representation by halving response times with having comparable accuracy. Although it has been stated by the authors that a lot of limitations have to be taken into consideration when interpreting their results (like varying the data density), these results look at least in parts promising for the usage of space-time cubes, at least for complex tasks and possibly in a combination with 2D maps which support the lower level tasks.

Similarly, Kjellin et al. [38] also reported promising results for the space-time cube, at least when used for tasks where the analysis of precise 3D-metric properties (which are perspectively skewed) does not play a crucial role, since in this case even *fairly unsophisticated 2D visualizations* support users better. Kjellin et al. [38] also pointed out that, in general, all graphical data expressions that can be used in a 2D visualization can also be integrated in a space-time cube, where the space-time cube has also the potential to add another additional dimension, allowing richer communication of data and gaining a detailed mental image to better understand the data [23].

Having said that, both Kristennsen et al. [41] and Kjellin et al. [38] recommend using 2D charts as a first choice, mainly due to the space-time cube's weaknesses for some tasks and economic considerations, since the usage of space-time cubes may be like breaking a butterfly on the wheel. Further it has to be noted that some points of discussion on 3D visualizations have not been addressed in the above presented studies to a satisfying extent so far, for instance usability issues when using 3D, which will be also a topic in Sect. 3.1.5.

As mentioned above, another way to visualize geo-spatial information is the usage of small multiples. Empirical research showed some limitations for juxtaposing maps (cf. b in Fig. 5). First of all, tasks concerning longer time-periods are better supported by this layout [12, 25], but sudden changes or subtle movements of data clusters are often not detected [28]. Further, findings from Dransch et al. [23] showed that juxtaposition is only appropriate for a limited number of maps. For large datasets this can lead to a coarse temporal resolution, where short-time anomalies or single missing data points can easily be overseen, since they might not pop-out saliently in a view where data was averaged over a time interval. These limitations can be seen as the result of the general disadvantage of small multiples failing to represent continuous temporal data efficiently, since every small view has to be extracted at a sacrifice (either by slicing or by averaging over a time interval) out of the temporal data stream.

3.1.5 Relational Data

Relational data, typically represented as node-link diagrams, often become more complex than visualization with maps, because even the topology or *the content of a map* is dynamic. But similar to layouts for space-time visualizations of maps, among the most prominent layouts for dynamic relational data are once more small multiples, superimposition as overlay of trajectories and the space-time cube, as depicted in Fig. 5 (Windhager et al. [80], see also Weaver et al. [79] for different layout variants).

Since the content of the time-slices changes in small multiples, it is the question of how much change should be shown. If too much change is shown, the problem of how the mental map may be preserved between time-slices may arise. Aside from graph theoretical views which layout algorithm for network graphs can support the maintenance of the mental map (see Purchase and Samara [55] or Federico et al. [27], for a discussion of actual developments), the empirical works of Purchase and Samara [55] as well as Saffrey and Purchase [62] lead to a differentiated view on how to preserve the user's mental maps. In the study from Saffrey and Purchase [62], it turned out that highly restricting the layout – and therefore highly preserving the mental map – produced most errors or no mental map was produced at all, probably since node overlapping makes tasks highly demanding and can lead to poor layouts. They showed that restricted node movement alone would not help preserving the mental map and also the overall relation of nodes and edges should be addressed for layout. In a user study by Purchase and Samara [55], the authors reported that the trade-off between stability (keep the position of a node) and consistency (make a correct computation of the layout in each slice) does not lie somewhere in the middle, since this compromise leads to the worst performance of the users. The authors therefore recommended – in case changes in their connectivity are of interest – to use layouts that show significant changes or movements, although individual nodes may be harder to track.

Saraiya et al. [65] reported results of a user study on a rather special case where small multiples are used and the network topology does not change over time. In this case, small multiples showed their strengths for (static) topology and temporal analysis tasks, but performed bad for single node analysis. Another outcome was that single node analysis is supported by a static node-link layout where the activities of each node are represented in a heatmap-overlay. In a user study about dynamic node-link diagrams from Smuc et al. [70], small multiples, two variants of superimposition (cf. a in Fig. 5) and a 2.5D-visualization were compared. Although the superimposition had some advantages in task-completion speed over small multiples, it suffered from visual clutter. Especially novice users reported about difficulties on how to read the graph. If the superimposition layout was color-coded using different colors for new, stable and outdated nodes and edges, many users reported to have problems decoding the graph. The best superimposition layout turned out to be the one where edges are hidden at first sight to reduce clutter. The edges could be turned visible interactively. Layouts with small multiples performed well and were preferred by many users, especially novice users.

For 2D layouts some usability issues have to be reported: the legibility of labels turned out to be a problem when time-slices are perspectively skewed. The 3D navigation of the space-time cube can be an additionally demanding task for some users and should be implemented with much care regarding controls and reactivity. Since visual clutter and overlapping of visual elements are problems for all of the three layouts as soon as trajectories are used to support single node tracking between time-slices, repeated user interaction is needed to analyze these visualizations in detail.

3.1.6 Interactive Interfaces

Navigation and interactive interfaces play an important role for the analysis of time-dependent data and is often one of the main topics when users are asked to suggest improvements (see Weaver et al. [79] for an example of exhaustive user suggestions). For example, Zhao et al. [83] showed that interactive interfaces enriched the visualization in order to support the user in exploratory analysis of time-dependent data, e.g. for the identification of trends or anomalies. Timelines turned out to be a good application area for interactivity in multiple view settings, since they provide overview (see, e.g., [4, 21, 22]). Furthermore, interactivity can be helpful to learn and understand visualization approaches especially if users are not familiar with the visualization [13].

However, a high demand for interactivity can have also its drawbacks if users feel inconvenient with a visualization (see Lam et al. [43]). For time sliders it is essential that the feedback about the degree of change is clear, that prominent timescales are used and that the other panels are clearly connected [4].

3.1.7   2D Versus 3D

Whether 2D or 3D visualizations for time-dependent data are more or less effective, is intensely discussed in the community. Similar to the usage of animation (as is discussed in Sect. 3.2), 3D visualization is not equally suitable for everybody. On the one hand the cognitive load and the user's mental effort to correctly interpret the represented data is higher with a 3D representation [47]. One the other hand, the performance depends strongly on the tasks (see, e.g., [38, 41]) and the interaction with 3D representation can be challenging (see, e.g., [48]). Furthermore, the users' experience with the usage of 3D representation often plays an important role and influences their preferences to use 2D or 3D visualizations. For example, Nakakoji et al. [52] found out that the participant who was an expert liked the 3D representation to get a feeling of changes in data. Similar insights were gained by Tekušová and Schreck [73] in their evaluation study. They found out that participants' preferences in regard to use 3D visualizations for time-oriented data depended strongly how familiar they were with such representations. They pointed out that the interpretation of data was more difficult for users who had no or less experiences with 3D visualizations than for users who already had experience with such representation. Furthermore, participants who already had experience with 3D representations preferred the 3D representation, because it allowed them to analyze more complicated relationships. This is also confirmed by Kristensson et al. [41] that 3D representation can support users to analyze complex patterns in datasets.

3.1.8   Views for Space Organization

The results of the evaluation studies show us that single views as well as multiple views have advantages but also weak points. For example, Saraiya et al. [64] found out that single views are better for topological information. Huang et al. [34] showed that using an overlapping presentation is more effective to see the differences in data (e.g., how the data changed between two time points) as with the corresponding side-by-side presentation. Furthermore, juxtaposition presentation is only appropriate for a limited number of views (see, e.g., [23]). In contrast to the use of single views, Aigner and Miksch [2] pointed out that participants found different views very helpful, e.g., to see the structure and detail information of the data. Boyandin et al. [12] found out that switching from one view to another can have a beneficial effect in order to gain insights about the datasets.

3.2   Animation

As mentioned in [39], there already exists a large number of research studies in cognitive psychology on the perception of movement and change which are relevant for the explanation of the usage of animation techniques in order to visualize time-dependent data. Examples are the phenomenon of change blindness as discussed in [58], that humans can concentrate only on restricted areas in their field of vision, and the ability of humans to only track a limited number of moving targets (e.g., Pylyshyn [56] found out that humans can track approximately four similar objects).

In this section, we analyze evaluation studies from the literature with a focus on animation approaches to present time-dependent data in information visualization.

3.2.1   General Aspects

Based on existing empirical research, animation was used as additional support for users to improve the handling and perception of changes in data in order to analyze data that evolves over time (see, e.g., [10, 15, 28, 66]).

Evaluations of animation have yielded mixed results. Although it was often suggested to use animation carefully and recommendations exist to use animation only for case where a static representation is not possible (see, e.g., [30]), animation seems to be beneficial under certain conditions. Animation techniques was often said to be very helpful for tasks related to the dynamic development in order to grasp changes of data and to understand how the data evolve over time (e.g., [6, 26, 49, 52, 61, 71–73]). A possible reason is that animation can provide micro steps that convey more information than for the case of an analysis of time-dependent data with small static multiple graphics [69]. McGrath and Blythe [49] also pointed out that the usage of animation in combination with space metaphor like graph representation is a good way to communicate different properties of data.

However, studies also found out that although a well-designed animation is a powerful technique, it is not suitable for everybody and can be confusing (cf. [82]). A possible reason is that it is only possible to concentrate on a small part of information and therefore it can happen that users miss important information for the further analysis.

For an effective usage of animation as additional support it is also necessary to consider the speed of animation (see Sect. 3.2.2) as well as interactivity (see Sect. 3.2.3) and to use additional indicators (see Sect. 3.2.4) in order to handle the problems like disappearance, attention, confidence, and complexity. It also depends on the flow of the animation in regard to order and synchronization [45]. For example, Midtbø et al. [50] found out that the usage of animation types depends on the used time span. The participants in their study preferred circular animations for time spans with repeating patterns and linear animations for time spans which had non-cyclical character.

Fig. 6 Example for video cassette recording (VCR)-style control elements

3.2.2 Animation Speed

As mentioned in [39], the speed of the animation influences the analysis of time-dependent data, e.g., to identify changes in the animation. Therefore, it is recommended that speed should be slow and clear enough to perceive movements and changes [78]. However, to find the optimal speed of animations is often a big challenge. For example, if the animation is too fast, users don't have sufficient time to realize changes and miss information (see the results of the evaluation studies by Griffin et al. [28] and Tekušová et al. [72]). But if the animation is too slow, problems with principles of grouping can happen, e.g., because users have problem to see the motion.

According to the results by Griffin et al. [28], Harrower and Fabrikant [31], and Kjellin et al. [38] the speed of the animation depends on different aspects like the distance of the moved objects, the duration of the animation and the frame rate (or also known as frame frequency) in the animation. Kjellin et al. [38] noted that the phenomenon of change blindness played also an important role. They found out that visual disruption (e.g., because a saccade was made) was a reason that the participants had problems to detect large changes.

Therefore, often a default animation speed was chosen which the user could change during they interactive with the animation in order to get control over the speed with the aim to spend more time on interesting parts [51]. Studies showed that the possibility to control the speed of animation was beneficial in order to support the creation of users' mental model and users were frustrated more easily if they had no control over the animation speed (see, e.g., [30, 52, 61]). Control elements to manipulate the speed and the duration of the animation are presented in Sect. 3.2.3.

3.2.3 Interactivity

Only to analyze data by the observation of the animation is in most cases not sufficient in order to support users to get answers to questions which they developed during their observation of the animated data [60]. Interaction possibilities allow users to control the animation which can improve the visual exploration better than when the users only watch an animation passively [11, 74].

Most approaches (see, e.g., [6, 10, 20, 26, 50, 52, 59, 61, 71]) consider well-known and widely understood standard Video Cassette Recording (VCR)-style control elements like play, pause, (fast) forwarding, and (fast) rewind in combination with a slider in order to control the speed (see Fig. 6 for an example).

User interactions are analyzed in several studies (see, e.g., [44, 52, 74]) with the insights that (a) interactivity can be a good solution to overcome the difficulties of perception and (b) the comprehension in animations helps users to get a feeling of the data. The evaluation studies (see, e.g., [6, 10, 12, 26, 52, 61]) also analyzed the usage of interaction elements during the analysis of the datasets. For example, it was observed that users preferred first to scan the whole animation in order to get a big picture and as second step they used the control elements to jump to positions of interest for a detail analysis, e.g., to compare the different frames. Because of the additional interaction possibility and the fact that the participants need more steps for the analysis, they need more time with an animation than with a static representation [15, 26, 61]. For example, in the study [12] they interacted for a longer time with the animation approach than with small multiples for the same number of findings. Therefore it is recommended to use animation if accuracy is more important than response time [6]. Further findings from comparison studies between animation and static representations are discussed in Sect. 3.3.

The findings of research studies in cognitive psychology in regard to perception of movement and change, showed that users can only follow four to five different objects at the same time [7, 78] which was mentioned as possible reason that participants got lost because the number of objects were too large and hence distracted their attention (cf. [26, 52, 61]). Therefore additional to the control elements further interaction features like a selection, filter, and search function are necessary in order to reduce complexity and information overload [30, 71]. For example, filter functions can be helpful to select specific animation parts of interest [26, 52, 57]. Tekušová and Kohlhammer [71] noted that zoom or detail on demand functions would be useful to control the number of displayed information at the same time in order to show or hide more or less information about the observed data.

3.2.4 Additional Indicators

In addition to the interaction possibilities in order to influence the flow of an animation, a combination of animation with further indicators like colors, traces or sound is recommended in the evaluation studies (see, e.g., [10, 15, 49, 50, 59, 66, 71, 82]). For example, indicators like colors or sound can be used to lead users' attention to specific parts in the animation or to specific animated objects. Traces can be used to visualize the history of the data points in order to support users to follow changes in data over the time. However there exist different opinions about the benefit of the usage of such traces. On the one hand, results showed that participants analyzed trends faster with traces than without the path information (see, e.g., [61]), but on the other hand it was noted as confusing especially for the visualization of traces for many datasets (cf. [59, 61, 71]).

3.3   Comparison Studies

The comparison studies between different visualization techniques showed that the usage of different techniques within the same system can have the benefit to support flexibility in regard to users' needs and tasks (see, e.g., [12, 18]). In this section we concentrate on findings of evaluation studies which include comparison studies between animation and static representation using the space metaphor.

The aim of such comparison studies was to identify if animation or the static representation were more effective for the analysis of time-dependent data (see, e.g., [6, 12, 26, 28, 52, 61, 64]). For the comparison studies, measurements of user performances like task completion time, error rates, accuracy rates were often combined with the evaluation of user preferences.

3.3.1   Animation Versus Static Representation

When animations or static representations like small multiples are compared whether they are more effective, it seems that it is dependent on which type of tasks is used. For presentation tasks, the reaction to the usage of animation was predominantly positive in regard to certain aspects of data and their changes over time (e.g., to identify trends in data), because the corresponding static representation was often noted as too overloaded or for the case of small multiples the views were too small and too many for comparison of the time-dependent data (see, e.g., [12, 15, 25, 52, 61]). Furthermore, it was easier for the participants to see how the data changed over time with an animation and to follow changes in data with the help of motion than with the comparison of before and after static representations (cf. [49, 71]).

The usage of animation is also useful to identify patterns. For example, although Griffin et al. [28] noted that the identifications of patterns depended on how visible the moving patterns were, they found out that the participants found more and in less time patterns with the help of animation than with small multiples. This was also confirmed by Boyandin et al. [12]. They noted that the animation can be an effective solution in order to detect local patterns and sudden changes [12].

For analysis tasks, studies like [6, 28, 61] showed that static representation appeared more effective than animation, e.g., to analyze topological information or to compare information at a specific time point.

Although there was no main difference in performance, Kjellin et al. [38] found out in their study that the learning effect of the participants was higher with static representation as with animation. However they also pointed out that further investigations would be necessary.

3.3.2 Motion as Additional Indicator

Bartram et al. [8] pointed out that motion as additional indicator for static representation (e.g., as a notification mechanism) would have its advantages. For example, the usage of motion in combination with static representation was for the participants easier to recognize and grabbed their attention than with the usage of other static visual variables like colors and shapes. This also confirmed by MacEachren [45].

3.3.3 Size of Datasets

Often the size of the represented datesets played an important role in the comparison studies in regard to how well the approaches supported the participants in their tasks. For example, Robertson et al. [61] noted that although the participants found the use of animation was more fun and exciting than the static representation, the participants were more successful with the animation for smaller datasets and with the static representation for larger datasets.

4 Recommendations

In the following, we will present the most important recommendations resulting from our overview of the literature:

- Linear charts are the most popular way to present time (cf. Sect. 3.1.1). Sometimes such charts become cluttered when to many lines are combined in one diagram. Such a visualization is still advantageous for comparison tasks with a local visual span (e.g., one point in time).
- Small multiples help to identify single time series better than linear charts.
- Horizon graphs which are designed carefully can increase estimation accuracy for larger datasets.
- Color coding is beneficial for simple tasks and for getting an overview.
- Cyclical layouts (cf. Sect. 3.1.2) are helpful for detecting periodic patterns. Users sometimes have to get used to this form of visualization.
- Visualization of time intervals helps users to see duration.
- 2D representations are less error-prone for simple tasks, 2.5D is better for more complex tasks. 2.5D also allows to get a deeper understanding of the data. It is more advantageous for experts than for novices.
- Small multiples are good for geo-spatial information under some conditions (e.g., long time periods).
- When representing development over time in node-link diagrams it is recommended to show significant changes between consecutive frames.
- Animation is beneficial under certain conditions (e.g., data showing dynamic development).

- Speed of animations should be slow and clear, but not too slow (cf. Sect. 3.2.2). Users should be able to control the speed.
- Interactivity (cf. Sect. 3.2.3) usually is beneficial for animation (e.g., jumping to a specific point in time, filtering, . . .).
- Animation versus static representations (cf. Sect. 3.3.1): This choice should depend on which type of task has to be solved.

We would like to point out that several of the recommendations depend on the context in which they are applied. Sometimes, it is therefore to be recommended to read the original paper where the empirical investigations were described.

5   Conclusions

In this chapter we gave an overview of empirical research concerning the evaluation of information visualizations of time-dependent data. We assume that such an overview can help to inform the design of future visualizations in that area. We derived several tentative recommendations for such a design process. It should be mentioned, however, that there is still too little research in that area to lay a firm foundation for general guidelines. Future research will help to develop a more comprehensive framework of guidelines in that area.

Acknowledgements This work is conducted in the context of the CVAST – Centre of Visual Analytics Science and Technology project. It is funded by the Austrian Federal Ministry of Economy, Family and Youth in the exceptional Laura Bassi Centres of Excellence initiative and also within the EXPAND – EXploratory Visualization of PAtent Network Dynamics project, supported by the program FIT-IT/BMVIT of the Federal Ministry of Transport, Innovation and Technology, Austria (Project number: 2883373).

References

1. Aigner, W., Bertone, A., Miksch, S., Tominski, C., Schumann, H.: Towards a conceptual framework for visual analytics of time and time-oriented data. In: Proceedings of the 39th Conference on Winter Simulation: 40 years! The best is yet to come, WSC '07, pp. 721–729. IEEE Press (2007)
2. Aigner, W., Miksch, S.: CareVis: Integrated visualization of computerized protocols and temporal patient data. Artif. Intell. Med. **37**(3), 203–218 (2006)
3. Aigner, W., Miksch, S., Schumann, H., Tominski, C.: Visualization of Time-Oriented Data. Human-Computer Interaction Series. Springer (2011)
4. André, P., Wilson, M.L., Russell, A., Smith, D.A., Owens, A., schraefel, m.: Continuum: designing timelines for hierarchies, relationships and scale. In: Proceedings of the 20th Symposium on User Interface Software and Technology, UIST '07, pp. 101–110. ACM Press (2007)
5. Andrienko, N., Andrienko, G.: Exploratory Analysis of Spatial and Temporal Data: A Systematic Approach. Springer (2005)

6. Archambault, D., Purchase, H., Pinaud, B.: Animation, small multiples, and the effect of mental map preservation in dynamic graphs. IEEE Transactions on Visualization and Computer Graphics **17**, 539–552 (2011)
7. Bartram, L.: Perceptual and interpretative properties of motion for information visualization. In: Proceedings of the Workshop on New Paradigms in Information Visualization and Manipulation, NPIV '97, pp. 3–7. ACM Press (1997)
8. Bartram, L., Ware, C., Calvert, T.: Moticons: detection, distraction and task. Int. J. Hum.-Comput. Stud. **58**(5), 515–545 (2003)
9. Bertin, J.: Semiology of Graphics: Diagrams, Networks, Maps. University of Wisconsin Press (1983). Translated by William J. Berg (French edn. 1967)
10. Bezerianos, A., Dragicevic, P., Balakrishnan, R.: Mnemonic rendering: an image-based approach for exposing hidden changes in dynamic displays. In: Proceeding of the 19th Annual ACM Symposium on User Interface Software and Technology, pp. 159–168. ACM Press (2006)
11. Blok, C.: Monitoring change: Characteristics of dynamic geo-spatial phenomena for visual exploration. In: Spatial Cognition II, Integrating Abstract Theories, Empirical Studies, Formal Methods, and Practical Applications, pp. 16–30. Springer (2000)
12. Boyandin, I., Bertini, E., Lalanne, D.: A qualitative study on the exploration of temporal changes in flow maps with animation and small-multiples. Comp. Graph. Forum **31**(3pt2), 1005–1014 (2012)
13. Campbell, J.D., Ganesan, A.B., Gotow, B., Kavulya, S.P., Mulholland, J., Narasimhan, P., Ramasubramanian, S., Shuster, M., Tan, J.: Understanding and improving the diagnostic workflow of MapReduce users. In: Proceedings of the 5th ACM Symposium on Computer Human Interaction for Management of Information Technology, CHIMIT '11, pp. 1:1–1:10. ACM Press (2011)
14. Carlis, J.V., Konstan, J.A.: Interactive visualization of serial periodic data. In: Proceedings of the 11th Annual ACM Symposium on User Interface Software and Technology, UIST '98, pp. 29–38. ACM Press (1998)
15. Chevalier, F., Dragicevic, P., Bezerianos, A., Fekete, J.D.: Using text animated transitions to support navigation in document histories. In: Proceeding of the 28th International Conference on Human Factors in Computing Systems, pp. 683–692. ACM Press (2010)
16. Chin, G., Singhal, M., Nakamura, G., Gurumoorthi, V., Freeman-Cadoret, N.: Visual analysis of dynamic data streams. Information Visualization **8**(3), 212–229 (2009)
17. Cleveland, W.S.: Visualizing Data. Hobart Press (1993)
18. Conati, C., Maclaren, H.: Exploring the role of individual differences in information visualization. In: Proceedings of the International Working Conference on Advanced Visual Interfaces, AVI '08, pp. 199–206. ACM Press (2008)
19. Correll, M., Albers, D., Franconeri, S., Gleicher, M.: Comparing averages in time series data. In: Proceedings of the SIGCHI Conference on Human Factors in Computing Systems, CHI '12, pp. 1095–1104. ACM Press (2012)
20. Craig, P., Kennedy, J., Cumming, A.: Animated interval scatter-plot views for the exploratory analysis of large-scale microarray time-course data. Information Visualization **4**, 149–163 (2005)
21. Dias, R., Fonseca, M.J., Gonçalves, D.: Interactive exploration of music listening histories. In: Proceedings of the International Working Conference on Advanced Visual Interfaces, AVI '12, pp. 415–422. ACM Press (2012)
22. Dias, R., Fonseca, M.J., Gonçalves, D.: Music listening history explorer: an alternative approach for browsing music listening history habits. In: Proceedings of the International Conference on Intelligent User Interfaces, IUI '12, pp. 261–264. ACM Press (2012)
23. Dransch, D., Kothur, P., Schulte, S., Klemann, V., Dobslaw, H.: Assessing the quality of geoscientific simulation models with visual analytics methods-a design study. Int. J. Geogr. Inf. Sci. **24**(10), 1459–1479 (2010)
24. Elmqvist, N., Fekete, J.D.: Hierarchical aggregation for information visualization: Overview, techniques, and design guidelines. IEEE Transactions on Visualization and Computer Graphics **16**(3), 439–454 (2010)

25. Fabrikant, S.I., Rebich-Hespanha, S., Andrienko, N., Andrienko, G., Montello, D.R.: Novel method to measure inference affordance in static small-multiple map displays representing dynamic processes. The Cartographic Journal 45, 201–215 (2008)
26. Farrugia, M., Quigley, A.J.: Effective temporal graph layout: a comparative study of animation versus static display methods. Information Visualization 10(1), 47–64 (2011)
27. Federico, P., Aigner, W., Miksch, S., Windhager, F., Zenk, L.: A visual analytics approach to dynamic social networks. In: Proceedings of the 11th International Conference on Knowledge Management and Knowledge Technologies, i-KNOW '11. ACM Press (2011)
28. Griffin, A.L., MacEachren, A.M., Hardisty, F., Steiner, E., Li, B.: A comparison of animated maps with static small-multiple maps for visually identifying space-time clusters. Annals of the Association of American Geographers 96(4), 740–753 (2006)
29. Hägerstrand, T.: Space, time and human conditions. In: A. Karlqvist, L. Lundqvist, F. Snickars (eds.) Dynamic allocation of urban space. Saxon House & Lexington Books (1975)
30. Harrower, M.: Tips for designing effective animated maps. Cartographic Perspectives 44, 63–65 (2003)
31. Harrower, M., Fabrikant, S.I.: The role of map animation in geographic visualization. In: M. Dodge, M. McDerby, T. M. (eds.) Geographic Visualization: Concepts, Tools and Applications, pp. 49–65. Wiley (2008)
32. Heer, J., Kong, N., Agrawala, M.: Sizing the horizon: the effects of chart size and layering on the graphical perception of time series visualizations. In: Proceedings of the SIGCHI Conference on Human Factors in Computing Systems, CHI '09, pp. 1303–1312. ACM Press (2009)
33. Hochheiser, H., Shneiderman, B.: Dynamic query tools for time series data sets: timebox widgets for interactive exploration. Information Visualization 3(1), 1–18 (2004)
34. Huang, D., Tory, M., Staub-French, S., Pottinger, R.: Visualization techniques for schedule comparison. Comput. Graph. Forum 28(3), 951–958 (2009)
35. Javed, W., McDonnel, B., Elmqvist, N.: Graphical perception of multiple time series. IEEE Transactions on Visualization and Computer Graphics 16(6), 927–934 (2010)
36. Johnson-Laird, P.N.: Space to think. In: P. Bloom, M.A. Peterson, L. Nadel, M.F. Garrett (eds.) Language and Space, pp. 437–462. MIT Press (1996)
37. Keim, D.A., Kohlhammer, J., Ellis, G., Mansmann, F. (eds.): Mastering The Information Age – Solving Problems with Visual Analytics. Eurographics (2010)
38. Kjellin, A., Pettersson, L.W., Seipel, S., Lind, M.: Evaluating 2D and 3D visualizations of spatiotemporal information. ACM Trans. Appl. Percept. 7(3), 19:1–19:23 (2008)
39. Kriglstein, S., Pohl, M., Stachl, C.: Animation for time-oriented data: An overview of empirical research. In: Proceedings of the 16th International Conference on Information Visualisation, IV '12, pp. 30–35. IEEE Computer Society (2012)
40. Kriglstein, S., Wallner, G.: Human centered design in practice: A case study with the ontology visualization tool Knoocks. In: G. Csurka, M. Kraus, L. Mestetskiy, P. Richard, J. Braz (eds.) Computer Vision, Imaging and Computer Graphics. Theory and Applications, CCIS, vol. 274, pp. 123–141. Springer (2013)
41. Kristensson, P.O., Dahlbäck, N., Anundi, D., Björnstad, M., Gillberg, H., Haraldsson, J., Mårtensson, I., Nordvall, M., Ståhl, J.: An evaluation of space time cube representation of spatiotemporal patterns. IEEE Transactions on Visualization and Computer Graphics 15(4), 696–702 (2009)
42. Kulyk, O.A., Kosara, R., Urquiza-Fuentes, J., Wassink, I.H.C.: Human-centered aspects. In: A.E. A. Kerren, J. Meyer (eds.) Human-Centered Visualization Environments, pp. 13–75. Springer (2006)
43. Lam, H., Munzner, T., Kincaid, R.: Overview use in multiple visual information resolution interfaces. IEEE Transactions on Visualization and Computer Graphics 13(6), 1278–1285 (2007)
44. Lowe, R.: Learning from animation. Where to look, when to look. In: R. Lowe, S. Wolfgang (eds.) Learning with Animation – Research Implications for Design, pp. 49–68. Cambridge University Press (2008)

45. MacEachren, A.M.: How maps work: representation, visualization, and design. The Guilford Press (2004)
46. Mackinlay, J.: Automating the design of graphical presentations of relational information. ACM Trans. Graph. 5(2), 110–141 (1986)
47. Mazza, R.: Introduction to Information Visualization. Springer (2009)
48. Mazza, R., Dimitrova, V.: Visualising student tracking data to support instructors in web-based distance education. In: Proceedings of the 13th International World Wide Web Conference on Alternate Track Papers & Posters, WWW Alt. '04, pp. 154–161. ACM Press (2004)
49. McGrath, C., Blythe, J.: Do you see what I want you to see? The effects of motion and spatial layout on viewers perceptions of graph structure. Information Sciences 5(2) (2004)
50. Midtbø, T., Clarke, K.C., Fabrikant, S.I.: Human interaction with animated maps: the portrayal of the passage of time. Proceedings of the 11th Scandinavian Research Conference on Geographical Information Science pp. 45–60 (2007)
51. Müller, W., Schumann, H.: Visualization for modeling and simulation: visualization methods for time-dependent data – an overview. In: Proceeding of the 35th Conference on Winter Simulation: driving innovation, WSC '03, pp. 737–745 (2003)
52. Nakakoji, K., Takashima, A., Yamamoto, Y.: Cognitive effects of animated visualization in exploratory visual data analysis. In: Proceeding of the 5th International Conference on Information Visualisation, pp. 77–84. IEEE Computer Society (2001)
53. Ordóñez, P., desJardins, M., Lombardi, M., Lehmann, C.U., Fackler, J.: An animated multi-variate visualization for physiological and clinical data in the ICU. In: Proceedings of the 1st ACM International Health Informatics Symposium, IHI '10, pp. 771–779. ACM Press (2010)
54. Pohl, M., Wiltner, S., Miksch, S.: Exploring information visualization: describing different interaction patterns. In: Proceedings of the 3rd Workshop: BEyond time and errors: novel evaLuation methods for Information Visualization, BELIV '10, pp. 16–23. ACM Press (2010)
55. Purchase, H.C., Samra, A.: Extremes are better: Investigating mental map preservation in dynamic graphs. In: G. Stapleton, J. Howse, J. Lee (eds.) Diagrammatic Representation and Inference, LNCS, vol. 5223, pp. 60–73. Springer (2008)
56. Pylyshyn, Z.W.: Things and Places: How the Mind Connects with the World. The MIT Press (2007)
57. Rensink, R.A.: Internal vs. external information in visual perception. In: Proceedings of the 2nd International Symposium on Smart Graphics, SMARTGRAPH '02, pp. 63–70. ACM Press (2002)
58. Rensink, R.A., O'Regan, J.K., Clark, J.J.: To see or not to see: the need for attention to perceive changes in scenes. Psychological Science 8(5), 368–373 (1997)
59. Rind, A., Aigner, W., Miksch, S., Wiltner, S., Pohl, M., Drexler, F., Neubauer, B., Suchy, N.: Visually exploring multivariate trends in patient cohorts using animated scatter plots. In: Proceeding of the International Conference on Ergonomics and Health Aspects of Work with Computers held as part of HCI International 2011, pp. 139–148. Springer (2011)
60. Rob, M.J.K., Edsall, R., Maceachren, A.M.: Cartographic animation and legends for temporal maps: Exploration and or interaction. In: Proceedings of the 18th International Cartographic Conference, pp. 23–27 (1997)
61. Robertson, G., Fernandez, R., Fisher, D., Lee, B., Stasko, J.: Effectiveness of animation in trend visualization. IEEE Transactions on Visualization and Computer Graphics 14, 1325–1332 (2008)
62. Saffrey, P., Purchase, H.: The "mental map" versus "static aesthetic" compromise in dynamic graphs: a user study. In: Proceedings of the 9th Conference on Australasian User Interface, AUIC '08, pp. 85–93. Australian Computer Society, Inc. (2008)
63. Saito, T., Miyamura, H.N., Yamamoto, M., Saito, H., Hoshiya, Y., Kaseda, T.: Two-tone pseudo coloring: Compact visualization for one-dimensional data. In: Proceedings of the IEEE Symposium on Information Visualization, INFOVIS '05. IEEE Computer Society (2005)
64. Saraiya, P., Lee, P., North, C.: Visualization of graphs with associated timeseries data. In: Proceedings of the IEEE Symposium on Information Visualization, INFOVIS '05. IEEE Computer Society (2005)

65. Saraiya, P., North, C., Duca, K.: Comparing benchmark task and insight evaluation methods on timeseries graph visualizations. In: Proceedings of the 3rd BELIV'10 Workshop: BEyond time and errors: novel evaLuation methods for Information Visualization, BELIV '10, pp. 55–62. ACM Press (2010)
66. Schlienger, C., Conversy, S., Chatty, S., Anquetil, M., Mertz, C.: Improving users' comprehension of changes with animation and sound: an empirical assessment. In: Proceeding of the 11th International Conference on Human-Computer Interaction, pp. 207–220. Springer (2007)
67. Shah, P., Freedman, E.G., Vekiri, I.: The comprehension of quantitative information in graphical displays. In: The Cambridge Handbook of Visuospatial Thinking, pp. 426–476. Cambridge University Press (2005)
68. Simkin, D., Hastie, R.: An Information-Processing Analysis of Graph Perception. Journal of the American Statistical Association **82**, 454–465 (1987)
69. Slocum, T.A., Block, C., Jiang, B., Koussoulakou, A., Montello, D.R., Furhmann, S., Hedley, N.R.: Cognitive and usability issues in geovisualization. Cartography and Geographic Information Science **28**(1), 61–75 (2001)
70. Smuc, M., Federico, P., Windhager, F., Aigner, W., Zenk, L., Miksch, S.: How do you connect moving dots? Insights from user studies on dynamic network visualizations. In: W. Huang (ed.) Human Centric Visualizations: Theories, Methodologies and Case Studies. Springer (2013)
71. Tekušová, T., Kohlhammer, J.: Applying animation to the visual analysis of financial time-dependent data. In: Proceeding of the 11th International Conference Information Visualization, pp. 101–108. IEEE Computer Society (2007)
72. Tekušová, T., Kohlhammer, J., Skwarek, S.J., Paramei, G.V.: Perception of direction changes in animated data visualization. In: Proceeding of the 5th Symposium on Applied perception in graphics and visualization, pp. 205–205. ACM Press (2008)
73. Tekušová, T., Schreck, T.: Visualizing time-dependent data in multivariate hierarchic plots – design and evaluation of an economic application. In: Proceedings of the 12th International Conference Information Visualisation, IV '08, pp. 143–150. IEEE Computer Society (2008)
74. Tversky, B., Morrison, J.B., Betrancourt, M.: Animation: can it facilitate? International Journal of Human-Computer Studies **57**, 247–262 (2002)
75. Wallner, G., Kriglstein, S.: DOGeometry: teaching geometry through play. In: Proceedings of the 4th International Conference on Fun and Games, FnG '12, pp. 11–18. ACM Press (2012)
76. Wallner, G., Kriglstein, S.: A spatiotemporal visualization approach for the analysis of gameplay data. In: Proceedings of the SIGCHI Conference on Human Factors in Computing Systems, CHI '12, pp. 1115–1124. ACM Press (2012)
77. Wang, T.D., Plaisant, C., Quinn, A.J., Stanchak, R., Murphy, S., Shneiderman, B.: Aligning temporal data by sentinel events: discovering patterns in electronic health records. In: Proceedings of the SIGCHI Conference on Human Factors in Computing Systems, CHI '08, pp. 457–466. ACM Press (2008)
78. Ware, C.: Information Visualization: Perception for Design, 3rd edn. Morgan Kaufmann Publishers Inc. (2012)
79. Weaver, C., Fyfe, D., Robinson, A., Holdsworth, D., Peuquet, D., MacEachren, A.M.: Visual exploration and analysis of historic hotel visits. Information Visualization **6**(1), 89–103 (2007)
80. Windhager, F., Zenk, L., Federico, P.: Visual enterprise network analytics – visualizing organizational change. Procedia – Social and Behavioral Sciences **22**, 59–68 (2011)
81. Wongsuphasawat, K., Guerra Gómez, J.A., Plaisant, C., Wang, T.D., Taieb-Maimon, M., Shneiderman, B.: Lifeflow: visualizing an overview of event sequences. In: Proceedings of the SIGCHI Conference on Human Factors in Computing Systems, CHI '11, pp. 1747–1756. ACM Press (2011)
82. Zaman, L., Kalra, A., Stuerzlinger, W.: The effect of animation, dual view, difference layers, and relative re-layout in hierarchical diagram differencing. In: Proceedings of Graphics Interface, GI '11, pp. 183–190. Canadian Human-Computer Communications Society (2011)
83. Zhao, J., Chevalier, F., Balakrishnan, R.: KronoMiner: using multi-foci navigation for the visual exploration of time-series data. In: Proceedings of the SIGCHI Conference on Human Factors in Computing Systems, CHI '11, pp. 1737–1746. ACM Press (2011)

Using Textbook Illustrations to Extract Design Principles for Algorithm Visualizations

J. Ángel Velázquez-Iturbide

Abstract The literature on algorithm visualizations addresses a number of important issues for educational use, such as instructional uses, graphical formats, effort of adoption, etc. However, there is a lack of clear principles to guide the construction of educationally effective visualizations. We have addressed an analysis of visualizations concerning three basic algorithm design techniques (divide and conquer, backtracking and dynamic programming). The material was the illustrations found in a number of prestigious algorithm textbooks, which prove to be high-quality sources. One contribution of this chapter is the final list of fields used to characterize visualizations, given that they embody the key features of illustrations. A second contribution is an outline of the findings of our analysis, which are a step toward stating design principles for algorithm visualizations.

1 Introduction

Algorithm visualization is one of the main approaches addressed in past 40 years to overcome novices' problems for effectively learning computer programming [1, 2]. During the eighties and first nineties of last century, the main contributions to the field of algorithm visualization were technical, but in the last decade the emphasis shifted to pedagogical issues. A key landmark was the metastudy conducted by Hundhausen et al. [3] on the features of 24 controlled experiments of educational effectiveness. Their main conclusion was that it is more important how students use visualizations than what students see. Based on this study, a working group at the ITiCSE'02 conference proposed a number of educational uses of visualizations,

J.Á. Velázquez-Iturbide (✉)
Departamento de Lenguajes y Sistemas Informáticos I, Escuela Técnica Superior de Ingeniería Informática, Universidad Rey Juan Carlos, C/ Tulipán s/n, 28933 Móstoles, Madrid, Spain
e-mail: angel.velazquez@urjc.es

W. Huang (ed.), *Handbook of Human Centric Visualization*, 227
DOI 10.1007/978-1-4614-7485-2_9, © Springer Science+Business Media New York 2014

known as the engagement taxonomy [4]. They suggested that the higher the students' engagement with animations, the better the learning outcomes.

Following Hundhausen et al.'s rationale [3], cognitive constructivism is a theory of learning that seems to be more educationally effective than others, in particular epistemic fidelity, according to the results of controlled evaluations of visualizations. Constructivism [5] claims that students construct knowledge rather than merely receive and store knowledge transmitted by the teacher, while epistemic fidelity [6] gives emphasis on the encoding of instructors' information to be transmitted to students. Consequently, the results of Hundhausen et al. produced a shift in emphasis from the visualizations themselves to their instruction. This concern on the educational use of visualizations led to further investigations on dissemination of visualization systems [7], factors that demand an effort from instructors to adopt a system [8], attitudes of instructors with respect to the systems [9], etc.

However, we cannot ignore the importance of algorithm visualizations themselves. Hundhausen et al. acknowledge that well-designed visualization contribute to learning. Other studies also have identified this as a key factor in educational success (e.g. [10] with respect to good examples). The surprising fact is the subsequent lack of systematic research on the representations themselves used in algorithm visualizations.

Obviously, a visualization must at least be able to "represent adequately" an algorithm behaviour. From a constructivist point of view, an algorithm visualization is a conceptual model built to communicate an algorithm to students; therefore, it must be precise, complete and consistent [11]. On the other side, a mental model is the representation that a student builds of a conceptual model, i.e. his/her understanding of it; consequently, a mental model often is partial and ambiguous [11]. We do not speak about correct mental models because they vary from one person to another. Rather, we speak about viable or unviable models, depending on whether they allow explaining or understanding the phenomena in observation.

In summary, the instructor should schedule his/her course with activities that help students in building viable mental models. However, he/she must also address the features of the conceptual models to be delivered, in particular of algorithm visualizations. A typical scenario we imagine is an instructor or a system constructor who wants to construct a visualization or set of visualizations for a course. What features should exhibit these visualizations? This is the concern we addressed and present in this chapter.

In this chapter we present a study of the characteristics of visualizations of algorithms designed according to three basic design techniques: divide and conquer, backtracking, and dynamic programming. The study was conducted by analyzing the illustrations included in prestigious textbooks of algorithms. Although our study is focused on visualizations manually constructed by instructors, we want to remark the potential importance of these results for constructors of algorithm visualizations. The features extracted could be a basis to state design principles that may characterize the visualizations constructed by current or future visualization systems.

The structure of the paper follows. In Sect. 2 the background of our work is presented. Section 3 describes the procedure followed to analyze the illustrations and the categories identified, while Sect. 4 presents the results of our analysis. The two last sections contain a discussion of the results, and our conclusions and plans for future work.

2 Background

In this section we review different recommendations on the features of algorithm visualizations, as given by researchers in the area or deduced from existing systems or from their evaluations. We go from the general to the specific recommendations, making clear on the way the kind of recommendations that are most interesting for our concern. We also sketch founded approaches to the design of visualizations in other fields.

2.1 Recommendations on Algorithm Visualization Systems

We find a great variety of recommendations on software visualization systems, including recommendations on the user interface, user functions, etc. The metastudy by Hundhausen et al. [3] and the engagement taxonomy [4] identified a set of functions that seemed to help in keeping the user active and therefore in increasing the educational effectiveness of visualization systems. These functions would allow users:

- Provide their own input data.
- Answer questions about the visualizations.
- Construct their own visualizations.
- Solve programming exercises.
- Present a visualization to an audience.

In subsequent studies or experiences, we find these or related recommendations again (e.g. [12, 13]). In addition, new engaging activities have been proposed and successfully addressed in these years (e.g. peer review of visualizations).

The list of functions that a visualization system could incorporate or support can be enlarged by accommodating more educational issues. Thus, Ihantola et al. [8] performed a comprehensive analysis of the characteristics that software visualization systems must take into account for their successful adoption and usage in educational contexts. They distinguished three categories: scope of visualizations, integration into courses, and interaction with the user.

However, informative as they are, these features do not give much information about the contents of the visualizations themselves. This concern can be better appreciated by analyzing the "ten commandments of algorithm animation" by

Gloor [14]. They are very general requirements that any animation builder could use as a checklist:

1. Be consistent.
2. Be interactive.
3. Be clear and concise.
4. Be forgiving to the user.
5. Adapt to the knowledge level of the user.
6. Emphasize the visual component.
7. Keep the user interested.
8. Incorporate both symbolic and iconic representations.
9. Include analysis and comparisons.
10. Include execution history.

In accordance with constructivism, requirements (1), (3) and (5) are extensible to any conceptual model, and requirements (2), (4) and (7) are important for the students' construction of their mental models. Requirements (6) and (8) address visual design. Finally, requirements (9) and (10) give general guidance on the algorithmic contents of visualizations.

We may throw some light on this diversity of recommendations by classifying them. We address this task in the following subsection.

2.2 Classes of Recommendations on Algorithm Visualization Systems

A number of taxonomies have been proposed that allow characterizing visualizations and visualization systems. A brief overview of eight taxonomies can be found in Hundhausen et al. [3]. For our purpose, we adopt the taxonomy by Price et al. [15] for classifying visualization systems because of its comprehensiveness. (Another comprehensive taxonomy, on algorithm animation languages [16], has similarities with Price at al.'s taxonomy.)

Price et al.'s taxonomy comprises six categories: scope, content, form, method, interaction, and effectiveness. We may analyze which of these categories are the most closely related to our concern, namely the contents of visualizations. Each category is characterized with the assistance of a question [15]:

1. Scope: What range of programs can the system take as input for visualization?
2. Content: What subset of information about the software is visualized by the system?
3. Form: What are the characteristics of the visualization?
4. Method: How is the visualization specified?
5. Interaction: How does the user of the visualization interact with and control it?
6. Effectiveness: How well does the system communicate information to the user?

Price et al. identify four potential roles for the persons involved with visualizations: system developer, "visualizer" (the person who specifies or constructs the visualization), programmer (the person who constructs the program to visualize), and user (the person who watches the visualization). We are interested in the role played by an instructor who wishes to construct a visualization, i.e. the visualizer:

1. Scope: This category only is relevant for the initial stages of the system's adoption. We assume here that the instructor has adopted a system appropriate for his/her course.
2. Content: This information is clearly relevant, especially the subcategory Algorithm. This category considers the visualization of instructions and data of an algorithm.
3. Form: In general, we do not consider highly relevant the information strictly related to visual design. Of course, it is important for the final look of the visualization but it does not play an important role for the choice of contents. However, some subcategories included here are related to contents, in the sense they are related to the visualization structure. This is the case of subcategories Granularity, Multiple views, and Program synchronization (i.e. multiple, synchronized algorithms).
4. Method: This category is the most important for the system developer and does not have a direct relation to the visualization contents.
5. Interaction: Our analysis for this category is similar to the analysis for the Form category. Some subcategories may interest the instructor to the extent they are related to the structure of the visualization, namely Navigation (and its subcategories Elision control and Temporal control).
6. Effectiveness: This category includes the functions of a visualization system that have been identified in empirical evaluations as important for educational success. However, it does not have a direct relation to the visualization contents.

In summary, categories Content and, to some extent, Form and Interaction deal with contents and structure of visualizations. In the next section, we review specific contributions to these issues.

2.3 Contents and Structure of Algorithm Visualizations

The contents of a visualization is a subset of the information present in an algorithm. This information can be gathered directly from the algorithm or it can be computed from the data gathered [17]. This information is very variegated and has different dimensions: instructions vs. data, static vs. dynamic information, etc. We are interested in the structure and contents of algorithm visualizations, therefore program visualizations closely related to the underlying programming paradigm are not considered here. Some surveys for specific programming paradigms can be found elsewhere [18, 19].

Detailed explanations of specific, well-designed animations can be very valuable to the interested instructor, e.g. as found in [20, 21]. However, we seek general advice, hints or guidelines about how to design visualizations. A few authors have given some guidance on the display of recursive algorithms. Stern and Naish [22] present a classification of recursive algorithms into three classes or recursive algorithms and they present visualizations that, according to their experience, are adequate for each class. Their classification is based on the manipulation operations performed on the data structures. Velázquez-Iturbide et al. [23] present and justify three graphical representations, which are implemented in the SRec system, for divide and conquer algorithms. The proposals combine in different ways the algorithm flow of control and the main data structure.

We also find some works addressing the structure of visualizations. Structuring takes several forms, which we may classify into: temporal structuring of animations, levels of abstraction, and multiple visualizations.

Temporal structure determines the events to display in an animation. Gloor [14] argues for giving a relevant role to the logical events of each algorithm. This relevant role may even be implemented as particular animation buttons. Velázquez-Iturbide et al. [24] present a detailed "instructor's guide" to identify the states of an algorithm execution that might be displayed. Their proposal was based on two classes of empirical studies conducted: analysis of existing animations and evaluation of animations by expert instructors. The resulting guide gives specific recommendations on the following issues: number of animations to illustrate a specific algorithm, structure and size of an animation, and input data values to be used for an animation.

A few systems have structured their visualizations at different levels of abstraction. In the HalVis system [25], every algorithm supported is animated at three levels of abstraction: an everyday metaphor of the algorithm, an animation of an algorithm's code, and an animation involving large input data. The "Algorithms in Action" system [26] supports up to three levels of pseudocode, each one coordinated with a visualization adapted to its degree of abstraction. The GreedEx system [27] distinguishes up to four levels of abstraction in the experimentation with and comparison of optimization algorithms. The levels of abstraction depend on considering one or several input data sets, and one or several simultaneous algorithms.

Finally, it is also common to provide multiple simultaneous visualizations. They may take different forms, depending on the factor for which several visualizations are simultaneously displayed [28]. The most common form is the simultaneous display of several states for the same algorithm (i.e. an animation). However, other forms of simultaneity also are relatively common: different algorithms, different views of the same algorithm, and different input data for the same algorithm.

2.4 Characterization of Visualization Contents in Other Fields

Some fields provide clear and explicit criteria about how to graphically display their objects of study, some of them as well-known as graphs [29, 30] or statistics [31, 32]. In other fields, these criteria do not exist, but there are antecedents about how to find them.

The works by Maneesh Agrawala et al. [33] are a remarkable effort for as different domains as route maps or assembling instructions. Agrawala et al. propose a methodology that consists of three steps: identification of design principles, implementation of the visualization system and evaluation of the system. "Design principles" are high-level rules that describe the features of the most effective visualizations in a given domain. For fields where principles have not been explicitly stated, Agrawala et al. consider several strategies to find them out: analysis of manual visualizations, usage of previous knowledge on perception and cognition, or to conduct perception and cognition experiments. In subsequent steps, the design principles would be implemented into a visualization system and its merits would be evaluated.

3 Description of the Study

As we claimed above, the goal of our research was the identification of structure and contents features exhibited by algorithm visualizations. Several years ago we analyzed the illustrations contained [34] in textbooks for the four basic algorithm techniques, namely divide and conquer, greedy algorithms, backtracking and dynamic programming. A surprising finding was that the scarce illustrations of greedy algorithms did not seem to adhere to a common scheme, therefore we may omit this technique if we seek general principles. In addition, the study was only focused on a few features, therefore it is insufficient to identify detailed features.

The study we present in this chapter was influenced by the empirical approach by Agrawala et al. [33], cited in the previous section. Following the steps proposed by Agrawala et al., we were lucky to have available "manual visualizations" to analyze in the first phase: the illustrations contained in algorithm textbooks. We focused on figures corresponding to algorithms designed according to the design techniques divide-and-conquer, backtracking and dynamic programming. We had several reasons for this choice of "manual visualizations". Firstly, we wanted to have the opinion of instructors, rather than the constructors of visualization systems. Textbooks are high quality sources directly delivered by instructors. Secondly, we wanted to extract features that could be shared with a large community. Textbooks are more widely disseminated and used on algorithm courses than visualization systems. Thirdly, some of the best known algorithms fit one of the three selected design techniques, e.g. quicksort or Floyd's algorithm. Finally, we wanted to extract features common to many algorithms, not only features of particular algorithms. The three design techniques represent a high percentage of algorithms used in CS education.

In the three subsections we describe the gathering of illustrations, the analysis procedure conducted, and the categories identified for an effective cataloguing of illustrations.

3.1 Gathering of Illustrations

The first step consisted in the gathering of illustrations. Most of the textbooks selected were prestigious books written in English, but we also analyzed two algorithm textbooks written by Spanish authors.

One researcher gathered figures for divide-and-conquer algorithms and other two researchers did the same task, several months later, for the backtracking and dynamic programming techniques. This temporal distance had an influence on the choice of textbooks selected. We analyzed 20 different books, only 8 of them being in common to the three techniques. Fourteen books were analyzed for divide and conquer [30, 35–47], and also 14 for both backtracking and dynamic programming [35, 37–39, 42, 45–53].

Each of these three researchers gathered illustrations independently. If a book had a chapter devoted to the corresponding design technique, that chapter was exhaustively analyzed. Otherwise, we looked in the term index for problems or algorithms representative of the technique (e.g. quicksort for divide and conquer). We did not argue about the adequacy of the algorithms included by any author in the different design techniques (although we had to later differentiate several classes of algorithms on analyzing the divide-and-conquer technique).

Then, each of these three researchers scanned the illustrations selected and elaborated a technical report [54–56]. Each report grouped the problems into classes; given the specific features of each design technique, these classes are different. For each problem, we collected its associated bibliographic data (e.g. chapter and pages), the problem statement, author terminology and the book illustrations. These data were also summarized in one or several tables.

3.2 Procedure for the Analysis

The analysis followed an iterative pattern. We did not start from a fixed set of categories to classify the illustrations, but they were refined as the analysis proceeded, according to the principles of grounded theory [57].

Once the illustration gathering process was finished, the fourth researcher read the reports and made an exploratory analysis of the illustrations. As a result, he wrote a working document with desirable features, illustrated with selected figures, and elaborated a set of interesting features that deserved further analysis. A second analysis was done again for the three design techniques. The classes of problems defined for each technique were revised, reducing their number, and the relationships among graphical representations, classes of problems, and illustration goals were analyzed.

After this second round, the results were shared, identifying common trends. We also identified what aspects had been analyzed for some technique but not for all of them. For the divide and conquer technique, it became evident a need to classify the algorithms collected. Two additional rounds were held where the cataloguing criteria were refined and the results of the three analyses were homogenized. In order to increase the reliability of our cataloguing, the fourth researcher met separately with each of the three first researchers, revising in pairs the three catalogues. Preliminary results after these phases have been presented elsewhere [58].

Finally, a fifth round was conducted by the fourth researcher and author of this chapter, where a final definition of cataloguing criteria was stated.

3.3   Cataloguing Criteria

Our final characterization of simple illustrations uses a number of criteria that we explain in this section as follows. We first discuss what we considered to be an illustration. Then, we present the most relevant cataloguing criteria for our analysis. Trivial cataloguing criteria, such as author or book page are not enumerated here. We only include seven criteria: kind of problem, kind of algorithm, graphical representation, learning objective, algorithmic objective, generality of the illustration, and treatment of time.

3.3.1   What an Illustration Is

An underlying issue is the definition of "illustration". Obviously, we considered that an illustration is what an author published in his/her textbook, unless it was irrelevant for our analysis (e.g. a figure containing source code). However, we found many compound illustrations and, conversely, very similar, usually consecutive illustrations. The former kind of illustrations could not be uniquely characterized, while we felt that the second kind of illustrations distorted a descriptive statistical analysis.

We faced this situation by distinguishing the concepts of illustration and simple illustration. An illustration is a figure as found in a textbook, but simple illustrations were the result of processing the illustrations as follows. If an illustration consists of several parts with different formats or objectives, we considered it a compound illustration and it was split into several simple illustrations. Conversely, several illustrations or parts of illustrations were joined into a single simple illustration in any of the following cases:

- Different cases with the same format and objective, e.g. several input examples or cases for which the algorithm made different treatments.
- Consecutive illustrations with the same format and objective, e.g. a sequence of states in a execution.
- Different illustrations which were a part of a hypothetical global illustration.

Finally, any other illustration was considered a simple illustration.

3.3.2 Kinds of Problems

Our classification was pragmatic, identifying classes of problems that "reasonably" group the problems found in every design technique. We finally used the following classification of problems:

- Game problems, e.g. the n-queens problem.
- Decision problems. They are combinatorial problems where a valid solution is sought, e.g. a contest schedule.
- Optimization problems. They are combinatorial problems where an optimal, valid solution is sought, e.g. the 0/1 knapsack problem.
- Graph problems, e.g. find a Hamiltonian cycle in a graph.
- Tree problems, e.g. the optimal binary search tree problem.
- String problems, e.g. find an optimal alignment of two sequences.
- Matrix problems, e.g. optimal chained multiplication of matrices.
- Array problems, e.g. find the maximum element of an array.
- Mathematical problems, e.g. Fibonacci numbers.
- Geometrical problems, e.g. determine the maximal elements in a set of points.

3.3.3 Kinds of Algorithms

We classified algorithms according to their underlying algorithm design technique, namely divide and conquer, backtracking or dynamic programming. However, we found that algorithms classified in the textbooks as divide-and-conquer had very different natures. We examined the illustrations and we came to the conclusion that we had to further differentiate four kinds of algorithms:

- Proper divide-and-conquer algorithms, e.g. mergesort or quicksort.
- Algorithms which are auxiliary to divide-and-conquer algorithms, e.g. merging for mergesort or splitting for quicksort.
- Lineal recursive algorithms. Algorithms commonly classified as divide-and-conquer exhibit multiple recursion. However, some authors also classify into this class of algorithms to lineal recursive algorithms that substantially reduce the size of data in successive recursive calls, e.g. binary search. Our concern here is that lineal and multiple recursions have very different behaviors and therefore the illustrations of their behavior also are very different. Consequently, we differentiated between both classes of algorithms.

- Iterative algorithms. Some authors include iterative algorithms to remark the distinguishing features of divide-and-conquer algorithms.

3.3.4 Graphical Representation

We include in this category the graphical representations used in the illustrations to display the variables and data structures relevant to understand an algorithm. Again, our approach was pragmatic, thus we distinguished as many kinds of graphical representations as necessary. A number of representations are well-known in computer science, e.g. arrays, trees or graphs. However, other representations are domain-specific, e.g. geometric diagrams or game boards.

3.3.5 Learning Objectives

Algorithm animations show the execution of an algorithm for given input data. This objective is shared with many textbooks illustrations. However, we found that textbooks authors used visualizations for additional purposes:

- Declaration. Simple problems are easy to understand, but more difficult problems require one or several examples to explain the problem statement. In the case of problems of a graphical nature, these examples are illustrated with figures. Furthermore, some problems require additional definitions of concepts which also may require illustrations, e.g. a Hamiltonian cycle in a graph.
- Design decisions. Algorithms of the three design techniques are based on some key decision. A divide-and-conquer algorithm is characterized by how a problem is divided into subproblems and how the subresults are combined into a result. The key decision in a backtracking algorithm is the organization of states in a search space. Dynamic programming algorithms use a table, with a specific number of dimensions and index values in each dimension, to store intermediate values. Furthermore, some algorithms rely on additional properties in the problem domain. All of these design decisions are sometimes easier to explain with the assistance of a figure.
- Analysis of an algorithm. An algorithm can be analyzed with respect to different criteria, such as correctness or efficiency. Again, a figure may assist in understanding the property under study.

3.3.6 Algorithmic Objectives

The learning objective of an illustration gives us information about the intents of the textbook author. Given this main objective, we sometimes want to be more specific about the particular issue that is being illustrated. We call algorithmic objective to this objective.

A rich set of algorithmic objectives are found in figures illustrating declarations. We find the following set of cases:

- An input of the algorithm. For some abstract data types, it is common to provide an abstract representation of input, e.g. a graph, and its implementation, e.g. an adjacency matrix. For decision problems, it is common not only to give examples, but also to give examples for which no solution exists.
- An output. For many problems, the problem statement determines uniquely the output for a given input. However, in combinatorial problems we find other situations which are sensible with respect to potential solutions: one valid or invalid solution, one partial solution, an optimal solution, and several or all the valid solutions.
- One input and its corresponding output.

Similarly, different illustrations of the design of divide-and-conquer algorithms show different parts of their inductive definition:

- Partition of the problem into subproblems.
- Partition into subproblems and recursive solving of them.
- Combination of solutions to the subproblems (i.e. subsolutions) into a global solution.
- Partition into subproblems and combination of subsolutions.
- Recursive solving of subproblems and combination of subsolutions.
- The three components of partition, recursion and combination.

The list of specific objectives is large, so we do not include here the full list.

3.3.7 Generality of the Illustration

There are different degrees of generality in an illustration:

- Generic illustration. It shows a general formulation or case, often in terms of some variable n, e.g. an array of n cells.
- Semigeneric illustration. It shows a general case of a specific size, e.g. an array of length 5 but undefined contents.
- Simplified illustration. It shows a particular case, but with some of its values removed. This is typical in geometric diagrams, where specific points are plotted in a plane, but their coordinates are omitted.
- Concrete illustration. It shows a particular case.

3.3.8 Treatment of Time

The treatment of time is very important in algorithm visualizations. In fact, algorithm visualization systems always provide dynamic visualizations (i.e. animations).

This is in contrast to visualization in other fields where other kinds of interaction are more important [59].

We could think that illustrations of textbooks, given their static nature, have very few ways of dealing with time. However, there is a great diversity of representations:

- One, two or, in general, a sequence of states of the algorithm execution. Each state is represented with the values of a given set of variables.
- Overlapping of states. Sometimes, several states can be collapsed into a single illustration where the changes in the algorithm state are represented with visual effects, e.g. highlighting. Sometimes, we also find a combination of sequence and overlapping consecutive states. In this case, overlapping acts as a kind of "common factor" to collapse several consecutive states into one.
- An illustration representing the history of the algorithm execution. A well-known example is recursion trees, which represent all the recursive calls invoked by an algorithm for given input data. Other history representations are dependence graphs (in multiple, redundant recursive algorithms) or search trees (in backtracking algorithms).

We should note that it does not make much sense to speak about time for some kinds of illustrations. For instance, a definition often has a static nature. In these cases, we discard the respective illustrations from our analysis.

4 Results

In this section we present the results of our analysis. For the sake of brevity, we only include in detail the results of analyzing divide-and-conquer algorithms. For the other two algorithm design techniques, we show some specific, interesting results.

4.1 Results for Divide-and-Conquer Algorithms

We found 19 problems to be solved by divide-and-conquer in the 14 textbooks analyzed, 3 of them without any illustration. For the remaining 16 problems, we found 62 illustrations. We found convenient to divide 4 illustrations into two simple illustrations each, and to group 13 illustrations into 4 simple illustrations. Consequently, we obtained 57 simple illustrations.

A first surprising fact was the high diversity of illustrations for this technique. More detailed analysis showed that it was partly due to the fact that a number of different algorithms were being illustrated under this epigraph, as explained in Sect. 3.3.3. Table 1 shows the number and percentage of illustrations of each kind. In this study we only focus on the 40 simple illustrations corresponding to the two first kinds of algorithms, which truly correspond to the divide-and-conquer technique.

Table 1 Kinds of illustrations under the epigraph "divide and conquer"

Design technique	# illustrations	% illustrations
Divide and conquer	38	67%
Divide and conquer, and auxiliary algorithm	2	4%
Auxiliary algorithm	6	11%
Lineal recursive algorithm	9	16%
Iterative algorithm	2	4%

Table 2 Kinds of problems illustrated and designed with the divide-and-conquer technique

Kinds of problems	# illustrations	% illustrations
Mathematical	3	8%
Over arrays	16	40%
Games	3	8%
Geometric	17	43%
Decision	1	3%

Table 3 Kinds of learning objectives of the illustrations

Learning objectives	# illustrations	% illustrations
Declaration	7	18%
Design	13	33%
Execution	19	48%
Analysis	1	3%

The 40 illustrations considered correspond to 12 problems of five classes, as shown in Table 2.

Notice that most illustrations correspond to array or geometric problems. In the rest of the section, global analyses will be made using all the illustrations, but those analysis related to the kind of problems will only focus on these two classes of problems, given that the coverage of other kinds of problems is marginal.

There is a clear relationship between the kind of problem and the graphical representation of the algorithm state. Mathematical problems solved by divide and conquer represent numbers as arrays, problems over arrays display arrays, games show boards, geometric problems use geometric diagrams, and the decision problem displays a table.

The figures illustrate different learning objectives, as shown in Table 3.

Notice that most illustrations show either some key aspect of the algorithm design or an execution of the algorithm.

If we analyze the learning objectives by kinds of problems, we find that most algorithms defined over arrays trace an execution ($14/16 = 87\%$). However, geometric algorithms are more diverse since most figures illustrate a design ($9/17 = 53\%$), but we also find a number of figures for declarations ($5/17 = 19\%$) or executions ($3/17 = 18\%$).

An analysis of algorithmic objectives is contained in Table 4.

Relating Tables 2, 3, and 4 allow knowing the educational intended usage of these algorithmic objectives. We find five input/out specifications, two concept definitions, and five auxiliary properties that are used in the design of some algorithm.

Table 4 Kinds of algorithmic objectives of the illustrations

Algorithmic objectives	# illustrations	% illustrations
Input/output specification	5	13%
Concept or property definition	7	18%
Algorithm main steps	27	68%
Complexity analysis	1	3%

Table 5 Kinds of algorithmic key steps explained in the illustrations

Kind of key step	# illustrations	% illustrations
Partition	4	15%
Partition and combination	3	11%
Partition and recursion	3	11%
Recursion and combination	1	4%
Partition, recursion and combination	16	59%

Table 6 Kinds of generality of the illustrations

Generality	# illustrations	% illustrations
Generic	12	30%
Semigeneric	2	5%
Simplified	15	38%
Concrete	11	28%

Four of the five input/output specifications, and all the concept definitions and property statements are found in geometric problems.

Notice that the majority of illustrations (all the executions and 8 design decisions) show the key steps of divide-and-conquer algorithms. If we analyze the specific key steps shown in these figures, we obtain the results of Table 5. Many illustrations show partial parts of the divide-and-conquer principle but the majority $(16/27 = 59\%)$ show the three key steps.

An analysis of generality is included in Table 6.

Notice that the different degrees of generality coexist. If we relate generality to the previous fields, we find the following. Most illustrations targeted to declarations contain simplified cases $(6/7 = 87\%)$. Illustrations of design are mostly generic $(10/13 = 77\%)$ and they never include semigeneric or concrete cases. Illustrations of executions mostly refer to concrete cases $(14/19 = 74\%)$ and never contain generic cases.

If we relate this category to the kind of problem, we find that most illustrations of array algorithms contain concrete instances $(11/16 = 69\%)$. However, most geometric illustrations contain simplified cases $(10/17 = 59\%)$.

Let us consider now the treatment of time, see Table 7.

Notice the high diversity of representations of time. In the seven figures catalogued as 'others', it does not make sense trying to identify treatment of time. This is the case, for instance, of input examples or definitions.

We restrict now to the first five treatments of time in the table, i.e. to 33 illustrations. For illustrations of design, the most outstanding treatment of time

Table 7 Treatment of time in the illustrations

Time	# illustrations	% illustrations
One or two states	5	13%
Sequence of states	5	13%
Overlapping states	10	25%
Sequence and overlapping of states	1	3%
Recursion tree	12	30%
Others	7	18%

is overlapping states (6/7 = 86%). However, executions are more often illustrated with recursion trees (11/19 = 58%) and also with sequences of states (6/19 = 32%). Sequences of states and all the recursion trees but one are used for executions.

If we refer to the kind of problem, most illustrations for array problems are recursion trees (12/16 = 75%) and almost all the illustrations of geometric problems display an overlapping (10/11 = 91%).

It is interesting to notice that almost all the recursion trees depicted (11/12 = 92%) are complete, and the only recursion tree partially displayed shows the higher levels of the tree for a visual complexity analysis. Two thirds of the sequences of states are displayed complete (4/6 = 67%), while one third are displayed abbreviated to the key algorithmic steps, i.e. they mirror the inductive definition of the algorithm.

From the findings given in the previous two paragraphs, notice that most illustrations of array algorithms contain a complete trace of an execution (either as a recursion tree or as a sequence of states) while most illustrations of geometric algorithms mirror the inductive definition of the algorithm (either as a partial trace or as a display of the definition).

4.2 Results for Other Algorithm Design Techniques

We gathered 55 illustrations of the backtracking technique for 23 problems. Seven of them were double, thus the total number of simple illustrations is 62. We also gathered 132 illustrations for 36 problems to be solved by dynamic programming. Some of them were decomposed, resulting in 148 simple illustrations.

Our analysis of illustrations of backtracking or dynamic programming algorithms shows in many cases trends similar to divide-and-conquer algorithms. Therefore, we do not include here detailed results, but we only focus on the findings that give us a wider vision. In order to focus the presentation of the results, they are shown in parallel for the two algorithm design techniques.

Let us start noting that the kinds of problems solved vary much with the algorithm design technique. Although all the problems are combinatorial or optimization problems, some of them can be more easily characterized by their domain (e.g.

Table 8 Kinds of problems illustrated and designed with the backtracking and dynamic programming techniques

Kinds of problems	Backtracking		Dynamic programming	
	# illustrations	% illustrations	# illustrations	% illustrations
Games	25	40%	—	—
Graphs	18	29%	34	25%
Strings	1	2%	14	10%
Trees	—	—	25	18%
Networks	—	—	3	2%
Mathematical	—	—	16	12%
Other decision problems	10	16%	1	1%
Other optimization problems	8	13%	44	32%
Total	**62**	**100%**	**137**	**100%**

Table 9 Kinds of learning objectives of the three algorithm design techniques

Learning objectives	% divide and conquer	# backtracking	% dynamic programming
Declaration	18%	32%	32%
Design	33%	42%	24%
Execution	48%	26%	44%
Analysis	3%	0%	0%

games) or main data structure (e.g. graphs). Table 8 shows the kinds of problems found in the illustrations of backtracking and dynamic programming.

We notice a great variation in the learning objectives of the three algorithm design techniques, see Table 9. Notice the increase in the percentage of illustrations devoted to explain declarations and. In backtracking algorithms, there is an increase of illustrations of the algorithm design and a decrease in the percentage of execution figures.

These percentages also vary for each kind of problem. Thus, the percentage of declarations is increased in some kinds of problems up to 56% (i.e. graphs solved by backtracking), whereas in other kinds of problems declarations illustrations even disappear (e.g. other decision or optimization problems).

In dynamic programming algorithms, we found that many graph algorithms combined abstract and concrete representations of graphs. This is due to the fact that the adjacency matrix is used itself as the table for the dynamic programming algorithm.

The most common algorithmic objective still is the illustration of the algorithm key steps. However, it is now performed almost exclusively with either search trees (in backtracking algorithms) or tables (in dynamic programming). For some kinds of problems (e.g. other decision or optimization problems in backtracking), search trees are the only graphical representation used.

We also notice an increase in the percentage of concrete examples and in the simplification of the representation of states (up to 47% in backtracking). This is very common in search trees, where the shape of the tree is remarked by removing the state associated to each node or by substituting it with a label (e.g. the order of visit in the search process).

Simplification also is used with respect to time. The percentage of simplified search trees is one third. It is also interesting to note that search trees are displayed either complete or simplified for design goals, but they are almost exclusively displayed complete to illustrate an execution. Many of these simplifications resemble the effect of focus + context techniques in information visualization [60].

5   Discussion

There are two important issues in the illustrations analysis presented above: the final set of cataloguing criteria, and the analysis results. Both are briefly discussed in this section.

5.1   Cataloguing Criteria

The cataloguing criteria were proposed and refined in the successive rounds of analysis conducted. Some of them did not change much or they changed in not very relevant directions, e.g. kinds of problems or graphical representations. However, other criteria suffered more drastic changes:

- Kinds of algorithms. The need to differentiate several classes of algorithms under the label "divide and conquer" raises some issues, mainly that of deciding whether lineal recursive algorithms are truly divide-and-conquer algorithms.
- Learning objectives. The obvious use of visualizations to illustrate algorithm executions has been widened to other educational uses, mainly input/output problem specifications, auxiliary definitions, and design decisions. This brings two important considerations. The first one is which of these learning objectives can be supported by visualization systems and, in the affirmative case, how and with what restrictions. The second consideration is the blurred frontier between some of these categories. In order to catalogue adequately, it is important to take into consideration the context of the illustration. For instance, an initial and a final state may be used for an input/output specification in a context where the problem is stated and no algorithm has been identified yet. However, in the context of an algorithm, they may represent two states of execution. Similar considerations apply to the frontier between design and execution.
- Generality of the illustration. Some values of this cataloguing criterion are related to the learning objectives. Thus, we have shown that illustrations intended for design usually use generic values while illustrations intended for executions

usually use concrete values. The domain also is important, especially for the display of simplified input data.

- Treatment of time. This was the most surprising criterion to us. Although in previous works, we were aware of the importance of time for algorithm visualization [28], we did not consider the need to state it as an explicit criterion until the last analysis round. We expected to find recursion trees and also sequences of states, but it also was surprising to us the extensive use of overlapping states.

We consider that the identification of a set of criteria that successfully allow characterizing visualizations is an important contribution itself. They can be used (and perhaps refined) in future analyses of illustrations or visualizations.

We examined the aspects of illustrations that we considered the most relevant in relation to their contents. However, in retrospective, the most important criteria found do not strictly correspond to those characteristics of the Contents and Form categories of Price et al.'s taxonomy [15] that we had anticipated (see Sect. 2.2). Thus, algorithmic objective, generality of the illustration and treatment of time are related to Contents, and graphical representation clearly fits the Form category. However, kind of problem and of algorithm belong to the Scope category, and learning objective to the Effectiveness category. Although they were not strictly related to contents, they allowed creating a context for our analysis.

An issue that we did not address is the identification of useful interaction actions other than animation. The common appearance of simplified data suggests a need for filtering facilities. This and other interactions may prove to be very useful for a flexible visualization system, as shown in the SRec system [61].

5.2 Main Trends

The results shown in Sect. 4 still do not allow us to state clear design principles for algorithm visualizations but do allow identifying trends:

- The problem domain imposes restrictions on the graphical representations. Problems that can be easily represented with formats common in computer science may benefit from them (e.g. array problems). However, other problems require custom visualization formats (e.g. geometric problems).
- For nontrivial problems, it is important to give examples of the input/output problem specification. This is especially important for problems of a visual nature, such as graph problems. For decision or optimization problems, the range of useful solutions is widened to consider input values without a solution and also invalid and suboptimal solutions. Furthermore, it is common for some data structures (e.g. graphs) to show several representations with different degree of abstraction.

- Definitions, properties and analysis of algorithm properties also are common. For less familiar domains, such as geometric problems, the two former classes of learning goals are necessary.
- The blurred distinction between design and execution may open a door to the use of execution visualizations to illustrate design decisions.
- Complex problems, involving large data or many execution states, present the problem of the representation scale. Consequently, visualizations of design are preferred over visualizations of execution. This problem also is solved by displaying fewer execution states or by simplifying design illustrations.
- The most obvious and universal representation of time is as a sequence of states (as in conventional animations), but the use of overlapping and history representations (e.g. recursion or search trees) must be considered for some domains or design techniques.

6   Conclusions and Future Work

We have conducted an analysis of illustrations contained in textbooks regarding three basic algorithm design techniques. The chapter contains two main contributions. Firstly, a set of criteria are identified for the characterization of illustrations. Secondly, the analysis of textbook illustrations has allowed us to outline some trends in their design. However, the current state of our research is not as advanced as to state design principles. Currently, we have found common trends or even guidelines, but their formulation as design principles deserves further study.

In the short term, we plan to gather some more illustrations about the divide-and-conquer design technique, given its higher variability. The analysis presented here will be updated with these figures and a first draft of design principles will be delivered. In a second phase, we plan to gather illustrations of other design techniques, such as greedy or approximate algorithms. This new set of illustrations will be used as a benchmark for testing the design principles drafted and we hope that will allow us to state a set of design principles.

Acknowledgments I want to thank Natalia Esteban-Sánchez, Antonio Pérez-Carrasco and Belén Sáenz-Rubio for their collaboration in previous phases of this research. This work was supported by research grant TIN2011-29542-C02-01 of the Spanish Ministry of Economy and Competitiveness.

References

1. S. Fincher and M. Petre. *Computer Science Education Research*. London, UK: Routledge, 2004.
2. R. Lister. The naughties in CSEd research: A retrospective. In *SIGCSE Inroads*, volume 1, no. 1, pp. 22–24, March 2010.

3. C. Hundhausen, S. Douglas and J. Stasko. A meta-study of algorithm visualization effectiveness. In *Journal of Visual Languages and Computing*, volume 13, no. 3, pp. 259–290, June 2002.

4. T. Naps, G. Roessling, V. Almstrum, W. Dann, R. Fleischer, C. Hundhausen, A. Korhonen, L. Malmi, M. McNally, S. Rodger and J. Á. Velázquez-Iturbide. Exploring the role of visualization and engagement in computer science education. In *SIGCSE Bulletin*, volume 35, no. 2, pp. 131–152, June 2003.

5. M. Ben-Ari. Constructivism in computer science education. In *Journal of Computers in Mathematics and Science Teaching*, volume 20, no. 1, pp. 45–73, 2001.

6. E. Wenger. *Intelligent Tutoring Systems*. Los Altos, CA: Morgan Kaufmann, 1987.

7. T. L. Naps, G. Roessling, J. Anderson, S. Cooper, W. Dann, R. Fleischer, B. Koldehofe, A. Korhonen, M. Kuittinen, C. Leska, L. Malmi, M. McNally, J. Rantakokko and R. J. Ross. Evaluating the educational impact of visualization. In *SIGCSE Bulletin*, volume 35, no. 4, pp. 124–136, December 2003.

8. P. Ihantola, V. Karavirta, A. Korhonen and J. Nikander. Taxonomy of effortless creation of algorithm visualization. *Proceedings of the 2005 International Workshop on Computing Education Research, ICER 2005*, pp. 123–133.

9. R. Ben-Bassat Levy and M. Ben-Ari. Perceived behavior control and its influence on the adoption of software tools. In *Proceedings of the 13th Annual Conference on Innovation and Technology in Computer Science education, ITiCSE 2008*, pp. 169–173.

10. P. Saraiya, C. A. Shaffer, D. S. McCrickard and C. North. Effective features of algorithm visualizations. In *Proceedings of the 35th Technical Symposium on Computer Science Education, SIGCSE 2004*, pp. 382–386.

11. D. Norman. Some observations on mental models. In *Mental Models*, D. Gentner and A. Stevens, Eds. Hillsdale, NJ: Erlbaum, 1983, pp. 7–14.

12. G. Roessling and T. L. Naps. A testbed for pedagogical requirements in algorithm visualizations. In *Proceedings of the 7th Annual Conference on Innovation and Technology in Computer Science education, ITiCSE 2002*, pp. 96–100.

13. S. Pollack and M. Ben-Ari. Selecting a visualization system. In *Proceedings of the Third Program Visualization Workshop, PVW 2004*, pp. 134–140.

14. P. A. Gloor. User interface issues for algorithm animation. In *Software Visualization*, J. Stasko, J. Domingue, M. H. Brown and B. A. Price, Eds. Cambridge, MA: MIT Press. 1998, pp. 145–152.

15. B. A. Price, R. Baecker and I. Small. An introduction to software visualization. In *Software Visualization*, J. Stasko, J. Domingue, M. H. Brown and B. A. Price, Eds. Cambridge, MA: MIT Press. 1998, pp. 3–27.

16. V. Karavirta, A. Korhonen, L. Malmi and T. Naps. A comprehensive taxonomy of algorithm animation languages. In *Journal of Visual Languages and Computing*, volume 21, pp. 1–22, 2010.

17. M. H. Brown. A taxonomy of algorithm animation displays. In *Software Visualization*, J. Stasko, J. Domingue, M. H. Brown and B. A. Price, Eds. Cambridge, MA: MIT Press. 1998, pp. 35–42.

18. J. Urquiza-Fuentes and J. Á. Velázquez-Iturbide. A survey of program visualizations for the functional paradigm. In *Proceedings of the Third Program Visualization Workshop, PVW 2004*, pp. 2–9.

19. P. Romero, R. Cox, B. du Boulay and R. Lutz. A survey of external representations employed in object-oriented programming environments. In *Journal of Visual Languages and Computing*, volume 14, pp. 387–419, 2003.

20. M. H. Brown and J. Hershberger. Fundamental techniques for algorithm animation displays. In *Software Visualization*, J. Stasko, J. Domingue, M. H. Brown and B. A. Price, Eds. Cambridge, MA: MIT Press. 1998, pp. 81–89.

21. M. H. Brown and M. A. Najork. Algorithm animation using interactive 3D graphics. In *Software Visualization*, J. Stasko, J. Domingue, M. H. Brown and B. A. Price, Eds. Cambridge, MA: MIT Press. 1998, pp. 119–135.

22. L. Stern and L. Naish. Visual representations for recursive algorithms. In *Proceedings of the 33th SIGCSE Technical Symposium on Science Education, SIGCSE 2002*, pp. 196–200.
23. J. Á. Velázquez-Iturbide, A. Pérez-Carrasco and J. Urquiza-Fuentes. A design of automatic visualizations for divide-and-conquer algorithms. In *Electronic Notes in Theoretical Computer Science*, no. 224, pp. 113–120, January 2009.
24. J. Á. Velázquez-Iturbide, D. Redondo-Martín, C. Pareja-Flores and J. Urquiza-Fuentes. An instructor's guide to design web-based algorithm animations. In *Advances in Web-Based Learning – ICWL 2007*, LNCS 4823, Springer-Verlag. 2008, pp. 440–451.
25. S. Hansen, D. Schrimpsher and N. H. Narayanan. Designing educationally effective algorithm animations. In *Journal of Visual Languages and Computing*, volume 13, pp. 291–317, 2002.
26. L. Stern, H. Sondergaard and L. Naish. A strategy for managing content complexity in algorithm animation. In *Proceedings of the 4th Annual Conference on Innovation and Technology in Computer Science Education, ITiCSE 1999*, pp. 127–130.
27. J. Á. Velázquez-Iturbide, O. Debdi, N. Esteban-Sánchez and C. Pizarro. GreedEx: A visualization tool for experimentation and discovery learning of greedy algorithms. In *IEEE Transactions on Learning Technologies*, in press.
28. J. Á. Velázquez-Iturbide. Characterizing time and interaction in a space of software visualizations. In *Proceedings of the Sixth Program Visualization Workshop, PVW 2011*, pp. 43–51.
29. K. Sugiyama. *Graph Drawing and Applications for Software and Knowledge Engineers*. Singapore: World Scientific, 2002.
30. P. di Batista, G. Eades, T. Tamassia and I. Tollis. *Graph Drawing: Algorithms for the Visualization of Graphs*. Prentice-Hall, 1999.
31. J. Bertin. *Semiology of Graphics*. Madison, WI: University of Wisconsin Press, 1983.
32. W. Cleveland. *Visualizing Data*. Mummit, NJ: Hobart Press, 1993.
33. M. Agrawala, W. Li and F. Berthouzoz. Design principles for visual communication. In *Communications of the ACM*, volume 54, no. 4, pp. 60–69, April 2011.
34. L. Fernández-Muñoz and J. Á. Velázquez-Iturbide. Estudio sobre la visualización de las técnicas de diseño de algoritmos. In *Interacción'06: Actas del VII Congreso Internacional de Interacción Persona-Ordenador*, pp. 315–324.
35. M. H. Alsuwaiyel. *Algorithms, Design Techniques and Analysis*. World Scientific, 1999.
36. G. Brassard and P. Bratley. *Algorithmics: Theory and Practice*. Prentice-Hall, 1988.
37. G. Brassard and P. Bratley. *Fundamentals of Algorithmics*. Prentice-Hall, 1996.
38. T. H. Cormen, C. E. Leiserson and R. L. Rivest. *Introduction to algorithms*. Cambridge, MA: MIT Press, 2nd ed., 2001.
39. J. Gonzalo-Arroyo and M. Rodríguez-Artacho. *Esquemas algorítmicos: enfoque metodológico y problemas resueltos*. Madrid, Spain: Universidad Nacional de Educación a Distancia, 1997.
40. R. Johnsonbaugh and M. Schaefer. *Algorithms*. Pearson Education, 2004.
41. J. Kleinberg and É. Tardos. *Algorithm Design*. Pearson Addison-Wesley, 2006.
42. R.C.T., Lee, S.S., Tseng, R.C., Chang and Y.T. Tsai. *Introduction to the Design and Analysis of Algorithms*. Singapore: McGraw-Hill, 2005.
43. A. Levitin. *The Design of Analysis of Algorithms*. Addison-Wesley. 2003.
44. N. Martí-Oliet and Ortega and J. A. Verdejo. *Estructuras de datos y métodos algorítmicos ejercicios resueltos*. Madrid, Spain: Pearson, 2004.
45. I. Parberry. *Problems on Algorithms*. Prentice-Hall, 2002.
46. S. Sahni. *Data Structures, Algorithms and Applications in Java*. Summit, NJ: Silicon Press, 2005.
47. M. A. Weiss. *Data Structures and Algorithms Analysis*. Addison-Wesley, 1992.
48. S. Baase and A. Van Gelder. *Computer Algorithms: Introduction to Design and Analysis*. Addison-Wesley Longman, 2000.
49. M. T. Goodrich and R. Tamassia. *Data Structures and Algorithms in Java*. John Wiley & Sons, 2nd ed., 2001.
50. E. Horowitz and S. Sahni. *Fundamentals of Computer Algorithms*. Pitman, 1978.
51. R. Neapolitan and K. Naimipour. *Foundations of Algorithms*. Jones and Bartlett, 1997.
52. R. Sedgewick. *Algorithms in Java*. Addison-Wesley, 2002.

53. S. Skiena. *The Algorithm Design Manual*. Berlin, Germany: Springer-Verlag, 1998.
54. A. Pérez-Carrasco, J. Á. Velázquez-Iturbide and F. Almeida-Martínez. Revisión bibliográfica de la representación de problemas de la técnica «divide y vencerás». In *Serie de Informes Técnicos DLSI1-URJC*, Universidad Rey Juan Carlos, Spain, no. 2012–02, 2012.
55. N. Esteban-Sánchez and J. Á. Velázquez-Iturbide. Revisión bibliográfica de problemas resolubles por la técnica de vuelta atrás. In *Serie de Informes Técnicos DLSI1-URJC*, Universidad Rey Juan Carlos, Spain, no. 2012–03, 2012.
56. B. Sáenz-Rubio and J. Á. Velázquez-Iturbide. Revisión bibliográfica de algoritmos de programación dinámica. In *Serie de Informes Técnicos DLSI1-URJC*, Universidad Rey Juan Carlos, Spain, no. 2012–04, 2012.
57. B. Glaser and A. Strauss. *The Discovery of Grounded Theory: Strategies for Qualitative Research*. Aldine, 1967.
58. N. Esteban Sánchez, A. Pérez Carrasco, B. Sáenz Rubio and J. Á. Velázquez Iturbide. Towards the identification of graphical principles for visualizing algorithm design techniques. In *Proceedings of the 2012 International Symposium on Computers in Education, SIIE 2012*, 5 pp.
59. J.S. Yi, a. Kang, J.T. Stasko and J.A. Jacko. Toward a deeper understanding of the role of interaction in information visualization. In *IEEE Transactions on Visualization and Computer Graphics*, volume 13, no. 6, pp. 1.224—1.231, November/December 2007.
60. Y. K. Leung and M. D. Apperley. A review and taxonomy of distortion-oriented presentation techniques. In *ACM Transactions on Computer-Human Interaction*, volume 1, no. 2, pp. 126–160. June 1994.
61. J. Á. Velázquez-Iturbide and A. Pérez-Carrasco. InfoVis interaction techniques in animation of recursive programs. In *Algorithms*, volume 3, no. 1, pp. 76–91, March 2010.

Part IV
Methods

Conceptual Design for Sensemaking

Ann Blandford, Sarah Faisal, and Simon Attfield

Abstract The focus of sensemaking research is often on process and resources such as "schemas" and "frames". Less attention has been paid to the conceptual structures that make up the schema or frame, or how visualisations can be designed to support users' conceptual structures. In this chapter, we present an approach to gathering user requirements based on the conceptual structures that people are working with when making sense of a domain. We illustrate the approach with examples drawn from our own experience of designing, prototyping and testing an interactive visualisation tool for making sense of academic literature and of studies of sensemaking by lawyers and journalists. We discuss how to move from requirements to design, drawing on a classification of visualisations that highlights their principal conceptual structuring basis. Since each individual *makes sense* in their own way, it is beneficial to include features that enable people to work with a representation in their own way; for this, appropriation tools are helpful. We discuss the design of such features. Finally, we present an approach to evaluating interactive visualisations in terms of their support for sensemaking, focusing on the quality of the fit between users and system.

1 Introduction

One of the important roles for visualisations is to support sensemaking, as people can often assimilate information much more rapidly through visualisations than through text [39]. In designing visualisations for sensemaking, however, relatively

A. Blandford (✉) • S. Faisal
UCL Interaction Centre, University College London, Gower Street, London, WC1E 6BT, UK
e-mail: a.blandford@ucl.ac.uk; sarah.fba@gmail.com

S. Attfield
School of Engineering and Information Sciences, Middlesex University, The Burroughs,
Hendon, London, NW4 4BT, UK
e-mail: S.Attfield@mdx.ac.uk

W. Huang (ed.), *Handbook of Human Centric Visualization*,   253
DOI 10.1007/978-1-4614-7485-2_10, © Springer Science+Business Media New York 2014

little attention has been paid to users' conceptual structures, or how visualisations can be designed to support those structures. Similarly, sensemaking research and practice has largely been on process and resources: on the pattern of activities involved in sensemaking and on the kinds of intermediate representations that people typically work with. The focus of this chapter is on how an understanding of users' conceptual structures can inform the design and evaluation of visualisations to support sensemaking.

We present an approach to gathering user requirements based on constructing an understanding of the conceptual structures that people are working with when making sense of a domain. In the following sections, we discuss how to gather user requirements for sensemaking visualisations, how to design suitable visualisations, and then how to evaluate those visualisations with users. This is grounded in our own experience of designing, prototyping and testing ALVi, an interactive visualisation tool for making sense of academic literature [17, 18] and of studies of sensemaking by lawyers and journalists [3].

2 Background

Information visualisations are tools that interactively generate and display visual representations of abstract domains. As users interact with the tool, "ah HA!" moments arise [33]. Recognising the power of externalisations, Ware [39] proposes that visualisation has moved out of the mind and onto the computer screen. This may be true from the designers' perspective, but for users, sensemaking is very much in the mind. As users interact with the externalisations, internal conceptualisations of the domain are created, updated and used [30]. The challenge for design is to create interactive visualisations that really support their users in making sense of the domain in question.

When it comes to the design of information visualisation tools, guidance exists for determining the visual attribute (e.g., color, shape, or size) that best communicates specific types of information (e.g., variance, type and extent) [5, 37]. There are also guidelines such as Shneiderman's [31] visual information-seeking mantra and its associated design guidelines: overview, zoom and filter, details on demand, view relationships, history and extract. These address the functionalities supporting the exploration of the visualisation. In addition, Craft and Cairns [12] list resources that information visualisation designers often rely on, namely design examples, taxonomies, guidelines and reference models. When it comes to designing the visualisation as a whole, though, there is little clear guidance on how to create the best possible conceptual structure; this is the focus of this chapter.

The core conceptual structure of the design is the same for all users, but there is also scope for creating individualisation features to accommodate individual differences. A further angle to our story is that not every user of a visualisation understands it or uses it in exactly the same way, so it is also important to support these individual differences as far as possible. The approach that we propose

Fig. 1 The Information
Journey

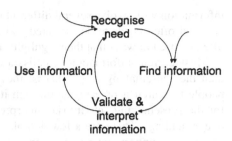

for doing this is by creating tools for appropriation, allowing the user to 'take ownership' of the visualisation to make it work for their own purposes.

To set the scene for this chapter, we briefly summarise relevant background work on sensemaking, on conceptual structures, and on appropriation.

2.1 Sensemaking

Sensemaking is generally regarded as a process through which an individual makes sense, or constructs understanding, of a domain or topic. The literature on sensemaking locates it within active information seeking and interpretation. Pirolli and Card [29] describe it as involving two interconnected loops of activity: information foraging and sensemaking. The foraging loop involves seeking, filtering and extracting information, while the sensemaking loop involves constructing a conceptualisation that best fits the evidence at hand. This process is individual and subjective. Using different language to describe very similar ideas, Klein et al. [24] explain sensemaking in terms of a data-frame theory. In their view, when engaged in sensemaking, people explain elements by fitting them into a frame which links them to other elements that have resulted from their past experiences. These representations, whether referred to as frames [24] or schemas [30], are subjective lenses through which people view, filter and structure the data.

The classical views of sensemaking have focused on the finding, organisation and synthesis of information, and somewhat overlooked the use of the resulting understanding—for example, in preparing a report or making a diagnosis. Blandford and Attfield [7] extend this to account (at least at a high level) for the use that is made of the information, through an Information Journey (see Fig. 1). An information journey typically starts with either identifying a need (a gap in knowledge) or encountering some information that addresses a latent need or interest. When working with a visualisation tool, the tool use might be instigated by a recognised need, or the user might come across some information (or construct some understanding) that makes them want to know more. When an explicit need has been identified, a way to address that need has to be determined and acted on. This might involve asking the person at the next desk, looking "in the world" or interacting with a visualisation tool; very often, for a complex need, finding information will involve several different resources and activities. Information that is found often needs to be validated and interpreted (made sense of). And it will often be used in ways that then highlight further information needs.

Most current information visualisation tools address the foraging loop more than the sensemaking loop—that is: they make information available, and allow people to search it or navigate through it, but they provide little explicit support for the sensemaking activity (i.e. interpretation of information relative to current understanding). There are a few notable exceptions to this, including Jigsaw [34, 35], which supports investigative analysis in fields such as law enforcement and intelligence, ALVi [17, 18], an academic literature visualisation tool, and Aruvi [32], a scatterplot visualisation that supports analytical reasoning.

Jigsaw [34, 35] supports investigative analysts making sense of collections of documents that comprise evidence in an investigation. The design is based on the sensemaking literature (e.g. [29]) and has a focus on explicitly addressing user requirements. Jigsaw uses information extraction to identify entities in raw texts and then show connections between entities across documents. Users can view chronologies, relationship diagrams and groupings of information. An evaluative study of investigative analysis [23], which included Jigsaw as the most sophisticated of four analysis tools, provides some evidence that Jigsaw is effective, but the focus of the study was on the investigative process (for which the tools provided support) rather than on explicitly evaluating Jigsaw.

ALVi is an interactive visualisation of academic literature (see Fig. 2 for an example screen) that is designed to support researchers in making sense of their literature domain [16]. The user can explore the literature using four linked views that focus on authors, author citations, publications, and publication citations. Users are also able to highlight items using a marking tool. The prototype has been implemented using the infoViz '04 dataset [19], and evaluated with researchers with an interest in information visualisation [18].

Unlike the other two visualisations described, Aruvi [32] is intended as a generic tool to support analytical reasoning (without targeting a particular user population). The design is based on both the sensemaking literature and also the cognitive reasoning literature. The prototype implementation provides three different views: a data view (including interactive visualisation tools); a knowledge view, which allows the user to make and manage notes on the analysis; and a navigation view, which maintains a history trace of the visualisations explored in the data view. A formative evaluation, in which four participants used Aruvi on their own datasets, highlighted the value of the visualisation.

All three examples illustrate a design approach based on understanding users' needs and the ways that they think about their tasks and activities, so that the systems that were implemented support, as well as possible, users' sensemaking activities. They all implement conceptual structures appropriate to the domain, as discussed in the next section.

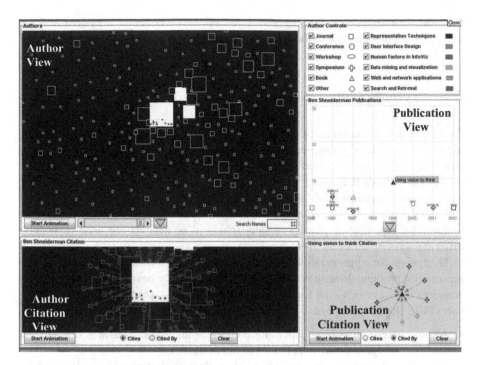

Fig. 2 Screenshot of ALVi

2.2 Conceptual Structures

Every domain has a structure [4] which influences the conceptualisations that users generate. Although there has been limited study of this topic over the years (e.g. [26, 28]), there has been surprisingly little work on how the conceptual structure of a design supports or hinders people working with a system. Norman [27] argued that it was the designer's responsibility to make the conceptual structure of the design clear to the user through the interface, but the converse argument—that the designer should aim to understand and actively support pre-existing conceptual structures—has rarely been made explicitly.

Johnson and Henderson [22] echo Norman's [27] view, explaining that "A conceptual model describes how designers want users to think about the application" (p. 18). It is implicit in their writing that they recognise the importance of implementing systems that support users' conceptual structures consistently and effectively. However their focus is on how to move from the conceptual model to the implementation, rather than on how users' pre-existing conceptual models might inform the designer's conceptual model.

The question of how to identify the user's conceptual model and how to assess the quality of fit between the user's conceptual model and that implemented within the system is addressed by Blandford et al. [9]. They present an evaluation method,

CASSM, that supports reasoning about conceptual fit between a system and its users. This approach is described in more detail below.

2.3 *Appropriation*

As well as depending on the suitability of generic conceptual structures, sensemaking also relies on the strategies that people adopt to generate personal conceptual structures. Since each individual makes sense in their own way, it is beneficial to include features that enable people to work with a representation in their own way; for this, appropriation tools are helpful. By *appropriation* we mean making technology 'one's own' in ways that the designer could not anticipate; for example, features for labelling or placing objects (physical or digital) are used in widely differing ways by different individuals and groups.

Designing for appropriation [13] allows people to express ownership of the represented problem domain in ways not anticipated by the designers. Appropriation concerns the ways in which technologies are adopted, adapted and shaped by their users [10]. Dix [13] proposes several guidelines for appropriation:

- Allow interpretation (e.g. providing colours or annotation features that enable people to assign their own meanings to the marking).
- Provide visibility: making the system state clear so that people can infer when they need to find a work-around (e.g. because an important feature is not functioning properly).
- Expose intentions: making the designers' intentions clear to the users so that they better understand when workarounds are and are not appropriate. The example Dix gives is of login procedures, which might be to provide security (which could be provided some other way) or to provide personalisation features (so that login is about identification, not just authorisation).
- Support not control: making it possible for people to vary routines in common tasks while nevertheless keeping them efficient.
- Plugability and configuration: creating systems that can easily be reconfigured by users.
- Encourage sharing: providing ways for end users to share tips on how to use a system.
- Learn from appropriation: develop new generations of technology that better support the new use (Carroll and Rosson [11] discuss this as a "task artefact cycle").

In the context of visualisations to support sensemaking, probably the most relevant and valuable of these guidelines are providing support for interpretation (as implemented, for example, in Aruvi's annotation feature and ALVi's marking tool) and ensuring that the system provides support, not control, so that people can explore data in ways that suit them.

3   Gathering Requirements

In order to understand how people think about a domain (the world of which they are trying to make sense), it is necessary to work with those people, listening to their descriptions of their world. For example, in our studies of academics' interactions with their literature [16], we found that *ideas* and *authors* (i.e. other researchers) are more important than particular papers; in a study of lawyers [3], we found that people and events were central to the building of chronologies (or causal narratives), which had to be backed up by evidence (typically in the form of documents and records such as telephone call logs).

There is often a tension between the way a user naturally thinks about their problem and concepts that can be implemented in a computer system. For example, in ALVi it was straightforward to represent the concept of *authors* of papers, but much more difficult to represent the concept of *ideas*, which are meaningful to users but are not directly represented within the text of a paper. Conversely, in implementing a system it may be possible to make new concepts available to the user that are actually valuable to them, and that extend their space of thinking. For example, when the first author started blogging, she had not considered what factors might make one post more or less popular than another, but the visualisations of statistics made available through the blogging tool encouraged consideration of how to publicise the blog posts, when people typically access the posts, and why some posts might be much more popular than others which, over time, informed her approach to writing and publishing posts.

There are many ways of gathering user requirements, but at the centre is a need to engage with the potential users of a system, to get a deep understanding of their goals and activities, and the ways that they think about their sensemaking challenges. There are entire textbooks on gathering user requirements (e.g. [6, 20, 25]), and it is not possible to do justice to all possible approaches within one chapter, so here we focus on techniques that are particularly pertinent for designing sensemaking visualisations. In particular, to gather information about people's conceptual structures, it is necessary to gather verbal data. So we focus on interviews and Contextual Inquiry as ways of gathering data, and then briefly outline one qualitative approach to analysing that data.

3.1   Interviews

Interviews may be more or less structured. For gathering requirements, semi-structured interviews are generally the most effective. Such an interview allows the interviewer to ensure that they cover the important questions (e.g. about what people are trying to achieve, what they do, what is important to them) while also having the opportunity to pursue unexpected interesting avenues as they arise.

Key steps to planning interviews are:

- Planning questions
- Identifying and recruiting participants
- Deciding on practicalities such as locations, times and lengths of interviews
- Deciding how to record and analyse data

When planning questions for designing visualisations for sensemaking, the heart of the interview is likely to be around how people make sense of the domain of interest, and what questions they have about it. For example, when designing ALVi, we wanted to know how young researchers went about identifying and familiarising themselves with the research literature in their field, and how more senior researchers thought about the literature in their field, and how they might go about learning about a new literature (e.g. if shifting fields).

Every interview has a beginning, a middle and an end. The aim at the beginning is to set the scene, to put the participant at ease (e.g. that there are no wrong answers, that they are the experts in the domain, that their participation is valuable and will inform the design of the proposed system), and to elicit basic background information (e.g. what their job function is, how long they have been active in this area). The aim of the middle is to build a rich understanding of their work and their sensemaking activities, while the aim of the end is to tie up any loose ends, thank the interviewee and close the interview.

Interview questions should, as far as possible, be short and to-the-point, using familiar language. They should not be leading (i.e. suggesting a particular answer), but should be as open and unbiased as possible. During the interview, it is often helpful to reflect back to the participant your understanding of what they have told you. Probably the most important questions, though, are ones that elicit detail: not general "what do you do?" or "what are the key factors that you take into account when ... ?", but "can you talk me through the last time you did ... ?" or "was there a particularly memorable occasion when ... ? Tell me about it." The details of a particular incident will bring the subject to life in a way that no amount of generalisation ever can.

When determining who to interview and how to approach them, there are both ideal and pragmatic considerations. Participants should be representative of the population for whom you are designing; maybe they are existing users of a related technology or potential users of the proposed visualisation tool. There may be typical users or users who are extraordinary in some way (e.g. recognised experts or people who struggle to do the job). There may be sub-populations within the overall population that need to be catered for. There may be people who are articulate or opinionated or quiet. In an ideal world, participants will be broadly representative of the intended user population. In the real world, where it might be difficult to recruit the ideal set of participants, it is sometimes necessary to work with more of a "convenience sample" of people who are willing and able to take part; in this case, it is important to be alert to possible biases in the data, and to find ways to limit them, and to validate the findings.

There are many possible ways of recruiting participants: within an organisation, this might be largely by word-of-mouth, or internal advertisements; externally, it might involve working through special interest groups, using a specialist recruitment agency or using social media to solicit participation.

People may participate for many different reasons: maybe you are offering a material reward (participation fee, a small gift, or entry into a prize draw); maybe they are inherently interested in the topic and value the chance to discuss it and reflect on it; maybe it is important to them that they have a chance to influence a new design; or maybe they have been asked to participate by their line manager (and are therefore doing it as part of their work). It is important to recognise and work with these alternative motivations in recruiting participants and in valuing their contributions.

You need to plan where and when to interview people. Is it important that you are physically co-located, or can the interview be conducted equally well by phone or over the internet? There are potential advantages to being co-located, including ease of establishing rapport and of sharing externalisations (e.g. sketching together or being shown relevant artefacts). Conversely, meeting up often requires that one person travel and that a suitable meeting space be available. The space should be comfortable, quiet and private enough to conduct the interview without significant distraction. Interviews in the place where people normally engage in the sensemaking activity may allow participants to illustrate activities or use resources from the environment to animate the discussion.

How to record data? This may depend on factors such as noise, confidentiality and privacy, and the forms of analysis that are planned. Sometimes, just pen and paper (or typed notes) are sufficient, or all that is possible in the circumstances; this limits the amount of subsequent transcription or analysis needed or possible. Going to the other extreme, a full video recording may capture details that would otherwise be missed, but setting up and using a video camera can be intrusive and distracting, and the benefits rarely outweigh the costs unless there is important action to record. Most often, an audio recording (which can be fully or selectively transcribed), and optionally still photos or sketches of key artefacts, achieves the best balance between richness of data and cumbersomeness of gathering and analysis.

Interviews are best conducted as conversations: asking questions naturally, and in a logical sequence (which may not be the order planned), showing an interest in the participant's accounts and stories, and valuing their expertise and insights.

One great advantage of interviews over observations is that they elicit people's understanding of the task or activity, which may be partly decoupled from the way they use a particular system to perform it. For example, someone talking about planning a holiday will probably talk about many factors that they use to decide where and when to go (the likely weather, the activities or attractions, the ease of getting there, etc.); and observation might focus on how they use flight booking websites, which will naturally highlight certain factors (e.g. cost of flights) while omitting others.

Conversely, interviews elicit perceptions, and can be unreliable for eliciting certain kinds of facts. For example, people's accounts of the steps they take to perform a task may omit steps that are so "obvious" that they are overlooked, and some information may be so taken-for-granted that it is never mentioned.

There are two widely used approaches to gathering observational data that also includes people talking about their activity. One is think-aloud [14], which can support incremental redesign (fixing limitations of an existing system), but is less suitable for identify user requirements for a novel system; we discuss think-aloud in Sect. 5. The second is Contextual Inquiry, which focuses more on the broader activity than on the details of an interaction, and hence is better suited to gathering requirements.

3.2   Contextual Inquiry

Contextual Inquiry (CI) is described in detail by Beyer and Holtzbatt [6], so here we provide just a brief overview of the approach. It is a form of data gathering and analysis that takes place in the work setting (or wherever the activity naturally takes place). It involves observation of the activities of interest, and questioning about those activities.

Just as, when discussing interviews, we emphasised the importance of eliciting concrete examples, not just generalisations, so in CI there is a focus on concrete data: on real artefacts that support the activity, and events that occur. Similarly, just as interviews should value the expertise of the interviewee, so CI values the expertise of participants in doing their work, and data gathering involves working with people on understanding their work experience. In most contexts, CI is conducted by a team of people, each gathering some data and then sharing insights and reaching a shared understanding; for the purposes of this chapter, we focus on what each individual analyst does within that process.

At the heart of CI is observation followed by questioning, to develop a rich understanding of what people are doing, why, and how they think about their activities. For sensemaking visualisations, key questions will include what questions people have, what information they draw on to address those questions, and how they think about the topics they are working on.

Many of the decisions (e.g. about who to work with and how to record) are very similar to those involved when planning interviews. One important difference is that CI should always take place in the workplace (or other place where the activity is performed). There may be more compelling reasons for having a video record for CI than for an interview, but costs and benefits still need to be traded off against each other.

One challenge is sometimes to persuade people to continue working as naturally as possible, as some prefer to stop and chat about work rather than performing work (which makes data gathering into an unstructured interview rather than a CI). It may also be important to consider the timing of questions to minimise disruption. A further consideration in the workplace is how to fit in: are there particular dress codes? How to take breaks? How to introduce yourself to "third parties", particularly if your participant has a role where they interact with many other people? There is not a single answer to such questions, which will depend on the situation, but they should be considered carefully.

Once data has been gathered, a full CI analysis involves representing the findings in terms of five models:

- A flow model, which focuses on the flow of communications around the organisation (between people mediated by artefacts).
- A sequence model, which focuses on the order of events.
- An artefact model, describing the physical structure and noting reasons for that structure.
- A cultural model, noting roles and relationships.
- A physical model, noting influences of the environment on performance.

The process of creating representations, as well as the results, provide a focus for thinking about new design solutions. However, when it comes to designing sensemaking visualisations, much of the information represented in the recommended models is of limited value and relevance, so in the next section we present an alternative approach to analysis that focuses on conceptual structures.

3.3   *Identifying Users' Conceptual Structures*

As discussed above, there are many kinds of user requirements, and in this chapter we are focusing on understanding the conceptual structures people work with, and how to translate that understanding into requirements for design. When evaluating a particular system, one can go further and assess the quality of the conceptual fit between user and system—a topic to which we return below.

A *concept* is a "thing" or a "property" that the user works with while making sense of information. These can be divided into *entities* and *attributes*, although early on in analysis, this distinction may not be important. An *entity* is often something that can be created or deleted within the system. Sometimes, entities are things that are there all the time, but that have attributes that can be changed. In ALVi, entities include authors and publications. An *attribute* is a property of an entity. For example, an author has a number of publications, signified in ALVi by the size of the rectangle representing the author. There may also be *relationships* between entities—for example, that one paper cites another paper.

Having gathered user data as described above, one way to conduct an analysis is to go through the words (e.g. transcription of users talking or documentation) highlighting nouns and adjectives, then deciding which of those words represent core concepts within the user's conceptualisation of the domain.

Depending on what matters most, the analyst might distinguish between entities and attributes to achieve clarity in the model. Concepts might also be grouped into related ones that function together, or that might be displayed together. Relationships between concepts also need to be noted.

To illustrate the approach, we take brief quotations from the requirements study that led to the development of ALVi and highlight key concepts and relationships. This analysis is described more fully by Faisal et al. [16].

Participant 3 said: "... there'll be a core body of people who are aligned with particular kinds of ideas". This highlights two key ideas: a *body of people* (which in turn comprises individual *people*) and *ideas*. It is unclear from the way participant 3 talks whether the *body of people* might also be called a *community*, or whether it is better regarded as a group within a community. Further analysing the data, we decided that this distinction did not matter, as the concept of a community is only loosely defined anyway.

Participant 6 also highlighted the importance of *ideas*: "... I think I would go for ideas ... what it means actually it is not the paper but the ideas". Ideas are presented in *papers*. In turn, *papers* are written by *authors*, as participant 2 stated: "It is hard to separate that [papers] from authors, 'cause ultimately they were written by authors". Note that different terms (people, authors, also researchers) were used by participants in the study to refer to the same concept (namely the people who do research and then write it up).

Participant 3 identified an attribute of a paper that was important, namely its *influence*: "I could name you papers that have been influential ... in terms of changing thinking in a particular area". As well as particular papers, there are also authors who are influential within a community, as participant 5 noted:"... you have to read what they [influential authors] are doing even if you don't agree"

The word "influential" was used in two different senses. As well as being recognised as being influential by a community, some participants recognised a more personal kind of influence. For example participant 3 said: "... there have been papers that have been influential ... actually changed the way I have thought of my work". Similarly, participant 4 noted that: "I suppose when you say influential I consider it to be influential to my own ideas". This difference between influence on the community and personal influence appeared to be important. The first might be represented indirectly by number of citations, whereas the second is internal to the user, and might be externalised (e.g. through the use of the marking feature mentioned earlier).

In summary, from these selected extracts we have identified important user entities of *person* (with attribute *community influence*), *community, idea, paper* (with attributes *community influence* and *personal influence*), and relationships that person is-member-of community; paper is-written-by author; paper encapsulates idea.

In this particular domain, relationships are important, and chronology (how a field develops over time) may also matter, so visualisations have to be designed to allow people to explore relationships and chronology. The design of ALVi is based on these principles. Other domains have different central properties that should be designed for, as discussed in the next section.

4   From Requirements to Design: A Space of Sensemaking Visualisations

In moving from requirements to design, it is necessary to determine what kinds of visualisations will best support people's sensemaking. Here, we propose and illustrate a classification of visualisations that highlights their principal conceptual structuring basis, based on work reported by Faisal et al. [15]. This classification (into spatial, sequential, networks, hierarchical, argumentation and faceted) reflects the different conceptual structures of different sensemaking domains.

For each visualisation structure, we propose a scenario of use for which that structure would be appropriate, and highlight key properties of the structure.

4.1   Spatial

Spatial representations depict objects and their spatial relationships. In such a representation, elements such as orientation, distance and location are critical components. Visualisations based on maps, such as geographical information systems, are typical examples of spatial visualizations. They support planning that is based on spatial attributes (such as planning a trip), and reasoning about properties that are based on space or geography, such as the spread of disease or the distribution of wildlife across a region.

4.1.1   Scenario: Planning a Trip

Samantha is planning a trip from London (UK) to South Africa. She has decided to visit Cape Town, Pretoria and Johannesburg. She checks out a travel website and determines that both Cape Town and Johannesburg have international airports, but that Pretoria does not. Her task now is to determine a traveling route. In order to do so she consults a map as she has only a vague memory of the layout of South African cities. She sees that Pretoria and Johannesburg are towards the north east whilst Cape Town is further south. She also notes that the distance between Cape Town and Johannesburg is too great to be covered in a day except by flying, whereas Johannesburg and Pretoria are within comfortable driving distance. From her previous spatial knowledge, she knows that London is far north. It will take her longer to travel from London to Cape Town. She decides that she would prefer to make the longer journey on her way to South Africa rather than on her way back to the UK. Therefore, she settles on the following travel route: fly from London to Cape Town and from Cape Town to Johannesburg, drive to Pretoria then back to Johannesburg in order to catch a plane back to London.

Fig. 3 Spatial visualisation

A simple design example to demonstrate the spatial representation is Google Maps (Fig. 3). As part of this application users are able to locate themselves in terms of the map's spatial layout, get directions to wherever they want to go and determine areas with high traffic which assists then in planning their journey.

4.1.2 Properties

Two and three dimensional spatial representations are generated in order to make sense of physical layouts. Information such as the relative and absolute locations of different kinds of objects is crucial and so these are generally mapped proportionally within the representation. Spatial representation can also be used to represent a domain metaphorically. For example, our own discussions with a programmer indicate that the way he makes sense of code is by conceptualizing it spatially. Spatial representation offers a means of depicting complex abstract relationships in a simplified way.

4.2   Sequential

Sequential representations depict movement through a series of elements based on a predefined order such as time. Chronologies are a common example of sequential representation, exemplified for example in legal investigation support software such as LexisNexis CaseMap (www.casesoft.com), in which elements are events connected within a time-series. Previous field research by Attfield and Blandford [3], which looked at collaborative sensemaking activities during large corporate investigations, found that chronologies, painstakingly constructed by large teams of lawyers, provided central visual representations for sensemaking. Legal investigations extend over time, are resource intensive, and require the sifting and re-representation of very large collections of electronic evidence. Using the chronologies that they created, teams of lawyers could review the underlying narrative of their investigated domain, identify periods of key concern or activities of protagonists that seemed odd and potentially suspicious and, using this, refine their investigation questions and searches in ways that were more focused and tractable. This could be understood as comprising phases of *data focusing* (identifying and structuring information to draw out facts relevant to a given set of investigation issues), followed by *issue focusing* (revising the issues and questions in the light of new insights).

4.2.1   Scenario

Carol is a legal investigator leading a team in an investigation into potential hidden liabilities within a large multinational organisation. Carol's team has been tasked with making sense of activities surrounding a particular contract that the company had and a particular individual who led the contract. As they review hundreds of recovered emails that have been returned by searches they have constructed, they select those that seem most relevant and use these to construct a timeline of how work on the contract unfolded. They look in detail at the run up to the contract bid submission and see something unexpected. During this period which, as one would expect, is particularly busy, the manager with responsibility for the bid sends an email saying that he will be flying out to a foreign capital for a short period and flying back home again. There is no explanation for why he is going. This is odd since the lawyers would expect that during the bid preparation he would remain close to the team he is leading. This appears potentially suspicions. These scenarios typically require 'all hands to the pump'. Carol's team mention what they have found at a daily investigation review meeting and ask whether anyone has anything that may shed light on it. Another team involved in the investigation say that they have evidence of activity in the same foreign capital on that same day. The two groups align their chronologies. Carol's team see that a representative of the company visited an overseas airport that day, signed a contract, and then returned directly home. They search for the contract and find that one of the signatories was the manager in question.

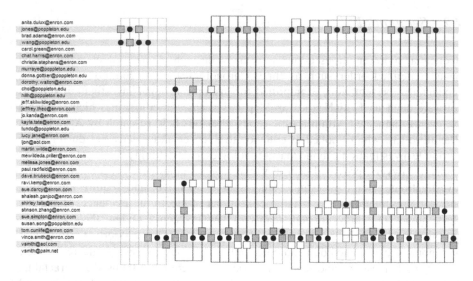

Fig. 4 ThreadsVI [2], a temporally organised representation of protagonist activity within an email collection

Legal investigations are one scenario in which temporal representations are particularly important. Other representation are used, but chronologies often provide a central representation for reasoning. Since lawyers expend considerable energy in filtering, selecting and structuring otherwise unstructured evidence collections (such as emails) into temporally ordered narratives, they would benefit from visualisations that can automatically perform that kind of visual structuring for them. This was the rationale behind a representation called ThreadsVI [2] which aimed to associate emails in a meaningful way by representing them visually in the context of their discussion threads and in terms of senders and receiver such that the user could gain informal visual impressions of levels of activity (similar to centrality) within a semantically filtered social network over time (Fig. 4).

4.2.2 Properties

Sequential temporal representations lay events or entities in order of, or in proportion to, separation on a temporal variable. Since sequence may demand only one dimension within a 2D or 3D visual representation, further dimensions may be used to represent other parameters in multivariate data that are salient to the usage scenario. Colour and shape can support the encoding of further variables.

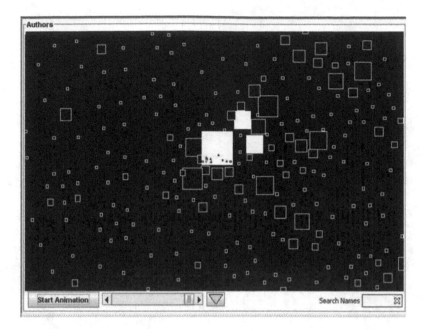

Fig. 5 ALVi authors' view

4.3 Networks

Network structures are relational structures where items may be linked to an arbitrary number of other items in many-to-many relationships [31]. These representations can, for example, support the understanding of complex social structures, e.g. Vizster [21], a visualization that allows for the exploration of online communities. ALVi's author view is another example, in which authors are clustered based on the collaboration relationship. Authors that have collaborated together will be seen as being grouped in a cluster. Based on such clustering needs, the layout of the authors view takes the form of a force-directed layout graph (Fig. 5).

4.3.1 Scenario

Tessa is new to social media networks. She recently created an account on Facebook and added her friend Mia as a friend. Tessa and Mia have been friends since high school. They also went to college together. Hence, they have a lot of friends in common. Through Mia, Tessa started to identify friends and add them to her list of friends. Similarly through those friends she started to identify more friends and acquaintances and started adding them. Slowly Tessa was creating a network of friends whereby each person on that list was connected to her and to other people in her list of friends. As she was doing that she started to realize that she

Fig. 6 Friend Wheel

had several networks being created. For example, a network that contained mainly school friends and another that contained work friends. As she was browsing her list of friends she realized that Anna a colleague of hers at work had a common friend, Emma. Emma was Tessa's friend in school. When Tessa asked Emma about this she realized that Anna and Emma had taken a summer course together and had been friends ever since.

Figure 6 shows a Facebook tool that visualises a friends network. It takes all the friends, for example Tessa's friends, then links and groups them together to form a colourful image. Each node represents a friend. When a line links two friends, it means that they are friends with each other, as per the example above, Anna and Mia.

Fig. 7 Hierarchical
categorization

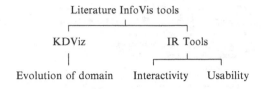

4.3.2 Properties

Network representations represent the interconnection of various items. They have a general graph structure with nodes and edges. Each edge connects two nodes. Nodes in a network can be connected by an unlimited number of edges, representing many-to-many relationships. Network representations are applied to a variety of domains including, but not limited to, social networks.

4.4 Hierarchical

Hierarchical representations model a domain by organizing elements according to asymmetric, one-to-many relations.

4.4.1 Scenario: Categorizing Research Areas

This scenario is based on the second author's experiences. Sarah is working in the area of academic literature visualization and would like to develop a scheme that she can use to structure her literature review. She works with a number of ideas which relate the papers in different ways, but there is no organization that seems to include all the work she wants to include in a neat way that she feels she can structure a narrative around. She considers different facets that seem to distinguish the papers. She realizes that you can describe all of the information visualization tools she has read about as falling into one of two categories: knowledge domain visualizations (KDViz) and Information Retrieval (IR) tools. This strikes her as a candidate for her high level organization. Then, as she looks more closely at the papers that fit into the IR tools' category, she sees that roughly half are concerned with interactivity and half are concerned with usability. She settles on this as her first plan for an organizational scheme (shown in Fig. 7).

An example of an online visualization is provided by Microsoft Academic Search's representation of supervisor-student relationships, which show "genealogies" of researchers, as illustrated in Fig. 8.

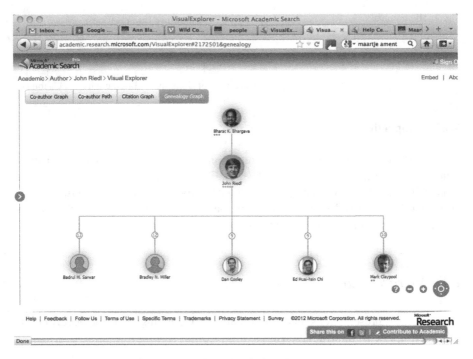

Fig. 8 Example hierarchy from Microsoft Academic Search

4.4.2 Properties

Hierarchical representations may be represented as trees [31], where each item has a link to a parent item except for the root. The relations used in hierarchical representations are often taxonomic. However, they can also be used to represent non-taxonomic relationships such as *part-whole*, or *parent–child*.

4.5 Argumentation Structures

Argumentation representations relate multiple propositions or ideas together through argumentation operators in a way that makes inferential relationships explicit. Visual representations of arguments have a long history. For example, in the early twentieth century, Wigmore developed a visual language for representing and analyzing arguments in legal cases [40]. The Wigmore diagrammatic convention was designed specifically for representing competing arguments in a contentious, legal setting. The unit of analysis in a Wigmore diagram is the evidential proposition or statement, with these laid out in relations of inferential support. Notably, Wigmore included conventions for representing aspects of two competing arguments which you would expect to be salient to a legal mind trying to make sense of them, evaluate their relative merits, and potentially develop a legal strategy. For example, evidence is visually coded as to whether it is testimonial (stated by a witness), circumstantial (requires inference), explanatory (reduces impact of testimonial or circumstantial evidence) or corroborative (supports testimonial or circumstantial evidence). Evidence is also represented in terms of the side who offered it (prosecution or defense). A strength of Wigmore's scheme is the representation of the perceived strength of elements of an argument and of the argument as a whole.

Whereas Wigmore was concerned with a specific domain of activity, namely legal cases, Toulmin later developed a diagrammatic convention for the representation of everyday arguments [36]. A philosopher, Toulmin was concerned that everyday persuasive arguments rarely correspond to classical models of inference, such as the syllogism. What he felt was needed was some way of representing the form of every day arguments so, for example, these could be reviewed in terms of validity. In Toulmin's scheme there are three parts to an argument: a claim (the conclusion of an argument); data (evidential support for a claim); and a warrant (a generalised assumption on which a link between claim and evidence depends). In everyday arguments warrants are often implicit. An advantage of the Toulmin approach was that by making them explicit they become amenable to consideration.

4.5.1 Scenario: Conducting a Literature Review

Paul is an HCI researcher. He is reading around the topic of information seeking looking for ideas for tools and functionalities that might usefully augment digital library systems. He is looking at studies of information behavior in order to gain insights about the things that people naturally do with paper documents in order to trigger ideas about how people might wish to interact with digital documents. Paul notes a study (a) which reports that people use physical piles of paper as a way of informally organizing task related information. This reminds him of a study (b) which reported on the way that paper documents on the desk are sometimes used as reminders for action. He notes that digital libraries don't provide tools that support these kinds of behaviors. Later he reads a paper (c) which describes a spatial hypertext system and how such systems allow users to create informal, visual document arrangements which can persist across sessions. Paul uses these claims to construct an argument which acts as motivation for his new idea of augmenting digital libraries with spatial hypertext functionality.

ClaiMapper [38] is an example of an argumentational representation. It allows users to create concept maps of literature by manually dragging and dropping concepts and building relationships between them. This is done with the aim of creating coherent arguments by sketching out rough structures as informally as required.

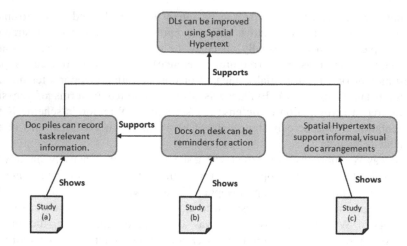

Fig. 9 The Spatial Hypertext example shown in a form of argument map. Rounded rectangles represent propositions and arcs either represent relations of inferential support or links with source study reports

4.5.2 Properties

Argumentation representations are formed around the integration of a series of claims from which a conclusion is inferred. As a result, the associated tools need to allow users to build dependency connections between different parts of the data, whether these are explicitly expressed by the data or inferred (Fig. 9).

4.6 Faceted

Faceted representations show a set of entities within a domain in terms of a set of properties. These will often also exploit other visual representations such as spatial (e.g. [41]) or temporal (e.g. [1]).

4.6.1 Scenario: Planning a trip to the cinema

Claudia and Jeff want to watch a film. They are not in the mood for anything too serious. A romantic comedy would be ideal. However, they do not have a particular movie in mind. They start browsing online to identify a title that interests them. There are too many to choose from. Hence, they decide to select something that is quite recent. They look for films that have been released in the past 2 years. Lots of titles come up which sound interesting. However, when looking at the reviews, some have low and average ratings. Consequently, they decide to look for a film

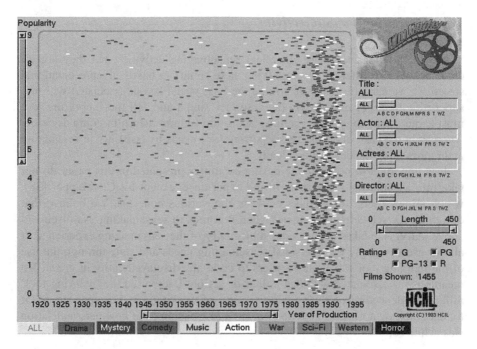

Fig. 10 FilmFinder

that has good reviews. This narrows down their choices. They settle on "Midnight in Paris" as it was a romantic comedy, released in 2011, with excellent reviews. This scenario involves relating film selection to an evolving set of facets derived by the user's changing knowledge of the various possible film criteria.

FilmFinder [1] (Fig. 10) is a visualization tool that represents films in terms of (and filtered by) their properties. In this application films are represented as colored squares, where color encodes genre (e.g. horror, comedy, science fiction, etc.). The films are laid out in a scatter plot, in which horizontal position indicates year of production and vertical position indicates popularity. Sliders allow users to filter the database by different properties, e.g. title, actor, etc. The FilmFinder application can be considered as an example that supports the facets representation allowing users to compare objects according to properties that best represent their interests.

4.6.2 Properties

As part of the faceted representation, entities within a domain are shown in terms of a set of facets. The best possible match is used to either make a choice or provide an interpretation of a specific situation.

4.7 Designing for Subjectivity

The scenarios presented here all reflect sensemaking activities, but they are all characteristically different. By classifying these situations based on representational structures we can effectively determine the best visual layout associated with the supporting tool. For example, in the case of Claudia and Jeff the supporting tool must capture and represent the various facets that support their film-going decision, whereas in the case of Sarah the supporting tool needs to assist her in organizing and representing the data hierarchically.

It is important to note that the design examples presented in this paper do not take into account the subjectivity of the experience, except for ClaiMapper [38], which allows users to add links between the various concepts. We believe that designing for subjectivity is crucial when it comes to sensemaking. As discussed above, ALVi also takes into account subjectivity of the sensemaking experience. It allows users to apply their personal sensemaking needs and style through the use of the marking tool which allows users to code any displayed entity in green. No meaning was associated with that colour or action. Users used the 'marking tool' differently depending on their goals and ways of working, i.e. they appropriated it to meet their own needs. The following are a few examples of how personalisation was achieved:

- *Subjectively filtering the data—U1 said*: "having a sort of representation of what responded to a query term and then being able to go through them and put my own, so it's like a two step filter so the system filters and then I filter that was really useful." *The user used the word "my own" to refer to his experience in using the marking tool. He indicated that he was able to further filter the data by overlaying his filtering scheme over the system's filtering by marking some of the results of the system's filtered outputs.*
- *Setting landmarks—U6 said in relation to the marking tool's benefits*: "I won't feel lost I can go back". *Similarly, U11 said*: "I started marking because I looked at stuff and then I moved on and then I came back to it and I realized I actually read that but because I hadn't marked it I did not realize I had read it".
- *Keeping track of the amount of work—U10 said*: "I have a sufficient amount of papers actually from it being highlighted so I know that I possibly have a sufficient background for this particular reason".
- *Generating personal overviews—U8 expressed it*: "I remember um searching names, marking the authors and then towards the end I remembered that you could mark all the associated authors which is a really handy little thing if you want to get a grips with much of the overview much more quickly".

From these examples it can be seen that the users were able to associate different meanings that were personal to them and that fitted their own sensemaking strategies with the marking colour. This feature was successful in allowing users to implement their personal sensemaking strategies.

5 Evaluating the Appropriateness of Conceptual Structures

So far, we have discussed gathering user requirements and selecting appropriate conceptual representations for the design of interactive visualisations. The third key step is evaluation of the resulting interactive visualisation in terms of its support for sensemaking. There are many possible aspects to evaluation, as discussed by Kang et al. [23] and Faisal et al. [18]. For the purpose of this chapter, we continue our focus on the quality of the fit between users and system [9].

In an earlier section, we discussed how to gather user requirements and identify key concepts. Evaluation builds on those techniques, comparing the users' conceptual structures with those now implemented within the visualisation tool. This approach can be used to evaluate an existing implementation (that was not developed by starting with user concepts); in that case, it is necessary to gather user concepts as described above, even if they are being used principally for evaluation rather than for design.

As well as the interview and Contextual Inquiry techniques described above, it may also be appropriate to make use of Think Aloud protocol at the evaluation stage, as this can highlight additional concepts that emerge in the interaction between user and visualisation system.

At this stage, we are considering conceptual fit between a user and the system: the user might have some concepts that are not represented within the system ([a] in Fig. 11); conversely, there may be system concepts that the user is not familiar with ([c] in Fig. 11). The concepts that are shared between system and user ([b] in Fig. 11) are a conceptually good fit.

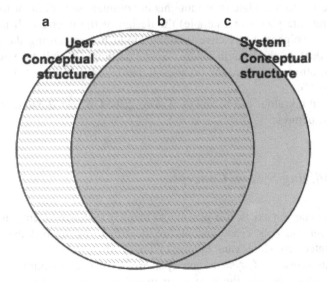

Fig. 11 User-system fit

5.1 Think Aloud Protocol

Think-aloud involves the user of a system articulating their thoughts out loud as they work with a system; it creates data on how people think about their activity and about the system they are working with. Think-aloud [14] involves recording and analysing people's thoughts about the activity they are currently engaged in, and typically focuses on the interaction with a current interface (and hence is well suited to identifying strengths and limitations of that interface).

If think-aloud data is being gathered to support analysis, based on how people use a current system, then it is important that the tasks given to study participants are domain-relevant, and give the participants scope for discussing domain concepts. For example, the analyst evaluating ALVi would get little useful data if the user task were given as "use ALVi to find out who has cited the work of Matthew Ward". The task description we used for evaluating ALVi was based on our original requirements study, and intended to represent a typical sensemaking task: "At this point of your research you need to examine the concept of "Dynamic Queries"; you do not know where to start. A colleague of yours has given you a paper reference as a good starting point: a paper written by Shneiderman in 1996 and titled "Incremental ..."" Your goal is to identify key researchers and publications that target this area and identify any commonalities or differences between these groups of people.." This version of the task description allowed participants to show how they thought about the literature domain, as well as how they used the tool to support their reasoning; these are the kinds of issue that matter for a CSII analysis.

When instructing people on how to think aloud, it is important to emphasise that they are thinking aloud, not providing an explanation to the analyst. So people are not expected to articulate their thoughts in complete sentences, or to provide a running commentary on (for example) the design of the interface. It is generally helpful for participants to practice thinking aloud before starting the activity, if that is practical in the circumstances. Some participants will take to thinking aloud easily; others may need the occasional gentle reminder (e.g. "What are your thoughts now?").

Data from think-aloud can be analysed as described in Sect. 3.3 to identify users' conceptual structures.

5.2 Identifying System Concepts

Once a system design exists, it is possible to identify system concepts and compare them with user concepts. User concepts are identified as described above (through analysis of interview, Contextual Inquiry or Think Aloud data).

The main sources of system concepts are system descriptions and maybe a running system. Again, the data is analysed in whatever ways are possible (depending on what data sources are available) to identify core system concepts.

In doing this analysis, one thing to avoid is extensive descriptions of interface widgets: rather, the analysis should focus on the underlying system representation. Interface widgets are a means to an end, not an end in themselves. For example, an analyst describing ALVi would focus on authors, publications, citations and the other concepts that users can access through the visualisations, not on the scrollbars, tick boxes or search boxes.

5.3   Identifying Misfits

Once suitable data has been gathered, misfits can be identified. The first step is simply to identify and compare system and user concepts; a second stage of analysis considers what actions are needed to change the system, and whether there are problems with actions. Misfits between user and system are the most important information-related misfits. These misfits fit into three classes, illustrated in Fig. 11:

User concepts that are not represented within the system, and hence cannot be directly manipulated by the user ([a] in Fig. 11). A very simple example related to ALVi is the concept of *an idea*. For example, when doing an evaluation study on ALVi, a participant gave this comment when asked whether publishing an influential article makes its author influential " . . . *I think I would go for* **ideas** . . . *what it means actually it is* **not the paper but the ideas**".

User concepts that are not represented in the system often force users to introduce workarounds, as users are unable to express exactly what they need to, and therefore use the system in a way it wasn't designed for. In these situations, users may adopt complementary tools, such as bits of paper or electronic notepads, to support their sensemaking. A well designed appropriation tool (such as those discussed above) can make such workarounds more seamless.

System concepts that the user has to know about but that are not naturally part of their initial understanding, and therefore need to be learned ([c] in Fig. 11). An example involving information structures might be the ways that information is organised in a hierarchical classification system.

For users, these misfits may involve simply learning a new concept, or they may involve the users constantly tracking the state of something that has little significance to them.

An example of this was seen when we were evaluating ALVi. A few of the study participants conceptualised their literature domains differently from the majority (for whom the tool was tailored). One of these users was trying to adjust his personal sensemaking strategies to fit the tool's design. This user thought in terms of papers/titles and not in terms of authors; as a result, he kept getting confused as ALVi was designed with the authors at the centre of the sensemaking strategy. This user commented: "I am treating this as a visualization of papers rather than a visualization of authors".

User- and system concepts that are similar but non-identical, and which are often referred to by the same terms. This could be considered as an amalgamation of the two categories above (a user concept that the system doesn't represent and a system concept that the user has to know about), but has a particular set of implications, in terms of how the user has to mould their understanding to the system. For example, Blandford and Green [8] found that users of an electronic diary system had to adjust their concept of a "meeting" from being a relatively informal agreement between people to congregate in a particular place (which might involve a separate activity of booking a room) at a particular time to discuss agreed topics to being a more structured concept that typically did not include the purpose of the meeting, but did include an expectation of inviting rooms and other resources to participate in the meeting.

These misfits may cause difficulties because the user has to constantly map his / her natural understanding of the concept onto the one represented within the system, which may have a subtly different set of attributes that the user then has to work with.

As well as concepts being absent, some may be available but present some kind of difficulty. For users 'difficult' concepts are most commonly ones that are *implicit*— ideas they are aware of if asked but not ones they expect to work with. An example, for many people, is the end time of a meeting: in people's paper diaries, many engagements have start times (though these are often flagged as 'approximate'— e.g. '2ish') but few have end times, whereas electronic diaries (which are sold as diaries, but are better described as scheduling systems) force every event to have an end time (or a duration, depending on how you look at it). This forces users to make explicit information that they might not choose to. Of course, there are (typically busy) people for whom the "scheduling" nature of electronic diaries suits them better than the relatively imprecise structure of paper diaries [8], but these are a minority of users.

5.4   Adding in Information About Actions

The analyst can define how actions change the existence of entities or the values of attributes as a further step of analysis. The key question for sensemaking visualisation is whether or not the actions supported within the system are easily discovered and interpreted by users, whether they have unexpected side effects, and whether there are important action sequences that are longer or more tedious than necessary.

6   Discussion

This chapter has focused on the design of sensemaking representations in terms of conceptual structures. Some sections (notably those on data gathering) address topics that are covered in more detail in other texts, and are included here for completeness. The heart of our argument is that sensemaking tools should where possible be transparent to their users, working as far as possible with people's existing conceptual representations, but that there is often additional benefit in introducing new, important and powerful concepts to people that allow them to think about the domain in a new way. Of course, to be included in a sensemaking tool, *someone* must have identified them as potentially valuable concepts to include. The sections on requirements and evaluation have focused particularly on conceptual fit between a typical user and a system.

Recognising that every user has individual interests and ways of thinking about a domain, we have also argued that appropriation tools such as features for annotating and marking representations can help individual users to make sense of a domain on their own terms, and to track their own developing understanding.

In the design section, we have proposed a set of representation types that might support users' sensemaking in different domains. We do not claim that this set is comprehensive, and we look forward to discovering (or generating) novel representation types for new sensemaking problems. This set has been proposed based on a review of the interactive visualisation literature and our own experience. A particular sensemaking tool may bring together multiple representations.

Having an appropriate conceptual structure is not the only success criterion for an interactive visualisation. Many other usability and usefulness criteria will also come into play. Conceptual structure is, nevertheless, an important contributor to usability and utility that has, historically, received insufficient attention in design and evaluation of interactive visualisations for sensemaking.

References

1. Ahlberg, C., & Shneiderman, B. (1994). Visual Information Seeking Using the FilmFinder. CHI'94.
2. Attfield S. & Blandford A. (2010) Discovery-led Refinement in e-Discovery Investigations: Sensemaking, Cognitive Ergonomics and System Design. Artificial Intelligence and Law - Special Issue on e-Discovery, 18.4, 387–412.
3. Attfield, S. & Blandford, A. (2011) Making sense of digital footprints in team-based legal investigations: The acquisition of focus. Human–Computer Interaction Journal. 26.1&2, 38–71.
4. Barone, R., & Cheng, P. C. H. (2004). Representations for Problem Solving: On the Benefits of Integrated Structure. In Proc. IV04. IEEE Computer Society (pp. 575–580)
5. Bertin, J. (1983). Semiology of Graphics: Diagrams, Networks, Maps. University of Wisconsin Press
6. Beyer, H., Holtzblatt, K. (1998) Contextual Design. San Francisco : Morgan Kaufmann.
7. Blandford, A. & Attfield, S. (2010) Interacting with information. Morgan & Claypool. http://www.morganclaypool.com/doi/abs/10.2200/S00227ED1V01Y200911HCI006
8. Blandford, A. E. & Green, T. R. G. (2001) Group and individual time management tools: what you get is not what you need. Personal and Ubiquitous Computing. 5.4, 213–230.
9. Blandford, A., Green, T. R. G., Furniss, D. & Makri, S. (2008) Evaluating system utility and conceptual fit using CASSM. International Journal of Human–Computer Studies. 66. 393–409.

10. Carroll, J., Howard, S., Vetere, F., Peck, J., & Murphy, J. (2002). Just what do the youth of today want? Technology appropriation by young people. In Proc. 35th Annual Hawaii International Conference, 1777–1785.
11. Carroll, J. M. & Rosson, M. B. (1992) Getting around the task-artifact cycle: how to make claims and design by scenario. *ACM Transactions on Information Systems*, 10.2, 181–21.
12. Craft, B., & Cairns, P. (2008). Directions for Methodological Research in Information Visualisation. In Information Visualisation, 2008. 44–50.
13. Dix, A. (2007). Designing for Appropriation. In Proc. 21st British HCI Group Conference (Vol. 2, pp. 28–30). Lancaster, UK
14. Ericsson, K.A. & Simon, H.A. (1984) Protocol analysis: Verbal reports as data. Cambridge, MA: MIT Press.
15. Faisal, S., Attfield, S. & Blandford, A. (2009) A Classification of Sensemaking Representations. CHI 2009 Workshop on Sensemaking.
16. Faisal, S., Cairns, P. & Blandford, A. (2006) Developing User Requirements for Visualizations of Literature Knowledge Domains. In *Proc. IV06*.
17. Faisal, S., Cairns, P. & Blandford, A. (2007) Building for Users not for Experts: Designing a Visualization of the Literature Domain. In Proc. *IV07*.
18. Faisal, S., Cairns, P. & Blandford, A. (submitted) Designing for seamless interaction with visual analytic tools. Submitted for journal publication.
19. Fekete, J., Grinstein, G., & Plaisant, C. (2004). IEEE InfoVis 2004 Contest, the history of InfoVis. Retrieved from www.cs.umd.edu/hcil/iv04contest (2004).
20. Hackos, J. & Redish, J. (1998) User and Task Analysis for Interface Design. John Wiley & Sons
21. Heer, J., & Boyd, D. (2005). Vizster: Visualizing Online Social Networks. InfoVis'05.IEEE
22. Johnson, J. & Henderson, A. (2012) Conceptual Models: Core to Good Design. Synthesis Lectures on Human-Centered Informatics. Morgan & Claypool Publishers. doi: 10.2200/S00391ED1V01Y201111HCI012
23. Kang, Y., Görg, C. & Stasko, J. (2011) How Can Visual Analytics Assist Investigative Analysis? Design Implications from an Evaluation, IEEE Transactions on Visualization and Computer Graphics, 17.5, 570–583.
24. Klein, G., Phillips, J. K., Rall, E. L., & Peluso, D. A. (2007). A Data-Frame Theory of Sensemaking. In R. Hoffman (Ed.), Expertise Out of Context: Proc. Int. Conf. on Naturalistic Decision Making. Lawrence Erlbaum Assoc. Inc, US, 113–155.
25. Kuniavsky, M. (2003). Observing the user experience: a practitioner's guide to user research, Morgan Kaufmann, San Francisco.
26. Moran, T.P. (1983) Getting Into a System: External-internal Task Mapping Analysis, in A.Janda (ed.), Human Factors in Computing Systems, pp. 45–49. ACM SIGCHI and Human Factors Society conference proceedings, Boston, December 1983. New York: ACM Press.
27. Norman, D.A. (1986) Cognitive Engineering. In D.A. Norman and S.W. Draper, Eds. User Centered System Design, Hillsdale NJ: Lawrence Erlbaum, 31–62
28. Payne, S.J., Squibb, H.R. & Howes, A. (1990) The nature of device models: The yoked state space hypothesis, and some experiments with text editors. Human-Computer Interaction, 5, 415–444.
29. Pirolli P. and Card S., (2005) The Sensemaking Process and Leverage Points for Analyst Technology as Identified Through Cognitive Task Analysis. In: Proc. International Conference on Intelligence Analysis (McLean, VA, May 2–6, 2005).
30. Russell, D. M., Stefik, M., Pirolli, P., & Card, S. (1993). The Cost Structure of Sensemaking. In Proc. INTERACT '93 and CHI '93 (pp. 269–276). Amsterdam, The Netherlands: ACM
31. Shneiderman, B. (1996). The Eyes Have It: A Task by Data Type Taxonomy for Information Visualisations. In Proceedings of the IEEE Symposium on Visual Languages (p. 336–343). IEEE Computer Society
32. Shrinivasan, Y. B., & Wijk, J. J. V. (2008). Supporting the analytical reasoning process in information visualisation. In Proc. CHI (pp. 1237–1246). ACM

33. Spence, R. (2007). Information Visualisation: Design for Interaction (2nd ed.). ACM Press Books
34. Stasko, J., Görg, C. & Liu, Z. (2008) Jigsaw: Supporting Investigative Analysis through Interactive Visualization, Information Visualization, 7.2, 118–132.
35. Stasko, J., Gorg, C., Liu, Z., & Singhal, K. (2008). Jigsaw: supporting investigative analysis through interactive visualisation. Information Visualisation, IEEE, 7(2), 118
36. Toulmin, S. (1958). The Uses of Argument, Cambridge, England: Cambridge University Press.
37. Tufte, E. R. (1986). The Visual Display of Quantitative Information. Graphics Press Cheshire, CT, USA
38. Uren, V., Buckingham Shum, S., Bachler, M., & Li, G. (2006). Sensemaking tools for understanding research literatures: Design, implementation and user evaluation. *IJHCS, 64*(5), 420–445.
39. Ware, C. (2004). Information Visualisation: Perception for Design. Morgan Kaufmann
40. Wigmore, J.H. (1931). The Principles of Judicial Proof. Little, Brown & Co.
41. Williamson, C. & Shneiderman, B. (1992) The Dynamic HomeFinder: Evaluating dynamic queries in a real-estate information exploration system, Proc. ACM SIGIR'92. 338–346.

An Introduction and Guide to Evaluation of Visualization Techniques Through User Studies

Camilla Forsell and Matthew Cooper

Abstract The objective of this chapter is to increase awareness of what constitutes a sound scientific approach to evaluation through user studies in visualization, and to provide basic knowledge of current research practice relating to usability and evaluation. The content covers the most fundamental and relevant issues to consider during the different phases of an evaluation: planning, design, execution, analysis of results and reporting. It outlines how to proceed to achieve high quality results and points out common mistakes during the different phases and how these could be avoided. The chapter can be used as a guide when planning to conduct an evaluation study. The reader will also learn to better judge the relevance and quality of a publication presenting an evaluation when reviewing such work, since the same guidelines apply.

1 Introduction

User-centred evaluation is often described by visualization researchers as being a prohibitively difficult task, and it certainly is a difficult area to master and to carry out consistently well. The value of it, however, cannot be overestimated. Well tested results are meaningful, positive results providing a basis on which the next generation of development can be built with greater confidence, while negative results provide useful knowledge which can help to refine the future work to ensure

C. Forsell (✉)
Associate Professor, Department of Science and Technology, C-research, Linköping University, Linköping, Sweden
e-mail: camilla.forsell@liu.se

M. Cooper
Associate Professor Department of Science and Technology, C-Research, Linköping University, Linköping, Sweden
e-mail: matthew.cooper@liu.se

W. Huang (ed.), *Handbook of Human Centric Visualization*,
DOI 10.1007/978-1-4614-7485-2_11, © Springer Science+Business Media New York 2014

that it moves in positive directions. Today user-centred evaluation is a key research challenge in visualization [8, 33] and researchers submitting their work to journals and conferences are often encouraged to conduct such studies to investigate and report on the usability of their proposed techniques. The difficulty in conducting scientifically sound studies is acknowledged, however, and inexperienced authors are often encouraged to partner with colleagues with greater knowledge to ensure high quality studies [1, 27]. The reality, however, is that many techniques are never subject to any evaluation study or, if they are, then often the approach used is questionable thus making it hard to trust the results and conclusions and so gain insight from them.

In this chapter we aim to present knowledge and guidance on sound scientific evaluation in an accessible form, and so outline how to proceed to achieve high quality results. The chapter covers the most fundamental and relevant issues to consider during different phases of an evaluation study. It points out common mistakes which many researchers may make during the different phases, and provides guidance and examples of how these can be avoided.

2 Background and Motivation

The goal of visualization is to provide techniques and tools that promote insight and understanding of a data set; to support and amplify cognition [6]. Today one cannot fail to be impressed by the vitality of the field in developing impressive images and structures of data, and complex interactive tools to support exploration, analysis and decision-making. The creativity of the thinking behind the new approaches that are proposed, the new applications to which methods are put, and the new analytical tools which underpin them, are outstanding. This is evident when, for example, attending the annual IEEE VisWeek (from 2013 renamed to VIS) [26], the triple-event covering Scientific and Information Visualization, and Visual Analytics, which is regarded as the most important conference in visualization in the world. The other side to this, however, is that from all of this creativity very few of the techniques and tools ever achieve a high uptake by users. Many of the proposed methods risk become only interesting oddities, which are disregarded outside of academic research. One reason for this is the lack of scientific evaluation to support and demonstrate their actual usability. To develop successful techniques we need to assess their merits and disadvantages and the value of evaluation cannot be overestimated. Lack of evaluation, on the other hand, allows promising and potentially useful ideas to fail to be expanded upon by other researchers, or adopted by users, since evidence of usability and measurable merits are not presented [36]. A number of observers have commented upon the scarcity of evaluation of the techniques shown by presenters at the major conferences in the world. For example Lam [31] showed that in the symposium on Information Visualization at IEEE VisWeek, between 1995 and 2007 only about one third of the papers (105 out of 283) included any kind of evaluation and only 39 out of these involved users and tasks. In later years the number has grown but still rigorous evaluations are scarce. In Scientific Visualization such evaluations have been even less common [21].

From our discussions with developers, many do indeed understand the need for evaluation and do want to evaluate their developed tools and techniques but find the complexities of the process daunting, which may mean they simply never begin. For the inexperienced person (as mentioned in the introduction) cross-disciplinary collaboration or consulting with an expert on user study design and data analysis can be an excellent means to overcome this problem. This is also reflected in the call for papers when submitting work to conferences (see, for example, [27] or [1]) that have previously included the statement:

> We do suggest that potential authors who have not had formal training in the design of experiments involving human subjects may wish to partner with a colleague from an area such as psychology or human-computer interaction who has experience with designing rigorous experimental protocols and statistical analysis of the resulting data.

However this is not always possible to achieve, which is unfortunate because many developers just do not have the skills needed to perform sound evaluation. This statement might seem dismissive of our fellow visualization researchers but it is merely the reality and it is our belief that most would agree. Many of us, after all, are technologists rather than from domains where we are taught experimental psychology and methodology. The result is that it is quite common to see publications presenting evaluations where the outcome is, in the worst case, meaningless since flaws in the method applied and/or the analysis of data makes it impossible to draw useful and scientific insights from the results. This is very unfortunate since the authors have invested a lot of time and effort into a study where the outcome is not as informative and valuable as could have been. Many of these flaws are easily avoided once the experimenter is aware of them and of how to avoid having them confound the study. In other cases it is the actual reporting of the study that is insufficient, not providing sufficient detail for the reader to gain insight and trust in the work. This could be prevented by better knowledge of what information to include and how to present it, and how to avoid common omissions and ambiguities.

Based on our experience as authors, reviewers, papers chairs, symposium chairs and tutorial presenters we believe this chapter will be valuable to all visualization researchers, developers and academic reviewers who do not have training or experience of evaluation methodology.

Parts of the chapter are based on previous publications [15, 17] and tutorials [16] which received very positive responses and where attendees expressed the great need for this type of easily accessible knowledge and guidance. This can be illustrated by the following quote from one tutorial attendee:

> After this tutorial I feel much more confident in actually doing another user study and now I fully agree with the reviewers and their comments when they rejected another I did.

The reader of this chapter is assumed to be aware of visualization as a methodology but not to be an expert in the field of evaluation, and so a basic tone is adopted. Our aim is not that such a reader will finish studying this chapter as someone now fully capable of designing and conducting an evaluation, analysing its results, and reporting it appropriately. However, the reader should come away knowing what factors are important to consider and some of the basics of how it should and should not be done. Hence, the main contribution of this chapter should be its value as a guide when planning to conduct and document an evaluation study. Also, the same rules that guide a sound scientific approach to conducting an evaluation also apply when reviewing the work of others. This means that the reader will also be better able to identify potential problems in an evaluation, see when studies have been poorly designed, identify basic problems with statistical methods used for analysis, and be able to ask questions about why evaluations have been carried out in an unsatisfactory way. The reader will also be provided with good references for further study in the area of evaluation in order to expand their knowledge and skills appropriately. The main focus of this chapter is on quantitative evaluation studies, often called experiments. Such studies are characterized by the need for a rather high degree of mechanisms for controlling the quality of the output. However, the general knowledge provided by reading this chapter applies to all kinds of studies. For example in a qualitative study, conducting an interview or applying a 'think-aloud' protocol to collect data, the level of control is not comparable with a controlled experiment, but the evaluator may still want to make sure that all participants are asked the same questions, or perform the same actions, in the same order and that the evaluator does not influence their answers or activities while interacting with them. Hence, most studies involve some precision and control and will benefit from considering, identifying and avoiding any factors that could confound the outcome.

The remainder of this chapter is structured as follows. Section 3 discusses different goals within scientific evaluation processes and the importance of a sound approach. Section 4 describes the different phases of an evaluation study (planning, design, execution, and analysis of the results) and how to proceed to achieve high quality results. This involves ensuring reliability, validity and consideration of any confounding factors that may bias the outcome and the analysis. Section 5 describes the importance of reporting and how to proceed to ensure readability, provide details relevant to the outcome, and enable replication. In Section 6 we present a catalogue of common problems and discuss when and why evaluation studies may fail. Finally, Section 7 concludes the chapter.

3   Scientific Evaluation

In visualization today, usability studies and user-centred evaluation are key challenges and considered important parts of research when developing techniques. Evaluation is, at its most basic, about testing something in some way in order to determine its value. In [37, p. 2] the author argues that there are two types of tests: formal comparative experiments and exploratory usability evaluations. The difference is explained in terms of the purpose for which the test is being done. In an experiment the aim is to collect data to demonstrate the relative worth of the subject under evaluation (in our case a visualization technique). Typically this is done comparing two or more ideas. An evaluation on the other hand aims at assessing worth by use and to gather information to inform design and implementation in an iterative development process [ibid]. In this chapter we consider evaluation to be either of these two activities.

'Usability' is a term which has been defined in a variety of ways. One commonly adhered to is the following, given in ISO 9241-11 [2], which defines usability to be:

The extent to which a product can be used by specified users to achieve specified goals with effectiveness, efficiency and satisfaction in a specified context of use.

Using this definition one can examine three dimensions of usability. Effectiveness assesses to what extent users can achieve their objectives with completeness and accuracy while efficiency refers to the amount of effort (for example time spent or cognitive load) expended in relation to the accuracy and completeness. Satisfaction is a subjective dimension referring to users' opinions, attitudes and preferences. When assessing usability it is, in many cases, valuable to address all three dimensions. Also, note that all three can be addressed using both objective and subjective measures, thus a subjective measure is not limited to satisfaction and objective measures can also be evaluated subjectively [24]. For example, objective time such as recorded response time for completing one task may be higher or lower than the subjectively experienced elapsed time while doing it [ibid]. This also shows that there exists no simple correlation between these dimensions. Thus they should be considered independent and investigated separately when necessary, since they may lead to different conclusions [20]. However, the more data that must be collected, the more complex and time consuming the evaluation and analysis will be, so it is necessary to decide what is sufficient to produce a meaningful answer to the research question.

What, then, is the purpose of evaluation studies in visualization? What kinds of questions are we trying to answer? In [30, p. 20] the authors have classified some general aims into the following common thematic areas:

- To evaluate strengths and weaknesses of different techniques.
- To seek insight into why and for what type of task a particular technique gives good performance.
- To demonstrate that a new technique is useful in a practical sense, according to some objective criteria for a specific task.
- To demonstrate that a new technique is better than an existing one according to some objective criteria for a specific task.
- To investigate whether theoretical principles from other disciplines apply under certain practical conditions (for example results from psychophysics or vision research may not extend to visualization).

Based on aims such as those listed above, one can develop a general research question to begin with. This is then refined into more specific problem statements, or hypotheses. Next it is possible to consider what methodological approach will be appropriate to yield robust, reliable answers. This is not always an easy and straightforward process. This chapter does not provide a literature review or clear classification of available evaluation methods or guidance on how to chose between methods. This would go beyond the scope possible in this short chapter. Instead, this chapter focuses on how to proceed once the method is chosen, and provides some specific examples, mainly from quantitative experimental research. We therefore refer the reader to publications such as [7] and other chapters in this book for such knowledge and guidance. We also strongly recommend the recent publication "Empirical Studies in Information Visualization: Seven Scenarios" [32]. In that work the authors, based on a comprehensive literature review covering over 800 publications, present a scenario-based classification of evaluation studies. The scenarios they derive are: evaluating visual data analysis and reasoning, evaluating user performance, evaluating user experience, evaluating environments and work practices, evaluating communication through visualization, evaluating visualization algorithms, and evaluating collaborative data analysis. Within each scenario the authors present different research goals that can be achieved, what types of questions can be investigated and possible output in terms of collected data, for example quantitative and qualitative metrics and subjective feedback. They also suggest possible methods for how to approach these goals and questions, and they use published work to illustrate and exemplify different approaches.

A clear research question is crucial since it both specifies what to investigate and it also largely determines how the research should be carried out in terms of the level of constraints, the method for collecting data, and the data analysis. Level of constraints refers to the extent to which we structure, limit or control any part of an evaluation study [22, p. 43]. These levels range from low-constraint research, as in naturalistic observation in a work context, to very high constraint, as in strictly controlled studies such as laboratory experiments. All levels of constraint are scientific when used correctly, and it is important to acknowledge that they have to map onto the question(s) to be answered [ibid].

In experimental research we investigate and compare participants' responses under different conditions. We do this by manipulating one (or more) factor or variable (the independent variable) to investigate its effect on one (or more) other factor (the dependent variable). We could, for example, compare performance using a 2D visual representation and a 3D visual representation for some task. The independent variable here is then the visual representation method (having two levels: 2D and 3D) which results in two experimental conditions. The dependent variable is performance which could be measured by, for example, accuracy or time to solution. This measure is then analysed using statistical tests to investigate whether there was any experimental effect, that is a difference between the two conditions (2D and 3D) or not. The aim is to control everything and vary only the independent variable(s). Then any effect observed can be attributed to the independent variable and not to any confounding factors or variables that might affect the dependent measure (accuracy or time in the previous example). Confounding factors (also called extraneous factors or variables) are factors that are not associated with the specified independent variable(s) but which are not controlled and thus may distort any (potential) effect [22].

A well thought-out design, evaluation procedure and analysis of the results will answer your research question clearly with well argued results and conclusions. A poor design, on the other hand, may render the conclusions invalid and the work wasted. There are many factors to consider when selecting specific methodological choices and it is often not possible to predict and control for everything that might have an impact on the results. However, there are many actions the evaluator can take to make sure that the obtained results will be scientifically sound. This involves ensuring reliability, validity and the consideration or elimination of known confounding factors that are threats to these concepts.

Reliability refers to repeatability, or the consistency of something from one time to another. It could be how consistently the different observers or scorers give judgements of the same aspect, the consistency of a measure from one time to another, or consistency in the behaviour of a person moderating an evaluation from participant to participant. Validity relates to soundness or quality, which is whether or not a study can scientifically answer the questions it is intended to answer. This includes [22]:

- Construct validity: Does the evaluation investigate and assess what was intended according to the theory behind the research question?
- Internal validity: Is the result based on the design, measures, setting and procedures or due to any confounding factors?
- External (ecological) validity. Is it possible to generalize the results to different people, contexts or places or are they limited to this particular evaluation?
- Statistical validity: Are the conclusions from statistical testing sound and justified from using the appropriate test?

Please note that [22, p. 81]:

A measure cannot be valid unless it is reliable, but a measure can be reliable without being a valid measure of the variable of interest.

A major concern when carrying out an evaluation is the reliability and validity of the measures, procedures and conclusions. We should always strive for theoretical and methodological appropriateness or soundness, this is how we can obtain high quality results having great value to the community. Issues related to the reliability, validity and generalizability of the outcome are often questioned by reviewers and it is important to ensure that these issues are considered and reported in any publication to support trust in the results. Unfortunately there are many issues that, if not adequately addressed, can compromise a study, or make it difficult to draw useful insights from the results. The remainder of this chapter outlines the most important issues to consider and provides guidance on how to address them through best practice.

4 Planning and Designing an Evaluation

The research question of an evaluation study and the design of the evaluation study are, to a high degree, interdependent. It is difficult to actually settle upon a research question, or any specific hypothesis, without having some clear idea about what study design will be used. The design will determine what sort of question(s) can be answered with the resulting data and how well this can be done. A well thought-out design and analysis of the outcome will answer your research question clearly with sound data and well argued results and conclusions. A poor design on the other hand may render the outcome invalid and the work wasted. When settling on a design there are many factors to consider. You need to answer such questions as:

- What type of participants do I need and how many should be engaged?
- Should I obtain data from one group of participants or more than one?
- What activities should the participants do and how should I present these appropriately?
- How am I going to instruct and train the participants before the evaluation session?
- How should data be recorded, compiled and analysed?

In the following subsections we provide answers to these questions and to other common and important matters to deal with. We also provide references to more detailed publications.

4.1 One Group of Participants or More

One important decision is whether the study should examine performance (or some other metric) of the same participants or compare data between two or more groups of participants. Both approaches have their strengths and weaknesses and, when deciding on which to choose, these need to be weighed and compromises may be necessary. In the first approach the same participants take part in all conditions of the study and any comparison takes place within the same group of participants with the data from each participant being related. Each participant is being compared to him- or herself and serves as their own control. Any individual difference having an effect on the outcome (such as experience, high or low motivation, age) is thus the same across all conditions [14] and cannot be attributed to differences between participants. Designs using a single group are also known to be more sensitive to small differences between conditions and so are more likely to detect differences if they exist (there is generally less variance within a single participant than across a group). In the literature this design is called a 'related' design, a 'same subjects' design, a 'within-subjects' design, or a 'repeated measures' design.

When the comparison of results is made between groups of participants their scores are unrelated [ibid]. This is usually referred to as an 'unrelated design', a 'between-subjects' design or a 'between groups' design. Hence, there are several different names referring to the same type of design. This is important to know, for example, if you need to analyse the data statistically since it is reflected in the names of the statistical tests associated to the designs. For instance, if the evaluation has examined how one group of participants has performed on a variable having two levels (such as a task being simple or complex) then it is appropriate to use a 'related t-test' [13, 14].

The decision of what design to apply depends on several factors, see for example [14, 37, 44]. Some important factors are:

1. *The aim of the study*: If the purpose is to investigate whether there is a difference between groups of people, for example two types of professions using the same visualization technique for work tasks, then it is fundamental to the research question that the evaluation compares performance between these two groups of participants. If, on the other hand, the purpose is to determine how easy it is to perceive different elements on a display then it is not as important where the participants come from and the same participants could be tested under all conditions. Performance on such low-level perceptual judgement tasks is regarded to involve a universal perceptual ability and so expertise is not a factor.
2. *Availability of participants*: A between-subjects design typically requires a larger number of participants since each person only provides data for one of the conditions. When it is hard to recruit, this option might simply not be possible. It may also be impractical due to limitations of other resources such as time and space.
3. *Confounding factors*: If it is suspected that participation in one condition of a study (solving tasks using visual representation A) will affect the result in the following condition (solving tasks using visual representation B) due to an unwanted learning effect, then it is a requirement to use a between-groups design to avoid such carry-over effects. Sometimes this can be avoided by controlling the order of presentation of conditions (see Sect. 4.2.1 on participants and assignment) but in other cases that may not help. Participation time is another major consideration. If each participant must spend a long time completing the study then it might be appropriate to choose a between-group design in order to reduce the number of trials that each participant must complete, and so avoid fatigue (negative practice effect) and decreasing motivation. Allowing rest periods can also help to control the fatigue but increases the overall time required. When using a between-groups design for these kinds of reasons it is necessary to ensure that there are no significant differences between the groups in the beginning. For example if the objective of the evaluation is to investigate how novices learn to use a visualization method then it is necessary to make sure that all participants are novices to begin with and don't differ in their level of previous experience.

As evident from the descriptions above there are many issues to consider when choosing an appropriate design and it is not always perfectly clear what would be the right one. Often the result is that a 'mixed-design' may be the best solution since it is common that neither a within- nor a between-subjects design meets all the requirements. This design has multiple variables some classified as within-subject variables and some as between-group variables. For example you could use a mixed design to examine whether there is a difference between men and women (two different groups) using a specific visualization technique (both groups take part in this condition).

4.2 Participants

The first critical task when it comes to recruiting participants to take part in an evaluation is to select them: a process decided by factors such as appropriateness, study goal(s), availability and cost. As mentioned previously, it is important to be able to generalize the findings so the participants should be representative of a larger group of people (the population) and not just the participants in the actual study (the sample), and they should match the target audience. Thus it is necessary to find a sample that correctly reflects the properties of the population [22]. This is harder than it sounds. Many visualization tools are intended for expert users (perhaps neurosurgeons) and such persons can be difficult to engage for a sufficient period of time and in sufficient numbers. Therefore students are often used instead since they are easier to access and often motivated by earning study credits or some small reward. When such persons are not considered representative enough you have to be aware of the limitations of the collected data and the conclusions that can be drawn based on these data. There is also a risk that the study will draw a great deal of criticism from persons reviewing the study. When appropriate, for example in investigating low-level perceptual tasks where expert users are not needed, it is possible to recruit large numbers of participants using a crowdsourcing platform such as Amazon Mechanical Turk [25]. Here you can effectively collect a great deal of data in little time and at low cost but it has several known limitations: the evaluator has no control over the hardware or the context in which the evaluation is done, participants cannot ask questions to resolve any confusion and there are issues with 'random clickers' [10]: participants who carry out the task without care for the accuracy of the results they enter. The discussion above directly raises the next critical question: how many participants are required to conduct an evaluation? This is one of the most commonly asked questions and also one for which it is very difficult to provide a simple answer. As stated in [41, p. 17]:

> There's no rule that says if you don't have at least x number of participants in a study, the data won't be valid.

The appropriate sample size depends on the aim of the study, the design used, the planned analysis of the resulting data and the tolerance for a margin of error. When doing quantitative studies, where data is analysed by statistical testing, there is a concern about 'statistical power'. This power refers to the sensitivity of a statistical test for detecting a significant experimental effect (a real difference) assuming it is present. When studies do not have enough power they don't include enough participants to be sure of detecting the effect, leading to the risk of missing an actual difference [13]. The power is described by a number between 0 and 1, and 80% or 0.8 is usually considered an acceptable value. The traditional way to increase power is to increase the number of participants. The size needed for a specific level of power can be computed, and there are software packages available for doing this (see for example [5]). The calculations do, however, require specification of a number of criteria about the evaluation that can be difficult to anticipate. Also, the result is often that a surprisingly large number of participants are required, so it is often necessary to compromise and accept lower statistical power. This can be particularly true when the experiment requires participants that are hard to recruit, or when time is short. Hence we have to make an educated decision from study to study. A good guideline is that for quantitative studies the more participants the better the result. This is also true when calculating confidence intervals, which indicates a range of values within which the population (all intended target users) mean is likely to fall based on the results from the sample (the participants in the evaluation). Regarding how much error is acceptable, confidence intervals change as a function of sample size. Thus, the larger the sample size, the more confidence one can have that the findings are representative of the population [14]. It is rare to see studies from trustworthy authors with less than about 10 or 12 participants per group. Hence, if the experimental design involves comparing 3 different groups, a minimum of 30 participants will be required. Good guidelines for appropriate sample sizes can be found in recognized published work, reviewing their choices and justifications.

One common misconception is that "for all user studies 5 participants is enough". This statement originates from Jakob Nielsen's work on Heuristic Evaluation [35] and is, as he clearly stated, only valid when the study involves using a homogeneous group of evaluators (usability experts in the original form of the method) to find usability problems in an application. Then, based on a mathematical formula, the gain of adding more evaluators is not in relation to the findings since five of them will typically find about 80% of the problems [35, p. 33]. It is recognized that these studies are not perfect and that they will not discover all the problems actually present. Instead they are based on a discount usability philosophy, and are considered good enough for debugging in iterative design processes. In no other case can this number, five, be recommended with confidence. There are also studies in the area of human-computer interaction that present the benefits of a larger number and how it may vary with what is evaluated (for example tool diversity) and what types of problems are found (complexity of tasks) [12, 45]. Most importantly, there is as yet no empirical evidence as to whether any of these simple recommendations regarding sample sizes will transfer and thus be applicable for evaluation studies in visualization.

4.2.1   Allocating Participants to Groups and Conditions

The next important task is to allocate the participants to the different groups and conditions in the study. Obviously the first issue is not relevant when using a within-subjects design but a between-subjects design means that each participant must be allocated to one or another of the groups. The best way to do this is through random assignment unless the study requires that the properties of each group be different. Random assignment on a sufficiently large group should ensure that the characteristics of the different groups will be approximately the same and therefore any difference (experimental effect) observed between the groups can be attributed to the effect of what is being studied and not to any characteristic of the individuals in the groups [13, 22]. It is important to note though that purely random assignment never guarantees that the different groups are equivalent, only that any differences are due to chance. Sometimes it is the case that the participants cannot be considered totally equivalent: perhaps 24 persons have been recruited and four of them are more experienced than the others. If these more experienced individuals can be identified in advance (which they should have been after having provided demographic information, see Sect. 4.4.) then, having two groups in the design, it is possible to randomly assign two of these more experienced persons to each group.

Randomization can also be used when it comes to the assignment of order of presentation of conditions or tasks [13,22]. For instance, if the objective of a study is to evaluate the performance in terms of accuracy and response time when identifying two types of glyph (A and B) using two different display techniques, perhaps 2D and 3D display. This design has two variables with each having two levels, display technique (2D and 3D) and type of glyph (A and B), and thus yields four different conditions: 2D glyph A, 2D glyph B, 3D glyph A and 3D glyph B. Using a within-subjects design all participants will take part in all conditions. Remembering the discussion about carry-over effects in Sect. 4.1, it is important to realize that the order of presentation of these four conditions is a major concern. Consider, for example, a participant who starts by performing tasks with the 2D glyph A condition and then proceeds to do an equivalent task in the 3D glyph A condition. This second condition might be found to be easier, with both better accuracy and better response times, and these results will appear in the obtained data. It might be the case that the 3D version of the glyph is more intuitive to perceive but it might also be the case that having previously used the 2D counterpart makes it easier to understand the 3D version. It might also be because the participant has simply become more familiar with the overall situation. Consequently the order of presentation has to be considered and handled appropriately. There are several techniques one can use to achieve a sound sequencing and the logic behind them is to control for order effects by having these contribute equally to each condition in the study [22]. First, we can randomly assign each participant to a different order of the four conditions by simply shuffling different orders. Any differences observed between conditions can then be attributed to a true difference between visualization methods and/or glyphs since any potential order effects can be assumed to have been evened out between them. A second approach is to counterbalance, which means that the presentation order of conditions is systematically varied. With complete counterbalancing an equal number of participants are assigned to each of the possible orders, and the number of participants needed can be calculated as $nX!$, where X is the number of conditions [22, p. 226]. This procedure is only useful when the evaluation includes few conditions, however, since the factorial expands too quickly for general use. In the previous example, with four conditions, there are 24 orders, which may well be manageable, but with six conditions there will be 720 orders and a full counterbalancing will almost certainly be impossible. In this case the best thing to do is to use a partial counterbalancing where, if the experiment uses a participant group of 12 persons then 12 out of the 24 (or 720) possible orders are selected and the participants are randomly assigned to these selected orders [22, p. 226]. Another method for counterbalancing is the 'Latin Square', where the order of presentation is determined using a matrix where the conditions are arranged in rows and columns so that each condition appears only once in each column and once in each row. In a partial Latin square one order can appear more times than another. There are also complete ones where all conditions appear in each position in the sequence an equal number of times, and all conditions follow each other condition an equal number of times [22, p. 226]. Using this complete version, again a large number of participants may be required.

To conclude, if there is any risk of an order effect and randomization and/or balancing is not employed in the allocation of participants to groups and conditions, properly, the outcome of the study will be meaningless and no statistical analysis of the data can overcome this flaw. However, there are cases when the order does not need to be controlled. For example, when conditions are unrelated to each other and so learning how to perform tasks in one of them will not help in the other. Thus any learning effects biasing the results are unlikely. In some situations it is actually inappropriate to control the order. This happens when a specific order of tasks or questions is necessary since they naturally are presented in that way, or need to be sequenced that way just to make sense. Here any order effects should be regarded as a general learning process and not as a result of a poorly designed evaluation [37].

4.3 Designing Tasks and Data

In an evaluation study participants are asked to perform one or more tasks, a specific activity, used to examine the effects, if any, of the independent variable(s) on the dependent variable(s).

The definition and selection of tasks is sometimes predetermined due to the visualization technique being developed with a specific target area of application in mind, for a specific group of users, to support specific tasks etc. In such a case it is clear which activities should be included. For example in [39] a study was performed with the objective to evaluate a novel interaction concept in visualization inspired by real-world behaviour of people comparing information printed on paper. Hence the tasks were chosen to mimic these real-world tasks of comparison: side-by-side, folding (browsing) and fading (shine-through). At other times the aim might be to evaluate a novel visualization concept, or to investigate a certain phenomenon, in which case appropriate tasks need to be invented. Here one option is to review known taxonomies in the literature and choose relevant tasks suggested by those examples. See for example [3, 4, 34, 42].

The task chosen have a tremendous impact on the outcome of an evaluation, on what aspects can be identified and to what extent any findings can be generalized. A task needs to be designed to meet a number of criteria such as: it should be well defined, unambiguous and straightforward to give a response to. It should be simple enough to allow a controlled comparison between conditions yet pinpoint the relative differences between them. Further, it should allow equivalent answers over all conditions thus not favour one condition over another or lead to a comparison between "apples and oranges" [37].

When selecting tasks there are two very important issues to consider. (1) The task (and its associated metrics for assessment) has to be appropriate and have construct validity. This implies that it is supported by both theory and empirical work related to the research question: that it assesses what it is supposed to assess. (2) The task also needs to be representative, meaning that it should be characteristic for tasks in the intended application domain, for the intended users etc. Ensuring this allows for the generalization of the results outside of the specific context of the evaluation [22]. For example, testing participants' performance for the identification of a specific relationship (pattern) in data when using a parallel coordinates visualization with a data set made up of 75 data items may not to be considered as representative for a practical usage situation where a data set might include 15,000 items. In [38], Stasko phrased it like this: "When some of the tasks being assessed are not exactly the kind of tasks a system was built for, then the evaluation simply will not be balanced and fair."

A task is what participants should do in a study, for example (as described above) to identify patterns in a data set. The concrete representation of the task in each condition is called a stimulus and is thus how variations in the independent variable(s) are presented to the participants. Each occasion a participant performs the task using a stimulus is called a trial (usually one of a number of repetitions of a task). For example, a study having two types of task executed 20 times each per participant yields an experimental design having 40 trials in total. In a study it is often good to collect as much data as possible and the more trials carried out, the more data can be aggregated for each task. Also, depending on the nature of the individual task and the study, it may be possible to aggregate data over tasks or they may need to be treated and analysed separately.

In [43] we evaluated a visual analysis tool representing spatio-temporal data. The aim was to compare performance between using a 2D and a 3D representation of such data. This visual analysis tool supports a variety of exploration tasks and we chose the following representative one: identify how many activities are concurrently performed by a group of individuals over a period of 24 h. This type of analysis is often done on groups of people, for example families, so a representative number of individuals to compare would be between two and eight. Here, one approach would be to choose one number from this range and create a selection of stimuli representing activity paths from that number of individuals. However this would result in no variation of the task. On the other hand, including all possibilities would result in seven different family types, leading to a very complex design and subsequent analysis. Hence we chose two variants: families having two and having four individuals, respectively. One option would be to aggregate data from all trials over these two variations by considering the task only as a means for investigating performance between the 2D and 3D visual representation, as long as we assume no relevant difference between the actual variations in the stimuli (in terms of the number of individuals represented). Instead, we considered the task as an independent variable having two levels (two individuals vs. four individuals) and treated data from each of these conditions separately. We could then also establish that there actually was a difference in complexity, which also led to follow up experiments including more family types. Also, if the conclusions from a study are found to hold for a number of tasks (or variations in task complexity) then they do support generalization of a phenomenon to a greater extent [37]. Hence, the number of tasks, and how these are realized as stimuli, may have an impact on the analysis and interpretation of the results. Also, the total number of tasks and trials in an evaluation affects total participation time and the nature of them (cognitively demanding, tiring, boring etc.) influences the overall experience. One should always aim for the evaluation to fit within a reasonable time-frame, not forgetting to include time for introduction, practice runs and so on. Then it will be a pleasant experience for the participants and thus minimize the risk of bias in the result due to fatigue and declining motivation towards the end. We typically aim to keep this below 1 h but, here, common sense, experience and pilot studies are the most valuable tools to aid your design decisions.

Another critical issue, and one upon which reviewers often comment, is whether to use a real or a synthetic data set, and opinions here are strongly divided. Generally it is advisable to strive to use real data, or at least a data set that is representative of such in terms of size and complexity, since this will lend greater credibility to the results. There are, however, many occasions where the study requires the use of a synthetic data set since it is only by using such data that it is possible to fully specify and control the structures in the data which can be necessary to ensure that participants can execute the tasks without interference from other factors in a data set, and for performance to be measured [29]. For example, in [18] the ability to discriminate between five different patterns in data presented in a parallel coordinates display was evaluated, and in [28] the threshold levels for perceiving noise in data were investigated. For both these studies it was crucial to use synthetic data to ensure that only the features being sought were present, and that the participants were not reacting to other features present in the data.

Finally, any research question(s) investigated in an evaluation could be addressed in several ways and it is common that reviewers complain about tasks being either too abstract or too specific, that they should have been performed with real data rather than the data set you have constructed, that they do not allow a fair comparison between different conditions etc. This is to be expected since judging the work of others is sometimes a rather subjective process. All you can do is to take the advice presented in this section in to consideration and present clear motivations for your design decisions and justify the reasons behind them. For a more elaborate discussion and advice on tasks and data we recommend for example Helen Purchase's recent publication [37].

4.4 Moderating an Evaluation Session

This section provides some guidance on how to moderate an evaluation session, an assignment that includes many aspects. Here we cover the most fundamental and important ones for a successful evaluation session and that, when neglected, can lead to some common mistakes. For further details we refer the reader to, for example, [11].

It is important to collect relevant background or demographic information from the participants as, in any study, the persons taking part should be reported somewhere in the text, typically in a subsection of the "Method" section called "Participants" (see Sect. 5 in this chapter for an example). For us this description often includes data such as age, sex, education, occupation, optical acuity, and experience in visualization. Second, demographic information can be very valuable when analysing and interpreting the results from the actual evaluation session. It may, for example, enable the evaluator to identify causes of participants' unusually good, bad or ambiguous results. No matter how carefully you select your participants, unexpected properties such as previous experience may be present and having an impact. Sometimes it is also necessary to test participants for specific conditions, such as colour blindness, eye dominance, or stereo vision ability, prior to the evaluation session. Even if participants are asked in advance about any deficiencies, they may not be aware of having them since many people function perfectly well with quite significant disabilities in their vision.

Participants need to be appropriately informed and trained in all parts of the evaluation. For some studies it is required (and for some only recommended) to obtain a signed informed consent form each participant. The content of this form varies, see for example [11] for details, but it serves the legal purpose of documenting their agreement to participate and for their data to be used, that they were informed about the study and about their rights. The purpose of the study should be explained to them prior to the session beginning, as long as it will not jeopardize the result. In some cases details cannot be revealed until afterwards, then this should be clearly explained to each participant it is often best explained by including this aspect in the consent form when such is provided.

It is recommended that written instructions be provided for the participants to review before the session starts, and illustrations are often useful for clarity. This material should explain the entire procedure including: tasks to be performed, questionnaires to fill in, how to give responses, whether questions are allowed during the session or not, breaks, total participation time that is expected etc. It is also recommended that the moderator of the evaluation session follow a written protocol for how to moderate each part of the study. These uniform procedures will ensure that all participants receive equal information and are treated the same way. Also, it allows for different people to carry out the moderation uniformly if a single person cannot deal with all of the participants taking part in the study.

In most studies the participants should perform training tasks although, when the aim is to investigate how intuitive a technique is, practice may not be appropriate. If, on the other hand, the participants need to familiarize themselves with a novel task, or pass an initial learning period to avoid confounding learning effects then training is an absolute requirement. Training is important since it enables the moderator to make sure that each participant has really grasped what they need to know to conduct the trials. This implies that it is necessary to check the accuracy (in most cases) by some means, and not just ask the participants, after the training, if they know what to do and how to do it. To motivate participants to engage in the training it is possible to present these tasks as part of the real session, without telling that they are just for practice, and then just exclude this data from the final analysis [37]. However, in this case checking for accuracy and correcting participants might be difficult. Finally, if the study is not self-paced, with the participant controlling when each trial begins, then it may be necessary to build in rest phases to avoid fatigue.

Another issue to consider is whether the moderator should monitor the session or not. In many cases it is best for the evaluator to leave the room and let the participant work alone. This reduces stress on the participants since they will not have the feeling of being observed. Also, this avoids a participant asking questions during the session, which may bias the results. The presence of an observer may encourage them to do that even if it has been stated that it is not allowed, which might be difficult to handle. The drawback of not being present is that the moderator has no control over what happens. In some evaluations observation is a requirement, in this case either a one-way window or a video camera may be a sufficient substitute for the presence of the observer. If the moderator is present, or is able to monitor the session from outside the room, it is useful to take observational notes. This includes noting any difficulties related to executing a task, impressions of the participants commitment and motivation, signs of fatigue or frustration etc. If there are concerns, then the participant can be questioned after the session and this information can be useful when looking at the quantitative results. For example a participant might admit that they got tired and just answered randomly in the end to finish as quickly as possible, then their data has to be rejected from the subsequent analysis.

Finally, it is highly recommended to conduct one or more pilot studies before the real evaluation takes place. It is exceptional that an evaluation runs correctly the first time it is done. All relevant aspects of it, such as equipment, viewing condition, response procedure, instruction material, orderings, tasks for familiar-ization/training, how to answer questions from participants, timing etc. could and should be tested. This is invaluable in refining and finalizing the study and will help to identify factors that have initially been overlooked, miscalculated or perhaps simply incorrectly designed. Any confounding factor(s) not controlled for will bias the outcome and, importantly, time for participants and others involved in the evaluation is a very valuable resource and should be handled with great care. By piloting you can ensure that the study and its procedures are adequate for both the research question and the persons taking part.

4.5 Results

For any outcome of an evaluation, some analysis process and interpretation are necessary to make inferences. It is critical to make the findings clear and motivated and here a systematic approach, both in terms of analysis and reporting, is crucial. The remainder of this section focuses on the analysis of quantitative data by means of using inferential statistics: significance testing [13]. Reporting is discussed in Sect. 5.

The choice of which statistical test is suitable for analysis of the significance of the data follows from the evaluation design chosen to investigate the research questions or hypotheses. Once the decisions have been made with respect to the design of the study which test to apply, will, in most cases, be automatically selected. In fact, the choice of test depends on only a very few decisions [23]. In textbooks on statistical analysis, such as [23] and [13], there are decision charts to aid the selection. You follow a path trough the chart and, depending on your answers along the way, you are directed to the test most appropriate to your particular design.

However, the test you choose must also be appropriate to the type of data obtained and here you (the analyst) do not have all the answers until the actual data have been obtained. Therefore, the first step in analysing the data should always be to explore it to get a first impression and overview of it: screen the data, look at descriptive statistics and plot graphs to support exploration and finally check some basic assumptions (the assumptions are described below) [13]. These activities will aid in the final decision of which test(s) to use to ensure statistical validity and thus credibility of the obtained results. Descriptive statistics summarize and describe data with just a few numbers showing measures of central tendency (for example the mean or median), measures of variability (such as range, variance, and standard deviation) and measures of shape (skewness and kurtosis). These statistics tell a lot about the nature of the data and facilitate comparisons, so obviously they are important to include when reporting the results. They also provide the basis for further analysis of the data using statistical tests and to help interpret the overall meaning of the results [13].

Statistical tests are classified into parametric and non-parametric tests [13, 14]. In non-parametric tests the data points are transformed into rank values and statistics are calculated on these ranks instead of on the actual scores. Most parametric tests have four basic assumptions that must be fulfilled for a test to give an accurate result whereas non-parametric tests do not; they are assumption-free [13]. The assumptions for parametric tests are: normally distributed data, homogeneity of variance, data are measured at least on an interval level, and independence. The last two assumptions are only subject to testing by means of common sense and, it is debatable whether it is appropriate or not when rating scale data, data measured at an ordinal level, is tested by parametric tests. Homogeneity of variance can be tested in different ways depending on the nature of the study and appropriate tests will be suggested by the statistical software (for example SPSS [13]) used for doing the analysis. The assumption of normal distribution of the data is often simply explored by plotting the data and just looking at it. However, this procedure is highly subjective and one should also quantify the normality with numbers. The statistical skewness (the symmetry of the distribution) and kurtosis (the peakedness of the distribution) are both 0 in a perfect normal distribution, so the further away from 0, the more non-normal the distribution is. There is no exact rule for how much deviation is acceptable so the best way to decide is to apply an objective test in the statistical analysis software. The Kolmogorov-Smirnov or Shapiro-Wilk tests [ibid] are two such tests that take into account both skewness and kurtosis simultaneously.

If the data violates any of the assumptions then the analysis should be done using a non-parametric test instead, to ensure statistical validity. Here you should also be aware that the assumptions that must be met for using a parametric test relate to the dependent variables separately [13]. Therefore it is commonly seen that authors apply different types of tests for different dependent variables in their analysis, for example they may use a parametric test for analysing the accuracy data while a non-parametric test is used for analysing the response times.

Apart from using the wrong test in relation to the data type there is another unfortunate mistake that often appears in the literature. The basis of all statistical tests of significance is to calculate the percentage probability, the level of significance, of obtaining a difference (an effect) in the scores if the scores are in fact occurring on a random basis rather than being the result of an experimental effect [23, p. 23]. This means that if the probability is very low, then the null hypothesis, that any differences between values in the evaluation are likely to be random, can be rejected and, instead, the hypothesis that the results are significant [ibid] can be accepted. Often a level of significance, referred to as the alpha level (written α), of 0.05 is used, which means that there is a 95% probability that the results are significant against 5% (1 in 20) that they are the result of random variation. This is referred to as having a p-value of 0.05. Hence, every time a test is conducted it is important to be aware that we might accept an effect in data that is not actually present. If many tests are conducted on the same data these errors accumulate since repeated testing increases the probability of finding a significant difference in the result [14, p. 172]. For instance, often you need to compare several means (not just two resulting in one test) in an evaluation. For example you might need to analyse values from three groups: A, B and C. Or you might have a study where one group of participants have performed within four different conditions and then these four values, for example the mean values for each condition, must be compared. It is not unusual to see publications where t-tests are used for this kind of analysis. However, a t-test compares only two means at a time and so, in the above examples, three tests are required in the first case and six in the second case to include all pairwise comparisons and, as stated above, the probability of error accumulates with each test. It should be noted that there are published papers including far more than six comparisons! To avoid this accumulation of probabilities when many comparisons must be made, the p-value (for example 0.05) used for accepting them as statistically significant must be reduced by the number of t-tests performed. This is done by dividing the probability value by the number of tests. If the procedure includes five tests the new value becomes $0.05/5 = 0.01$ [14].

When many comparisons must be made, tests should be used which, instead, look for an overall effect between several means (a difference between them) at once while maintaining the 5% level of significance. Analysis of variance (ANOVA) is a well-known example of such a test and ANOVA is also used for analysis when there is more than one independent variable. If there is an overall effect the next step is to examine where the difference exists, between which specific means, using so called post-hoc tests. These tests compare every mean with every other one in a manner that can be thought of as doing multiple t-tests but where these tests are calculated in a way such that the overall 5% level of significance is maintained. Bonferroni correction is one example of such a test that is often used in literature and this correction is done by dividing the probability value by the number of tests, as described above [14, pp. 173–174].

Finally, in recent years, several researchers have emphasized the value of going a step beyond statistical significance, and complementing p-values by also reporting the effect size of a result. The motivation for this is that, although a study has yielded a statistically significant difference effect, this does not automatically mean that it is important or meaningful [13, p. 56]. The magnitude, and thus the importance of an obtained effect, on the other hand, is an objective and standardized measure that can be used to compare findings over different studies (with different variables, different scales of measurement etc.). The effect size (small, medium or large) is commonly expressed in standard deviation units, Cohen's d [9], or correlation using Pearson's correlation coefficient, r [13]. It is still unusual to see authors reporting the effect size but it is becoming more common and should be encouraged.

To conclude this section, to learn more about statistical analysis the reader is strongly recommended to examine the very accessible books by Field [13, 14].

5 Reporting an Evaluation

The process of sound scientific evaluation also encompasses sound reporting. The aim should always be for the description to be as complete and clear as possible. It should be possible to clearly understand every aspect of the experiment that has been conducted, and it is the authors' job to ensure it is presented sufficiently well. This enables the reader to assess the reliability, validity and generalizability of a study, and allow it to be replicated by others to verify its results.

Reviewers of papers submitted to a conference or journal may have a variety of concerns and criticisms and it may be impossible to satisfy all desires and requests. However, we hope that following the advice given in this section will leave the person reporting a study well prepared to predict and address the most common ones. It is also very important to remember not to describe only what was done you also need to describe why it was done, and motivate certain choices, decisions and actions taken in the design and execution. Including such information increases the readers' understanding, and often your chances of convincing them about the validity of your approach increases dramatically.

When reporting an evaluation study some introduction and an overall description of the study itself will, of course be required. This includes the research question and the motivation for addressing it, and a thorough description of the visualization technique or system under investigation etc. This can be organized in several ways and will typically vary greatly depending on the type of publication and study, the area of interest for the study itself and any page limitations. The remainder of this section will focus more specifically on what can be regarded as the 'method' and 'results' sections of an evaluation publication. These sections should describe all relevant facts and details of what has been done within the study, how it was carried out and how it was analysed. Considering this section as a recipe, it should be possible to replicate the study exactly by following the description.

The information provided should be presented in a way that is scientific, unambiguous and useful. Here terminology is very important but often, unfortunately, there is no "universal code" to apply. Several words can be used to refer to the same thing. For example, "a trial" is one of a number of repetitions in an experiment, but it is sometimes used to refer to an entire experiment. Also, one of a number of all repetitions can be called a task or a case. Publications do appear where several different words have been used within the paper to refer to the same thing, making the description very hard to follow. In each place in the text where an aspect is first raised, that is when it is first given an operational definition, it should be quite clear what is meant and the term used should then be used consistently throughout the text. Good operational definitions define and describe variables and procedures so that they cannot be misunderstood and so that other researchers can replicate them by following the descriptions [22, p. 75].

There are several fundamental issues that need to be covered and commonly the method section is divided into different subsections for structure and clarity. How to name and organize these subsections is quite flexible, and depends on what will constitute a logical order of presentation both to enhance readability and to avoid repetition. Naturally, depending on the precise nature of the study and the publication, some subsections can be collapsed or perhaps some of their content would make more sense if placed elsewhere. Below some typical subsections are described. Some example sentences taken from [18] and [28] and have been inserted in some of these as examples of how to describe certain aspects. For greater detail we recommend reviewing these publications. Other examples on how to write a method section can be found in [14, 44].

Task. The actual task(s) used in the evaluation should be thoroughly explained. This includes the exact nature of the task, why it was chosen, how it is representative for the research question and why and to what extent it could or could not be generalized beyond the context of the evaluation. The task is often closely related to the main objective and aim of the study. This means that when the reader has reached the method section, and the subsection describing the task, information like that listed above has often already been described and discussed in previous sections, at least in general terms. This is especially true if the publication is a pure evaluation paper (see for example [18, 19, 28]), in which case this subsection needs to cover very detailed information related to how the task should be correctly completed in the evaluation. Sometimes this information can be distributed in other sections, it is a matter of what constitutes a logical order of reporting and what makes the best presentation in terms of avoiding repetition and redundant information. It is, however, essential that this information is included in a clear and concise manner.

Stimuli (or material). This section presents details about all the stimuli and materials used in the evaluation. You need to provide a full description of each stimulus or other material to allow a clear understanding of their appearance (exactly what it looked like to the participant) and content. For example a part of a description of a visual stimulus could read: "Each stimulus display was comprised of a 12×12 matrix of grid cells creating a total of 144 grid positions with a square size of 0.8×0.8 cm". Any buttons for interacting with the software, for giving responses or controlling the sequence of trials etc., which appear on the visual stimuli should be described. Also, any information provided to help participants keep track of the evaluation session, such as "5/25" in the right upper corner of the display to indicate where in the sequence of trials the participant is at a particular moment should be described. Although the descriptions should be comprehensive, more general issues (about the type of stimuli and/or the visualization technique when applicable) could be explained elsewhere in the paper for clarity and then this section need cover only what the actual images presented on the display looked like. If not explained previously in the paper, this is the section that should include a description of the data set(s) used. Where it has been explained previously, a reference to that section should be included. Descriptions of material such as questionnaires or interview guides should also be reported in this section. It is important that in this section you don't include too much detail about how the stimulus and materials etc. were used during the evaluation, as this should be left to the procedure section. When space is an issue a full description of techniques, for example of the details of the data sets etc., might not be feasible. In such a case, it is possible to refer to previous publications or to on-line material and only include the most crucial details in the current reporting.

Apparatus. This is where you describe the equipment used and its configuration for the evaluation, together with the usage conditions. This includes type of computer, monitor, response apparatus and other hardware, and also what software was used to present the stimuli and to record responses. Even details such as the name of the manufacturer's products, perhaps the type of graphics card used, may sometimes be needed. Anything relevant to the physical configuration, such as viewing condition, lighting, or any potentially disturbing factors such as noise should also be reported.

Participants. The role of this section should describe essential information about the people that took part in the evaluation. One important characteristic is the number of participants. Another important factor is the nature of the participants, this includes sex and age and here we advise to state the age range, and then median age instead of mean age since that is more informative. Further characteristics are education, occupation, level of experience with visualization, optical acuity, and, sometimes, even nationality. It could also be described how participants were recruited (students or experts for example) and it should be stated whether or not they received any compensation for taking part.

Study design. In the design section a logical outline of the evaluation is provided. A description of the structure should be presented: what design was used, what were the independent variables and how many levels did they each have. Further, what were the dependent variables and what, if any, procedures applied to assign orders should be described. For example, "The study was designed as a two variable, mixed design with group of target patterns (pattern-group A and B) as within-subject variable and the order of presentation of the two groups as between-subject variable". Another example could read like this: "The presentation order was balanced using a Latin-square procedure".

In this section it is also a good place to state how many trials in total the design yielded per participant (total number of repetitions of all tasks). For example "Each staircase was covered by 75 trials and the three patterns included appeared 25 times each in a randomized order of presentation. This design yielded a total of 150 trials per participant (75 trials $\times$ 2 pattern-groups)". The description should be clear and concise and not include too much detail of the actual procedure used. To avoid unnecessary repetition such details are presented in the procedure description below.

Procedure. The objective of this description is to give details as to how the study was conducted in a practical sense. Thus it is a narrative of the evaluation and tells what happened from start to finish. This includes instruction, training and task to be performed and how this was done. Other important facts are whether feedback was provided or not, the response procedure, whether the sequence of trials was self-paced or not, and how the equipment was operated. For example, "Participants reviewed written instruction material and completed a block of practice trials to learn the concept and usage of the visualization and the two types of task to be performed", and "Stimuli were displayed until a response was given". Timing information should also be presented regarding the total participation time as well as any constraints on individual task completion times, whether due to some objective usability criteria or to keep the evaluation session to a reasonable duration.

Results. This section presents the findings: the results from applying the method. When the results are reported any treatments of data should be included in the description. For example "We employed a logarithmic transformation of the data before further statistical testing". It should also be stated what test was used and how the data fitted into this procedure: "Group mean values were calculated and a between-subject ANOVA was carried out using a decision criterion of 0.05.

Variables were visualization type (2Dm vs. 3Dm vs. 2Da) and sequence of task type". It is also usually required to state the finding to which the test relates, report the test statistic, usually with its degrees of freedom, the probability value and associated descriptive statistic. For example, "There was a significant effect of visualization type $F(2, 24) = 5.528, p < 0.01$. The response times for the 3Dm visualization were significantly faster than for the 2Dm visualization ($T = 2.4891, n = 10, p < 0.05$). The group mean value for the search times with 3Dm was 23.9 s with a standard deviation of 1.35 and 37.2 s, std 1.61 with 2Dm".

Sometimes it is beneficial to include a short discussion of the results at the end of this section, or in a subsequent one, especially if the results lead to follow up studies that are also described in the same publication. In the majority of cases, however, the interpretations and discussions of the results should come in the sections covering a general discussion and conclusions. The results should stand for themselves and one should separate facts from opinions: that is, a result could be accurate whereas a conclusion based on it may not be.

To conclude, to write the method section in an evaluation publication is not an easy task and a great deal of detail needs to be included. A good description of a study thus also requires a considerable amount of space, which can be difficult when faced with a page limit, and there is a trade-off between including irrelevant information and leaving relevant aspects out. A good approach is to have someone without prior knowledge about the study provide feedback about what is missing, since there is always the risk of leaving information out that may seem obvious to the author. You will also get feedback on parts that could be left out since they are repetitive or even irrelevant. When space is critical it is more important than ever to focus on the details you think are most important to the outcome of the evaluation and for replication.

6 Catalogue of Standard Problems

The authors of this work are experienced reviewers of research papers in the field of visualization and have seen many examples of evaluation papers where the results, as documented in the publications, do not support the conclusions. There are many reasons for these failures but they can be found in all aspects of the evaluation process described above. Here we summarize a few examples to help the reader in determining when they are seeing good or bad examples of evaluation in publications.

Assertion of benefit. This is the most common example of a flawed evaluation in the sense that it does not contain any user-centred evaluation at all. Developers of new visualization methods and tools often spend significant amounts of time and effort analysing a visualization problem and designing a new approach to it which they then spend a great deal of time implementing and optimizing. They then write a paper describing the method and finish with the conclusion that this is a wonderful new approach to the problem. This assertion that the method is beneficial is, however, quite hollow. While in some cases it is quite believable that the method might have benefits, it is often fairly likely that the new approach may have no overall benefit over existing approaches and may even have negative effects that the developers have not foreseen. Hence asserting benefit is not a scientific approach and any claims not supported by scientifically sound conducted evaluation should be regarded with great scepticism by the reader. On occasion the assertion of benefit is accompanied by woolly statements to the effect that some target users have found the tool to be beneficial. Statements such as these add little to the argument since, without knowing about how the users reached that determination the reader cannot be certain that any valuable contribution is made by it.

Case studies. It has become quite common in recent years for visualization publications to include a case study to support the assertion of benefit by the developers of a new visualization technique or tool (not to be confused with the method called 'expert review' [40]). While one or two examples where the tool seems to be useful (at least when used by the developer) can certainly add a little credibility to the idea that the tool has been well designed, they should not be regarded as a valid substitute for a well-designed and conducted evaluation. Specific problems chosen by the developer to show off their tool or technique at its best can be misleading and one should also consider that, when the tool is used by someone other than the developer, it may be rather less intuitive than the developer believes. The reader should also consider how that result might translate to other cases where the data is not as clean, has more or less dimensions, has greater or smaller size or any other changes which might make the method work less well. These are the type of questions that can only be answered through a rigorous user-centred evaluation.

Poor study design. As discussed above there are so many choices which the evaluator must make in the design that there are many ways in which they can make mistakes, leading to an evaluation which does not assess what it set out to assess. Considering a few examples from recent papers which we have seen, either in review or in publication, we could describe a study where the experiment uses two different types of equipment to present two conditions. In another the two conditions are tested using different experimental conduct. In a third different settings are employed in the visual representation between two conditions but the changes are not included in the definition of the conditions. In a fourth the order of presentation of the different conditions is not considered or reported. In each of these examples the evaluation results in measurements from which the authors attempt to draw conclusions about the two conditions without taking any account of the different equipment, different experimental conduct, or extraneous and unrecorded factors employed in each condition or the order of presentation. The conclusions, therefore, are quite likely to be meaningless, regardless of the level of sophistication of the statistical analysis applied, and in any event cannot be trusted by the reader of these publications.

Poor statistical analysis. Statistical analysis is a complex area in its own right and the opportunities to undo any amount of careful evaluation design are many, hence this is a common place to see evaluation papers introduce problems which render the interpretation of their results impossible. As described above the most common problems lie in the size and number of groups used, the determination of the validity of the results (the probability that the 'effect' is actually a random fluctuation) and the management of the statistical errors through good choices of statistical methods. Further, it is rare that we see authors describing having applied any tests to check whether the assumptions for using parametric tests have been met or not, and it is common to see that repeated tests are applied without correcting the probability level accordingly. We even see reports which present results in the form of p-values without stating which tests have been applied to produce these statistical probabilities.

The majority of visualization researchers are not as well schooled in statistical methods as they should be to undertake this area of the work and we would, once again, recommend teaming up with experts in the field of evaluation to help to ensure that this is not the part of the process that undermines the whole effort.

Poor reporting. Thorough evaluations (qualitative studies as well as quantitative) are typically very complex, and to enable a clear understanding and replication of such a study exactly, the reporting of the methods and techniques used must be very detailed. A common mistake authors make is that they do not provide enough information or that the information is ambiguous. Of course the pressures of page limits and limited time make it difficult to reproduce the evaluation in sufficient detail and this can also undermine the value of the work. When the reader cannot understand exactly how an important part of the process has been carried out, the sceptical reader must assume the worst, the conclusions are likely to be considered invalid, and the value of the evaluation is lost.

Ignoring the results. It is surprisingly common to see the authors of a publication, despite including an apparently well conducted evaluation, ignore the results when drawing their conclusions when the results have failed to go their way. We see papers where the authors report that the hypothesis that their proposed method is superior to the alternatives has not been supported by the experimental data but then continue to claim that their method is clearly superior in spite of the cold, hard facts that they have found. This is a human failing, and in many ways understandable: a method that they have spent time developing and refining, and which they obviously believe to be a good idea, has not been shown to work and its value has been substantially diminished. It is also the case that the negative result is much harder to get published even though, as scientists, we know that the negative result can be as valuable as the positive. The 'publish or perish' mentality in modern academia makes this negative result even more problematic.

Instead, the authors should take the negative result and attempt to understand that. What is the problem with the method? Is it actually not as effective as was thought, or is it the type of task to which it was being applied? Was it a problem in the experimental design which might invalidate the result? If it is concluded that the result is correct and the method ineffective, then what is the new information about the user and their particular needs that can be extracted from this negative result? It is, after all, only by knowing both what does and what does not work, and why, that we will advance visualization as a scientific discipline.

7 Conclusion

As stated in the introduction, user-centred evaluation is an enormously complex topic and is far too large to cover in detail in a single chapter such as this. We have attempted, therefore, to give the reader an introduction to some of the mysteries of this area in the hope that, first, it will convince them that evaluation is very important for the future of research in visualization, second that they can learn to do evaluation for themselves even if, for the time being, they are better off teaming up with experimental evaluation experts from other fields such as HCI or psychology, and third to help to arm the visualization researcher against 'snake-oil' in the form of publications with absent, badly designed, badly conducted and badly analysed evaluations, which can fool the less experienced reader. By considering the information detailed in this chapter, and applying a healthy level of scepticism and common sense, the reader should now be able to understand some of the problems with research publications that they read in the journals and see at conferences they attend.

There is a huge need for evaluation in visualization. Without it there can be little meaningful progress in the field, particularly when dealing with abstract visual representations of the large, multidimensional and time-varying data which are becoming the standard fare of visualization research today. We hope that this chapter will have helped make this clear to the reader and have encouraged them to bring such scientific evaluation methodology to their own work and attempt to measure the usability and worth of such applications in the future.

References

1. The eurograpics conference on visualization. accessed 2013-01-29. URL http://www.eurovis2013.de/content/full-paper-submission
2. Iso 9241-11: Ergonomics requirements for office work with visual display terminals, part 11. guidance on usability (1998)
3. Amar, R., Eagan, J., Stasko, J.: Low-level components of analytic activity in information visualization. In: Proceedings of the IEEE Symposium on Information Visualization, InfoVis'05, pp. 111–147 (2005)
4. Amar, R., Stasko, J.: Knowledge task-based framework for design and evaluation of information visualizations. In: Proceedings of the IEEE Symposium on Information Visualization, InfoVis'04, pp. 143–149 (2004)

5. Buchner, A., Erdfelder, E., Faul, F.: How to use g*power (1997). URL http://www.psycho.uni-duesseldorf.de/aap/projects/gpower/how_to_use_gpower.html
6. Card, S.K., Mackinlay, J.D., Shneiderman, B.: Readings in information visualization: using vision to think. Morgan Kaufmann (1999)
7. Carpendale, S.: Evaluating Information Visualizations., pp. 19–45. Springer (2008)
8. Chen, C.: Top 10 unsolved information visualization problems. IEEE computer graphics and applications 25(4), 12–16 (2005)
9. Cohen, J.: A power primer. Psychological Bulletin 112(1), 155–159 (1992)
10. Downs, J.S., Holbrook, M.B., Sheng, S., Cranor, L.F.: Are your participants gaming the system?: screening mechanical turk workers. In: Proceedings of the SIGCHI Conference on Human Factors in Computing Systems, CHI '10, pp. 2399–2402 (2010)
11. Dumas, J., Loring, B.: Moderating Usability Tests: Principles & Practices for Interacting. Morgan Kaufman (2008)
12. Faulkner, L.: Beyond the five-user assumption: Benefits of increased sample sizes in usability testing. Behavior Research methods, Instruments and Computers 35, 379–383 (2003)
13. Field, A.P.: Discovering statistics using SPSS. Sage Publications Limited (2009)
14. Field, A.P., Hole, G.: How to Design and Report Experiments. Sage Publications Ltd (2003)
15. Forsell, C.: A guide to scientific evaluation in visualization. In: Proceedings of the 14th International Conference Information Visualisation (IV'10), pp. 162–169. IEEE (2010)
16. Forsell, C., Cooper, M.D.: Scientific evaluation in visualization. In: Eurographics 2011-Tutorials, p. T6. The Eurographics Association (2011)
17. Forsell, C., Cooper, M.D.: A guide to reporting scientific evaluation in visualization. In: Proceedings of the International Working Conference on Advanced Visual Interfaces, pp. 608–611. ACM (2012)
18. Forsell, C., Johansson, J.: Task-based evaluation of multirelational 3d and standard 2d parallel coordinates. In: Proceedings of SPIE 2007 - Visualization and Data Analysis. The International Society for Optical Engineering (2007)
19. Forsell, C., Seipel, S., Lind, M.: Surface glyphs for efficient visualization of spatial multivariate data. Information Visualization 5(2), 112–124 (2006)
20. Frøkjær, E., Hertzum, M., Hornbæk, K.: Measuring usability: are effectiveness, efficiency and satisfaction really correlated? In: In Proceedings of the SIGCHI Conference on Human Factors in Computing Systems (CHI'00), pp. 345–352. The Hague, Netherlands (2000)
21. Gabbard Jr., J.L., Swan II, J.E., North, C.: Quantitative and qualitative methods for human-subject visualization experiments. Tutorial presented at VisWeek 2011 Providence, R.I., USA (2011)
22. Graziano, A.M., Raulin, M.L.: Research methods: A process of inquiry (2nd ed.). Harper-Collins College Publishers, New York, NY, US (1993)
23. Greene, J., D'Oliviera, M.: Learning to use statisitical tests in psychology, 2nd edition. Open University Press, Philadelphia (2001)
24. Hornbæk, K.: Current practice in measuring usability: challenges to usability studies and research. International Journal of Human Computer Studies 64(2), 79–102 (2006)
25. https://www.mturk.com:Amazonmechanicalturk.accessed2013-01-29
26. IEEE: Vis. accessed 2013-01-29. URL http://ieeevis.org/
27. IEEE: Visweek. paper submission guidelines. accessed 2013-01-29. URL http://visweek.vgtc.org/year/2012/info/call-participation/paper-submission-guidelines
28. Johansson, J., Forsell, C., Lind, M., Cooper, M.: Perceiving patterns in parallel coordinates: Determining thresholds for identification of relationships. Information Visualization 7(2), 152–162 (2008)
29. Keim, D.A., Bergeron, R.D., Pickett, R.M.: Test Data Sets for Evaluating Data Visualization Techniques, pp. 9–22. Springer Verlag (1995)
30. Kosara, R., Healey, C.G., Interrante, V., Laidlaw, D.H., Ware, C.: Thoughts on User Studies: Why, How, and When. Computer Graphics and Applications 23(4), 20–25 (2003)
31. Lam, H.: A framework of interaction costs in information visualization. IEEE Transactions on Visualization and Computer Graphics 14, 1149–1156 (2008)

32. Lam, H., Bertini, E., Isenberg, P., Catherine, P., Carpendale, S.: Empirical studies in information visualization: Seven scenarios. IEEE Transactions on Visualization and Computer Graphics 18(9), 1520–1536 (2012)
33. Laramee, R.S., Kosara, R.: Challenges and unsolved problems. In: A. Kerren, A. Ebert, J. Meyer (eds.) Human-Centered Visualization Environments, chap. 5, pp. 231–254. Springer Lecture Notes in Computer Science, Volume 4417: GI-Dagstuhl Research Seminar Dagstuhl Castle, Germany, March 5–8, 2006 Revised Lectures. (2007)
34. Morse, E., Lewis, M.: Evaluating visualizations: using a taxonomic guide. International Journal of Human-Computer Studies 53, 637–662 (2000)
35. Nielsen, J.: Heuristic Evaluation. In: J. Nielsen, R.L. Mack (eds.) Usability Inspection Methods. Wiley & Sons, New York, NY, US (1994)
36. Plaisant, C.: The challenge of information visualization evaluation. In: Proceedings of the Working Conference on Advanced Visual Interfaces - AVI '04, pp. 109–116. ACM Press, New York, New York, USA (2004). DOI 10.1145/989863.989880. URL http://portal.acm.org/citation.cfm?doid=989863.989880
37. Purchase, H.C.: Experimental Human-Computer Interaction: A Practical Guide with Visual Examples. Cambridge University Press (2009)
38. Stasko, J.: Evaluating information visualizations: Issues and opportunities (a position statement). In: Proceedings of BEyond time and errors: novel evaLuation methods for Information Visualization (BELIV06). Venice, Italy (2006)
39. Tominski, C., Forsell, C., Johansson, J.: Interaction support for visual comparison inspired by natural behavior. IEEE Transactions on Visualization and Computer Graphics 18(12), 2719–2728 (2012). DOI 10.1109/TVCG.2012.237
40. Tory, M., Möller, T.: Evaluating visualizations: Do expert reviews work. IEEE Computer Graphics and Applications 25, 8–11 (2005)
41. Tullis, T., Albert, B.: Measuring the User Experience. Collecting, Analyzing, and Presenting Usability Metrics. Morgan Kaufmann (2008)
42. Valiati, E., Pimenta, M., Freitas, C.: A taxonomy of tasks for guiding the evaluation of multidimensional visualizations. In: Proceedings of BEyond time and errors: novel evaLuation methods for Information Visualization (BELIV06), pp. 1–6 (2006)
43. Vrotsou, K., Forsell, C., Cooper, M.D.: 2D and 3D Representations for Feature Recognition in Time Geographical Diary Data. Information Visualization 9(4), 263–276 (2010)
44. Wood, C., Giles, D., Percy, C.: Your Psychology Project Handbook: Becoming a Researcher. Pearson (2009)
45. Woolrych, A., Cockton, G.: Why and when five test users aren't enough. In: In Proceedings of the IHM-HCI 2001 Conference, pp. 105–108 (2001)

User-Centered Evaluation of Information Visualization Techniques: Making the HCI-InfoVis Connection Explicit

Carla M.D. S. Freitas, Marcelo S. Pimenta, and Dominique L. Scapin

Abstract In the last decade, the growing interest in evaluation of information visualization techniques is a clear indication that usability and user experience are very important quality criteria in this context. However, beyond this level of agreement there is much room for discussion about how to extend the variety of usability evaluation approaches for assessing information visualization techniques, and how to determine which ones are the most effective, and in what ways and for what purposes. In this chapter we take a user centered, Human-Computer Interaction-based perspective to discuss usability evaluation of information visualization techniques. We begin by presenting a singular view of the evolution of visualization techniques evaluation, briefly summarizing the main contributions of several works in this area since its humble beginning as a collateral activity until the recent growth of interest. Then, we focus on current issues related to such evaluations, particularly concerning the way they are designed and conducted, taking into account a background of well-known usability evaluation methods from HCI to help understanding why there are still open problems. A set of guidelines for a (more) user-centered usability evaluation of information visualization techniques is proposed and discussed. Our ultimate goal is to provide some insight regarding if and how sound ergonomic user-centered knowledge can be transferred to the information visualization context.

C.M.D. S. Freitas (✉) • M.S. Pimenta
Institute of Informatics, Federal University of Rio Grande do Sul (UFRGS),
Porto Alegre, RS, 91.501-970 Brazil
e-mail: carla@inf.ufrgs.br; mpimenta@inf.ufrgs.br

D.L. Scapin
INRIA-Rocquencourt, B.P. 105, Le Chesnay Cedex, 78153 France
e-mail: Dominique.Scapin@inria.fr

W. Huang (ed.), *Handbook of Human Centric Visualization*,
DOI 10.1007/978-1-4614-7485-2_12, © Springer Science+Business Media New York 2014

1 Introduction

In the last decade, the growing interest in evaluation of information visualization (InfoVis) techniques [6–9, 19, 41] has become a clear indication that usability and user experience (UX) are important quality criteria for such techniques.

Following the same path of desktop graphical user interfaces and Web-based interfaces, the use of information visualization (InfoVis) techniques depend on their usability. Whereas the first information visualization techniques were presented without thorough evaluation studies (see, for example, the classical paper on the hyperbolic tree by Lamping and Rao [38]), along the years researchers have become aware of the importance of evaluation approaches, reporting from simple comparison of techniques [74] to a whole set of experiences [6–8] based on Human-Computer Interaction (HCI) concepts and methods. However, despite an evident progress towards "good practices" for design and usability evaluation of such techniques, several aspects related to a user-centered perspective for InfoVis techniques remain as open issues. Indeed, beyond some level of agreement, there is much room for discussion about how to extend the variety of usability evaluation approaches adopted for InfoVis evaluation, and how to determine which approaches are the most effective, in what ways and for what purposes.

User-centered perspectives have their origin in User Centered Design (UCD). The principles now accepted as the basis for UCD—early focus on users and tasks, empirical measurement, iterative design [52]—basically involve making real users and their goals as the driving forces behind software development. Thus, a user-centered perspective tries to optimize the user interface around how people can, want, or need to work, rather than forcing the users to change how they work to accommodate the system or function. In short, evaluation with a user-centered perspective is an evaluation based on the needs of the user and, for that, we need to know the users, their goals and their tasks.

Although there are a great variety of techniques for information visualization, there is not yet a consensus about their usability evaluation: what is the meaning of *usability* for such techniques? Which characteristics do we have to evaluate, and how can we do so? In addition, which aspects are generic to all types of interactive systems and which are specific to InfoVis techniques? What are the most relevant issues for their evaluation?

The HCI domain has already solid principles and methods for interaction design and evaluation of interactive systems, and the adoption of this already existent basis of theory and practice not only saves some effort and makes InfoVis technique design and evaluation more efficient, but also provides a common working method for the area.

In the last years, we have been particularly interested in usability evaluation [57], evaluation of InfoVis techniques [26, 65, 70, 71], UX [56], and also in making the InfoVis-HCI connection more explicit [27]. In this chapter, our aim is to discuss how InfoVis could follow this HCI perspective and how HCI concepts, principles, methods and techniques were adapted and used in the evaluation of information visualization techniques.

Our goal is twofold: first, to summarize current problems related to the usability evaluation of InfoVis techniques, including the way these evaluations are designed and performed, and second, to propose a set of guidelines for a (more) user-centered usability evaluation of information visualization techniques. Our final intention is to provide some insight about if and how sound ergonomic user-centered knowledge from HCI domain [55] can be transferred to the information visualization context, particularly adopting the Human-Centered Perspective, and to contribute for overcoming the limitations stressed by Kasper Hornbæk in his talk at BELIV 2010 [8]: *To keep advancing InfoViz, I believe we need to address two limitations of our evaluations. On the one hand, few empirical studies are motivated by theory or are comparing equally plausible hypotheses. Mostly, the InfoViz literature proposes radical innovations (in the terms of William Newman) and does little to develop and test concepts. On the other hand, many of the practical, low-level decisions in InfoViz evaluations are problematic. Like most HCI researchers, we evaluate our own interfaces, use mostly simple outcome measures, rarely study the process of interaction, and select tasks somewhat randomly.*[1]

The chapter is structured as follows. Section 2 presents our view on the evolution of evaluation methods adopted by the InfoVis community, and Sect. 3 summarizes current problems related to the usability evaluation of InfoVis techniques, especially how these evaluations are designed and performed. The main contribution of this chapter is in Sect. 4, where we propose a set of guidelines for a (more) user-centered usability evaluation of information visualization techniques. Finally, in Sect. 5 we present some final comments.

2 Evaluation of InfoVis Techniques

In this section we present our view on the evolution of evaluation methods adopted for assessing InfoVis techniques, briefly summarizing the main contributions of several works, from the first reports about evaluation as a collateral activity to the current growth of interest.

The first information visualization techniques were developed almost in an ad-hoc way, without adopting any systematic evaluation approach. However, during the subsequent years, different questions related to the evaluation of these techniques have become research issues in information visualization [19, 22, 36, 49, 66] and lately in visual analytics [1, 58]. Among many open problems, we are particularly interested in finding answers to the following question: *How do we know if information visualization tools are useful and usable for real users performing real visualization tasks?* In fact, for effective and well-accepted adoption of information visualization tools, it is essential that these tools be effective, efficient and satisfying the intended users' needs.

---

[1]"Conceptual and Practical Challenges in InfoViz Evaluations" (http://www.beliv.org).

As new applications are more often related to larger and more complex datasets, the challenges of information visualization and visual analytics involves not only the selection of typical datasets and tasks but defining evaluation methodologies and finding the appropriate case studies and users.

2.1 First Evaluation Studies

The first well-organized report on the evaluation of information visualization techniques[2] included a meta-analysis of empirical studies of information visualization [19]. The authors restricted their meta-analysis to studies where information structures were visualized in the form of trees and networks.

From a first set of 35 papers reporting experimental studies in which independent variables were related to one of three variables (users, tasks and tools), they ended up analyzing six studies. They compared the results of these studies in terms of effect sizes and significance levels, namely, effects of individual differences on accuracy, effects of users' cognitive abilities on efficiency, and effects of visualization (compared to traditional interfaces) on accuracy.

The conclusions they reached are still mostly valid today: *in order to improve the quality, clarity and comparability of experimental studies of information visualizations, future experimental studies of information visualizations should carefully take into account 6 aspects of an experimental design: the use of standardized testing information; the clarity of descriptions of visual-spatial properties of information visualizations; the use of standardized task taxonomies for activities such as visual information retrieval, data exploration and data analysis; the focus on the task-feature binding to be investigated in experimental studies; the use of standardized cognitive ability tests; the level of details in reporting statistical results.*

Although the work of Chen and Yu [19] focused on evaluation based on experiments with users, reflecting the common practice that was being established, some early works addressed other approaches [13, 48].

Brath [13] proposed a set of quantitative metrics for evaluating the efficiency of static visual representations, basically plots. For each display he measured the number of data points (data density), number of dimensions (cognitive complexity), occlusion rate and the number of identifiable data points. The investigation of quantitative and qualitative metrics to assess visual representations has been research theme along the years—see, for example, [10, 28, 40, 46, 58, 75]. The results reported so far point that there is no generic set of metrics: which metrics to use depends on users, tasks, and context.

---

[2]The special issue of the International Journal on Human-Computer Studies (volume 53, no. 5, 2000).

Pirolli and Rao [48] pioneered the use of GOMS [15] to evaluate visualization techniques. They reported the evaluation of Table Lens [51] in two typical tasks of exploratory visual analysis: assessing the properties of a set of data (for example, range, central values, dispersion and symmetry), and finding the relationships among variables (for example, correlation). Methods for performing these tasks with Table Lens (and another tool named Splus) were described using GOMS, and they obtained useful results regarding possible refinements of Table Lens to increase its effectiveness in exploratory data analysis tasks.

Low-level action analyses similar to those that support GOMS are still valid today due to the availability of different technologies [20, 28]. Goldberg and Helfman [28] discuss advantages and disadvantages of eye tracking methods and related metrics, and use the empirical results obtained from experiments to suggest a set of design heuristics for graph visualizations, since eye tracking can help understanding the human visual scanning behavior associated to different layouts. In the present book, the same authors provide a thorough and deeper discussion about eye tracking on visualizations [29], aiming at guiding a researcher to understand how eye-tracking data can be used for visual comparison tasks.

The first main issue, which drove evaluation of information visualization techniques for many years, has been proving that a new technique being described would make users perform some tasks efficiently, and preferably even better than with other techniques. This approach was evident in early works (see, for example, [17, 18, 48]), and eventually became an expected standard for papers describing new techniques, which from that point on should ideally present some sort of comparative evaluation.

Such use of evaluation might have led to the understanding that evaluation of information visualization techniques was primarily summative. However, as pointed out by Ellis and Dix [22], looking at the discussions of many papers reporting such evaluations one finds suggestions for improvements, making the use of evaluation formative.

2.2 Heuristics and Tasks Taxonomies in Evaluation of Visualization Techniques

Although controlled experiments with users proved to be the most performed evaluation method, other methods have also been employed for evaluating visualizations.

Heuristic evaluation [42], for example, has been largely used due to its rich outcome: a group of evaluators using a set of heuristics can find most of the usability problems of an interface. In evaluating visualizations, the heuristics have been usually a list of tasks that a user should be able to perform with the technique. Research on task taxonomies (for visualization) has its origins in an early work by Wherend and Lewis [72] followed by Springmeyer et al. [63]. But it was only after Shneiderman's now classic information-seeking mantra [60] that visualization evaluation started to be based on heuristics evaluation. The tasks specified by Shneiderman were: overview, zoom, filter, details-on-demand, relate, history and extract.

Later on, Zhou and Feiner [77] introduced another categorization of tasks. They separated *presentation intents* (goals a user has when using a visual representation) from low-level *visual techniques* (the exact operation performed on a given object presented in the display) by means of an intermediate level, the *visual tasks. Visual tasks* can be considered abstract visual techniques, since they indicate a desired visual effect in the representation while a visual technique is a *way* to achieve that desired effect, either by the user or the system. Zhou and Feiner characterized visual tasks along two dimensions: *visual accomplishments* and *visual implications*. Visual accomplishments correspond to the presentation intents a visual task is supposed to support while visual implications specify the visual techniques that could be used to fulfill the visual task. Regarding visual accomplishments, two classes of visual tasks can be identified: *inform* and *enable. Inform tasks* can be further distinguished as *elaborate* and *summarize* tasks, while *enable* tasks can be divided in *explore* tasks and *compute* tasks. At the bottom level of this hierarchy of abstract visual tasks one still has tasks like *categorize, cluster, compare, correlate, identify*, etc., i.e., generic operations like those identified by Weherend and Lewis [72].

Based on the observation of principles for visual perception and cognition, Zhou and Feiner also established a link from the visual tasks to the adequate visual techniques. For example, to identify a piece of information, one can give its name, point at it in the display, or give a range of attributes as a profile, all of these implying certain concrete visual tasks like name input, mouse pointing and filtering.

Amar and Stasko [3] discuss the notion of analytic gap, representing the obstacles presented by visualization systems in facilitating high-level analytical tasks, such as domain learning and decision making under uncertainty, which are usually not covered by the existing works in design and evaluation of information visualization systems. The authors claim that, although Wehrend and Lewis's and Zhou and Feiner's low-level tasks are essential, they do not offer a consistent basis to fill the analytic gaps. In [3] they also propose a new taxonomy, with higher-level tasks that can provide a better support to visualization systems designers and evaluators. Limitations of existing visualization systems (at that time) were grouped into two major categories: the Rationale Gap and the Worldview Gap. The first one was defined as the gap between perceiving a relationship and actually being able to explain confidence in that relationship, as well as its usefulness. In fact, users need to be able to relate data sets to the realms in which decisions are being made. The second one is defined as the gap between what is being shown and what actually needs to be shown to draw a straightforward representational conclusion for making a decision. In fact, users need to be able to formulate a strategy for browsing a visualization, and for creating, acquiring and transferring knowledge or metadata about important domain parameters within a data set.

In a subsequent work, Amar et al. [4] proposed a taxonomy of 10 low level tasks based on 196 analytic questions found by students when analyzing data with commercial visualization systems: Retrieve Value, Filter, Compute Derived Value, Find Extremum, Sort, Determine Range, Characterize Distribution, Find Anomalies, Cluster, Correlate.

The works of both Zhou & Feiner and Amar & Stasko were the basis of another task taxonomy by Valiati et al. [70]. These authors proposed five analytical, high-level tasks (identify, infer, determine, compare, and locate) and two more operational, intermediate-level tasks (configure and visualize). At that same year and venue [6], Lee et al. [39] presented a task taxonomy for graph visualization, which served as heuristics in the evaluation of a new graph visualization tool [64, 65].

Other works discussing heuristics evaluation in the realm of information visualization are by Tory and Moeller [68] and Zuk et al. [79]. Tory and Moeller [68] used heuristics evaluation for assessing two visualization systems, concluding that experts and users should be included in evaluation processes. Zuk et al. [79] used three sets of heuristics, which are actually task taxonomies by Zuk and Carpendale [78], Shneiderman [60] and Amar and Stasko [3], in a case study where evaluators assessed two specific views provided by a system. In a meta-analysis performed with the results from the evaluations they found characteristics of each set of heuristics, including the usefulness of visualization-specific heuristics, which allowed capturing problems not discovered by usability heuristics.

Heuristics evaluation is appropriate as a formative (or analytical) method, and so a combination of usability and visualization-specific heuristics would provide useful design guidelines for visualization developers. A recent work by Forsell and Johansson [25] reported the review of six sets of heuristics suggested for evaluating information visualization techniques (Amar and Stasko [3], Freitas et al. [26], Scapin and Bastien [55], Shneiderman [60], Tory and Moeller [67], Zuk and Carpendale [78] and Nielsen's 10 usability heuristics). After an extensive case study with 6 evaluators using the 63 heuristics in a total of 4,662 judgments for each participant, Forsell and Johansson proposed a set of 10 heuristics, claiming that they provide the best explanatory coverage out of all possible combinations of the 63 candidate heuristics: information coding, minimal actions, flexibility, orientation and help, spatial organization, consistency, recognition rather than recall, prompting, remove the extraneous, and data set reduction.

The results of Forsell and Johansson's study was used in a more recent work by Scholtz [58], which proposed guidelines for assessing visual analytics environments based on the reviews performed for the 2009 Visual Analytics Science and Technology (VAST) Symposium challenge and from a small user study with three intelligence analysts analyzing five videos from the VAST 2009 Challenge.

Few authors have selected other HCI methods for evaluation purposes. For example, Allendoerfer et al. [2] used cognitive walkthrough for evaluating the usability of CiteSpace social network visualizations, and Elmqvist et al. [23] performed an informal scenario-based evaluation as a case study on the use of ScatterDice.

2.3 Empirical Methods for Evaluating Information Visualizations

As it is well known from the HCI literature, empirical methods have been used for years in the evaluation of interactive techniques [52], with controlled experiments, observation, field studies, interviews and questionnaires being the most used ones.

Interviews and questionnaires are often used as complementary methods for collecting data before and after the other methods have been employed.

Although there is a sound knowledge from HCI regarding controlled experiments, results from different reports reviewing experimental studies suggest that developers still do not perform rigorous scientific evaluation procedures [19, 22, 24, 37].

Based on the analyses of 19 quantitative empirical user studies, Lam and Munzner [37] recommend improvements both in experiment design and report of the outcomes of experiments: (1) use comparable interfaces in terms of visual elements, information content and amount displayed, levels of data organization displayed, and interaction complexity; (2) capture usage patterns in addition to overall performance measurements to better identify design tradeoffs; (3) isolate and study interface factors instead of overall interface performance; and (4) report more study details, either within the publications, or as supplementary materials. While with their recommendations Lam and Munzner aimed at facilitating meta-analysis of experimental studies, Forsell [24] focused on the phases of an experimental study, discussing specific issues related to experiment design, definition and selection of tasks, selection of participants and assignment of tasks to participants, performing the evaluation, analyzing data and reporting the results.

While an important issue in an experimental study is the set of tasks used for the evaluation, it is also important to notice that in exploratory data analysis and visual analytics, the main results from tasks are the insights users may achieve while observing visual representations and interacting with the visualization. The first reported attempt to quantify and categorize insights in information visualization was presented by Saraiya et al. [53]. From a pilot experiment, the authors quantified certain characteristics of insights: observation (the process of finding itself), time, domain value, hypotheses, directed versus unexpected, correctness, breadth versus depth and category. The authors conducted an experiment with 30 users, which were given questions regarding three data sets they had to explore with five popular visualization tools. Think-aloud protocol and video were used to complement data collected from the experiment. Although, the study allowed interesting observations, it was limited in time and, in a subsequent work [54], the authors explicitly stated that *the study failed to address the most important factor—motivation—that drives a data analyst to spend days and often months analyzing a particular data set. Also, the study did not capture the ability of a data analyst to judge the significance of reported insights, which is usually based on users' domain knowledge and familiarity with the data background and the experimental context.* The authors performed a longitudinal study that allowed them to attest three important functions of such studies: provide a deeper understanding about the actual process of data analysts; guide visualization designers in constructing tools that fulfill the needs of the analytic process; and guide evaluators in designing studies that assess the effect of visualization tools on visual analytics.

Some other works also discussed insight-based methods in visualization evaluation [44, 50, 62, 76]. North et al. [45] performed a comparison between benchmark-task and insight-based methods. As initial data for comparison purposes, the authors collected overall performance in terms of time and accuracy, and performance for individual tasks using alternative visualizations in the task-based method. As for the insight-based method, the authors collected overall performance in terms of time and accuracy, and performance based on insight categories. An interesting result from this study was that the benchmark-task method measures how efficiently a visualization supports a task, while the insight-based method measures how much a visualization promotes a given task to users. Of course, insight occurs when a task is efficiently supported by a visualization technique. So, one might conclude that a combination of both evaluation methods is more interesting, confirming what Carpendale [16] suggested about applying a greater variety of evaluation methodologies.

In this scenario, longitudinal studies play an important role. Gonzáles and Kobsa [30] conducted a longitudinal case study along 6 weeks, involving five data analysts, aiming at verifying which were the factors related to the adoption of commercial information visualization systems for visualizing administrative data. Hetzler and Turner [32] reported some lessons learned from an observational study involving a group of expert users, using a visualization system for analyzing some data sets.

Shneiderman and Plaisant [61] introduced MILCS—"Multi-dimensional In-depth Long-term Case Studies" for the evaluation of information visualization techniques. *Multi-dimensional* makes reference to the usage of different methods (for example, interviews, observations, etc.) for evaluating user performance, efficacy and utility of an interface. *In-depth* refers to intensive engagement of the researcher with the expert users to the point of becoming a partner or assistant. *Long-term* refers to the execution of longitudinal studies with users who are experts in a specific domain, and finally, *case studies* refers to detailed reporting about a small group of users (typically from 3 to 5 domain experts) doing real tasks in normal work conditions and environment, and using visualization for analyzing their own data.

Seo and Shneiderman [59] conducted longitudinal case studies comprising 6 weeks of participatory observation and interviews, with three expert users; they used the *Hierarchical Clustering Explorer*—HCE tool to analyze their database. Perer and Shneiderman [47] used four long-term case studies to prove the power of integrating statistics and visual exploratory features in *SocialAction*, a tool for social network analysis. Employing this tool, four experts explored datasets making discoveries that would not be possible without the integration of statistics and visualization. Valiati et al. [71] reported their experience in conducting 7 MILCS (7 expert users, 5 different domains, 13 data sets and 5 visualization tools). Although two of the case studies were still ongoing at that time, the experience allowed them to obtain high-quality understanding and significant results with respect to information visualization systems usage, in addition to the preliminary results described in [70].

3 Problems with Usability Evaluation of InfoVis Techniques

In this section we focus on some current problems related to the usability evaluation of InfoVis techniques, and in the way these evaluations are designed and performed. We will take into account the well-known usability evaluation methods employed in HCI to help understanding four main problems:

Problem I—The diversity of methods used for evaluating information visualization techniques is quite limited

First of all, few methods, from those used in traditional interaction evaluation, are actually used for evaluating InfoVis techniques. In practice, most evaluations in InfoVis are oriented towards user-testing methods. These methods aim both at checking if usability goals are reached and identifying usability problems by conducting experiments, in which users often try to solve realistic tasks involving the understanding of vast amounts of dynamically changing data. The general idea of such tasks is that users can interact with (parts of the) visualization to find out more about certain data elements and/or about the overall structure of an entire data set. The dependent variables usually measured in this process are task duration time and task accuracy. The effects of the modification on one or several independent variables are collected or measured, and data is analyzed statistically, in order to capture the central usability measures, i.e., effectiveness, efficiency, and user satisfaction.

However, user testing is not always the best choice. It is a very time consuming process, with high costs. In addition, although allowing to integrate real tasks and excluding possible interfering variables such as system crashes or incomplete functionality, it does not provide much support for improving a product, since finding answers to usability flaws is not the main focus.

Clearly, evaluation is not restricted to empirically based user testing, although the importance of active user involvement is heavily stressed. Sometimes, instead of empirical methods like user testing—which can only be used after some form of interaction design is available, and for which direct access to end-users is required— it may be interesting to explore other methods in an attempt to bring down the cost and time requirements of traditional user testing, adopting some analytical methods that could be used earlier when just a preliminary design is available, like expert evaluation (inspection based solely on the evaluator's knowledge and experience), document-based evaluation (inspection based on some guiding documents, at various degrees of precision) or even model-based methods (inspection based on some theoretical models, usually cognitive). These methods are particularly useful when it is not possible to collect data directly from users; but also, they can simply be a useful first step to uncover some major usability flaws before investigating more complex problems.

In order to increase the diversity of methods to be adopted, we need to understand why there is such diversity. As a common characteristic, every method, when applied to an InfoVis Technique (shortly IVT), produces a list of (potential, observed or detected) usability problems as its output. Despite this common characteristic—that in theory should make the various methods comparable—there is in practice a set of criteria we can use to guide the choice of which IVT usability evaluation method is more adequate in a situation and why [31].

We assume the selection of the method to be adopted depends on:

(a) The evaluation objective;
(b) The software development stage when evaluation occurs; and
(c) What is evaluated.

Criteria (a) and (b) are important to determine which type of evaluation methods one can choose. In general, formative evaluation methods aim at determining usability problems that need to be eliminated through redesign e.g., any evaluation performed during development to improve a design is a formative evaluation. Summative evaluation, on the other hand, is used to determine the efficacy of the final design or to compare competing design alternatives in terms of usability. It is to be performed after development to assess a design.

As for criterion (c), an orthogonal perspective is used to distinguish evaluation methods in terms of what is evaluated—a representation of a design result (like, for example, a model, a diagram, a prototype) or the design result itself. Hix and Hartson [33] describe two kinds of evaluations: analytic and empirical. Analytic evaluation is based on the analysis of the characteristics of a design, through the examination of a design representation, prototype, or implementation. Empirical evaluation is based on observation of the performance of the design in use.

Clearly, we can choose distinct methods for IVTs evaluation according to the (a), (b) and (c) criteria above, but it is recognized that today the majority of evaluation of IVTs are summative and empirical. However, a variety of usability evaluation methods can be employed in a wide spectrum of IVTs contexts by designers depending on different goals and needs.

Having diverse usability evaluation methods is a good way to allow selecting an appropriate method to meet evaluation requirements and constraints. It is also an important step that leads to useful evaluation outcomes and presumably effective redesign of IVTs.

In this context, two additional problems arise. *Evaluation happens too late when employing only user testing*, because such testing is mainly applied in later stages of development, it requires a working, running system and in general, the *evaluation process does not follow a general usability evaluation methodology*. These problems will be further discussed.

Problem II—As a consequence of Problem I, evaluation happens too late, because user testing is mainly applied in later stages of development: to test, a running system is mandatory

User testing can only be used after some form of interaction design is available, and for which direct access to end-users is required. Sometimes, instead of empirical methods like user testing, it may be interesting to adopt some analytical methods like the well-known expert-evaluation, document-based evaluation or even model-based evaluation methods. These methods are particularly useful when it is not possible to collect data directly from users. Moreover, they can be simply a useful first step to uncover some major usability flaws before further investigating more complex problems. In addition, because usability testing often occurs late in the design process, developers can be motivated to look at methods that could be used earlier when only an immature design is available.

Problem III—In general, an evaluation process does not follow a general usability evaluation methodology

In general, a usability evaluation process includes at least four activities:

1. Definition of evaluation goals and context of use;
2. Selection of variables to measure, i.e., which variables are taken into account for evaluation;
3. Selection of evaluation methods, i.e., which methods are considered appropriate for the selected variables;
4. Application of the selected methods.

In Activity 1 we are concerned with the appropriate identification of aspects of the context of use and the goals to be reached (e.g., why evaluate?). Since evaluation is not independent of the context of use for which one is designing, many conditions have to be taken into account, such as the characteristics of the user population, the nature of the task demands, or even the technical environment. As can be observed [34], the concept of usability places a considerable emphasis on the need of identifying the context of use. For usability evaluation to achieve meaningful results, it is essential that the context used for the evaluation match as closely as possible the intended or actual context of use. In practice, a context of use includes:

  (i) Representative users, preferably sharing the same knowledge, previous experience and training;
 (ii) Realistic and representative tasks;
(iii) A simulated physical and social environment.

Among the four main purposes of usability evaluation pointed by Bevan [11], two are more oriented to quality of processes (either development or organizational)—and thus out of the scope of our concerns –, but other two purposes are related to "quality" in a sense very close to the concepts we are interested in:

(a) Evaluation of "quality in use": evaluation of user performance and satisfaction when using the product or system in a real or simulated working environment;

(b) Evaluation of "product quality": evaluation of the characteristics of the interactive product, tasks, users and working environment to identify any obstacles to usability.

In our point of view, IVTs evaluation is currently mainly focused on the first type above (a). Indeed, although there is an evident progress in recent years to provide answers to some issues along the first purpose (a), very little attention has been paid to the second one (b). We assume the second type (b) is very suitable to provide a richer outcome. But for that, we need a better definition of properties (or variables) related to IVT's usability and we need answers to the following basic questions:

- What is the meaning of usability for InfoVis techniques?
- Which characteristics should be evaluated?
- Which aspects are generic to all types of interactive systems and which are specific to IVTs?
- Which of these issues are particularly relevant to the evaluation of IVTs?
- How do we apply usability inspection methods (e.g., heuristics evaluation) to IVTs?

In Activity 2 and Activity 3, we are focusing on the assessment of appropriate usability methods. Methods for evaluating usability are described in terms of the variables that should be measured and in the way the evaluation itself is conducted. The detailed description of a method to be used is related to a narrowly defined set of user goals, in a specified task domain, with limited metrics.

The ISO/TR 16982:2002 standard [35] provides descriptions of individual methods for collecting data as part of usability activities within design, and provides a brief assessment of the advantages and limitations/constraints of each method. They may be classified in relation to their relative suitability for use at different stages of the lifecycle process, suitability for use depending on the constraints of the project environment, access to users, the nature of the task and the availability of the skills and expertise needed to use them.

The purpose of evaluation may be primarily formative, to identify and fix any obstacles for effective usability, or summative, to validate the usability of a system. A problem that might arise is that once someone has had success with a particular method, it may be used routinely for all projects irrespective of its appropriateness. There are many categories of methods, from those based on observing users of a real or simulated system, to methods based on models or based on expert assessment of the characteristics of a system.

It is not our intention here to make explicit recommendations about specific methods, although clearly one may draw conclusions about which methods are most appropriate for use in given circumstances.

In Activity 4, we are focusing on the capability of conducting the selected methods in an appropriate manner. Since user testing is the most commonly selected method, phases 1–3 are usually not performed. The evaluators have only to adequately prepare the application of user testing (the way the evaluation is conducted). It should also be noted that agreement on the "walk up and use" test method is achieved more easily because the practical demands of testing an appropriate number of people create less serious barriers than adopting other usability evaluation methods. However, it is difficult—even for technically trained staff—to plan how to correctly apply the methods. Some advocate that only skilled individuals should carry out the usability work [12], others that development staff can easily be trained to take on routine work [11].

Problem IV—InfoVis techniques are usually developed (and evaluated) following a technology-oriented perspective rather than a user-centered perspective.

IVTs are usually developed following a technology-oriented perspective rather than a user-centered perspective. IVTs must be considered by their capacity to support users in achieving specified task goals, with efficiency, effectiveness and satisfaction. IVTs should be more flexible when applied to a real world situation and thus they need to be more adequate and compatible with users' daily routines, providing support to the different phases of the data analysis process in users' work environment. An IVT will fail if it does not fulfill what a user needs. This is a complex endeavor because there is a huge variety of users and tasks. In order to assess an IVT, besides knowing who are the users, what are their goals, and what tasks they need to perform, one needs to characterize what steps users need to take, create scenarios of actual use and finally decide which users and tasks to include in the evaluation.

Evaluating the usability of information visualization techniques has still many open research questions, and we are convinced that methodologies based on a user-centered perspective may yield significant results.

4   Guidelines for a User-Centered Evaluation of InfoVis Techniques

Ideally, the evaluation of visualization techniques must be able to:

- Identify user goals and verify if the user can reach them with an application which implements the information visualization technique;
- Identify which interaction mechanisms made available to the user by the visualization technique are useful to accomplish the user tasks;
- Identify the graphical rendering functions that have been employed by the visualization techniques to show information;
- Relate user goals, interaction mechanisms and graphical rendering.

In a user-centered approach, the requirements stated above can be achieved by understanding users, their tasks, and the context of that task. In this section, a set of guidelines for a (more) user-centered usability evaluation of information visualization techniques is proposed and discussed. In particular, during an evaluation, we need to take into account:

- Which IVT would be suitable to help users understand the data being visualized;
- Which IVT would best support the users' activities.

These guidelines are based on accumulated experience of our research group, but the current set of guidelines is mostly inspired by several case studies conducted by Valiati et al. [71] following the MILCs approach [61].

After a process of intensive analysis, interpretation and discussion about the data collected in these case studies, we devised the following four guidelines:

1. The context of usage for evaluation must be defined before the beginning of evaluation;
2. Evaluation needs to know who the users are, of what are their goals, and to decide which users to support;
3. Evaluation needs to understand which tasks users need to perform and their characteristics (steps, constraints, and other tasks attributes like frequency, priority, etc.) and to decide which tasks to support;
4. Evaluation should be performed earlier in the design-development cycle.

We think it is important to use these guidelines in combination rather than focusing on only one. More detailed explanation of each guideline follows.

Guideline#1: The context of usage for evaluation must be defined before the beginning of evaluation

The standard 9241 Part 11 Guidance on Usability [34] defines usability as *The extent to which a product can be used by specified users to achieve specified goals with effectiveness, efficiency and satisfaction in a specified context of use.*

According to this definition usability is only meaningful in a specified context of use. In practice, "context" is very difficult to define and most general-purpose definitions are inadequate in Computer Science [5]. In fact, there is a specific definition of context in the literature of each of the many fields, such as AI, Software Engineering, CSCW, etc.

HCI's point of view on context is based on a social science investigation of everyday activity (e.g., see [21]). Within this approach, one of the central concerns is the question of "how and why, in the course of their interactions, do people achieve and maintain a mutual understanding of the context for their actions?"

In an IVT evaluation, "context" identifies the IVTs users, the goals of the use of IVTs, the nature of IVT tasks, and the environment within which use occurs. In effect, it asks for the identification of the characteristics and range of the independent variables to which the measured outcomes will apply.

The evaluator has responsibility for the identification of the relevant aspects of the context of use and the identification of the appropriate measures, and the criteria to be applied, always determined within the particular design context. There is no prescription for how different measures are weighted against one another, and while it seems realistic to argue that one cannot achieve usability without achieving some level of effectiveness, it is possible to envisage situations where efficiency is not very important, or indeed situations where satisfaction is the dominant factor. Whatever particular measures are chosen, the approach to measurement for the purposes of specification and evaluation is empirical and based on user testing.

Guideline#2: For evaluating IVT one needs to know who the users are, and to decide which users to support

From an HCI perspective, any design should start with a study aiming at identifying and knowing the users and their goals, as well as the tasks they need to perform to achieve the goals. Thus, a good design needs a better understanding of users.

Indeed, the two first steps of the six basic steps in a UCD process are:

1. Get to know the users;
2. Analyze user tasks and goals.

Typical visualization applications are normally used by a wide variety of people with different skills, background and interaction preferences.

A straightforward way of user description is user profiling, i.e. the definition of user profiles (the literature also uses respectively the terms "user modeling", for the first one, and "user models", for the second one). A user profile is a group of attributes whose values characterize the user in a distinct way. The attributes forming the user profile can change according application domain, application goal, etc. The most common attributes related to user modeling are individual characteristics (gender, age, etc.) but some attributes are particularly interesting for evaluation purposes like user skills, education level, previous experience in using visualization tools, etc.

Having a user profile is a sound way to make an InfoVis application more responsive to user needs and adapted to user profile characteristics.

In short, a good evaluation process should take advantage of the knowledge about users (possibly in the form of user profiles) and of the understanding (and representation) of their tasks, performed to accomplish their goals using an InfoVis technique.

Guideline#3: For evaluating IVTs one needs to understand which tasks users have to perform and their characteristics (steps, constraints, and other task attributes like frequency, priority, etc.) and to decide which tasks to support

Usability testing of visualization techniques also needs the definition of user tasks. Clearly, an understanding of user tasks is a necessary condition for any statement of usability goals. Task Analysis has emerged from ergonomics as an important aid to user-centered approaches because it is an empirical method to understand how people carry out the tasks necessary to carry out their job [14]. A task analysis produces an explicit model of tasks in a domain, a Task Model. In classical task models, we can use different levels of abstraction to represent the complete hierarchical structure of tasks for an entire application: the highest level contains the tasks; these high level tasks are split in subtasks or actions at intermediary levels until the lowest level (also called articulatory level), which contains the physical non-decomposable actions directly associated with devices. Here, we are concerned with the description of device-independent tasks (and actions) corresponding to the fundamental aspects of the user activity.

The understanding and representation of tasks that a user performs while analyzing data analysis are essential for an effective evaluation of information visualization systems.

The relationships between tasks and goals are clear as described in Norman's theory of action [43]. The user's behavior during interaction with a system corresponds to a 7-stages cycle: the user has goals; formulates intents; verifies possible actions and selects the most appropriate one according to their intentions; executes the chosen action; perceives, interprets and evaluates the system's results until completion of the task. Sometimes, a user goal can be mapped to only one task, but often it may require the coordinated execution of several. Clearly, users' needs and goals must ideally be taken into account throughout design and development. Evaluation is the process responsible for validating: (a) how a system cover efficiently users' goals, and (b) how users' tasks using (tasks of a) system meet the exact user goals in an effective, efficient, safe and satisfying way [52].

Indeed, the identification and understanding of the nature of the users' tasks in the process of acquiring knowledge from visual representations of data has received attention in information visualization research.

However, few authors explicitly explore the set of user tasks for InfoVis evaluation purposes. In particular, in this guideline we are interested in bringing attention to how the understanding and representation of tasks (e.g., the well-known HCI task model) can be used for InfoVis evaluation purposes [73]. Some works have proposed InfoVis task taxonomies as reviewed in Sect. 2, but further investigations are still needed.

Guideline#4: Evaluating early

IVT developers often ask for assistance regarding usability testing only when they realize late in development that their techniques are difficult to use.

We suggest that IVT development needs to adopt formative evaluations in order to allow a presumably more effective IVT design and redesign.

We aim also to stimulate the use of expert-based evaluation method as a way to evaluate earlier and without user involvement, mainly when the design process is not yet finished or only an incomplete design is available. Such evaluation methods (or in fact any evaluation method without user involvement) are useful to detect some basic usability problems, which have to be fixed before the beginning of more sophisticated and complex user testing activities.

5 Final Comments

In this chapter we summarized part of the literature on evaluation of information visualization techniques, focusing on the identification of appropriate usability methods. Methods for evaluating usability are described in terms of the characteristics that should be present in the way the evaluation is conducted. The selection of an evaluation method depends on the set of user goals, in a specific task domain, and involves limited metrics. Much work is needed to extend the scope of current evaluation methods to cope with the many possible dimensions of InfoVis techniques. For example, how do we effectively apply usability inspection methods (like heuristics evaluation, for example) to InfoVis techniques? Or how do we guide development of such techniques with these usability considerations?

Consequently, some interesting questions are still open for discussion: (1) can sound ergonomic knowledge (style guides, architectures, and design and evaluation methods that have been proved adequate for GUIs and Web-based contexts) be transferred to the context of IVTs? If so, how do we deal with the idiosyncrasies of InfoVis techniques: in general, or in a way more customized to IVT-specificities? How do we ensure user involvement in usability evaluation? In fact, user involvement is a direct way to accelerate the process of improving usability evaluation of InfoVis techniques.

Our experience [26, 27, 65, 70, 71] and that of others [69] allow us to conclude that doing user testing earlier results in usability knowledge being gained rapidly, rather than having simply the technology perfected without considering the users.

The guidelines presented here focus on a user-centered point of view for the usability of IVTs. With such perspective, IVTs must be considered from their capacity to support users in reaching specified task goals with efficiency, effectiveness and satisfaction. Clearly, one should not "re-invent" methods, but apply as much as possible sound user-centered knowledge from HCI and Ergonomics that can be transferred to IVT environments.

Clearly, these guidelines are not exhaustive: user-centered perspective may contribute, but sometimes we will need other complementary perspectives—e.g., user experience concepts like enjoyment, satisfaction, and even fun—in order to help the choice of an evaluation technique or method. Different goals need different concepts, techniques and assessment methods.

As a contribution of the chapter, we argue that evaluating the usability of information visualization techniques still has open research questions, and we are convinced that methodologies based on a user-centered perspective can lead to significant results.

Acknowledgements This research has been supported by the Brazilian funding agencies CNPq and CAPES to the first and second authors. We acknowledge the work of former students E.A.Valiati, P.R.G. Luzzardi and R.A. Cava, who helped to shape our experience, and are deeply grateful to M. Winckler, L. Nedel and A. Spritzer for the fruitful discussions on the theme along the last years.

References

1. G. Albuquerque, M. Eisemann, and M. Marcus. Perception-Based Visual Quality Measures. *Proceedings of IEEE Conference on Visual Analytics Science and Technology* (Providence, RI, USA, October 23–28, 2011), IEEE, pp. 13–20, 2011.
2. K. Allendoerfer, S. Aluker, G. Panjwani, J. Proctor, D. Sturtz, M. Vukovic, and C. Chen. Adapting the Cognitive Walkthrough Method to Assess the Usability of a Knowledge Domain Visualization. *Proceedings of IEEE Symposium on Information Visualization* (Minneapolis, MN, USA, October 23–25, 2005), IEEE, pp. 26, 2005.

3. R. Amar and J. Stasko. A knowledge task-based framework for the design and evaluation of information visualizations. *Proceedings of IEEE Symposium on Information Visualization* (Austin, TX, USA, October 10–12, 2004), IEEE, pp. 143–149, 2004.
4. R. Amar, J. Eagan and J. Stasko. Low-Level Components of Analytic Activity in Information Visualization. *Proceedings of IEEE Symposium on Information Visualization* (Minneapolis, Minnesota, USA, October 23–25, 2005), IEEE, pp. 111–147, 2005
5. M. Bazire and P. Brézillon. Understanding Context Before Using It. *Proceedings of the 5th international conference on Modeling and Using Context* (CONTEXT'05), Springer-Verlag, Berlin, Heidelberg, pp. 29–40, 2005.
6. BELIV: Beyond Time and Error: Novel Evaluation Methods for Information Visualization. *Proceedings of Workshop of the Advanced Visual Interfaces Conference* (Venice, Italy, May 23, 2006), ACM Digital Library, Eds. E. Bertini; C. Plaisant; G. Santucci, 2006
7. BELIV: Beyond Time and Error: Novel Evaluation Methods for Information Visualization. *Proceedings of Workshop of the Conference on Human Factors in Computing Systems* (Florence, Italy, April 5, 2008), ACM Digital Library, Eds. E. Bertini; C. Plaisant; G. Santucci, 2008.
8. BELIV: Beyond Time and Error: Novel Evaluation Methods for Information Visualization. Proceedings of Workshop of the Conference on Human Factors in Computing Systems (Atlanta, USA, April 10–11, 2010), ACM Digital Library, 2010, Eds. E. Bertini; A. Perer; H. Lam, 2010.
9. E. Bertini, H. Lam, and A. Perer. Special Issue on Evaluation for Information Visualization. In *Information Visualization*, volume 10, no. 3, July 2011. (online version available at http://ivi. sagepub.com/content/10/3)
10. E. Bertini and G. Santucci. Visual Quality Metrics. *Proceedings of BELIV'06* (Venice, Italy, May 23, 2006), ACM Digital Library, pp. 44–48, 2006.
11. N. Bevan. Cost benefits framework and case studies. In R.G. Bias and D.J. Mayhew (eds): *Cost-Justifying Usability: An Update for the Internet Age*. Morgan Kaufmann, 2005.
12. R.G. Bias, A. Druin, B.M. Wildemut, and S. Hirsch. Usability in practice: Avoiding pitfalls and seizing opportunities. In *Proceedings of the Annual Meeting of the American Society of Information Science & Technology*, volume 40, no. 1, pp. 432–433, 2003.
13. R. Brath. Concept Demonstration: Metrics for Effective Information Visualization. *Proceedings of the IEEE Symposium on Information Visualization* (Phoenix, AZ, USA, October 1997), IEEE, pp. 108–111, 1997.
14. S. Caffiau, D. L. Scapin, P. Girard, M. Baron, and F. Jambon. Increasing the expressive power of task analysis: systematic comparison and empirical assessment of tool-supported task models. In *Interacting with Computers*, volume 22, no. 6, pp. 569–593, November 2010.
15. S. K. Card, T. P. Moran, and A. Newell. *The psychology of human-computer interaction.* Hillsdale, NJ: Lawrence Erlbaum Associates, 1993.
16. S. Carpendale. Evaluating Information Visualizations. In A. Kerren, J. Stasko, J-D. Fekete, C. North (eds.) *Information Visualization,* Lecture Notes in Computer Science, volume 4950, Berlin-Heilderberg, pp. 19–45, 2008.
17. J. Carrièrre and R. Kazman. Interacting with Hierarchies: Beyond Cone Trees. In *Proceedings of Information Visualization Symposium* (Atlanta, GA, USA, October 30–31, 1995), IEEE, pp. 74–81, 1995.
18. R. Cava, P. R. G. Luzzardi, and C. D. S. Freitas. The Bifocal Tree: a Technique for the Visualization of Hierarchical Information Structures. *Proceedings of IHC 2002 - 5th Workshop on Human Factors in Computer Systems* (Fortaleza, CE, Brazil, October 2002). SBC/UFCe, 2002
19. C. Chen and Y. Yu. Empirical studies of information visualization: A meta-analysis. In *International Journal of Human-Computer Studies*, volume 53, no. 5, pp. 851–866. 2000.
20. S. Conversy, S. Chatty, and C. Hurte. Visual scanning as a reference framework for interactive representation design. In *Information Visualization*, volume 10, no. 3, pp. 196–211, 2011.
21. P. Dourish. What we talk about when we talk about context. In *Personal and Ubiquitous Computing*, volume 8, no. 1, pp. 19–30, 2004.

22. G. Ellis and A. Dix. (2006) An Explorative Analysis of User Evaluation Studies. *Proceedings of BELIV'06* (Venice, Italy, May 23, 2006), ACM Digital Library, pp. 15–20, 2006.

23. N. Elmqvist, P. Dragicevic, and J-D. Fekete. Rolling the Dice: Multidimensional Visual Exploration using Scatterplot Matrix Navigation. In *IEEE Transactions on Visualization and Computer Graphics*, volume 14, no.6, pp. 1141–1148, 2008.

24. C. Forsell. A Guide to Scientific Evaluation in Information Visualization. *Proceedings of the 14th International Conference on Information Visualisation* (July 27–29, London, UK), IEEE, pp. 162–169, 2010.

25. C. Forsell and J. Johansson. An heuristic set for evaluation in information visualization. *Proceedings of AVI 2010* (Rome, Italy, May 25–29, 2010), ACM Digital Library, pp. 199–206, 2010.

26. C.M.D.S. Freitas, P.R.G. Luzzardi, R.A. Cava, M.A.A. Winckler, M.S. Pimenta, and L.P. Nedel. Evaluating Usability of Information Visualization Techniques. In *Proceedings of IHC 2002 - 5th Workshop on Human Factors in Computer Systems* (Fortaleza, CE, Brazil, October 2002). SBC/UFCe, pp. 40–51, 2002

27. C.M.D.S. Freitas, M.S. Pimenta, and D. Scapin. User-Centered Evaluation of Information Visualization Techniques: Issues and Perspectives In *Brazil/INRIA Colloquium: Cooperation, Advances and Challenges* (Bento Gonçalves, RS, Brazil, July 2009), INF/UFRGS, pp. 199–202, 2009.

28. J. Goldberg and J. Helfman. Eye tracking for visualization evaluation: Reading values on linear versus radial graphs. In *Information Visualization*, volume 10, no. 3, pp. 182–195, 2011.

29. J. Goldberg, J. Helfman. Eye Tracking on Visualizations: Progressive Extraction of Scanning Strategies. *This book, chapter 13.*

30. V. Gonzales and A. Kobsa. A Workplace Study of the Adoption of Information Visualization Systems. *Proceedings of I-KNOW'03: Third Int'l Conf. Knowledge Management* (2003), pp. 96–102, 2003.

31. H.R. Hartson, T.S. Andre and R.C. Williges. Criteria for evaluating usability evaluation methods. In *International Journal of Human-Computer Interaction,* volume13, pp. 373–410, 2001

32. E. Hetzler and A. Turner. Analysis Experiences Using Information Visualization. In *IEEE Computer Graphics and Applications*, volume 24, no. 5, pp. 22–26, 2004.

33. D. Hix and H.R. Hartson. *Developing User Interfaces: Ensuring Usability through Product & Process.* John Wiley & Sons Inc, 1993.

34. ISO 9241–11: 1998. Ergonomic requirements for office work with visual display terminals (VDTs)—Part 11: Guidance on usability, 1998.

35. ISO/TR 16982: 2002. *Ergonomics of human system interaction – Usability methods supporting human centered design,* 2002.

36. O. Kulyk, R. Kosara, J. Urquiza, and I. Wassink. Human-Centered Aspects. In A. Kerren et al. (eds.): *Human-Centered Visualization Environments*, LNCS 4417, Springer-Verlag, Berlin Heidelberg, pp. 13–75, 2007.

37. H. Lam and T. Munzner. Increasing the utility of quantitative empirical studies for meta-analysis. *Proceedings of BELIV '08* (Florence, Italy, April 5, 2008), ACM Digital Library, Article 2, 7 pages, 2008.

38. J. Lamping, R. Rao, and P. Pirolli. A focus + context technique based on hyperbolic geometry for visualizing large hierarchies. *Proceedings of the* CHI '95 (Denver, CO, May 07–11, 1995) ACM Press/Addison-Wesley Publishing Co., New York, USA, pp. 401–408, 1995.

39. B. Lee, C. Plaisant and C.S. Parr. Task Taxonomy for Graph Visualization. Proceedings of BELIV'06 (Venice, Italy, May 23, 2006), ACM Digital Library, pp. 82–86, 2006.

40. S. M. McNee and B. Arnette. Productivity as a Metric for Visual Analytics: Reflections on E-Discovery. *Proceedings of BELIV '08* (Florence, Italy, April 5, 2008), ACM Digital Library, Article 1, 2008.

41. T. Munzner. A Nested Model for Visualization Design and Validation. In *IEEE Transactions on Visualization and Computer Graphics*, volume 15, no. 6, pp. 921–928, 2006.

42. J. Nielsen. Heuristic Evaluation. In J. Nielsen and R.L. Mack (eds) *Usability Inspection Methods.* John Wiley & Sons, pp. 25–62, 1994.

43. D. Norman. Psychology of Everyday Action. In *The Design of Everyday Things*. New York: Basic Book, 1988.
44. C. North. Toward Measuring Visualization Insight. In *IEEE Computer Graphics and Applications*, volume 26, no. 3, pp. 6–9, 2006.
45. C. North, P. Saraiya, and K. Duca. A comparison of benchmark task and insight evaluation methods for information visualization. In *Information Visualization*, volume 10, no. 3, pp. 161–181, 2011
46. W. Peng, M.O. Ward, and E.A. Rundensteiner. Clutter Reduction in Multi-Dimensional Data Visualization Using Dimension Reordering. *Proceedings of IEEE Symposium on Information Visualization* (Austin, TX, USA, October 10–12, 2004), IEEE, pp. 89–96, 2004.
47. A. Perer and B. Shneiderman. Integrating Statistics and Visualization: Case Studies of Gaining Clarity during Exploratory Data Analysis. *Proceedings of CHI 2008* (Florence, Italy, April 05–10, 2008) ACM Press, pp. 265–274, 2008.
48. P. Pirolli and R. Rao. Table lens as a tool for making sense of data. *Proceedings of the AVI'96* (Gubbio, Italy, May 27–29, 1996), ACM, New York, USA, pp. 67–80, 1996.
49. C. Plaisant. The Challenge of Information Visualization Evaluation. *Proceedings of AVI'04* (Gallipoli, Italy, May 25–28, 2004), ACM Press, pp. 109–116, 2004.
50. C. Plaisant, J-D. Fekete, and G. Grinstein. Promoting Insight-Based Evaluation of Visualizations: From Contest to Benchmark Repository. In *IEEE Transactions on Visualization and Computer Graphics*, volume 14, no. 1, pp. 120–134, 2008.
51. R. Rao and S. Card. The Table Lens: Merging Graphical and Symbolic Representations in an Interactive Focus + Context Visualization for Tabular Information. *Proceedings of CHI'94* (Boston, Massachusetts, April 24–28, 1994), ACM Press, pp. 111–117, 1994.
52. Y. Rogers, H. Sharp, J. Preece. *Interaction Design: Beyond Human-Computer Interaction*. John-Wiley, 3rd edition, 2011.
53. P. Saraiya, C. North, and K. Duca. An insight-based methodology for evaluating bioinformatics visualizations. In *IEEE Transactions on Visualization and Computer Graphics*, volume 11, no. 4, pp. 443–456, July-August 2005
54. P. Saraiya, C. North, V. Lam, and K.A. Duca. An Insight-Based Longitudinal Study of Visual Analytics. In *IEEE Transactions on Visualization and Computer Graphics*, volume 12, no. 6, pp. 1511–1522, November/December 2006
55. D. L. Scapin and J.M.C. Bastien. Ergonomic Criteria for Evaluating the Ergonomic Quality of Interactive Systems. In *Behaviour and Information Technology*, volume 16, no. 4, pp. 220–231, 1997.
56. D. L. Scapin, B. Senach, B. Trousse, and M. Pallot. User Experience: Buzzword or New Paradigm? Proceedings of ACHI 2012 - The Fifth International Conference on Advances in Computer Human Interactions (Valencia, Spain, Jan 30–Feb 4, 2012).
57. D. L. Scapin, E. Law, and N. Bevan. Review, Report and Refine Usability Evaluation Methods. In *3rd COST294-MAUSE Project International Workshop* (Athens, March 5, 2007). Technical Report available at http://141.115.28.2/cost294/upload/522.pdf
58. J. Scholtz. Developing guidelines for assessing visual analytics environments. In *Information Visualization*, volume 10, no. 3, pp. 212–231, 2011.
59. J. Seo and B. Shneiderman. Knowledge discovery in high dimensional data: Case studies and a user survey for the rank-by-feature framework. In *IEEE Transactions on Visualization and Computer Graphics,* volume12, no. 3, pp. 311–322, May/June, 2006.
60. B. Shneiderman. The Eyes Have It: A Task by Data Type Taxonomy for Information Visualization. *Proceedings of the IEEE Symposium on Visual Languages* (Boulder, Co., Sept. 3–6, 1996), IEEE, pp. 326–343, 1996.
61. B. Shneiderman and C. Plaisant. Strategies for Evaluating Information Visualization Tools: Multi-dimensional In-depth Long-term Case Studies *Proceedings of BELIV'06* (Venice, Italy, May 23, 2006), ACM Digital Library, pp. 38–43, 2006.
62. M. Smuc, E. Mayr, T. Lammarsch, W. Aigner, S. Miksch, and J. Gartner. To score or not to score? Tripling insights for participatory design. In *IEEE Computer Graphics and Applications*, volume 29, no. 3, pp. 29–38, 2009.

63. R. Springmeyer, M. Blattner and N. L. Max. A characterization of the scientific data analysis process. In *Proceedings of IEEE Visualization* (San Francisco, October 23–26, 1990), pp. 235–242, 1990.

64. A.S. Spritzer and C.M.D.S. Freitas. A Physics-based Approach for Interactive Manipulation of Graph Visualizations. *Proceedings of AVI'08* (Napoli, Italy, May 2008), ACM Press, pp. 271–278, 2008.

65. A.S. Spritzer and C.M.D.S. Freitas. Design and Evaluation of MagnetViz - a Graph Visualization Tool. In *IEEE Transactions on Visualization and Computer Graphics*, volume 18, no. 5, pp. 822–835, 2012.

66. J. Stasko. Evaluating Information Visualizations: Issues and Opportunities. *Proceedings of Workshop of the Advanced Visual Interfaces Conference* (Venice, Italy, May 23, 2006), ACM Digital Library, pp. 5–8, 2006.

67. M. Tory and T. Möller. Human Factors in Visualization Research. In *IEEE Transactions on Visualization and Computer Graphics*, volume 10, no.1, pp. 72–84, 2004.

68. M. Tory and T. Moeller. Evaluating Visualizations: Do Expert Reviews Work? In *IEEE Computer Graphics and Applications*, volume 25, no. 5, pp. 8–11, 2005.

69. M. Tory. User Studies in Visualization: A Reflection on Methods. *This book, chapter 16*

70. E.R.A. Valiati, C.M.D.S. Freitas, and M.S. Pimenta. A Taxonomy of Tasks for Guiding the Evaluation of Multidimensional Visualizations. *Proceedings of BELIV'06* (Venice, Italy, May 23, 2006), ACM Digital Library, pp. 86–91, 2006

71. E.R.A. Valiati, M.S. Pimenta, and C.M.D.S. Freitas. Using Multi-dimensional In-depth Long Term Case Studies for Information Visualization Evaluation. *Proceedings of BELIV'08* (Florence, Italy, April 5). ACM Digital Library, Article 9, 2008.

72. S. Wehrend and C. Lewis. A Problem-oriented Classification of Visualization Technique. *Proceedings of IEEE Visualization* (San Francisco, October 23–26, 1990), pp. 139–143, 1990.

73. M. Winckler, C.M.D.S. Freitas, and P. Palanque. Tasks and Scenario-based Evaluation of Information Visualization Techniques. *Proceedings of 3rd International Workshop on Task Models and Diagrams for User Interface Design - TAMODIA'2004* (Prague, Czech Republic, 2004), ACM, pp. 165–172, 2004.

74. U. Wiss and D.A. Carr. An Empirical Study of Task Support in 3D Information Visualizations. *Proceedings of International Conference on Information Visualisation,* (London, England, 1999), IEEE, pp. 392–399, 1999.

75. J. Yang-Peláez and W.C. Flowers. Information Content Measures of Visual Displays. *Proceedings of IEEE Symposium on Information Visualization* (Salt Lake City, Utah, 2000) IEEE, pp. 99–104, 2000.

76. J.S. Yi, Y. Kang, J. Stasko, and J. Jacko. Understanding and Characterizing Insights: How Do People Gain Insights Using Information Visualization? *Proceedings of BELIV'08* (Florence, Italy, April 5). ACM Digital Library, Article 4, 2008.

77. M. Zhou and S.K. Feiner. Visual Task Characterization for Automated Visual Discourse Synthesis. *Proceedings of CHI' 98* (Los Angeles, USA, April 18–23), ACM Press, pp. 392–399, 1998.

78. T. Zuk and S. Carpendale. Theoretical analysis of uncertainty visualizations. *Proceedings of SPIE-IS&T Electronic Imaging, Vol. 6060: Visualization and Data Analysis 2006.* (Robert F. Erbacher and Jonathan C. Roberts and Matti T. Gröhn and Katy Börner, Eds.) SPIE, page 606007, 2006.

79. T. Zuk, L. Schleiser, P. Neumann, M.S. Hancock, and S. Carpendale. Heuristics for Information Visualization Evaluation. *Proceedings of BELIV'06* (Venice, Italy, May 23, 2006), ACM Digital Library, pp. 55–60, 2006.

Eye Tracking on Visualizations: Progressive Extraction of Scanning Strategies

Joseph H. Goldberg and Jonathan I. Helfman

Abstract Eye tracking methods can be used to help understand what people think about as they search for or compare visual information, such as data displayed in graphs, charts, and other visualizations. This chapter, intended for the visualization researcher who is new to eye tracking, focuses on decision making during data analysis. It steps through the progressive data abstraction that is required during analysis, describing both resources needed and results found. Examples are taken from a study that presented two graph types (linear, radial) in three styles (bar, line, area) to 32 participants. Tasks comprised comparisons between two data series on each graph, across 1–8 categorical dimensions. Analysis proceeded from creating visualizations (Level 0), to statistical factor effects (Level 1), to adding Areas of Interest (Level 2), to comparing scanning sequences (Level 3). Objective, metric-based analysis is advocated throughout, but deeper analyses (Level 3) can be tedious and resource-demanding. Research and new product needs are highlighted.

1 Introduction

Visualizations such as bar charts, timelines, and treemaps are routinely exploited by both consumer and enterprise users to find and compare values, and to understand trends across multiple values. Design guidance for these and other visualizations is typically based upon heuristics, as opposed to empirical data. Eye tracking is

J.H. Goldberg (✉)
Applications UX, Oracle America, 500 Oracle Parkway, MS 3op310,
Redwood Shores, CA 94065, USA
e-mail: joe.goldberg@oracle.com

J.I. Helfman
Measurement Research Laboratory, Agilent Technologies, 5301 Stevens Creek Blvd.,
Santa Clara, CA 95051, USA
e-mail: jonathan_helfman@agilent.com

W. Huang (ed.), *Handbook of Human Centric Visualization*,
DOI 10.1007/978-1-4614-7485-2_13, © Springer Science+Business Media New York 2014

presented here as one empirical approach that can help the investigator to understand visual scanning strategies on many types of visualizations. While neither a panacea nor completely automated, eye tracking can provide insight not available by other methods.

Eye tracking is a well accepted methodology to understand certain characteristics of observers' scanning behavior on visual scenes. Reliable, though still expensive hardware can be obtained from many vendors, and software is available to design, control, and analyze studies. Primers are available to help novices understand and use these systems and methodologies [e.g., 4, 20, 23], and results from extensive web evaluations have been published [34]. Many of the technical aspects of defining fixations and visualizations of eye tracking data have been clearly explained [8]; in particular, a comprehensive explanation of metrics has been published [23]. Current research is reported in at least two dedicated conference series [11, 12], as well two online journals [28, 29].

Studies using eye tracking can be subjective or objective. Subjective approaches can provide very rapid design feedback from a few participants, at the potential risk of over-interpreting individual participant trends. Objective approaches can find subtle differences between conditions, at the expense of requiring substantial resources to interpret and study large datasets. The present chapter illustrates the objective approach with examples from a user study that has dependent and independent variables, factor manipulations, counterbalancing, participant sampling, and other carefully designed criteria.

The real challenge to using eye tracking methodologies is not hardware-related; it is in selecting and interpreting appropriate analyses for specific questions and hypotheses. The cost of eye tracking hardware and associated software is rapidly declining and commoditizing. Eye tracking will soon appear in laptop computers and cell phones. New open source eye tracking solutions use webcams to detect pupil and gaze direction. Calibration-free solutions have been demonstrated. A tremendous amount of data is available within seconds, but generalized approaches to organizing this data have not been presented.

The goal of the present chapter is to provide guidance in the objective analysis of eye tracking data for visual comparison tasks. This is accomplished by stepping through a series of analyses on a dataset, ranging from broad factor comparisons to highly focused sequential understanding of scanning strategies. Throughout, emphasis is placed on an objective understanding of aggregate (as opposed to individual) behaviors, and on understanding the relative resource demands at each level of analysis. We illustrate typical analyses at progressively deeper levels of understanding, but do not present all analysis possibilities. Emphasis is placed upon determining what answers have been provided, and what questions remain, following each level of analysis. This chapter should help the visualization researcher answer: (1) What information can be obtained by adding an eye tracking methodology to an analysis of visualization usage? (2) Should an objective understanding of visual scanning behaviors be pursued? (3) How can empirical answers obtained from one level of analysis drive subsequent questions for deeper analyses? (4) What level of understanding is needed to answer empirical questions?

2 Eye Tracking Background

2.1 Basic Eye Tracking Method

2.1.1 Hardware

The most popular eye tracking approaches bathe the eyes in infrared (IR) illumination, causing the pupils to appear white to one or more IR cameras. Cameras may track one or both eyes by modelling vectors from the pupil centers to small light glints on the surface of the corneas. The direction and length of these vectors are reasonably robust indicators of gaze direction, and are somewhat independent of head or body movement by the participant. Successful eye tracking does, however, require a clear view of the pupil center and of the corneal glint. Extraneous point light sources reflecting in the cornea, for example, can confuse the eye tracker, resulting in poor data quality.

Eye tracking cameras sample gaze location at a specified frequency, typically 30–240 Hz (or higher). Slower cameras can capture average gaze locations reliably enough for usability purposes, but higher frequencies (some >1,000 Hz) can capture micro-fluctuations while fixating, adjustments during eye movements, and movement velocity envelopes while initiating and stopping eye movements. Typical systems for usability studies record at 30, 50 or 60 Hz. See [8, 23] for further discussion of eye tracking hardware.

2.1.2 Preparation

As for any study, careful preparation is crucial to obtaining answers to questions that are posed. While questions regarding *where* people look may only require a single visual target, questions regarding *how* people solve tasks and identify trends require multiple stimuli with multiple conditions or treatments. Static screens are more easily prepared than web pages with dynamic or animated content. Eye tracking can also be conducted on video presentations, but objective, empirical analysis is presently extremely tedious.

Eye tracking studies often use verbal instructions and spoken output to avoid mouse usage, which may cause a subject's eyes to track the cursor, rather than search for desired visual targets. Verbally-presented task instructions will typically invoke eye movements during their presentation, so it is recommended to provide these before viewing a stimulus screen. Also, a time lag may be present before a web-presented screen is fully refreshed.

2.1.3 Selection

Certain participant-specific factors can limit the ability of eye tracking systems to resolve gaze locations accurately [23]. Some of these factors include ocular dominance, droopy eyelids, allergies, and use of mascara, eyeglasses, and contact lenses. If the study design or resolution of analysis is sufficiently sensitive to data loss, selection or restriction of participants during recruitment is recommended. The selection criteria must be carefully documented in subsequent reports and publications.

2.1.4 Calibration

Participants typically complete a 2–16-point calibration, where they must actively look at (or follow) a small visual target on a display. This procedure is often automated, relying on a location-variance calculation about each point. The eye locations at the visual plane are then mapped, via kinematic equations, to actual screen locations [8]. The calibration may be redone during the study to ensure accuracy. Careful attention to good calibration can, in general, help prevent data loss.

2.1.5 Analysis

With a completed dataset, analysis can be conducted subjectively or objectively. Subjective analysis emphasizes real-time playback of gaze locations and viewing attention maps of gaze locations during task completion. The latter, often termed *heatmaps*, are discussed further below. Participants are often encouraged to make comments verbally while watching replays of their gaze locations. These retrospective verbal protocols can help elucidate participant solution strategies, but have been criticized for potentially conveying what a participant believes they're thinking, as opposed to what they're actually thinking at the time [21, 41].

Objective analyses rely upon metrics generated by commercial, open-source, or home-built software [e.g., 14, 39]. Popular metrics include fixation duration, number of fixations, and time to first target fixation, but many other metrics are possible [23, 40]. Data export to other software packages may be required to fully analyze these data.

2.2 Progressive Data Abstraction

For eye tracking to help us understand how individuals complete tasks, a tremendous amount of very low-level data must be filtered and aggregated, allowing the comprehension of data trends at progressively higher levels of abstraction.

2.2.1 Gaze Samples

A study that records 10 min of eye tracking data at a 60 Hz sampling rate gathers more than 36,000 samples of gaze locations. Some of these samples may occur during movement periods, while others occur while the eye is stationary. Many of these are also filtered out as invalid, due to various system glitches. These data are typically abstracted to meaningful units, to better understand larger search patterns.

2.2.2 Fixations

Occurring at about 3 Hz, fixations are short dwells where the eye can process information without blur. Algorithms to define the initiation and termination of fixations have long been compared [8, 23, 37] and it is generally agreed that longer fixations are associated with greater visual and/or cognitive complexity. Since visual targets are often fixated prior to observer awareness, the elapsed time to first fixating the target may be a more sensitive indicator of saliency than task completion time [23].

Fixation algorithms are based upon some combination of gaze sample dispersion, minimum number of gaze samples, and eye movement velocity. While the investigator can typically choose which fixation algorithm to use for analysis: (1) the same algorithm must be used throughout the analysis, (2) the algorithm should match that used by other studies if valid comparisons are to be made, and (3) the algorithm used should be stated in the analysis report.

2.2.3 Scanpaths

A string of temporally related fixations can be combined into a scanpath, which describes the visual scanning path over some time period. Many visualizations and metrics can be defined using scanpaths; some of these are described further below. Example metrics include scanpath length and summed angles or complexity. Scanpaths can provide a temporal or sequential understanding of a single observer's scanning strategy, but scanpath aggregation methods across multiple observers is a current topic of research.

2.2.4 Areas of Interest

Visual targets of arbitrary shape can be marked as 'Areas of Interest' (AOIs). These are typically described manually for each visual target, by dragging rectangles or other shapes. However, there are also methods that assign AOIs automatically from clusters of attention in an eye movement dataset [35]. Heuristics abound for AOI specifications, such as minimum size or minimum margins around visual targets. Software packages may use hierarchically-nested groups of AOIs, allowing the investigator to study AOI visits between groups at multiple levels. Typical analyses include time to first fixation in an AOI, number of fixations in AOIs, total AOI visit times, and transition probabilities between AOIs. The latter has led to sophisticated pattern analysis, using for example, Hidden Markov Models [36]. In addition, the aggregated order of viewing AOIs on a visual scene can be valuable for understanding the flow of visual attention [34].

The above abstractions and metrics can also be combined with other behavioural metrics, such as locations and mouse-clicks (when used). Metrics provide useful snapshots or comparisons of areas of visual attention within and between page elements, but they cannot fully transcribe individual and/or group scanning strategies while completing tasks. A more sequential understanding of visual scanning is needed.

2.2.5 Metrics

Objective analysis of eye tracking data must rely upon objective metrics or measures that can be interpreted graphically and statistically, to understand the impact of specific treatments. Metrics are defined to answer particular questions, such as viewing order of AOIs, or viewing coverage area [20]. Holmqvist et al. [23] provides an organized compendium of eye tracking metrics, from basic fixation counts to those requiring advanced transformations and pattern analysis. This organized framework can be used for defining standards to support cross-study comparisons. Fixation-related metrics (e.g., number, counts, locations, and durations) are often used because of a long research history that has shown that nearly all visual information processing occurs when the eye is stationary, avoiding scene blurring. Metrics that involve saccadic response (e.g., peak velocity, latency, acceleration) have been posed for answering questions such as fatigue countermeasures, or ability to track animated objects. Scanpath-related metrics (e.g., length, complexity, similarity, coverage) can be tedious computationally, but can provide answers to questions that are oriented toward the solution strategy for a task.

Many questions and solutions in eye tracking mimic those of other domains, such as optimized route planning, expected process completion times, or sales territory coverage. Operations research provides many potential metrics of coverage area and routing [19]. Other metrics of search can rely upon the directionality of scanpaths, by summing and aggregating the cumulative angles between fixations [17, 19].

2.3 Eye Tracking on Visualizations

Eye tracking has augmented studies of solution strategies while using visualizations for more than two decades. Prominent visualizations that have been studied include network graphs (or node-link diagrams) and business graphs.

Huang and Eades [24, 25] had participants count how many nodes separated two individuals in social network graphs. Although eye tracking results were used only subjectively, they determined that observers followed edges that lead to target nodes, and that denser areas containing many nodes and edges are avoided. Fairly sophisticated models of graph comprehension with encoding, pattern interpretation, and integration sub-processes have since been validated by eye tracking methods [6].

Business graphs, such as bar and radial graphs, provide rich opportunity for eye tracking methods. Lohse [32] developed a model to predict the perceptual and cognitive time required to read information in tables, bar, and line graphs. Eye tracking helped to validate the model, by decoding the time and order in which graphical elements were observed. Task types included reading data, comparing series, and understand larger trends. When looking up values on line graphs, Shah [38] found that most eye movements were from relating information from data lines to labels, rather than viewing the lines themselves.

Goldberg and Helfman [18] validated aspects of prior graph models [32, 38], using a rigorous eye tracking methodology. Participants were presented short value lookup tasks, then received a specific graph type presented in a specific style. Eight graphs were defined: types included linear and radial graphs, and styles included bar, line, area, and scatter. Radial graphs can be difficult to use; ability to use radial graphs easily could be driven by specific individual differences or characteristics [7, 13]. The hypothesized model for value lookup included three stages: (1) find dimension, (2) find associated datapoint, and (3) find datapoint value. Although a relatively simple stage model, the intricacies of providing experimental control, defining areas of interest, and aggregating metrics can make analysis quite tedious. Metrics, obtained from eye movements, included response time, first target fixation time, and computed stage times. Linear graphs provided faster lookup for these tasks, mainly due to difficulties in tracking concentric rings in radial graphs. Radial graphs with concentric rings that are too closely spaced can cause inadvertent lookup errors, because eye movements can traverse to the wrong ring. Confusion also increases towards the origin of radial graphs, due to smaller available space.

Business graphs routinely support much more complex tasks than just value lookup. Readers may compare values within one or more dimensions (e.g., is stock price X greater than stock price Y?), or they may try to understand larger trends (e.g., is stock X going to be worth more next year?). The present study considers some of these more complex questions as a background for illustrating the flow of eye tracking data analysis.

2.4 Visualizations of Eye Tracking Data

Visualizations of eye tracking data are frequently used to promote a high level view of scanning on an image or page. Heat or attention maps [43] are a popular eye tracking visualization that generally show all fixations within a time window as either multi-colored or transparent areas that are superimposed on the stimulus image. Redder or more transparent areas have a greater numbers of fixations or fixations of longer duration (e.g., Fig. 3). Fixation metrics could also indicate the

Fig. 1 Radial graph
visualizations of prototypical
scanning trends

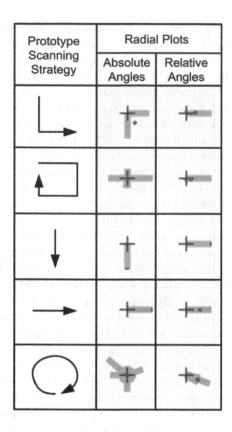

Prototype Scanning Strategy	Radial Plots	
	Absolute Angles	Relative Angles

number of 'raw' gaze samples, the number of individuals who fixated that area at
least once, and possibly other metrics. Many eye tracking researchers don't know
or report which metric is used [1, 5]. Heatmaps have been promoted as a method to
aggregate the strategies of multiple observers [34], but have also been questioned
for not conveying sequential understanding of scanning strategies [5, 16, 33].

Many other static visualizations of eye tracking data have also been presented.
Gaze plots show scanpaths of individual observers during a time period, with longer
fixations denoted by larger circles (e.g., Fig. 4, left side). Gaze plots may show
sequential fixations as numbered circles connected with lines. But gaze plots do
not aggregate multiple sequences well and become overwhelmed with data from
multiple observers (e.g., Fig. 4, right). Some static visualizations may convey
diagnostic information about scanning strategies. Goldberg and Helfman [17], for
example, developed radial histograms that convey high level scanning information
by reporting the relative frequency of saccades that travel in a specified orientation.
(Radial graphs are not advocated here for routine use [13], but rather are presented as
an alternative data visualization for studying differences in scanning strategies.) As
illustrated in Fig. 1, the shape of radial graphs can signify high level scanning trends
if they are sufficiently similar to prototype sequences, such as clockwise scanning,
or scanning down and to the right [17]. Each radial plot can be constructed using

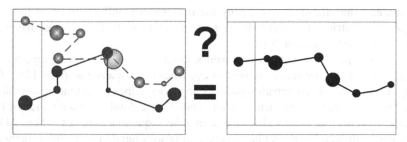

Fig. 2 Understanding the central tendency of multiple scanpaths is a complex, yet important problem

either angles that are *absolute*, measured with respect to the stimulus image, or *relative*, measured with respect to the current direction of gaze movement. Also, a small dot is provided to summarize the average spatial angle of travel. Separation of the shape of these plots could potentially be diagnostic, providing a high-level description of a scanning strategy.

2.5 Understanding Sequences and Strategies

Comparing sequential scanning strategies between individuals can provide a deeper understanding of group and individual scanning strategies than metrics such as aggregated times and durations. Sequential analysis can help the investigator to detect differences between predefined datasets, determine within versus between-group strategy differences, and find clusters of common scanning strategies within a dataset [33]. The fundamental challenge, illustrated in Fig. 2, is to understand the central tendency of multiple scanpaths, which include (1) different numbers of nodes, (2) different durations of nodes, and (3) different lengths of edges. Recent developments in this area, now called the "Average Scanpath Problem" [22], are described below at a summary level.

When each fixation in a scanpath is mapped to the name of the AOI that contains it, the sequence of AOI names can be used to represent the scanpath as a sequence or string. String editing methods compare scanning sequences by calculating the minimum number of editing steps necessary to transform one scanpath into another. The Levenshtein Distance, in particular, counts the number of insertion, deletion, and substitution steps necessary for transforming one sequence (e.g., a sequence of AOI names) into another sequence [30]. Note that sequential order is the only temporal difference modelled by such an analysis. Temporal differences such as saccadic velocity, saccade length, or fixation durations, are ignored. Still, string editing comparisons have been useful in examining real scanning strategies on a variety of stimuli [e.g., 27]. An alternate way of comparing scanpaths is to map each fixation in one string to that in another string, and vice-versa [33], based upon Euclidean distances. The overall similarity metric is then the sum of the two sets of summed value distances, divided by the maximum length of nodes of the two strings. This metric was found to be more sensitive than the Levenshtein distance, when normalized for length [33].

Dot-matrix plots (or dotplots) are well known in certain domains (e.g., bioinformatics [42]), but are relatively new to eye tracking sequence analysis [15, 16]. Strings of concatenated scanpath sequences are cross mapped to a matrix, with dots placed in cells where the sequence tokens match. The dotplot is then searched for collinear patterns that indicate forward or inverse sequential matches, where AOIs associated with matches do not have to be contiguous neighbors. Dotplot methods are further presented later in this chapter.

Scanpaths can also be represented as a series of vectors, not just as edges and nodes, with each vector representing the amplitude and direction of subsequent saccades [26]. In this method, scanpaths are simplified, temporally aligned, then compared by averaging differences in saccade lengths, fixation distances, saccade angles, and fixation durations. While this method has the potential to be very diagnostic, it has not been shown to be more sensitive or reliable than comparison using Levenshtein Distance. However, aspects of shape, position, or length can potentially be compared in detail.

A tree visualization of scanpath AOIs can also be used to compare scanpath similarity. In eSeeTrack [40], the analyst selects a set of concatenated fixations from a timeline. A second window displays a tree diagram showing AOI transitions from the selected root AOI. The shape of the tree diagrams can then be compared across observers and conditions, to indicate similarities and differences.

3 Methods: Scanning Business Graphs

Discussion of the practice and analysis of eye tracking is easier to follow if accompanied by a running example. Data are presented here from a study in which business graphs were scanned to find answers to presented task questions [18]. The present unpublished data represent the second half of a larger study, in which participants compared two data series along one or more dimensions, using a variety of graph types. The presented tasks were representative of realistic comparisons that could be conducted using these visualizations.

3.1 Participants

As reported previously, 32 professional colleagues from Oracle Corporation (18 male, 14 female) participated [18]. Their age range was 24–61 years with mean (median) of 36.5 (37.0) years. Each was screened for normal color vision using six Ishihara color plates on an Optec 2000P vision tester. Each reported significant experience in reading and comparing data series in business graphs.

3.2 Apparatus and Procedure

Each participant was individually tested over a 30-min period. The height and distance of the eye tracker was adjusted, and a 5-point calibration was conducted. The following scenario was read aloud: *You are a corporate manager, reviewing employee candidates for a job opening. The following graphs represent 8 job competencies (such as leadership, teamwork, communication, and motivation), labelled a-h. On each graph, BLUE represents minimum job competency, and ORANGE represents the job candidate.* Five practice tasks were completed, followed by 12 experimental trials. Each of these was initiated with a screen-presented task, which the participant read aloud. The participant could ask any questions for clarification, then a graph was presented. The graph was searched, and answer(s) given. The next trial ensued, following a pause and blank screen. No manual input device was used by participants.

Eye tracking data were recorded using a Tobii T60 eye tracker and Tobii Studio v.3.0 control and analysis software [39]. This binocular eye tracker uses infrared cameras and lights to model the gaze location of the eyes, without any external attachment. The software allows experimental preparation, fixation definition, data visualization, and data exporting.

3.3 Stimuli

Graph stimuli, examples shown in Table 1, were displayed at $1,280 \times 1,024$ resolution. Two data series were presented on each of two graph types (radial, linear) by three graph styles (bar, line, area) by two datasets (ab, cd). Each graph had eight hypothetical abscissa dimensions (or categories), labelled a-h, and six unlabeled ordinate values, marked by light gray linear or radial grid lines. The two series that were presented on each graph each plotted eight values, one value per dimension. Two sets of dual-series datasets were defined (ab, cd), each of which had mean = median = 3.0, and had variance values of 3.0–4.0. Due to space limitations, Table 1 only shows dataset ab for bar and line graphs, and dataset cd for area graphs. The orange and blue series had complementary colors, and were rendered with sufficient transparency to clearly see underlying values.

3.4 Tasks

All 12 tasks required relative comparisons within and among the orange and blue plotted series on each graph. These comparisons required either: (1) both series on 1 dimension, (2) 2 dimensions on one or two series, or (3) All 8 dimensions on both series. Table 2 shows selected tasks by each of these categories. The difficulty of

Table 1 Graph stimuli from study, including type, style, and dataset

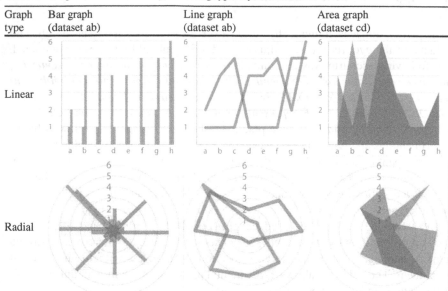

Graph type	Bar graph (dataset ab)	Line graph (dataset ab)	Area graph (dataset cd)
Linear			
Radial			

Table 2 Selected tasks, by difficulty categories

Task type	Presented task
1 dimension, 2 series	Which series has a greater value for dimension A: orange or blue?
	Considering only dimension F, which series had a greater value: orange or blue?
	Which series (orange or blue) had a greater value on dimension E?
2 dimensions, 1 or 2 series	For the job candidate represented by the orange series, which value was larger: dimension B or dimension H?
	Assume larger values are more desirable for dimensions A and C. Based only upon these two dimensions, which candidate (orange or blue) is better?
8 dimensions, 2 series	Which dimension(s) shared the same value for the orange and blue series?
	Across all dimensions, how many times were the data points equal between the blue and orange series?
	On which dimension(s) was the value of the orange series greater than the value of the blue series?

these tasks can be roughly characterized as a function of the number of dimensions required in series comparisons. Answering single dimension task questions required three stages: (1) locate proper dimension, (2) find relative values on blue and orange series, and (3) compare these values. Answering dual dimension task questions required scanning relative series values for two dimensions, while answering eight dimension tasks required scanning values for each of 8 dimensions.

3.5   Experimental Design

Each participant received a series of 12 tasks, with each task assigned to one of the aforementioned 12 graphs. Counterbalancing minimized learning and fatigue effects by presenting tasks and graphs in different orders across the participants. Each participant received one of the 12 task orders, and one of the 12 graph orders. Although not a formal Graeco-Latin square design, subsequent analyses confirmed that there were sufficient replicate task and graph orders to minimize learning and fatigue effects across participants.

Designed independent factors included: Graph Type (linear, radial) × Graph Style (bar, line, area) × Task (1, 2, 8 dimensions). Dependent measures included: Completion time, Number of fixations, Fixation duration, AOI visit duration, AOI transitions, and other measures from sequential analysis.

4   Results

Analysis of eye tracking data can proceed at many levels, and a multi-step decision making approach to these levels of analysis is advocated here. While successively narrower questions can be answered at each level, additional time and resources are generally required. These analysis levels are: (0) Visualizations that provide an initial view of data, (1) Times and durations of tasks and fixations, providing statistical comparisons of effects, (2) AOI partitioning, in which specific areas or elements of visual scenes are considered, and (3) Comparing sequential scanning strategies between observers and conditions. This section considers decision and results at each of these analysis levels, using the present dataset.

Data was recorded at 60 Hz, and fixations defined using an I-VT filter [37] in Tobii Studio 3.0 software [39]. A Fixation was defined by a minimum duration of 60 ms and eye movement velocity threshold of 30°/s. Adjacent fixations were merged if they were separated by less than 75 ms. Data quality was high for these participants, with less than 10 of gaze samples dropped (e.g., due to blinking).

4.1   Results (Level 0): Visualizations

4.1.1   Heatmaps and Gazeplots

Visualizations offer a quick way to aggregate data across participants, but are limited in their ability to explain larger data trends. Heatmaps, in particular, have become a very popular way to summarize eye tracking data, but can vary widely depending on fixation criteria, and modelling assumptions [3]. Figure 3 presents heatmaps from two different 8-dimension tasks, using linear area (left) and radial area graphs

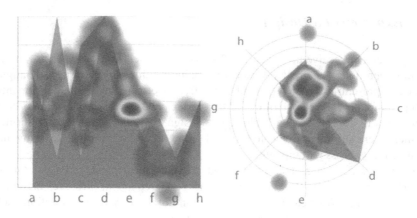

Fig. 3 Example heatmaps on linear area (*left*) and radial area (*right*) graphs

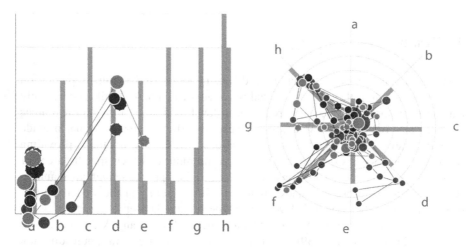

Fig. 4 The same three participants scanning linear bar graph (*left*) and radial bar graph (*right*)

(right). These are computed from the number of fixations at each location, as opposed to duration of fixations. Data from the same three participants is aggregated on each graph, but the tasks used different graph dimensions. At a high level, it appears that fixations are much more spread out in the linear graph, than on the more compact radial graph. Also, there are few fixations on dimension labels in either case, as the tasks required relative comparisons between the two series. Observers scanned mostly along the orange series in the linear graph (indicated here by the 2nd bar in each cluster), but viewed both series in the radial graph.

Compare the heatmaps of Fig. 3 with the gazeplots in Fig. 4, which show three scanpaths from a specific task. Figure 4 shows typical scanpaths from the linear (left) and radial bar (right) graphs. The gaze plots retain the sequential information discarded by the heatmaps, but how can scanpaths from multiple participants be compared? One approach is to carefully study each scanpath, then aggregate these into larger trends [34]. This approach is tedious and can overweight individual participant strategies. In simple tasks, however, scanning strategies may be close enough that an aggregated understanding of strategy is possible. In Fig. 4 (left), all three participants immediately located the label for the "a" dimension, then scanned upwards to read its value. None of these observers appeared to have any hesitation or confusion over the labelling of the dimension. More complex tasks, as in Fig. 4 (right), may be solved in many more ways, little overlap may exist between scanpaths, and it becomes much harder to develop an aggregated understanding of strategy. Also, for radial graphs, scanpaths may cross the graph to get to another dimension, as opposed to following a concentric ring, further complicating the discovery and grouping of strategies. Crossing through the center of a radial graph may be harder to detect with a heatmap (e.g. in Fig. 3 right, where the hotspot is mainly aligned with dimension "a", although this could indicate that crossing through the center was not required due to the high saliency of the dimension).

4.1.2 Reflection

While traditional visualizations of eye tracking data can help us understand where people look over some time period, it is hard to develop an aggregated understanding of strategies used to solve tasks. These visualizations provide an understanding of spatial aggregation, *where* people looked, but not *when* they looked there. Any understanding obtained of timing, order, or sequential scanning strategies is subjective and may overweight individual participant strategies in some cases. A more objective understanding of sequential factors and effects may be obtained through analysis of appropriate metrics.

4.2 Results (Level 1): Times and Durations

In our example study, data exported from the eye tracker's software into Excel included task durations, number of fixations, fixation durations, participant number, trial order, task, stimulus, graph type, and graph style. The tasks were then aggregated into difficulty levels (1, 2, or 8 required dimensions). The dataset was also exported to a statistics application for further analyses.

4.2.1 Task Completion Times

The completion time within each graph condition by task was determined from the time difference between the graph appearance and the initiation of the participant's verbal response. As a starting point, factor influences on completion time can be further studied using various metrics from eye tracking. A three-factor ANOVA,

Table 3 Summary of 3-factor ANOVAs on time and fixation measures

| | | F-statistic | | |
Source of variation	DF	Task times	No. fixations	Fix. duration
Graph type	1	1.33	**7.34****	**19.69*****
Graph style	2	**3.17***	2.16	1.98
No. task dimensions	2	1.37	1.80	0.44
Type × style	2	1.87	1.30	1.33
Type × dims	2	1.20	2.28	0.41
Style × dims	4	**2.87***	1.56	1.97
Type × style × dims	4	0.31	0.54	1.67
Within	366			
Total	383			

*p<.05; **p<.01; ***p<.001

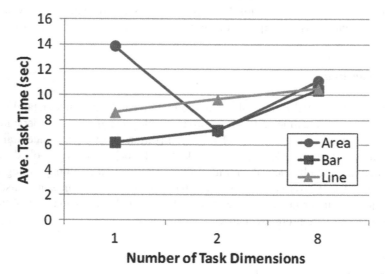

Fig. 5 Average task completion time by number of required task dimensions

summarized in Table 3, was computed on completion time, with graph type, graph style, and number of task dimensions serving as independent factors. Graph style influenced completion times ($F_{2,366} = 3.2$, $p < 0.05$), but neither the graph type nor task dimension main effects were significant ($p > 0.10$). Using pairwise Bonferroni comparisons ($\alpha < 0.05$), the area graphs produced significantly longer completion times than the bar graphs (Fig. 5). A significant interaction between graph style and number of dimensions in task time ($F_{4,366} = 2.9$, $p < 0.05$) was due to area graphs requiring about 7 s more time than the other graph styles when tasks required only a single dimension. Area graphs also required more time than other graph types in general.

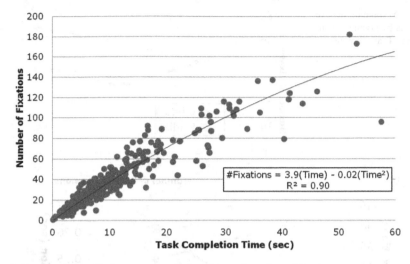

Fig. 6 Strong relationship between completion time and number of fixations

4.2.2 Number and Duration of Fixations

Task completion time is usually strongly correlated with total number of fixations during a task. (In search tasks for a single visual target, note that completion time is also correlated with the number of fixations required to *initially* locate the visual target [e.g., 18]). In short tasks, this relationship is highly linear because attention is highly focused, and the number of fixations can serve as a proxy for completion time. As tasks become longer than a few seconds, attentional dissociation can occur [20], and fixation durations can become more variable. In general, it is good practice to check the correlation of completion time with total number of fixations for each study conducted.

The present relationship between completion time and number of fixations is shown in Fig. 6. The relationship is quite linear up to about 30 s, after which it appears to become asymptotic. A quadratic regression has $R^2 = 0.90$, and a linear slope that is close to four fixations/s. It is likely that attentional dissociation started to occur at about 30 s, and that longer searches may have included some element of frustration or other cognitive processing.

Graph type strongly influenced both number of fixations and fixation durations ($F_{1,366} > 7$, $p < 0.01$). Participants averaged 40 fixations on the linear graphs, but only 32 fixations on the radial graphs (Fig. 7). Although the graph type × number of dimensions interaction on number of fixations wasn't significant ($p = 0.10$), note that 1-dimension tasks didn't differ, whereas 2 and 8-dimension tasks showed an approximate 10 fixation advantage for radial graphs.

Average fixation durations were significantly longer on radial than on the linear graphs ($F_{1,366} = 19.7$, $p < 0.001$), as shown in Fig. 8. This could potentially signal

Fig. 7 Average number of
fixations by graph type and
number of task dimensions

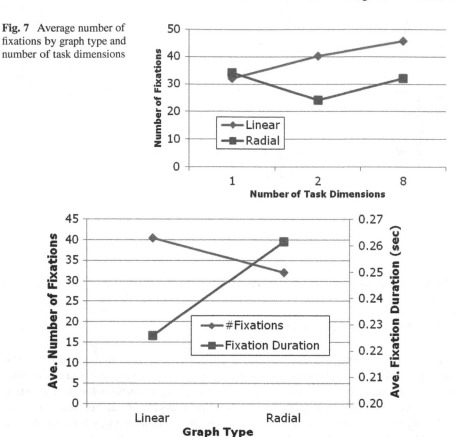

Fig. 8 Average number and duration of fixations by graph type

greater perceived complexity on radial, compared to linear graphs [20]. Note that
longer fixation durations are also commonly accompanied by fewer fixations, if all
other factors are held constant.

4.2.3   Reflection

Linear graphs can support comparison tasks with about the same completion time
as radial graphs, but more, shorter, fixations are made on linear graphs, while
fewer, longer, fixations are made on radial graphs. Greater numbers of fixations
occur on tasks that require comparison of more than one dimension. Longer
fixations generally denote periods when a user is interpreting, thinking, processing
information, or is possibly confused in some way. In addition, for comparison tasks
in general, both linear and radial area graphs require more time than those presented
on bar or line graphs.

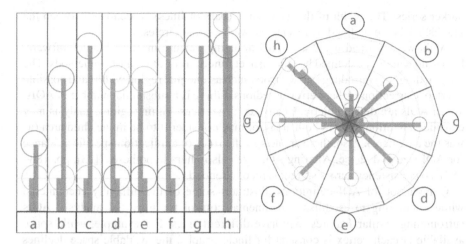

Fig. 9 Linear bar (*left*) and radial bar (*right*) graphs, illustrating AOIs: dimension labels (*indicated by boxes or circles surrounding letters*), dimension areas (*indicated by rectangles or wedges surrounding the data bars*), and series vertices (*indicated by a circle at the end of each data bar*)

The level of analysis provided by a statistical understanding of differences among graphs and tasks was not possible by inspecting heatmaps and gaze plots. Appropriate metrics had to first be selected, requiring familiarity with both eye tracking methods and the specific problem domain. Selection of appropriate metrics is an important stage in objective analyses, considering the very large set of possibilities [19, 23]. Data preparation for this step was not overly tedious, but required in-depth knowledge of spreadsheet software and statistics, as well as careful checking to ensure that no errors were introduced.

Several questions loom as a result of these analyses. Why were more fixations made on linear than radial graphs, as the number of required task dimensions increased? Why were fixation durations longer on radial than on linear graphs? Why did area graphs require more time to complete tasks than the other graph types? To help address these questions, we now introduce an additional abstraction layer: Areas of Interest.

4.3 Results (Level 2): Areas of Interest

4.3.1 AOI Definitions

In the quest to address specific questions and hypotheses about number and duration of fixations between linear and radial graphs, AOIs can be a beneficial transformation to help understand higher level task solution strategies. Figure 9 shows AOIs defined for the linear and radial graphs in this study; linear bar (left side) and radial bar (right side) are plotted with identical datasets as lighter and darker series. The width of the data marks (bars or lines) is identical between the graphs. AOIs are outlined by rectangles, wedges, and circles.

AOIs were defined using a manual dragging gesture in Tobii Studio software, but could equally be defined by spreadsheet functions or other ad-hoc methods. The manual dragging introduces small errors, such as unequal region widths and possible overlap. Correction of these errors is tedious if there is a substantial number of AOIs. When AOIs are defined hierarchically, as for vertices within regions, analysis may consider a fixation to be in both areas, or may consider it to be in the child area (as was the case here). Also, there are no specific rules or application-specific guidance for AOI size and shape. Altering AOI sizes also alters hypotheses ([23], pp. 188–189), so consistency across stimuli is recommended.

Comparison of AOI definitions illustrates some of the difficulty encountered while attempting to establish experimental control for impartial analysis. AOIs surrounding similar features may have different sizes. For example, while space available at each vertex is constant for linear graphs, the available space declines towards the origin in radial graphs. This makes it harder to capture fixations accurately towards the origin of radial graphs, than for either linear or the outer portions of radial graphs. In radial graphs, fixations that are assigned to region AOIs, may actually be in vertex AOIs. Similarly, AOIs for radial dimensions are pie-shaped, but are rectangular for linear graphs. The number of expected fixations in regions could depend on the relative areas of these shapes. Direct comparison between linear and radial graphs should ideally control for these size differences in metrics that use the AOIs. Without such control, however, it is expected that when solving identical tasks, more AOIs would be visited in radial than linear graphs, due to their proximity near the graph's origin. Similarly, the time spent looking at specific region AOIs should be shorter in radial than in linear graphs.

Typical eye tracking software can re-interpret collected data with respect to new AOI definitions, re-associating fixations with AOIs. Here, fixation and AOI visit durations (including AOIs, participants, tasks, and graphs) were calculated and exported from Tobii Studio into an Excel pivot table and a statistics package. When many AOIs are defined, they may be aggregated within groups, such as Dimension Labels, Dimension Regions, and Vertices. (Other group aggregations were certainly possible, depending on analysis questions. For example, AOIs could be grouped as 'targets,' those *required* to complete each task, versus non-targets.)

Figure 10 shows the first 12 sequential fixations on a linear area graph, with overlaid AOIs. Although difficult to clearly see in the present grayscale images of the color stimuli, the initial fixation was on a vertex of the darker series (in dimension g), then proceeded to the dimension labels, b and a. Next, the scanpath travelled to the darker series vertex at dimension a, scanned to label c, then up to the darker series vertex c. The AOIs are hierarchical, so viewing a vertex also accumulated time within its region. For p1's total scanpath, four fixations were on the dimension labels, eight were in dimension regions, and four of these were also in vertices. Many synthetic measures can be computed, based upon these values. For example, a high ratio of numbers of fixations in dimension labels to those in regions could indicate difficulty in finding proper vertices when comparing series.

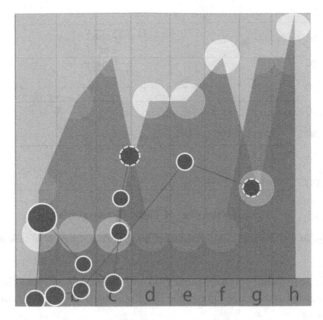

Fig. 10 Scanpath from p1 on linear area graph with superimposed AOIs

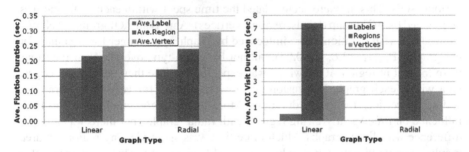

Fig. 11 Average fixation duration (*left*) and AOI visit duration (*right*) as a function of graph type and defined AOI grouping

4.3.2 Fixation and Visit Durations

Three-factor ANOVAs were conducted on average fixation durations, as a function of graph type (linear, radial), graph style (area, bar, line), and the AOI group (labels, dimensions, vertices). Fixation durations (Fig. 11, left) were significantly longer in radial than in linear graphs ($F_{1,940} = 21$, $p < 0.001$) as was previously found, and the AOI group was also very influential ($F_{2,940} = 56$, $p < 0.001$). Note that, within each graph data cluster in this figure, the left bar indicates Labels, the middle bar Regions, and the right bar, Vertices. While the graph type × AOI group interaction wasn't significant, pairwise comparisons on the AOI group main effect showed that fixations were longest on vertices (0.274 s), shorter in dimension regions (0.230 s), and shortest on dimension labels (0.176 s). Further ANOVAs pinpointed the fixation

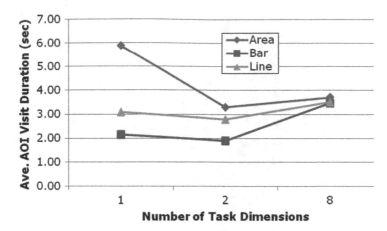

Fig. 12 Average AOI visit duration, as a function of graph style and number of task dimensions

duration increase in radial graphs to the region and vertex AOI groups. This suggests that finding regions and vertices may have been more difficult on the radial than on the linear graphs.

AOI visit duration (Fig. 11, right) defined the amount of time spent in the various AOIs. This measure accumulated the time spent within each AOI, and was aggregated by AOI group (labels, regions, vertices). Across all AOI groups, ANOVA showed that visit durations were influenced by graph style (longer for area than for bar graphs, $F_{2,1134} = 3.6$, $p < 0.05$), but not by graph type overall. Visit durations were longest in region AOIs, which was expected because they occupy the largest area within each graph. Dimension labels were viewed for very short durations, and series vertices for longer durations. More interestingly, graph style had little influence on time spent viewing regions with radial graphs, but there was a larger difference with linear graphs. Much more time was spent viewing regions in area graphs, compared with bar graphs ($F_{2,366} = 4.0$, $p < 0.05$). This same trend also occurred, to a lesser extent, when viewing vertices.

Similar to the task completion time results (Fig. 5), the AOI visit duration × graph style interaction was significant ($F_{4,1134} = 3.3$, $p < 0.01$), with single dimension area graphs showing longer times than the other graph style × task dimension combination (Fig. 12). The increase in task time for 1-dimension tasks and area graphs was due to longer visits within AOIs.

4.3.3 AOI Transitions

The aggregated transitions from one AOI to another AOI were computed, then compared between graph types and styles. These data provide unconditional probabilities of looking at an area on the graph, starting within a particular area. We consider 2-state transitions, but analyzing 3-state and longer transition strings can serve as a basis for Markov models to predict sequences and transition times. Transition analysis is another tool to help describe the aggregated scanning strategy

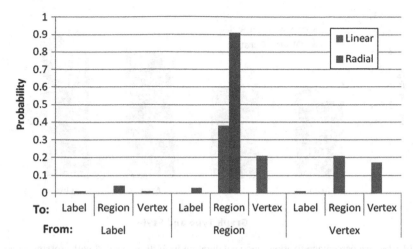

Fig. 13 Unconditional transition probabilities between AOI groups, using p1, p13, and p25, and both graph types

of a group of observers, for a specific task. Figure 13 shows these probabilities from the bottom row regions to the upper row regions in the graph. Linear graphs are noted by the darker, blue bars, and Radial graphs are noted by the lighter, red bars. Very few transitions were made to or from dimension labels in either graph type. Most transitions within the radial graphs were made from region to region, as vertices were so small, close to the origin of the graphs. The linear graphs showed much greater tendency (about 40%) to transition from region to vertex and from vertex to region. Although these graph type differences were most likely due to differences in AOI sizes, this analysis illustrates how analysis of AOI transitions can help to bridge from a factor-effect understanding, to a sequential understanding of scanning differences between graphs.

4.3.4 Scanning to False Vertices

The gaze plot visualization that was shown in Fig. 10 highlighted potential confusion by dual-series crossings that do not actually represent data values. For example, in Fig. 10, these series crossings were evident between dimensions c and d, and between dimensions f and g. Indeed the scanpath of Fig. 10 ended with a fixation on the c-d crossing point. The question naturally arises, whether this confusion was equally likely between graph types (linear, radial), styles (area, line), and datasets. (Note that no such crossing points occur for bar graphs.)

To investigate this post-hoc question, we assigned AOIs on top of the crossing points between dimensions c-d and f-g for the first dataset (ab), and between dimensions a-b and b-c for the second dataset (cd). Two metrics were summed for each of these: number of fixations and visit duration. Because these metrics differed as a function of graph type and style, we normalized each by the total task time and

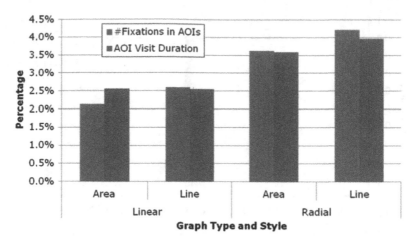

Fig. 14 Relative percentage fixations and visit duration in AOIs associated with crossing series

fixations. Three-factor fully-crossed ANOVAs were computed as a function of graph type, graph style, and dataset. Both the proportion of fixations in these "crossing" AOIs ($F_{1,248} = 9.2$, $p < 0.01$) and proportion visit duration ($F_{1,248} = 4.6$, $p < 0.05$) were greater for radial than linear graphs, but no other factors were significant. Figure 14 shows that twice as many fixations and twice as much time emerged from the "crossing" AOIs in radial graphs, compared to those in the linear.

4.3.5 Reflection

Longer visits were made to AOIs in linear, compared to radial graphs. This was perhaps due to greater required searching for vertices in linear area graphs, where data values may be harder to ascertain than in radial graphs. Another possibility, however, is that region AOIs were very close together near the origin of radial graphs, meaning that one scanpath might rapidly traverse several different regions. The analysis showed that AOI visits were nearly twice as long in area graphs compared to other graph types, for single-dimension tasks, but this difference disappeared for more difficult tasks. It is possible that the solid areas were confusing in tasks that only involved a single dimension. The finding of mostly region-region AOI transitions in radial graphs was most likely due to very small vertices towards the origin of these graphs, compared to linear graphs. This is evidence that the area of AOIs should be included as covariates in ANOVAs; however, substantial resources are needed to measure these pixel areas. An example sub-analysis considered the notion of false vertices that are created inadvertently by crossing of multiple series. Adding additional post-hoc AOIs allowed us to discover that both more fixations and longer visits are made on these false vertices in radial than in linear graphs. Overall, the AOI-level analysis provided a moderate level of new information; AOIs, however, are also a requirement for deeper analysis of scanning sequences.

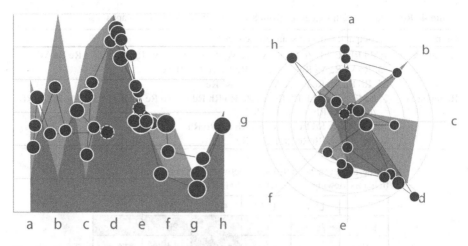

Fig. 15 Scanpaths from same task; linear (*left*) and radial (*right*) area graphs

4.4 Results (Level 3): Sequence Analysis

Analysis of scanpath sequences is becoming popular to understand subtle differences in scanning strategies by those completing carefully controlled tasks. Fixation sequences can be exported and analyzed to determine (1) if different observers scan in a similar way, and (2) how close observers are to optimal scanning, or matching a reference scanpath. Sequence analysis starts by obtaining a scanpath sequence from each observer conducting a specified task. Although many comparisons are possible, for present illustration purposes we will compare scanning strategies using radial versus linear area graphs, completing an 8-dimension task (*Across all dimensions, how many times were the data points equal between the blue and orange series?*). Two typical scanpaths from this task are shown in Fig. 15. Note that the scanpath on the linear area graph (left) tended mostly to follow vertices of the two series, whereas that on the radial area graph (right) crossed frequently through the origin of the graph.

4.4.1 Sequence Reduction

The scanpath sequences were coded and analyzed for similarity. The coding schema used here for each visited AOI was "Xy", where:

"X" = {R if region, L if label, and V if vertex}
"y" = {a-h, indicating the series dimension of the graph}

Reduced strings, shown in Table 4, were created by removing duplicate, contiguous AOIs. String reduction is commonly done to emphasize first entries into each AOI ([23], p. 193), because contiguous fixations in the same AOI are often

Table 4 Reduced scanpath sequences from 8-dimension tasks on area graphs

Graph	Participant: reduced scanpath sequence
Linear area	1: Rd Rb Ra Rb Rc Rd Re Ve Re Ve Re Vd Re Ve Vf Rg Rh Vg Vf Re Rd Ve
	13: Rc Va Ra Va Rb Vc Rc Vd Re Rf Ve Rh Rf Ve
	25: Rb Ra Rb Rc Re Ve Vf Vg Vf Rf Re
Radial area	2: Rc Rb Ra La Lb Rb Ra Rd Re Rb Rh Rd Ra Rb Rc Ra Rh Lh Rh Rg Rf Re Rf
	Re Rh Ra Rb Rh Rc Rd Re
	14: Ra Rh Rc Rd Re Rf Rc Rh Rb Ld Re Rh
	26: Ra Rh Ra Rb Rc Rd Re Rf Rc

Table 5 Levenshtein distances between sequence pairs, with participants viewing linear (L) or radial (R) graphs

	L13	L25	R2	R14	R26
L1	25	24	32	26	30
L13		13	41	14	14
L25			45	11	11
R2				43	45
R14					12

considered part of the same AOI visit. In the present task, an optimal scanpath would traverse the vertices in a linear or circular pattern (e.g., Va,Vb, . . . ,Vh). The actual vertices with equal values were Vd, Ve, Vg, and Vh. Note that vertices were organized within dimension AOI regions. Fixations on vertices were only counted within vertices; those within regions were not considered within a vertex.

4.4.2 Sequence Matching Using String Editing

The Levenshtein Distance is frequently used to describe the similarity/difference between strings, regardless of length or length differences [23, 27, 30]. It is based upon the summed number of insertion/deletion/substitution operations needed to transform one string to another string. A smaller number is associated with strings of greater similarity.

Three sequence strings were available from the linear area graph, and three from the radial area graph, each solving the same 8-dimension task using the same dataset. Reduced strings were used to emphasize transitions among different AOIs, as opposed to transitions within the same AOIs. Table 5 shows the Levenshtein Distances between the reduced sequences that were shown in Table 4. The average distance among linear graphs was 20.7 operations, and 33.3 among radial graphs. The average difference *between* Linear and Radial graphs was 24.9 substitutions. This tells us that, for these tasks, scanning was more similar between participants for linear, than for radial graphs. This should not come as a great surprise, as scanning strategies for linear graphs are much more practiced than for radial graphs. Whereas

Table 6 Timeline visualization of scanpath sequences, by graph type and participant

Graph	Participant	Scanpath Sequence
Linear	1	Rd Rb Ra Rb Rc Rd Re Ve Re Ve Re Vd Re Ve Vf Rg Rh Vg Vf Re Rd Ve
Linear	13	Rc Va Ra Va Rb Vc Rc Vd Re Rf Ve Rh Rf Ve
Linear	25	Rb Ra Rb Rc Re Ve Vf Vg Vf Rf Re
Radial	2	Rc Rb Ra La Lb Rb Ra Rd Re Rb Rh Rd Ra Rb Rc Ra Rh Lh Rh Rg Rf Re Rf Re Rh Ra Rb Rh Rc Rd Re
Radial	14	Ra Rh Rc Rd Re Rf Rc Rh Rb Ld Re Rh
Radial	26	Ra Rh Ra Rb Rc Rd Re Rf Rc

scanning on linear graphs proceeded horizontally between regions, scanning on radial graphs could cross in many directions, as well as around the circumference. This was seen in Fig. 15, with the radial graph exhibiting more haphazard scanning than the linear graph equivalent condition.

4.4.3 Timeline Visualization of Sequences

Visualizations of scanpath sequences are a quick way to highlight regular patterns that might exist between AOIs. In Table 6, timelines were constructed from reduced scanpath sequences by shading each AOI a different color. The present color scale assigns blue to red from dimension a-h; regions are solid and vertices are graduated. Relative duration of each scanpath element was not considered, but could be coded by length of each cell using Gantt charts or stacked bar charts. Here, Participant 1 had many repeated scanning sequences between regions and their vertices. Also, scanpaths from the radial graphs show many instances of directly scanning from region a to h, which was not possible when scanning linear graphs.

4.4.4 Sequence Matching Using Dotplots

A more formalized approach to finding matching sequences uses dotplots. This scalable method allows quick identification of matching substring sequences that are contained within longer, more complex strings [15, 23]. Matches are identified despite any number of intervening AOI visits, allowing improved identification of matching sequences. The method of dotplot extraction of matching sequences in a linear area graph (using reduced sequences for an identical 8-dimension task) is illustrated in Fig. 16; the sequences are from p1 and p13 in Table 4. Normally, this analysis is conducted at a much larger data scale, using the concatenated scanpath sequence from all participants in a study. Each set of collinear dots can be modelled using linear regressions. The significance of each regression can serve as a threshold towards determining whether matching sequences exist [15, 16]. Negative slopes indicate sequential matches; positive slopes indicate 'reverse' matches in the opposite sequential order. While the present example promotes understanding of the method, automated analysis methods scale better to larger datasets encountered with eye tracking strings [15]. In this case, the solid (red) line matches the sequence {Ra Rb Rc Re Ve Ve}, which is contained within both scanpaths. This indicates a left to right scan across regions, until region e, where vertex e is found.

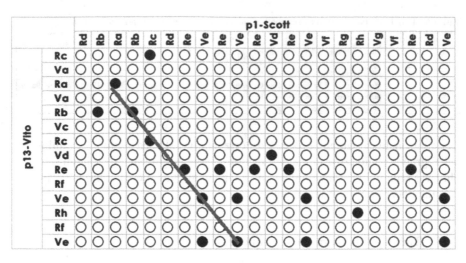

Fig. 16 Dotplot illustration of matching AOI sequence between p13 and p1. As indicated, one matching sequence is: {Ra Rb Rc Re Ve Ve}

An automated sequence matching tool first concatenates all AOI sequences, then generates and statistically analyses dotplots for collinear sequences. Multiple regression lines, associated with multiple matching sequence strings, are allowed, and regressions can be tuned to make it easier or harder to include AOIs in matching strings [15]. Scanning sequence data was input from three participants (p1, p13, p25), who scanned the linear area graph using the same 8-dimension task. The regression characteristics to be considered collinear (shown in Fig. 16) were tuned to three sequence matching criteria: (1) Points within $\pm (\mu + \sigma)$ from the least-squares fitted regression line, (2) Minimum three points per line, and (3) Minimum model $r^2 > 0.1$. Figure 17 also shows how a 'sketch query' can be performed by dragging the mouse cursor over a thumbnail stimulus image; the mouse gesture samples are treated as fixations in a synthetic scanpath that is matched against the actual scanpaths in the dataset. The query shown in Fig. 17 models a left-to-right scanning strategy and is labelled as 'query' in Figs. 18 and 19.

The resulting dotplot (Fig. 18, left) indicates significance of regression-based matching with shading. Darker regions correspond to stronger matches, based upon the regression model fit. The tooltip in Fig. 18 (right) indicates the matching scanning sequence from p1 to the strategy: {Ra Rb Rc Re Vg Rh}. A matching sequential cluster was found between the three participants. The cluster's aggregate scanpath is identified as −3 and indicated in red at the top and left of Fig. 19. Pairwise matching AOI sequences were: P1, P25: {Rb Ra Rb Rc Re Ve Vg Vf Rf Re}; P1, P13: {Rc Rb Rc Vd Re Ve Vf Ve Re}; P13, P25: {Rc Rb Rc Re Ve Re}.

In each case, scanning traversed from region b or c, rightwards, ultimately landing in region or vertex e. Additional clustering of scanning strategies could be investigated with a greater number of participants completing the same task on

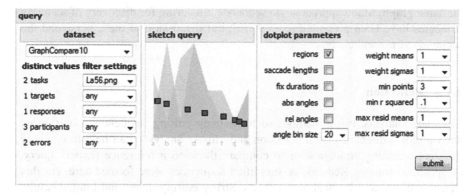

Fig. 17 Sequence analysis tool query panel, with regression criteria parameters

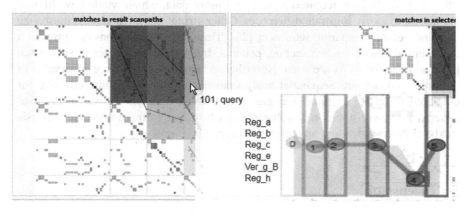

Fig. 18 Dotplot with scanning strategies; tooltip shows matching strategy between two participants

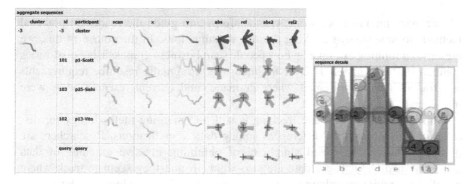

Fig. 19 Sequence cluster and associated aggregate scanpath (−3), showing matches with three participants, and the input 'sketch query'. The aggregate scanning strategy for the cluster is shown as a connected (*red*) sequence on the linear area graph

the same graph. Many approaches to sequence clustering for discovery of scanning strategies have been presented [e.g., 9, 16], but discussion of these is beyond the scope of the present chapter.

4.4.5 Reflection

Sequence analysis, using methods such as string editing, sequence visualization, and dotplot sequential match detection and cluster analysis, can help to compare different scanning strategies, or to compare these to a reference (expert, query, or optimal) strategy. Reduced or simplified sequences were formed here, but this step is not required by analysis methods. String editing, here, had limited results because comparisons were done at an AOI group level, rather than individual AOIs. The latter would have required a great deal more data, which would have limited our ability to easily illustrate differences. Other string edit methods first normalize or align, before comparing sequences [23]. Timeline visualizations can provide a quick initial analysis when seeking patterns, but trends are harder to spot when large numbers of AOIs are used. Nevertheless, some patterns were observable in the presented dataset. Sequential analysis using dotplots was quite effective for extracting matching scanning strategies, but did require custom software to discover patterns. Even with appropriate software tools, however, formatting data for analysis can be tedious, depending on required data specifications.

5   Discussion

5.1   Objective, Aggregated Data Comprehension

The primary intent of our work was to inform and guide the reader on practical methods to analyze eye tracking data on visualizations. In the course of this, we presented a running example on how people scan business graphs, using dual-series comparison tasks across 1, 2, or 8-dimensions. Pragmatic resource requirements and tradeoffs were presented where germane, and current research areas were introduced.

Although well-known, eye tracking is still an emerging methodology for discovering subtle usability differences among data visualizations. Researchers are impressed by the ease and simplicity of obtaining massive amounts of data almost immediately. These data are also valid, in that they seem to track where people look while completing tasks and search for objects. However, developing an understanding of data trends across multiple observers and/or visualization conditions can be tedious at best, and nearly impossible at worst. This paper has highlighted methods for developing an aggregated, objective understanding of data trends from eye tracking in relevant tasks on visualizations.

Table 7 Cost-benefit comparison at each analysis level

Level	Primary Results (Benefit)	Relative Effort (Cost)
0: Visualization	**General**: Spatial aggregation can provide a basis for question and hypothesis generation. **Here**: Scanning radial graphs was less organized than on linear graphs	**Easy**: Built into tool, setting specifications.
1: Factors & Effects	**General**: Substantial objective understanding of factors and interactions. **Here**: More fixations (but longer) on linear than radial as difficulty increased; area graphs required more time than other graphs.	**Medium**: Exporting and data manipulation; loading in statistical tool; statistical interpretation and plotting of results.
2: AOIs	**General**: Understanding of visits/transitions within and between AOIs. **Here**: Fixation durations longest in vertices, shortest in labels, longer in radial regions and vertices, than linear; AOI visits longer for area than bar graphs, particularly in easier tasks, otherwise little influence of graph style. Most AOI transitions in radial graphs were between regions, and were more spread out in linear graphs. More fixations and longer visits in radial graph series crossing points than linear graphs, but no difference between area and line graph.	**High**: Precisely defining and copying areas; controlling AOI sizes; defining metrics; exporting data; statistical analysis; analyzing transitions.
3: Sequences	**General**: Understanding and clustering of scanning strategy differences between tasks and stimuli. **Here**: Scanning was more similar between linear than between radial graphs. Pattern similarities noted from visualization; formal analysis located specific matching sequences for a subset of users; e.g., left-to-right scanning on linear area graph.	**High**: Sequences simplified, then compared via string editing, visualization, and dotplot analysis. Additional tools may be required.

We believe, along with many other applied eye tracking researchers, that subjective analysis of scanning trends can serve as a basis for forming questions and hypotheses, but that aggregated data are needed to separate influences of design factors in visualizations objectively. Methods for aggregating eye tracking data that retain the temporal or sequential aspects of the data can reveal behaviors and strategies that are repeated across participants.

5.2 Choosing Appropriate Data Abstraction

Analysis of eye tracking data can be considered a progressive abstraction of a dataset. This is done through sequential application of filters and transformations, until posed questions can be answered, and more focused questions can be asked. The pragmatic question of "How much abstraction is required, given current resources?" is considered here. Table 7 compares the relative type of results (benefit) to the relative effort (cost) required at each level of analysis. The analysis traversed from visualization (Level 0) to basic consideration of factor effects (Level 1) to AOIs (Level 2) to sequence comparisons (Level 3). One distinction between the last two levels is that Level 2 focuses on expected transitions within or between one or more AOIs, whereas Level 3 analysis focuses on finding matching sequences of any length across many scanpaths. Each successive level requires additional resources, such as specialized statistical or sequence analysis software. Visualizations (like heatmaps and gazeplots) play an important role in eye tracking analysis: they form a basic representation for question and hypothesis generation. Heatmaps and gaze plots provide a basis for question and hypothesis generation, but provide limited objective, aggregated understanding of task strategies.

When generating a dataset for statistical analysis, we highly recommend initially including many potential eye tracking measures. These include, for example, AOI visited, visit duration, fixation count, fixation duration, time to first fixation, etc. This is because many post-hoc analyses will rely on new or synthetic metrics that were unexpected at the start of the study. An example of a post-hoc metric that was used here was "Percentage of fixations in crossing AOIs." Because post-hoc analyses can require substantial spreadsheet work, we recommend exploiting metrics implemented by vendors' eye tracking analysis software wherever possible.

The choice of how "deep" to extend analyses *should* be based upon the questions asked, or problem to be solved. Heatmaps (Level 0), for example, are sufficient for sequence independent questions like, "Where did they look?", or "What was the relative difference in spread of visual attention between conditions?" A more objective understanding of factor effects (Level 1) is required when subtle differences are sought between various experimental treatments, such as the influence of changing line widths on completion time and errors in value lookup on bar graphs. In carefully defined studies, significant multi-factor interactions can provide serendipitous guidance for subsequent analyses and next steps. In our study, we found that radial graphs caused longer fixations, but fewer fixations than linear graphs, and that area graphs required much more time to complete single-dimension tasks, than multiple dimension comparison tasks. Such findings led us to drill further on area graph analysis. As one example of a deeper analysis, we asked whether there were graph type and style influences on fixation durations, visit times, and transitions, as a function of AOI groups (Level 2), labels, regions, and vertices. The question of whether "crossing AOIs" were viewed more with linear or radial area versus line graphs was another example where AOIs were defined within eye tracking analysis software, and measures statistically compared. Delving deeper into sequence analysis (Level 3) enabled us to find matching AOI sequences among participants, and to compare these with specific hypotheses, such as whether scanning is left-to-right. Although seemingly trivial in our example, the potential to find matching strategy clusters and to compare empirical with hypothesized strategies is extremely powerful.

5.3 Multivariate Understanding

Rather than traditional univariate analysis approaches, eye tracking is increasingly being considered as a multivariate concept, with vectors of independent and dependent factors. While more than 120 eye tracking-related measures can be defined [23], many of these have correlated and/or hierarchical relationships. They also differ in validity, reliability, and usefulness. An important trend is the broader consideration of multiple measures, which has the potential to provide a better understanding of scanning on visualizations.

Recent research has extended eye tracking analysis into the realm of machine learning and classification. For example, a task-independent framework has recently been developed for predicting where an individual will look, and for determining their intent [1]. This machine learning approach built feature sets from fixation-related, saccadic eye movement, and pupil diameter measures. The system could successfully discriminate between intentional and non-intentional feature vectors. This technology could adapt aspects of visualizations to real-time search behaviour; an example would be a user interface with views that zoomed automatically based upon an understanding of user intention. Another example of a machine learning approach is a classifier for discriminating reading from skimming text [2]. This real-time classifier uses features such as average forward saccadic velocity and angularity. With reported classification accuracy of 86%, this system could potentially segregate browsing from reading data in visualizations. Classifying expertise while using visualizations is another area that has benefited from a multivariate machine learning approach [31]. Using a hierarchical probabilistic framework (e.g. Hidden Markov Models) to extract clusters of eye movement patterns, subtle differences between scanning behaviors of experts and novices can be uncovered, then subsequently applied to training methods or adaptive visualizations.

5.4 Call for Research

The present paper, coupled with recent work [18], is only one step towards understanding how visualizations are scanned while solving relevant tasks. Our intent was to provide the visualization researcher with "how to" guidance toward objective analysis of eye tracking data. Statistical analysis of trends can be overwhelming to researchers who are new to eye tracking, due to large amount of required data abstraction, and the need to understand intricacies of eye movements. This guidance included decisions and analyses from an unexplored dataset in which participants compared dual data series on eight different graphs. There are, however, still several areas that can greatly benefit from additional research.

Methods to promote fair, standardized, less tedious assignment of AOIs to visual stimuli are woefully needed. There are minimum required sizes and margins due to technological hardware and display constraints. Automated methods that are based on aspects of tasks and stimuli have been demonstrated [e.g., 10], but are not yet standardized or widely used. While AOIs may be assigned hierarchically, new methods are needed to flexibly define which level(s) to consider in analyses. Methods to assign AOIs automatically to dynamic content, such as popup contextual panels, are needed; currently, manual assignment of AOIs to these regions is quite tedious. The lack of standards (or even heuristics) to assign AOIs has also meant that scanpaths (made from AOI sequences) cannot be compared across studies.

Commercial eye tracking software may not adequately support flexible stimulus assignment that is needed to ensure careful experimental control. For example, the current study relied on counterbalancing both stimulus order and task-stimulus assignments across participants. To implement counterbalancing, 12 different test sequences had to be defined in the software [39], which severely limited our ability to use the integrated software to create visualizations and tables of metrics that included all participants—multiple tables had to be assembled in Excel for subsequent metric computations. Requiring such resource-demanding operations will limit researcher's use of more objective analyses in general.

Comparison of sequences, like with the dotplot-based method used here, should be routinely included as part of vendors' eye tracking analysis tools. Much like defining AOIs on static media, the capability is needed to: (1) select or define one or more pre-existing scanpath sequences, then find matching sequences in the rest of the database, and (2) find sequential clusters of scanning strategies without defining initial sequences [9, 15, 16]. Providing these capabilities within reference analysis tools [e.g., 39] will alleviate much of the tedium of transforming data toward the Level 3 analysis advocated here.

Eye tracking will continue to provide objective answers to questions regarding appropriate design of visualizations. While hardware has evolved to become extremely usable and useful, software for objective analysis has lagged. We hope that embedded analysis solutions for AOI creation, sequence comparison, sequential clustering, reading versus scanning, and other innovations will soon become part of the applied eye tracking, standard toolkit.

References

1. Bednarik, R., Vrzakova, H., and Hradis, M. (2012), What do you want to do next: A novel approach for intent prediction in gaze-based interaction, *ETRA 2012, ACM Symposium on Eye Tracking Research & Applications*, ACM Press, pp. 83–90.
2. Biedert, R., Hees, J., Dengel, A., and Buscher, G. (2012), A Robust Realtime Reading-Skimming Classifier, *ETRA 2012, ACM Symposium on Eye Tracking Research & Applications*, ACM Press, pp. 123–130.
3. Bojko, A. (2009), Informative or Misleading? Heatmaps deconstructed. *Proc. HCII 2009*, Springer-Verlag, LNCS 5610, 30–39.
4. Bojko, A. (2012, in press), Eye Tracking the User Experience, Rosenfeld Media.
5. Bojko, A. and Adamczyk, K.A. (2010) More Than Just Eye Candy, *User Experience*, 9(3): 4–8.

6. Carpenter, P.A., and Shah, P. (1998), A Model of the Perceptual and Conceptual Processes in Graph Comprehension, *J. Exp. Psych. Appl.*, 4(2): 75–100.
7. Conati, C., and Maclaren, H. (2008), Exploring the Role of Individual Differences in Information Visualization, *Proc. AVI 2008*, Napoli, Italy, ACM Press, pp. 199–206.
8. Duchowski, A.T. (2007) *Eye Tracking Methodology*, 2nd Ed., London: Springer-Verlag.
9. Duchowski, A.T., Driver, J., Jolaoso, S., Tan, W., Ramey, B., and Robbins, A. (2010), Scanpath Comparison Revisited, *ETRA 2010, ACM Symposium on Eye Tracking Research & Applications*, ACM Press, pp. 219–226.
10. Egawa, A., and Shirayama, S. (2012), A Method to Construct an Importance Map of an Image Using the Saliency Map Model and Eye Movement Analysis, *ETRA 2012, ACM Symposium on Eye Tracking Research & Applications*, ACM Press, pp. 21–28.
11. ECEM Conference Series, European Conference on Eye Movements, Biennial independent meeting of eye tracking researchers (European locations).
12. ETRA Conference Series, ACM Symposium on Eye Tracking Research & Applications, Biennial ACM-sponsored meeting of eye tracking researchers (U.S. locations).
13. Few, S. (2005) Keep Radar Graphs Below the Radar – Far Below. http://www.perceptualedge.com/articles/dmreview/radar_graphs.pdf.
14. Gazetracker. Eye tracking analysis software, Applied Science Laboratories, Bedford, MA, www.asltracking.com.
15. Goldberg, J.H., and Helfman, J. (2010), Scanpath Clustering and Aggregation, Proceedings of *ETRA 2010, ACM Symposium on Eye Tracking Research & Applications*, ACM Press, pp. 227–234.
16. Goldberg, J.H., and Helfman, J.I. (2010), Identifying Aggregate Scanning Strategies to Improve Usability Evaluations, *54th Annual Meeting of the Human Factors and Ergonomics Society*, pp. 590–594.
17. Goldberg, J.H., and Helfman, J.I. (2010), Visual Scanpath Representation, *ETRA 2010, ACM Symposium on Eye Tracking Research & Applications*, ACM Press, pp. 203–210.
18. Goldberg, J., and Helfman, J. (2011), Eye Tracking for Visualization Evaluation: Reading Values on Linear versus Radial Graphs, *Information Visualization*, 10(3): 182–195.
19. Goldberg, J.H., and Kotval, X.P. (1999), Computer Interface Evaluation Using Eye Movements: Methods and Constructs, *Int. J. Industrial Ergonomics*, 24(6): 631–645.
20. Goldberg, J.H., and Wichansky, A.M. (2003), "Eye Tracking in Usability Evaluation: A Practitioner's Guide," in Hyona, J., Radach, R., and Deubel, H. (Eds.), *The Mind's Eyes: Cognitive and Applied Aspects of Eye Movements*, Elsevier Science Publishers, pp. 493–516.
21. Guan, Z., Lee, S., Cuddihy, E., and Ramey, J. (2006), The Validity of the Stimulated Retrospective Think-aloud Method as Measured by Eye Tracking, *CHI 2006, ACM/SIGCHI Conference on Human Factors in Computing Systems, Montreal*, Canada, ACM Press, pp. 1253–1262.
22. Hembrooke, H., Feusnder, M., and Gay, G. (2006), Averaging Scan Patterns and What They Can Tell Us., *ETRA 2006, ACM Symposium on Eye Tracking Research & Applications*, ACM Press, 41.
23. Holmqvist, K., Nystrom, M., Andersson, R., Dewhurst, R., Jarodzka, H., and van de Weijer, J. (2011), *Eye Tracking, A Comprehensive Guide to Methods and Measures*, Oxford Univ. Press.
24. Huang, W., and Eades, P. (2005), How People Read Graphs, *Proc. Asia Pacific Symp. Info. Vis. (APVIS 2005)*, in Hong, S. (Ed.), *Conf. on Research and Practice in Info. Technology*, Vol 45, pp. 51–58.
25. Huang, W., Eades, P. and Hong, S. (2009) Measuring effectiveness of graph visualizations: A cognitive load perspective. *Information Visualization* 8: 139–152.
26. Jarodzka, H., Holmqvist, K., and Nystronm, M. (2010), A Vector-based, Multidimensional Scanpath Similarity Measure, *ETRA 2010, ACM Symposium on Eye Tracking Research & Applications*, ACM Press, 211–218.
27. Josepheson, S., and Holmes, M.E. (2002), Visual Attention to Repeated Internet Images: Testing the Scanpath Theory on the World Wide Web, *ETRA 2002, ACM Symposium on Eye Tracking Research & Applications*, ACM Press, 43–49.

28. Journal of Eye Movement Research, http://www.jemr.org.
29. Journal of Eye Tracking, Visual Cognition and Emotion, http://revistas.ulusofona.pt/index.php/JETVCE/index.
30. Levenshtein, V.I. (1966), Binary Codes Capable of Correcting Deletions, Insertions, and Reversals, *Soviet Physics Doklady*, 10(8): 707–710.
31. Li, R., Pelz, J., Shi, P., Alm, C., and Haake, A. (2012), Learning Eye Movement Patterns for Characterization of Perceptual Expertise, *ETRA 2012, ACM Symposium on Eye Tracking Research & Applications,* ACM Press, pp. 393–396.
32. Lohse, J. (1991), A Cognitive Model for the Perception and Understanding of Graphs, *CHI 1991, ACM/SIGCHI Conference on Human Factors in Computing Systems,* ACM Press, pp. 137–144.
33. Mathot, S., Cristino, F., Gilchrist, I.D., and Theeuwes, J. (2012), A Simple Way to Estimate Similarity Between Pairs of Eye Movement Sequences, *J. Eye Movement Research*, 5(1): 1–15.
34. Nielsen, J. and Pernice, K. (2010), *Eyetracking Web Usability*, Berkeley, CA, New Riders Press.
35. Papenmeier, F., and Huff, M. (2010), DynAOI: A Tool for Matching Eye-Movement Data with Dynamic Areas of Interest in Animations and Movies, *Behavior Research Methods*, 42(1): 179–187.
36. Salvucci, D. (1999), Inferring Intent in Eye-based Interfaces: Tracing Eye Movements with Process Models, *CHI 1999, ACM/SIGCHI Conference on Human Factors in Computing Systems*, ACM Press, pp. 254–261.
37. Salvucci, D., and Goldberg, J.H. (2000), Identifying Fixations and Saccades in Eyetracking Protocols, *ETRA 2000, ACM Symposium on Eye Tracking Research & Applications,* ACM Press, pp. 71–78.
38. Shah, P. (1997), A Model of the Cognitive and Perceptual Processes in Graphical Display Comprehension, *Proc. Am. Assoc. Art. Intel., AAAI Technical Report FS-97-03*, Stanford Univ., CA.
39. Tobii Studio. Eye tracking analysis software, Tobii Systems, Inc., Danderyd, Sweden. http://www.tobii.com.
40. Tsang, H.Y., Tory, M., and Swindells, C. (2012), eSeeTrack—Visualizing Sequential Fixation Patterns, *IEEE Trans. Visualization and Computer Graphics*, 16(6): 953–962.
41. Van Gog., T., Paas, F., and Van Merrienboer, J. (2005), Uncovering Expertise-related Differences in Troubleshooting Performance: Combining Aye Movement and Concurrent Verbal Protocol Data, *Applied Cognitive Psychology*, 19, pp. 205–221.
42. Wikipedia Bioinformatics, http://en.wikipedia.org/wiki/Dot_plot_%28bioinformatics%.
43. Wooding, D.S. (2002), Fixation Maps: Quantifying Eye-Movement Traces, *ETRA 2002, ACM Symposium on Eye Tracking Research & Applications* ACM Press, pp. 518–528.

Evaluating Overall Quality of Graph Visualizations Indirectly and Directly

Weidong Huang

Abstract Visualization is one of the popular methods that are used to explore and communicate complex non-visual data. However, representing non-visual data in a visual form does not automatically make the process of exploration and communication effective. The same data can be visualized in many different ways and different visualizations affect the process differently. Therefore, it is important to have the resultant visualizations evaluated so that their quality in conveying the embedded information to the end users can be understood. In designing an evaluation study, at least three issues need to be addressed: what kind of quality is to be evaluated? What methods are to be used? And what measures are to be used? A range of methods and measurements have been used to evaluate visualizations in the literature. Overall quality is often considered as a multidimensional construct and the elements of the construct have limitations in evaluating overall quality. In this chapter, we introduce two one-dimensional measures. The first one is an indirect measure called visualization efficiency that is based on task performance and mental effort measures, while the second is a direct measure that is based on aesthetic criteria. These new measures take into consideration the elements of its corresponding multidimensional construct and combine them into a single value. We review related work, explain how these measures work and discuss user studies that were conducted to validate them.

1 Introduction

Non-visual data can sometimes be difficult to make sense of in their original format. Visualization is one of the popular methods that have been used for the purposes of exploration and communication of these data. Visualization takes advantage of

W. Huang (✉)
CSIRO ICT Centre, Corner Vimiera and Pembroke Roads, Marsfield, NSW 2122, Australia
e-mail: tony.huang@csiro.au

W. Huang (ed.), *Handbook of Human Centric Visualization*,
DOI 10.1007/978-1-4614-7485-2_14, © Springer Science+Business Media New York 2014

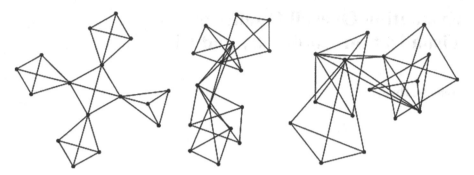

Fig. 1 Three drawings of a graph

the powerful human visual system and can help turn a complex cognitive process into simple perceptional operations [26]. However, representing non-visual data in a visual form does not automatically make the process of exploration and communication effective. This is particularly true for graph visualization. Graph data are often visualized as node-link diagrams and an issue with node-link diagrams is that the mapping between a graph and its node-link diagram representation is not unique. Figure 1 gives an example showing three different drawings of the same graph. Among these drawings, the only difference is layout. Looking at these drawings from a user point of view, it is clear that relatively, the left one is the best visualization (although it can be further improved) since it clearly shows the structural characteristics of the underlying graph. On the other hand, the other two drawings may take viewers some effort to make sense. Indeed, empirical research has shown that layout affects how users perceive the underlying data (e.g., [10, 15]).

Given the fact that quality varies with layout, it is important for visualization designers to be able to evaluate the layout and understand the quality of the drawings in consideration so that the right visualization can be determined. To do evaluation, at least three issues need to be addressed: what kind of quality is to be evaluated? What methods are to be used to evaluate the quality? And what measures are to be used to measure the quality? A range of methods and measurements have been proposed and used to evaluate visualizations in the literature [4, 5, 31]. In this chapter, we focus on overall quality definition and the measurement of it.

Visualization quality can be defined according to the specific purposes of the visualization. For example, visualization can be tested based on how much insight it can provide [30], how well the layout displays the graph structure [17], or whether it has the intended layout feature such as how well the Euclidean distance between any two nodes represents their graph-theoretic distance [29]. Traditionally the quality of graph visualization is understood as a multidimensional construct and measured in two ways. As shown in Table 1, the empirical construct is task performance based. That is, a drawing is better if for example, users spend a shorter time to complete a task with fewer errors. The computational construct is aesthetics based. That is, a

Table 1 Multidimensional constructs of overall quality

	Overall quality
Computational construct	Fewer crossings, larger crossing angles, higher angular resolution . . .
Empirical construct	Less time, more accurate . . .

drawing is better if for example, it has fewer edge crossings, larger crossing angles and higher vertex angular resolution.

The multidimensional measures enable us to evaluate the quality from different perspectives at the same time and have proved to be useful in their own right as demonstrated in the graph drawing literature in the past two decades [1]. However, they also have limitations. First, regarding the empirical construct, due to a widely acknowledged speed-accuracy trade-off, it is very likely that for example, a user, consciously or unconsciously, spends more time to complete a task with fewer errors for one visualization, and then takes less time with more errors with another visualization. This behavior can be more often observed when two visualizations are close in quality. On this occasion, the quality of these two visualizations can be tricky to decide.

Second, regarding the computational construct, it measures the visual properties of layout and does not involve human users. However, aesthetics measure different visual properties separately and to what extent this construct is related to human graph comprehension is not immediately known. Further, aesthetics are often conflicting with each other, which means that all aesthetics criteria cannot be achieved at the same time. A drawing may have fewer edge crossings, but have smaller crossing angles, while another drawing of the same graph may have more crossings, but have larger crossing angles. Again, in this situation, it is difficult to judge which one has better overall quality.

Therefore, there is a need for a measure that evaluates overall quality with a single numerical value. In an attempt to meet this need, two overall quality measures have been proposed and initial studies have been conducted to test their validity [12, 19, 20]. In this chapter we introduce these two measures. The first one is an indirect measure called visualization efficiency that is based on task performance measures and mental effort, while the second is a direct measure that is based on aesthetic criteria. Each of the new measures takes into consideration the elements of its corresponding multidimensional construct and combines them into a single number. In the rest of the chapter, we briefly review related work with examples, explain how these new measures work, and discuss two user studies that were conducted to validate them.

2 Related Work

In this section, we selectively review graph evaluations by giving examples, with a focus on task performance and aesthetic criteria.

2.1 Evaluating Quality Based on Aesthetics

Evaluating drawing quality based on aesthetics is mostly done for comparing graph drawing algorithms. Graph drawing algorithms are designed to draw graphs quickly, as well as satisfying some pre-specified aesthetic criteria. Therefore, the resulting drawings are evaluated to see to what extent the algorithm in consideration has met the criteria. The literature has also seen that different algorithms are evaluated by comparing the resulting drawings in terms of a set of pre-specified aesthetics. These aesthetics are not necessarily those considered as part of the algorithm design. For example, Hismsolt [25] compared 12 graph drawing algorithms. In this study, large sets of sample graphs were used and drawn with a graph drawing system GraphEd. Evaluations were conducted based on geometric properties, combined with subjective human judgments. It was shown that individual criteria might be not as important as they appeared and that "balance is often better than optimization." Di Battista et al. [23] conducted a study comparing four graph drawing algorithms. In this study, the test data were general graphs derived from application domains of software and databases. The quality measures included crossings, vertex distribution, bends, edge length and running time and it was found that there were trade-offs between these quality measures. Didimo et al. [24] conducted an experimental study that compared two new topology-driven heuristics with three existing graph drawing algorithms. The quality of the resultant drawings was compared based on the extent to which they conformed to each of a set of readability aesthetics. The aesthetic criteria used for comparison included the number of crossings, crossing angle resolution, geodesic edge tendency and vertex angle resolution. The results indicated that drawings of the topology-driven algorithms had better trade-offs between these criteria than others.

2.2 Evaluating Quality Based on Task Performance

There is a large body of empirical research that investigates how specific drawing features affect drawing quality in the literature pioneered by Purchase [1]. For example, Purchase et al. [9, 10] conducted a study validating aesthetics for their relevance to human graph comprehension. Graphs were drawn with high and low values for each of the aesthetics in consideration and users were asked to perform graph reading tasks with the resulting drawings. Using task performance as the indication of drawing quality, it was found that for example, drawings with fewer crossings had better quality. Using a similar approach, Korner and Albert [16] conducted a study on how visual properties of hierarchical graphs affected speed of comprehension and it was found that crossing of lines was the most influential factor affecting comprehension performance. Huang et al. [13, 18] conducted a study that evaluated the quality of sociograms (node-link diagrams of social networks). These diagrams were drawn based on different drawing conventions. Subjects were asked to both perform domain specific tasks and indicate their personal perceptions of quality in relation to a specific task. It was found that a perceived quality drawing and the best performed drawing may be different. Blythe et al. [22] also conducted a user study with sociograms and found that layout affected users' perception of actor centrality and existence of groups.

3 The Indirect Measure: Performance Based Visualization Efficiency

3.1 How It Works

Although performance measures, time and accuracy, have been widely used in visualization evaluation, these may not be enough. For example, suppose that we have two different visualizations. One could exert more effort with one visualization and perform equally well with the two. In this case, if only the performance measures are considered, we would not be able to tell the quality difference between them. Therefore it would be helpful if we take an additional measure, mental effort, into consideration for quality evaluation.

In addition, mental effort has been widely used in Cognitive Load Theory research [3, 33]. It is a measure of perceived cognitive load imposed on the users and is often obtained by asking users to indicate the level of the effort they devoted in performing a task based on a 9-point Likert scale [32]. In this community, a relative difference between test performance and effort is used to measure the efficiency of instructional methods [2]. More specifically, this measure normalizes test data into z scores and combines them into a single score as below:

$$E_{instructional\ efficiency} = \frac{Z_{test\ score} - Z_{test\ effort}}{\sqrt{2}}$$

Inspired by this practice, it was proposed to combine response time, accuracy and mental effort and measure overall quality of visualizations using the equation below:

$$E_{overall\ quality} = \frac{Z_{accuracy} - Z_{mental\ effort} - Z_{response\ time}}{\sqrt{3}}$$

In this equation, overall quality is defined as a difference between cognitive gain (performance accuracy) and cognitive cost (mental effort and response time), which we call *visualization efficiency*. In the context of this equation, a visualization is of high efficiency if high performance accuracy is achieved with low mental effort and less response time, and vice versa.

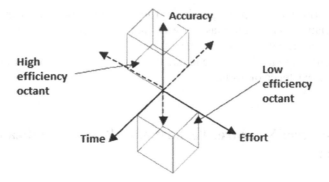

Fig. 2 Illustration of visualization efficiency in a 3d space (Modified from [2])

Please note that (1) normalizing data into z scores is to make them addable, since the original data of time, accuracy and effort are each measured on a different scale. (2) The computed overall quality can be negative or zero. If it is negative, it means that the cognitive cost outweighs the cognitive gain. If it is zero, it means that the cost and the gain are balanced. (3) Adding $\sqrt{3}$ to this equation enables us to graphically illustrate the equation in a three-dimensional space, with accuracy representing the vertical axis and the other two representing the horizontal plane, as shown in Fig. 2. If we plot the z scores of the three measures as a point in this space, then the overall quality represents the perpendicular distance from this point to the plane whose equation is:

$$Z_{accuracy} - Z_{mental\ effort} - Z_{response\ time} = 0$$

3.2 How z Scores Are Computed

In statistics, z score indicates the number of standard deviations an observation or score is below or above the mean of all observations [6]. It is obtained by subtracting the mean from the score and then dividing the difference by the standard deviation of all observations. Z score can be understood as a common yard stick for all types of data and has been used as a common statistical way of standardizing data on one scale so that data measured based on difference scales can be compared [7].

Take the time measure as an example for the purpose of demonstration. Suppose that there are m visualizations (conditions), and n subjects have performed a task on all visualizations. The time spent with each visualization is recorded. As a result, we have $m \times n$ data entries obtained for time. To standardize the time scores, we compute the mean (μ) and the standard deviation (σ) of all these data entries first. Then the z score of each time entry (t) can be computed by using the following equation:

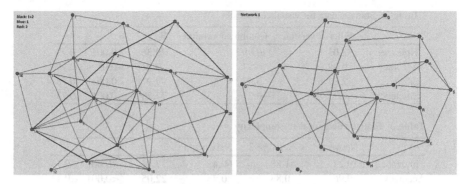

Fig. 3 A drawing of the combined version (*left*) and a drawing of filtered version (*right*) for a multi-relation network. Only the drawing of the relation relevant to the tasks is shown for the filtered version

$$Z_{response\ time} = \frac{t - \mu}{\sigma}$$

Z scores of mental effort and accuracy can be computed in the same way as that of response time.

3.3 A User Study

3.3.1 Method

The original study was conducted to validate the proposed interaction model of task performance, mental effort and cognitive load. However, it was also expected that the study provided evidence for the need of an overall quality measure and gave insights into validity of the proposed measurement. For more details, see [12].

The study used three multi-relation social networks of different sizes as graph data and drew them into socioagrams in two different ways: a combined version and a filtered version. The combined version represents the network in a single visualization in which different colors were used to represent different relations. The filtered version represents the network in multiple visualizations with each visualization representing a particular relation. An example of such drawings is shown in Fig. 3.

There were three tasks of different difficulties. They were find one neighbor of a given actor (task 1), find all common neighbors of two given actors (task 2) and find all triangle patterns (task 3).

Thirty subjects participated in the study. The experiment employed a within-subject design and was conducted on an individual basis. The diagrams were randomly displayed by a custom-built system. For each randomly ordered task, the

Table 2 Statistic data for visual complexity

	Filtered version	Combined version	F	p	η^2
Time (sec.)	32.03	48.99	79.36	<0.001	0.73
Accuracy	0.88	0.84	3.91	>0.05	0.12
Effort	2.82	3.53	99.91	<0.001	0.78
Efficiency	0.24	−0.24	77.20	<0.001	0.73

Table 3 Statistic data for data complexity

	Network 1	Network 2	Network 3	F	p	η^2
Time (sec.)	19.23	41.88	60.42	77.34	<0.001	0.73
Accuracy	0.96	0.83	0.78	22.05	<0.001	0.43
Effort	2.19	3.38	3.95	94.28	<0.001	0.77
Efficiency	0.71	−0.12	−0.59	202.47	<0.001	0.88

Table 4 Statistic data for task complexity

	Task 1	Task 2	Task 3	F	p	η^2
Time (sec.)	11.76	33.57	76.20	64.85	<0.001	0.69
Accuracy	0.98	0.81	0.78	22.53	<0.001	0.44
Effort	1.56	3.12	4.85	144.60	<0.001	0.83
Efficiency	0.01	0.00	−1.01	183.62	<0.001	0.86

subjects looked at the picture, wrote down the answer and indicated his mental effort on the answer sheet provided. Then a button on the screen was pressed to proceed to the next diagram. This procedure was repeated until all pictures were viewed. The time spent for each answer was recorded online by the system.

3.3.2 Results

The study had three independent variables: task with three levels, visualization version with two levels and graph data with three levels. The dependent variables were time, effort, accuracy and visualization efficiency. The efficiency data were computed based on the data of the other three dependent variables using the procedure described in Sect. 3.1.

We deliberately manipulated the complexity levels of the independent variables. We expected that all independent variables had significant effects on the dependent variables. More specifically, if the proposed measure was valid, we expected that visualization efficiency was able to detect the condition differences with good agreement of the corresponding complexity levels.

The mean values of all dependent variables in all conditions and the results of repeated ANOVA tests under each of the independent variables are summarized in Tables 2, 3, and 4. In these tables, along with mean values and F statistics, p values and eta squared (η^2) are also reported. Eta squared is a measure of effect size and indicates how large the effect of one variable on another is, while p tells us

Table 5 Statistic data for network 1 and task 1 on visual complexity

	Filtered version	Combined version	F	p	η^2
Time (sec.)	10.77	11.39	0.21	>0.05	0.01
Accuracy	1.00	0.97	1.00	>0.05	0.03
Effort	1.40	1.67	6.27	<0.05	0.18
Efficiency	1.09	0.95	3.16	>0.05	0.10

whether this effect is statistically significant or not at the significance level of 0.05. According to common rules of thumb, the effect is small if eta squared is smaller than 0.10, is large if eta squared is larger than 0.50 and is medium if eta squared is between 0.10 and 0.50.

As can be seen from these tables, on average, when the complexity of independent variables increased, the time spent and effort exerted by the subjects increased, while performance accuracy and efficiency decreased. These were in good agreement with what we expected. The repeated ANOVA tests revealed that all these trends were statistically significant, except the trend of accuracy for visual complexity.

Further, the benefits of considering mental effort in quality evaluation can be seen from Table 5. This table shows the data of the two visual versions when network 1 and task 1 were used. Overall, the size of the effect of visual complexity on time, accuracy and efficiency was small as shown in the values of eta squared. Performance measures (time and accuracy) found no significant difference between these two visual versions. However, if we look at the effort data, it is clear that there was a difference between the visualizations. That is, the same level of performance was achieved by devoting significantly more effort with the combined version, indicating that the quality of the combined version was not as good as the filtered version.

3.4 Discussion

In this study, performance measures and mental effort were consistent in showing the effects of the independent variables. Although possible tradeoffs between these individual measures were not obvious to us in the experimental data, it was demonstrated that the proposed efficiency measure was able to detect differences between the experimental conditions as expected.

It is worth mentioning that the proposed efficiency measures overall quality in terms of the cognitive gain relative to the cognitive cost. Although it has the advantage of giving a single numerical value, it can be misleading if only the efficiency measure is considered when it comes to the overall quality evaluation. In other words, visualization efficiency should be considered as a supplementary rather than an alternative measure of the traditional performance based measurements. After all, these are different measures emphasizing different aspects of visualization quality and findings obtained based on the overall efficiency measure and individual performance measures are not necessarily identical. Taking both efficiency and performance into consideration at the same time will enable visualization designers to make thoughtful decisions.

Further, due to various considerations, the measurements used for quality evaluation can be different in different experimental settings. The equation of visualization efficiency can be modified accordingly to be used in these changed situations. For example, user preference is often used instead of mental effort. In this case, visualization efficiency can be measured as follows:

$$E_{overall\ quality} = \frac{Z_{accuracy} + Z_{user\ preference} - Z_{response\ time}}{\sqrt{3}}$$

In the context of this equation, a visualization is of high quality if high performance accuracy is obtained in association with high user preference and a short response time. In addition, in a particular situation, one particular measure may be considered more important than others. In this case, this equation can be further modified by assigning different weights to corresponding measures according to the levels of their importance as follows:

$$E_{overall\ quality} = \frac{w_1 \times Z_{accuracy} + w_2 \times Z_{user\ preference} - w_3 \times Z_{response\ time}}{\sqrt{3}}$$

where $\sum_{i=1}^{3} w_i = 1$.

Although visualization efficiency can be helpful in evaluating overall quality, it requires conduction of user studies. User studies can be time-consuming and require recruitment of suitable participants, which can be costly to run. And mostly, to have a proper evaluation, summative studies are required and this type of evaluations can only be done after all visualizations have been developed. It would be more desirable if we have a one-dimensional measure that does not require user studies. This measure will help visualization designers to quickly evaluate the candidate drawings at hand and make decisions accordingly. This measure is introduced in the next section.

4 The Direct Measure: Aesthetics Based Overall Quality

4.1 How It Works

Aesthetics have been widely used as drawing rules or quality criteria to produce "good" drawings [28]. Some of them have been empirically validated having impact on human graph comprehension. However, meeting one aesthetic criterion can lead to changes in other aesthetics and the impact of one aesthetic can change when statuses of other aesthetics change. For example, reducing the number of crossings could make the vertex angular resolution poor. The impact of crossings can be more prominent when crossing angles are small, while the impact can be reduced if crossing angles are made larger [34]. Based on these facts, a theoretic measure for overall quality ($O_{overall\ quality}$) was proposed as a function of individual aesthetics (a_i) and their interactions as follows [11]:

$$O_{overall\ quality} = \sum_{i=1}^{n} f(a_i) + \sum_{i,j=1}^{n} f(a_i a_j) + e$$

To make it practically useful, this equation was simplified by measuring overall quality as an aggregation of z scores of the following four most discussed aesthetics for general random graphs [20]:

- Minimum number of edge crossings (cross number)
- Maximum crossing angle resolution (cross resolution)
- Maximum vertex angular resolution (angular resolution)
- Uniform edge lengths (edge length)

And the overall quality can be expressed as below:

$$O_{overall\ quality} = Z_{angular\ resolution} + Z_{cross\ resolution} - Z_{cross\ number} - Z_{edge\ length}$$

In this equation, angular resolution is measured as the minimum size of angles formed by any two neighboring edges [35]. Edge length is measured as the standard deviation of all edge lengths. Cross resolution is measured as the minimum size of all crossing angles [34]. In the context of this simplified equation, a smaller number of crossings, a smaller standard deviation of edge lengths, a larger angular resolution and a larger cross resolution will make overall quality better.

It is possible that the value of overall quality obtained this way is zero or negative. When this is the case, it does not necessarily mean that the visualization is bad. This only means that the negative aesthetic elements in the equation outweigh or are balanced with the positive elements.

4.2 How z Scores Are Computed

Normalizing aesthetics into z scores is to make them addable and comparable as these aesthetics are originally measured in different contexts. Further, to make the comparison meaningful, the scores of these aesthetics need to be normalized across conditions. It should be noted that here we follow a general statistical approach for data normalization; Purchase has proposed a different approach by normalizing aesthetics into a value between 0 and 1. For more details, see [8, 21].

To give an example, suppose that we would like to compare three different layouts of a graph and that these layouts have 4, 7, 10 crossings respectively. To compute z scores for the aesthetic of crossings, the mean (μ) and the standard deviation (σ) of these three numbers need to be computed and they are 7 and 3 respectively. Then the z scores of the cross numbers (x) for the three drawings can be computed as below:

$$Z_{cross\ number} = \frac{x - \mu}{\sigma}$$

and they are -1, 0 and 1, respectively. The z score of -1 means that the cross number of 4 is one standard deviation, which is 3, below the mean of 7. The z score of 1 means that the cross number of 10 is one standard deviation above the mean of 7. The other aesthetics can be standardized into z scores in the same way.

4.3 A User Study

4.3.1 Method

Suppose that we have a number of drawings of a graph with their relative quality known beforehand. If the proposed measure is valid, then the measured quality of all drawings should be consistent with their corresponding pre-known quality. In this study, 20 random graphs with each having 30 nodes and 40 edges were generated (see [20] for more details). These graphs were generated using a random graph algorithm, thus having similar structures. To produce drawings with pre-known relative quality, a force-directed graph drawing algorithm was used to draw these graphs. It is well known that force-directed algorithms draw graphs by starting from a random layout and applying forces on nodes and edges to move them around. This process is repeated until a stable converge is reached [27]. It is also known that each time the process is repeated, the layout is generally getting better. Therefore, for each graph, we recorded the layout when the process was repeated for 3,000, 6,000, 9,000 and 12,000 times. That is, four drawings were produced for each graph with their quality increasing from layout1, layout2, layout3 to layout4 (which corresponded to the repeated times of 2,000, 6,000, 9,000 and 12,000 respectively). As a result, we had four experimental conditions with each condition having 20 drawings. Figure 4 shows an example of four drawings of a graph used in the experiment.

As mentioned previously, task performance is often used as an indication of overall quality. To compare performance measures and the proposed direct measure for their ability in detecting quality difference between drawings, a user study was conducted with the drawings of all four conditions (this study also included extra drawings to test the predictability of the proposed measure for task performance which is not reported here). The task was to find the shortest path between two

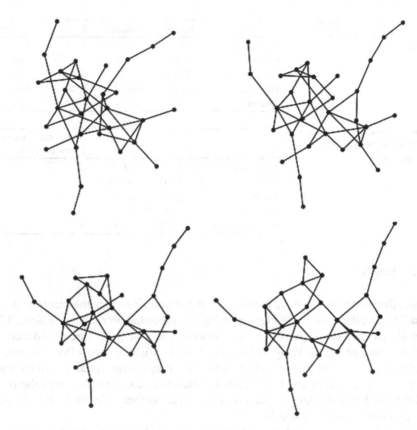

Fig. 4 Four drawings of a graph used in the experiment

pre-specified nodes. For each graph, the two nodes were randomly selected with constraints that there was only one shortest path between them and that the length of the path is between 3 and 5 inclusive. A custom-built system was used to display the drawings randomly and the pre-specified nodes were highlighted as red. To reduce the possible fatigue caused by too many drawings, only half of the drawings in each condition were viewed by each subject.

Thirty-five subjects participated in the study. After a short information and training session, the subjects were given time to ask questions and practice the system. Once ready, they started to run the system and perform the task online. For each drawing, once the answer was found, the subject pressed a key on the keyboard immediately and the time spent for that drawing was recorded. Then he/she continued to indicate the answer and mental effort exerted on the screen. To proceed to the next drawing, a key was hit and the process was repeated until all drawings were viewed. The subjects were asked to perform the task as quickly and as accurate as possible. The whole session took about 40 min for each subject on average.

Table 6 Means of dependent variables in the four layout conditions

	Layout1	Layout2	Layout3	Layout4
Time (sec.)	9.91	9.51	7.19	7.11
Effort	3.60	3.26	3.27	3.09
Accuracy	0.69	0.75	0.76	0.76
Efficiency	−0.74	−0.22	0.28	0.48
Overall quality	−2.14	0.04	1.02	1.08

Table 7 Results of repeated ANOVA tests with post-hoc comparisons

	F	p	# of condition pairs found different
Time (sec.)	2.804	<0.05	2
Effort	7.491	<0.001	3
Accuracy	2.118	>0.05	0
Efficiency	7.442	<0.001	3
Overall quality	28.596	<0.001	5

4.3.2 Results

Time, effort and accuracy were recorded by the system. Based on the recorded data, visualization efficiency was computed using the procedure introduced in Sect. 3.1. Overall quality was computed for each drawing using the procedure introduced in Sect. 4.1. As a result, the study had five dependent variables: time, effort, accuracy, efficiency and overall quality. There was one independent variable, which was layout. It was expected that the condition differences could be shown in the data of the dependent variables. The means of the dependent variables in the four layout conditions are shown in Table 6.

As can be seen from Table 6, with the increase in drawing quality from layout1 to layout4, the time spent and the effort exerted by the subjects decreased while the task accuracy and overall efficiency increased. The measured overall quality also followed the similar pattern as accuracy and efficiency did and increased with the drawing quality.

To test whether these trends were statistically significant, repeated ANOVA tests with post-hoc comparisons were conducted. The results are shown in Table 7.

As can be seen from Table 7, there was a significant layout effect on time, effort, accuracy and overall quality, but not on accuracy. The finding that there was no difference on accuracy across the conditions showed that the difficulty of the task was at the appropriate level and the subjects did not compromise accuracy for speed. The difference on the measured overall quality showed that the overall quality measure was able to detect the condition differences as expected.

Further, the post-hoc comparisons revealed that in this particular study, the dependent variables showed different levels of ability in detecting condition differences. In particular, time found 2 pairs of conditions different, effort found 3, accuracy found none and efficiency found 3. The measured overall quality found the largest number of condition pairs different, which was 5 out of 6. This indicated that the proposed measure was more sensitive than the other measures in detecting quality differences between drawings.

4.4 Discussion

The study showed that the aggregation of the four aesthetics can be useful for measuring overall quality of drawings, and that given a graph, the proposed measure was able to differentiate drawings based on overall quality. The experimental data further demonstrated that the proposed measure had better performance in detecting quality differences between drawings than other measures did including performance, mental effort and the proposed indirect quality measure, efficiency. This finding is encouring and makes the proposed direct measure more preferable for designers to use during the process of visualization design. However, it is worth noting that this finding should not be interpreted as evidence that the direct measure should be used as a substitute for other ones. This is mainly because these measures are essentially different kinds of measures serving for different purposes.

Further, the proposed measure treats its aesthetic elements equally. In practice, we may want to treat them differently according to their importance or priority in the final layout. This can be achieved by assigning different weights to the aesthetics with higher weights to more important ones. This helps to make more important aesthetics have larger effects on the final layout. Accordingly, the proposed measure can be refined as below:

$$O_{overall\ quality} = w_1 \times Z_{angular\ resolution} + w_2 \times Z_{cross\ resolution} - w_3$$
$$\times Z_{cross\ number} - w_4 \times Z_{edge\ length}.$$

where $\sum_{i=1}^{4} w_i = 1$.

One limitation of the proposed measure is that it assumes a linear relationship between overall quality and each of its aesthetic elements. This is an oversimplification of the reality. For example, Huang et al. [14] conducted a user study investigating the effect of crossing angles on human graph comprehension. In that study, the drawing quality was measured by task response time. The study revealed that there was a significant quadratic relationship between the size of crossing angles and the drawing quality as shown in Fig. 5. Although more studies are needed so that the actual relationships between them can be fully understood, the empirical evidence presented in our study has shown that the simplified version does give useful insights into the relative overall quality between drawings.

5 Concluding Remarks and Future Work

In this chapter, we introduced two measures that measure overall quality with a single value. Although they were proposed to overcome the limitations of the existing multidimensional measures, these newly proposed measures should not be used as the replacement of the existing ones. Both multidimensional and one-dimensional measures have their own advantages and disadvantages. Further, these

Fig. 5 A quadratic relationship between cross resolution and drawing quality

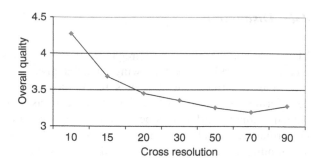

two new measurements measure overall quality based on individual elements of their corresponding multidimensional constructs. In other words, those individual elements need to be measured first before the proposed measures can be computed. Therefore it is always a good practice to evaluate overall quality based on both the multidimensional construct and its corresponding one-dimensional one. When the elements of a multidimensional construct give conflicting information, the proposed measures provide another perspective for us to look at the overall quality.

Two experiments were presented to demonstrate the validity of the quality measures. However, these experiments have limitations like any other laboratory based ones. Cautions should be taken for generalization. The work presented in this chapter has only made a promising start towards an one-dimensional quality measurement. More studies are needed in this area. It is hoped that the proposed measures can be further refined and improved in the future work based on rigorously designed and conducted empirical studies.

Acknowledgements The author acknowledges the contribution made by Peter Eades, Seokhee Hong, Mao Lin Huang and Chun-Cheng Lin for the research mentioned in the chapter. The author also thanks the study participants for their time and effort.

References

1. H. C. Purchase, Experimental Human-Computer Interaction: A Practical Guide with Visual Examples. Cambridge University Press; 1 edition (July 23, 2012).
2. Tuovinen, J. and Paas, F. (2004) Exploring Multidimensional Approaches to the Efficiency of Instructional Conditions. Instructional Science, 32: 133–152.
3. J. van Merrienboer and J. Sweller (2005) Cognitive Load Theory and Complex Learning: Recent Development and Future Directions.
4. C. Plaisant (2004) The Challenge of Information Visualization Evaluation. In Proc. the working conference on Advanced Visual Interfaces (AVI'04), 109–116.
5. C. Chen and M. Czerwinski (2000) Empirical Evaluation of Information Visualization: An Introduction. Int. J. Human-computer Studies, 53(5): 631–635.
6. Wikipedia. Standard score. http://en.wikipedia.org/wiki/Standard_score. Accessed on 1/9/2012.

7. Jeff Sauro. (2004)What's a Z-Score and Why Use it in Usability Testing? http://www. measuringusability.com/z.htm. Accessed on 1/9/2012.

8. H. C. Purchase, Metrics for graph drawing aesthetics, Journal of Visual Languages and Computing, vol. 13, no. 5, pp. 501–516, 2002.

9. H. C. Purchase, R. F. Cohen, and M. James, Validating graph drawing aesthetics, in Proceedings of the Symposium on Graph Drawing (GD'95), Springer-Verlag, 1995, pp. 435–446.

10. H. C. Purchase, D. A. Carrington, J.-A. Allder: Empirical Evaluation of Aesthetics-based Graph Layout. Empirical Software Engineering 7(3): 233–255 (2002)

11. W. Huang, S.-H. Hong, P. Eades, and C.-C. Lin, Improving Multiple Aesthetics Produces Better Graph Drawings. Journal of Visual Languages and Computing.

12. W. Huang, P. Eades, and S.-H. Hong, Measuring effectiveness of graph visualizations: a cognitive load perspective. Information Visualization, vol. 8, no. 3, pp. 139–152, 2009.

13. W. Huang, S.-H. Hong and P. Eades. 2006. How people read sociograms: a questionnaire study. In Proceedings of the 2006 Asia-Pacific Symposium on Information Visualisation (APVis'06), 199–206.

14. W. Huang, P. Eades, and S.-H. Hong, Effects of crossing angles. In Proceedings of the IEEE Pacific Visualization Symposium 2008 (PacificVis'08), pp. 41–46.

15. C. Ware, H. Purchase, L. Colpoys, and M. McGill, Cognitive measurements of graph aesthetics, Information Visualization, vol. 1, no. 2, pp. 103–110, 2002.

16. C. Korner and D. Albert, Speed of comprehension of visualized ordered sets, Journal of Experimental Psychology: Applied, 8:57–71, 2002.

17. S. Hachul and M. Junger, An experimental comparison of fast algorithms for drawing general large graphs, GD'05: 235–250.

18. W. Huang, S.-H. Hong and P. Eades: Layout effects: Comparison of sociogram drawing conventions. TR No.575, University of Sydney, 2005.

19. W. Huang, S.-H. Hong and P. Eades: Predicting graph reading performance: a cognitive approach. APVIS 2006: 207–216.

20. W. Huang, C.C. Lin and M.L. Huang, An Aggregation-Based Approach to Quality Evaluation of Graph Drawings. The International Symposium on Visual Information Communication and Interaction (VINCI12), 2012.

21. C. Dunne, and B. Shneiderman, Improving graph drawing readability by incorporating readability metrics, TR No. HCIL2009-13, University of Maryland, 2009.

22. J. Blythe, C. McGrath and D. Krackhardt: The effect of graph layout on inference from social network data. In Proceedings of the Symposium on Graph Drawing (GD'95), pp. 40–5 1995.

23. G. Di Battista, A. Garg, G. Liotta, R. Tamassia, E. Tassinari, and F. Vargiu. 1997. An experimental comparison of four graph drawing algorithms. Comput. Geom. Theory Appl. 7, 5–6 (April 1997), 303–325.

24. W. Didimo, G. Liotta, and S. A. Romeo. 2010. Topology-driven force-directed algorithms. In Proceedings of the 18th international conference on Graph drawing (GD'10), 165–176.

25. M. Himsolt: Comparing and Evaluating Layout Algorithms within GraphEd. J. Vis. Lang. Comput. 6(3): 255–273 (1995)

26. M. Tory and T. Möller. 2004. Human Factors in Visualization Research. IEEE Transactions on Visualization and Computer Graphics 10, 1 (January 2004), 72–84.

27. U. Brandes, 2001. Drawing on physical analogies. In: Kaufmann, M., Wagner, D. (Eds.), Drawing Graphs: Methods and Models. Vol. 2025 of LNCS. Springer-Verlag, pp. 71–86.

28. G. di Battista, P. Eades, R. Tamassia and I. Tollis 1998. Graph Drawing: Algorithms for the Visualization of Graphs. Prentice Hall, Upper Saddle River, New Jersey.

29. U. Brandes and C. Pich, 2009. An experimental study on distance-based graph drawing. In: Proc. of 16th International Symposium on Graph Drawing (GD 2008). Vol. 5417 of LNCS. Springer-Verlag, pp. 218–229.

30. P. Saraiya, C. North, V. Lam, K. Duca: An Insight-Based Longitudinal Study of Visual Analytics. IEEE Trans. Vis. Comput. Graph. 12(6): 1511–1522 (2006)

31. BELIV workshop. BEyond time and errors: novel evaLuation methods for Visualization. http://www.beliv.org/
32. F. Paas and J. Van Merrienboer (1993) The Efficiency of Instructional Condition: An Approach to Combine Mental Effort and Performance Measures. Human Factors, 35: 734–743.
33. F. Paas, J. Tuovinen, H. Tabbers and P. van Gerven (2003) Cognitive Load Measurement as a Means to Advance Cognitive Load Theory. Educational Psychologist, 38(1): 63–71.
34. W. Huang, P. Eades and S.-H. Hong (2013) Large crossing angles make graphs easier to read. Submitted.
35. W. Huang, C.C. Lin and M.L. Huang (2011) Aesthetic of angular resolution for node-link diagrams: Validation and algorithm. VL/HCC 2011: 213–216.

Visual Analysis of Eye Tracking Data

Michael Raschke, Tanja Blascheck, and Michael Burch

Abstract Eye tracking has become a valuable approach to evaluate visualization techniques in a user centered design process. Apart from just relying on task accuracies and completion times, eye movements can additionally be recorded to later study visual task solution strategies and the cognitive workload of study participants. During an eye tracking experiment many data sets are recorded. Standard techniques to analyze this eye tracking data are heat map and scan path visualizations. However, it still requires a high effort to analyze scan path trajectory data to find common task solution strategies among the study participants. In this chapter we discuss three existing methodologies for analyzing the vast amount of eye tracking data from a visualization and visual analytics perspective. These three approaches are a classical static visualization, visual analytics techniques and finally a software prototype, which helps the user to manage, view and analyze the recorded data in a simple interactive way.

1 Introduction

A key factor for the readability and thus, for the success of an information visualization is its ergonomics. To evaluate visualizations user experiments have to be performed to study the readability, efficiency and cognitive workload by controlled experiments, usability tests, longitudinal studies, heuristic evaluations, or cognitive walkthroughs [14, 22] with controlled experiments being the primary

M. Raschke (✉) • T. Blascheck
Institute for Visualization and Interactive Systems, University of Stuttgart, Stuttgart, Germany
e-mail: michael.raschke@vis.uni-stuttgart.de; tanja.blascheck@vis.uni-stuttgart.de

M. Burch
Postdoctoral Researcher of Computer Science, Visualization Research Center (VISUS),
University of Stuttgart, Stuttgart, Germany
e-mail: michael.burch@visus.uni-stuttgart.de

W. Huang (ed.), *Handbook of Human Centric Visualization*,
DOI 10.1007/978-1-4614-7485-2_15, © Springer Science+Business Media New York 2014

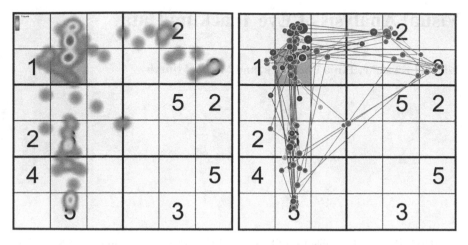

Fig. 1 Most prominent visualization techniques for eye tracking data: Heat maps are time aggregated density based representations (*left*), scan paths are line based visualizations suffering from overdrawing and visual clutter (*right*)

method for evaluating visualizations [7]. Since the recording of eye movements became easier during the last decade, many user study designers additionally use eye tracking techniques.

Standard metrics to evaluate visualizations are accuracy rates and completion times. Accuracy rates show how many correct answers have been given by a participant while carrying out one or more given tasks. The completion times indicate the time participants needed to perform a given task. Both metrics can be analyzed by standard statistics techniques such as t-tests, ANOVA, and the like.

Additionally, eye tracking techniques provide information about eye movements of a participant during a user experiment. In most cases the participants' fixation positions on the screen, the fixation durations and the scan path structure is of interest. Two very common visualization techniques for this spatio-temporal data are heat maps and scan paths, see Fig. 1. Both techniques are very common, but have some drawbacks concerning visual clutter (scan paths) and time aggregation (heat maps). Furthermore, detailed analysis of recorded eye tracking patterns requires a high effort. For example, a user experiment with 30 participants, three types of tasks and with 30 stimuli for each task leads to 2,700 scan paths in total. Each scan path typically consists of a high number of fixations. In this example, more then 10,000 extra data points have to be stored, formatted, and visualized to finally prove or disprove a hypothesis.

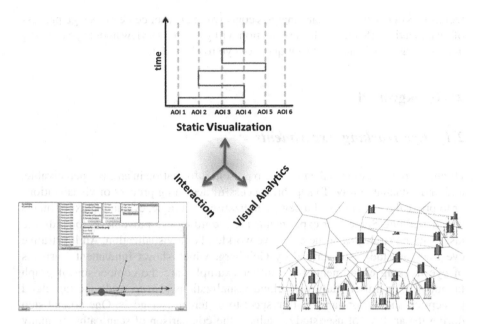

Static Visualization

Fig. 2 This chapter will present three visual analysis approaches to deal with the large amount of data recorded in an eye tracking experiment through classical static visualization, visual analytics, and a prototype to interactively manage, view, and analyze the recorded data

Eye tracking techniques provide additional information about the eye movements of a participant during a user experiment besides the standard metrics accuracy rates and completion times. However, a detailed analysis of recorded eye tracking patterns requires a high effort. This chapter discusses three approaches, which deal with the visual analysis of a large amount of data recorded in an eye tracking experiment: a classical *static visualization*, a *visual analytics* approach and a prototype to *interactively* manage, view and analyze the recorded data.

This chapter will discuss this issue and will present three approaches to deal with the large amount of data recorded in an eye tracking experiment (Sect. 3). These three approaches are (cf. Fig. 2): (1) a classical *static visualization* which shows important metrics of an eye tracking experiment at a glance (Sect. 4.1), (2) a *visual analytics* approach to analyze the statistic properties of the spatio temporal eye tracking data structure (Sect. 4.2), (3) a software prototype, which helps the user to manage, view and analyze the recorded data in a simple *interactive* way (Sect. 4.3). Finally, we will compare these three approaches with each other and with standard techniques of eye tracking data analysis concerning the user centered design process of visualizations (Sect. 5). The next section will give an overview about eye tracking and existing visual analysis techniques for eye tracking data.

2   Background

2.1   Eye Tracking Experiments

Often, the purpose of visualizations is to present information in an easy, perceivable, and understandable way. To support successful designing process of visualizations or to prove the readability of a new visualization technique, evaluation experiments are carried out. Some user experiments are recording eye movements to study the task solution strategies or the cognitive workload of a visualization. An illustrative eye tracking study is described by Goldberg, which shows fundamental aspects of eye tracking experiments [8]. Further examples are the comparison of graph layouts [5], or a user study with cloud visualizations [13] to study the tag cloud perception and performance with respect to different user goals. One crucial step during the analysis of user study results is the comparison of scan paths. In many cases, the analysis aims at finding common eye movement patterns, which can be interpreted as common cognitive strategies to perform a given task. Thereby, eye tracking metrics such as completion times, fixations, fixation rates, pupil sizes and others can be used [11, 16]. For a detail description of eye tracking experiments, please read the chapter by Goldberg in this book.

2.2   Visualization of Eye Tracking Data

Visualizations are used to study eye tracking data to find interesting scan path structures. Today, there is already a large number of visualization techniques for representing eye tracking data. But only a small number of them have been integrated in existing eye tracking devices. Heat maps [4] or scan paths [6] are the most prominent ones, that focus on a direct exploration of the recorded eye tracking data, see Fig. 3.

But these simple visualization approaches also have some drawbacks: Heat maps are problematic, because they only show a density based representation of the time varying trajectory data. Such a diagram only shows the hot spots where participants have focused on more frequently and where not. There is no information about a sequential order of visited areas of interest (AOIs) or points of interest (POIs). Hence, such heat maps belong to the class of time aggregated diagrams. Scan paths on the other hand allow the exploration of the time varying behavior by using a line based trajectory representation where all single trajectories for each participant are drawn on top of each other. This approach soon leads to the drawback known as visual clutter [19] as many trajectories have to be plotted in one diagram to compare them.

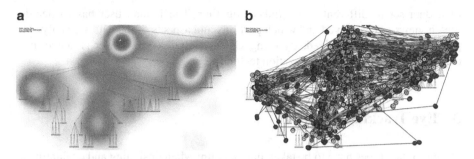

Fig. 3 Most prominent visualization techniques for eye tracking data: (**a**) Heat maps are time aggregated density based representations. (**b**) Scan paths are line based visualizations suffering from overdrawing and visual clutter

Gaze replays can be used in an animated fashion to show the sequence of eye fixations for a certain number of participants. The general drawback of animated diagrams is that the human's short term memory can only remember a few time steps of this animated gaze replay. Static diagrams or multiple representations will do a better job in many cases [24].

Another technique is used by eSeeTrack which combines a timeline and a tree structured visual representation to extend current eye tracking visualizations by extracting patterns of sequential gaze orderings. Displaying these patterns does not depend on the number of fixations on a scene [23]. Aula et al. present a non-overlapping scan paths technique [3]. Further techniques to analyze eye movement patterns are transition matrices [9, 15] which are based on areas of interest on the stimulus.

2.3 Visual Analytics

Visual analytics approaches benefit from visualization, interaction, applied algorithms, data mining, human reasoning, and the like [12, 21]. Applying visual analytics methodologies to eye tracking data has been proposed by Andrienko et al. [1] with the goal to achieve better visualizations for common visual task solution strategies. In the work of Andrienko et al., the authors investigated techniques, succesfully applied to geographic movement data, to analyze vast amounts of eye tracking data recorded during eye tracking experiments.

Since eye tracking data has a spatio temporal structure combined with other meta data about the participants as well as task accuracies and completion times, a single visualization technique such as a heat map or a scan path cannot succeed as a stand alone tool. Visual analytics methods with several views on the same data can be combined, which is known as Multiple Coordinated Views [18]. Interactive features [25] such as brushing and linking can be integrated to inspect aspects of a data set in different views linked together. The human user has to decide which views are displayed and which portions and aspects of the data are analyzed, preprocessed, and visualized. By using visual analytics, the viewer is hence, not depending on one single visualization technique.

3 Eye Tracking Data

Different data types have to be taken into account when designing and evaluating a user study. Psychology research differentiates between *independent* and *dependent variables*. These two types of evaluation variables are also valid for an eye tracking study. In eye tracking one can differentiate between the data which will be recorded by the eye tracker, the benchmarks completion times and accuracy rates, or meta data about the participant itself. This kind of data will be defined, discussed and classified in the following sections.

3.1 Variables

In psychological experiments one differentiates between *independent* and *depended variables*. Independent variables will be changed during the experiment to test a hypothesis. Depended variables will not be changed; they are used to investigate the effect the change of the independent variable has on them. Furthermore, there are *extraneous variables* which might also have an effect on the experiment and should be controlled. Some examples for extraneous variables are the age, gender, or the environmental testing situation. Those variables have to be controlled by either collecting the information, e.g., asking the participant about his or her age and gender, or by controlling it, e.g., all participants perform the experiment in the same room under the same conditions. Additionally, *distractor variables* are those, which should be removed or controlled to a high degree during an experiment because they can have an influence on the measured variables. For example, the behaviour of the operator can be a distractor variable because his or her behaviour varies over several run through of the experiment.

3.2 Recorded Data

What kind of data will be recorded during the experiment depends on the investigated hypotheses. If a hypothesis is to evaluate two visualizations (independent variable), the most obvious choice is to record the completion times and accuracy rates of the participants (dependent variables). This data will have to be statistically analyzed to prove or disprove the hypothesis. However, this data does not provide

Timestamp	Number	FixationIndex	GazePointX	GazePointY	Event	StimuliName	AoiNames
267	16	4	674	374		barchart.png	Content
284	17	4	678	379		barchart.png	Content
301	18	4	681	376		barchart.png	Content
317	19	4	679	359		barchart.png	Content
334	20	4	675	375		barchart.png	Content
351	21	4	676	365		barchart.png	Content
367	22	0	0	0		barchart.png	Content
384	23	5	677	380	KeyPressed	barchart.png	Content

Fig. 4 An eye tracking software records timestamps, fixations, pupil sizes and other information like events for key presses, mouse moves, and project management information such as stimuli file names and identifier for areas of interest

precise information why one visualization performs better than another one. To do so, "soft" metrics such as subjective affection about the visualizations could be recorded by questionnaires during the experiment. One great drawback of these metrics is that although they are quantitative, they represent a very subjective statement by the participants. A more precise metric, which can be used to answer the question why one visualization can be read better than another, are eye tracking metrics. They will be described in the next section.

3.3 Eye Tracking Data

In eye tracking experiments four main data types can be distinguished. A *gaze point* is a single eye gaze which has a specific position. Many gaze points will be grouped together as a *fixation*. Each fixation is defined by a two- or three-dimensional position, which is the averaged geometrical position of all gazes. Additionally, every fixation has a *fixation duration* at this position. If the eye tracker records pupil sizes, information about changes of the pupils is added to the fixation. Between every fixation the eyes change their focus. This focus change is called a *saccade*. During a saccade the vision is suppressed. The sequence of fixations and saccades are called a *scan path* (in some literature also called *gaze path* or *gaze trajectory*). Figure 4 shows an cutout of a CSV-file exported by the Tobii Studio 2.0 software.

3.4 Eye Tracking Trajectory Data

The analysis of eye tracking data is mostly concerned with obtaining insight from scan paths. In the context of this work, we will concentrate on two-dimensional scan paths. Thus, a scan path T of length $n - 1$ can be expressed as a sequence of n fixations with a two-dimensional position p_j

$$T := p_1 \rightarrow p_2 \rightarrow \ldots \rightarrow p_n \quad (p_j := (x_j, y_j) \in \mathbb{N} \times \mathbb{N}, 1 \leq j \leq n).$$

Each fixation position p_j contains a fixation duration $t_{d_j} \in \mathbb{N}, 1 \leq j \leq n$, expressing the time spent to inspect this point. The time it takes the eye to move from fixation point p_j to p_{j+1} is denoted by t_{m_j}. The total time to inspect a stimulus can be expressed by

$$t := \sum_{j=1}^{n} t_{d_j} + \sum_{j=1}^{n-1} t_{m_j}$$

4 Three Visual Analysis Approaches of Eye Tracking Data

The standard analysis method for eye tracking data are visualizations. An advantage of visualizations is that they show data structures in an accessible fashion by exploiting the benefits of visual perception and the strengths of the human visual system. However, static visualizations, in particular line based diagrams, have drawbacks known as visual clutter. To overcome this problem, statistical and interactive techniques can be added to the visualizations. Statistical techniques are used in visual analytics approaches. This section contributes an example for three approaches: (1) the *parallel scan path visualization technique*, which visualizes eye movements of many participants in a single static visualization with a parallel layout containing various levels of detail, such as fixations, gaze durations, eye shift frequencies and time at a glance, (2) a *visual analytics* approach to analyze the statistic properties of spatio temporal eye tracking data, and finally (3) *eTaddy*, which is a software prototype, that helps users to view and analyze their recorded eye tracking data in a simple and interactive way.

4.1 Parallel Scan Path Visualization

The first approach for a visual eye tracking data analysis is the parallel scan path visualization technique (in the following "PSP"). A key feature of the PSP technique is the visualization of eye movements of many participants on a single screen in a parallel layout. The visualization presents various properties of scan paths, such as fixations, gaze durations and eye shift frequencies at one glance. We have developed the PSPs because we wanted to answer the following questions in our eye tracking experiments: Are there any general eye movement patterns when working with simple visualization techniques? Can different user groups be distinguished based on their eye movement patterns? Answering these questions can help to adapt and optimize the layout of visualizations to a specific user group. Our new visualization technique should have a structured layout of scan path lines, that could also be used for a larger number of scan paths, displaying various properties of scan paths, supporting quick detection of common eye movement patterns and studying

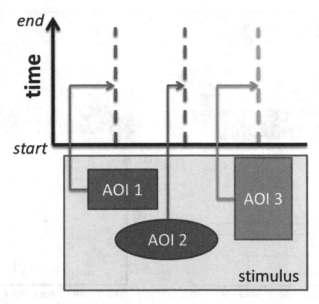

Fig. 5 Parallel scan path visualizations map areas of interest, which have been defined on the stimulus, to vertical coordinate axes. The *leftmost vertical axis* indicates time

temporal properties of scan path. All this information should be viewable at a glance. More details about the PSP technique can be found in [17].

> Key feature of the **parallel scan path visualization** is the visualization of eye movements of many participants in a single visualization with a parallel layout containing various levels of detail, such as fixations, gaze durations, eye shift frequencies and time.

The PSP visualization technique is based on areas of interest and maps gaze durations and fixations to vertical axes. Figure 5 shows a sketch of this approach, where the three areas of interest AOI1, AOI2, and AOI3 are mapped to three coordinate axes. The leftmost axis indicates time, starting from the bottom with the start time of the eye tracking measurement. The orientation of the parallel scan path visualization is arbitrarily represented. In the following we use a vertical time axis from bottom (start of the scan path recording) to top (end of the scan path recording) depending on user preferences. The horizontal axis displays all selected areas of interest as independent values. We have developed three types of parallel scan path visualizations, which we will explain in more detail in the following:

1. Gaze duration sequence diagram: Shows only scan paths.
2. Fixation point diagram: Shows scan paths together with fixations and frequencies of eye movement shifts between areas of interest.

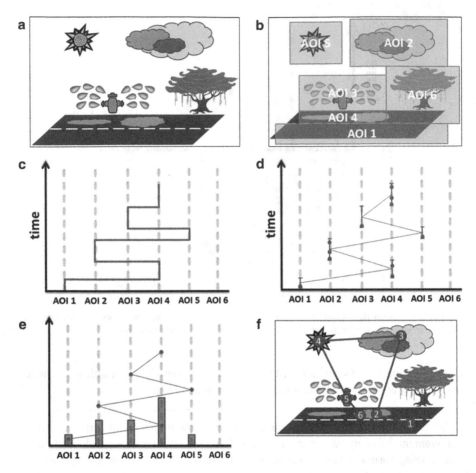

Fig. 6 Example stimulus (**a**) with areas of interest (**b**). Three visualization techniques have been developed which are using the same visualization concept with a parallel layout: gaze duration sequence diagram (**c**), fixation point diagram (**d**), and gaze duration distribution diagram (**e**). The scan paths of all three visualization techniques are plotted in (**f**)

3. Gaze duration distribution diagram: Shows scan paths together with information about frequencies of eye movement shifts between areas of interest and gaze durations in areas of interest.

4.1.1 Gaze Duration Sequence Diagram

Figure 6c shows a sketch of a gaze duration sequence diagram. This example shows the scan path of one participant. Every continuous line on a vertical axis represents a gaze duration inside an area of interest. Horizontal lines indicate a change of attention from one AOI to another. Figure 6c shows a scan path starting in AOI1, moving to AOI4, back to AOI2 and so on. Figure 6a shows the stimulus together with its areas of interest (Fig. 6b). By means of the time axis both gaze duration parameters like start and end times and the temporal sequence of changes between areas of interest can be identified. Changes of participant's attention can be studied by following the scan path line in the visualization.

4.1.2 Fixation Point Diagram

Fixation point diagrams additionally show single fixations. The scan path of one participant is shown in Fig. 6d. Now, every fixation on the stimulus inside an area of interest is additionally plotted as a filled circle. The gaze duration inside an area of interest is shown with a vertical continuous line on the corresponding area of interest axis. For each such group of fixations, the center of the respective lines are connected by ascending or descending lines. By using a fixation point diagram, characteristics of eye movements during a gaze duration such as the frequency and number of fixations can be studied.

4.1.3 Gaze Duration Distribution Diagram

Figure 6e shows a gaze duration distribution diagram. The gaze duration distribution diagram is based on eye tracking data visualizations for website analysis used in [10]. Scan paths through areas of interest are plotted similar to gaze duration sequence diagrams. However, this third type of a PSP visualization does not directly indicate the time of gaze durations. Now, time is shown by the midpoint of a gaze duration which is plotted as a filled circle. Lines between these circles display eye movements between areas of interest. A bar chart is overlaid showing the summation of the percentage of gaze durations in the areas of interest.

4.1.4 Implementation

All three visualization techniques have been implemented in a software tool together with classical heat map and scan paths, see Fig. 7. A simple project management supports users to select participants, participant groups, and areas of interest.[1]

---

[1]For further information about the software please contact the authors.

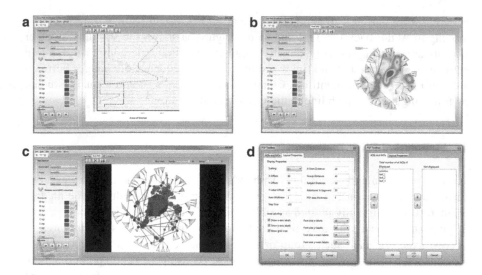

Fig. 7 The Parallel scan path visualization approach (**a**) was implemented in a software tool together with heat map visualizations (**b**) and classical scan paths (**c**). A project management allows users to easily select participants and groups of participants. Various layout properties can be set via option dialogues (**d**)

4.2 Visual Analytics for Eye Tracking Analysis

Due to the progress in hardware technology much bigger datasets can be generated and stored than some years ago. This observation also holds for eye tracking data. Though there is no technical limitation for producing vast amounts of eye tracking data, the analysis of the data mountains is the more challenging part than its storage.

Visualization in general and visual analytics in particular have been applied as a promising example that exploit the perceptual abilities of the human user with the goal to find visual patterns very fast. This concept helps when the problem at hand cannot be specified purely algorithmically and its solution cannot be computed by a simple algorithm. Oftentimes, a mixture consisting of algorithms and visualizations are the medium of choice with the human user in-between.

In particular, visualization techniques have also been applied for visually exploring eye tracking data, i.e., gaze trajectories, and have also been integrated into eye tracking systems in form of simple heat maps and gaze plots. The spatio temporal nature of the data makes it difficult to find insight in such data by just using simple visualization techniques. More exactly, heat maps are time aggregated representations which do not allow to examine the sequential order of fixations. Gaze plots on the other hand are line based diagrams that suffer from overplotting and visual clutter. This means common visual task solution strategies cannot be examined by just inspecting a static gaze plot.

This is exactly the point where visual analytics comes into play as a means to uncover insight in the spatio temporal data. Andrienko et al. [1] proposed a visual analytics methodology for analyzing vast amounts of trajectory data. In this work the focus is on deriving common task solution strategies for a given static stimulus shown to a certain number of participants.

> Visualization in general and *visual analytics* in particular have been applied as a promising example that exploit the perceptual abilities of the human user with the goal to find visual patterns very fast. Andrienko et al. [1] proposed a visual analytics methodology that have originally used for geographic data for analyzing vast amounts of eye tracking data.

A rich source of visualization techniques exists for analyzing geographic movement data. Since such data has some commonalities with eye tracking trajectories some of the existing visual analytics techniques can be adapted to this novel domain. From about 30 generic methods relevant to movement data [2], 23 methods were available and tested. From these, only six methods have been found ineffective and the rest was judged as useful. Andrienko et al. [1] discusses the selected methods and the eye movement analysis procedures in which these methods are used. The methods and procedures combine computational techniques for data transformation and analysis, visual displays, and interactive operations. Figures 8a–f illustrate some of the applied visualization techniques in this tool.

4.3   eTaddy: Interactive Eye Tracking Data Analysis

The last approach, which we would like to present, uses interaction to support the user during the analysis of eye tracking data. eTaddy (eyeTracking Analysis, conDuction and Designtool for user studYs) is an integrative framework for the creation, conduction, and analysis of eye tracking user studies.[2] Each step of a user study is represented with an own window in the framework. The data used in the framework, including eye tracking data, questionnaire answers, or participant data, is stored in a database. The framework can be extended via a plug-in system, to add own metrics, statistics, or visualizations. This framework allows users an efficient task oriented analysis process. In the following we will describe the architecture of eTaddy and the user interface.

---

[2]For further information about the software please contact the authors.

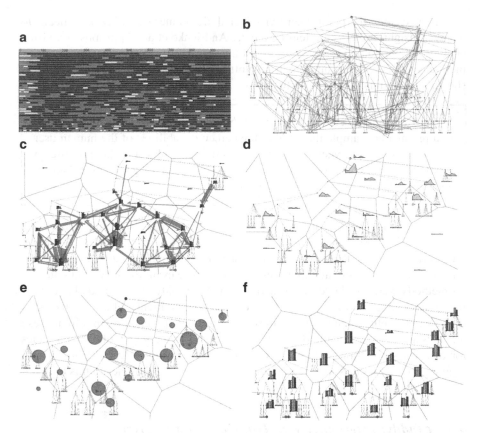

Fig. 8 A small set of the visualization techniques provided by the visual analytics tool [1]:
(**a**) Time varying distances to points of interest for all participants. (**b**) Line based trajectory
visualization aggregated over time for a specific time interval. (**c**) Time aggregated representation
showing number and direction of eye gazes. (**d**) Time varying histograms for fixation frequencies
in specific areas of interest. (**e**) Time aggregated fixation numbers in areas of interest. (**f**) Barchart
representation for different metrics in areas of interest

eTaddy is an integrative framework to design, conduct and analyze eye tracking
user studies. The framework supports the embedding of additional metrics,
statistical tests, and visualizations via a plug-in system. Both data from the user
experiment and data which is generated during the analysis process is stored in
a database.

4.3.1 Architecture

One drawback of commercial eye tracking software like BeeGaze (SMI) or Tobii
Studio (Tobii) is the limited flexibility during the analysis process. Usually the data
collected in the experiment is stored on the eye tracking computer. This means, that

only one person can conduct an eye tracking experiment at a time. Furthermore own visualizations and metrics cannot be added to the state-of-the-art eye tracking software suites. The data usually has to be exported and imported into statistic software like R or SPSS for the statistical evaluation of the hypotheses. To reduce the amount of time which is spent into this data management, we have implemented an integrative framework which is connected to a database for data storage.

The eTaddy framework consists of a plug-in system to include own metrics, visualizations or statistics and a database is connected for data storage. The plug-in system is implemented via an XML file, which will be parsed when starting the software in which new plug-ins can be added. The used database model defines classes for the user study itself, the different scenarios and tasks, answers from questionnaires, the results which are generated during the analysis process, and the eye tracking data from the eye tracker system. By using a database the complete analysis process can be done in the framework. There is no more need to import or export data, except of an initial import of the eye tracking data at the beginning.

4.3.2 User Interface

To analyze a user study, the first step using eTaddy is to create a new user study data set. The user can define information about the eye tracking system and can create scenarios of the user study. Each group of tasks from a scenario represents one hypothesis and consists of different stimuli. Besides the creation of scenarios the user can define the study procedure the participant will have to perform. This study procedure includes questionnaires, task descriptions, a vision test, or other text documents, which are needed for the user study conduction. The questionnaires can be created with eTaddy by using an integrated questionnaire editor.

Two separate computer screens can be used for the conduction of the user experiment: The first computer screen presents information for the moderator, the second computer screen presents information and stimuli for the participant. The participant will be presented tutorials, questionnaires, or vision tests consecutively. The main eye tracking recording is done with an standard eye tracking software. However, the recorded data is later used in the analysis process in eTaddy.

Figure 9 shows a screen shot of the implementation. To analyze the results of the eye tracking experiment the user first has to import the eye tracking data from the eye tracker software into the database (1). Then, all stimuli and participants will be displayed and the user can choose one or multiple stimuli and participants for the analysis (2). Next, the user can choose between different metrics, statistic analysis techniques and visualizations (3). The solution of such a calculation or visualization is displayed in the main panel of the window (4). This part is called the *history* and each calculation or visualization is displayed in a so called *history node* which contains information about the scenario, stimulus, chosen calculation, participants, time stamp and solution, respectively the visualization. Furthermore, eTaddy provides visualizations for the questionnaires and comes with a printing feature for a user study report.

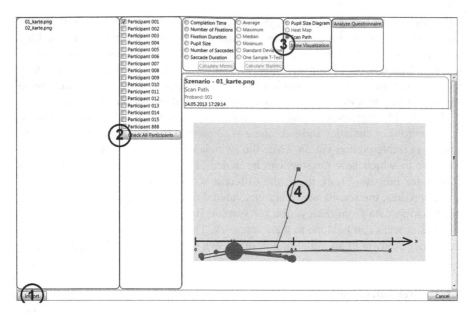

Fig. 9 Screen shot of the eTaddy prototype (in german). At first the eye tracking data has to be imported (1). Secondly, the user chooses one or multiple stimuli and participants (2). Then, she can choose between different metrics, statistics and visualizations (3). eTaddy presents the results of the calculations or of the visualizations in its main panel (4)

5   Discussion

Figure 10 shows an overview of the three visual analysis methodologies for eye tracking data, which have been presented in this chapter. This overview can help users to find the most appropriate visual analysis technique for their experimental results and research questions. Users start from the top of the decision tree diagram with the question whether they are interested in an overview about the recorded eye tracking data. If they are, they can use heat map visualizations or the presented approaches used in eTaddy. If they are interested in statistical information about fixations, fixation durations, frequencies and other metrics they also can use eTaddy. Next, the user would like to study the order in which items on the stimulus have been focused on. If areas of interest are not available, they can use traditional scan path visualizations. If areas of interest are available they can continue their analysis with methods from visual analytics as presented. If additionally a large number of participants has to be analyzed, they can either also use visual analytics techniques, eTaddy for an easy management of the recorded data, or the PSP visualizations. If they are interested in studying fixations, they can use fixation point diagrams. If they want to visualize gaze durations, they can proceed their analysis with gaze duration sequence diagrams or with gaze duration distribution diagrams in the case, if they wish to overlay summarized eye tracking information in the diagram.

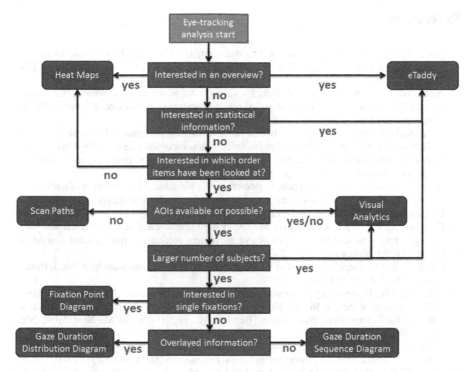

Fig. 10 Overview about the three presented approaches together with the classical scan path and heat map visualization techniques. This diagram can help users to plan their analysis process

6 Conclusion

In this chapter we have presented three approaches for a visual analysis on eye tracking data. Our three approaches deal with the large amount of data which is recorded during an eye tracking experiments. The parallel scan path visualization technique has the key feature, that it visualizes eye movements of many subjects in a single visualization with a parallel layout containing various levels of detail, such as fixations, gaze durations, eye shift frequencies and time. We have presented the visual analytics approach by [1] to use visual analytics techniques from geographical data for the analysis of eye tracking data. This is possible, because eye tracking data and geographic data have both a spatial temporal structure. The last approach, which we have presented, is based on the Information-Seeking mantra [20] and allows the user to analyze eye tracking data by the three steps overview, filtering and details-on-demand.

Depending on the user study tasks, one or more of these techniques can be used to efficiently analyze data from eye tracking experiment and thus, allows researchers to evaluate their new visualization techniques from a user centered perspective.

References

1. Andrienko, G., Andrienko, N., Burch, M., Weiskopf, D.: Visual analytics methodology for eye movement studies. Visualization and Computer Graphics, IEEE Transactions on **18**(12), 2889–2898 (2012)
2. Andrienko, G.L., Andrienko, N.V., Bak, P., Keim, D.A., Kisilevich, S., Wrobel, S.: A conceptual framework and taxonomy of techniques for analyzing movement. Journal of Visual Languages and Computing **22**(3), 213–232 (2011)
3. Aula, A., Majaranta, P., Raiha, K.J.: Eye-tracking reveals the personal styles for search result evaluation. Ifip International Federation For Information Processing **3585**, 1058–1061 (2005)
4. Bojko, A.: Informative or misleading? Heatmaps deconstructed. In: International Conference on Human-Computer Interaction, pp. 30–39 (2009)
5. Burch, M., Konevtsova, N., Heinrich, J., Hoeferlin, M., Weiskopf, D.: Evaluation of traditional, orthogonal, and radial tree diagrams by an eye tracking study. Visualization and Computer Graphics, IEEE Transactions on **17**(12), 2440–2448 (2011)
6. Çöltekin, A., Fabrikant, S.I., Lacayo, M.: Exploring the efficiency of users' visual analytics strategies based on sequence analysis of eye movement recordings. International Journal of Geographical Information Science **24**(10), 1559–1575 (2010)
7. Chen, C., Yu, Y.: Empirical studies of information visualization: a meta-analysis. Int. J. Hum.-Comput. Stud. **53**(5), 851–866 (2000)
8. Goldberg, J.H., Helfman, J.I.: Comparing information graphics: a critical look at eye tracking. In: Proceedings of the 3rd BELIV'10 Workshop: BEyond time and errors: novel evaLuation methods for Information Visualization, BELIV '10, pp. 71–78. ACM, New York, NY, USA (2010)
9. Goldberg, J.H., Kotval, X.P.: Computer interface evaluation using eye movements: methods and constructs. International Journal of Industrial Ergonomics **24**(6), 631–645 (1999)
10. Guan, Z., Cutrell, E.: An eye tracking study of the effect of target rank on web search. In: Proceedings of the SIGCHI conference on Human factors in computing systems, CHI '07, pp. 417–420. ACM, New York, NY, USA (2007)
11. Jacob, R.J.K., Karn, K.S.: Eye Tracking in Human-Computer Interaction and Usability Research: Ready to Deliver the Promises. The Mind's eye: Cognitive The Mind's Eye: Cognitive and Applied Aspects of Eye Movement Research pp. 573–603 (2003)
12. Keim, D.A., Mansmann, F., Schneidewind, J., Thomas, J., Ziegler, H.: Visual analytics: Scope and challenges. In: Visual Data Mining, pp. 76–90 (2008)
13. Lohmann, S., Ziegler, J., Tetzlaff, L.: Comparison of tag cloud layouts: Task-related performance and visual exploration. In: Proceedings of the 12th IFIP TC 13 International Conference on Human-Computer Interaction: Part I, INTERACT '09, pp. 392–404. Springer-Verlag, Berlin, Heidelberg (2009)
14. Plaisant, C.: The challenge of information visualization evaluation. In: Proceedings of the working conference on Advanced visual interfaces, AVI '04, pp. 109–116. ACM, New York, NY, USA (2004)
15. Ponsoda, V., Scott, D., Findlay, J.: A probability vector and transition matrix analysis of eye movements during visual search. Acta Psychologica **88**(2), 167–185 (1995)
16. Poole, A., Ball, L.J.: Eye Tracking in Human-Computer Interaction and Usability Research: Current Status and Future Prospects, vol. Encyclopedia of HCI. Pennsylvania: Idea Group, Inc (2005)
17. Raschke, M., Chen, X., Ertl, T.: Parallel scan-path visualization. In: Proceedings of the Symposium on Eye Tracking Research and Applications, ETRA '12, pp. 165–168. ACM, New York, NY, USA (2012)
18. Roberts, J.C.: State of the art: Coordinated & multiple views in exploratory visualization. In: International Conference on Coordinated and Multiple Views in Exploratory Visualization, pp. 61–71 (2007)

19. Rosenholtz, R., Li, Y., Mansfield, J., Jin, Z.: Feature congestion: a measure of display clutter. In: Proceedings of the Conference on Human Factors in Computing Systems (CHI), pp.761–770 (2005)
20. Shneiderman, B.: The eyes have it: A task by data type taxonomy for information visualizations. In: Proceedings of the 1996 IEEE Symposium on Visual Languages, pp. 336–. IEEE Computer Society, Washington, DC, USA (1996)
21. Thomas, J., Cook, K.: Illuminating the Path. IEEE Computer Society Press, Los Alamitos (2005)
22. Tory, M., Staub-French, S.: Qualitative analysis of visualization: a building design field study. In: Proceedings of the 2008 Workshop on BEyond time and errors: novel evaLuation methods for Information Visualization, BELIV '08, pp. 7:1–7:8. ACM, New York, NY, USA (2008)
23. Tsang, H.Y., Tory, M., Swindells, C.: eSeeTrack–visualizing sequential fixation patterns. IEEE Transactions on Visualization and Computer Graphics 16(6), 953–62 (2010)
24. Tversky, B., Morrison, J.B., Bétrancourt, M.: Animation: can it facilitate? International Journal of Human-Computer Studies 57(4), 247–262 (2002)
25. Yi, J.S., ah Kang, Y., Stasko, J.T., Jacko, J.A.: Toward a deeper understanding of the role of interaction in information visualization. IEEE Transactions on Visualization and Computer Graphics 13(6), 1224–1231 (2007)

User Studies in Visualization: A Reflection on Methods

Melanie Tory

Abstract In this chapter I will reflect on many years of running user studies in visualization, examining my experience with how effectively different methodological approaches worked for different goals. I first introduce my own categorization of user studies based on their major goals (*understanding* versus *evaluation*, each with specific subcategories) and common methodological approaches (*quantitative experiment, qualitative observational study, inspection,* and *usability study*), providing examples of each combination. I then use examples from my own experience to reflect upon the strengths and weaknesses of each methodological approach.

1 Introduction

User studies are becoming standard practice in visualization research and design, as a way of both understanding users and evaluating visualization methods and tools. But the "user study" is not one single research method; rather it encompasses a large collection of empirical methods involving human participants. Designing a user study can be a complex and challenging activity involving many difficult choices. Which methods should I consider? How should I choose which method to use? When is each type of study appropriate? What metrics can best answer my question? Answering these questions is often very difficult.

Moreover, the adoption of user study methods within visualization has taken some time because visualization has some unique challenges compared to other fields such as Human Computer Interaction (HCI). Many of these were documented by Plaisant [18]. For example, many real world scenarios involve experts who will use the visualization tools on an extended basis; yet it is hard to get much time with

M. Tory (✉)
Department of Computer Science, University of Victoria, 3800 Finnerty Road,
Victoria, BC, V8W 2Y2 Canada
e-mail: mtory@cs.uvic.ca

W. Huang (ed.), *Handbook of Human Centric Visualization*,
DOI 10.1007/978-1-4614-7485-2_16, © Springer Science+Business Media New York 2014

experts, and training novices for a long time period is impractical. Additionally, tasks in visualization studies often need to be oversimplified in order to make them measurable. There is also a trade-off between a desire to build generalizable tools and the need to make sure tools work in specific domains.

The purpose of this chapter is to share lessons learned from running user studies in visualization, and to provide a structure to help designers and researchers navigate the rather large space of methods involving human participants. For readers who are new to the field, the chapter may be used as an overview of empirical study approaches and when to use them, with references to follow for more detail on specific methods. For those who have done some work in the area, the chapter provides a conceptual structure to explain the design space of user studies, and some insight on the state of user study methods in visualization and where to go from here.

2   User Study Goals and Types

User studies have been categorized in numerous different ways. In this section, I focus on categorizing studies based on the researcher's primary goals.

Some researchers have classified user studies in visualization by doing surveys of research papers. For example, Komlodi et al. [11] identified four major types: *controlled experiments comparing design elements, usability evaluation of a tool, controlled experiments comparing two or more tools,* and *case studies of tools in realistic settings.* In a more recent review, Lam et al. [12] characterized seven different empirical study scenarios, divided at the highest level into *understanding data analysis* and *understanding visualizations.* Nearly all of these categories are captured within my own description of study types.

Munzner's [16] nested model of the visualization design and validation process identifies four nested layers of design: *domain problem characterization, data/operation abstraction design, encoding/interaction technique design,* and *algorithm design.* While her model does not directly focus on characterizing user study techniques, it makes the important points that evaluation needs to occur within each layer, and the evaluation method chosen needs to match the design contribution of the layer. As an extreme example, evaluating the speed of an algorithm does not verify that the visualization algorithm solves a specific domain problem. The high-level user study goals that I present below have some similarity to the goals of Munzner's design layers.

2.1   Types of Visualization Studies Based on High Level Goals

Over the last several years within the field of visualization the term *evaluation* has become synonymous with *user study.* This is rather unfortunate because there

Table 1 Study types based on the high level goals of the researcher. Study types are classified at the highest level into those with understanding-based goals and those with evaluation-based goals

Study goal		Description
Understanding	Understand perceptual and cognitive principles	Understand human perceptual or cognitive characteristics, often by measuring performance at abstract tasks
	Understand context	Understand the context in which a visualization or future visualization will be used, including user characteristics, tasks, environment, social context, work practices, communication practices, etc.
Evaluation	Compare visualization techniques, tools, or interaction techniques	Directly compare two or more approaches to identify strengths and weaknesses of each or to validate a hypothesized improvement over a baseline design
	Evaluate one visualization technique, tool, or interaction technique	Identify strengths, weaknesses, and/or limitations of one single approach

are other reasons to run users studies besides evaluating visualization tools and techniques. In particular, many user studies are designed to better understand human behaviour, in order to provide insight to design. Often these studies take place before a visualization system or technique has even been developed.

I propose a simplified categorization of user study types based loosely on the prior work described at the beginning of Sect. 2, but mostly based on classifying past research studies that I am familiar with. I divide user study goals into two high-level categories, *understanding* and *evaluation*, each with two lower-level categories, as shown in Table 1.

From a practical perspective, a user study designer can use these categories as the first step of study design. Choosing one of the study types can help the designer to articulate their goal, and can narrow down the choice of empirical methods (see Sect. 3 for how the study types relate to empirical methods).

2.1.1 Understanding Perceptual and Cognitive Principles

In this sort of study, the researcher wishes to learn something about human perception or cognition in order to inform design. Typically, the study employs similar methods to applied psychology studies, and aims to provide insights into human behaviour that can be used to formulate generalizable guidelines for visualization design. For example, Healey and Enns [6] described a series of experiments to better understand how humans perceive textural elements such as height, density, regularity and colour. They used the results of these experiments to inform the design of geographic representations of multivariate data.

2.1.2   Understanding Context of Use

In a context of use study, the researcher wishes to better understand the circumstances surrounding the use (or potential use) of a visualization tool, often for a specific domain. This could involve understanding the domain problem, user characteristics, tasks, environment, social context, work practices, collaborative or communication practices, analytical reasoning processes, etc. Typically a study would not investigate all of these, but would focus on the subset that is most relevant to the problem at hand. Most often, studies to understand context are done before initiating design, in order to inform the design process; such studies have been termed *pre-design studies* by Isenberg et al. [8]. However, this does not mean that visualization tools are never used in the studies. For example, people might already make use of visualization tools in their current work practices, and a study might assess how well the existing tools support the users' tasks, attempting to identify ways in which the tools might be modified or enhanced. Kang and Stasko [9] provide a recent example of a study to understand context. Through a longitudinal field study, they observed the work process of intelligence analysts to inform the design of analytics tools. Similarly, Sedlmair et al. [19] conducted extensive studies of automotive engineering prior to designing their visualization tool to support current work practices.

2.1.3   Compare Visualization Techniques, Tools, or Interaction Techniques

In a comparison study, the researcher directly compares two or more visualization techniques, complete visualization tools, or interaction techniques. Here *visualization technique* refers to a visual encoding method, *interaction technique* refers to a software and/or hardware interface for the user to interact with a visualization technique, and *tool* refers to a complete visualization system that would typically include multiple visualization and interaction techniques. The goal of a comparison study may be either to compare a prototype idea to a baseline 'best-of-breed' approach, to validate that the new technique is better, or to compare two or more competing ideas to identify their strengths and weaknesses. An example of technique comparison is Li and North [14]. They compared dynamic query sliders to brushing histograms (implemented in otherwise identical systems) for interacting with geographic visualizations. In contrast, Kobsa [10] conducted a study comparing complete systems for tree visualization.

2.1.4   Evaluate a Visualization Technique, Tool, or Interaction Technique

In a single approach evaluation study, the researcher typically wishes to assess the strengths, weaknesses, and/or limitations of a particular visualization technique, tool, or interaction technique, without making direct comparison to a competing method. This type of study might be used when a designer has developed an

Table 2 Example research question for each major study type

Understand perceptual and cognitive principles	Which type of motion (blink, shake, or expand) attracts attention in one's peripheral vision most quickly, assuming equal motion frequencies?
Understand context	What data analysis challenges do teams of intelligence analysts encounter during in-person team meetings?
Compare visualization techniques, tools, or interaction techniques	Can users navigate horizontal tree layouts more quickly with a fisheye lens or an overview + detail display?
Evaluate one visualization technique, tool, or interaction technique	Do side-by-side treemaps enable people to answer tree comparison questions? What design improvements are necessary to make these comparison questions easier?

approach for a specific problem and wants to identify ways to improve the design, or simply needs to validate that the approach does indeed support users to accomplish their tasks. It also might be used when the competing approaches are so primitive that comparison against them would be pointless. For example, in one study we wanted to evaluate a visualization-based photo browser designed for construction managers [27]. We chose to evaluate only the prototype interface rather than compare to a baseline system because the default interface used to accomplish the photo management tasks (a standard computer folder system) was clearly inadequate.

2.2 Specific Research Questions

Identifying the high-level study goal (Sect. 2.1) is only a first step in user study design. The next step is to identify a more specific set of research questions or objectives. These research questions will similarly help in identifying research methods and metrics. Example research questions for each of the four major study types are given in Table 2.

A major challenge with defining research questions is that often the first iteration of a research question is much too vague, and therefore cannot help the researcher in narrowing down the methods [4]. For example, consider the research question, "Do multiple tag clouds improve information finding as compared to a single tag cloud?". This only makes sense if we already know answers to a bunch of additional context details, like who (?) will be using the tag clouds, for what task (?), in what domain (?), what exactly does information finding mean (?), and how will we define when the information finding has been successful (?). Precisely defining the research question is an important first step to understanding what techniques and metrics will be appropriate. Note that the example questions in Table 2 are part way there, but each one still needs to have some additional context defined.

For studies with an understanding goal, research questions need to focus the study in order to make its scope manageable. For instance, in a perceptual study, the designer will typically choose a perceptual attribute (e.g. visual attention to motion in one's peripheral vision) and a metric (e.g. performance, specifically time to notice a motion event). The designer will then need to specify the conditions under which the phenomenon will be studied (e.g. young adults with normal vision, conducting a reading task under normal lighting conditions with no other distractions), which will help to identify how other factors will be controlled. For a context study, the designer would typically narrow down a focus area (e.g. understanding task requirements for a specific visualization problem, with a specific user population).

For studies with an evaluation goal, the research questions will similarly need to narrow down the context of the evaluation, including the precise users and tasks. Often one of the most difficult choices to make is what specific outcomes are most important to evaluate. Specific outcomes that might be considered are:

- Performance (efficiency, errors)
- Usability
- Learnability
- User experience/preference
- Utility of the feature set (e.g. which features are useful, not useful, missing, or require improvement)
- Support for insight generation
- Support for communication
- Support for learning
- Effect on the analytical reasoning process
- Effect on the collaboration process (including attributes such as efficiency/effectiveness/ease of group work, and support for information sharing and awareness, which in turn need to be precisely defined)

Typical studies focus on one or a small number of the above outcomes. Comparative evaluations will ask which technique/tool/interaction technique has the best result for the specific outcome and why. Single approach evaluations will typically ask whether the approach meets a threshold level of the outcome, or what the strengths and weaknesses of the approach are with respect to the outcome.

Early in a project, the research questions are often *exploratory* in nature. These types of questions help us to better understand the problem, and identify factors that are relevant. For example, exploratory questions might be worded as, "What factors are important for X?", "What is X like?", "How often does X happen?", or "What are the conditions under which X is needed?". *Understanding context* goals are nearly always exploratory in nature. At later stages of a project, we may ask *confirmatory* questions because we already have some specific hypotheses in mind. For example, we might ask, "Does Y cause X?" or "How much does X improve performance at task Y?" For a more detailed description of research question types and additional examples, see Easterbrook et al. [4].

3 Empirical Approaches and Methods

User studies can be categorized according to the general methodological approach taken by the researcher, as well as by the specific data collection methods used.

3.1 High Level Approaches

I classify empirical user study approaches into four major categories, as shown in Table 3. Note that this set of approaches is not meant to be exhaustive (e.g., it is possible to have a qualitative experiment or a quantitative observational study), but instead is meant to capture the most common methods used currently in visualization. Numerous authors in other fields have described research methods involving human subjects in much greater depth. See McGrath [15], Lazar et al. [13], Creswell [3], or Easterbrook et al. [4] for a more comprehensive introduction and to gain an understanding of the philosophical differences among research traditions. For example, McGrath [15] provides a particularly useful taxonomy of general research approaches and the tradeoffs among them in terms of realism, generalizability, and precision. In contrast, my aim in this section is to focus specifically on visualization, and provide a small number of categories that describe the most common methods currently used in practice.

Combinations of the above approaches are also possible. For example, a quantitative experiment might also include the collection of qualitative data to help explain

Table 3 Common empirical research approaches in visualization

Quantitative experiment	Makes a direct comparison between two or more controlled conditions and measures a quantitative difference between them. Results are analyzed using statistical hypothesis testing. Emphasis is on generic research results that can inform design of a whole class of visualization systems.
Qualitative observational study	Answers exploratory questions using mainly qualitative data gathered through techniques such as in-person or video observations, interviews, or journals. Groups may be compared but the partitioning into groups may be self-selected.
Inspection	A small number of experts inspect visualization tool(s), interface(s) or technique(s) using a pre-defined protocol and provide a report of their findings. Example inspection techniques include cognitive walkthroughs [17], heuristic evaluations [17], and abstract task evaluation [2].
Usability study	Users complete tasks with a visualization tool, technique, or interaction method to assess whether it meets specified criteria. Normally used to assess whether a tool/technique/interaction method meets specific desired usability outcomes such as those in Sect. 2.2. Emphasis is on generating design improvements for a specific tool rather than generalizable design insights.

the results. As another example, in [22] we conducted an experiment with primarily qualitative observations, but also analyzed some quantitative differences between conditions.

3.2 Specific Data Collection Methods

Each of the study approaches will require data collection. A variety of data collection methods can be used, and these often are applicable across multiple approaches. Specific data collection methods used in visualization include:

- Direct observation
- Performance measurements
- Questionnaires
- Interviews
- Eyetracking
- Journaling
- Log analysis

For more detail on Eyetracking, see the Eyetracking chapter in this book. For other data collection methods, see Lazar et al. [13].

4 Relationship Between Study Goals and Approaches

Table 4 illustrates the relationship between study goals and common research approaches, giving a short description and examples for each combination. The table reveals that all of the approaches can be used for multiple goals, and all of the goals can be achieved through more than one approach. However, not all combinations make sense. For example, Inspection is not possible unless there is a prototype system to evaluate, so it is not useful for the Understanding goals. Similarly, it would be difficult to understand the rich and complex nature of system use (Understanding Context) through a quantitative experiment since the approach is not exploratory. The text following the table describes advantages and challenges associated with each approach, with examples from my own experience.

4.1 Quantitative Experiments

I have conducted quantitative experiments for both understanding perceptual principles (e.g. to understand perceptual tasks with scatterplot and landscape visualizations of multidimensional data [24, 25]) and for evaluating visualization techniques (e.g. new approaches to schedule visualization [7]). An advantage of quantitative

Table 4 Relationship between research approaches and study goals. Text in each cell describes a typical study of that type, with citations to examples. X's mark combinations that are unlikely to make sense

	Understanding goals		Evaluation goals	
	Understand perceptual and cognitive principles	Understand context	Compare visualization techniques, tools, or interaction techniques	Evaluate a visualization technique, tool, or interaction technique
Quantitative experiment	Typically a lab experiment with quantitative measures. Emphasis on generic insight into human attributes such as perception of different visual encodings [6, 24, 25]	X	Lab or field experiment to compare two or more methods quantitatively (typically performance). Emphasis on results relevant to a specific domain or visualization type [7, 10, 14]	X
Qualitative observational study	Lab or field observations to understand cognitive processes in data analysis. Emphasis on generic understanding rather than domain-specific design knowledge [5]	Observations of naturalistic practices, often in field settings, and interviews, especially with domain experts [9, 19, 26]	Compare methods in a lab study or field deployment, with a focus on qualitative outcomes [22]	Observe use of a deployed system, ideally over a long time frame [19–21]
Inspection	X	X	Comparative evaluation by experts using a usability inspection approach [23]	Single system evaluation by experts using a usability inspection approach [28]
Usability study	X	Evaluate usability of existing systems to establish requirements for a new system	Usability study comparing multiple prototypes. Emphasis on identifying design improvements	Prospective users conduct tasks with a prototype system. Emphasis on design improvements for that tool [27]

studies is that they can provide convincing statistical evidence that one technique outperforms another. The downside is that they provide only a pinhole view since the study conditions must be strictly controlled. This typically means that the results are only reliable for the specific task and user population tested. As a result, a large collection of studies on the same topic is necessary in order to gain any holistic understanding. For example, our studies of landscape and scatterplot spatializations of multidimensional data sets revealed that scatterplots often outperformed landscape displays. However, we must note that our understanding of this topic is limited to two specific abstract tasks (memory of specific views, and finding regions within a specified value range) with an undergraduate student population. While we anticipate that these results are likely to be more generalizable than this, we cannot be certain that scatterplot views will outperform landscapes in all cases.

This example also brings up another important challenge: choosing an appropriate benchmark task. Because of the number of repeat trials required for statistical analysis, it is often impossible to test a large number of tasks within one study, so the set of tasks must be selected carefully. It can also be a challenge to choose tasks that are both ecologically valid (i.e. match what people do in the real world) but also easily measurable (i.e. they have a correct answer that can be validated, and ideally the measurements have some intermediate levels of granularity). Time is often a better primary metric than accuracy since it is measured on a continuous scale rather than a binary (correct/incorrect) one. However, even when time is the main metric, a clearly defined task is still necessary so that it is clear when a trial should end and the timer should stop. Abstraction versus domain specificity is another important factor to consider. In our landscape/scatterplot studies we chose tasks that were abstracted away from any particular domain, in order to increase generalizability of the results. This is both a benefit and a drawback: it is more *likely* that the results apply widely, but not *certain* that they apply in particular domain such as document analysis for intelligence analytics. For further reading, Lazar et al. [13] provide a good introduction to designing and analyzing a quantitative experiment.

4.2 Qualitative Observational Studies

Qualitative observational studies can be used for any of the goal types, as illustrated in Table 4. For example, we conducted a qualitative field study to understand the current work practices of interdisciplinary building design teams [26] (understanding context). In the laboratory, we used qualitative analysis to understand the process of visualization construction by novices [5] (understanding cognitive principles). In these projects, the research questions were exploratory (i.e. we did not know in advance which factors might be important for design). Qualitative methods allowed us to gain a holistic understanding of current practices. Qualitative analysis can also be useful in evaluation, to learn the complexities of when and why a visualization technique works/does not work.

Qualitative methods are very well established in the social sciences. The most common methods in Visualization are Grounded Theory (a general approach that involves building a theory to explain behaviour by observing undisturbed real world activities) and Content Analysis (a more specific method for analyzing the content of interviews, questionnaire responses, video observations, blog comments, etc.). These methods typically involve iteratively *coding* the material (i.e. assigning named categories to statements or observed events) until a cohesive description of the results emerges. Codes may be either based on existing knowledge and theory (fixed coding), or may emerge during the analysis (open coding). See Lazar et al. [13] or Creswell [3] for a more detailed introduction to qualitative methods and details on the coding process. In addition, Isenberg et al. [8] provide a visualization-specific perspective and Shneiderman and Plaisant [21] describe a specific approach for longitudinal evaluation of deployed visualization tools.

Despite their benefits, qualitative approaches do not come without costs. Because they are more susceptible to researcher bias than quantitative experiments, it is important to follow a structured and established method. Collecting the data can be time consuming, especially in a field study, because you cannot predict when interesting events will happen. Therefore you may have to observe many hours of repeated or irrelevant activities in order to capture the interesting and relevant ones. Analyzing the data is even more time consuming and it is not obvious at the outset what findings will be interesting and novel. For example it took me several months of analysis before the data from the building design field study began to make any sense, and many of the early "findings" turned out to have been already reported by other researchers in different domains. Therefore, it took several iterations to identify findings that were novel and interesting to the research community, and to structure those findings in a cohesive way. New researchers often feel lost in the analysis process; there is typically a stage where it feels as though you are drowning in data with no possible way of making sense of it all. If you find yourself in this state, do not despair! Persistence pays off.

4.3 Inspection

In HCI, usability inspection is generally thought of as a "discount" method of usability evaluation since asking a small number of experts (typically around five) to review a prototype typically takes less time than running a study with end users. Inspection techniques have been used successfully in visualization and are typically able to identify some strengths, problems, and limitations for a specific design. However, the quality of the results depends on several factors. Zuk et al. [28] reported that the findings of heuristic evaluation (one type of usability inspection, where the interface is evaluated with respect to several pre-defined heuristics) depended highly on the heuristics used and the types of evaluators that were chosen. In particular, they suggest using visualization-specific heuristics rather than just usability heuristics, and including experts from usability, visualization, and the application domain. We came to similar conclusions in our own previous research [23].

In general I have found that inspection techniques provide a quick way to identify likely problems with a visualization design. Conducting an inspection typically identifies many issues that were already known (e.g. missing functionality that was left out of a research prototype in the interest of time), but also nearly always reveals some problems that were not anticipated. However, the problems that are identified tend to be at a somewhat superficial and generic level. Unless some domain experts are included, and given realistic tasks for their domain, it is very hard to predict how well a system will perform in real use. Therefore this approach seems most useful for formative (early) evaluation of prototypes, where the emphasis is on finding problems and improving the design, rather than summative (late) evaluation, where the emphasis is on validation.

4.4 Usability Studies

A usability study is a study with end users that has a primary emphasis on identifying problems to directly improve the design of a next version, or to validate that users can accomplish certain tasks with a design. A usability study may use many of the same data collection approaches as quantitative experiments or qualitative studies, but is typically less formal and rigorous. For example, we took a usability study approach to evaluate the construction photo browser [27] described earlier because our primary objectives were to identify design problems and verify whether users could accomplish designated tasks with the interface.

Usability studies are generally quicker to set up and analyze as compared to more rigorous studies, yet provide better evidence of what will happen in real world use as compared to inspection. In visualization research, it is very difficult for results of a usability study to be a research contribution on their own due to the lack of rigor. For example, it may be hard to provide definitive evidence of which features are essential for each task. However, usability studies can provide insight to improve design, and evidence to validate design ideas. In non-research settings, iterative prototyping and evaluation via usability testing may be the most effective method to ensure a successful final design.

5   Lessons Learned

Over the years, I have learned numerous lessons about conducting user studies:

- **Know your goal, one goal per study**. Too many studies start with a very vague goal (e.g., "I need to evaluate my technique") or too many goals. Clearly defining the goal (and research question) at the start of user study planning helps to narrow down the research methods and metrics that will be useful. Similarly, trying to answer too many questions at once can make a study unwieldy. Often it is worth considering running a set of related studies, each one answering just one or two specific questions.

- **Choose tasks carefully**. One of the biggest challenges with study design in visualization is the choice of task. Most importantly, the task needs to be ecologically valid (i.e. relevant to real world use of the visualization). Particularly in controlled studies, it can be easy to get caught up in finding a task that is easily measurable, losing track of whether it represents a realistic scenario of use.

- **Pilot the instructions**. Particularly for open-ended tasks, the wording of the instructions can have a very strong impact on participant behaviour. For example, in a tabletop collaboration study [22], we observed that small wording changes or ordering of instructions impacted how much responsibility each participant took for their part of the work, how they prioritized their ultimate objectives, and how they chose to work together versus separately. Some bias of people's work practice is inevitable, but testing the instructions through pilot studies can help to minimize this effect.

- **Abstract away from the tool in quantitative experiments**. Quantitative experiments can compare either visualization or interface techniques, or complete visualization tools. Generally, those that compare complete tools have fewer generalizable outcomes, since it is impossible to isolate which features or design attributes made one tool better. It is often better to implement one system and then add and remove features of interest to understand their effects in isolation. Note, however, that *qualitative* evaluations of complete tools can provide some of this understanding.

- **User studies are not only for evaluation**. As illustrated by the examples in this chapter, user studies are not just for the purpose of evaluating visualizations. Understanding characteristics of users and the context surrounding the use of visualizations is just as important to achieving a successful design.

- **Use a variety of methods**. Qualitative and quantitative methods each have their place. For example, field studies are very realistic, but lack in precision, whereas lab experiments are very precise (in terms of measurements), but limited to specific (and often simplistic) tasks in a controlled environment and thus not realistic [15]. In my early work, I focused strictly on quantitative approaches, putting great value on the results being scientifically conclusive. However, I soon realized that my understanding was incomplete, because each study revealed only a small part of the picture. Qualitative methods can help to fill in the gaps and answer research questions with a more exploratory nature. Therefore, a good solution is to use either mixed-method designs or conduct multiple studies using complementary approaches.

- **Visualization is not that different**. Research methods for interacting with people have been around for a long time. Fields such as HCI, Psychology, and Social Sciences have documented these methods well, and many of the methods are very similar across disciplines. Rather than reinventing the wheel, we should aim to understand existing research approaches, adopt them for our purposes, and then tweak them where necessary. Visualization does have some specific challenges, especially the need to access expert users and examine how tools support very complex analysis tasks. However, these are not totally unique to visualization, and can be addressed by established methods such as longitudinal field studies.

6 Where to Go from Here?

Over the last decade, user studies have become the de facto standard for understanding users and evaluating the effectiveness of visualization tools and techniques. As a community, we have gradually adopted a greater variety of user study methods from HCI and other fields. By and large, these methods have been very effective and have not required substantial modifications to work for visualization. But the variety of adopted methods is not very large; for example, the vast majority of visualization user studies are still quantitative experiments. Use of qualitative and other approaches has only just begun. As a community, we should continue to explore a variety of research methods from other disciplines and bring them into our repertoire. One example of a method that might provide deeper understanding is ethnography. Among the few qualitative studies done in visualization, few could be considered a true ethnography, where the researcher is not only an observer of human activities but also an active participant who is immersed in the culture. Closest is probably Sedlmair et al.'s [19] research with automotive engineers, where the researchers worked closely with the target users over a three-year period. More work of this nature should be encouraged. More generally, the visualization community will benefit from a greater number of pre-design studies and long-term deployment studies that can provide a rich and detailed understanding of how visualizations can fit into work practices. In addition, online studies (e.g. through Mechanical Turk [1]) offer the potential to collect certain types of data more quickly, less expensively, and with more participants than laboratory studies. These should be considered as an alternative or complementary approach to laboratory studies with straightforward, controlled tasks.

References

1. Amazon Mechanical Turk [online]. Available: https://www.mturk.com/mturk/welcome [Accessed: June 15, 2012].
2. C. Ardito, P. Buono, M.F. Costabile, and R. Lanzilotti. Systematic Inspection of Information Visualization Systems. In *Proceedings of the 2006 AVI Workshop on Beyond Time and Errors: Novel Evaluation Methods for Information Visualization*, pp. 1–4, 2006.
3. J.W. Creswell. *Educational research: Planning, conducting, and evaluating quantitative and qualitative research*. Pearson/Merrill Prentice Hall, 2007.

4. S. Easterbrook, J. Singer, M.A. Storey, and D. Damian. Selecting Empirical Methods for Software Engineering Research. In Shull et al., eds., *Guide to Advanced Empirical Software Engineering*, Springer-Verlag London Limited, 2008, Section III, pp. 285–311.

5. L. Grammel, M. Tory, and M.A. Storey. How information visualization novices construct visualizations. In *IEEE Transactions on Visualization and Computer Graphics*, volume 16, no. 6, pp. 943–952, Nov./Dec. 2010.

6. C.G. Healey and J.T. Enns. Large Datasets at a Glance: Combining Textures and Colors in Scientific Visualization. In *IEEE Transactions on Visualization and Computer Graphics*, volume 5, no. 2, pp. 145–167, April-June 1999.

7. D. Huang, M. Tory, S. Staub-French, and R. Pottinger. Visualization Techniques for Schedule Comparison. In *Computer Graphics Forum*, volume 28, no. 3, pp. 951–958, June 2009.

8. P. Isenberg, T. Zuk, C. Collins, and S. Carpendale. Grounded Evaluation of Information Visualizations. In *BELIV 2008: BEyond time and errors: novel evaLuation methods for Information Visualization*, 2008.

9. Y. Kang and J. Stasko. Characterizing the Intelligence Analysis Process: Informing Visual Analytics Design through a Longitudinal Field Study. In *Proceedings of IEEE Visual Analytics Science and Technology*, pp. 21–30, October 2011.

10. A. Kobsa. User Experiments with Tree Visualization Systems. In *IEEE Symposium on Information Visualization*, pp. 9–16, 2004.

11. A. Komlodi, A. Sears, and E. Stanziola. Information Visualization Evaluation Review, *ISRC Tech. Report, Dept. of Information Systems, UMBC-ISRC-2004-1*, 2004. Reported in [18].

12. H. Lam, E. Bertini, P. Isenberg, C. Plaisant, and S. Carpendale. Empirical Studies in Information Visualization: Seven Scenarios. In *IEEE Transactions on Visualization and Computer Graphics*, 30 Nov. 2011, IEEE Computer Society Digital Library, Volume: 18, no. 9 pp. 1520–1536, Sept 2012.

13. J. Lazar, J.H. Feng, and H. Hochheiser. *Research Methods in Human-Computer Interaction*. United Kingdom: John Wiley & Sons Ltd., 2010. ISBN-10: 0470723378, ISBN-13: 978-0470723371.

14. Q. Li and C. North. Empirical Comparison of Dynamic Query Sliders and Brushing Histograms. In *IEEE Symposium on Information Visualization*, pp. 147–153, 2003.

15. J.E. McGrath. Methodology Matters: Doing Research in the behavioral and social sciences. in *Readings in Human-Computer Interaction: Toward the Year 2000*, R.M. Baecker, J. Grudin, and W.A.S. Buxton, eds. pp. 152–169, 1995.

16. T. Munzner. A Nested Model for Visualization Design and Validation. In *IEEE Transactions on Visualization and Computer Graphics*, volume 15, no. 6, pp. 921–928, November 2009.

17. J. Nielsen and R.L. Mack, eds. *Usability Inspection Methods*. New York: John Wiley & Sons, 1994, ISBN 0-471-01877-5.

18. C. Plaisant. The Challenge of Information Visualization Evaluation. In *Proceedings of the working conference on Advanced Visual Interfaces*, pp. 109–116, 2004.

19. M. Sedlmair, P. Isenberg, D. Baur, M. Mauerer, C. Pigorsch, and A. Butz. Cardiogram: Visual Analytics for Automotive Engineers. In *ACM Conference on Human Factors in Computing Systems*, pp. 1727–1736, May 2011.

20. J. Seo and B. Shneiderman. Knowledge discovery in high dimensional data: Case studies and a user survey for the rank-by-feature framework. In *IEEE Transactions on Visualization and Computer Graphics*, volume 12, no. 3, pp. 311–322, May/June 2006.

21. B. Shneiderman and C. Plaisant. Strategies for Evaluating Information Visualization Tools: Multi-dimensional In-depth Long-term Case Studies. In *BELIV 2006: BEyond time and errors: novel evaLuation methods for Information Visualization*, 2006.

22. A. Tang, M. Tory, B. Po, P. Neumann, and S. Carpendale. Collaborative Coupling over Tabletop Displays. *In ACM CHI 2006*, pp. 1181–1190, Apr. 2006.

23. M. Tory and T. Möller. Evaluating Visualizations: Do Expert Reviews Work. In *IEEE Computer Graphics and Applications*, volume 25, no. 5, pp. 8–11, Sept./Oct. 2005.

24. M. Tory, D.W. Sprague, F. Wu, W.Y. So, and T. Munzner. Spatialization Design: Comparing Points and Landscapes. In *IEEE Transactions on Visualization and Computer Graphics*, volume 13, no. 6, pp. 1262–1269, Nov./Dec. 2007.
25. M. Tory, C. Swindells, and R. Dreezer. Comparing Dot and Landscape Spatializations for Visual Memory Differences. In *IEEE Transactions on Visualization and Computer Graphics*, volume 16, no. 6, pp. 1033–1040, Nov. / Dec. 2009.
26. M. Tory, S. Staub-French, B. Po, and F. Wu. Physical and digital artifact-mediated coordination in building design. In *Journal of Computer Supported Cooperative Work*, volume 17, no. 4, pp. 311–351, Aug. 2008.
27. F. Wu and M. Tory, "PhotoScope: Visualizing Spatiotemporal Coverage of Photos for Construction Management," *ACM Conference on Human Factors in Computing Systems*, pp. 1103–1112, 2009.
28. T. Zuk, L. Schlesier, P. Neumann, M.S. Hancock, and S. Carpendale. Heuristics for Information Visualization Evaluation. In *BELIV 2006: BEyond time and errors: novel evaLuation methods for Information Visualization*, 2006.

Part V
Perception and Cognition

On the Benefits and Drawbacks of Radial Diagrams

Michael Burch and Daniel Weiskopf

Abstract Visualizations based on circular shapes are oftentimes referred to as radial representations in the literature. Radial diagrams have recently gained increasing popularity in many application domains. One reason for this popularity is the aesthetic look of those diagrams. However, there are advantages for corresponding Cartesian (rectangular) representations as well, including possible perceptual benefits, more efficient use of rectangular screen space, or ease of implementation due to simpler graphical primitives. We describe several examples of visualizations that exist in Cartesian or radial form, and we discuss their respective benefits and drawbacks. We focus on the evaluation of those diagrams by comparative user studies. In particular, we consider how quantitative and qualitative evaluations, including eye tracking, can be employed to analyze which of the two visual mapping approaches is more suitable for given datasets and tasks.

1 Introduction

When looking at infographics, diagrams, plots, charts, or visual depictions of data in general, we can observe two important types of shapes for visual data mappings: Cartesian and radial. The *Cartesian* representation is oriented along a Cartesian coordinate system, i.e. rectangular geometry. Alternatively, data can be visually encoded in circular regions on screen, leading to *radial* depiction of data.

M. Burch (✉)
Postdoctoral Researcher of Computer Science, Visualization Research Center (VISUS),
University of Stuttgart, Stuttgart, Germany
e-mail: michael.burch@visus.uni-stuttgart.de

D. Weiskopf
Professor of Computer Science, Visualization Research Center (VISUS), University of Stuttgart,
Stuttgart, Germany
e-mail: daniel.weiskopf@visus.uni-stuttgart.de

W. Huang (ed.), *Handbook of Human Centric Visualization*, 429
DOI 10.1007/978-1-4614-7485-2_17, © Springer Science+Business Media New York 2014

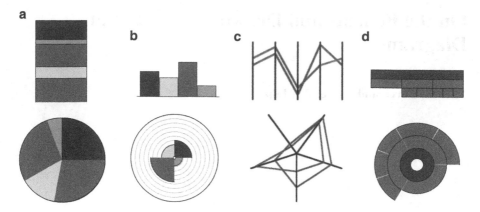

Fig. 1 Cartesian (*top row*) vs. radial (*bottom row*) diagrams: (**a**) Quantitative data visualization by Cartesian stacked bar chart and radial pie chart. (**b**) Quantitative data visualization by Cartesian bar chart and radial rose diagram. (**c**) Multivariate data visualization by parallel coordinates plot and radial star plot. (**d**) Hierarchical data visualization by Cartesian and radial layered icicle plots

Adopting the transformation from Cartesian to polar coordinates, almost any Cartesian diagram can also be represented in a radial way, see for example Fig. 1. However, from the perspective of visualization, the transformation from a non-radial to a radial representation only makes sense if it supports better readability of the diagram or visual analysis of the data. If aesthetics is the only reason for the radial diagram, then this choice of visualization is questionable. Figure 1a, b show simple, yet most common examples of Cartesian diagrams and their radial counterparts: two variants of bar charts, a pie chart, and a rose chart. For these visualization examples, it is known that bar charts outperform pie charts for most tasks related to analysis of quantitative data [13, 14]. Figure 1c, d show examples of more complex plots handling more complex data structures. Parallel coordinates plots that display multivariate data can be mapped to a circular shape, known as Kiviat diagrams, spider web diagrams, or star plots, see Fig. 1c. Also, hierarchical data representations can be displayed in both styles, see Fig. 1d for space-filling layered icicle representations. Another kind of hierarchy visualization is illustrated in Fig. 2: node-link diagrams in traditional top-down representation and with a radial organization.

Figures 1 and 2 show only a small selection of available Cartesian and radial diagrams. With the large number of visualization techniques present in the literature, the question arises which diagram type is most suited for a specific data visualization problem. We restrict ourselves to assessing the usefulness of radial vs. non-radial representations. We illustrate various general benefits and drawbacks of radial diagrams compared to their Cartesian counterparts, also in the context of different application domains and related visualization tasks. We discuss different kinds of aspects that play a role in this assessment: perceptual characteristics related to recognizing length, area, angle, or other quantities in diagrams, space efficiency of the visual representation, aesthetic appearance, computational costs associated with the visualization technique, and ease of implementation.

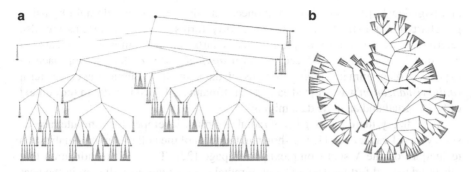

Fig. 2 Node-link tree diagrams where the large filled circle represents the root vertex of the hierarchy: (**a**) Traditional top-down representation. (**b**) Radial depiction of the same hierarchy

Despite those general assessment criteria and guidelines, the ultimate test of usefulness has to involve the human "consumer" of the diagram. Therefore, comparative user studies are critical for definitive answers to the problem of assessing those diagrams. We highlight a few representative comparative user studies. Some of those studies include eye tracking systems for data acquisition. The advantage of eye tracking is that gaze trajectories are recorded in addition to the standard measurements of completion time and accuracy for task performance. In this way, eye tracking data researchers expect to better understand the visual task solution strategies applied by the participants, which is a great benefit of eye tracking studies. However, analysis of eye movement data is challenging because of its high complexity and inherently spatio-temporal nature. Apart from the complicated analysis, the design and conduction of eye tracking studies already require much more time and knowledge than traditional controlled studies that only consider task performance.

The objective of this chapter is to illustrate the general benefits and drawbacks of radial diagrams. Furthermore, we discuss the approaches to user-study-based evaluation of this type of diagrams. In particular, we describe the difficulties that occur when investigating the strengths and weaknesses of radial visualizations by applying eye tracking techniques. This chapter reports on respective results from our research, complemented by a short discussion of related previous work from the literature. We focus on the use of visualization for interactive data analysis, following Card [9]: "Information visualizations should do for the mind what automobiles do for the feet".

2   Related Work

Radial depictions of data were used long before the invention of computers and have their origins in statistical graphics of the nineteenth-century [21]. Several examples of this old kind of diagrams can be found in the work of Playfair [28]

and Nightingale [27], who became pioneers in this visualization domain by using pie charts for statistical graphics. These early forms of radial diagrams are also referred to as polar area diagrams, rose charts, or coxcombs in the case of Nightingale. She also introduced the term *applied statistics* for the approach of exploiting diagrammatic representations of data. In the computer age, the term *radial visualization* was adopted early by Hoffmann et al. [18] in 1997 in the context of DNA visual and analytic data mining.

Accordingly, there is a long list of radial diagrams developed for computer use, as surveyed by Draper et al. [12] or shown in the form of the collection of visualization techniques on the Visual Complexity web page [22]. There is no indication that this trend toward further development of radial representations will stop in the near future. The most common radial graphic in use—although a very simple one— may be the traditional pie chart that is almost ubiquitous in our everyday life when reading newspapers or magazines. Several variables of quantitative nature are graphically displayed to provide the reader—typically a layman from a visualization perspective—with a fast overview of the ratio of several quantitative values that typically sum up to a 100%. As a drawback, even in this simple form, the radial depiction of data may be questionable due to issues related to visual perception: judging areas and angles is less accurate and slower than perceiving spatial lengths and positions; therefore, a bar chart would often do a better job than a pie chart [13, 14].

Cleveland and McGill [10] conducted a user experiment to explore the accuracy of six basic judgments of graphical perception. Among those six judgments, they investigated position along a common scale, position along identical but non-aligned scales, length, angle, slope, and area judgments. Translating that into the scenario of judging quantitative values from pie charts and bar charts leads to the hypothesis that radial representations (pie charts) are less effective for this task than Cartesian representations (bar charts). This hypothesis can be formulated because Cleveland and McGill found in their experiment that position and length judgments (required for bar charts) are more accurate than angle, slope, and area judgments (required for pie charts). These fundamental results concerning basic visualization techniques have already been extended to some degree to more complex types of datasets and also more complex visualization techniques that make use of either radial or non-radial visualization approaches.

An experiment investigating the user performance for a set of simple types of diagrams, supported by eye tracking, was conducted by Goldberg and Helfman [15]. In this study, linear and radial stimuli for bar, line, area, and scatter graphs were presented to the participants. An analysis concerning the gaze trajectory data uncovered a three-stage processing model, i.e. participants identified the desired data dimension, found its data point, and mapped the data point to its value. In this study, the mapping of a data point to its value was found to be slower in radial than in non-radial diagrams.

To the best of our knowledge, there are no comparative studies investigating the differences in user performance for complex radial and non-radial visualizations. In particular, the evaluation of the benefits and drawbacks of interactive features in both radial and Cartesian diagrams is challenging, leaving room for interesting future research. The recent survey by Draper et al. [12] provides a good overview of the various existing and most prominent radial visualizations, categorized into seven high-level design patterns. However, their survey does not cover qualitative or quantitative evaluation of the effectiveness of such diagrams.

Despite the results reported in the aforementioned papers, there is no global and general finding that radial visualizations outperform their Cartesian counterparts or vice versa. However, there is some evidence that for many typical tasks, the non-radial version performs better with respect to accuracy and completion time. There is also some evidence that radial visualizations look more aesthetically appealing than non-radial ones, which consequently have the benefit that they better attract the viewers' attention. One might argue that the aesthetics is one, maybe the most prominent, reason why radial diagrams are widely used despite their tendency toward lower user performance compared to their Cartesian counterparts.

In this chapter, we want to discuss the benefits and drawbacks of radial diagrams in more detail. In particular, we report that non-radial diagrams perform better than the radial depiction for certain tasks, but there are also some tasks where radial techniques have their strengths. To this end, we present several user study examples investigating typical data-analysis tasks that have to be answered in either Cartesian or radial representations. In addition, we discuss criteria that can be employed to assess the quality of visualizations. Such criteria include the amount of visual clutter (in the form of excess of items or their disorganization) [30], the avoidance of chart junk, the design of displays with good lie factor [31], and space efficiency of the representation on screen.

3 Cartesian and Radial Visualization Techniques

Many existing complex interactive tools are built on visual mappings that exploit radial diagrams, sometimes also only parts of it providing different views of the data connected by linking and brushing features. In this section, we demonstrate several example visualizations that exist in both Cartesian and radial styles; some of them are simple ones, whereas others are more complex. We also illustrate benefits and drawbacks by means of these examples.

3.1 Diagrams in Both Worlds

As already discussed in the introductory section, diagrams for simple kinds of data exist in either Cartesian or radial style. Most common examples are bar and pie charts. The drawbacks of pie charts have already been discussed. Therefore, one may be tempted to conclude that radial depictions of data are not useful in general. While this finding may hold for many examples of diagrams and tasks,

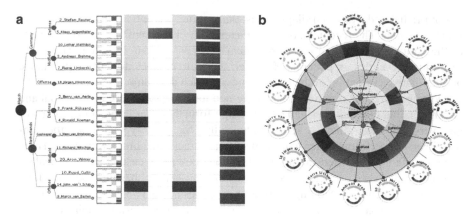

Fig. 3 Visualization of transaction sequences by using stacked graphical primitives [5]: (**a**) Cartesian Timeline Trees. (**b**) Radial TimeRadarTrees

the question arises if more complex radial visualizations only benefit from their aesthetic appearance or if they also benefit from other aspects. Indeed, there are radial visualizations that improve readability and accelerate task performance just by making things radial. In this section, we describe visualization techniques that exist in both Cartesian and radial form, along with a discussion of benefits and drawbacks.

For instance, Burch et al. developed visualization techniques for representing sequences of transactions (*Timeline Trees* [3]) and time-varying weighted directed compound graphs (*TimeRadarTrees* [6]), i.e. adjacency relations among leaf nodes of an information hierarchy. Both visualization methods are able to display the same dataset in either Cartesian or radial style, i.e. dynamic weighted transactions in this case. The techniques avoid visual clutter by exploiting stacked and aligned space-filling graphical primitives instead of sequences of node-link diagrams.

Figure 3a illustrates the Cartesian Timeline Trees approach for a small sequence of five transactions acquired from the common ball contacts of players during a move in a soccer match. Figure 3b shows the same dataset in the radial representation of the TimeRadarTrees. For the Cartesian Timeline Trees diagram, time starts at the left hand side following the prevalent reading direction in Western countries, whereas the radial TimeRadarTrees diagram uses a timeline that starts in the center of the circle and grows radially outward. This concept allows newer time steps to be placed closer to the circle circumference, which gives them more display space than older time steps and hence, is also an intuitive visualization concept for radial diagrams showing time-series data.

Both types of diagrams show the temporal evolution of the data in a static image, not by using animation. The great benefit of a static diagram for time-series data is the support for direct analysis of trends, countertrends, and anomalies: the simultaneously visible subsequence of time-varying elements leads to faster and more accurate comparisons than for animated diagrams since high cognitive load is associated with interpreting animation, see Tversky et al. [32]. Furthermore, the mental map [26] is preserved when inspecting several time steps of time-varying data in a static view, which also leads to a reduction of cognitive effort. Another benefit of a static diagram is that additional information such as a hierarchical organization in the case of Fig. 3 can be added easily and interactive features are more easily applied than in an animated visualization.

A user study on perceptual differences of Cartesian and radial variants of these visualizations [5] uncovered some benefits of the radial diagram variant. This eye tracking study showed that thumbnails are used to draw conclusions from the given transaction dataset and to solve the tasks in the study accurately and fast. An inspection of the gaze trajectories and heatmap representations revealed that the participants used the small thumbnail representations in the radial TimeRadarTrees more frequently than in the Cartesian Timeline Trees. The reason for this benefit of radial visualization is that different orientations, slopes, angles, and graphical primitives are generally used compared to the Cartesian visualization, where each rectangular object has the same size, orientation, slope, and angle. Here, only color coding and position can be used as a visual feature to draw conclusions from the diagram. A drawback of the radial representation is the smaller size of the graphical primitives (i.e. circle sectors) in the circle center. However, if the user wants to focus on newer time steps, this problem is no longer present because more display space is available close to the circle circumference.

Figure 4a shows a dynamic graph visualization by *parallel edge splatting* [8]. Figure 4b illustrates the radial variant of this visualization concept [4], showing the same dynamic graph dataset of the evolving call relations of an open source software project. In both types of visualization, sequences of node-link diagrams are shown side-by-side in a static display, allowing good comparisons between subsequent graphs and better support for mental map preservation [26] than it would be the case for an animated diagram. To achieve these goals each graph is mapped to a stripe either vertically or circularly. The graph edges are drawn as either straight links from left to right or curved links from the inside to the outside of the circle. Edge splatting is used to compute the edge coverage information and a density field which makes the cluttered link structure visible again.

In this visual mapping of time-varying graph data, the benefits of the radial variant over the Cartesian diagram can readily be observed. The diagram in Fig. 4a contains several long links crossing each of the vertical stripes, which cause a higher probability of visual clutter. By using the radial variant of this visualization technique for the same dataset, the long links disappear; see Fig. 4b. The reason for this crucial benefit for line graphics is that the two endpoints of the Cartesian diagrams are now connected in the radial diagrams, which consequently allows us to draw a link in one of two possible directions—clockwise and counter-clockwise. Visual clutter in these radial diagrams is reduced by choosing the direction that produces the shorter link. In the Cartesian diagram, one could also apply this concept of choosing the shortest link by leaving the display either on top or at the bottom and entering it again at the opposite side. This visualization strategy would also lead to shorter links similar to the radial diagram, but the

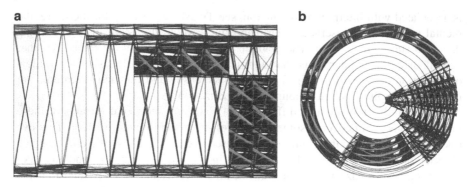

Fig. 4 Dynamic graph visualization by edge splatting: (**a**) Cartesian parallel edge splatting. (**b**) Radial edge splatting for the same dataset

human eye cannot easily follow such links that are broken by a gap and that start again somewhere else on the display. Gestalt principles [20] come into play here.

There is another interesting advantage of such a radial diagram. Since there are no endpoints on the circle circumference, vertices can be shifted in any direction and to any extent. If the order of the vertices is retained, the visual patterns will stay the same apart from the fact that they are moved around the circle by some angle. This shifting by an angle is not problematic for the human visual system. The Gestalt principle [20] of invariance says that geometric objects are still recognized independent of rotation, translation, and scale. This feature is not possible for the Cartesian diagram: shifting vertices in the same way would lead to totally differently looking visual patterns in many cases. Therefore, the radial visualization is invariant under the shifting operation, but the Cartesian is not.

Apart from these benefits, we can also identify drawbacks of such a radial depiction of dynamic graph data. For example, the circle annuli are much smaller than the vertical stripes of the Cartesian diagram, which leads to a more aggregated representation of this pixel-based heatmap-like diagram. Furthermore, curved links have to be used to avoid overlaps with different annuli but it is known that curved links are more difficult to follow by the human eye than traditional straight links.

There are many more complex visualization examples that exist in both worlds—the Cartesian and the radial one. This section showed only some benefits of radial visualizations apart from the aesthetically nice looking pictures.

3.2 Using Radial Diagrams Recursively

The concept of radial diagrams is oftentimes used in a recursive manner. This is possible if there are self-similar structures in a dataset such as in hierarchical data. The *bubble* or *balloon tree* diagrams [16, 23, 25] make use of this concept. The hierarchy elements on the first level are mapped to a circle circumference, whereas

Fig. 5 An example of a
bubble or balloon tree layout
of a hierarchy containing 31
vertices at a maximum
depth of 3

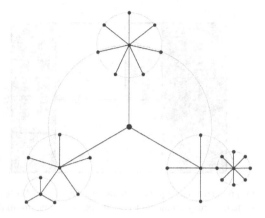

elements of each sub-hierarchy are visually encoded in smaller circles with the root node being placed on the circumference of the circle used to display the elements of the superordinate hierarchy level, see Fig. 5.

Applying this layout concept to node-link tree diagrams seems to be most natural. If trees in nature are inspected more closely and if we make a cut through the tree crown, such a bubble tree layout becomes visible.

Although the bubble layout seems to be intuitive and produces aesthetically pleasing tree diagrams, it has some serious drawbacks from a visualization perspective. Since hierarchies may grow very deep, such a tree layout soon reaches circles with very small radii for visually encoding deeper levels of a hierarchy and hence, such tree diagrams lack visual scalability. Furthermore, the applied layout algorithm is more complicated to implement than an algorithm for the Cartesian counterpart because such diagrams should be free of overlaps.

Similarly, *treemap* representations as developed in their original form by Johnson and Shneiderman [19] can be transformed to *circular treemaps* [36]. However, the radial version of the diagram has the drawback of many empty spaces in a display since circles cannot be packed without wasting space, see Fig. 6.

Figure 6a demonstrates a color-coded treemap representation that uses rectangular shapes only and is generated by applying the slice-and-dice technique. The radial counterpart for the same dataset in the same color coding is illustrated in Fig. 6b. Each of the nested circles contains many empty regions, wasting a lot of display space. Furthermore, text labels are very difficult to attach to the graphical elements.

On the positive side, a circular treemap makes the hierarchical structure much clearer to the viewer by allowing empty regions between single bubbles. These empty regions refer to the inner hierarchy vertices that must be displayed as nested rectangles in Cartesian treemaps. Using rectangular shapes only, i.e. horizontal and vertical border lines, makes Treemaps harder to interpret for the viewer.

In general, recursive algorithms are a powerful concept to process data by applying the same subroutines to smaller and smaller instances of the given problem until it becomes that simple causing a termination of the algorithm.

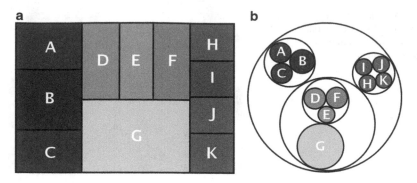

Fig. 6 Treemap representations: (**a**) Traditional Cartesian treemap. (**b**) Circular treemap of the same hierarchy containing 11 leaf nodes at a maximum depth of 3

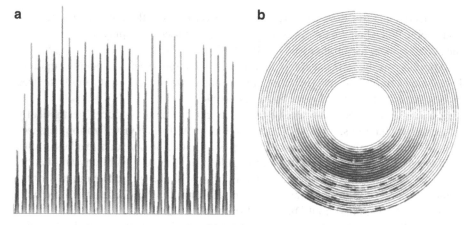

Fig. 7 Visualization of periodic data: (**a**) Traditional Cartesian timeline. (**b**) Periodic timeline resulting in a spiral diagram (©[2012] IEEE. Reprinted, with permission, from [35])

3.3 Spiral

Spiral representations are also some kind of radial representations of data. If there is a periodic behavior in the data such as seasonal trends, each season is typically mapped to one complete circle. To avoid jumps and discontinuous changes, the end of one circle representing a season and the starting point of the new season are connected, leading to a spiral representation.

The Archimedian or arithmetic spiral shown in Fig. 7b is the locus of points corresponding to the locations over time of a point that moves away from a fixed circle center with a constant speed along a line that rotates with constant angular velocity.

Spiral representations have the advantage that a periodic behavior in the data is visually represented by aligning the same time steps in each period to the same angle. This feature allows easy comparisons of the periodic behavior and explorations of seasonal trends over time. An example of such a dataset is the consumption of energy over time of a specific geographic region, which is typically low in the night and increases in the morning and in the evening because people switch lights and electronic devices on more frequently. These daily patterns can be observed by a spiral representation, but also seasonal effects can be inspected over 1 year due to different times for sunrises and sunsets in summer and winter seasons. Figure 7 illustrates the sunshine intensity, where the spiral diagram makes the exploration of daily periodic patterns much easier than the Cartesian diagram.

A major problem for the spiral diagram is the computation of a suitable periodicity in the data if it is not known beforehand, which requires a sophisticated algorithm. In the example of sunshine intensity, the periodicity is known beforehand and comes from the periodic time patterns. In some scenarios, the period in the data is even changing over time, which requires spiral representations different from the Archimedean or arithmetic spiral.

For spirals, the focus is on newer time steps that are located close to the circle circumference: graphical elements are larger close to the circumference than those close to the circle center.

3.4 Application Domains for Radial Diagrams

Radial diagrams have been introduced in various application domains. Software engineering, social networking, statistics, bioinformatics, or flow visualization are just a few examples of a rich source of application scenarios. In many cases, a developer of a radial technique may not be aware of the fact that the Cartesian counterpart would do a better job, although it might not look as aesthetically pleasing to a viewer as the radial counterpart.

Radial visualizations of abstract data such as graphs only make sense if these have a benefit for human perception compared to their Cartesian counterparts. In the visualization scenarios illustrated in Sect. 3.1, the advantage of the radial representations are for example the reduction of link lengths and the reduction of visual clutter. Furthermore, by the radialization process the visual patterns are invariant under the shifting operation.

A similar argument can be used to justify the usage of a radial diagram for the *Stargate visualization* of developer activity and code ownership in the domain of software engineering [24]. The hierarchical organization of a software project is abstract and there are no explicit end points in the data. Consequently, a radial layered icicle plot is exploited to show the hierarchy. The developer activity is displayed as trajectories along the locations of the arithmetic means of the locations of all files used in one transaction.

Fig. 8 Flow Radar Glyph:
(1) An oscillation phase in
downward direction occurs in
the beginning, (2) constant
behavior in downward
motion, (3) two cycles,
(4) direction of flow leads
downward with oscillations
of varying intensity

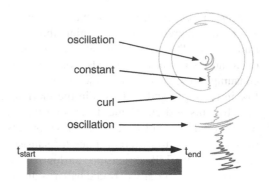

In the domain of 2D flow visualization, a radial approach has recently been used by Hlawatsch et al. [17]. The reason for relying on a radial depiction of flow data is the inherent nature of the data: a 2D vector is decomposed into a direction component and magnitude. Direction is then mapped to the angle in the radial representation, whereas radii show time. In this way, the flow at a given point is drawn as a curve that points radially outward. The time-varying flow can now easily be shown in a static radial diagram known as *Flow Radar Glyph*, see Fig. 8.

4 User Studies

We have already explained and discussed benefits and drawbacks of radial diagrams and illustrated the findings by means of example figures. To fully understand the relative performance of visualization techniques, comparative user studies have been designed and conducted. We cover both uncontrolled crowd-sourcing experiments, which have recently gained popularity, and controlled lab studies, which are a traditional approach to user-oriented testing of visualization.

Crowd-sourcing studies are typically conducted online, study participants are recruited via the World Wide Web by e-mail newsletters, website announcements, forum contributions, or providing a hyperlink to the study Web page via another frequently visited Web page. The main advantage of such a study is the large number of participants whose performance can be measured in a very short time frame. As downside, such experiments should be conducted in a short time (typically up to 3 min) so that participants are not discouraged. Therefore, only simple experiments can be conducted. If personal information is recorded, this can further discourage participants. Finally, the intrinsic problem of an uncontrolled study is that input cannot be guaranteed to be correct and the study conditions may vary without being considered in the evaluation of the results.

In contrast, a controlled laboratory user study has the advantage that an operator is present to control the situation and to guarantee reliable and correct measurements. Controlled experiments can last longer than crowd-sourcing experiments, allowing for the evaluation of more complex visualization methods; the typical length of a study session is around 1 h per participant. Another advantage is that more data can be collected from the participants. A good example is recording of eye gaze tracks by an eye tracking device. The main disadvantage of laboratory experiments is the labor and cost intensive execution, especially if a large group of participants is required for statistically significant results: suitable room and equipment for the study has to be set up, an operator has to be present, and participants are typically paid for their efforts.

In Sects. 4.2 and 4.3, we report on the results of different user studies that compare radial and non-radial visualization. The studies from Sect. 4.2 were conducted to investigate the memorization of positions of visual elements in Cartesian or radial coordinate systems. The memorization of visual elements is critical for exploring datasets by means of visualizations. This study is split into an uncontrolled crowd-sourcing study and a controlled study with far fewer participants. Section 4.3 reports on a study that investigates the readability of different types of node-link tree diagrams for hierarchies. This study was conducted by also exploiting an eye tracking system. The recording of participant's eye movements allowed us to identify different visual task solution strategies for radial and non-radial node-link tree diagrams.

4.1 Eye Tracking

Eye tracking refers to the process of recording the gaze positions over time and the sequential order in which specific locations or regions of interest are fixated and for how long. Such eye movement sequences are typically called gaze trajectories or scan paths. Eye tracking has emerged as a great tool to support visualization researchers in understanding the behavior of study participants. A high degree of control leads to reliable recorded data. Eye tracking data is typically complemented by additional data that is collected during the study. Prominent examples of such additional data are the traditional task performance indicators: accuracies and completion times. However, the large increase of the amount and complexity of recorded data makes the analysis or statistical evaluation of eye tracking experiments challenging.

Although eye tracking is the method of choice for exploring people's exploration behavior when using a visualization technique or interactive system, it comes with serious drawbacks that have to be considered before conducting such an experiment. Eye tracking systems are expensive devices and the design and conduction of a useful and reliable eye tracking study is complicated and time-consuming. If participants wear glasses or contact lenses, the eye tracking device may not work correctly or calibration is not possible. If the calibration of the eye tracking system is possible, it may take a lot of time, which may discourage the subject from participating in the study. The actual study has to be designed to avoid spurious eye gazes; for example, if the solution of a given task has to be confirmed by mouse click, there is an impact on the recorded scan paths because the participants have to find the mouse cursor on screen first before they can move it to the location of the correct answer. Furthermore, eye tracking might not fully capture what participants perceive and use for understanding the visualization; in particular, peripheral vision is not considered by eye tracking systems.

The standard visualizations for eye tracking data are heat map representations [2] and gaze plots. Heat maps show the distribution of eye movements aggregated over any number of participants as a color-coded density field, but the information about the order of visited points or areas of interest is lost in a heat map representation. The trajectory data, i.e. saccades and sequence of fixation points and fixation durations, is visualized in a gaze plot that suffers from a vast amount of visual clutter caused by many line crossings and much overplotting. Recently, a visual analytics methodology has been introduced for the better exploration of eye movement data, aiming at understanding applied visual task solution strategies of participants [1].

4.2   Study Results: Memorization of Positions

The results of our studies for memorizing positions [11] show interesting differences between Cartesian and radial visualizations. We identified memorizing positions as a generic task when using visual representations of data. The comparison of several visual patterns can only be solved reliably and effectively when the viewer is able to keep recognized patterns in mind for a while. To investigate the differences in error rates and completion times when memorizing visual objects in either Cartesian or radial grids we split the study on memorization into two parts. An uncontrolled online user study with 674 participants was conducted first. A second study with 21 participants served as a controlled experiment to explore further questions raised by the first study in more detail. Both studies are conducted by participants from Western civilized countries.

The comparison of two visualization techniques can be difficult because, often, these do not only differ in the property under study (independent variable), but also in other properties. Even if each tool is designed to be the direct counterpart of the other one, some secondary property (confounding variable) might inevitably differ. This problem becomes clear when thinking about the layout of labels in radial and non-radial visualization techniques without introducing a bias. To address this issue the study on memorizing objects in radial or Cartesian grids follows a reductionistic strategy by looking at simple and generic visualizations.

In the memorization studies, we chose an $n \times n$-matrix in its rectangular representation as the basis for the Cartesian part. Either one matrix cell is color-coded and then removed after a short time, or three matrix cells are color-coded and one of the three is removed after some time. We also checked the impact of background information as an additional independent variable in the controlled study part and show the color-coded matrix cells in either a blank grid pattern or by extra randomly black colored matrix cells. Apart from the Cartesian matrix, the

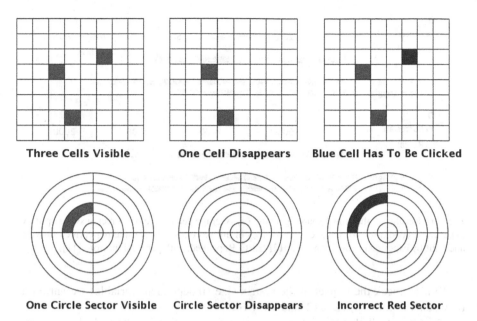

Three Cells Visible   One Cell Disappears   Blue Cell Has To Be Clicked

One Circle Sector Visible   Circle Sector Disappears   Incorrect Red Sector

Fig. 9 Cartesian stimulus: memorization of color-coded cells in a matrix and clicking at the corresponding cell where one of them disappeared (*top*). Radial stimulus: memorization of one circle sector and clicking on the position where it was shown (*bottom*)

memorization of visual objects in a radial grid pattern is investigated. For this, we transform the Cartesian matrix by using a polar coordinate system: the rows and columns are mapped to sectors and rings of a circle. Figure 9 illustrates example stimuli from the study in Cartesian or radial style. For more details about the study design, participants, experimental setup, statistical evaluations, and results we refer to Diehl et al. [11].

The tasks were designed to require the movement of the eye over longer distances on the display and hence, the memorization of positions of visual objects for a few seconds. The user's spatial memory and object location memory play a crucial role for the task performance. Research focusing on this topic in the field of cognitive science is very challenging. Furthermore, a controversial discussion on the details about such cognitive processes for the object location memory is illustrated by Wang et al. [33].

The eye tracking study [5] discussed in Sect. 3.1 serves as a preliminary study for the more in-depths studies that we describe here. The preliminary study showed that the Cartesian visualization outperformed the radial one for many tasks. For the tasks where the radial variant performed better, we expected that the memorization of visual objects is easier because radial visual patterns can be distinguished from each other much better than non-radial ones. We followed up on these initial results with the extended crowd-sourcing study.

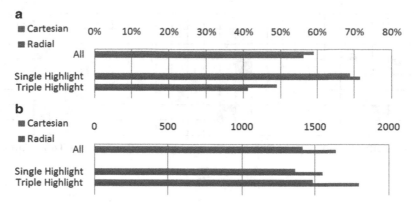

Fig. 10 Results for the crowd-sourcing study: (**a**) Average percentage of correct answers split by experiment condition. (**b**) Average answer times in milliseconds only considering correct answers and split by experiment condition (©[2012] IEEE. Reprinted, with permission, from [11])

To also check the impact of the matrix size (independent variable) we varied it between 4×4, 8×8, 12×12, and 16×16.

The uncontrolled study led to a couple of interesting results and observations related to task performance:

- **Result 1:** The error rate was significantly higher in the radial variants than in the Cartesian variants, see Fig. 10 (top).
- **Result 2:** The time taken by the users to answer correctly is significantly shorter in the Cartesian visualizations than in the radial counterparts, see Fig. 10 (bottom).
- **Observation 1:** Memorizing single cells is easier in radial coordinate systems, whereas memorizing three cells is easier in Cartesian coordinate systems.
- **Observation 2:** In radial visualizations, positions on rings about half way between center and circumference seem to be most difficult to memorize.
- **Observation 3:** There is an effect of reading direction for Cartesian coordinate systems: accuracy increases from left to right and from top to bottom.
- **Observation 4:** When depicting as many sectors as rings in a radial visualization, sector positions are easier to memorize than ring positions.

A second study was conducted to refine and complement the findings for memorization tests. Here, the experiment took much longer than the one used in the crowd-sourcing study. In this controlled study, we only used a matrix size of 20×20 cells. The most relevant results can be summarized as follows (Figs. 11 and 12):

- **Result 3:** Adding visual context in form of background patterns increases task performance significantly.
- **Result 4:** Providing visual context increases the performance more in Cartesian coordinate systems than in radial ones.

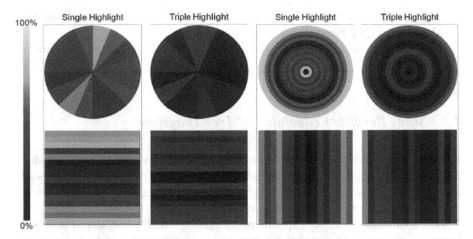

Fig. 11 Uncontrolled crowd-sourcing study: average percentage of correct answers summarized for radial visualizations by sector and ring, and for Cartesian visualizations by row and column (©[2012] IEEE. Reprinted, with permission, from [11])

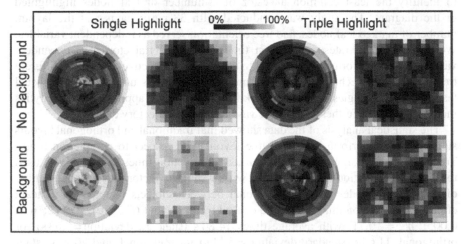

Fig. 12 Controlled study: percentage of correct answers per cell, smoothed and split by experiment conditions (©[2012] IEEE. Reprinted, with permission, from [11])

From both studies—uncontrolled crowd-sourcing and controlled laboratory study—our findings suggest some interesting implications:

- Use a Cartesian coordinate system unless there are clear reasons to favor a radial one (Results 1 and 2).
- Provide a recognizable local visual context, especially in Cartesian visualizations (Results 3 and 4).
- For radial visualizations, encode the more important dimension in sectors, not in rings (Observation 4).

- Omit the innermost ring in radial visualizations, but do not refrain from using the other inner rings (Observation 2).
- Consider the dominating reading direction in Cartesian visualizations (Observation 3).

4.3 Study Results: Readability of Tree Diagrams

The above studies directly address the memorization of visual objects in either Cartesian or radial grids. Let us now turn to another study [7] to investigate visual reading strategies for node-link tree diagrams. Here, visual memory also plays a role in the differences between the reading strategies.

This controlled laboratory experiment collected the traditional data of task performance (accuracy and completion times) and complemented it with eye tracking. Different node-link tree layouts were tested, among them a radial variant. To trigger visual reading of the trees, the following kind of task was used: participants had to identify the least common ancestor of a number of leaf nodes highlighted in the diagram. The study was conducted with 38 participants and the layout, number of marked leaf nodes, and tree orientation served as independent variables. We recorded error rates, completion times, and gaze trajectories as dependent variables. Task performance was used to compare the effectiveness of the different tree visualization techniques. The eye movement data was used to identify visual task solution strategies. In recent work, we additionally applied visual analytics methodologies for further analyze the vast amount of trajectory data [1].

The statistical analysis of the data showed that traditional and orthogonal layouts significantly outperform the radial tree layout with respect to completion times. Tree layout had a significant effect on task completion times. ANOVA was used to analyze completion times with two within-subjects factors layout and number of marked leaf nodes and one additional between-subjects factor. A significant effect on completion time was found caused by the layout ($F(2; 70) = 5.72$; $p < 0.005$; $\eta^2 = 0.71$) with average times 14.84 s (standard deviation = 9.63 s) for orthogonal, 11.65 s (standard deviation = 6.11 s) for traditional, and 20.95 s (standard deviation = 12.65 s) for radial layouts. Pairwise t-tests uncovered significant differences between radial and non-radial node-link tree layouts for the task of finding the least common ancestor of a number of highlighted leaf nodes ($p < 0.005$ Bonferroni-corrected for multiple comparisons). To fully understand the reason for the nearly twice as high task completion time the scan paths of our participants have to be analyzed as well, which is much more challenging than an evaluation of accuracies and completion times.

The analysis of the gaze data from eye tracking showed that it took participants more time to analyze radial diagrams because of the time taken to cross-check the potential solution. The visual task solution strategy of cross-checking the solution node in a tree diagram stimulus was also confirmed by a second more sophisticated analysis [1] based on the eye movement data (set of trajectories) itself instead of

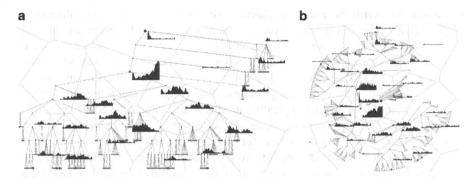

Fig. 13 Number of fixations over time shown as time-based histograms for regions of interest. (**a**) Traditional node-link tree diagram. (**b**) Radial node-link tree diagram

just task completion times, see Fig. 13a, b. From the more detailed trajectory data analysis, we also uncovered that the cross-checking phenomenon also appeared in the non-radial diagrams but not that often as in the radial counterparts. For more details about the eye tracking study and the results, we refer to Burch et al. [7] and Andrienko et al. [1].

We also recorded qualitative feedback and the subjective preferences of the participants. Feedback questionnaires had to be filled out after the actual eye tracking experiment. A Likert scale was used to analyze these preferences (1 means very good and 5 means very bad). All participants had to check a box to tell which layout they liked most.

As a result, participants preferred the traditional node-link tree layout with respect to three criteria: motivating (1.42), intuitive (1.42), and suitable (1.76). In contrast, orthogonal and radial layouts performed worse. The results for the orthogonal layout were: motivating (2.18), intuitive (2.55), and suitable (2.63). Finally, the radial layouts showed the worst results with respect to these points: motivating (3.39), intuitive (3.42), and suitable (3.32).

5 General Benefits and Drawbacks

Generally, radial visualizations have some benefits, but there is a long list of drawbacks that have to be taken into account. Apart from their more complex geometric forms and algorithms to be applied, perceptual challenges are the consequence and cognitive efforts are higher than for non-radial counterparts. Furthermore, due to their more complex forms interactive features are more difficult to develop and to apply. On the positive side, radial diagrams are said to look more aesthetically pleasing, although empirical studies on this topic are rare. If there is an inherent circular behavior in the data, radial depictions make sense but do not have not be the means of choice. In the following, we give a summary of points that have to be taken into account when visually encoding data by radial shapes.

- **Geometric Objects and Graphical Primitives:** In radial visualizations, geometric objects in use are typically some sort of circular elements. Instead of using rectangles and squares, the visualization designer is now confronted with more complex graphical primitives such as circle sectors, circle segments, or circle annuli. There is some kind of asymmetry in a radial display. Visual elements close to the center are given less display space than those close to the circle circumference which typically introduces a lie factor to the visualization. Sectors may all have the same shape but are differently oriented, although they may present the same quantity (see, for example, bar and pie charts).
- **Attached Detail Information:** If there is additional information to be attached to a radial visualization such as text labels, the question arises if these labels are radialized, too. Radial text labels are difficult to be generated and are hard to read, whereas non-radial labels in a radial diagram impair aesthetics.
- **Rendering Performance:** Mapping abstract data to circular shapes means using more complex graphical primitives for the geometric objects. Rectangular and quadratic shapes have to be transformed to circular objects such as circle sectors, segments, or annuli. Although many application programming interfaces (APIs) and graphics libraries provide good support for graphical objects, it tends to be more complicated to compute circular geometric primitives and their combination.
- **Perceptual Challenges and Cognitive Effort:** A radial representation places visual objects in the field of view and hence, it is in easy reach to the viewer. This means that corners on screen are seldom fixated by the eye in a Cartesian visualization. On the negative side, visual objects in radial diagrams are harder to compare because there is no straight guiding line that can be followed by the eye easily. As demonstrated by Cleveland and McGill [10], angle, slope, and area judgments (used in radial diagrams) are less accurate than position and length judgments (used more frequently in Cartesian diagrams).
- **Interactivity and Navigation:** Selecting circular regions in a radial visualization requires more complex algorithms. In a Cartesian diagram, a selected region is typically spanned by just dragging the mouse from a starting point and dropping it at the destination point. Radial distortion is used as an interactive feature to enlarge selected circle sectors. Shifting elements in a radial diagram typically does not destroy the visual pattern as it may happen in Cartesian counterparts. The visual pattern has just moved around for some angle, which means that a radial diagram is invariant under the shifting operation.
- **Aesthetics:** Aesthetics of diagrams is very hard to measure and depends on many factors. Male users typically have a different feeling for aesthetics than female users. Similar findings should be expected for diagrammatic representations of data. To the best of our knowledge there is no empirical evidence that radial depictions look more aesthetically appealing than non-radial ones for the same dataset. However, the design of a radial diagram is often justified by elegance of form, symmetries, wholeness, and closure.

   Aesthetics from a visualization perspective is typically interpreted in the sense that a diagram is useful, readable, and easily understandable. This means, a diagram belongs to the class of aesthetic representations if a viewer is able to efficiently derive insights from a dataset. In the field of graph visualization, for example, people call a node-link diagram aesthetically pleasing if the generated layout of a graph follows several aesthetic graph drawing criteria, such as the minimization of link crossings and overlaps, the maximization of symmetries, or the even distribution of graph vertices [29, 34].

- **Data Characteristics:** A radial depiction of abstract data makes sense if there is already an inherent circular or periodic behavior in the given dataset. Mapping this circular behavior to a radial diagram is an intuitive concept and mirrors the data appropriately. Radial diagrams use circular shapes in many orientations which consequently produce many differently looking graphical elements on screen. Cartesian representations do not benefit from this fact. If we have to deal with abstract data that has no spatial structure, the circular form benefits from the fact that it has no end points as a Cartesian representation.

6 Conclusion and Future Work

In this chapter, we have illustrated benefits and drawbacks of radial representations for data of different types and application domains. Apart from aesthetic reasons, there are some scenarios where a radial visualization is suitable, whereas other scenarios benefit from using non-radial, Cartesian diagrams. Comparative user studies need to be conducted to fully understand the benefits and difficulties of radial visualization. We presented a few of those studies. However, with the large design space of radial visualization, many open question remain that will have to be investigated in future studies.

Acknowledgements We would like to thank Stephan Diehl, Fabian Beck, Felix Bott, Rainer Lutz, Natalia Konevtsova, Julian Heinrich, Markus Höferlin, Gennady Andrienko, and Natalia Andrienko for the very pleasant and productive collaborations that led to the results reported in this chapter. We thank Marcel Hlawatsch for Fig. 8. Furthermore, we would like to thank the many study participants for their patience and their interest especially in the eye tracking study. Without their support, this work would not have been possible.

References

1. Andrienko, G., Andrienko, N., Burch, M., Weiskopf, D.: Visual analytics methodology for eye movement studies. Transactions on Visualization and Computer Graphics (2012)
2. Bojko, A.: Informative or misleading? Heatmaps deconstructed. In: International Conference on Human-Computer Interaction, pp. 30–39 (2009)
3. Burch, M., Beck, F., Diehl, S.: Timeline Trees: visualizing sequences of transactions in information hierarchies. In: Proceedings of Advanced Visual Interfaces (AVI '08), pp. 75–82 (2008)

4. Burch, M., Beck, F., Weiskopf, D.: Radial edge splatting for visualizing dynamic directed graphs. In: Proceedings of the International Conference on Information Visualization Theory and Applications, pp. 603–612 (2012)
5. Burch, M., Bott, F., Beck, F., Diehl, S.: Cartesian vs. radial – a comparative evaluation of two visualization tools. In: Proceedings of International Symposium on Visual Computing, pp. 151–160 (2008)
6. Burch, M., Diehl, S.: TimeRadarTrees: Visualizing dynamic compound digraphs. Computer Graphics Forum **27**(3), 823–830 (2008)
7. Burch, M., Konevtsova, N., Heinrich, J., Höferlin, M., Weiskopf, D.: Evaluation of traditional, orthogonal, and radial tree diagrams by an eye tracking study. IEEE Transactions on Visualization and Computer Graphics **17**(12), 2440–2448 (2011)
8. Burch, M., Vehlow, C., Beck, F., Diehl, S., Weiskopf, D.: Parallel edge splatting for scalable dynamic graph visualization. IEEE Transactions on Visualization and Computer Graphics **17**(12), 2344–2353 (2011)
9. Card, S.: Information visualization, in A. Sears and J.A. Jacko (eds.), The Human-Computer Interaction Handbook: Fundamentals, Evolving Technologies, and Emerging Applications. Lawrence Erlbaum Assoc Inc (2007)
10. Cleveland, W.S., McGill, R.: An experiment in graphical perception. International Journal of Man-Machine Studies **25**(5), 491–501 (1986)
11. Diehl, S., Beck, F., Burch, M.: Uncovering strengths and weaknesses of radial visualizations— an empirical approach. IEEE Transactions on Visualization and Computer Graphics **16**(6), 935–942 (2010)
12. Draper, G.M., Livnat, Y., Riesenfeld, R.F.: A survey of radial methods for information visualization. IEEE Transactions on Visualization and Computer Graphics **15**(5), 759–776 (2009)
13. Few, S.: Save the pies for dessert. In: Perceptual Edge Visual Business Intelligence Newsletter (2007)
14. Few, S.: Our irresistible fascination with all things circular. In: Perceptual Edge Visual Business Intelligence Newsletter (2010)
15. Goldberg, J.H., Helfman, J.: Eye tracking for visualization evaluation: Reading values on linear versus radial graphs. Information Visualization **10**(3), 182–195 (2011)
16. Grivet, S., Auber, D., Domenger, J., Melançon, G.: Bubble tree drawing algorithm. In: Proceedings of International Conference on Computer Vision and Graphics (ICCVG '04), pp. 633–641 (2004)
17. Hlawatsch, M., Leube, P., Nowak, W., Weiskopf, D.: Flow Radar Glyphs - static visualization of unsteady flow with uncertainty. IEEE Transactions on Visualization and Computer Graphics **17**(12), 1949–1958 (2011)
18. Hoffman, P., Grinstein, G.G., Marx, K.A., Grosse, I., Stanley, E.: DNA visual and analytic data mining. In: Proceedings of IEEE Visualization, pp. 437–442 (1997)
19. Johnson, B., Shneiderman, B.: Tree maps: A space-filling approach to the visualization of hierarchical information structures. In: Proceedings of IEEE Visualization, pp. 284–291 (1991)
20. Koffka, K.: Principles of Gestalt Psychology. Harcourt, Brace, New York (1935)
21. Lewi, P.: Speaking of graphics, http://www.datascope.be/sog.htm (2008)
22. Lima, M.: A visual exploration on mapping complex networks. http://www.visualcomplexity.com/vc/
23. Lin, C.C., Yen, H.C.: On balloon drawings of rooted trees. Graphics Algorithms and Applications **11**(2), 431–452 (2007)
24. Ma, K.L.: Stargate: A unified, interactive visualization of software projects. In: Proceedings of IEEE VGTC Pacific Visualization Symposium, pp. 191–198 (2008)
25. Melançon, G., Herman, I.: Circular drawings of rooted trees. Tech. rep., Amsterdam, The Netherlands (1998)
26. Misue, K., Eades, P., Lai, W., Sugiyama, K.: Layout adjustment and the mental map. Journal of Visual Languages and Computing **6**(2), 183–210 (1995)

27. Nightingale, F.: Notes on matters affecting the health, efficiency, and hospital administrations of the british army: Founded chiefly on the experience of the late war (1858)
28. Playfair, W.: The statistical breviary; shewing, on a principle entirely new, the resources of every state and kingdom in Europe (1801)
29. Purchase, H.C., Cohen, R.F., James, M.I.: Validating graph drawing aesthetics. In: Proceedings of Graph Drawing, pp. 435–446 (1995)
30. Rosenholtz, R., Li, Y., Mansfield, J., Jin, Z.: Feature congestion: a measure of display clutter. In: Proceedings of the Conference on Human Factors in Computing Systems (CHI), pp. 761–770 (2005)
31. Tufte, E.R.: The Visual Display of Quantitative Information. Cheshire, CT: Graphics Press (1983)
32. Tversky, B., Morrison, J.B., Bétrancourt, M.: Animation: can it facilitate? International Journal of Human-Computer Studies **57**(4), 247–262 (2002)
33. Wang, H., Johnson, T.R., Sun, Y., Zhang, J.: Object location memory: The interplay of multiple representations. Memory & Cognition **33**(7), 1147–1159 (2005)
34. Ware, C., Purchase, H.C., Colpoys, L., McGill, M.: Cognitive measurements of graph aesthetics. Information Visualization **1**(2), 103–110 (2002)
35. Weber, M., Alexa, M., Müller, W.: Visualizing time-series on spirals. In: Proceedings of IEEE Symposium on Information Visualization, pp. 7–14 (2001)
36. Wetzel, K.: Using circular treemaps to visualize disk usage. Available at http://lip.sourceforge.net/ctreemap.html

Measuring Memories for Objects and Their Locations in Immersive Virtual Environments: The Subjective Component of Memorial Experience

Matthew Coxon and Katerina Mania

Abstract The utility of Virtual Environment (VE) technologies for training systems is predicated upon the accuracy of the mental representation formed of the VE. It is therefore important to benchmark the use of these environments through the application of appropriate memory tests. In addition to measuring accuracy, a consideration for such tests is the quality of the memories being reported: Are these vivid memories where the person can mentally visualize the environment? Or are responses based on a strong feeling that happens to be correct? This chapter reviews what is currently known about these more subjective aspects of memories in respect to visualizations of immersive VEs. Current understanding about judgements of 'remembering' and 'knowing' will be summarised from the psychological literature. This will be related to recent attempts to measure these awareness states in VEs, and the consequences of these efforts. An information-processing approach is taken, interpreting the relationship between these experiences and potential influencing variables. As well as highlighting the importance of measuring the subjective component of memorial experience, a brief descriptive model is proposed to aid considerations of these issues in the design and implementation of such visualizations, and future research.

M. Coxon (✉)
Faculty of Health and Life Sciences, York St John University, Lord Mayor's Walk, York YO31 7EX, UK
e-mail: m.coxon@yorksj.ac.uk

K. Mania
Laboratory Of Distributed Multimedia Information Systems and Applications,
Department of Electronic and Computing Engineering, Technical University of Crete,
Chania, Crete, Greece
e-mail: k.mania@ced.tuc.gr

W. Huang (ed.), *Handbook of Human Centric Visualization*,
DOI 10.1007/978-1-4614-7485-2_18, © Springer Science+Business Media New York 2014

1 Introduction

The utility of Virtual Environment (VE) technologies for training systems, such as flight simulators, is predicated upon the accuracy and detail of the mental representation formed of the VE. A central research issue, therefore, is how an interactive synthetic scene is cognitively encoded and retrieved. Importantly, as the visualizations in immersive VEs become increasingly sophisticated it is possible to fill these environments with a range of realistic-looking objects placed in a wide variety of locations in a photo-realistically illuminated scene. For training or instructional purposes it can be important for people to remember details of these scenes, such as the type or location of objects within them, yet relatively little is known about the underlying cognitive representations of these virtual environments. Measurements of memory for these scenes can therefore be fundamental to the development process and are often incorporated in benchmarking processes when assessing the design of VE simulations created for training purposes [3, 24, 35].

Memory for synthetic environments, has generally been measured according to how much people remember correctly from, or about, these environments. A key limitation of such measurements is that they do not reflect the quality of the memories: Are these vivid memories where the person can mentally visualize the environment? Or are these responses based on a strong feeling that happens to be correct? The implications for training are significant as the designers of VEs for training may be intending for subjectively rich memories to be formed, although the extent to which this is achieved is rarely measured. In this regard, the field of cognitive psychology can play an important role in helping designers and developers understand the impact of their design decisions on the hidden mental processes of individuals in their simulations.

This chapter reviews what is currently known about these more subjective aspects of memory for the contents of immersive VEs. In particular, current understanding about the experiences of 'remembering' and 'knowing' will be summarised from the psychological literature. This will be related to recent attempts to measure these awareness states in immersive VEs, and the consequences of these efforts on our understanding of memory for such visualizations. An information-processing approach is taken, interpreting the relationship between these experiences and potential influencing variables. As well as highlighting the importance of measuring the subjective component of memorial experience, a descriptive model is proposed to aid future considerations of these issues in the design and implementation of such visualizations.

2 Measuring Memories in Virtual Environments

In general, what we know from cognitive psychology is that accurately remembering information over even short periods of time is peculiarly difficult. It has been firmly established that memory is a complex concept influenced by a wide range of factors, related in complex ways to many other cognitive processes (such as attention and perception), and involves many as of yet unspecified systems and processes with complex relationships between them. Memory is therefore not a unitary concept and cannot be studied as such. Nevertheless, psychological research on human memory can provide a number of important insights into potential processes and variables that influence someone's memorial experiences in immersive virtual environments.

An important measure for benchmarking virtual training environments is the extent to which participants can remember aspects of the environment to which they had recently been exposed. Measurements of memory in these scenarios tend to take one of two forms: a recall memory test; or a recognition memory test. Recall memory tests require the person taking the test to provide information about something that has happened in the past, such as what they saw or experienced whilst in the virtual environment. This can be in the absence of any cues that might facilitate the memories (free recall), although cues may sometimes be included (cued recall). For example, having spent 2 min in a virtual environment a participant could be asked to write down all of the objects they can remember seeing in that environment. This would constitute a free recall test as no additional cues or support are provided. In contrast, having spent 2 min in a virtual environment a participant could be asked to write down all of the objects they can remember seeing in that environment with a list of potential cues (i.e. what was on the table?), or asked to recall objects from specific parts of the room. This would constitute a cued recall test.

An alternative to the use of recall memory tests is the use of recognition memory tests. A typical recognition memory test involves the presentation of a list of things that have potentially been seen or experienced in the past. For example, having spent 2 min in a virtual environment a participant could be given a list of objects that could have been in the environment and asked which of them they recognise as having been there. The test list should be composed of things that were seen or experienced (old items) and things that were not (new items). Someone's ability to accurately remember on this test therefore involves correctly recognising the old items, but not the new items.

There are four possible responses to items on a recognition memory test. The first response is to indicate that an old item has been seen before, this is referred to in the psychological literature as a 'hit'. The second response is to indicate that an old item has not been seen before, this is referred to as a 'miss'. In terms of new items (things that weren't seen before), if the participant claims that they recognise it this is known as a 'false alarm'. Finally, if a participant claims they don't recognise a new item this is known as a 'correct rejection'. Whilst accuracy is intuitively considered as the number of 'hits' (correct recognition of old items) it can also be useful to take into account these other responses, particularly false alarms.

2.1 Measuring the Subjective Component

Both recall and recognition measurements provide an indication of how much a person can accurately remember. This is imperfect as it provides no indication as to the strength or the quality of the memories. A correct response on these measures may be indicative of a strong memorial experience in which the participant can actively visualise the thing being remembered, perhaps recalling additional details about it, or even what they were thinking about when they first perceived it. In contrast, a correct response may also be indicative of a weak memorial experience in which the participant has a vague feeling that they have seen something in the past, and remembers nothing else about it. Accuracy measures per se do not provide sufficient detail that would allow researchers or designers to differentiate between such different memorial experiences when creating training or simulations in virtual environments. Alterations to the technical specifications of an environment might drastically change the participants' memories from weak feelings of familiarity to strong recollective experiences with little change in overall accuracy on standard measures. It is therefore important that some indication of the strength or quality of the memorial experience is gathered.

One potential option is to ask participants to say out loud what they remember [26, 30]. In this scenario, participant's responses could be coded by the researchers according to criteria that are deemed appropriate. For example, the data could be coded as to whether the memories contain contextual details or not. Whilst this represents one solution, in practical terms this rich qualitative data cannot always be collected. Furthermore, it relies upon individual researcher's interpretations of the qualitative data, which in turn may be limited by a participant's ability to express themselves clearly, willingness to disclose their personal experiences, or willingness to deceive the experimenter for either personal gain or as a response to perceived expectations on the part of the experimenter.

One solution, is therefore to require participants to make this classification themselves, based upon their own personal experiences, before reporting them non-verbally such as writing down, or otherwise indicating, the classification of their experience from a choice of discrete responses. For example, in researching the processes of recognition memory, cognitive psychologists have required participants to distinguish between two types of memorial experience: remembering and knowing. Participants are only required to indicate whether their memory is associated with a 'remembering' experience or a 'knowing' experience without necessarily providing elaborated narratives that may be more vulnerable to bias.

Asking participants to provide these two types of responses, to capture the more subjective components of their memorial experience, was first proposed by Tulving [34]. Since then, self-reports of these distinct experiences have been used to investigate a wide range of psychological phenomena in respect to memory. Importantly, it has become well established that participants are able to make these reliable discriminations in a reliable way [13]. Some researchers have attempted to further divide these judgements into remember, know, familiar and guess [7] although this approach has not been widely adopted. It is however noteworthy that, in a further refinement of this technique, it has been suggested that the terminology of 'remember' and 'know' may be impacting upon the accuracy of the responses being given [25].

2.1.1 The Impact of Terminology

Whilst psychologists use 'remember' to refer to vivid recollective experiences and 'know' to refer to feelings of familiarity, these are not always common usages for these terms outside of this field. Despite receiving specific instructions as to what constitutes a 'remember' response or a 'know' response some participants may continue to apply, to some extent, their own understanding of these terms to their responses. This is of particular importance when researching the design of virtual environments because the origins of such research is globally diverse, with much of it conducted in countries in which English is not the first language. Understandings of the terms 'remember' and 'know' are therefore likely to vary between languages and cultural contexts.

One solution is to use more neutral terms to refer to these types of memory [4, 18, 36]. Consistent with this, recent empirical evidence suggests that participants make more reliable judgements if 'remembering' is instead presented to participants with the label 'Memory A', and if 'knowing' is presented to participants with the label 'Memory B' [25]. Each memory type is then associated with a standardised definition. To aid readers an example of these somewhat lengthy definitions are provided here, in the context of recognising words from a word list. Each definition can be adapted to suit the purposes of the research, although it is advisable that this consists of only minor alterations to the examples of stimuli that will be seen. The following definition of 'Memory A' has been suggested to enable more accurate measurement:

> When you see a word on the test, it may bring to mind the exact thought you had from when you first studied the word at the start of the experiment. If you can recall the exact thought you had from when you studied the word earlier you should press the A key to indicate a Type A response. Often when people give a Type A response it is because they can recall a personal association that came to mind when they first saw the word, or some other details about when they studied the word.
>
> For example, imagine you had studied the word BOOK earlier in the experiment. Imagine also that when you studied the word BOOK that you thought of the title of a book you have recently been reading. If you then saw the word BOOK on the test, and you recalled that when you were studying it you had thought about the title of the book you have been reading, then you would give a Type A response for the word BOOK. There are other details you may recall about studying a word that would lead you to give a Type A response, such as a particular feeling you had when you saw the word, or a mental image that came to mind while you were studying the word.
>
> You may also be able to recall that you associated the word with another word that you studied, or you may recall what the word looked like on the screen. If you can be sure you studied the word because you can recollect specific details about when you studied it, then press the A key to indicate a Type A response. [25, pp. 407–408]

The following definition of 'Memory B' has also been suggested:

> If you see a word on the test and you believe it was presented but you cannot recall any specific association that you made when you studied it, press B to indicate a Type B response. In other words, a Type B response means you "just know" you studied the word, even though you can not recall any details from when you studied it. [25, pp. 408]

Not only does this simplify the interpretation through reducing ambiguity it also leads to increased accuracy, in part through the reduction of false alarms with remember responses [25]. False alarms are responses in which a participant claims to have a memory which contains vivid recollective detail, but this memory is actually false because it is a memory for something that was never shown. Given that it is desirable to reduce such responses, future adoption of the labels 'Memory A' and 'Memory B' when measuring the subjective components of memorial experience is therefore advisable.

2.2 Remembering and Knowing in Virtual Environments

The rationale for measuring experiences of 'remembering' and 'knowing' when designing and developing virtual environments is simple: accuracy on memory tests per se provides an incomplete picture of the underlying cognitive representations which can at least in part be clarified through measuring the more subjective components of memory. Importantly, these subjective experiences have the potential to provide greater insight into the impact that technical and design changes are having on the richness of memories for those environments. It is therefore important that this information regarding quality is collected in addition to any concerns over accuracy. Collecting 'remember' and 'know' judgements, or 'Memory A' and 'Memory B' judgements, helps provide this important information.

However, this rationale would be irrelevant if technical changes within virtual training environments had no impact upon these measures. That is to say that the value of taking these additional measurements is dependent upon them highlighting systematic variations in the underlying representations. What follows is a brief review of investigations which have incorporated these additional measurements of 'remembering' and 'knowing' in immersive VEs. Each of these investigations has studied the impact of changing different aspects of the technical specifications upon participants' judgements of remembering and knowing. Crucially, it has been found that not only do these additional measurements help differentiate the underlying cognitive representations formed as a result of different environments, but they have given rise to a series of unanticipated outcomes with important practical implications.

2.2.1 Previous Research

To date measurements of the more subjective components of memorial experience in virtual training environments has remained rare. One of the first reports of using 'remember' and 'know' judgements in virtual training environments was published in 2001 by Mania and Chalmers [21]. They investigated whether there were differences in learning from a seminar if it was presented: in reality; using a 3D environment presented on a desktop display with audio (taken from the reality condition); using a 3D environment presented on a Head Mounted Display (HMD) with audio (taken from the reality condition);or the audio on its own. The seminar itself lasted 15 min and was on a non-science topic which none of the participants had any prior knowledge of.

Whilst the main focus of the reported study was investigations of feelings of presence, the researchers also took measurements concerned with memory and learning. Specifically, participants were tested on both their memory for the factual information presented in the seminar, and on their memory for the environment itself such as the objects in it and their locations. For both tests participants were able to choose from four possible answers, of which only one was correct. In addition, participants were required to report their confidence (five-point scale: no confidence to certain) and they were required to classify their judgements as being based on remembering, knowing, a feeling of familiarity or guessing. Participants were given instructions as to what each of these responses signified prior to responding. Measures of presence were also taken, alongside measures of simulator sickness where appropriate.

In terms of judgements of remembering and knowing, it was found that participants were more likely to provide correct responses accompanied with a 'remember' experience when they had been exposed to the seminar in the HMD. This numerical difference was statistically significant when comparing responses in the HMD to those in reality in particular. In the reality condition, accurate recognition memory was more often associated with feelings of familiarity ('familiar' responses). Interestingly, when questioned about aspects of the environment itself there were no overall differences in accuracy across the environments despite these differences in the qualitative experience. Although little discussion was made of it in this publication it was noteworthy that a less realistic environment, the display presented on a HMD, resulted in more memories based on the experience of 'remembering'. This becomes of increasing relevance when considered alongside further measurements that have been taken over the past 10 years.

In 2003, a second study was published in which measurements of 'remembering' and 'knowing' were taken in a virtual training environment [23]. In this investigation all participants were exposed to a realistic virtual room containing different objects in different locations. Participants in this study were exposed to a room either in the real-world, wearing custom goggles to restrict their field of vision, or as a photorealistic virtual environment on a desktop display or through a HMD (Fig. 1). Presentations within the HMD were made in the following conditions: monocular and head-tracked; stereo and head-tracked; monocular and mouse navigated. The real-world condition consisted of a room with the exact same geometry and features of the virtual environment.

After exposure, participants were tested with a blueprint diagram of the room which included numbered positions of objects in various locations. For each position participants were asked to indicate which of three possible objects they saw there, their confidence (on a five-point scale from no confidence to certain), and whether their judgement could be classified as 'remember', 'know', 'familiar' or 'guess'. Participants were given instructions as to what each of these responses signified prior to responding. Measures of presence and simulator sickness were also taken.

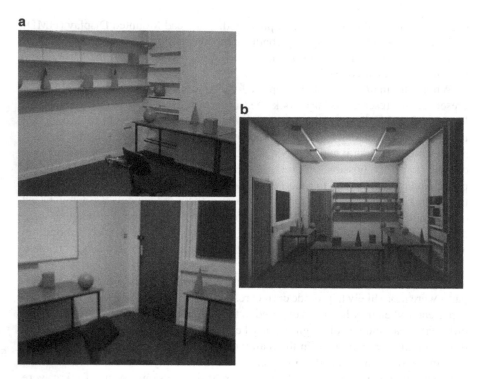

Fig. 1 Experimental space in the real-world (**a**) and in a virtual environment (**b**) (Adapted from [23])

It was noted by the authors that they had anticipated that the number of correct remember judgements would be higher in conditions that incorporated more naturalistic interfaces such as head tracking. However, in general, it was found that participants reported a significantly greater number of 'remember' judgements in the HMD mono mouse condition, than the HMD mono head tracked condition, and the HMD stereo head tracked condition. Participants therefore reported more vivid memorial experiences when the interactions with the virtual environment were less realistic. In comparison, there were no significant differences between the conditions on overall accuracy. Whilst participants were no less accurate across the different technical specifications there are significant impacts upon the strength or quality of the memory traces in unanticipated ways. This was therefore consistent with the results previously reported in 2001 [21].

Since these two studies, further investigations have revealed a similar pattern of results. These not only support the efficacy of gathering information on the more subjective components of memory in these virtual training environments but also supporting the finding that less 'real' or 'naturalistic' environments can lead to more vivid memorial experiences (remember responses). This is a particularly important finding given that the role of many virtual training environments may be to accurately reproduce reality as faithfully as possible.

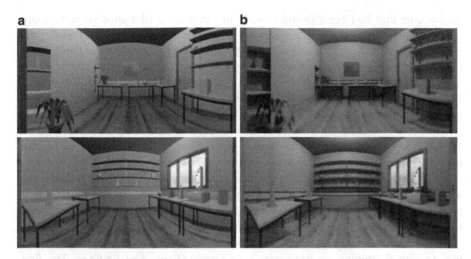

Fig. 2 Flat-shaded rendering (**a**) and radiosity rendering (**b**) of an experimental space consisting of two interconnected rooms (Taken from [24])

In 2006, it was found that changes in rendering (radiosity vs. flat-shaded) in a 3D room presented via a stereo-tracked HMD also led to differences in the distribution of 'remember', 'know', 'familiar' and 'guess' responses [24]. Participants were either exposed to a room with flat-shaded rendering or a room with radiosity rendering (Fig. 2). After exposure to the environment participants were again required to indicate the nature and location of objects within the room. No significant differences in overall accuracy were found between the two rendering conditions. However, participants were more likely to indicate that they had experienced the more vivid memorial experiences associated with 'remember' responses when the training environment had been flat-shaded rather than radiosity rendered. This finding was therefore consistent with those detailed above in which 'remember' and 'know' judgements differed depending upon technical differences in the interaction interfaces. This study extended that observation to variations in visual fidelity.

These findings were also latterly extended when it was investigated whether they transfer into a real-world environment following exposure in a virtual training environment [20]. Again, participants were exposed to a synthetic scene of a room via a stereo-tracked HMD. The visual fidelity of the room was manipulated with both a flat-shaded rendering and a radiosity rendering. After exposure participants were taken to a real version of the same room. In this real room they were required to choose from a range of objects and place them in the locations that they had seen them in during exposure in the virtual training environment.

As well as indicating their confidence on a five-point scale, participants were also required to classify their memorial experience into either remember, know, familiar or guess responses. Again, it was found that the viewing conditions did not impact upon overall measurements of accuracy. However, it was found that participants that had been in the low fidelity condition (flat-shaded) were more likely to associate their correct responses with 'remember' experiences. Participants in the less realistic version of the training environment therefore had more vivid recollective experiences when their memory was tested. Consistent with the research reported above, this study again indicates that the impact of technical differences can be seen in terms of the more subjective components of memory and that less realistic environments promote more vivid recollective experiences.

There are also further indications that variations in mono depth cues and stereo cues may influence remember responses when the type of objects are taken into account [2], and that this may also extend to variations in latency [28]. In terms of variations in latency, early indications are that these may influence both overall accuracy and the distribution of memory awareness states, unlike the research mentioned above. Furthermore, there are indications that this may again be related to the types of objects being remembered from the synthetic scenes.

In general, research that has differentiated between 'remembering' and 'knowing' in virtual training environments have provided two key findings. The first is that whilst some technical variations, such as changes in interaction or visual fidelity, may not influence overall memory measures of accuracy they are associated with changes in the more subjective experiences of 'remembering' and 'knowing'. Secondly, and counter intuitively, technical variations that increase the realism of the virtual training environments appear to decrease the number of remember experiences reported. Vivid memorial experiences of remembering are more heavily associated with less realistic training environments. But why might this happen?

3   Understanding Remembering and Knowing in VEs

It is important to try to understand why certain factors (stereoscopy, interaction mode, rendering) may influence these more subjective components of memory. The psychological factors that influence mental representations of virtual environments can be split into at least two broad categories: perceptual or sensory factors; and conceptual or contextual factors. Each of these broad categories represents a combination of multiple underlying processes and systems. With the limited data available, the interpretation presented here will not attempt to completely disentangle each of these processes and their relative contribution to the underlying mental representation of virtual environments. Instead, the level of analysis adopted here will be much broader providing initial suggestions as to some of the factors that influence information processing that may be of relevance.

To put the distinction between perceptual and conceptual factors into context imagine standing in a virtual flat. As you turn your head to look in the kitchen area you see a kettle and a trombone. In this simple scenario the perceptual or sensory factors include (but not exclusively) the naturalness of the motion of the scene as it is rendered and the lines, and the colours and shapes that define the objects and their surroundings. The conceptual or contextual factors in this scenario include (but not exclusively) expectations of the contents of a scene, and expectations of their relative position within that scene. For example, the kettle may be expected in a kitchen on a supported surface, but a trombone floating in the middle of the air may not be. From an information processing perspective a distinction can therefore be drawn between perceptual processing (processing of the perceptual aspects of a display) and conceptual processing (processing of the conceptual aspects of a display).

The distinction between perceptual processing and conceptual processing is important because theorists have suggested that increased levels of conceptual processing increases the proportion of 'know' judgements, reflecting increased feelings of familiarity in the absence of vivid recollection. In contrast, more salient perceptual processing has been associated with increases in the proportion of 'remember' judgements, potentially reflecting an increase in the number of recollective experiences [11, 12, 14].

One potential hypothesis is therefore that any aspects of a visualization's design that promote perceptual processing will lead to an increase in 'remember' judgements, reflecting more vivid or stronger memorial experiences. Conversely, any aspects that promote fluent conceptual processing (matching participants' expectations) may lead to an increase in 'know' judgements, reflecting less vivid or weaker memorial experiences. Although intuitive, this is inconsistent with the findings reported here in which the need for less perceptual processing (low rendering, limited movement) led to a greater proportion of remember judgements. This suggests that an alternative framework is needed to understand memorial experiences in virtual environments, compared to early suggestions from the cognitive psychology laboratories.

3.1 Distinctiveness, Remembering and Knowing

To interpret the available data it seems more appropriate to adopt the idea that the distinctiveness of processing is crucial for reports of remembering. It has been suggested by other theorists that experiences of 'remembering' are determined by the distinctiveness of the item (object/scene) being processed, such that distinctive experiences (e.g. a trombone in a kitchen) may lead to more vivid recollective experiences [31, 32]. Within this framework distinctiveness can be either conceptual or perceptual in nature, it is the property of 'distinctiveness' which is of importance. Distinctiveness is not necessarily considered a unitary concept that any single thing has irrespective of its context, but instead it is considered as a cognitive property that is also determined by its context, such that a trombone would be less distinctive in a music shop than in a kitchen. Conversely, in this framework, fluency of processing is thought to determine experiences of 'knowing'. Given that the effects evident within the literature noted above are centred on differences in 'remembering', notions of distinctiveness will remain the focus of this brief discussion.

In applying this framework to the current issue further assumptions and clarifications are necessary. Specifically, in defining distinctiveness it is important to disentangle those aspects that might be unique to a cognitive psychology lab to those that are of significance to an immersive virtual training environment. Importantly, conceptions of distinctiveness within the psychological literature often refer to distinctiveness within the laboratory experience. Traditionally, this may reflect distinctiveness within or across the stimuli participants are exposed to, which are often presented serially for short periods of time. In contrast, 'distinctiveness' in an immersive virtual environment that is designed to resemble reality (as virtual training environments are) may reflect a broader concept that captures sensorimotor interactions, as well as more global attributes of the experience, presented in parallel with one another. Participants are also exposed to these environments for much longer periods, and these environments are more perceptually and contextually detailed than the stimuli provided in your average psychology laboratory. The extension of these concepts from laboratory to virtual environments are therefore made cautiously with an awareness that there are several differences which be may prove to be of importance as our understanding develops.

For these purposes then, distinctiveness is better conceptualised as a 'variation from real'. The aim of a virtual training environment is naturally to replicate the real-world, to the extent that any learning that occurs within it can be usefully transferred into real-world environments. As such, the starting point is that the training environment reflects the real world, so by definition a distinctive aspect is one that varies from reality. By defining the construct in this way it can be proposed that the distinctiveness of variations from reality (either perceptual or conceptual) determines whether it promotes memorial experiences based on vivid recollections (remember responses). More distinctive variations result in more vivid memorial experiences.

There are indications that the availability of attentional resources may play a mediating role in this relationship. Previous psychological research has indicated that remember responses require more attentional processing in the first instance than those based on familiarity [6, 29]. There are also indications that more distinctive stimuli are more likely to recruit attention [5]. Therefore, by extension, virtual training environments that are distinctive may in turn recruit more attentional resources. This additional attentional processing may bring about a change in the encoding of the experience, which is reflected in more elaborate remembering when participants later recall the environment from the underlying cognitive representation.

The concept of distinctiveness is therefore an important one when considering the potential impact of technical variations in virtual training environments on the subsequent subjective components of memories for that environment. The extent to which this can be applied effectively to inform the design of such environments remains an open question to be explored.

3.2   Recollection and Familiarity

The initial measurement of 'remember' and 'know' experiences is therefore straightforward and introduces an additional source of information for designers of virtual training environments as part of their benchmarking and evaluation processes. In general, the previous research indicates that varying technical elements (mode of interaction, fidelity) can change the richness of participants' memories, often in the absence of overall changes to accuracy on memory tests. These increases in rich subjective experience have been associated with less 'realistic' or 'naturalistic' aspects of the virtual training environments, which speculatively may be due to the influence distinctiveness has upon our attentional processing.

Thus far we have focused on interpreting the differences that result from technical changes. However, it is also important to ask how we should best interpret the 'remember' and 'know' responses themselves independent of these experimental variations. The answer to this question is not straightforward and has been a matter of debate within cognitive psychology in recent decades. A very brief overview of this debate will be presented here to help inform any reader who decides to measure, and interpret, remember and know responses in the design of their virtual environments.

It is worth nothing that when the measurement of remember and know judgements was first proposed it was hypothesised that these measures reflected two separate processes in the mind: a recollection process; and a familiarity process [34]. 'Remember' responses were thought to be indicative of a process of recollection, whilst 'know' responses were thought to be indicative of a process of familiarity. This interpretation of the underlying information-processing is controversial with respect to the number of processes that are necessary, and (if two) how these relate to each other.

Some theorists have advocated two separate processes, with the subjective experience of our memories supported by one or the other of them [1, 17, 19, 39]. By way of contrast, other theorists have argued that relevant data can be accounted for by a single process and that there is no apriori need to distinguish between recollection and familiarity [8, 9, 16, 33]. Most recently hybrid models have been developed which consolidate these two approaches. These hybrid models advocate the existence of dual processes which interact to form a single memory trace upon which judgements of remembering and knowing are based [38]. The theoretical position that is adopted has important consequence for how 'remember' and 'know' responses are both analysed and interpreted.

3.2.1   Dual Processes Versus a Single Process

Dual-process accounts of recognition memory focus on two separate processes that operate in parallel: familiarity and recollection [1, 17, 19, 39]. Familiarity is conceived of as a continuous variable ranging from low to high. For any one thing (e.g. object, face, smell etc.) a person may experience a differing degree of familiarity for it on this scale. The degree to which it is judged to be familiar will, in part, determine if someone judges to have recognised it from the past (i.e. from memory). Researchers that advocate this model have argued that 'know' judgements provide an index of familiarity.

In models that contain two processes, judgements of familiarity are supplemented by feelings of recollection. Feelings of recollection are based on the recall of explicit details of the source of a memory. For example, in recognising a digital recreation of a kettle I may remember seeing it in a virtual environment I was in earlier because I thought of how I would like a cup of tea at the time. The explicit memory for these additional details about past occurrences is considered part of a recollective process. Within an information-processing approach, this process is therefore considered a categorical variable: it is either all (additional contextual details are recalled) or none (nothing additional is recollected). In measurement terms, this is argued to correspond to participants' 'remember' responses.

In these dual-process accounts recognising something from memory, i.e. recognising that one has encountered it in the past, is presumed to make use of the distinct but parallel processes of familiarity and recollection. Dual-process accounts draw on a wide-range of behavioural evidence which indicates that these measures of recollection and familiarity can be independently influenced by a range of variables [13].

In contrast, some researchers have argued that recognition memory can be accounted for in terms of a single process that represents the relative 'strength' of a memory [8, 9, 16]. Evidence for this perspective primarily comes from the analysis of receiver operating characteristics (ROCs). In brief, a ROC is a mathematical function that takes into account the proportional hit rate (items correctly recognised as previously seen) and its relationship to the proportional false alarm rate (items incorrectly recognised as previously seen). ROC curves, a plot of this relationship, can predominantly be explained by single process models without the need to postulate two separate processes [40]. Instead, 'remember' responses are taken to indicate a higher strength of memory, compared to 'know' responses on the same scale.

3.2.2 Hybrid Models

The underlying mental concepts of recollection and familiarity can therefore be conceived of as either two separate parallel processes, or as a single unitary process. However, more recent interpretations of remember/know judgements provide a more integrated hybrid approach. Whilst the details of different models continue to be debated the underlying principles, and their implications for measuring remembering and knowing, are of increasing importance. For example, one suggestion is that both recollection and familiarity are represented internally in a continuous way but are also aggregated into a single trace strength [38]. This hypothesis therefore suggests that there are degrees of recollection, and degrees of familiarity. This varies from the traditional dual process account in which recollection is categorical in nature, additional details can either be recollected or not. It also varies from a more unitary approach in that it suggests that dual parallel processes exist, but are later aggregated as a single strength variable to make remember/know judgements.

More specifically, within this example of a hybrid account it is hypothesised that a judgement regarding whether an item is old or new is based on a single combined memory strength signal. Nevertheless, subsequent 'remember' and 'know' judgements are thought to be based on two different dimensions of memory strength, recollection and familiarity, and participants can reliably report which their decisions are primarily based upon. Importantly, it is not assumed in this approach that these judgements are process pure. What is meant by this is that a 'remember' judgement may also be informed by high levels of familiarity (they are not mutually exclusive). Similarly, a 'know' judgement as a result of very high feelings of familiarity does not exclude the possibility of some recollective experience too. The consequences of this for the interpretation of remember/know judgements in virtual environments are important.

Current thinking suggests that we cannot assume that a remember judgement reflects a pure experience of recollection (in the absence of familiarity) or that a know judgement reflects a pure experience of familiarity (in the absence of any recollection) [10, 15]. Instead, this current theorising suggests that judgements instead reflect a memory's strength based upon the separate processes of recollection and familiarity. So what are the practical implications of this conceptual shift?

If the measurements of 'remember' and 'know' are to be interpreted as recollection or familiarity then we must take steps to ensure that these are as process pure as possible. One important step is to take into account confidence levels of the judgements being made. It is suggested that if confidence is not equated then 'remember' responses reflect stronger memories (not recollection), and 'know' responses reflect weaker memories (not familiarity) [38]. However, if confidence is equated across the measures then they are more likely to be process pure [34, 38]. This not only involves collecting both remember/know responses and participants confidence ratings but also using these in a combined way within analyses. In practical terms, confidence can be equated through only comparing remember and know judgements in our analyses when they both have a high level of confidence (or at least the same level of confidence). It is also important to equate accuracy in the same way. Once accuracy and confidence have been equated then it can be assumed that the responses being analysed are of equivalent strength, varying only on the dimension of recollection and familiarity.

It is also important to take into account differences between false alarm rates alongside differences in hit rates [37]. The difference between false alarm rates and hit rates in recognition memory measures was outlined earlier in this chapter. To re-cap, a hit is when someone correctly recognises an old item as something they have seen before. A false-alarm is when someone falsely claims to recognise a new item as something they have seen before. Whilst methods for controlling false alarm rates vary within the literature, the most common method is simply to subtract one from the other (hits minus false alarms). Through controlling for false alarm rates it is again possible to provide a more accurate index of the underlying cognitive processes rather than memory strength per se.

In each case, the purpose of making adjustments to the measures is to help us to accurately measure the mental processes of recollection and familiarity, so that we can ascribe our findings to specific types of mental events. If these steps are taken then any differences in 'remember' and 'know' responses are more likely to resemble underlying differences in these cognitive processes. However, in pragmatic terms, it may be sufficient when evaluating interactive synthetic scenes to simply acknowledge that remember/know judgements only index these two processes, and are not process pure.

4 Conclusions and a Descriptive Model

It is clear that taking measurements of 'remember' and 'know' responses, in addition to accuracy and confidence, may be beneficial when using recognition memory tests to benchmark virtual training environments. On an information-processing level, it is argued that these additional measurements may index the strength of a single memory trace which is based upon two dual continuous processes (recollection and familiarity), and is influenced by the distinctiveness of the virtual environment from reality. On the basis of the literature reviewed above, this final section summarises a brief descriptive model to aid future considerations of these issues in the design and implementation of immersive visualizations used for training purposes. The descriptive model put forward here is by its very nature ad hoc and is intended only to formalise the key messages emerging from the relevant literature across the discipline boundaries. More broadly, the intention is to capture both the existing data and current thinking in cognitive psychology to guide future approaches to benchmarking as well as future research efforts.

The model has the following three propositions:

(a) The subjective quality of memories for objects and their locations in synthetic scenes is determined by at least two processes (recollection and familiarity);
(b) Recollection processes are positively influenced by the extent to which the sensorimotor context or perceptual context differs from reality;
(c) These processes form part of a memory 'strength' which determines later subjective experience.

Given the ad-hoc and speculative nature of this model, each of these propositions is supported to a different extent by the existing literature on the processing of synthetic scenes. More specifically, proposition (a) is most highly supported by both the psychological literature (reviewed in this chapter) and the accompanying evidence that indicates participants in virtual environments cannot only make these judgements, but that these judgements are clearly influenced by design variables. Proposition (b) is supported by the accompanying literature reviewed here which suggests that the less 'real' a synthetic scene is, either perceptually or in terms of sensorimotor interactions, more detailed vivid memorial experiences are reported ('remember' responses). Proposition (c) is necessary given recent evidence which suggests that the processes of recollection and familiarity are not mutually exclusive. This has yet to be investigated in detail within synthetic scenes in virtual environments and it is only assumed that it extends to these.

4.1 Implications and Future Directions

The implications of this model for designers of virtual training environments are simple. It is important to measure the quality of memories as well as the quantity of memories to determine the impact that any design decisions may be having on both of these. If desirable, more vivid recollective experiences may be promoted through modes of interaction, or global perceptual computations, that are not characteristic of reality. Research indicates that this impact transfers beyond the virtual environment into real-world environments [20]. But, if measured, consideration should be given to controlling for both accuracy and confidence when interpreting the results [38].

In terms of further validating the model there are several important directions which need to be addressed. The analyses presented in the original literature on 'remember' and 'know' judgements in virtual training environments was predicated on the existence of two independent processes of recollection and familiarity. Although the results are nominally consistent with the model presented here it is appropriate for future analyses to also take into account both confidence and accuracy as control variables, as suggested [38]. In particular, these controls may resolve some of the minor anomalies and inconsistencies noted in the data. If not, then these aspects should be explored further with a view to developing and validating the model proposed.

In terms of developing the model further, it will be important to consider the impact of design factors upon the familiarity process as well as upon the recollection process. Inspection of the existing data suggests there is little impact upon 'knowing' in these environments but given this is a null result (no difference) further data is required before confidence in this can be increased. Equating confidence and accuracy in analyses may again be of importance here.

In addition, the impact of higher cognitive variables needs to be explored. Initial indications are that consistency of objects with more generalised mental representations of their contexts (schema) can positively impact upon overall accuracy under similar conditions under which 'remembering' is promoted [22, 27]. The extent to which these higher cognitive processes also impact upon the distribution of 'remembering' and 'knowing' experiences is therefore also worthy of investigation.

It is worth noting that the research evidence underpinning these differences within virtual training environments is based upon research conducted by a relatively small pool of researchers, predominantly including the authors of this chapter. It would therefore be beneficial if these more subjective components of memory for virtual training environments were replicated, or further explored, by an alternative research lab. In particular, there is a need for conceptual replications using alternative paradigms and/or alternative measures.

It has been noted that the definition of 'remember' and 'know' responses differs somewhat from some participants' common usage of these terms. It has been indicated that this may lead to inappropriate interpretations, or inaccurate judgements, of 'remember' and 'know' instructions on the part of the individual.

Given that many participants in the accompanying studies may not have spoken English as a first language, it is important to clarify whether similar effects can be found with alternative terms such as 'Memory A' and 'Memory B'. This is particularly important given that a variety of applications are being developed for training purposes across the globe. It is important to assess whether the impact of translation, and cultural contexts, on understanding of the instructions is likely to be lessened with the more neutral terms highlighted.

Above all, given that the utility of taking these additional memory measures has been satisfactorily demonstrated across several different implementations it is now necessary to increase the volume of research and data upon which this model can be developed whilst encouraging others to incorporate it into their benchmarking processes. As part of the development process it would clearly be beneficial to explore the impact of different technical aspects of virtual environments on the more subjective components of these memories. Ultimately, the aim of this knowledge development is to aid more efficient design and benchmarking of the visualizations of synthetic scenes created for training purposes.

References

1. Atkinson RC, Juola JF (1974) Search and decision processes in recognition memory. In Krantz DH, Atkinson RC, Suppes P (eds) Contemporary developments in mathematical psychology, Freeman, San Francisco.
2. Bennett A, Coxon M, Mania K (2010) The effect of stereo and context on memory and awareness states in immersive virtual environments. doi: 10.1145/1836248.1836275
3. Bliss JP, Tidwell PD, Guest MA (1997) The effectiveness of virtual reality for administering spatial navigation training to firefighters. Presence-Teleop Virt 6: 73–86.
4. Bowler DM, Gardiner JM, Gaigg SB (2007). Factors affecting conscious awareness in the recollective experience of adults with Asperger's syndrome. Conscious Cogn 16: 124–143.
5. Brandt KR, Gardiner JM, Macrae CN (2006) The distinctiveness effect in forenames: The role of subjective experiences and recognition memory. Brit J Psychol 97: 269–280.
6. Brandt KR, Macrae CN, Schloerscheidt AM, Milne AB (2003) Do I know you? Target typicality and recollective experience. Mem 11: 89–100.
7. Conway MA, Gardiner JM, Perfect TJ, Anderson SJ, Cohen, GM (1997) Changes in memory awareness during learning: the acquisition of knowledge by psychology undergraduates. J Exp Psychol Gen 126:393–413.
8. Donaldson W (1996) The role of decision processes in remembering and knowing. Mem Cognition 24: 523–533.
9. Dunn JC (2004) Remember-know: A matter of confidence. Psychol Rev 115: 426–446.
10. Eldridge LL, Engel SA, Zeineh MM, Bookheimer SY, Knowlton BJ (2005) A dissociation of encoding and retrieval processes in the human hippocampus. J Neurosci 25: 3280–3286.
11. Gardiner JM (1988) Functional aspects of recollective experience. Mem Cognition 16: 09–313.
12. Gardiner JM, Java RI (1990) Recollective experience in word and nonword recognition. Mem Cognition 18:23–30.
13. Gardiner J, Richardson-Klavehn A (2000) Remembering and knowing. In Tulving E, Craik FIM (eds) The Oxford handbook of memory. Oxford University Press, New York.
14. Gregg VH, Gardiner JM (1994) Recognition memory and awareness: A large cross-modal effect on "know" but not "remember" responses following a highly perceptual-orienting task. Eur J Cogn Psychol 6:131–147.

15. Hicks JL, Marsh RL, & Ritschel L (2002) The role of recollection and partial information in source monitoring. J Exp Psychol Learn 28: 503–508.
16. Hirshman E, Henzler A (1998) The role of decision processes in conscious recollection. Psychol Sci 9:61–65.
17. Jacoby LL (1991) A process dissociation framework: Separating automatic from intentional uses of memory. J Mem Lang 30: 513–541.
18. Levine B, Black SE, Cabeza R, Sinden M, McIntosh AR, Toth JP et al. (1998). Episodic memory and the self in a case of isolated retrograde amnesia. Brain 121: 1951–1973.
19. Mandler G (1981) Recognizing: The judgement of previous occurrence. Psychol Rev 87: 252–271.
20. Mania K, Badariah S, Coxon M, Watten P (2010) Cognitive transfer of spatial awareness states from immersive virtual environments to reality. ACM Trans Appl Percept 7: 9:1–9:14
21. Mania K, Chalmers A (2001) The effects of levels of immersion on memory and presence in virtual environments: A reality centred approach. Cyber Behav 4: 247–264.
22. Mania K, Robinson A, Brandt K (2005) The effect of memory schemas on object recognition in virtual environments. Presence-Teleop Virt 6: 73–86.
23. Mania K, Troscianko T, Hawkes R, Chalmers A (2003) Fidelity metrics for virtual environment simulations based on spatial memory awareness states. Presence-Teleop Virt 12: 296–310
24. Mania K, Woolridge D, Coxon M, Robinson A (2006) The effect of visual and interaction fidelity on spatial cognition in immersive virtual environments. IEEE T Vis Comput Gr 12: 396–404.
25. McCabe DP, Geraci LD (2009) The influence of instructions and terminology on the accuracy of remember-know judgements. Conscious Cogn 18: 401–413.
26. McCabe DP, Geraci L, Boman JK, Sensenig AE, Rhodes MG (2011) On the validity of remember-know judgements: evidence from think aloud protocols. Conscious Cogn 20: 1625–1633.
27. Mourkoussis N, Rivera F, Troscianko T, Dixon T, Hawkes R, Mania K (2010). ACM Trans Appl Percept 8: 2:1–2:21.
28. Papadakis, G., Mania, K., Coxon, M., & Koutroulis, E. (2011). The effect of tracking delay on spatial awareness states in immersive virtual environments: An initial exploration. doi:10.1145/2087756.2087848
29. Parkin AJ, Gardiner JM, Rosser R (1995) Functional aspects of recollective experience in face recognition. Conscious Cogn 4:387–398.
30. Perfect TJ, Dasgupta ZRR (1997). What underlies the deficit in reported recollective experience in old age? Mem Cognition 25: 849–858.
31. Rajaram S (1996) Perceptual effects on remembering: Recollective processes in picture recognition memory. J Exp Psychol Learn 22: 365–377.
32. Rajaram S (1998) The effects of conceptual salience and perceptual distinctiveness on conscious recollection. Psychon B Rev 5: 71–78.
33. Rotello CM, Zeng M (2008) Analysis of RT distributions in the remember-know paradigm. Psychon B Rev 15: 825–832.
34. Tulving, E (1985) Memory and consciousness. Can Psychol 26: 1–12.
35. Waller D, Hunt E, Knapp D (1998) The transfer of spatial knowledge in virtual environment training. Presence-Teleop Virt 7: 129–143.
36. Wheeler MA, Stuss DT (2003). Remembering and knowing in patients with frontal lobe injuries. Cortex 39: 827–846.
37. Wixted JT (2009) Remember/know judgements in cognitive neuroscience: An illustration of the underrepresented point of view. Learn Mem 16:406–412.
38. Wixted JT (2010) A continuous dual-process model of remember/know judgements. Psychol Rev 117: 1025–1054.
39. Yonelinas AP (1994) Receiver-operating characteristics in recognition memory: Evidence for a dual-process model. J Exp Psychol Learn 20: 1341–1354.
40. Yonelinas AP, Parks CM (2007) Receiver operating characteristics (ROCs) in recognition memory: A review. Psychol Bull 133:800–832.

Human-Centric Chronographics: Making Historical Time Memorable

Liliya Korallo, Stephen Boyd Davis, Nigel Foreman, and Magnus Moar

Abstract A series of experiments is described, evaluating user recall of visualisations of historical chronology. Such visualisations are widely created but have not hitherto been evaluated. Users were tested on their ability to learn a sequence of historical events presented in a virtual environment (VE) flythrough visualisation, compared with the learning of equivalent material in other formats that are sequential but lack the 3D spatial aspect. Memorability is a particularly important function of visualisation in education. The measures used during evaluation are enumerated and discussed. The majority of the experiments reported compared three conditions, one using a virtual environment visualisation with a significant spatial element, one using a serial on-screen presentation in PowerPoint, and one using serial presentation on paper. Some aspects were trialled with groups having contrasting prior experience of computers, in the UK and Ukraine. Evidence suggests that a more complex environment including animations and sounds or music, intended to engage users and reinforce memorability, were in fact distracting. Findings are reported in relation to the age of the participants, suggesting that children at 11–14 years benefit less from, or are even disadvantaged by, VE visualisations when compared with 7–9 year olds or undergraduates. Finally, results suggest that VE visualisations offering a 'landscape' of information are more memorable than those based on a linear model.

L. Korallo (✉) • N. Foreman
Psychology Department, Middlesex University, The Burroughs, Hendon, London NW4 4BT, UK
e-mail: lkorallo@yahoo.co.uk; n.foreman@mdx.ac.uk

S. Boyd Davis
School of Design, Royal College of Art, Kensington Gore, London SW7 2EU, UK
e-mail: stephen.boyd-davis@rca.ac.uk

M. Moar
School of Art and Design, Middlesex University, The Burroughs, Hendon,
London NW4 4BT, UK
e-mail: m.moar@mdx.ac.uk

W. Huang (ed.), *Handbook of Human Centric Visualization*,
DOI 10.1007/978-1-4614-7485-2_19, © Springer Science+Business Media New York 2014

1 Introduction

Our work is concerned with *chronographics*—the visualisation of chronology, especially that of history. The approach is human-centric in two respects. We have undertaken extensive user-testing, comprising 12 experiments involving a cumulative total of 512 participants, the results of which are summarised and discussed in this chapter. While many chronographic visualisations have been created in recent years, none has been evaluated experimentally until now. Our investigation focuses in particular on questions of memorability. The second human-centric aspect of the work is that the user is literally placed at the centre of our visualisations using virtual environment (VE) technologies, positioned so as to take egocentric views on time past, to undertake personal explorations of 'history-space' looking through time and, in our most recent work, looking 'across' time too, rather as though exploring a landscape. We hoped that the use of such an embedded, spatialised user view would produce particular benefits.

Although our application was the learning of history and especially the recall of chronology, our findings have broad relevance. We report on surprising differences in the effectiveness of VE visualisations for different age groups, on some effects of multimedia and other components which are not strictly functional in expressing chronological information, and in our most recent work, suggestions that exploitation of two dimensional 'landscapes' of information are more effective than those that are essentially linear.

1.1 Chronographic Visualisation: The Timeline

In what follows we use the word *timeline* frequently, denoting a graphic layout where time is mapped to a surface or space. The word first appears in its modern sense in William James' Principles of Psychology of 1890 [1], in relation to recording experimental data against time. More than a century earlier there had been a shift from typographic, tabular layouts of historical events to truly graphical time-maps inspired by the ideas of Descartes and Newton [2]. For centuries prior to that, historical events had only been organised into lists and tables. In the mid-eighteenth century, French and then English pioneers began instead to map events in a linear, graphical way. One example was a printed paper chart 16.5 m (54 ft) long, attempting to encompass all history since the biblical Creation. The idea of situating the user within a dynamic representation of historical time was already claimed as a benefit: the timeline was described by its author Barbeu-Dubourg as 'a moving, living tableau, through which pass in review all the ages of the world [...] where the rise and fall of Empires are acted out in visible form' [3], and in fact this particular example was available in a 'machine' where time could be scrolled back and forth by turning handles [4], a surprising anticipation of modern digital approaches to navigating history.

Many of the early aspirations for chronographic visualisations are still with us now. A recurrent theme was memorability, the focus of the present chapter. Le Sage, for example, asked, 'Why is it that an object in geography communicates an idea that is so precise and so specific, and leaves such lasting traces, while a moment in history, by contrast, sinks into nothingness, leaving behind nothing but fleeting impressions? [...]: simply that the knowledge of geography is engraved in our mind by *images*, while that of history is only arrived at by *words*.' [5] (original emphasis). These alleged advantages of visualisation were of course based on intuition and assumptions, not experimental evidence. There was no way to judge whether one visualisation was more successful for the user than another.

Currently digital timelines proliferate, especially on the Web. Often the term is used just to mean a time-ordered list, but many truly graphical examples also exist, plotting time horizontally, vertically or in virtual depth. Sometimes events are attached to a single line, as in most of our examples discussed below, or to multiple lines or a time 'surface'. Different degrees of interactivity are made available, above all scrolling and zooming and related forms of navigation. But again any form of user-centric evaluation is noticeable by its absence.

1.2   A Problem and a Possible Solution: Adopting VE Technologies

We originally set out to address a problem in the learning of history, particularly within school education. An important aspect of historical knowledge is the *framework* of events: both sequence in time, and synchronism of contemporaneous events, perhaps in different fields. History only 'makes sense' when events can be fitted into a framework of this kind. Yet historical time and sequences of historical events are difficult concepts for children to acquire and comprehend. In schools, children usually learn about such abstract concepts by relying on semantic information most often provided on printed worksheets. To learn dates of events, for example, children have no option but to memorize them laboriously, which imparts little understanding of meaningful historical relations. Responding to a questionnaire conducted by the present authors, history teachers reported having used timelines to make history 'less kaleidoscopical and more coherent' [6]. The timeline is the most popular classroom tool to assist children in understanding chronology [7–9].

We wanted to know whether locating the user *within* such a visualisation, using Virtual Environment (VE) technologies to construct a three-dimension time-space which the user could navigate, would make a difference in particular to the memorability of the information it contained. No timeline visualisations have previously been subjected to this kind of research. Our findings do not offer an unequivocal answer, but our most recent experiments suggest the most promising routes to follow.

It is important to note that VE technologies have been extensively applied to history, but generally with the aim of recreating historical sites and artefacts. Our work instead visualises historical time itself, positioning visual markers such as paintings, photographs or objects representing events in a three-dimensional space, of which one dimension represents time. One of the most striking uses of VE technologies for a three-dimensional timeline is Kullberg's 1995 M.Sc. project representing the history of photography [10]. The user could navigate among photographs attached to lines representing the lives of individual photographers, travelling in different directions, and had a choice of either obtaining further information about a selected photograph (by clicking on a relevant icon) or moving on in time to further items. It offered an overview of the environment (from an elevated virtual viewpoint) making it potentially easier for the user to establish spatial relationships—to establish an effective cognitive 'map' [11] amongst places/images—that may subsequently improve recall of the information. However, Kullberg's project included no user-evaluation.

1.3 The Rationale for Using Virtual Environments

One might expect that VE presentations of historical data would have all the standard benefits of visualisation when compared with memorising lists of names and dates. In addition, by situating the user in a time-space we hoped to harness spatial memory rather than semantic memory, in particular since spatial memory is not obviously limited in terms of capacity [12]. Although participants could in principle remember a simple verbal nine-item list, it was hoped that spatial memory would be employed preferentially. In previous studies, for example in which participants experienced rows of shops in a VE, they quickly acquired a good spatial memory for the layout of the shopping mall and for positions of individual shops [13, 14]. After a short period exploring a VE, a participant can make spatial judgments that could only be made using a cognitive "map", such as pointing in the direction of currently-not-visible landmarks [15, 16]. Ours is the most recent incarnation of a long tradition of using physical spaces as mnemonic aids, often referred to as the 'Theatre of Memory', for example in Yates's seminal study The Art of Memory [17].

2 Overview of the Series of Experiments

In all, twelve experiments took place. We do not describe each experiment in detail but rather focus on illustrative examples and on the accumulated findings and discussion. The reader is referred to our other publications for more detailed accounts [12, 18–20]. The purpose here is to give sufficient information to indicate the general nature of the investigation, the characteristics of the different participant groups, and to indicate some firm and some more tentative findings relevant to user-centric visualisation. The aggregated findings on gender effects over all the experiments were inconclusive, so this aspect is omitted.

In our studies, except where otherwise noted, nine historical events were presented as images in a chronological sequence in three conditions, each condition experienced by independent experimental groups.

In many of the experiments described below, two screen-based conditions were evaluated. One was a VE visualisation, in which the user navigated a simple 3D space, so that it seems as though the user travelled in both space and historical time. The other used PowerPoint to sequence a series of images and associated text. In the VE condition, we did not take advantage of the immersive effects of head-mounted displays and stereoscopic vision, principally because our target users were mainly school-children for whom such facilities would currently remain inaccessible. Our use of VE technologies was therefore limited to the construction and delivery of time-spaces which were subsequently displayed and navigated on conventional computer displays.

In the simplest format used for most experiments, pictures or virtual objects representing events were positioned along a line in the virtual space, with successive images appearing alternately on the left or right of the axial line representing time. The user navigated along this timeline sagittally, that is orthogonally to the surface of the screen (for a discussion of the use of the three cardinal dimensions for time, see [21]). Clearly in the case of both screen conditions, the image is in reality two dimensional; however the design of the VE condition using perspectival cues and movement creates an impression that the user moves through a time-space rather than simply seeing a sequence of images.

In the case of classroom studies, efforts were made with the help of teachers to ensure that the comparison groups were equally capable in terms of their previous classroom performance in history lessons, as reflected in standard classroom assessments.

Some aspects of the experiments were modified in the light of experience. Early experiments simply exposed participants to the timeline material. As this produced generally poor results, an element of challenge was introduced into the exploration. These changes are described in more detail below. Other differences between experiments occurred through adaptation to local circumstances in the United Kingdom and in the Ukraine.

The size of the groups used in the experiments ranges from 10 to 20 participants per condition. From a practical point of view, conducting experiments using VEs in schools, it is difficult to access larger populations. The size of the groups was equivalent to those in previous studies of spatial learning conducted using VEs [13, 22–24].

Virtools Dev 3.0 educational version software (www.virtools.com) was used to create the virtual fly-throughs as a Virtools Player File. This was run in the Virtools Player in a standard browser on desktop computers with graphics cards sufficient to deliver smooth full-screen animation and, where necessary, synchronised audio.

Participating schools and teachers were told that the purpose of the research was to attempt to discover means to assist history teaching and learning, so that some benefits might accrue to the school (and other schools) in the medium term. After completion of the studies, children were presented with a simplified version of the results, and teachers were also debriefed. Staff were told their assistance would be acknowledged in any publications. No other incentives were offered. Consent forms were signed and returned by parents in conformity with ethical requirements. With regard to the studies conducted in schools in Ukraine, two separate ethical approvals were obtained: from Middlesex University and from local education authorities in Ukraine.

2.1 Scoring Methods for Experiments

In all 12 experiments, a score was allocated to the degree of error per item in each participant's performance when attempting to place items in the correct sequence, and to the number of correct answers in allocating the items to sequenced slots. A number of other measures were used depending on the focus of each experiment. These are summarised in Table 1.

1. REM score (i.e., "REMOVED" score—how far a picture was placed from its correct position in the sequence; see [12]). For instance, for a particular picture that ought to be placed in position 3, but was placed in position 6, the REM score would be $6 - 3 = 3$. Correspondingly, a correctly placed picture would obtain a score of 0. Each list constructed by a child was given an overall REM score by totalling the REM scores for each of the nine items in the list. This measure was used in all experiments.
2. REM1 the same score as REM, but analysed after a delay period (variously 2–6 weeks). This measure was used in Experiments three, four, six, seven, eight and nine.
3. REM2 was calculated by subtracting from the total Removed Score, the score that was ascribed to the highest-scoring picture. In a nutshell, the difference between REM and REM2 lies in the fact that the former indicates overall accuracy of ordering the pictures, whereas the latter avoids very high scores due to the very bad placement of a single item, despite the overall sequences of the nine pictures being generally well remembered (perhaps all otherwise correctly remembered). This measure was used in Experiments two, three.
4. Correct Order measurement (Corr) indicated how many of the nine pictures were placed correctly in their true list positions in the initial testing phase; participants were given nine slots on a page, as successive dotted lines and labelled 1–9; they therefore placed as many items as possible in the correct numbered slot. This measure was used in all experiments.
5. Correct Order 1 (Corr1), the same as Correct order but measured after delays. This measure was used in Experiments six, seven, eight, nine.

Table 1 Summary of the measures applied in each experiment. See text for an explanation of the abbreviations. The accumulated N for all experiments is 512. Experiments reported in detail elsewhere are indicated in the right-hand column (F07: [12]; K12a: [18]; K12b: [19])

Exp	REM	REM1	REM2	Corr	Corr1	SPE	Qs	Tries	TotErr	Location	Ages	N	Pub'd
1	•			•						UK	U/grad	45	
Environment complexity and 'decoration' has no effect on recall for undergraduate students													
2	•		•	•		•	•			UK	18-22	39	F07:1
VE visualisation enhanced recall compared with two paper-based conditions for undergraduate students													
3	•	•	•	•	•	•	•			UK	11-14	62	F07:2
VE produced no benefits in recall compared with paper-based or PowerPoint conditions for middle-school children													
4	•	•		•						UK	7-9	72	F07:3
VE impeded recall for primary school children; multimedia effects seemed counter-productive													
5	•			•		•	•			UK	18-27	36	
VE with integrated challenge enhanced recall compared with paper-based and PowerPoint conditions for u-grad students													
6	•	•		•	•	•		•	•	UK	8	52	K12a:1
VE produced no benefits in recall compared with paper-based or PowerPoint conditions for primary school children													
7	•	•		•	•	•		•	•	UK	8-9	45	K12a:2
VE enhanced recall compared with paper-based or PowerPoint conditions for primary school children, given more extensive exposure to the material													
8	•	•		•	•			•	•	Ukraine	7-8	30	K12a:3
VE enhanced recall (but not long-term) compared with paper-based or PowerPoint conditions for primary school children, as Exp7, in an alternative context.													
9	•	•		•	•			•	•	Ukraine	mean 12	30	
VE produced no benefits in recall compared with paper-based or PowerPoint conditions for middle-school children, as Exp3													
10	•			•				•	•	UK	Middle School	49	
VE did not benefit recall compared with paper-based or PowerPoint conditions for middle-schoolchildren, despite the use of perhaps more engaging material													
11	•			•				•	•	UK	mean 25	25	
The addition of music synchronised to events located in the VE seemed to be counterproductive													
12	•			•				•	•	U/grad		27	K12b:1
The use of three parallel timelines to create a VE 'landscape' benefitted recall in undergraduate students													

6. SPE: serial position effects. It was of interest to know whether, after experiencing a series of locations laid out in a sequence in space, information would be remembered best (or selectively lost) at the start (primacy) or middle, or end (recency) of the list. The number of items correctly remembered and placed in list positions 1–3, 4–6 and 7–9 was therefore recorded. This measure was used in Experiments two, three, five, six and seven.

7. Qs: Use of a set of questions in the form "Did X come before Y?" Not all studies were designed to explore this variable. Used in Experiments two, three and five.

Measures eight and nine were used when challenge was introduced into the protocol, as described below (Experiment five onwards):

8. Tries: The number of passes through the experiment that participants required to meet the researcher's criterion of two successive passes without error in the training phase. This measure was used in Experiments six, seven, eight, nine, ten, eleven and twelve.

9. TotErr: A total error score, i.e., how many errors were made throughout all passes prior to reaching criterion in the training phase. This measure was used in Experiments six, seven, eight, nine, ten, eleven and twelve.

10. In Experiment twelve, where multiple timelines were used in parallel, additional variables were introduced.

3 The Experiments

We present a sequence of 12 experiments, each of which contributes to one or more of our main findings overall. Two interim discussions are offered, while overall conclusions and discussion end the chapter.

3.1 Experiment One: A Comparison of Historical Chronological Learning from Three Complexities of VE

We describe this experiment in some detail in order to indicate the kinds of VE visualisation created and the experimental methods used. The specific question in the first experiment was whether, in order to be effective for recall, an environment should include non-functional environmental features, imparting some sense of visual and experiential realism, or whether simpler 'diagrammatic' characteristics should be preferred.

Forty-five participants took part in the experiment (9M, 36F). The participants were selected pseudorandomly from within a university student population. The subject domain was the history of art. All participants confirmed that they had no formal art education and were unfamiliar with most art works presented to them during testing. It was established that they were unaware of the chronological ordering of the paintings or the specific year when any one was painted.

Nine paintings were included in the timeline. Within the environment, each painting was inscribed with its title, author and date. Participants were pseudorandomly allocated to one of the three conditions: high, medium and low VE complexity: one (basic or low complexity) was a featureless corridor, one (medium complexity; Fig. 1) modelled a real corridor with windows and other features, and a third (high complexity) allowed user manoeuvres, i.e., using a lift between floors and going upstairs and downstairs.

Participants could move at a constant velocity forward through the virtual space by depressing a key. Other movements were disabled—we had discovered during

Fig. 1 Art History represented in a medium-complexity VE. Nine pictures from the history of art were located in a virtual corridor

a previous pilot phase that users could become disoriented and travel backwards in time while believing themselves to be travelling forwards. The speed of movement gave participants time to read the name of the artist, title, and year of each painting. Participants could also pause in their journey. Passing through the VE took typically 5–6 min. Participants passed through five times, after which they moved on to testing, being given a set of the nine images that they had seen in the environment (minus the inscribed text), printed on A4 sheets of paper. They were asked to place these in the order in which they had seen them on the computer. When they had completed the task, the order of their placed pictures was recorded and scored.

Two dependent variables were analysed: the number of pictures placed in their correct positions in the sequence (Number Correct), and REM (removed scores) using a one way independent ANOVA. The result showed that there was no significant difference obtained between the three conditions on either the REM scores or the Number Correct variable, $F(2,42) = .388$, $p > .05$ and $F(2,42) = .691$, $p > .05$, respectively.

The results showed that the effectiveness of the environment did not depend on its complexity or the inclusion of potentially distracting details. Statistical analysis revealed that participants retained the same amount of information irrespective of the complexity of the environment they experienced. However, later experiments (Four, Eleven) cast additional light on the possible effects of VE complexity, particularly when these use multiple media.

3.2   Experiment Two: Memory for Imaginary Historical Information Acquired from a VE, a 'Washing-Line', Text Alone

In this study, undergraduate students were tested using a nine-item series of historical events that depicted the chronological history of an imaginary planet. A 'washing-line' condition, described below, was introduced because this is a popular way of conveying chronology in school class rooms [7, 9, 25]. A verbal/text protocol was used as the control condition, its presentation using only semantic information being familiar from conventional teaching without visualisations.

A group of 39 undergraduate students (15M and 24F, aged 18–22 years), was pseudorandomly chosen from among the university student population and was pseudorandomly allocated to one of three groups, no specific attention being paid to their prowess in history classes in school. None was a history specialist.

A set of nine images comprising pictures and dates was created, each representing an event in the history of the imaginary planet. These were positioned as successive objects in a VE timeline. Participants could fly through the environment using forward movement only but with full control over their velocity. For the 'washing-line' control condition, the same pictures (with captions and dates) were printed on nine A4 sheets which were then pegged along a string across one wall of the room. For the printed verbal/text control condition, the procedure was the same except that the nine images plus event name and dates were printed, three per page, on three successive A4 sheets in portrait orientation.

The participants were allowed to spend as much time as they wished in each pass-through of the VE (the total time required at maximum velocity being 67 s). After each fly-through, an on-screen dialog prompted them to return to the beginning of the sequence.

In the washing-line condition, participants were asked to scan slowly along the line from left to right. In the verbal/text condition they were asked to look at the three A4 sheets. In all three conditions, the participants were asked to attempt to memorise the history of the planet represented.

All participants, in all three conditions, passed through the materials five times, taking roughly the same length of time to complete the exercise.

The test had two parts: a questionnaire that posed nine questions of the form "Did X come before Y?" requiring true/false responses; and a task to place the nine pictures in their correct chronological order. No time limit was imposed but on average, participants did not spend any longer than 4–6 min doing this.

The following measurements were taken: (1) "Correct number" was the number of pictures placed in their original places in the one to nine sequence; (2) the second was the number of questions correctly answered (out of nine) on the questionnaire; (3) the REM or "Removed" score assessed how far each picture was placed from its correct position; an additional score, Removed2 or REM2 was used, when testing was repeated after an interval. In order to examine serial position effects in the data (SPEs), the number of items placed correctly in list positions 1–3, 4–6, and 7–9

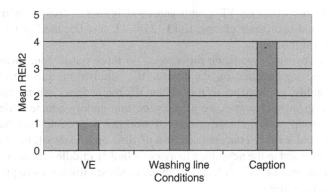

Fig. 2 Mean REM2 scores for the three groups (Experiment Two). The lower the score, the better the performance

were recorded separately for each participant. All data were normally distributed, allowing the use of parametric testing. Post-hoc group comparisons were made using the Least Significant Differences test (with two-tailed probabilities, unless otherwise specified) following the main analysis, when effects were found to be significant. There was a group significant difference in placing the pictures in their correct position $F(2,33) = 4.41$, $p < .05$. Participants in the VE group performed significantly better than participants in the two other groups (either washing-line, $p < .05$, or verbal/text, $p < .05$, groups). Further analysis showed that there was no significant difference found between washing-line and verbal/text groups, $p > .05$. For the number of questions answered correctly, the analysis revealed no significant difference, though the result bordered on significance, $F(2,33) = 2.99$, $p = .06$. Mean scores for the VE group indicated that the number of errors committed in this condition was arithmetically less than the numbers of errors made in the other two groups.

The third variable investigated was the Removed scores. There was a highly significant difference among groups, $F(2,33) = 5.95$, $p < .05$. The participants in the VE condition performed significantly better than those in the other two groups, washing-line and text (p's $< .05$ and .003 respectively). No significance was found between washing-line and text groups, $p = .19$. An additional variable was investigated, Removed2 scores, which revealed the same tendency (Fig. 2), ANOVA indicating that the three groups differed, $F = (2, 33) = 4.64$, $p < .05$. The VE group performed significantly better than the washing-line and caption/paper groups, p's $< .05$ and .005 respectively but no significance emerged between the latter groups, $p > .05$.

Since data variances were not homogeneous for the SPE measure, this was analysed by employing a non-parametric test, the Kruskal-Wallis test, to conduct a one way independent groups' analysis on each successive serial block. Group differences were then examined using the Mann–Whitney U-test. The result showed that there was a group difference in the middle block only (position block 4–6),

$X2(2) = 5.91$, $p < .05$. The VE group achieved higher scores compared to the washing-line and text/paper groups, $U(13,13) = 42$, $p < .05$; the latter two groups failed to differ significantly.

This study revealed significant differences on three out of four measures, and almost reached significance on the fourth, the number of questions answered correctly. Notably, participants who used the VE made fewer errors than those in the other two groups. However, we were aware that undergraduate students might be a special group, with more experience of working with computers than school-age children, and a fuller conceptualization of time and history [26]. With this in mind, a VE was used in the next study to assess whether middle school pupils would benefit from the use of VEs in learning about medieval history as required by the UK national curriculum.

3.3 Experiment Three: The Use of VE Fly-Throughs as Adjuncts to National Curriculum History at Middle School Level

Sixty-two children in two North London schools (29M, 33F, age 11–14 years) were pseudorandomly allocated by class teachers into two groups in one school and three groups in the other (Paper: $N = 24$, 17F, 7M; VE: $N = 26$, 9F, 17M; PowerPoint: $N = 13$, 7F,6M). The data were subsequently combined. The teachers were asked to equally distribute their children across the groups, taking into account their classroom performance in history.

The material, this time on medieval history, was presented in a similar manner to Experiment Two. An innovation was to introduce into Sub-study two ($N = 13$; 6M, 7F) a PowerPoint condition as a second control variable. The visual materials used in PowerPoint were identical to those used in the other two conditions. In order to move on to the next image, a key was pressed. At the end of a session of nine images, an additional screen would appear to invite the participants either to continue with the training task by returning to the starting point (as in the VE) or to proceed to a testing stage. The time taken to pass through the nine items was paced such that it was similar to that in the VE condition. The Paper Condition ($N = 24$; 7M, 17F) involved the children looking through the images presented by the researcher. The pictures would appear in the predetermined correct order, the time taken to pass through all nine being similar to that in the VE condition. As usual, the sequence of nine events was passed through five times in all three conditions.

Each participant was tested individually. The interval between exposure and testing was 48 h. The researcher explained the task by showing nine images presented in an A4 format and asking the children to place the nine images in the correct order. After this, the children were asked to complete a questionnaire, in which they were required to answer questions of the form "Did X come before Y?" To test the hypothesis that VE materials have a greater durability compared with the

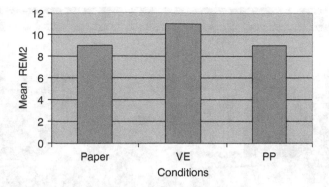

Fig. 3 The mean scores for REM2 for the three conditions (Experiment Three)

materials used in other conditions, a further test session was carried out, comparing a sample of the Paper Group (13 participants) with a sample of the VE Group (13 participants), two months after the original training and testing was completed.

Data were analysed in the same manner as in the previous experiments. In terms of picture ordering, there was no significant effect of condition; $F(2,57) = 1.12$, $p > .05$. When the number of questions answered correctly was analysed, the same pattern emerged, there being no significant differences found. Removed and Removed 2 Scores (Fig. 3) also failed to show any significant result. There was no significant result observed between groups in terms of primacy, middle or recency position blocks, $X2(2) = 1.03$, 1.18 and 1.53 respectively; p's $> .05$.

After a two month interval, there was no difference obtained when two subgroups were compared. Children in the VE condition were not better able to remember the items than those in the Paper condition. Further analysis revealed that there was a high correlation between the picture ordering score in the first round of testing and in testing after the delay, $r(24) = .7$, $p < .001$, suggesting that the measure used was sensitive and reliable.

The results from this experiment were disappointing: the VE presentation was not successful in promoting good scores as seen with undergraduate participants in the previous study. Indeed, participants showed no benefit on any measure from using the VE format in learning the sequence of historical events, and there was no benefit of using a VE in terms of the longevity of memory.

3.4   Experiment Four: The Use of VEs in the Teaching of Primary Level History

The next study involved younger children (primary school participants) who worked with material that had not yet been taught to them in the classroom.

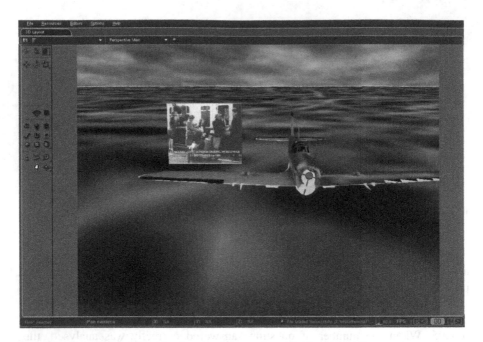

Fig. 4 The multimedia VE with animation and sound used in Experiment Four

Seventy two primary school children took part in this experiment (39M, 33F), 35 children in year 3 (19M, 16F, 7–8 years) and 37 children in year 4 (20M, 17F, 8–9 years). All children had at least some regular classroom experience of operating a computer keyboard.

A set of nine images was used as in the previous studies. A new, multimedia VE format was used, incorporating some animations and sounds such as a French battle cry accompanying the depiction of the battle of Hastings, a rolling Viking boat, and a noisy Hurricane aircraft flying over a depiction of evacuees in World War II (Fig. 4). Movement through the environment was controlled by depressing the space bar. The PowerPoint condition materials were presented as sequences of slides, without any auditory material, using the same computer as the VE condition. The same nine images were used in the Paper condition.

The participants were divided into three separate groups with equivalent numbers of boys and girls and ability range in each condition (VE condition N = 24; 13M, 11F; PowerPoint N = 23; 12 M, 11F; Paper N = 25; 14M, 11M). Testing took place two days after exposure. Each participant was tested individually and spent about 57 min completing the testing task, placing the nine images in order. Subsequently nine yes/no questions were posed in the form "Did X come before Y?"

With regard to the task in which the participants had to place pictures correctly, ANOVA revealed that the three conditions failed to show any difference, $F(2,66) = 1.38$, $p > .05$.

For the number of questions answered correctly, the three conditions differed significantly, $F(2,66) = 3.86$, $p < .05$. The Paper condition was significantly better compared to two other conditions, PowerPoint and VE, p's $< .05$ respectively, these latter groups failing to show any significant difference from one another. Teachers' prior ratings of ability correlated significantly with the questionnaire performance (Spearman's rho$[N = 72] = .22$; p [one-tailed] $= 03$).

When the difference between Removed and Removed1 was analysed statistically, there was no significant difference, $F(2,66) = 1.8$ and 1.4; p's $> .05$. The teachers' ratings of ability were significantly correlated with the participants' performances on both Removed and Removed1 scores, rho$[N = 72] = .19$ and $.20$, p's [one-tailed] $.05$.

Serial order effects were analysed. The Kruskal-Wallis test showed that there was no significant difference when comparisons were made within individual serial position blocks. When data from the first two blocks were combined, however, placement accuracy in these list positions (1–6) was significantly better in the Paper group than in the two computer groups combined, $U(25,47) = 423$, $p < .05$.

The results showed a *dis*advantage of using a VE. The detrimental effect was especially evident when scores for items in early/intermediate positions were analysed.

4 Interim Discussion: Introducing Challenge and Pre-Training VE Experience

At this point in our research, it was clear that our VE chronological visualisations were not universally useful, and could even be counter-productive. Although undergraduate students seemed to benefit from using the VE format, other age groups did not. Middle school children failed to recall more than from other media. Moreover, primary school children actually performed worse compared to control conditions, though they were perhaps distracted by the animations and sounds used in the version of the VE visualisation they experienced.

Other issues might include a lack of engagement with the environment which could affect how much information participants could retrieve when tested since they had experienced it only passively: the only activities available to them were to look and to move their position, far less then, for example, when playing a computer game [16].

In light of out generally disappointing results from simple navigation of a VE visualisation, we next experimented with a more game-like format, in which successive representations of events (paintings, representing epochs of art history) had to be memorized and anticipated during use. As in a computer game, participants' scores were displayed in the upper right corner of the computer screen.

In the interests of brevity, details of the statistical analysis are omitted for the remaining experiments, the conclusions being summarised. Further details of Experiments Six, Seven, Eight and Twelve are available in [12, 18, 19]

4.1 Experiment Five: Introducing Challenge Into the Interaction

Thirty six undergraduate students (18M, 18F) were pseudorandomly drawn from an undergraduate population. They were aged 18–27 years.

The environment used was as in the studies above. The nine pictures were displayed as successive objects in the space with the title, name of the artist, and date of the painting displayed in the upper right corner of each picture. The viewpoint was held stationary while participants guessed what the up-coming image would be. The PowerPoint and Paper conditions were as previously.

All participants were trained individually. For the VE condition, the participants were instructed to observe the environment carefully while depressing the forward arrow key to move through the environment. They were told to look at the pictures and try to remember the order of the pictures, if necessary using terms such as "blue flowers" as descriptors. No attempt was made to draw their attention to specific elements depicted within each picture. The same initial procedure as in Experiment Four was applied. However, on the second fly-through, at the point when the next picture became visible, it was always blank (Fig. 5) and the viewpoint was held stationary by the experimenter. The participant had to describe the still invisible picture; if the answer was correct, the experimenter would click on the screen to display the hidden picture, after which the participant was free to move forward to repeat the procedure with the next image, and the score would increase by one. If the participant described the picture incorrectly, he/she was asked to choose again and an error was recorded. At the beginning of each pass through the environment, the screen counter was reset to zero. The experimenter noted all errors and continued until the participant achieved two successive error-free fly-throughs. In the PowerPoint condition, the same images were displayed as in the VE condition, using full screen images. For the training procedure with challenge a blank screen was displayed and replaced when the image was correctly anticipated. For the Verbal/Text condition, participants were tested with semantic information provided on each plain sheet of paper only (the artist's name, as text, the picture's name and the date it was painted). Following training, after an interval of 5 min participants were assessed using three tests:

1. The numbers one to nine were listed vertically down a test sheet and users were asked to fill in as much information as they could recall about the nine successive pictures, if possible providing the painter's name and the picture's title and date. Then the sheet was removed.

Fig. 5 The VE timeline with following images blanked, waiting for the user to attempt to recall what the next image will be (Experiment Five)

2. A list of nine questions about picture order, of the form: "Did Kandinsky come before Matisse?" was then posed, answerable with "yes" or "no".
3. Finally, a set of nine pictures were placed pseudorandomly (without names or dates) and participants were asked to order them correctly, i.e., to reproduce the order in which they were shown in the training stage. No time limit was imposed, though on average 8 min were spent completing the three tasks.

The dependent measures were: a) During use of the VE visualisation: (1) the number of passes required, excluding the first, for users to achieve two successive error-free passes; (2) the total number of errors made before criterion was reached. b) During post-use testing: (1) the amount of information provided correctly in the first test (nine pictures each having three items of information: painter, picture title, date) so a possible maximum of 27 items; (2) the number of questions answered correctly out of a total of nine; (3) the number of pictures placed in correct order, calculated using a REM score procedure as previously.

Analysis of error positions was conducted by totalling the number of errors made by each participant in training within three successive list position blocks, representing list position blocks 1–3, 4–6 and 7–9, respectively. Note that the VE condition showed almost error free learning, and therefore median scores for all

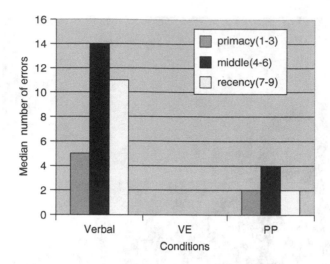

Fig. 6 Median number of errors made in the three conditions (Experiment Five)

blocks were zero, while the PowerPoint group made errors most frequently in the middle list position, and for controls a large number occurred in middle list positions (Fig. 6).

When participants were assessed for their ability to remember information about the pictures, the VE group differed significantly both from the PowerPoint and Verbal/Text controls. When REM score (reflecting the ability to place the pictures in correct chronological order) was assessed, the VE group's performance was entirely error free. As for the PowerPoint condition, two participants made two errors each, while 11 out of the 12 controls made between 2 and 4 errors (overall group mean was 2.5 errors).

This study showed a strong advantage of using a VE format compared with the PowerPoint condition (cf. [27, 28]. Participants in the verbal control group performed especially poorly. During the training procedure, it was evident that participants from the VE condition learned more, and more quickly, compared to the two other conditions; the latter two groups also showed poor retention when tested afterwards. This accords with the study by Hartley et al. [29], who claimed that the spatial relationship between objects is durable and can remain stable over a long time.

It appears that the verbal control group concentrated more on particular items (the picture name) while the experience of each picture with its accompanying textual information enabled the VE participants to absorb more of all kinds of information provided in the environment (spatial sequential and associated verbal). Interestingly, although the amount of information recalled (out of a maximum of 27 items) far exceeded the $7+/-2$ items associated with short term memory [30], suggesting that participants were using a memory store with a limit greater than that traditionally regarded as the short term memory store used for the learning of simple lists of items. On the other hand, the VE group was far from perfect, and their results revealed that they could remember only half of the total information presented.

4.2 Experiment Six: The Use of Challenge in Enhancing Learning in Primary History Teaching

In an earlier study with primary children (Experiment Four, above), nine sequential images were presented chronologically in a VE, depicting eras of history from ancient Greece to World War II. It was found that children in this primary group did not benefit from exploring historical events in the VE format. In fact, they performed significantly poorer in the VE condition than pupils given the same information sequentially on paper (Paper condition) or as a non-spatial sequence displayed sequentially on a computer monitor (PowerPoint condition). The present experiment was designed to improve upon the earlier study by encouraging children to anticipate what was going to appear next, at each sequential choice point. When they anticipated correctly they scored a point (their score being displayed on the screen). This format, therefore, involved more active participation of children in the task and moderate challenge, rather like many computer games. Besides, children were asked to think carefully about historical events presented to them. This adapted protocol might also help to overcome another disadvantage. In the previous experiment (Experiment Four above), children were apparently overexcited by the animations used in the environment and perhaps concentrated less on the main task as a result. By introducing challenge (requiring anticipation, and displaying their score on the screen), children were arguably more concentrated on the main task in this experiment. It was hypothesised that children in this study would perform considerably better than those in the earlier study. Further, the environment itself was designed not to feature any elements that could be considered distracting.

Participants were 52 children (32M, 20F) drawn from a primary school in North-East London, UK. The children were from a single class, the average age being 8 years and 6 months at the time of testing.

A set of nine pictures was used, representing historical epochs, the same set as used in the previous study with primary age children (Experiment Four above). Each picture was dated. A brief description of the picture was added to each; for instance, to represent the ancient Egyptian era, a picture was used which depicted pyramids, with label and date added conspicuously in white lettering to the upper part of the figure. Features were 3D-modelled and introduced into the VE to help to make the child feel "located" in space rather than just viewing a picture. For example, models of Egyptian pyramids were located around the picture of pyramids (Fig. 7), and a virtual Hurricane aircraft flew overhead as the participant approached the evacuation picture. As before, participants could proceed in a forward direction only, achieved by depressing the space bar.

Fig. 7 Images in the VE are supplemented by relevant 3D objects to create a sense of inhabiting the space (Experiment Six)

For the PowerPoint condition during the training phase, the pictures were separated from each other by using a blank PowerPoint screen, displayed for approximately the same length of time (8 s) as participants took to move from one picture to the next in the VE.

In the control Paper condition, the same material was used as above, the pictures being printed on A4 sheets and presented to the child in landscape orientation with text added as in other conditions. Intervening blank pages were shown for 8 s each.

Children were pseudorandomly divided into three groups on the advice of teaching staff, to encompass a similar range of ability in history in each group. These were a Paper group (N = 16; 8M, 8F), a PowerPoint group (N = 18; 12M, 6F), and a VE group (N = 20; 14M, 6F). As in Experiment Five, participants were introduced to the VE with all pictures initially visible, followed by additional passes in which the anticipation (challenge) element was introduced. Scoring was as before. On average, participants required four fly-throughs to achieve criterion. In the Text condition, the nine pictures were presented to the participants by the experimenter. The same anticipation routine was used in all three conditions.

When training was complete, the participant was taken to an adjacent set of desks on which were placed the nine test items, in pseudorandom order. The participant had to place these in the correct chronological order.

Five scores were obtained. Two were during initial training: (1) the number of passes to criterion and (2) a total error score, summing all errors committed prior to reaching criterion. Two further measures were obtained from the initial post-criterion testing: (3) REM score, and (4) Correct Order. A further two scores were obtained when testing was repeated 3 weeks after the original training and testing phases: (5) REM1, and (6) Correct Order1 scores, measures (5) and (6) being calculated in the same ways as (3) and (4).

The results showed that there was no significant difference between groups on any measure, though on total errors, a group effect approached significance. Participants trained in the Text condition were found to have made fewer errors than the PowerPoint group, a result that approached significance. The data showed that even the introduction of challenge into a VE visualisation of historical chronology is not sufficient to ensure effective recall. Indeed, in terms of total errors committed prior to achieving criterion, the Text condition made arithmetically fewer errors than those in the PowerPoint condition to an extent approaching significance, but there was no hint of significance between VE and PowerPoint conditions. Other measures showed no significant differences. The results reinforced earlier findings, that PowerPoint seems to be an especially ineffective medium for conveying chronologically sequenced information [31], and indicated that children of primary school age appear not to experience the kind of benefit from using VEs in the history context that characterizes an older, undergraduate sample.

Some children commented that they did not have computers at home and that they found the task rather difficult to perform, so the poor results might have arisen partly from participants' lack of familiarity in using computers generally. Therefore in Experiment Seven, the same basic protocol was used as in the first study, but children were given extra experience with the environment and input device control before full training commenced.

4.3   Experiment Seven: Challenge and Pre-Training Experience in the Use of VEs to Teach Historical Chronology at Primary Level

This experiment was as the previous one, but the children were given more time to explore the VE. It was hypothesized that by making this modification, ensuring adequate familiarity with the medium, error free learning would be achieved.

Forty-five primary school children (32M, 13F) were pseudorandomly drawn from the population of a school in north-east London by the teaching staff. The numbers for three conditions were: Paper N = 15; 11M, 4F; VE N = 15; 10M, 5F; PowerPoint condition N = 15; 10M, 5F. The children in this study had approximately 5–10 min of extra exposure compared with Experiment Six. Children in PowerPoint and Paper conditions were also given an extra pass through the materials which was adjusted to take approximately the same time as the extra VE training. All other procedures were followed as in Experiment Six.

The same six measures were taken as in Experiment Six. When the number of rounds/passes to criterion was analysed the result was highly significant, showing substantial differences between groups. Post-hoc tests revealed that participants in the VE condition required fewer trials to meet the criterion than in Paper and PowerPoint conditions while there was no significant difference between Paper and PowerPoint conditions. The Kruskal-Wallis Test was used to compare the three conditions, VE, PowerPoint, and Text. The result obtained was significant, $X2(2) = 6.2$, $p < .05$. The Mann–Whitney U-test showed that the VE group was significantly superior to the Paper group on the REM variable, $U(N1 = N2 = 15) = 64.00$, $p < .05$ (two-tailed) and that VE trained participants performed better than PowerPoint participants U $(N1 = N2 = 15) = 70.5$, $p < .05$ (two-tailed) while there was no significant difference found between Paper and PowerPoint groups, $U(N1 = N2 = 15) = 108.5$, $p > .05$ (two-tailed). Clearly from these results, PowerPoint presentation was not as ineffective as suggested by the results of the earlier studies (above). Post-hoc tests failed to reveal significant differences between VE and Paper or VE and PP groups. Other measures showed no significant group differences.

The results were compared with the previous findings from primary school children who did not have challenge incorporated in the protocol, and who performed particularly poorly (Experiment Four, above). Comparability between schools is complicated by differences that may arise from differences in curriculum, computer use and teaching strategies, though this of course applies equally to all of the experimental conditions in which the children were tested. Control groups (both Paper and PowerPoint) did not differ between the experiments but those who used VEs did perform significantly better on the REM and Correct Order variables than those using VEs previously.

Compared with the previous study, the addition of extra pre-training for VE participants clearly improved retention of the historical materials. Significantly better learning was reflected in the lower number of trials needed to reach performance criterion in training and by significantly better Correct Order and REM scores at test. Indeed, performance was error free for all VE participants and thus substantially better than for VEs in the previous Experiment Six and in earlier work.

5   Can VEs Benefit Children's Learning of Historical Chronology in a Culture Where Computer Experience Is More Limited?

The studies reported in above chapters were all conducted in schools in a culture, the UK, where most pupils reported using computers on a regular basis. This might influence results in at least two ways: computer familiarity might make it easier for pupils to use VEs, and navigate more naturally and freely, leading to good retention of historical materials. On the other hand there was evidence from one study

(Experiment Four) that primary children (with limited knowledge of computers, on account of their age) were apparently overawed by the computer experience, leading to especially poor retention. Therefore two studies were conducted, in primary and secondary schools in Sumy, Eastern Ukraine, to examine the effects of using VEs in a country where children have much lower levels of computer familiarity. Challenge was incorporated, as above, by having the children anticipate up-coming images, plus prolonged pre-training, since this combination proved effective in the present experiment. The same comparisons were made among conditions as in UK samples.

5.1 Experiment Eight: Use of a VE to Enhance the Learning of Ukrainian History in a Primary School in Eastern Ukraine

Thirty pupils (14M, 16F, aged 7–8 years old) from school number N.23 in Sumy in the Eastern part of Ukraine took part in the experiment. Children were randomly selected and equally divided into three conditions by the teachers: PowerPoint (N = 10, 4M, 6F), VE (N = 10, 5M, 5F) and Paper (N = 10, 5M, 5F). Teachers asked pupils for details of their typical daily computer use, which was found to be an average of 2.5 h per week. This compares with 10.5 h per week in Experiment Six and 13.8 h per week for Experiment Seven, both conducted in the UK. Unfortunately the VE visualisations used in Ukraine were not identical in form to those used in the UK because of the preferences of the teachers concerning the design.

Nine pictures representing significant events in Ukrainian history were selected with the assistance of teachers, based on the materials used to teach history in the classroom to this age group and representing events considered important for children to remember chronologically. A new VE format was designed (based on teachers' requests) that consisted of four gallery rooms located on two floors in a virtual gallery, similar to those that pupils might visit on school excursions. Each floor consisted of two rooms of the same size. On level one a first room contained two pictures, on opposite walls, while another had two on adjacent walls. The same room layout was replicated on a second floor, in which three pictures were placed in one room and two pictures placed in the final room. In order to move from the first to the second floor a child was required to call a lift, from which the participant was required to go along a short corridor, leading to the first of the level two rooms, after which they could pass across the corridor to the final room. In the training stage, all pictures were dated and named. The PowerPoint condition was conducted using the same materials but as a succession of single screen displays with dates and text; the Paper/Text condition used A4 pictures with dates and text, so replicating the conditions used in Experiments Six and Seven.

As before, teachers were asked to select the children randomly and children were assigned to the VE, PowerPoint or Text conditions. Children in the VE group were asked to look at the VE visualisation together with the researcher, who explained how the environment worked. As in the preceding UK experiment, the experimenter went through the environment with the participant reading and explaining all information depicted on each picture. They were told to try to remember the order of the pictures, plus dates and titles. Participants were then invited to explore the environment by themselves until they were comfortable to move to another stage of the training phase. At this point challenge was introduced (as in Experiments Six and Seven, above) so that participants had to guess which picture would be displayed next, using the same protocol as previously. On average, children required two to three passes to reach criterion. The same was conducted with the other conditions, moving between PowerPoint slides or between successive sheets of A4 paper with printed images, in all cases having the same labels and dates as displayed in the VE condition. After reaching criterion, all children were required to place the images (provided on A4 sheets, but without dates) in correct chronological order. On average 3–4 min were spent completing this task. Overall, children spent 7–10 min carrying out the whole experimental procedure. After a 2 week interval, the testing was repeated.

Six dependent variables (as in Experiments Six and Seven) were analysed. Post-hoc Tukey analyses revealed that for Total Errors, the computer groups (VE and PowerPoint) made more errors than the Paper group, p's $< .05$. On the REM variable the VE group performed much better than the PowerPoint group, $p < .05$, but there was no significant difference between VE and Text groups, $p > .05$. On the Correct order variable, the VE group answered more questions correctly than PowerPoint, $p < .05$ but there was no significant difference between VE and Text groups, $p > .05$. On the Correct1 variable (2 weeks after initial training and testing), the VE group gave fewer correct answers compared to the Text group, p .05.

The main result from this study showed that even among pupils who do not use computers as often as those in the UK, and do not have the same degree of computer familiarity, when challenge is incorporated there is some benefit in using a VE visualisation to acquire historical chronological information. This further reinforces the conclusions from previous studies, showing the benefits of active involvement [22, 32, 33]. Interestingly, however, children in the VE condition here answered fewer questions correctly than in the Paper condition when they were retested after a 2 week interval, which suggests that there was no benefit of using VE presentations in terms of the longer-term retention of information.

Another controversial aspect of the findings was that participants in the VE group during the training phase made more errors in the course of the trials required to meet the "two successive correct passes" criterion, compared to the Paper condition. This is not consistent with the findings from previous experiments (Experiment Six and Seven), in which VE participants made fewer errors in the course of training trials. It is unclear why children did better in the VE group when tested straight after the training phase, but failed to show any significant effect after 2 weeks. This is in conflict with the finding [29] that spatial memory remains stable over a long period of time.

It is important to reiterate that the VE used in this study was different from the environments employed in the research described above. Despite the complexity of the environment that required additional mental effort (using left/right turns, manoeuvring up and down the lift) primary school children did benefit significantly from the VE experiences, although this advantage was no longer evident at follow-up testing, and so there was no lasting effect.

5.2   Experiment Nine: Use of a VE to Enhance the Learning of Ukrainian History in a Middle School, Eastern Ukraine

Having achieved generally disappointing results from the middle school in the UK, but a significant benefit of using VEs with challenge when children were adequately pre-exposed to the medium, the aim of this study was to see whether this would apply equally to a group of children of the same age in the Ukraine, but having much less experience of computer use. Challenge was again incorporated during training, and children were introduced to the VE format individually by the experimenter and given time to explore the environment, to familiarize themselves with the medium prior to beginning the experiment per se.

Thirty (15M, 15F) pupils from a Ukrainian middle school were tested in the experiment. The group was a year group, the average age being 12 years. Typical daily computer use was found to be an average of 1.5 h per week. Ten out of 30 participants did not have any access to computers.

The same materials were used in the experiment as described in the previous study with primary children in Ukraine. The same VE layout was employed. However, three new pictures were added to the existing environment to match the learning material covered by teachers in classroom lessons. All pictures were named and dated as previously.

Children were in three groups: a Paper N = 10; 4M, 6F; PowerPoint group N = 10; 5M, 5F; and a VE N = 10; 5M, 5F, with a similar range of ability in history in each group. The protocol followed was as before. On average, participants required four fly-throughs to achieve criterion.

As in the UK sample (Experiment Three) middle school children showed no benefit from VE training. Most of the variables explored did not show any significant differences. Participants from the PowerPoint group required more trials in order to remember all historical events.

5.3   Experiment Ten: Use of VEs to Enhance Historical Understanding Amongst Middle School Children in the UK

In this experiment, a second exposure to the VE visualisation was included, separated by a period of time from the first. While no immediate beneficial effect of using VE visualisations with Middle School pupils had been found (Experiment Three),

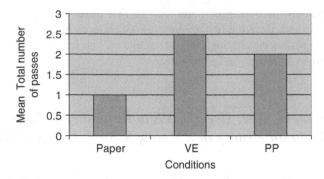

Fig. 8 Mean Total number of passes to meet criteria (Experiment Ten, Part 1)

such might become evident were participants to revisit the same environment after an interval, relearning the same materials and perhaps reinforcing retention.

It was hoped that the introduction of new materials, selected by teaching staff, in this experiment might also encourage children to be more engaged with the environment. In the previous experiment (Experiment Three) with the same age group, where performance was rather poor, it was speculated that this may have occurred because participants were asked to learn medieval materials that they had also been taught about in the classroom, which might have affected their enthusiasm for the experiment.

Forty-nine middle school pupils from North London were randomly selected by teaching staff (26M, 23F). Typical computer use 6.5 h per week.

As usual pictures and labels were the same in all conditions. Images were selected with the assistance of a teacher, who advocated using a horrific image of the victims of the Holocaust with the intention of evoking sympathy and engaging the 11–14 year-old pupils with the content. Other images represented discoveries regarded by history staff as being especially significant. Items were thus selected on the basis of the teacher's assessment of their apparent interest, rather on UK National Curriculum requirements.

The same procedure was applied as in Experiment Eight. Children were randomly divided into the same three conditions, a Paper condition (N = 16; 9M, 7F), a PowerPoint condition (N = 15; 7M, 8F) and a VE condition (N = 18; 10M, 8F).

As usual, initial exposure to the full set of materials was followed by a challenge in which users had to anticipate the next, invisible, item. The difference here was that after a 1 month interval the participants were asked to undergo the same experiment, in which they were asked to go through the training followed by the testing stages, exactly the same procedure being applied as in Part 1 of the experiment.

Five variables were analysed in the initial phase: Total number of passes required to meet criterion, Errors to criterion, REM, Correct order and Serial Order. The Total number of trials differed highly significantly $F(2,46) = 10.35$, $p < .001$ (Fig. 8). A further post-hoc test revealed that participants trained in the VE and PowerPoint conditions required more passes to meet criterion than in the Paper condition (both p's < .001).

The additional measures taken at retraining and retesting were Total errors1, Total number of trials1, REM1, Correct order1 and Serial order effect1. Participants who were trained in the PowerPoint condition tended to place more items correctly than the participants trained in the VE condition. The other three variables did not yield any statistical differences.

We found that VE participants, far from benefiting, required more passes through the environment to meet the experimental criterion. The Correct order 2 variable showed an interesting feature, insofar as participants who were trained in the Power Point condition (contrary to the previous findings above, in which the PowerPoint failed to deliver effective learning) placed significantly more items correctly than VE participants. On the other hand, the Serial Order Effect, when further analysed, showed that the participants who were trained in the VE condition placed more items correctly in the early list positions 1–3 than their counterparts in the PowerPoint condition. Despite the fact that the participants were exposed to the same training and testing stages twice, so that there was plenty of opportunity for any benefits of VEs to emerge, the results did not show any such effect. Throughout all of the above studies with middle school pupils, using different materials, different formats and with different nationalities, the absence of any advantage from using VEs (with or without challenge) was consistent and repeated, in contrast to the benefits that were observed with other age groups when equivalent training procedures were adopted.

6   Second Interim Discussion

The foregoing studies produced interesting data insofar as they demonstrated that VEs might not be effective as memorable media for chronological materials for all age groups, and especially not with middle school children. Despite the fact that other age groups profited from the use of VEs once challenge and familiarity with the medium were incorporated (see Experiments Seven and Eight), children aged 11–13 years old were found consistently not to benefit. In the second study the participants were allowed to explore the environment longer by being trained and tested twice after a short interval. The same strategy has been employed in classroom for children using 2-D timelines [25]. Still, the result showed that even extra exposure did not provoke the participants to perform better in the VE. An additional variable was tested, exploring the lasting effect of the use of VEs.

Although most of the results were non-significant, Experiment Ten demonstrated that children in the second part of the experiment showed a better grasp of materials learnt in a PowerPoint format. The present findings are therefore in disagreement with previous results consistently demonstrating the ineffectiveness of PowerPoint learning (Experiments Six, Seven and Eight).

There are several possible explanations for the fact that middle school children consistently failed to benefit from VE use. First, as suggested above, they may suffer an overload of information, which could be related to rapid biological/hormonal changes that may indirectly affect their ability to concentrate on a task or remember any new materials.

The changes that children experience in their lives at this age should not be underestimated; they experience novel activities that require independent thinking and are encouraged to take full responsibility for their actions (such as travelling to school independently, and learning new routes and strategies). This may reflect changes taking place in their cognitive styles and skills. Studies (see [34–36]) have previously argued that this is a stage at which children's spatial memory is undergoing important changes. In the context of the present studies this is important, since it means that children approach the test situation with immature structures and strategies that might be expected to make high demands on working memory. In other words, they perhaps have greater difficulty than other age groups, in employing the necessary strategies to encode materials in chronological time-space.

From the previous studies, it is evident that the use of a VE format to present sequential historical material for retention might not be beneficial for all ages, especially for middle school children. Undergraduate participants did retain more historical information from VE exposure compared to the other conditions. This can be explained in terms of their familiarity with computers and computer games, though it could also reflect better developed spatial capacities. When challenge and pre-training experience were introduced, undergraduates showed virtually error-free learning, but children at primary level also substantially improved their performance in retaining historical chronological materials. It seems that a computer "game" format might be effective in the teaching of historical chronology when using a VE as it allowed active participation and engagement, and introduced challenge that encouraged participants to be more motivated and try harder. In addition, most of the studies showed that a PowerPoint presentation might not be effective; participants tended to retain less historical information after PowerPoimt experience when compared to the other two conditions.

6.1 Experiment Eleven: Use of VEs in Conveying Parallel Timelines: Art and Music

Following the earlier success in studies carried out with undergraduate students working with VE visualisation, a new study was designed in a similar fashion, using the same paradigm as previously with the same age group, but including an additional domain of information, combining art and music. The number of items presented in each timeline was again nine. While our experiments have dealt with nine-item timelines using a single line, we also want to know whether this number can be exceeded using a more complex VE visualisation. If spatial memory is harnessed in the recall of VE visualisations, we can take advantage of the high capacity of human spatial recall. This should allow us to far exceed Miller's $7 +/- 2$ [30], but if the short term memory buffer is the limiting factor, and it becomes overloaded as successive items are remembered, art information will be dislodged by musical information, so that the total items remembered from the display may total nine but will not exceed it.

In an initial study designed to investigate the storage capacity for materials learned from a VE, the new study used a single timeline but with both art and musical materials presented simultaneously. In this case, a single timeline was used but it incorporated two domains of information—music was played as a line of successive pictures was viewed. The situation replicated what is sometimes reported anecdotally: that a particular piece of music can help spatial recall of a place, or that returning to a place might evoke a memory of music previously heard there. Examples of evocative paintings would seem particularly appropriate to this purpose. The use of spatial memory would be indicated were the amount of information recalled from this timeline greatly exceed nine. For both art and music events, the name of the picture and the tune, the name of the artist and the composer, and the year/period in which they were both created were presented in combination, so that a total of 45 items of information were presented in the course of a participant's passing from the start to the end of the VE visualisation. It was hypothesized that (1) after several successive exploratory trips through the VE, the total information remembered would exceed nine, and (2) a greater proportion of these 45 items of information would be recalled after exploring the timeline in a VE format than by either hearing the extracts of music while viewing the linked pictures as PowerPoint screens or while viewing them conventionally printed on sheets of paper.

Twenty-five undergraduates (9M, 16F) took part in the experiment: VE (N = 7), Power Point (N = 11) and Paper (N = 7). The average age was 25 years.

The nine images were placed in correct chronological order with the title of the paintings, the name of the artist and the year in which the painting was produced superimposed on the picture. A text adjacent to the picture gave details of the concurrent music (name of composer, title of the extract, and year in which it was composed). The music and paintings were selected and paired in such a way that they were chronologically matched.

The extract was programmed to begin playing as the corresponding picture was approached. Challenge was introduced into the three environments. A pair of headphones was used to allow participants to listen to the music excerpts in the VE condition. In a second condition PowerPoint was used, the same protocol being used as in the VE condition, such that participants viewed the same paintings along with the music excerpts. Similarly, the music details as well as the paintings' details were also shown on the screen. In the third, Paper condition, the painting was provided on a plain piece of paper with the name of the artist and the title of the painting. In contrast to the other two conditions, the music was not played at the same time as the pictures were shown, but the details of the music were displayed.

Individual training was provided for each participant in the VE condition. Participants were told that their task was to remember the order and details of each painting as it appeared on the screen. Participants could only move forward as in previous experiments. At the same time as the painting was displayed the music was played, matching the duration of time with the painting displayed. After this, participants were told to move forward to reach another painting; the same protocol was used throughout the environment. The participants were instructed to look at each painting along with the details of the music. Also they were told that they would be later asked to anticipate which painting was going to appear next. To meet a criterion, the participants had to guess nine paintings correctly twice in a succession. After completing the first fly-through, they were asked whether they felt confident enough to complete a task i.e. to recall the images in correct order. If they did not feel confident enough, the experimenter would reset the environment from start point until the participant completed the task successfully. The participant had to recall each painting by saying the name of the artists, the title of the painting or by describing the themes of the images.

For the first test, all nine paintings including the artists' name and the title were presented randomly on a desk. The participants were asked to place them in historical chronological order, the order in which they were displayed during training. The participant was then asked to place the name and the title of the corresponding music in chronological order on a desk. For the final part of the experiment, the experimenter instructed the participants to match the music details along with the name and the title of the images. The experimenter marked the order of the music as well as the paintings. There was no time limit to perform this task. The whole procedure would typically take about 4–5 min to complete.

The dependent measures of the present study were: the number of correct images placed in chronological order; the total number of error made; the number of passes until the learning criterion was met; the amount of information remembered in three testing conditions (correct chronological order in music, placing paintings and matching music and paintings together); ability to place items in an orderly chronological sequence, assessed using a REM score.

Analysis of the variables showed that the REM picture measurement was significant, $F(2,22) = 3.98$, $p < .05$. The REM music variable showed no significant difference between conditions, $p > .05$. REM music and pictures also showed that there was no significant difference between control and experimental groups. However, the Tries variable revealed that condition differed significantly, $F(2,22) = 7.087$, $p < .05$. The REM picture variable showed that there was a significant difference obtained between VE and PowerPoint groups suggesting that VE trained participants made fewer errors when they were tested on placing pictures in order, $p < .05$.

Contrary to hypothesis, when the additional variable was added—music—participants' performances varied but were not universally enhanced. Not all information was equally well remembered. Clearly, the addition of music might have distracted and detracted from the learning of the art materials. While placing pictures in order benefitted, other variables failed to yield any significant differences. A very surprising aspect of the study was that participants who were exposed to the VE condition required more passes compared to the PowerPoint condition to reach criterion.

6.2 Experiment Twelve: Can Undergraduate Students Acquire Knowledge Effectively in Three Domains Simultaneously Using a VE with Three Parallel Timelines?

The final experiment dealt with a two dimensional time structure situated in time-space, rather than a mere line with attached objects. Spatial memory systems are distinct but interacting [37]. Multiple cues and landmarks can be used as navigational aids that allow the formation of organizational relationships with other points in space [38]. Thus people acquire knowledge about a route by seeing objects sequentially [37], that can be encoded in relation to other locations rather than from a particular stand point [29, 39]. The spatial relationship between objects is durable and can remain stable over a long period of time; it can encompass large complex, vivid and detailed spaces [29, 40–42]. We wanted to know whether presenting events in a triple timeline structure would take better advantage of spatial memory than did a single line.

A previous study using a nine item fly-through showed that undergraduate participants benefited significantly from learning about the history of an imaginary planet by using VEs, when exposed (without challenge) to just one timeline (Experiment Two, above). A further series of studies working with primary school children also showed the benefit of using VEs, especially when children had adequate time to explore the environment and when challenged by using a game format. In the present study a different form of environment was used, incorporating 12 items in three different timelines, history of psychology, general history, and art. Participants were given more time to explore the VE (over a 2 week period) after which they were asked to return and participate in a series of tests. From previous research, and experiments above, it was evident that longer exposure to the environment improves participants' performance in the short term; despite some authors having exposed participants to virtual environments for only a few minutes [43] the acquisition of spatial information from very large scale virtual environments has been said by others to require a considerable period of time [44].

Twenty-seven participants (21F, 6M) took part, fourteen in the VE group (4M, 10F) and thirteen in the Booklet group (2M, 11F). They were randomly selected from a Year 1 university student population. It was ascertained that they did not have specialist knowledge in advance of any area covered by the timelines beyond a Year 1 knowledge of Psychology. Their average age was 24 years.

A triple timeline VE visualisation was used. The same materials (images and information) were used to produce three booklets (in A4 format with coloured images) were produced. Events in the three domains—psychology, general history, and art—were matched according to the year in which they occurred (Fig. 9).

Participants were asked to read a brief introduction to the study which specified what they needed to do. Participants were randomly divided into two groups, one (experimental group) that was exposed to the VE and another (control group) who worked with a paper version of the environment designed in a booklet format. The VE group received training that consisted of passing through the environment

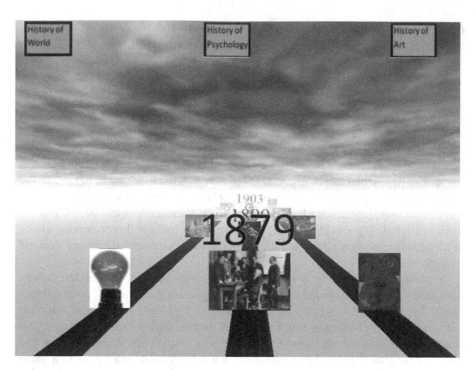

Fig. 9 The three-timeline environment representing History of Psychology, History of Art and General History (Experiment Twelve)

together with the researcher, who ensured that the participant knew how to operate (load, run and fly through) the environment. After the training procedure was complete, the participant was asked to take the environment home (or they were sent it as an e-mail attachment) where he/she could explore it in greater detail at their leisure. The latter was strongly emphasized by the researcher. Also, the researcher pointed out that all information presented in the environment should be considered, as if the participant was being asked to revise for an examination. The control (booklet) group was effectively given the same task, but asked to learn the materials in the three timelines by using three separate booklets depicting the same historical events as presented in the VE. A similar amount of time was spent with controls, explaining the booklets and required procedure, as was spent with the VE group explaining the fly-through. All participants were provided with a chart, on which they had to log the number of hours they had spent working with the materials (VE or booklet). The participants were asked to return after 2 weeks for a testing stage, although the objectives of the test were not disclosed in advance. The testing stage, for both groups, consisted of four parts. In Test 1, the participants had to recall the items learnt in their condition, but not in any particular order. In Test 2, they had to place events presented in a selected timeline in the correct chronological order. The same procedure was repeated for each component timeline. In Test 3,

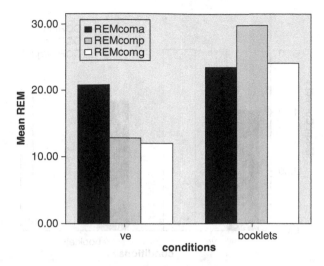

Fig. 10 REMcom: Mean REM scores for each domain/timeline when the three were tested together (*a* art, *p* psychology, *g* general history) (Experiment Twelve)

participants had to place together the events that took place in the three domains, i.e., History of Art, History of Psychology and General History, simultaneously. Finally, a questionnaire was designed to investigate whether participants could relate one timeline to another, and whether simultaneity could be identified between the events in the timelines. For instance, one of the questions asked: "What happened in the History of Psychology when event X occurred in the History of Art?"

The independent variables in the present study were the domain (art, psychology, general history), condition (Virtual Environment versus Booklets), and the gender of participants.

Fourteen dependent variables were measured: six revealed a significant statistical difference between the two groups. The VE group performed better in terms of correct recall of list positions for all three timelines (Figs. 10, 11, and 12). Participants from the Virtual group could also answer more questions correctly than controls. Total items correctly remembered, across all three timelines, approached significance. The VE group performed much better overall than controls in terms of their ability to relate together the events occurring simultaneously in the three timelines.

There was no difference between the groups in terms of the amount of time they spent in studying the materials, either reading the booklets or learning the materials from the VE. On average the two groups spent 3 h on the activities prescribed by the researcher.

This study differed from its predecessors in that a VE group was compared only with a group learning from a booklet, though using a booklet to learn historical materials is a suitable control since it resembles the materials often used in teaching. Participants were given much longer familiarization periods, to encourage the use of spatial encoding and the memorizing of materials rather like learning the layout of a small town when making daily trips through its streets.

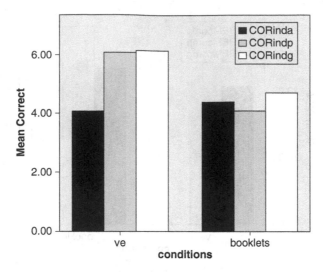

Fig. 11 Mean number of correctly placed items for each domain/timeline when tested independently (*a* art, *p* psychology, *g* general history) (Experiment Twelve)

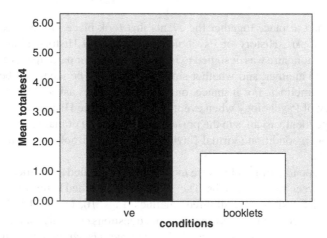

Fig. 12 The total number of items recalled across all three timelines in Test 4 by two groups (Experiment Twelve)

The results suggest that this protocol was successful. They consistently showed that using a VE gave significantly better performance than learning from a booklet. Six variables showed a statistically significant difference between the two groups, demonstrating the effectiveness of a VE, at least for this undergraduate age group. The amount of information to be remembered was substantial, in total 36 items to remember, with dates and textual information, yet still VE participants remembered more than their counterparts using booklets. According to the verbal reports of VE participants, the most important factor that helped them was their ability to connect events with each other—to see a structure and a point of reference, being able to look across the three timelines, suggesting in turn that they were genuinely using a "survey" form of cognitive representation (c.f. [11, 38]). This may be explained in terms of the fact that initially VE participants, unlike booklet participants, were exposed to the three timelines simultaneously, movement in virtual space giving them a better experience of time-space, allowing them to change their position fluently in relation to landmarks, historical images, and facts. As for the control group, they were limited in performing these activities in the sense that they could not easily visualise which historical facts happened simultaneously. For them the information that they were asked to memorize was presented with items isolated from each other, so lacking a sense of historical coherence, structure and organization. The result of the present study strongly suggests that VE visualisations of time have potential for further investigation, because although only three timelines were used in the present study, there is no reason why a more elaborate spatial environment could not encompass many timelines and a quantity of information similar to that remembered (as buildings, streets, squares) in a familiar town.

7 Concluding Discussion

This research has generated interesting but challenging results. It started from a naïve hypothesis, that just moving past events presented as pictures and other markers placed spatially in a virtual environment would instil these as places in users' spatial-temporal memory and make them memorable in correct order. The hypothesis, although naïve, still proved to be a good starting point, being plausible from previous work in which VEs have successfully conveyed spatial information. The results overall have suggested that the technology could be developed in such a way as to be valuable for specific purposes if used in the correct ways, but there are still questions over the effect of the users' age. Indeed the most striking finding, one which may turn out to be applicable to other forms of visualisation, concerns the effect of the age of the users. As we have seen, this is not a story of increase or decrease with age: the youngest and oldest participants performed better than those in between.

As we move around a real world environment, even at first visit to a new location, it seems as though we more or less effortlessly store some model of the place/space, and can remember other information with the assistance of that model. From our experiments, it seems that simply encountering events in time modelled as a spatialised environment does not have the same 'automatic' benefits. We have shown that we often had to cajole our participants using in-built game-like challenges (though these were also introduced across the other conditions) in order to produce significant gain.

An unresolved question is whether the spatial visualisation we were using carried all the potential benefits of experienced spatiality. This might be part of the explanation of our failure to get the effects associated with learning the layout of a real place. Although earlier work had shown that VE models of physical places seem to be learned in similar ways to real places, important cues may be missing when the user merely sits at a monitor navigating a virtual space. Further work is needed to discover what benefits for a similar domain might arise from (1) using immersive VE technologies in place of the screen and (2) harnessing the physicality of movement and proprioception as discussed for example by Price and Rogers [45].

It might be expected that an environment having a variety of engaging and 'realistic' features would promote the greatest learning, especially where the motivation of young children is concerned. However, when a more complex environment was used that included many animations and sounds, primary school children appeared to be distracted and consequently did not gain as much historical information as expected and, when tested, they showed no improvement in retaining information compared to other conditions. There is thus no evidence of benefit from 'decorative' motivational objects and experiences in the environment. Indeed, deciding for any given visualisation which aspects are 'necessary' or functional in itself requires investigation. In the domain of charts and similar visualisations there is unresolved controversy over the question of 'chart-junk' [46, 47], a concept analogous to some of the features we introduced into timelines here.

Some prominent questions raised therefore include:

1. To what extent are the findings of the research reported here, in particular relating to age-related difference, generalizable to other domains, users and formats?
2. What more can be discovered about the benefits and drawbacks of non-functional elements in diagrammatic visualisations?
3. More fundamentally, how can we define the borderline between those aspects of a visualisation which are strictly functional and those which are 'decorative'?
4. Would the results be different if the users' encounter with the VEs was immersive, using, for example, head-mounted displays with stereoscopic viewing?

Further work is also needed on the dynamic relation of the user to the visualisation. We explained that we constrained the movement of the user following the pilot studies. Work is needed to explore the most favourable kinds of allowable movement and constraints.

Acknowledgements The authors would like to thank David Newson for creating the VEs; headteachers, teachers and pupils at Alexandra Park School, North London, Hoddesdon School in Hertfordshire, Northumberland Park School in Tottenham, North London (especially Peter Molife), Worcesters and George Spicer Schools in Enfield, London, and School No. 23 in Sumy, Ukraine for their participation; and the Leverhulme Trust for financial support (grant number F/00 765/D).

References

1. W. James. *The Principles of Psychology*. (2vols.) New York: Henry Holt, p. 86, 1890.
2. S. Boyd Davis, E. Bevan and A. Kudikov. Just In Time: defining historical chronographics. *Proc. Electronic Visualisation and the Arts: EVA London 2010*. 5th-7th July 2010, British Computer Society, London. 355–362. [Online]. Available: http://www.bcs.org/content/conWebDoc/36111 (Accessed 30 July 2012), 2010.

3. J. Barbeu-Dubourg. *Chronographie, ou Description des Tems* (Chronography, or Depiction of Time). Pamphlet accompanying Chronographic Chart. Paris. (With a translation by S. Boyd Davis, 2009). [Online]. Available at http://goo.gl/vNhN (Accessed 30 July 2012), p.8, 1753.
4. S. Ferguson. The 1753 Carte Chronographique of Jacques Barbeu-Dubourg. *Princeton University Library Chronicle*. (Winter 1991). [Online]. Available at http://www.princeton.edu/~ferguson/ (Accessed 30 July 2012), 1991.
5. A. Le Sage. [Emmanuel, comte de Las Cases]. *Genealogical, Chronological, Historical and Geographical Atlas by Mr Le Sage*. London: J. Barfield. 1801.
6. L. Korallo. *Use of virtual reality environments to improve the learning of historical chronology*. Unpublished Ph.D. thesis, Middlesex University, p. 40, 2010.
7. A. Hodkinson. Historical time and the national curriculum. *Teaching History*, 79 (April 1995), pp. 18–20. 1995.
8. J. West. Young children's awareness of the past. *Trends in Education*, 1, pp. 8–15, 1978.
9. S. Wood. Developing an understanding of time-sequencing issues. *Teaching History*, 79 (April 1995), pp. 11–14, 1995.
10. R.L. Kullberg. Dynamic Timelines: visualizing historical information in three dimensions. MSc Dissertation, Massachusetts Institute of Technology, September 1995. [Online]. http://citeseerx.ist.psu.edu/viewdoc/download?doi=10.1.1.51.5278&rep=rep1&type=pdf (Accessed: 1 August 2012). 1995.
11. J. O'Keefe and L. Nadel. *The hippocampus as a cognitive map*. Oxford: Clarendon Press. 1978.
12. N. Foreman, S. Boyd Davis, M. Moar, L. Korallo and E. Chappell. 2007. Can Virtual Environments Enhance the Learning of Historical Chronology? *Instructional Science* 36(2). ISSN 0020-4277, pp. 155–173, 2007.
13. N. Foreman, P. Wilson, D. Stanton, H. Duffy and R. Parnell. Transfer of spatial knowledge to a two-level shopping mall in older people, following virtual exploration. *Environment and Behaviour*, 37, pp. 275–292, 2005.
14. M. Tlauka, P. Donaldson and D. Wilson. Forgetting in spatial memories acquired in a virtual environment. *Applied Cognitive Psychology*, 22, pp. 69–84, 2008.
15. N. Foreman, D. Stanton, P.N. Wilson and H. Duffy. Spatial knowledge of a real school environment acquired from virtual or physical models by able-bodied children and children with learning disabilities. *Journal of Experimental Psychology: Applied*, 9, pp. 67–74, 2003.
16. G. Sandamas and N. Foreman. Spatial reconstruction following virtual exploration in children aged 5–9 years: effects of age, gender and activity-passivity. *Journal of Environmental Psychology*, 27(2), pp. 126–134, 2007.
17. F.A. Yates. *The Art of Memory*. Chicago: University of Chicago Press, 1966.
18. L. Korallo, N. Foreman, S. Boyd Davis, M. Moar and M. Coulson. 'Can Multiple 'Spatial' Virtual Timelines Convey the relatedness of Chronological Knowledge across Parallel Domains?' *Computers & Education* 58(2), February 2012, pp. 856–862, 2012.
19. L. Korallo, N. Foreman, S. Boyd Davis, M. Moar and M. Coulson. Do challenge, task experience or computer familiarity influence the learning of historical chronology from virtual environments in 8–9 year old children? *Computers and Education* 58(4). May 2012, pp. 1106–1116, 2012.
20. L. Korallo. Using VEs to enhance chronological understanding of history. *Research Studies, Literature Reviews and Perspectives on Psychological Sciences*. Edited by Thanos Patelis. Athens Institute Of Education and Research, pp. 113–143. In press.
21. S. Boyd Davis. History on the Line: time as dimension. *Design Issues* 28(4). September 2012, pp. 4–17, 2012.
22. J. Pedley, L. Camfield and N. Foreman. Navigating memories. in B. Ahrends and D. Thackara (Eds.), *Experiment: Conversations in Art and Science*. London: Wellcome Trust, pp.173–235, 2003.
23. A. Cockburn and B. McKenzie,. Evaluating spatial memory in two and three dimensions. *International Journal of Human-Computer Studies*, 61, pp. 359–373, 2004.

24. R.A. Ruddle, S.J. Payne and D.M. Jones. Navigating buildings in "desk-top" virtual environments: Experimental investigations using extended navigational experience. *Journal of Experimental Psychology: Applied*, 3, pp. 143–159, 1997.
25. I. Dawson. Time for Chronology? Ideas for developing chronological understanding. *Teaching History* 117, pp. 14–22, 2004.
26. J. Howson. Is it the Tuarts and then the Studors or the other way round? The importance of developing a usable big picture of the past. *Teaching History*, 127, pp. 40–7, 2007.
27. L. Smart. *Using IT in primary school history*. Cassell, London. 1996.
28. J.D.M. Underwood and G. Underwood. *Computers and Learning*. Basil Blackwell Ltd., Oxford. 1990.
29. T. Hartley, I. Trinkler and N. Burgess. Geometric determinants of human spatial memory. *Cognition*, 94, pp. 39–75, 2004.
30. G.A. Miller. The magical number seven plus or minus two: Some limits on our capacity for processing information. *Psychological Review*, 63, pp. 81–97, 1956.
31. T. Haydn. Multimedia, Interactivity and Learning: some lessons from the United Kingdom. Current Developments in Technology Assisted Education (2006), *Proceedings of m-ICTE2006*, vol. 1, 110–114. University of Seville, Spain, 22–25 September 2006, pp. 110–114, 2006.
32. P.N. Wilson. Use of virtual reality computing in spatial learning research. In N. Foreman & R. Gillett (Eds.), *Handbook of spatial research paradigms and methodologies: Volume 1, Spatial cognition in the child and adult* Hove: Psychology Press, pp. 181–206. 1997.
33. G. Wallet, H. Sauzeon, A. Prashant and B. N'kaoua. Benefit from an active exploration on the transfer of spatial knowledge: Impact of graphic richness. *13th International Conference on Human-Computer Interaction*. 19–24 July, San Diego, 2009.
34. A. Flickinger and K.J. Rehage. Building time and space concepts. *Twentieth yearbook*. National Council for the Social Studies. Menasha, WI: George Banta Publishing, pp. 107–116, 1949.
35. G. Jahoda. Children's concepts of Time and History. *Education Review*, 15 (February 1963), pp. 87–104, 1963.
36. W.J. Friedman. Conventional time concepts and children's structuring of time. In W.J. Friedman (Ed.) *The Developmental Psychology of Time*. London: Academic Press, pp. 171–206, 1982.
37. G. Jansen. Memory for object location and route direction in virtual large-scale space. *The Quarterly Journal of Experimental Psychology* 59(3), pp. 493–508, 2006.
38. P. Jansen-Osmann and G. Wiedenbauer. The representation of landmarks and routes in children and adult: a studying a virtual environment. *Journal of Environmental Psychology* 24, pp. 347–357, 2004.
39. S. Moffat, A. Zonderman and S. Resnick. Age differences in spatial memory in a virtual environment navigation task. *Neurobiology of Aging* 22, pp. 787–796, 2001.
40. E. Maquire, R. Nannery and H. Spiers. Navigation around London by a taxi driver with bilateral hippocampal lesions. *Brain* 128, pp. 2894–2907. 2006.
41. M. Moscovitch, L. Nadel, G. Winocur, A. Gilboa and R.S. Rosenbaum. The cognitive neuroscience of remote memory: a focus on functional neuroimaging. *Current Opinion: Neurobiology* 16, pp. 179–190, 2006.
42. R.S. Rosenbaum, S. Priselac, S. Kohler, S.E. Black, F.Q. Gao, L. Nadel and M. Moscovitch. Studies of remote spatial memory in an amnesic person with extensive bilateral hippocampal lesions. *Nature Science*, 3, pp. 1044–1048, 2000.
43. D. Waller. Individual differences in spatial learning from computer-generated environments. *Journal of Experimental Psychology: Applied*. 6, pp. 307–321. 2000.
44. R. Darken and J. Silbert. Navigating large virtual spaces. *International Journal of Human-Computer Interaction* 8, pp. 49–71, 1996.
45. S. Price and Y. Rogers. Let's get physical: the learning benefits of interacting in digitally augmented physical spaces. *Computers & Education* 43(1–2). August 2004, pp. 137–151, 2004.

46. E. Tufte. *Envisioning Information*. Cheshire, Conn.: Graphics Press, pp. 34–35, 1990.
47. S. Bateman, R.L. Mandryk, C. Gutwin, A. Genest, D. McDine and C. Brooks. Useful junk?: the effects of visual embellishment on comprehension and memorability of charts. *Proceedings of the 28th international conference on Human factors in computing systems (CHI '10)*. ACM, New York, NY, USA, pp. 2573–2582, 2010.

Visualizing Multiple Levels and Dimensions of Social Network Properties

Cathleen McGrath, Jim Blythe, and David Krackhardt

Abstract In this chapter, we develop a framework for understanding how multidimensional, multilevel data is most effectively conveyed in social network diagrams. We build on work begun in 1994, with a series of explorations of social network visualization with the theme of helping viewers make accurate judgments about network properties. In contrast to contemporary work on layout aesthetics, we were interested in helping viewers with questions that may lie beyond the mathematical graph to the social group that it represented, and that may not be readily available by inspection, unlike path length or degree. We found that different layouts of the same graph had appreciable effects on viewer's judgment of the importance of individuals with the network individuals or the number of subgroups. In addition, we found that the use of motion to link alternative views of a network had different effects depending on the layouts used. In this chapter we describe and organize some of these results with respect to the long-term goal of aiding human judgment over complex structured data.

1 Introduction

Social network visualizations have become ubiquitous from boardrooms to Facebook (see for example, Paul Butlers note on Visualizing Friendship among a very

C. McGrath
Loyola Marymount University, Los Angeles, CA, USA
e-mail: cmcgrath@lmu.edu

J. Blythe (✉)
Information Sciences Institute, University of Southern California, Los Angeles, CA, USA
e-mail: blythe@isi.edu

D. Krackhardt
Heinz School of Public Policy and Management, Carnegie Mellon University, Pittsburgh, PA, USA

W. Huang (ed.), *Handbook of Human Centric Visualization*, 513
DOI 10.1007/978-1-4614-7485-2_20, © Springer Science+Business Media New York 2014

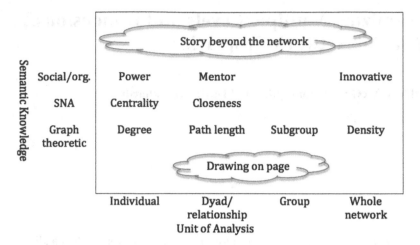

Fig. 1 Levels and dimensions of social network properties

large set of Facebook users [3]). Individuals can use these visualizations to assess the general social landscape, for example, or see where they fit in a complex social web. As tools to provide visualizations have become increasingly sophisticated they have equally become more accessible. That is why we need to understand how the presentation of social network information can support the way it is understood. There is so much use of social network visualization that it is more important than ever to take time to understand how and how much information is communicated through social network visualization.

We review a series of experiments that we began in 1994 to answer this question. Our first set of studies attempted to strip a visualization to its barest bones of a black and white display of nodes and edges in two dimensions to explore the impact of spatial positioning of nodes on viewers' inferences. We proceed to a study that is based on real organizational data and uses motion and differential graph layout rules.

When we use graphs to represent people and organizations, we are communicating multi-dimensional knowledge, and when individuals consider a visualization, they are typically reasoning about the underlying social structure. There is usually more information known about the structure than is explicitly represented in the visualization. Semantic knowledge about the graph includes the graph theoretical dimension, social network concepts, and information about the social and organizational dynamics of the group. At the same time, social network visualizations present information about different levels of the graph. When viewing a social network, one can learn information about individuals, dyads, groups and the whole network. Figure 1 gives some examples of types of information of different levels and relating to different groupings that may be conveyed in a social network graph.

Through our work on KrackPlot we recognized the role of the graph as a means of communication of information about social groups [8]. The success of the communication depends not only on the way researchers choose to encode information through graph representation techniques such as graph layout, animation, and user interaction, but also on the way the graph is perceived and information is decoded by the viewer of the graph. Our work began to uncover the fact that the viewer is likely to make inferences about the rich social structure underlying the group at multiple levels, even if the viewer had not intended to convey this information explicitly.

We can separate the semantic content of a network into (1) the structure of the data to be displayed and some display goal, (2) the background/general information about the domain that the observer has, that affects what they draw from the diagram and (3) the set of display conventions that are being used by the observer in interpreting the network. Note that the display conventions observed may be inadvertent, perhaps unavoidable, artifacts of the display process. Partly because of the background knowledge we explicitly provide in these studies and partly from the background of the classes that subjects were taking, these factors and the interplay between them play a role in all our experiments, but most of all in the project involving motion. It is also important to distinguish whether the experiment involves solving a problem that relies on background information beyond what can be observed, or if it involves estimating a network property from the diagram. Problems involving background knowledge are closer to the real-world uses of network visualizations in management and other fields. One effect of a more complex problem is to amplify the semantics involved, another is to introduce a utility for the inferences, so that the observers beliefs about the network matter to them and may be tested more completely.

For each study, we begin with the assumption that the structural characteristics of the network itself does impact viewers perceptions because of the reality of the structural properties and also because of the constraints that network structure places on how graphs can be drawn. However, we find evidence of the impact of graph presentation that is independent of graph structure. In describing each of the studies below, we begin with the main network property to be communicated to observers and the geometric properties that we hypothesize may be used to convey them. We describe the experimental setup and in particular whether it introduces background semantics or a decision framework beyond the graph-theoretical network. Finally we cover how we analyzed the results to investigate both the communication of the main network property and determine whether observers are making secondary inferences from the graph.

In all studies, we attempt to follow Tufte's admonition to make graph visualizations as efficient as possible [11]. At times we have manipulated placement and edge crossings in ways that are inconsistent with efficiency, but necessary for manipulating experimental conditions. We also adhere to standards of graph drawing aesthetics, such as described by Davidson and Harel [4].

We must note that these aesthetics were created for a particular type of graph for which there is a physical meaning to the edges. As is the case with a network that represents an electronic circuit. In the case of social network graphs, the connection between real world physical constraints on the relationships and the graph representation is less explicit than in some other network drawing contexts.

In the next section we briefly introduce KrackPlot, the software platform used in the studies we describe. Then we describe each of the three studies. We end with a discussion of lessons learned and pointers to future work.

1.1 Krackplot

The experiments described below were either conducted with layouts generated by the visualization tool KrackPlot [7], or they used KrackPlot directly for animation or instrumenting observer responses. Here we provide an overview of the tool, which was evolving during these experiments and has continued to evolve. KrackPlot supports a number of layout routines but the most common, used in each of the experiments below, is a simulated annealing approach based on Davidson and Harel [4]. This optimizes an energy function that sums weighted terms for the inverse distance between nodes and the length of an edge. In addition, terms are optionally included for the number of edge crossings, the closest distance between edges and the orientation of directed edges. The last term can be used to prefer diagrams that reflect a DAG structure on the y axis. Some effects of this optimization expression are described in [9].

Over a series of investigations we have found this optimization approach to be very flexible. While standard force-directed approaches require a differentiable function, the simulated annealing method admits extra terms that can be computed via any procedure, and may for example be a function of discrete node or link attributes. The approach was used to create animations of graphs over time by optimizing layouts at each time point combined with a separate optimization step that aligned snapshots by scaling and rotation. It was also combined with constraints on node position to create readable timeline layouts as we describe below.

KrackPlot 2 was written in Fortran and was not interactive. KrackPlot 3 was written in C++ as a Microsoft Windows program [7], and KrackPlot 4 was written in Java for availability on many platforms and as a web applet.

2 Perceptions of Grouping and Centrality

In the first study, we investigated the effect of spatial arrangement on observers inferences about the existence of subgroups and the relative centrality of actors [1, 10].

2.1 Layout Hypothesis

We tested the hypothesis that, while the structural properties of the graph would be the most important factor in observer decisions, certain spatial relations among the nodes would largely predict any deviations from the most appropriate judgment. We developed simple spatial measures that might influence observations. First, we tested the influence of the distance from the center of the network on perceptions of centrality of nodes. Second, we tested the influence of an angular measure of betweenness on perceptions of bridging. Third, we tested the influence of spatial clustering of nodes on perceptions of subgroups.

2.2 Experimental Setup

Eighty graduate students participated in the study, each viewing three of five layouts of one network of 12 actors and 48 ties. The layouts varied the proximity of nodes to each other and their positioning toward center or periphery of graph. Subjects were asked about their perceptions of individual nodes. Figure 2 shows the layouts that were used and the five nodes for which subjects responded.

The perceptions of groups within a network, of bridging and of the centrality of nodes are graph theoretic in the sense that they do not require any additional knowledge of entities that the nodes and links may represent beyond the graph. However they all typically involve judgment calls by the observer. Standard graph-theoretic metrics are not used since, for example, large, completely connected subgraphs are rarely found in social networks. Subjects were told that the graph represented communications observed in different merger and acquisitions teams of a merchant banking firm.

2.3 Experiment Analysis

Mean-centered scores were used for prominence and bridging in order to adjust for individual tendencies to score high or low. Analysis of variance and regression tests were used to compare accuracy across layouts and test the impact of geometric features on perceptions.

Positioning at the center of the graph had limited influence on perceptions of node prominence, especially for nodes that were structurally prominent. Positioning nodes between groups highlights their role as bridges.

Arrangement 2 has the highest proportion of correctly ordered responses for both prominence and bridging. Arrangement 5, the circle, had the highest proportion of responses saying that every node had the same bridging or prominence – 28% for bridging, while no other arrangement had more than 4% of responses scoring all nodes the same. We note, however, that the circle layout did not optimize the ordering of nodes to highlight graph-theoretic or higher-level features.

From this starting point we can see that, holding everything else constant, simply changing spatial position of nodes in a layout can have a great influence on the inferences viewers draw from the graph. This is true even in networks of such small

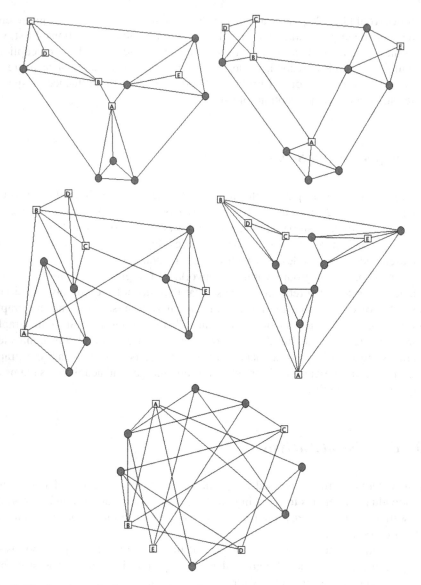

Fig. 2 The five alternative layouts, re-drawn from [10]

size as those used here, and one might expect the effect to be more pronounced in larger networks, as the viewer has less opportunity to inspect the diagram directly for properties of interest and relies more on the visual cues that are provided.

3   Layout and Group Membership

We extended the work on understanding how user perceptions of individual nodes depended on layout to examine perceptions of subgroups within the graph [9]. Observers were asked to assign nodes to at most five different groups by clicking on the nodes and labeling them. They used an instrumented version of KrackPlot that recorded not only the clicks and their orderings but also their timings. The main network property we investigated in this study goes beyond the number of groups to gather group attribution for individual nodes. From this information we begin to explore the level of consensus that a node belongs to a group and to relate this both to structural and layout properties.

3.1   Layout Hypothesis

We hypothesized that the most important layout feature for grouping would be spatial proximity to other members of the group, and that nodes placed more between the centroids of different groups would have a lower overall consensus of belonging to their main group. However since the notion of groups in this case is an inferred property, we frame the hypothesis and statistical testing in terms of the co-attribution of nodes to groups. In other words, we ask how frequently two nodes are assigned to the same group by observers.

3.2   Experimental Setup and Problem Semantics

Sixty-one graduate students were each shown one of five different orderings of five layouts of the same graph and asked to assign the nodes to groups using the interactive tool. The networks were presented as small organizations, but no other semantics were given explicitly.

The graph itself contains two clear groups according to metrics that compare internal and external links, one with 10 nodes and one with 16 nodes, with two links between these groups. We compared responses to three layouts shown in Fig. 3. In the first, nodes are arranged automatically with simulated annealing according to Harel-Davidson heuristics, and are in general close to their neighbors. In the second layout, one bridging node and one node from the larger clique are deliberately moved to disguise their properties. In the third, nodes are arranged spatially into three groups, but with many more links between them than within them, reflecting the fact that these groups are anti-correlated with clusters in the graph.

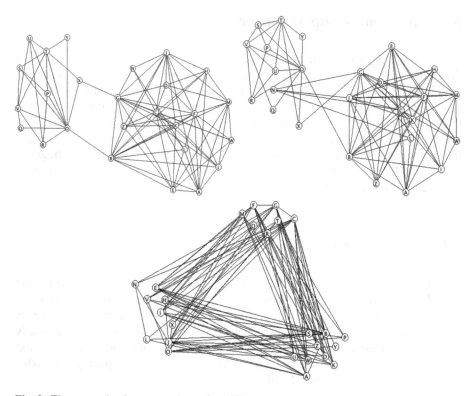

Fig. 3 Three grouping layouts, re-drawn from [9]

3.3 Experiment Analysis

We examined timing data, the number of groups chosen by users, a multi-dimensional scaling of average co-assignments chosen and a QAP analysis comparing co-assignments with both graph and spatial properties. Together, they tell an interesting story of the primacy of spatial distance for selecting groups and also of clearer understanding when the spatial properties are aligned with graph properties, as they should be. Individuals made assignment decisions significantly more quickly in the first layout in 72.5 s on average compared with 100.7 s in the second layout and 132.8 s in the third. The mode average number of groups was two in the first two layouts but three in the third, and with a higher consensus.

We performed multi-dimensional scaling (MDS) of the nodes based on the frequency with which observers placed nodes in the same group. This technique highlights the general groupings found by observers, but can also make apparent the relative degrees to which nodes are perceived to belong to their group or are connected to others. Evidence that viewers were responding to secondary properties such as bridging could be found in the MDS for the first layout, but not for the second or third [9]. In the MDS based on the first layout, the nodes that participate in the cross-group links appear with their groups but somewhat closer to the other groups, indicating that occasionally respondents placed them in those groups. In the second layout, the node manipulated to appear in the smaller group appears between the groups, indicating that respondents are ambiguous about its membership. The bridging effect was destroyed not only for the node that was manipulated but also for the other three nodes, although all remained spatially between the two groups. It seems that the greater difficulty associated with interpreting this graph has removed to tendency to recognize these nodes.

Finally the QAP analysis showed that Euclidean distance is negatively correlated with co-grouping in all three layouts, confirming our hypothesis. Graph adjacency and path distance were correlated in the first two but not in the third. This, combined with the polarization of the number of groups in the third layout, suggests that Euclidean distance is not only important for observers when structural information can easily be recognized in the layout, but it is commonly used as a fall-back metric when structural information is harder to see.

4 Exploring Motion and Semantics

The third study tested the link between the use of smooth motion and the observer's perception of change between two graphs. It also tested a layout algorithm specifically designed to convey one node feature, 'status', against a standard, Davidson-Harel layout. Finally, it uncovered insights into how layout combines with the background knowledge that observers use in interpreting the diagram to create an overall impression of the graph [8].

4.1 Layout Hypotheses

We have seen already that layout has a significant impact on observer ability to understand properties of a graph and that this can sometimes be attributed to geometric properties that relate to features of interest. A natural question is whether layouts that embody pre-determined geometric properties are effective at conveying the corresponding graph properties. For example, Brandes [2] proposed a layout to convey the 'Katz status' of nodes in a graph in which the status was mapped linearly to the height of the node in the layout, while the x axis was allowed to vary in order to create a readable layout. Our study compared the effectiveness of such layouts, which we termed "spatial hierarchy" layouts, with the effectiveness of a Davidson-Harel layout, which we termed "spatial centrality". Specifically we hypothesized that spatial hierarchy would be more effective than spatial centrality for conveying the status of nodes.

We were also interested in how to convey the differences between two networks, in this case with the same nodes but some changes to the edges, to maximize observer accuracy. One possibility is to use a layout where the edge differences will cause nodes to be placed differently, combined with smooth motion so that attention is drawn to the nodes that move and so edge deletions and additions are more likely to be noticed. We tested the hypothesis that smooth motion between such layouts would be more effective than static before-and-after pictures.

The two networks we compare represent the formal hierarchy of an organization and an informal advice network for the same organization. A standard hierarchy layout also uses the height of a node to convey importance, and so we developed a third hypothesis that the motion between formal and advice network would be more effective when the advice network was displayed with a spatial hierarchy, because in this case change in height specifically reflects change in status between the two graphs.

4.2 Experimental Setup and Problem Semantics

We created a classroom case exercise based on a change management case described by Krackhardt [6]. We presented a story of a small (14 person) auditing group that is part of a large organization. Viewers first saw the formal organizational chart and then saw the advice network. We manipulated the layout of the informal advice network using the two different approaches, spatial hierarch and spatial centrality. We also varied whether the two networks were seen with smooth motion or as static before-and-after pictures. Each of 133 subjects saw both the spatial hierarchy and the spatial centrality layout for the advice network, presented as two different groups. The graphs are shown in Fig. 4. Since the original article appears in an online journal [8], the effects of smooth motion can be viewed at that site with a java-enabled browser.

The graphs were presented in the context of a class discussing the use of social networks as a managerial tool. It was emphasized to subjects that although organizational restructuring often focusses on the formal relationships, both formal and informal networks are affected. We asked subjects to identify the four most powerful people in each of the formal and informal networks and estimate the amount of change undergone by each individual between the two networks. After showing both networks, we asked which team they felt would be "more likely to successfully implement a change in the auditing procedure" and gathered responses as free text.

4.3 Experiment Analysis

We used analysis of variance to compare the accuracy of subjects in estimating the change of status when using the spatial hierarchy or spatial centrality model. Although we hypothesized that the spatial hierarchy model would lead to better

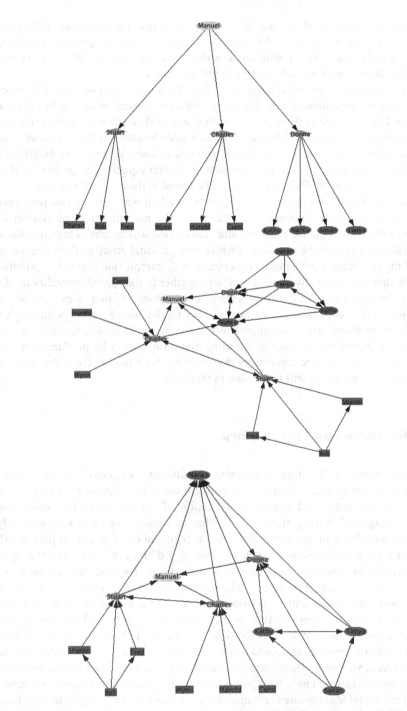

Fig. 4 The authority and advice networks, with spatial centrality in the center and spatial hierarchy at the bottom, copied with permission from [8]

accuracy independent of the use of motion, we found no significant difference when there was no motion. However the spatial hierarchy approach produced a significantly better result with motion than without motion. With the spatial centrality layout, motion made no significant difference.

When we compared the text responses for which team (layout) would be more better able to implement a new procedure, observers were evenly split between the two. However, the reasons given for choosing each team were quite different. Observers picking the spatial hierarchy team tended to attribute their success to its simpler information flow, its central authority and efficiency. Observers picking the spatial centrality team cited its informal nature, greater equality and greater level of overall communication. Full responses can be found at the original web site.

Recalling that these layouts show the same graph, it was striking that they gave such different overall impressions of the team. We note that a large number of networks shown in organizational literature tend to use a hierarchical layout to show a formal structure, and the association with a less informal group probably becomes part of the graphical lexicon that an observer uses to interpret the graph. It is possible that the shorter edges between nodes closer together in the spatial centrality model convey the impression of closer or greater communication that is evident in the responses. It is possible that the spatial centrality model, where arrows point roughly equally up or down, gives the impression that a hierarchical arrangement, where bulk of the arrows would point in the same direction, cannot be produced for this graph. However, these are conjectures that have not been tested. They illustrate the scope of the work that is still to be done in this area.

5   Discussion and Future Work

The experiments we described represent three initial explorations of the interplay of geometric features, graph features and background knowledge in predicting what observers will understand from a visualization of network data. Following this work, Huang et al. have performed a number of studies that more systematically examine the effects of geometric features on perceptions of graph properties [5]. However, much still needs to be done to develop and broaden our understanding of this interplay. In order to make progress, for example, we focused on small networks depicting organizations that might model an observers everyday interactions, rather than thousands of nodes extracted from online data. One line of future work would examine the effect of network size on these results. We also used the same graph for each layout in these studies, to focus on the effects of alternative visualizations. Another line of future work would investigate manipulating graphical structure as well as layout variables. There are a number of ways that one could experimentally control the level of semantic background that is introduced about networks during visualization and also capture the semantics that observers are bringing to their task.

References

1. Blythe, J., McGrath, C., Krackhardt, D.: The effect of graph layout on inference from social network data. In: Graph Drawing, pp. 40–51 (1995)
2. Brandes, U., Raab, J., Wagner, D.: Exploratory network visualization: Simultaneous display of actor status and connections (2001)
3. Butler, P.: Visualizing friendships (2010). URL https://www.facebook.com/notes/facebook-engineering/visualizing-friendships/469716398919
4. Davidson, R., Harel, D.: Drawing graphs nicely using simulated annealing. ACM Trans. Graph. 15(4), 301–331 (1996)
5. Huang, W., Eades, P., Hong, S.: Measuring effectiveness of graph visualizations: A cognitive load perspective. Information Visualization 8(3), 139 (2009)
6. Krackhardt, D.: Social networks and the liability of newness for managers. Trends in Organizational Behavior 3, 159–173 (1996)
7. Krackhardt, D., Blythe, J., McGrath, C.: Krackplot 3.0: An improved netword drawing program. Connections 17(2), 53–55 (1994)
8. McGrath, C., Blythe, J.: Do you see what i want you to see? the effects of motion and spatial layout on viewers' perceptions of graph structure. Journal of Social Structure 5(2) (2004). URL http://www.cmu.edu/joss/content/articles/volume5/McGrathBlythe/
9. McGrath, C., Blythe, J., Krackhardt, D.: Seeing groups in graph layouts1. Connections 19(2), 22–29 (1996)
10. McGrath, C., Blythe, J., Krackhardt, D.: The effect of spatial arrangement on judgments and errors in interpreting graphs. Social Networks 19(3), 223–242 (1997)
11. Tufte, E.: Envisioning Information. Graphics Press (1990)

Part VI
Dynamic Visualization

Adaptive Diagrams: A Research Agenda to Explore How Learners Can Manipulate Online Diagrams to Self-Manage Cognitive Load

Shirley Agostinho, Sharon Tindall-Ford, and Sahar Bokosmaty

Abstract This chapter presents an emerging research agenda focused on empowering learners to apply well-known instructional design principles, reserved mainly for application by instructional designers, to the design of diagrams to support their learning. Significant advances have been made in terms of developing design principles that can be applied to the design of diagrams to facilitate the efficient learning of diagrammatic information. However, little is known about how these design principles can be applied by learners themselves. In a technologically rich environment where learners can access a range of online diagrammatic information, we argue that it is imperative that learners' are equipped with strategies on how to physically manipulate digital diagrams in ways that optimise their learning. This can be considered an example of human-centric visualisation. The chapter explains the theoretical basis for our research, presents two empirical studies and concludes with a discussion of our ideas to build on our current work as a future research agenda.

1 Introduction

Information and communication technologies have essentially transformed how we learn. Today, learners have at their fingertips an abundant amount of information related to any topic of interest. The learning potential of such an online environment is vast, as learners can access and download a myriad of online content as well as create, upload and thus share information online. Learners, however, can

S. Agostinho (✉)
Interdisciplinary Educational Research Institute, University of Wollongong,
Wollongong NSW 2522, Australia
e-mail: shirleya@uow.edu.au

S. Tindall-Ford • S. Bokosmaty
University of Wollongong, Wollongong NSW 2522, Australia
e-mail: sharontf@uow.edu.au; saharb@uow.edu.au

W. Huang (ed.), *Handbook of Human Centric Visualization*,
DOI 10.1007/978-1-4614-7485-2_21, © Springer Science+Business Media New York 2014

become overwhelmed and cognitively overloaded by the information they access and consequently can learn very little [1]. Research in educational psychology and cognitive science has contributed significantly to better understanding the learning process and thus to the development of effective instructional strategies that instructional designers can implement to assist learners to increase learning performance [2, 3]. Whilst this work has been of great significance for the effective design of instruction, its limitation has been a reliance on the instructor/teacher to create high quality learning materials for learners. Yet in an environment where learners can access a range of online learning materials of varying quality, namely, materials that have not been designed based on cognitive science principles, we claim that it is important that learners are equipped with strategies to manage their own cognitive load when exposed to these kinds of materials.

For example, consider the evidence base for the design of diagrams. There has been extensive empirical research that has shown that if learners are presented with static diagrams that are formatted in a way where explanatory text about the diagram is spatially positioned close to the relevant part of the diagram, then efficient learning occurs [4]. This is because the two sources of information, that is, the text and the relevant parts of the diagram are integrated. This allows the learner to focus on understanding the information. If the diagram and explanatory text are separated, the learner would need to firstly engage in the cognitive process of splitting their attention between two separated sources of information by reading the text and then searching the diagram to see how the text relates to the diagram, and then focus on understanding the information. This process of searching and matching to mentally integrate textual information with diagrammatic information is not directly related to learning and imposes an unnecessary cognitive load on working memory; this burden on limited working memory hinders learning. Cognitive load refers to the amount of working memory resources a learner has to allocate to deal with learning information. According to cognitive load theory, the example presented is referred to as the split-attention effect [4].

What is not known is what effect on learning occurs if learners are shown how to manage their own cognitive load by being taught to self-manage split-attention by physically manipulating digital diagrammatic information. Physical manipulation would involve moving text explanations about the diagram closer to the relevant parts of a diagram, that is, adapting a diagram. Research on this kind of diagram design principle (split-attention effect) has been premised on the teacher or instructor managing the cognitive load for the learner by providing the learner with a visualization that integrates the textual information with the diagrammatic information. We propose that in the current educational context where learners can access a vast amount of online information, as well as create, upload and thus share information online, there is less likelihood that learners will always access well designed visualizations; therefore there is an increasing need to assist learners to engage in more self-directed strategies where they can adapt a diagram themselves to encourage sense-making [5].

We argue that it is important that learners are equipped with strategies to adapt diagrams for themselves so that they *self-manage their cognitive load*. The focus of our research is to explore how learners can physically manipulate visualizations, that is, adapt a diagram by positioning digital text objects closer to the relevant parts of a diagram, so that the learner does not need to engage in the search and match process, which is superfluous to learning. The result of the learner self-managing their cognitive load would be increased working memory resources available for learning the textual and diagrammatic information.

The structure of this chapter is as follows: the next section introduces theoretical underpinning of our research—cognitive load theory. We then present our research that is exploring how learners can adapt diagrams by self-managing the split-attention effect. In the final section we discuss our ideas as a future research agenda.

2 Background

Much of the cognition based research that has informed the development of effective learning environments has focused on modifying the instructional environment to take into account the underlying cognitive processes of the learner. Cognitive load theory (CLT) has been at the international forefront of this work [2, 6, 7].

2.1 Overview of Cognitive Load Theory

CLT is concerned with how humans process information and how instruction can be best designed to aid learners to efficiently and effectively learn new information. CLT suggests that only when the conditions of learning are aligned with human cognitive architecture does learning take place efficiently [8–12]. Human cognitive architecture assumes that information is inputted into sensory memory, processed in working memory and then can be stored and later retrieved from long-term memory.

CLT's particular focus is on the critical relations between working memory load and instructional design. It suggests that our cognitive architecture includes as one of its critical components, working memory, which is the structure that is used to hold and process information we are provided with [13, 14]. Working memory is related to consciousness. We are only conscious of what is held in working memory. On one hand, working memory has limited capacity when dealing with novel material, but on the other hand it can process intricate formerly learned information [12].

2.1.1 Working Memory

The most important two features of working memory are: (1) its limited capacity, as concluded by Miller [15] and (2) its limited duration, discussed by Peterson and Peterson [16]. Material that is learned well and understood does not suffer from either of these limitations [14].

Information enters working memory via two routes: from long-term memory if it has been previously learned or from sensory memory if the information is new [17]. How to process the information in working memory depends on the source and this leads to instructional design issues. Based on what Peterson and Peterson [16] found, when learners are presented with new information, instruction has to be designed to compensate for the limited duration of working memory otherwise it will be lost within seconds. In addition, according to Miller [15], working memory can hold between five and nine elements of novel unfamiliar information, or even less depending on the nature of processing (e.g., if some information must be contrasted or combined) [17]. Instructions need to take working memory limitations into account.

Information that enters working memory from long-term memory has different characteristics to information entering working memory from the environment. There are no known limits for the amount of information that working memory can process if it comes from long-term memory. Sweller [17] gives an example of the word restaurant that is stored in long-term memory. The information related to this word includes food, the building, service, tables, chairs etc. This information can be moved from long-term memory to working memory without overloading it and be processed as a single element.

2.1.2 Long-Term Memory

It is common knowledge that we all possess a long term-memory since we are able to recall things learned a long time ago. Its importance to cognitive functioning has been clarified in the last few decades. De Groot [18] studied long-term memory in higher cognitive functioning. He showed that expert chess players rely on previously learned moves when encountering similar conditions and configurations in new games. They store those moves in long-term memory and this is how they defeat beginner or weekend players. He found that the skills of master chess players have nothing to do with thinking ahead and considering more moves than beginner players. Upon giving less able players a few seconds to reproduce a board configuration taken from a real game, they did not perform as well as master players who could usually place most pieces correctly.

Knowledge of a large number of moves that are stored in long-term memory as *schemas* changes the characteristics of working memory. Research conducted by Chase and Simon [19] confirmed that the difference between experts and novices was not in their working memory capacity, since the board recall results were the same for novices and experts when random configurations were used. Simon and Gilmartin [20] estimated that chess grand masters could learn up to 100,000 configurations. Consequently, they can reproduce configurations that they are familiar with but they do not perform any better than beginner players when dealing with unfamiliar, random configurations. This knowledge of chess moves is stored in long-term memory after years of practice leading to high levels of expertise. It might be the only learnable factor contributing to differences in levels of skill among players [21].

The same results have been obtained in other complex tasks. Egan and Schwartz [22] displayed electronic wiring diagrams for a short period of time to expert and beginner electronic technicians who were asked to reproduce the same diagrams. The performance of expert technicians was better than that of the novice. However, upon repeating the experiment using random diagrams, the difference faded. Chiesi et al. [23] provided students with some prose about baseball. Learners with some knowledge of baseball performed better at recalling details of baseball games than those with less knowledge.

These experiments and studies have shown that the difference between expert problem solvers and beginners is not knowledge of refined strategies, but exposure to and knowledge of a huge number of different problem states and their associated moves [12]. Long-term memory allows us to solve problems, perceive, and think efficiently. Deliberate practice and rehearsal leads to high levels of intellectual performance [21].

The above characteristics of human cognitive architecture have significant implications for instructional design issues. Unlimited amounts of complex information can be stored in the human cognitive system. Long-term memory can store very complex, intricate procedures and facts. Human cognitive skills do not come only from the ability to perform complex reasoning activities in working memory but from stored knowledge in long-term memory. The finding that working memory has limited capacity and duration suggests that humans are able to deal with intricate reasoning only when the information they are presented with includes elements that are stored in long-term memory. Only then can they perform well. As a result, instructional designs that require learners to engage in complex reasoning processes that deal with unfamiliar elements are ineffective [7].

2.1.3 Schema Development

When information is stored in long-term memory, it is categorized according to how it is going to be used in schematic form [24]. By definition a schema is a cognitive structure that allows us to consider several elements as a single element that is categorized according to how it will be used [12]. When one sees a tree, one immediately perceives it as a tree even though each tree is different from every other tree in colour, number, shape, branches. A tree schema, stored in long-term memory, allows us to categorize this information according to how it is used and treated as a single element [7]. It is schemas held in long-term memory that define the learning process and outcomes. Our ability to read is possible because of the enormous number of schemas stored in long-term memory. Regardless of the text, we can recognize an infinite variety of shapes as the letter 'a'. Combinations of letters that form different words and combination of words that form phrases and letters are stored in higher-order schemas. Consequently when we read, we can ignore all the other details and focus on the meaning [12].

Only since the 1980s have schemas become important to modern cognitive theory and in particular to problem solving theories. Due to the studies conducted by

Larkin et al. [25], and Chi et al. [24] it is evident that schemas provide learners with the ability to classify many elements of information as one element, resulting in increased working memory capacity being available for learning. The cited research demonstrated the critical role of schemas in expert problem solving. Tens of thousands of schemas permit expert problem solvers to recognize certain problem situations in relation to suitable moves. Hence schema theory suggests that in order to be skilled in any domain, one has to acquire specific schemas and store them in long-term memory. The tens of thousands of configurations stored in the form of schemas in long-term memory allow chess masters to defeat novice players upon recalling problem states and the corresponding moves [18]. The same mechanism applies to all areas of expertise [12].

Decreasing the load on working memory is another essential function for schemas, in addition to storing and organizing information. In spite of the fact that working memory has a limited capacity in the sense that the number of elements that can be process is limited, the size, complexity, and sophistication is not [17]. Stored schemas may include a huge amount of information. The restaurant schema mentioned previously is a good example. It is held as a single entity, but it includes wide knowledge, everything from food to the structure of a building. Though the number of elements (or schemas) that working memory can process is limited, there are no limits on the size of an element. As a result, the two functions of schemas can be summarized as the storage and organization of information and a decrease of working memory load.

2.1.4 Automation

Information can be processed consciously or automatically. Conscious processing of information has the characteristics described previously. However, automatic processing circumvents working memory [26, 27]. With practice, knowledge may become automated and less conscious effort is required for information to be processed. A clear example is related to reading text. When reading, a competent reader does not consider the individual letters that make up the text. Processing letters becomes automated in childhood. However, processing each letter consciously is essential when young children are learning to read [7].

Thus, automation has the same consequence for working memory as schema acquisition; they both reduce the load on working memory. Kotovsky et al. [28] indicated that when rules are memorized to the extent that they can be repeated easily, then problem solution becomes easier since the rules are not processed consciously and planning a solution takes place in what is now a working memory with a reduced load. When rules are not automated, effort is exerted in working memory to recall them and reaching solutions becomes difficult [7, 29]. The experiments conducted by Kotovsky et al. [28] reflected the essential role of automation in problem solving.

Working memory capacity increases when a learner has a more automated schema. For example, when the schemas related to letters, words, and phrases are automated, the capacity of working memory is used to comprehend the text. In contrast, less proficient readers whose schemas are not fully automated, may be able to read the text fluently but they may not comprehend the text fully because they do not have enough working memory capacity to derive meaning from it [7]. This implies that instructional designs should not just focus on the construction of schemas and storing of information, but also on the automation of these schemas that supports problem solving [30].

2.1.5 Summary

To sum up, a powerful long-term memory, a limited working memory, and learning mechanisms that involve schema construction and automation are the constituents of our cognitive system [7]. Furthermore, recent CLT research argues about an extension of the human cognitive architecture that incorporates an evolutionary view [31]. For example, biologically primary knowledge, which encompasses skills that are acquired effortlessly such as recognising human faces and learning to speak, may be less affected by working memory limitations than biologically secondary knowledge, which are skills that require more conscious effort such as understanding mathematical concepts. Based on the evolutionary educational psychology view [32], it is suggested that human movement can be considered a biologically primary skill. Recent research based on the theoretical framework of grounded or embodied cognition has shown a link between visual and motor processes in the brain when cognitive tasks such as reading, comprehension, mental arithmetic, and problem solving are performed [e.g., 33]. Thus our research that is exploring how learners can adapt digital diagrammatic materials to manage their own cognitive load, may also contribute to the investigation of whether human movements, related to moving digital text objects, can be considered a biologically primary activity that can be used to self-manage cognitive load and therefore facilitate the acquisition of biologically secondary knowledge.

2.2 Contribution of CLT to the Design of Instruction

CLT has a rigorous empirical base spanning the last three decades which has led to a number of instructional design principles based on our understanding of the human brain and how learners process information [9].

CLT suggests that efficient learning requires limited working memory resources to be used effectively during learning by: (1) reduction of irrelevant/unproductive cognitive load (extraneous), (2) an increase in productive/relevant cognitive load (germane) and (3) management of intrinsic cognitive load (the complexity inherent in the information to-be learned) [34]. Extensive empirical research has found that when learners engage with instructional materials that comply with CLT design principles, efficient learning occurs [35]. Some of the major design principles derived as listed below. For a complete summary, refer to Sweller et al. [36].

- Worked example effect. This design principle states that problems are more effectively explained when the solution is explained out step-by-step for learners [37–39]
- Split-attention effect. This design principle advises to replace multiple sources of information with a single, integrated source of information [4, 6, 40]
- Modality effect. This design principles suggests to replace a written explanatory text and another source of visual information with a spoken explanatory text and a visual source of information [41]
- Redundancy effect. This effect has shown that it is better to avoid presenting the same information, which can be understood independently, in different modalities as this is redundant to learning [42]
- Expertise reversal effect. This design principles advises to tailor instruction based on levels of learner expertise [43]; and more recently, the
- Human movement effect has shown that it is better to use animation rather than statics to teach cognitive tasks involving human movement [44, 45].

The design principle of focus in this chapter is the split-attention effect. The next section summarises how the split-attention effect was derived and outlines the key split-attention effect research.

2.3 Effective Visualisations: Reducing the Split-Attention Effect

When the worked example effect was derived from research, it was noted that some worked examples did not take into consideration the limitations of working memory by, for example, requiring learners to split their attention between many sources of information [46]. Some of the materials that were given to learners may have included a picture and written information that was positioned either above, below or to one side of the diagram. It was found that such instruction led to a split in attention as the learner had to mentally integrate both sources of information (text and diagram) in order to process and comprehend the material [47].

Diagrammatic information is often designed with a split-attention layout, and this adds to the extraneous cognitive load if weighed against layouts that are physically incorporated. An example of a geometry worked example with evident split-attention is having a diagram with a list of steps under the diagram leading to the solution of the problem Sweller [34]. If one of the steps suggest that, Angle ABC = Angle XYZ, students have to locate the two angles. The interacting elements related to Angle ABC are the statement "Angle ABC" and almost all the angles on the diagram. The learner must go over all the angles till he/she locates the correct one. In order not to check angles that have been checked before, the learner needs to keep in mind the ones he or she checked before. The same applies to Angle XYZ. When the learner locates both, he or she has to work on proving why they are equal. This process entails extraneous cognitive load and places a burden on limited working memory resources. CLT research suggested that extraneous cognitive load related to searching for the angle should be removed [34]. This could be done by incorporating the statement "Angle ABC = Angle XYZ" within the diagram. It should be positioned in a suitable place on the diagram instead of below the diagram. Integrating text and diagram can also be supported by the use of arrows to reduce the search and match process.

Sweller et al. [48] demonstrated that incorporating text into diagrams aids learning. Two groups of students were involved in these experiments. The first group was given separate text and diagrams while the other group was given the same problem in which the text was incorporated into diagrams. The experiment was divided into two stages. During the first stage, students were given either the conventional or the modified instruction and they were allowed to use up as much time as needed to understand it. The second stage was the assessment stage. The results showed that the group that was presented with the incorporated information performed better in both stages. They spent less time understanding the material, which suggests a lower cognitive load due to the incorporated material, which in turn made learning easier since they performed better than the other group on the later tests. Presenting the solution on the diagram facilitates learning and reduces split attention since learners are presented with only one source of information.

The split-attention effect does not just relate to text and diagrams. It is present whenever there are multiple sources of information that learners have to make use of at the same time. This includes two or more sources of textual information. Chandler and Sweller [4] demonstrated this effect with the traditional format of educational psychology research papers in which the outcomes of the experiments and the discussions are conventionally presented in separate sections even though readers have to refer to both in order to comprehend the results thoroughly. The split attention was reduced when the results and the discussion were integrated into a single entity. Chandler and Sweller [40] described a split-attention situation created by referring to software and a hardcopy of a manual in order to understand how the software functions. The best alternative is not to use the computer when learning but rather to refer to diagrams in the manual. Chandler and Sweller [40] found that learners who first studied the manual without the presence of a computer did better than those who simultaneously used both the computer and the hardcopy manual.

The focus of the split-attention effect research over the last two decades in terms of diagram design, has been on developing visualizations that physically locate related information and joins them together in order to avoid extensive search and match behaviour, the result being a reducing of extraneous load. It has been demonstrated in a multitude of learning domains that providing visualizations that include physically integrated text in the diagram rather than presenting the text and diagram separately in a traditional split source format, reduces cognitive load and enhances students learning [49]. This is the most efficient method to date for dealing with split-attention when designing diagrams.

The line of research presented in this chapter, however, proposes a different perspective that has not been investigated for dealing with split-attention with digital diagrams. This new perspective involves allowing the learners themselves to manage split-attention by adapting a diagram (positioning digital text objects closer to the relevant parts of a diagram). The research work being carried out in this new line of research is discussed below.

3 Emerging Evidence of the Effectiveness of Adaptive Diagrams

The first research study to explore self-management of split-attention was recently conducted by Roodenrys et al. [50]. They investigated how university students can be guided to self-manage cognitive load with print-based split-attention materials. A key finding of this study was that learners who physically manipulated paper-based split-attention instructional materials by making connections between the text and diagram (by drawing circles around text, drawing lines and arrows from the text to the diagram and highlighting text) showed a positive effect on learning. Three instructional conditions were compared; regular split-attention instructions, split-attention instructions, which incorporated self-management strategies, and an integrated condition. As predicted, the regular split-attention condition performed poorly relative to the other two conditions. The most interesting contrast was between the integrated condition and the self-managed split-attention condition. On a near transfer task, that is, the application of acquired knowledge in new problems within the same domain, learners who self-managed the split-attention effect performed just as well as learners who studied the integrated instructional materials. This was a significant finding as it demonstrated that very subtle manipulations to everyday split-attention instructions performed by learners themselves had the same effect on learning than instructor manipulated techniques through integrated instruction.

The two studies explained below build on this initial evidence base for the effectiveness of adaptive diagrams by examining how learners can manipulate digital diagrams to reduce split attention by moving digital text objects closer to the relevant parts of the diagram.

3.1 Study 1: Exploring How University Students Adapted a Digital Diagram to Self-Manage Split Attention

The previous research discussed how students self-managed split attention materials for paper-based instructions. Following this research, a series of two experiments investigated how university students who were exposed to split source materials in an online environment could reduce split attention using digital tools. The aim of the research was to ascertain if students who were provided with split-attention materials and were guided on how, and the reasons why to self-manage split attention by moving textual information to relevant parts of the diagrammatic information, may out perform students who were provided the same split-attention materials but were not able or instructed to self-manage split attention.

3.1.1 Experiment 1: Research Design, Method and Participants

Participants were undertaking an undergraduate subject focused on how information and communications technologies (ICT) can be used for teaching and learning. The instructional materials presented to the students, the Cognitive Theory of Multimedia Learning [51], was a component of their area of study. As part of their studies, all students were familiar with the interactive whiteboard software used in the research and were able to competently move text objects using the software on a computer. As per the previous research discussed, there were three groups: the first group was provided online split attention materials, where students could not move textual information and consequently were not taught that self integrating text to pertinent parts of diagram may enhance learning. The second group had the ability to move textual information to relevant parts of the diagram, and were explained the benefits of self-managing information to reduce split attention. The third group were provided exactly the same online instructional content, but an instructor had integrated the textual information to relevant parts of the diagram; the textual information was not moveable. Figure 1 is an example of the online traditional static split source materials presented to students in the first group.

Students in the second group were provided exactly the same textual information as the first group, but had the ability to self-adapt the materials by moving text to relevant parts of the diagram to reduce split attention. Figure 2 provides an example of the adaptive instructional materials presented to Group 2.

The third group was provided exactly the same information as previous groups however the textual information was integrated to relevant parts of the diagram, to reduce split attention and cognitive load. As stated previously, students in this group were unable to move text. Example of the instructional material for Group 3 is provided in Fig. 3.

Students were provided a short period of time to study the diagram. It was predicted that students in Group 1 presented with a static split source diagram, would expend cognitive resources by mentally integrating text and diagram. This is an extraneous task to learning and would not support students' understanding the model of multi media. It was hypothesised that Group 2 provided with an adaptive diagram where text could be moved to relevant parts on the diagram would out perform Group 1 due to their ability to self manage load by reducing the search and match process. It was envisaged that Group 3 where text was meaningfully integrated to relevant sections on the diagram would out perform Group 1 and confirm the split attention effect.

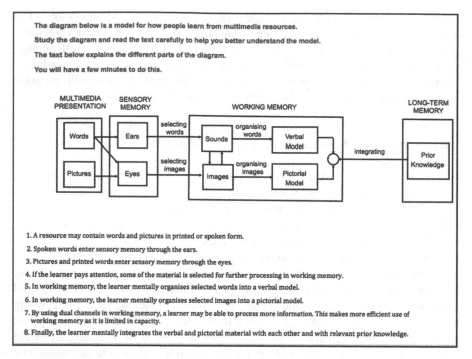

The diagram below is a model for how people learn from multimedia resources.

Study the diagram and read the text carefully to help you better understand the model.

The text below explains the different parts of the diagram.

You will have a few minutes to do this.

1. A resource may contain words and pictures in printed or spoken form.

2. Spoken words enter sensory memory through the ears.

3. Pictures and printed words enter sensory memory through the eyes.

4. If the learner pays attention, some of the material is selected for further processing in working memory.

5. In working memory, the learner mentally organises selected words into a verbal model.

6. In working memory, the learner mentally organises selected images into a pictorial model.

7. By using dual channels in working memory, a learner may be able to process more information. This makes more efficient use of working memory as it is limited in capacity.

8. Finally, the learner mentally integrates the verbal and pictorial material with each other and with relevant prior knowledge.

Fig. 1 Group 1: Static split-attention materials

Students were given a three minute learning phase to carefully follow the instructions provided and study (Group 1 & 3) or move text (Group 2) to understand the model of multi media. After completing this learning phase students were asked to rate on a nine-point likert scale the metal effort expended while trying to understand the model of multi media. The concept of mental effort was explained prior to commencing the experiment, with the scale based on previously developed mental effort scales [37]. Following the learning phase and mental effort rating, students were asked a series of paper-based questions, which tested their understanding of how the model worked.

3.1.2 Experiment 1: Results

The graph presented in Fig. 4 provides the overall test performance scores for the three groups. Results from the experiment confirmed a split attention effect, with Group 3 significantly outperforming Group 1 on test scores. Group 2 performed slightly better than Group 1 but not at a statistically significant level. Although Group 2 was asked to perform an extra activity, that is moving textual information to relevant parts of the diagram, they did perform slightly better than Group 1 but did not perform as well as Group 3, those students who were presented with instructor managed rather than self managed integrated materials.

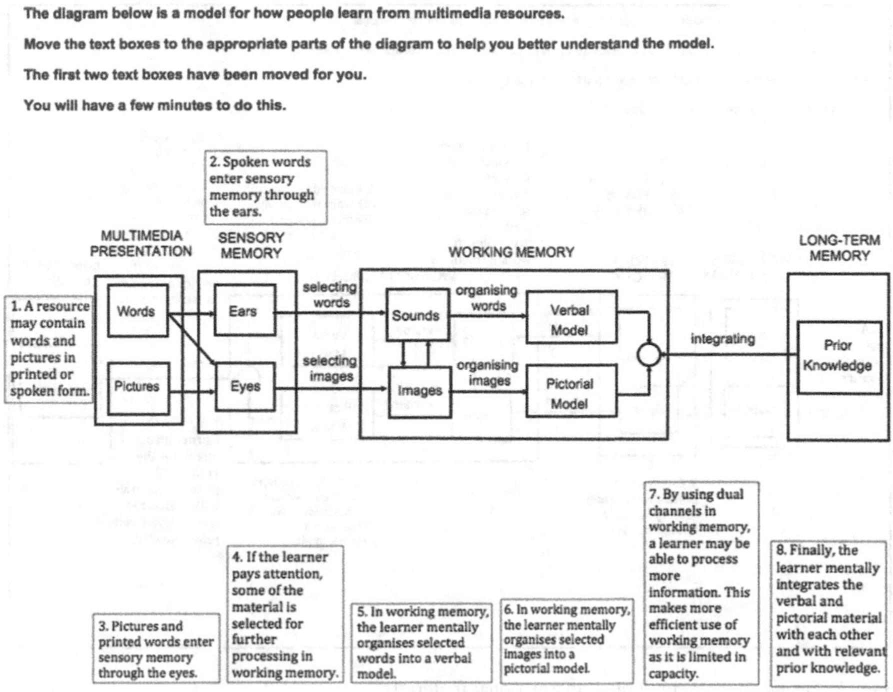

The diagram below is a model for how people learn from multimedia resources.

Move the text boxes to the appropriate parts of the diagram to help you better understand the model.

The first two text boxes have been moved for you.

You will have a few minutes to do this.

Fig. 2 Group 2: Adaptive split-attention instructional materials

Mental Effort ratings for the three groups demonstrated that the extra task of moving text led Group 2 to report a higher mental effort rating compared to the other two groups. The graph in Fig. 5 provides the three groups average mental effort recorded for the instructions. There was no statistical difference between the three groups mental effort ratings.

The results provide further evidence that instructions presented in integrated format are superior to instructions presented in a split source format. The research also suggests that instructor managed integrated instructions may be superior to student managed integrated instructions. This may be expected as instructor-integrated instructions, are developed by an expert, who understands the critical links between text and diagram and integrates accordingly. In contrast students who are novices and do not have the required prior knowledge, schemas, to make informed decisions regarding explicit connections between diagram and text may not always self integrate in a meaningful manner. Analysis of the recorded online files indicate that when Group 2 students integrated text to diagram there was a tendency to integrate based on superficial links between text and diagram for example the same words rather than conceptual connections. To understand the results from this experiment and the cognitive processes students undertake while self-integrating text with diagram, a second experiment was undertaken.

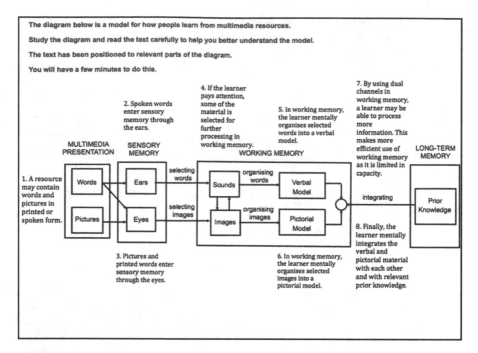

Fig. 3 Group 3: Static integrated instructional materials

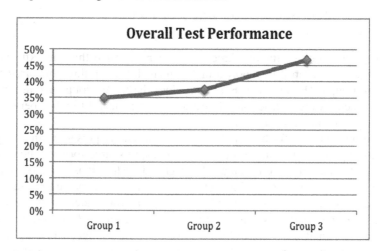

Fig. 4 Means for overall test performance

3.1.3 Experiment 2: Research Design, Method and Participants

The aim of the second experiment was to collect qualitative data, verbal protocols, to provide an understanding into the cognitive processes learners undertake while studying split source materials (Group 1), integrated materials (Group 3) or when

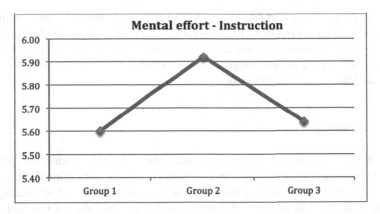

Fig. 5 Means for overall mental effort

students are required to self integrate materials (Group 2). The second experiment was conducted in exactly the same manner as the first experiment, with the same instructional materials, mental effort rating scale and test materials. The only difference was when the participants were studying instructions (Group 1 and 3) or when self-integrating text to diagram (Group 2) they were asked to "think aloud" the thoughts and processes they were undertaking. The technique used to collect of verbal protocols followed that of Ericsson and Simon [52].

3.1.4 Experiment 2: Results

The transcribed and analysed "think alouds" provided interesting insights into the thought processes of the three groups. Group 1 verbal protocols suggest that when students are provided split source materials the focus is on reading text, memorising key aspects of the text and then matching it with parts of the diagram. Statements like; "I am looking for symbols and for dual channel and how they relate to words and images" and "I am reading about working memory" suggest an intent to remember rather than understand how the model of multi media works. Analysis of Group 2 verbal protocols suggest that the focus for this group was on matching text with diagram rather than meaningful integration of text with diagram. This group used words like "match" and made comments "pictorial model" so I put the text under "pictorial" on the diagram. Few students undertook meaningful and deliberate integration of text and diagram to try and understand how the model of multi media worked. This suggests that explicit training is required for self-integration to be successfully undertaken and possibly some form of understanding and expertise of the basic content is required for students to meaningfully engage with the self-management process. Group 3 initial comments indicated being overwhelmed with the nature of the instructional format but by the end of the three-minute study time more affirmative language was expressed. Comments like "Bit confused.. "

to "oh yeah fairly comfortable .... " to "think I understand it" and comments like; "just reading the points of this diagram and trying to understand it" and " the text helped me understand the model". Consistently verbal protocols from Group 3 demonstrated a more focused engagement with the materials and actively trying to understand the model of multi media.

3.1.5 Discussion

These two experiments like previous research about the split-attention effect validated the superiority of integrated instructions compared to traditional split source materials. However, this was the first study of its kind that focused on investigating how learners can self-manage split-attention by adapting digital materials. The findings provided an insight into the efficacy of the self-management of split attention as an alternative to instructor managed integrated instructions. The two experiments suggest that for self-management to be successful the following may need to be undertaken. Firstly students need to be explicitly taught how to self-manage split-attention and the reasons why self-integrating text within a diagram in a meaningful way may enhance learning. The self-management of split-attention needs to be viewed by the learner as not only about physically moving text to diagram but more importantly that the movement of text is purposefully undertaken to support understanding. For the latter to happen it may require learners to have some level of expertise, that is, some form of schema development so that they can understand conceptual links between text and diagram rather than possibly superficial ones.

3.2 Study 2: Exploring How High School Students Adapt Diagrams to Self-Manage Split Attention When Learning Mathematics

A study currently in progress that builds on the previous research study is investigating how high school students can adapt digital diagrams to learn mathematical concepts about parallel lines. The digital instructional materials are presented using interactive whiteboard software on a computer. Participants learn about corresponding and alternate angles and are required to complete questions on their understanding of the materials.

A similar research design as per the previous study has been adopted. The study is conducted on an individual basis and there are three instructional conditions: conventional split-attention instructions (Group 1), split-attention instructions that incorporate self-management strategies (Group 2), and an integrated condition (Group 3). Only the self-managed split-attention instructional condition can move textboxes closer to the diagram.

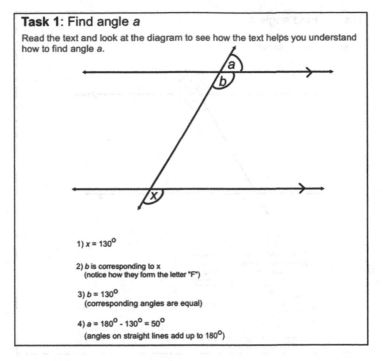

Task 1: Find angle *a*

Read the text and look at the diagram to see how the text helps you understand how to find angle *a*.

1) $x = 130^{\circ}$

2) *b* is corresponding to x
   (notice how they form the letter "F")

3) $b = 130^{\circ}$
   (corresponding angles are equal)

4) $a = 180^{\circ} - 130^{\circ} = 50^{\circ}$
   (angles on straight lines add up to 180°)

Fig. 6 Group 1: Static split-attention materials

Participants are asked to explain what they are thinking whilst studying the diagrammatic information, as with the previous study discussed, verbal protocols aim to provide insights into learners' cognitive processes when interacting with the instructions. The experiment comprises three phases. The first phase focuses on revising pre-requisite knowledge required for the experiment such as parallel lines and straight lines. The second phase is the training phase where each condition is instructed on what is required during the next phase—the acquisition phase. For example, participants in Group 1 are instructed to read the textual information carefully and identify how the textual information matches with the diagram. Participants in Group 2 are explicitly taught how to reduce split-attention by moving text objects closer to the relevant parts of the diagram. Participants in Group 3 are instructed to read the textual information that is integrated within the diagram carefully to understand how the text relates to the diagram. The acquisition phase comprises four tasks that participants complete using the study strategies shown in the learning phase. An example of a task in the acquisition phase for each condition is provided in Figs. 6, 7 and 8 below. The final phase is the test phase where participants complete a paper-based test. Preliminary analysis of verbal protocols from this study indicates that when students are explicitly taught self-management strategies, like moving text to appropriate places of a diagram, students are motivated to manipulate diagram and text to make it more meaningful for them.

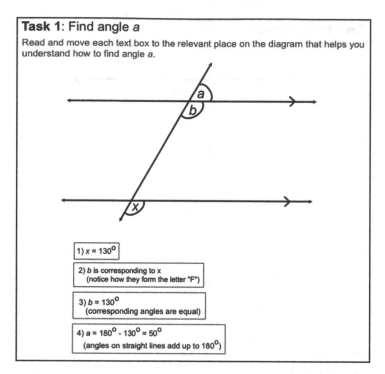

Fig. 7 Group 2: Adaptive split-attention instructional materials

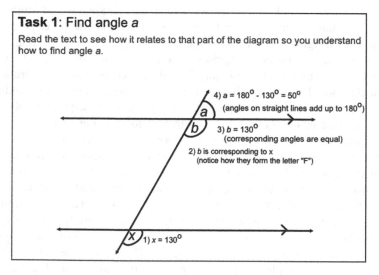

Fig. 8 Group 3: Static integrated instructional materials

Learners' comments suggest some form of prior experience and knowledge of the content, for example familiarity with parallel lines and angles support integration of text with diagram in a meaningful manner. Learners with less exposure to core content of parallel lines and angles found difficulty in self-management of split source materials.

The hypotheses for this study are that on test items the integrated condition (Group 3) would be superior to the split-attention condition (Group 1), validating the split-attention effect. It is anticipated that the self-managed split-attention condition (Group 2) outperforms conventional split-attention instructions (Group 1) suggesting that the self-managed split-attention condition is superior to the split-attention condition. It is also hypothesised that Group 2 performs as well as Group 3 demonstrating that self-management of split-attention can be as effective as instructor managed integrated condition.

4 Future Research Agenda

Once Study 2 is complete, the next step in our research is to build on this work by investigating the effect the physical movement involved in adapting diagrams has on learning. This will contribute to an emerging body of research in evolutionary educational psychology [31, 32] that proposes there is a link between cognition and the human motor system. This research would lead to the possibility that small human movements such as, moving text boxes on a touch screen like an interactive whiteboard, may decrease the load on working memory and thus support learning. A similar study as per Study 2 will be conducted but extended to include a fourth experimental condition to isolate the effect of human movement. This fourth group would watch how textual information is integrated rather than actually physically undertaking the movement themselves. For this future study, diagrammatic information will be presented on an interactive whiteboard (instead of a computer) and the additional condition will be similar in format to the split-attention format; participants will receive guidance on how to self-manage split-attention but instead of implementing this guidance themselves, they will observe the physical movement of text objects in animated form.

As today's learners now access and develop their own learning materials online, our research into how learners can control their own learning by adapting digital diagrams using evidenced-based design principles is timely and can be considered an example of human centric visualisation. Whilst our current research utilises interactive whiteboard software, another possible research area involves investigating how the principles of adapting digital diagrams can be implemented using eReader functionality. Online textbooks are proliferating the education context, yet there is little empirical research that explores how online textbook functionality can influence learning. Our future research can contribute to this empirical base and is timely as much of the research about eReader technology highlights a current lack of functionality for a learner to interact with text [e.g., 53].

5   Conclusion

This chapter has discussed an emerging research agenda that explores how learners can adapt diagrams using well-known design principles to aid their learning. The research is aligned with other investigations exploring instructional practices that take into account how humans process visual and textual information. The theoretical framework for this research is cognitive load theory, the research discussed utilises this theory to explore the efficacy of learners self-managing their cognitive load. The chapter provided an overview of critical aspects of cognitive load theory that underpins the research, and then summarised a series of paper-based and computer-based experiments that investigated self-management of split attention materials. Evidence from this preliminary research suggests that self-management of split source learning materials is a technique that can empower learners to take control of their learning and be a viable alternative to instructor managed instructional materials. The efficacy of self-management of split attention and other possible cognitive load theory effects in educational technologies like eReaders and interactive whiteboards is viewed as areas for future research. Clearly to optimize learning, educators and instructional designers need to support learners to efficiently and effectively engage with learning materials and enabling learners to adapt diagrams using evidence-based principles—a form of human centric visualization, is one such strategy.

References

1. P. Chandler. The crucial role of cognitive processes in the design of dynamic visualisations. In *Learning and Instruction*, volume 14, no. 3, pp. 353–357. 2004.
2. F. Paas, T. Van Gog, and J. Sweller. Cognitive load theory: New conceptualizations, specifications and integrated research perspectives. In *Educational Psychology Review*, volume 22, pp. 115–121. 2010.
3. J. Sweller, P. Ayres, and S. Kalyuga. *Cognitive Load Theory*. New York: Springer. 2011.
4. P. Chandler, and J. Sweller. The split-attention effect as a factor in the design of instruction. In *British Journal of Educational Psychology*, volume 62, pp. 233–246. 1992.
5. NMC Horizon Project Preview: 2012 K-12 Edition. The New Media Consortium. 2012.
6. P. Chandler, and J. Sweller. Cognitive load theory and the format of instruction. In *Cognition and Instruction*, volume 8, pp. 293–332. 1991.
7. J. Sweller, J. J. G. Van Merrienboer and F. Paas. Cognitive architecture and instructional design. In *Educational Psychology Review*, volume 10, pp. 251–296. 1998.
8. F. Paas, J.Tuovinen, H.Tabbers, and P.Van Gerven. Cognitive load measurement as a means to advance cognitive load theory. In *Educational Psychologist*, volume 38, pp. 63–71. 2003
9. F. Paas, A. Renkl, and J. Sweller. Cognitive load theory and instructional design: Recent developments. In *Educational Psychologist*, volume 38, pp. 1–4. 2003
10. F. Paas, A. Renkl, and J. Sweller. Cognitive load theory: Instructional implications of the interaction between information structures and cognitive architecture. In *Instructional Science*, volume 32, pp. 1–8. 2004.
11. J. Sweller. Cognitive load during problem solving: Effects on learning. In *Cognitive Science*, volume 12, pp. 257–285. 1988.

12. J. Sweller. *Instructional design in technical areas.* Melbourne: ACER Press. 1999.

13. A.D. Baddeley. Working memory. In *Science,* volume 255, pp. 556–559. 1992

14. K.A. Ericsson, and W. Kintsch. Long-term working memory. In *Psychological Review,* volume 102, pp. 211–245. 1995.

15. G. Miller, G. The magical number seven, plus or minus two: Some limits on our capacity for processing information. In *Psychological Review,* volume 63, pp. 81–97. 1956.

16. L.R. Peterson, and M.J. Peterson. Short-term retention of individual verbal items. In *Journal of Experimental Psychology, volume* 58, pp. 193–8. 1959.

17. J.Sweller. Instructional design consequences of an analogy between evolution by natural selection and human cognitive architecture. In *Instructional Science,* volume 321, pp. 9–31. 2004.

18. A. De Groot. *Thought and choice in chess:* The Hague, Netherlands: Mouton. (Original work published in 1964). 1965.

19. W.G. Chase, and H.A. Simon. Perception in chess. In *Cognitive Psychology,* volume 4, pp. 55–81. 1973.

20. H. Simon, and K. Gilmartin. A simulation of memory for chess positions. In *Cognitive Psychology,* volume 1, pp. 29–46. 1973.

21. K. A. Ericsson, and N. Charness. Expert performance: Its structure and acquisition. In *American Psychologist,* volume 49, pp. 725–747. 1994.

22. D.E. Egan, and B.J. Schwartz. Chunking in recall of symbolic drawings. In *Memory and Cognition,* volume 7, pp. 149–158. 1979.

23. H. Chiesi, G. Spilich, and J.F. Voss. Acquisition of domain-related information in relation to high and low domain knowledge. In *Journal of Verbal learning and Verbal Behaviour,* volume 1, pp. 257–273. 1979.

24. M. Chi, R. Glaser, and E. Rees. Expertise in problem solving. In R. Stenberg (Ed.), *Advances in psychology of human intelligence,* pp. 7–75. Hillsdale, NJ: Erlbaum. 1982.

25. J. Larkin, J.R. McDermott, and D. Simon, & H. Simon. Models of competence in solving physics problems. In *Cognitive Science,* volume *4,* pp. 317–348. 1980.

26. W. Schneider, and R.Shiffrin, R. Controlled and automatic human information processing: Detection, search and attention. In *Psychological Review,* volume 84, pp. 1–66. 1977.

27. R. Shiffrin, and W.Schneider. Controlled and automatic human information processing: II. Perceptual learning, automatic attending, and a general theory. In *Psychological Review,* volume *84,* pp. 127–190. 1977.

28. K. Kotovsky, J.R. Hayes, and H.A. Simon. Why are some problems hard? Evidence from Tower of Hanoi. In *Cognitive Psychology,* volume *17,* pp. 248–294. 1985.

29. F. Paas, and J.J.G. Van Merriënboer. Instructional control of cognitive load in the training of complex cognitive tasks. In *Educational Psychology Review,* volume *6,* pp. 51–71. 1994.

30. J.J.G. Van Merriënboer. *Training Complex Cognitive Skills: A Four-Component Instructional Design Model for Technical Training.* Englewood Cliffs, NJ: Educational Technology Publications. 1997.

31. F. Paas, and J. Sweller. An evolutionary upgrade of cognitive load theory: Using the human motor system and collaboration to support the learning of complex cognitive tasks. In *Educational Psychology Review,* volume *24,* pp. 27–45. 2012.

32. D.C. Geary. An evolutionarily informed education science. In *Educational Psychologist,* volume *43,* pp. 179–195. 2008

33. S. W. Cook, Z. Mitchell, and S. Goldin-Meadow .Gesture makes learning last. In *Cognition,* volume *106,* pp. 1047–1058. 2008.

34. J. Sweller, J. Element interactivity and intrinsic, extraneous, and germane cognitive load. In *Educational Psychology Review,* volume 22, 123–138. 2010.

35. R. Clark, F. Ngyuen, and J. Sweller. *Efficiency in Learning: Evidence Based Guidelines to manage Cognitive Load.* San Francisco: Pfeiffer. 2006.

36. J. Sweller, P. Ayres, and S. Kalyuga. *Cognitive Load Theory.* New York: Springer. 2011.

37. F. Paas. Training strategies for attaining transfer of problem-solving skill in statistics: A cognitive-load approach. In *Journal of Educational Psychology,* volume *84,* pp. 429–434. 1992.

38. F. Paas, and T. Van Gog. Optimising worked example instruction: Different ways to increase germane cognitive load. In *Learning and Instruction,* volume 16, pp. 87–91. 2006.
39. F. Paas, and J.J.G. Van Merriënboer. Variability of worked examples and transfer of geometrical problem-solving skills: A cognitive-load approach. In *Journal of Educational Psychology,* volume *86*, pp. 122–133. 1994.
40. P. Chandler, and J. Sweller, J. Cognitive load while learning to use a computer program. In *Applied Cognitive Psychology,* volume 10, pp. 151–170. 1996.
41. S. Tindall-Ford, P. Chandler, and J. Sweller, J. When two sensory modes are better than one. *Journal of Experimental Psychology: Applied*, volume 3, pp. 257–287. 1997.
42. S. Kalyuga, P. Chandler, and J. Sweller, J. Managing split-attention and redundancy in multimedia instruction. *Applied Cognitive Psychology,* volume 13, pp. 351–371. 1999.
43. S. Kalyuga, P. Ayres, P. Chandler, and J. Sweller. The expertise reversal effect. In *Educational Psychologist*, volume 38, no. 1, pp. 23–31. 2003.
44. T. Van Gog, T., F. Paas, N. Marcus, P. Ayres, and J. Sweller. The mirror-neuron system and observational learning: Implications for the effectiveness of dynamic visualizations. In *Educational Psychology Review,* volume 21, pp. 21–30. 2009.
45. A. Wong, N. Marcus, L. Smith, G.A. Cooper, P. Ayres, F. Paas, et al. Instructional animations can be superior to statics when learning human motor skills. In *Computers in Human Behavior,* volume *25*, pp. 339–347. 2009.
46. J. Sweller. Evolution of human cognitive architecture. In *The Psychology of Learning and Motivation,* volume 43, pp. 215–266. 2003.
47. G. Cooper, G. *Research into Cognitive Load Theory and Instructional Design at UNSW.* Retrieved from http://www.arts.unsw.edu.au/education/CLT_NET_Aug_97.HTML. 1998.
48. J. Sweller, P. Chandler, P. Tierney, and M. Cooper. Cognitive load and selective attention as factors in the structuring of technical material. In *Journal of Experimental Psychology,* volume 119, pp. 176–192. 1990.
49. P. Ginns. Integrating information: A meta-analysis of the spatial contiguity and temporal contiguity effects. In *Learning and Instruction,* volume 16, pp. 511–525. 2006.
50. K. Roodenrys, S. Agostinho, S. Roodenrys, and P. Chandler. Managing one's own cognitive load when evidence of split attention is present. *Applied Cognitive Psychology*, volume 26, no. 6, pp. 878–886. 2012.
51. R. Mayer, and R. Moreno, Techniques that reduce extraneous cognitive load and manage intrinsic cognitive load during multimedia learning. In J. Plass, R. Moreno, and R. Brünken. (Eds.), *Cognitive load theory,* pp. 131–152. New York: Cambridge University Press. 2010.
52. K.A. Ericsson, and H.A. Simon. *Protocol analysis: Verbal reports as data.* MIT Press, Cambridge, MA. 1993.
53. A. Behler, and B. Lush. Are you ready for e-readers? In *The Reference Librarian*, volume 52, pp. 75–87. 2011.

Dynamic Visualisations and Motor Skills

Juan Cristobal Castro-Alonso, Paul Ayres, and Fred Paas

Abstract Due to their popularity, dynamic visualisations (e.g. video, animation) seem attractive educational resources. However, in the design of any instructional material, not only must the appealing factor be acknowledged, but also the cognitive limitations. To consider the limitations of human cognitive architecture when designing instructional resources has been the *leitmotif* of cognitive load theory (CLT). CLT research has shown that the transitory nature of dynamic visualisations imposes such a high working memory load that, in many cases, these depictions are no more effective for learning than static visualisations. However, dynamic visualisations have been shown to be superior to static visualisations when the depiction involves human motor skills, a special case which might be explained by the mirror neuron system (MNS) aiding working memory to cope with transitory information.

We will begin this chapter by presenting instructional properties of dynamic visualisations. Next, we will discuss the main differences between dynamic and static visualisations and how each can affect learning. Then, we will describe briefly CLT to give a more detailed account on instructional strategies to improve learning from dynamic visualisations. Next, we will summarise video modelling of motor skills. To end this chapter, we will focus on the MNS and how it aids humans to learn motor skills through observation.

J.C. Castro-Alonso (✉) • P. Ayres
School of Education, University of New South Wales, Sydney, Australia
e-mail: j.c.castroalonso@unsw.edu.au; p.ayres@unsw.edu.au

F. Paas
Institute of Psychology, Erasmus University Rotterdam, Rotterdam, The Netherlands
e-mail: paas@fsw.eur.nl

W. Huang (ed.), *Handbook of Human Centric Visualization*,
DOI 10.1007/978-1-4614-7485-2_22, © Springer Science+Business Media New York 2014

1 Dynamic Visualisations and Learning

1.1 General Properties of Dynamic Visualisations

The words *picture*, *visualisation*, *image*, *depiction* are employed in education and psychology research interchangeably to describe representations that can convey stationary or moving elements. In the latter case, the representations are generally called *dynamic pictures*, *dynamic visualisations*, or—the shorter and preferred form—*animations*. So, what do education and psychology researchers mean by this word? According to Roncarrelli [84], animation can be defined as:

> Producing the illusion of movement in a film/video by photographing, or otherwise recording, a series of single frames, each showing incremental changes. . . , which when shown in sequence, at high speed, give the illusion of movement. The individual frames can be produced by a variety of techniques from computer generated images, to hand-drawn cells. (p. 8)

One important aspect to highlight in this definition is that animations are composed of a series of single frames (i.e. static pictures). The quantity of these still-frames changing per unit of time can be defined as *temporal granularity* [88]. An animation with many frames per unit of time would have higher temporal granularity than one with a lower number of frames in the same amount of time. Figure 1 shows two animations with different temporal granularities, 12 frames per second (fps) versus 24 fps.

Following mainstream conventions of temporal granularity, we can mention that the animated cartoon standard of 12 fps is a condition of lower temporal granularity than the film industry norm (24 fps), or the US analog television and video convention (NTSC, 30 fps).

The standard of 12 fps has been used extensively in animated cartoons [cf. 109] because it is sufficient to achieve the illusion of movement—due to persistence of vision—that is mentioned by Roncarrelli. As it can be inferred, when the pace

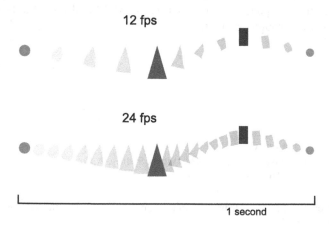

Fig. 1 Comparison between two animations with different temporal granularities

of the dynamic visualisation is progressively reduced, the illusion of animated frames gradually disappears, until they are perceived as static pictures. On the contrary, in typical conditions of watching animations (≥ 12 fps), the individual static frames are hardly noticed, and, more importantly, each frame is visible for only an instant ($\leq 1/12$ s). This means that one distinguishing property of animations is their *transiency*.

A second important aspect to highlight from the definition above is that the approach employed to produce the animations, whether vector or raster graphics, is irrelevant. On the one hand, *vector graphics*, the branch of design that illustrates line drawings and shapes, is associated with the traditional concept of animation and cartoons; on the other hand, *raster* or *bitmap graphics*, the branch of production that records audiovisual frames, is associated with the traditional concept of film and video [84]. As an example, consider a chemistry multimedia that includes three different dynamic visualisations: (1) a video (bitmap) of the procedures to mix substances in the laboratory; (2) a symbolic representation (2D vector) of the equations describing reactants and products; and (3) a molecular depiction (3D vector) of the changes in chemical bonds occurring [cf. 54]. Although these three screens show very different visualisations, which were produced by different methods, according to the definition used here we can consider all of them as *animations*.

Then, why not distinguish dynamic depictions according to these subcategories? Although we acknowledge that vector and bitmap graphics have obvious visual differences [cf. 43], we want to focus on the *similarity* of animated cartoons, 3D depictions, films, videos, television, etc., which is their transiency. Thus, any transient visualisation will be defined as *animation* or *dynamic visualisation* henceforth. As it can be inferred, this transient property of animations affects learning, what we describe next.

1.2 Relevance of Dynamic Visualisations to Learning

Animations are popular and their influence transcends educational boundaries. In fact, dynamic visualisations are so ubiquitous that viewers may confound their commercial-recreational or informational-educational roles [cf. 102]. For example, Wineburg (2000) noted that both students and their parents made references to fictional movies such as Forrest Gump and Schindler's List to support claims in their arguments about historical events [as cited in 108]. Also, Lewis et al. (1997) showed that students employed detective series, science fiction movies, and television programmes as their main sources for science and genetics knowledge [as cited in 52]. Similarly, a study with 218 subjects (12–13 year-old) that learned from a material consisting of 16 short animations and 16 corresponding short text pieces, showed that the students spent more time learning from the animated information. This preference for dynamic visualisation over text was also unexpectedly observed in the condition where subjects were notified that they would be assessed later, although it had been predicted that they would choose text, as this representation is the conventional for school evaluations [13]. These three examples show that dynamic visualisations are so appealing that they are preferred for informational-educational purposes over other instructional materials that might be more suitable for that role.

Interestingly, this preference can be exploited by using animations as motivational enhancers of other educational media or materials. This is not the same as stating that the only educational purpose of dynamic visualisations is a boost in motivation. Above all, researchers agree that the design of instructional animations must be aimed beyond motivation per se. In this sense, dynamic visualisations can give students opportunities beyond the spatial or temporal limitations of learning from static materials, as the following examples show:

- To shorten, via fast motion, the duration of phenomena excessively slow to be observed in real time [102]. Also, the method of fast motion can be used to analyse dynamic recordings that have extended for lengthy periods [29].
- On the contrary, to study the details of movement via slow motion, irrespective of the original speed of the phenomenon [102]. For example, this technique is very important for studying motor tasks, facial expressions of emotion, and nonverbal communication [29].
- To zoom at microscopic details or at far away phenomena in outer space [102].
- To study events that occur in unapproachable places, such as inside the human body, underwater, in extreme temperature conditions [102], or in situations that a live observer would rather avoid [29].
- To watch otherwise invisible events, by using X-rays, infrared, gamma, ultraviolet, holographic techniques, and many others [102]. For example, the assistance of X-rays can be used in medical diagnosis [29]. Otherwise, dynamic visualisations can assist to explain not inherently visual phenomena, such as forces, energy or electrical circulation [12].
- To analyse the same phenomenon multiple times [102]. Also, the analysis can be performed by many reviewers, what can benefit interjudge reliability [29].
- To compare simultaneously two or more events by splitting the screen accordingly. This can be a valuable tool in sports training, where a close-up view can be juxtaposed to a wider setting [31].

According to these capabilities, and inspired by research into explanations and maps, Tversky et al. [104] gave the following summary on how animations might promote learning:

> Explaining a process can be supplemented with analogies to other processes, with enlargements of subprocesses. . . . Just like good explanations and good maps, animations can include other views, other scales, other examples, other processes; they can use language and extra-pictorial devices to connect them. Animations can exaggerate and minimize and distort and highlight information. (p. 281–282).

Because of this potential, an enormous variety of animations, videos and films have been designed to assist learning, including school-age instruction, university settings, advanced and continuing education, job training, health education, safety education, medical treatment preparation, and special population instruction [e.g. 30]. However, educational animations motivated by different aims than students' learning, chiefly for technology showcase, may prove ineffective. This focus exhibits what Mayer [65] criticised as a *technology-centred approach*, which forces learners to adapt to technologies, instead of shifting to the desirable opposite direction with a *learner-centred approach*. Hence, for animations to be effective learning tools they must be designed according to the cognitive architecture of the learner.

In conclusion, animations may have educational relevance and potential, provided that limits of the human cognitive architecture are considered when designing these dynamic visualisations. Acknowledging these limitations is also fundamental to designing other materials for learning, for example, the instructional static visualisations that we incorporate in the next discussion.

1.3 Learning Outcomes with Dynamic Versus Static Visualisations

When learning a dynamic system from static visualisations, the student must deduce the movement of the components involved, by a process described as *mental animation* [39]. In contrast, when learning the same dynamic system from an animation, the movement of the components is shown directly [40, 68]. If we consider that it may be more difficult to deduce than to observe directly the dynamics of the moving system, there appears to be a cognitive advantage of a dynamic visualisations as compared to its static equivalents [40, 68]. One example of better learning with animation versus static pictures is an experiment with 119 fourth and fifth graders from an urban elementary school, who were given the topic of Newton's laws of motion. In this study, children in the animated conditions scored significantly higher on the posttest than those in the static pictures conditions [82]. Another experiment with 160 first year university students that learned from two multimedia instructions of astronomy and geology, revealed that the dynamic condition (12 animated steps of each phenomenon) outperformed the static version (12 key frames of each) in tests of retention. Similarly, the researchers observed that, when the learning setting was in dyads of students, the animation groups presented higher transfer scores than the static conditions [81]. These findings tend to support that better learning is achieved after watching an animation of a dynamic system rather than after inferring mentally the dynamics from static visualisations.

However, the fact that a dynamic depiction shows directly the motions of the components of the system, can also be a cognitive disadvantage, since it may promote "mental laziness" in the learners. According to this view, if learners are able to invest cognitive effort to perform a mental animation when given static pictures, giving them the already animated visualisation may impede this form of active generative processing that should enhance learning [40, 89].

Whether an animation fosters either a learning advantage or a disadvantage over static visualisations has not been resolved by research [e.g. 78]. For example, after reviewing studies of chemistry multimedia, Kozma and Russell [54] concluded: "We are not able to say. . . for which topics or students it is best to use animations versus still pictures or models. Nor can we say how these various media can be used together. . ." (p. 424). Another example that fails to show a significant advantage of animation or static pictures is an experiment with 82 students learning how surfactants dissolve dirt from a surface, in which the authors concluded that "neither animations nor static pictures are generally superior" [44]. To add to this inconclusiveness, there are some voices, for example Schnotz and Lowe [88], who consider it irrelevant to compare statics versus dynamics:

> In our everyday lives, we do not continually compartmentalize our environment into static and dynamic parts. . . Even when we move through a static environment, our visual field is continuously changing. . . On this basis, it seems difficult to justify a sharp distinction between learning from animated and learning from static pictures, and there appears to be little reason to assume that animations are necessarily easier or more demanding than static pictures. (p. 305).

In addition, to complicate the dynamic versus static comparisons further, there is research that goes beyond these two extremes to include middle ground conditions, which either: (1) lower the temporal granularity of the animation, or (2) show only some important frames of the animation. In one example of the first situation, where the learning content was the kangaroo hopping cycle, participants watching the condition of 1 fps outperformed another group that observed the same depiction at the higher granularity of 8 fps [59]. In one example of the second scenario, better learning outputs resulted when the participants watched either an animation of a flushing cistern or three key frames, as compared to a single static image [41].

It is important to mention that the lack of definitive conclusions to support dynamic or static visualisations may depend on the fact that some studies fail to control all the experimental variables that distinguish both representations. In line with this, a review on animation found that the seeming victories of the dynamic over the still pictures were mainly because the formers presented more or better information, or included facilitators of learning such as interactivity [103]. Take the case of a study with 415 university students who were given the topic of electrochemical features of a flashlight in a lecture supplemented with either dynamic or static visualisations [112]. As the researchers discussed, the better learning of the class exposed to animations could also be explained by alternative factors: (1) *multiple exposures* (in the animation class the depictions were played several times), or (2) *motivation on attention* (the animations were coloured, but the static pictures were black and white). Nonetheless, this does not mean that there is no evidence for better outcomes of dynamic versus static pictures in studies that did control these extra variables. For example, a study with 112 university participants, found a significant medium-to-large effect favourable to animations in the retention test of concepts about the rock cycle, even when the interactivity in both conditions was equivalent [55].

Fostering static visualisations, there is accumulating evidence that shows better learning from static pictures instead of animations. For example, in a study that compared *illustrations plus text* versus *narrated animations*, Mayer et al. [68] found that the groups with static pictures outperformed the groups with dynamic visualisations. Although it can be argued that this investigation did not control the different media employed (the comparisons were between *paper* illustrations versus *computer* animations), the contents depicted were varied enough—lightning, ocean waves, toilet tanks and car's brakes—to at least show an important tendency to promote static pictures. More evidence comes from a review of animation in e-learning, where Clark [21] concluded: "these studies provide preliminary evidence that at least for relatively simple procedures and processes, nonanimated treatments that communicate motion such as line drawings with motion indicators can result in learning equivalent to that resulting from animations". Similarly, Koroghlanian and Klein [53], who found in an experiment with biology students that animation implied more instructional time than static pictures, without corresponding improvement, concluded that "[w]hether to include animation or not in multimedia. . . is still a matter of instinct, not research, and the final decision may be dictated by pragmatic concerns such as budget or time" (p. 40).

So what is the explanation of these results in favour of static visualisations? One advantage of processing static as opposed to dynamic pictures is that, generally, static depictions focus on the fundamental steps of the processes, thus allowing learners to study exclusively from the most important pieces of information [68]. Other cognitive advantage of statics over dynamics is that, learners can control the speed of presentation of the static visualisations in order to meet their cognitive capacities [68]. When studying static pictures, as learners can observe permanently available depictions, they can accumulate information without speed limits via consecutive eye fixations; on the contrary, when studying dynamic pictures, information that was available at one instant disappears a moment later, obstructing the accumulation of information to be processed [40, 88]. In cognitive load theory, this phenomenon has been termed the *transitory effect of animations* [4], which we describe in a following section.

On the whole, it seems that there is not a unique answer to whether animation or still pictures is a better learning tool. In line with this, a number of moderating variables might be considered, such as learner characteristics. For example, in an experiment with participants with a cognitive style of either *visualisers* or *verbalisers*, it was shown that visualisers performed better with animation and that verbalisers showed a trend (not significant) to perform better with static pictures [44]. A more important learner characteristic is most likely spatial ability. A meta-analysis of 27 different experiments that involved learning from visualisations revealed that high-spatial-ability learners were significantly more superior to low-spatial-ability learners when learning with static pictures instead of dynamic pictures. In other words, although high-spatial learners still learned better with animations than low-spatial learners, the difference was reduced as compared to learning with statics. The researchers remarked that, because many studies of dynamic versus static visualisations had not controlled for spatial ability of the subjects, this variable could explain the heterogeneous findings regarding comparisons between both instructional depictions [42].

Another moderating variable besides learner characteristics is depiction characteristics. One of these is the topic to be learned. In fact, there are some contents that seem better portrayed as dynamic rather than static visualisations. One first example where animation shows an advantage is in topics that require students to visualise motion and trajectory attributes [82]. A second example is in recognising faces or facial expressions [1, 11]. However, more empirical evidence for a favourable learning with animations can be found in manipulative-procedural tasks. Correspondingly, a review by Park and Hopkins [78] aimed at finding instructional applications for animations, concluded that the dynamic visualisations were most effectively applied for "demonstrating sequential actions in a procedural task" (p. 443). In a similar way, a more recent meta-analysis by Höffler and Leutner [43] showed that the largest mean effect size in favour of dynamic as compared to static visualisations was found when procedural-motor knowledge was depicted.

Why is learning from either dynamic or static visualisations influenced by the learning topic or task? To understand this, we need to consider the evolution of our cognitive architecture, which has been highly influential in recent developments of cognitive load theory.

2 Cognitive Load Theory and Dynamic Visualisations

2.1 General Aspects of Cognitive Load Theory

Key components of humans' cognitive architecture are two memory subsystems called *long-term memory*, and *working memory*. Long-term memory can sustain large amounts of elements over long periods of time. Unlike long-term memory, working memory (WM) has processing limitations in both the quantity of elements to manage [22, 71], and the retention of these elements [79]. As a result, WM is a limited processor in managing cognitive load when learning.

The total cognitive load imposed on WM when learning can be divided into two different categories [97]: (1) *intrinsic cognitive load*, and (2) *extraneous cognitive load*. As its name implies, intrinsic cognitive load cannot be modified substantially by instructional interventions because it is intrinsic to the material being learnt [98]. A learning material with a high intrinsic cognitive load comprises many elements related to one another, so they must by processed simultaneously in WM; by contrast, there is low intrinsic load when the learning material consists of elements that can be processed independently, so they are not to rather than simultaneously in WM [19, 98]. As its name implies, too, and opposed to intrinsic load, extraneous cognitive load is not related to the complexity of the content to be learnt but to the way that this content has been instructionally designed. In other words, extraneous load is a typical problem of poorly designed educational materials. An obvious conclusion is that extraneous load must be minimised when designing instruction, as it diverts cognitive processes from the goal of learning [48].

That conclusion can be regarded as the *leitmotif* of cognitive load theory [for a recent review, see 99]. In brief, cognitive load theory (CLT) is a learning and instruction theory, which considers the human cognitive architecture in order to prescribe ways to manage cognitive load in learning events [47]. As reviewed by Paas et al. [77], the potential of CLT has been empirically tested in fields such as mathematics, statistics, physics, computer programming, electrical engineering, paper folding, and with different types of participants (children, young adults and older adults). In all these experiments, compared to conventional designs that did not consider the limitations of WM, the designs based on CLT required less training time and less mental effort to reach the same or significantly higher learning and transfer performances.

To summarise, instructional methods that consider the limitations of WM will be more effective than approaches that do not. Then, why do apparently difficult tasks, such as speaking or gesturing, seem not to be constrained by these limits? An explanation to this dilemma can be better understood through the lens of David Geary's *evolutionary educational psychology*. Next, we explain this relatively new addition to CLT.

2.2 Cognitive Load Theory and Evolutionary Educational Psychology

Most humans effortlessly learn tasks such as speaking, recognising faces, and gesturing. As Sweller [95] argues, because these tasks are easily learnt before going to school, it is tempting to suggest that *going to school* is the problem. In other words, the pedagogy of schools—namely, guided methods or explicit instructions—should be blamed. Additionally, it is also tempting to claim that the learning methods of pre-school children—such as pure discovery approaches—should be taught in schools, because they are easily adopted, even by juveniles. However, most of the evidence based on randomised and controlled experiments points in the opposite direction and supports the need for explicit instructions in schools [for reviews, see 50, 61].

This apparent paradox may be solved by not focusing on the *methodology* of the task to teach but in the *task* itself. For example, learning to gesture is easy not because it is taught without educational guidance, but because it is an easy task to learn. In other words, although gesturing may involve the coordination of many processes of the motor repertoire that could overload WM, nonetheless it is an easy skill for humans [95]. As evolutionary educational psychology portrays, tasks such as understanding gesturing and nonverbal behaviour, using physical materials as tools, or learning to speak are all effortless for humans because they have evolved those skills over many generations [34].

The *Homo sapiens* species had many years, and thus opportunities, to evolve or refine, for example, the skill of gesturing. Using a theory of evolution's term, the refinement to gesture was driven by the *struggle for existence*. As Darwin (1859) explained, struggling for existence means that different species compete for limited resources, with the final goal of perpetuating their own kind [as cited in 80]. Therefore, learning to gesture was a skill that evolved because it helped the human species to establish beneficial relationships for accessing essential resources, what to that supported the species in its survival or struggle for existence [34]. In the same vein, every other skill that has benefited humans in their survival should have evolved, and thus be in the species' motivational and cognitive disposition [34]. As such, these evolved skills are learned easily, without conscious effort and without explicit education [96].

Geary has named these evolved effortless skills as *biologically primary abilities*, which are grouped under the term *folk knowledge*. This primary knowledge can be subdivided into three main categories of human survival: (1) *folk psychology* considers social abilities for survival, such as competing for mates, recognising facial expressions or understanding gesturing and nonverbal behaviour; (2) *folk biology* is related to the knowledge of local flora and fauna that assists in survival, such as discriminating food from poison; and (3) *folk physics* includes abilities such as manipulating physical materials as tools [34].

In contrast to the primary abilities, we can mention *biologically secondary abilities* or secondary knowledge, which is not evolved and thus effortful. As its name implies, this secondary knowledge emerges from the primary knowledge, as the secondary scientific and academic domains converge around the primary areas of folk psychology, biology, and physics [34].

Stated another way, since species evolutionary adaptations have a fundamentally slower pace than intellectual and academic shifts, evolution has only been capable of equipping humans to manage effortlessly biologically primary but not secondary knowledge [95]. This has two implications: (1) we can use the relatively easy acquisition of primary tasks to assist students to learn the more difficult secondary knowledge skills [34], as we shall see in last section; and (2) students need guided methods and explicit instruction that consider their limited WM to learn secondary tasks [95].

Restating the last implication, CLT and resultant strategies to manage cognitive load apply mainly to biologically secondary knowledge [95]. For that reason, the instructional methods for dynamic visualisations that we present in next section are generally aimed at learning tasks or topics that deal with secondary as opposed to primary knowledge.

2.3 Instructional Strategies to Manage Cognitive Load in Dynamic Visualisations

In order to manage cognitive load when designing educational materials of biologically secondary knowledge, CLT has fostered diverse strategies. They are generally referred to as *cognitive load effects* [e.g. 97], but some are sometimes called *multimedia learning principles*, especially in the context of Mayer's cognitive theory of multimedia learning [for a review, see 65].

Building on those strategies, CLT has opened research directions aimed at methods for designing better dynamic visualisations. In line with this, one appealing new field of study is the *transitory effect of animations* [4]. This effect deals with the fact that some educational animations can be highly transitory, meaning that as the animation progresses, the elements depicted continually disappear. As a result, this transitoriness produces a high extraneous cognitive load in learners, because they are forced to perform three cognitive tasks simultaneously in WM: (1) process the current visible information [4], (2) remember the previous elements that are no longer visible [4], and (3) integrate these two streams of information in order to comprehend the material [106]. To help students cope with these demanding mental processes, CLT has provided various instructional strategies.

Next, we will describe the following methods to manage the transitory problem of dynamic visualisations: (1) *pace-control*, (2) *modality*, and (3) *attention cueing*.

2.3.1 Pace-Control

In a *pacing-controlled*–also named *stepwise, self-pacing,* or *learner-controlled*– animation, the learner can control the incremental progress of the visualisation, as opposed to a *continuous* or *system* animation where the depiction controls how to run the display [89]. Hence, the *pace-control effect* occurs when learners have control over the speed of the animation, and thus they can manage its transitoriness in accordance with the learners' capacities [66]. As a result, a CLT prediction is that animations may be more effective learning materials if the students have control over the pace of the presentation [3].

Supporting evidence for this effect in animations is an experiment about historical inquiry that was presented in a multimedia tutorial of 11 min. After this study, it was concluded that providing a pace-control feature could facilitate basic (retention) and deep (transfer) knowledge acquisition [60]. Another experiment to support the pace-control effect was conducted with 82 university students, where the instructional content was the chemical process of dirt removal from a surface [44]. In this study, it was shown that the self-pacing groups outperformed the system-pacing conditions, and that the former also reported a lower subjective cognitive load. Extending from the latter results, we could assume that both examples of a better performance in the self-pacing conditions might have been caused by a reduction in the cognitive load of the instructional material.

Related to the pace-control effect is the *segmenting effect* or *principle*, which advocates segmenting whole animations into shorter sections that do not overload the learners' WM capacity [72]. An obvious CLT prediction of the segmenting or segmentation effect is that animations may be more effective learning tools if they are divided into smaller segments [3]. Figure 2 illustrates the application of either pace-control or segmenting strategies to a whole-continuous animation.

We have included segmenting techniques inside this pace-control discussion because both strategies: (1) share the basic action of controlling the transitoriness of a lengthy animation, and (2) have been used interchangeably or combined in the literature [e.g. 60, 64, 67]. Detailing the second point, we can mention a study consisting of two experiments in which university students received narrated

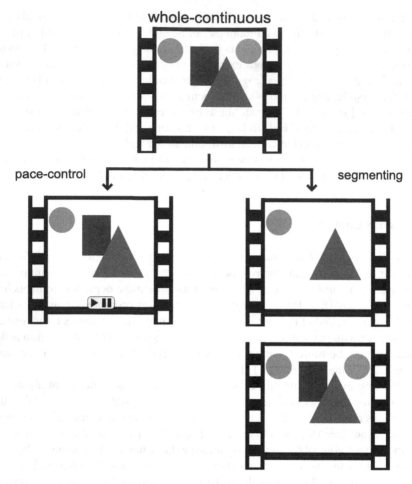

Fig. 2 Pace-control or segmenting strategies to a whole-continuous animation

animations in formats that combined *partial student-paced* (*P*) with *whole system-paced* (*W*) visualisations. In this work, Experiment 1 compared PW (partial student-paced followed by whole system-paced) with WP conditions, whereas Experiment 2 contrasted PP against WW. Taking together the two experiments, it was consistently shown that groups that firstly received the P format outperformed in transfer tests those students who firstly watched the W design. This result has a CLT explanation: starting by studying a short segment of an animation places less unnecessary load on WM than attempting to study the whole longer animation firstly [67]. Although the results support the CLT explanation, this study did not isolate segmenting from pace-control effects. In other words, the experimental conditions differed by two variables instead of one: (1) *animation's extension* (whole versus short segments), and (2) *controller agent* (system versus learner paced). However, a later study that considered this confounding variable problem was conducted with 72 male primary school students (ages 9–11) who watched a multimedia presentation that explained the causes of day and night. In this experiment, the participants who were assigned to the conditions that followed either the pace-control or the segmentation methods outscored the conditions of system-controlled pace, in the more complex questions [38]. In addition, since the pace-control and the segmentation conditions did not differ between them, these results support use of either or both of these two strategies in designing better instructional animations.

In conclusion, the rationale that both pace-control and segmentation strategies share, is to reduce the transitoriness of a whole animation, thus allowing pauses in the visualisation, which are needed to process both current and previous information shown. Consistent with this reasoning is what has been termed the *piecemeal hypothesis of mental animation*, where mentally animating a dynamic system is supposed to consist of animating the individual components one by one, in a causal chain of events, rather than mentally animating the whole system in a single process, which should be more cognitive demanding [39]. Although the piecemeal hypothesis fosters divisions of whole *systems*, whereas pace-control and segmentation effects are related to divisions of whole *animations* both approaches imply reduction of cognitive load generated by the transitoriness of the visualisations.

Recognising the similarity between pace-control and segmentation techniques, there is a critical difference between them, dependent upon the agent who segments the animation. In this case, the depiction might be segmented by either: (1) the *learner*, which comprises the pace-control strategy, or (2) the *instructional designer*, which comprises the segmenting method. As a result, each of these agents influences two factors differently: (1) *interactivity*, and (2) *segment boundaries*.

Firstly, if learners segment the visualisation via pace-control, they incur an interaction with the material—chiefly through *delivery learner control* [47]—, which is generally absent in the segmentation strategy. Although delivery control does not use all of the interactivity potential, it is enough to incorporate learning variables that fall beyond the scope of this chapter, such as motivation and adaptation [e.g. 87]. Since these "interactivity variables" are moderators of the pace-control effect, they should be considered with respect to the studies that employ this strategy [For overviews of interaction in animations, see 74, 111].

Secondly, if designers divide the visualisation via segmentation, they can present the information in meaningful narrative pieces, which may help to better understand the depiction [94]. Put differently, rather than relying via pace-control on students' capacity to determine where to split the animation's sequence into meaningful substeps, this decision of finding narrative boundaries is achieved for the students via segmentation. Therefore this boundaries factor should also be considered as a moderator, especially for the segmentation effect [94].

2.3.2 Modality

It is commonly accepted that WM consists of two relatively independent channels or modalities: visual and auditory [6, 7, 20]. As a consequence, it may be more

Fig. 3 Modality strategy
to an animation with
on-screen text

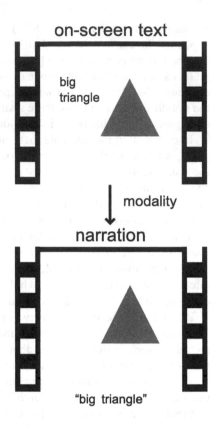

efficient to use both channels simultaneously than to use either independently, especially when the learning situation involves trying to deal with the visual load of an animation that contains written descriptions. In this scenario, learners may manage the transiency of the visualisation better if they are given the descriptions in narrated format, where they can employ the underused auditory channel, while the visual channel processes the dynamic frames. This use of both channels is known as the modality effect, which postulates that it is better to learn from animation and concurrent narration rather than from animation and on-screen written text [58, 65, 66]. Figure 3 shows an example of the modality strategy.

The above explanation of the modality effect can be referred to as the *visuo-spatial load hypothesis* [86] or the *hypothesis of expansion of effective working memory capacity* [98]. In support of this effect we mention two experiments with dual-task methodology, where the primary or learning tasks were computer-based training animations about the cardiovascular system or the city of Florence [17]. In each experiment, the participants had a better secondary task performance when the primary task was designed with lower visual load (narration) rather than higher visual load (on-screen text). Since the secondary or monitoring task was a visual-load task (noticing when a letter on the screen changed), these results tend to validate that there are less visual resources available to study when watching an animation with on-screen text versus a narrated condition.

Additional evidence in support of the dual-mode instructional strategy was found in an experiment that employed computer-controlled animation on theoretical aspects of heat soldering. It was found in this study that participants in the narration group showed lower subjective ratings of cognitive load and higher test scores compared to the on-screen text trainees [48]. In another example, researchers compared retention and transfer scores for students who watched either a narrated multimedia or the same depictions with on-screen text. In these studies about lightning storms and car breaking systems, the average modality gain was 33 % for retention and 80 % on transfer scores [65]. Similarly, a related study with animated pedagogical agents showed that the learners achieved better transfer performance with spoken narrative as compared to written texts [75].

In addition to the argument that effective working memory capacity is expanded, we will describe two other hypotheses that may explain the modality effect: (a) *social-cue hypothesis*, (b) *simpleness of the orality*. It should be noted that all these hypotheses may not be mutually exclusive.

Firstly, the social-cue hypothesis posits that the modality effect results from the additional interest that a voice can bring to a visualisation, compared to the presence of on-screen text [73, 75]. This hypothesis can be included in the broader *social agency theory*, which states that the social responses that personality cues prompt in students, motivate them to engage in greater learning [63]. In line with this theory, it has been shown that students can achieve higher transfer scores when narrations are in a more dialogical rather than formal style—*personalisation principle*—or when the words are not spoken in a foreign-accented or a machine-generated voice—*voice principle*—[63]. In addition, as the *speaker/gender effect* shows, female speakers are not only perceived as being more appealing than male speakers, but also when the feminine preference is presented there is a higher problem-solving performance. These results cannot be explained from a purely cognitive view because the processing of either a male or a female voice should require the same amount of mental resources [56]. In this way, the social-cue hypothesis, or other similar social-motivational perspectives, can be considered alternative or supplementary explanations for the modality effect.

Secondly, according to Liberman (1995), the modality effect can also be ascribed to the simpleness of the orality, shown in evidence such as: (a) speech appears before written words both in the history of each individual and in the records of humankind, (b) literacy is not as universal as orality, (c) speech is learned more easily than reading and writing, (d) although text can be overlooked when not attending to the written words, the persistent property of sound hinders speech to be missed [as cited in 73]. In the same line, we can add the described distinction between primary versus secondary knowledge, as humans may have evolved the primary skill of listening, but did not evolve the secondary skill of reading [76].

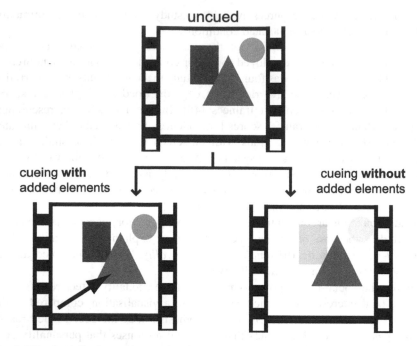

Fig. 4 Two strategies for cueing an animation

2.3.3 Attention Cueing

The *attention cueing effect*—also referred to as the *signalling* [62] or the *attention-guiding principle* [12]—states that students learn more deeply when cues or signals are added to a dynamic visualisation in order to highlight the important information and, therefore, to direct the learner's attention towards it [62]. As a result, a CLT prediction is that animations may be more effective learning materials if the key information is cued or signalled [3].

Cueing techniques can be divided into two broad groups: (1) *with* added elements, and (2) *without* added elements. Examples of the former are signalling with arrows or other pointing elements, lines, thick frames, symbols, texts, etc. Examples of the latter can be colours or patterns, dynamic contrasts, transparency shifts, flashing, zooming, panning, exaggeration and simplification [88]. It seems that the current trend is to prefer the latter method without added visual elements, possibly to avoid adding extraneous cognitive load by crowding the visualisation [cf. 25]. Figure 4 illustrates the application of either cueing techniques to an animation.

Empirical support for the strategy of attention cueing with added elements was found in an experiment with 129 French primary school participants studying animations of different gear systems. The study showed that students with low expertise levels benefited from attention cueing in the forms of arrows, dots and words [15]. Another example is an experiment that compared cueing (on-screen technical text and colour coding) versus no cueing in a sample of 83 biology college students. In this study, by adding these cues to a system-paced animation of the structure and function of an enzyme, Huk et al. [45] found a medium effect size for retention scores, and also that students reported fewer problems in comprehension. This cueing effect was replicated—but with a smaller effect size—in a more ecologically valid setting of a classroom scenario where the animation was learner-paced and part of a multimedia learning environment [45]. Interestingly, the cueing effects were observed both in 2-D and 3-D representation formats of the enzyme ATP-Synthase animations. More support in favour of this strategy, was found in a study which compared learning about an earth science topic for conditions that included red arrow cueing versus conditions without these added elements. Students in the conditions with signalling had significantly higher learning efficiency gains. However, cueing failed to show an effect in the retention tests, both for concept and process questions [55]. The lack of a more robust effect of cues may be due to the signalling technique used, which employed the extra elements of red arrows that could produce overcrowding of the visualisation.

In line with this, although the outcomes of attention cueing with added elements are encouraging, a more efficient signalling strategy might be one that does not add extra elements to the display, but, for example, increases the salience or visual contrast between the key elements and the secondary components [25]. One example of these salience techniques is a study with 73 psychology undergraduates who were required to learn about the cardiovascular system either from cued or uncued animations. In this example of *spotlight-cueing*, all components in the animation had their luminance decreased except for the cued key elements. It was found that the cued version resulted in better performances on retention, inference and transfer questions than the animation without this type of signalling [26]. Other example of attention cueing without added elements is an experiment with 102 undergraduate students who learned from different cueing conditions of animation about the cerebral base of language production. In this study, two methods to increase visual contrast or salience where employed, namely, change of colour and sudden appearance. It was found that salient colouring was significantly better than no salient colouring for the tasks *diagram completion*, *process retention*, *function retention*, as well as the *perceived ease of use scale*. Additionally, it was found that the students who learned with the cueing of sudden appearance performed significantly higher than those without this signalling, in the tasks of *diagram completion* and *function retention* [46]. More direct evidence favouring attention cueing strategies without extra elements was found in a study that coloured dynamically important parts of a piano animation. Results indicated better comprehension of the piano mechanism with the colouring method, rather than the use of external pointing in the form of arrows [16].

To conclude, pace-control, modality, and attention cueing are three examples of strategies that can be considered when designing educational animations of biologically secondary topics. For dynamic visualisations of primary knowledge, such as motor skills, we present the following section.

3 Learning Motor Skills Through Dynamic Visualisations

3.1 Video Modelling of Motor Skills

Modelling—also referred to as *observational learning*—is the learning process by which an individual (the model) demonstrates actions that can be imitated by another individual (the observer) [33]. Models may be live enactments, recorded in diverse ways, described in many different forms or even imagined [33]. Moreover, although we will focus on *observation* in this chapter, it should be noted that *listening* and other forms of perception have been employed in modelling methods [e.g. 28].

From this broad area, we will describe modelling via observation of dynamic visualisations, more specifically, *video modelling*. Two powerful advantages of video recordings over live enactments are: (1) video can incorporate many different situations and types of models, and (2) video can focus attention on certain aspects of the depiction by use of the camera or editing tools [33]. However, consistent with all learning environments, quality products must be pursued. For example, producing videos for modelling of professional skills may involve all the following actions: (1) structuring the content, in preproduction; (2) monitoring the conditions of recording, in production; (3) finding the most instructional efficient depictions, in postproduction; and (4) considering alternatives in all these steps depending on the intended learning outcomes of the trainees [modified from 33].

We will next centre the discussion on video modelling about tasks that involve human motor skills, for example, manipulative-procedural tasks. To illustrate a few examples, we present the following list that shows some video modelled motor skills and their respective references:

* Whole-Body Skills

    – Executing foul basketball shots with perfect form [36].
    – Following effective behaviours of lecturing in a lesson [92].
    – Performing ballet routines [35].
    – Practising complex movements of a modern dance sequence [23].

* Manipulative- Procedural Skills

    – Tying nautical knots [91]. Tying a series of 3 Scoubidou knots [5].
    – Learning paper-folding tasks [110].
    – Performing individual assembly tasks, like folding cardboard boxes [32].
    – Disassembling a machine gun [93].
    – Assembling an abstract and novel machine device [107].
    – Building a model helicopter with 54 pieces of an assembly kit [8].

* Other Hands or Arms Skills

    – Displacing wooden barriers following a pattern [28].
    – Following a percussionist's hitting pattern through wooden barriers [14].

- Following an action pattern of moving a lightweight paddle [18].
- Practicing different first-aid procedures [2].
- Performing dart throwing subskills [51].

So, what does video modelling of motor skills entail? Using the first levels of the *social cognitive model of sequential skill acquisition* [51, 90], we divide the learning process of video modelling of motor skills into two sequential steps: (1) *observation*, where modelling experiences gives the observer a representation of an accurate way to perform the motor skills, and (2) *emulation*, where modelling experiences continue to be important but now the corporal experiences (i.e. practice) of the observer becomes relevant [51].

As stated, video modelling of motor skills involves both observation and practice. Approaches that consider one step, for example emulation only, may be less effective. For instance, an experiment about 5 dart throwing skills with 60 high school girls allocated to conditions *observation plus practise* compared to *practise alone*, found higher levels of dart skill, intrinsic interest and self-efficacy perceptions in the groups exposed to the combined observation and practice condition [51]. A similar result was found in an experiment with undergraduate students following a sequence to displace seven vertical wooden barriers in a fixed time. In this study, a *sole practice* group forgot significantly more of the task after a period of 48 h, than a combined *observation plus practise* group [85]. Other studies have compared *observation plus practice* versus *observation alone*, showing that the combination is more effective. For example, in an experiment where the task was to replicate a videorecorded percussionist's hitting pattern, the authors concluded that observation alone may produce approximate reproductions of the motor task, but measurable improvement was only achieved through physical practice [14].

A distinction to make in the first step of video modelling of motor skills concerns the type of model that the student observes, which can be: (1) *coping model*, who starts performing with errors but gradually corrects them; or (2) *mastery model*, who always performs in a flawless way. Evidence that supports the coping model is the mentioned study about dart skills, where the participants who watched the coping model displayed higher performance than the group observing the mastery model [51]. In a related vein, a study that contrasted these types of model in 36 snake-anxious female undergraduates, showed that participants exposed to films with coping models—who initially showed fear, but later dominated it—displayed significantly more snake-approach behaviour than subjects watching mastery models—who demonstrated complete fearlessness [70]. Although this experiment involves behavioural change rather than motor learning, it adds empirical evidence in favour of coping modelling. An explanation for the better learning generally reported in coping modelling when compared to mastery modelling, might be that the former tends to demonstrate the desired skill in attainable steps rather than an unrealistic target performance [33]. This explanation can be framed in a CLT interpretation: since the coping model shows achievable steps, each of them contains few key elements; thus, every step in the coping model contains a number of new elements that do not exceed the processing capacity of the learner. On the contrary, a mastery model may exhibit in a single step a greater number of new elements than those the observer can handle.

To conclude, according to the social cognitive model of sequential skill acquisition, modelling experiences may involve not only observation of a model's performance, but also listening to the performance or watching its outcomes [51]. As all three different modelling experiences—observe actions, listen to actions, and observe results—assist learning, an important question is how does this actually happen? The answers may be at least partially explained by the brain's *mirror neuron system*, which we describe next.

3.2   The Mirror Neuron System

Initially mirror neurons were thought to be a class of visuomotor nervous cells that activate when individuals complete a particular motor action and also when they observe other individuals doing a similar action [for a review, see 83]. More recent data has expanded that view indicating that mirror neurons can also be fired when an individual either: (1) listens to actions related to motor skills [e.g. 101], (2) imagines his own actions being performed [e.g. 23], or even (3) observes the result of an action that can be linked to a motor skill [e.g. 57].

Mirror neurons were firstly identified in area F5 of the premotor cortex of the brain in macaque monkeys (*Macaca nemestrina*). The original findings described that in a macaque's inferior area 6, termed sector F5, there were neurons that fired during particular goal-directed hand movements, such as holding, tearing or grasping [27]. Furthermore, for actions like grasping with the mouth or the hands, and rotating and manipulating objects, similar F5 neuronal discharges were recorded during direct execution of the actions by the monkey and during observation of these actions performed by the researchers. It was also found that these mirror neurons participated in action prediction. Notably, some neurons where activated both when the macaque performed an action (for example, grasping an object) and when it observed the experimenters performing a related but preparatory action (placing the object near the animal). Thus, mirror neurons were found to play an important role in the rich social interactions (understanding direct actions and intentions) of macaques [27].

In addition to these results in monkeys, a similar mirroring phenomenon has been reported in humans. It has been found that the precentral motor cortex was not only activated when humans manipulated a small object but also when they observed another individual executing the same action. Although the activation observed in the motor cortex was weaker during observation as compared to direct execution, these findings revealed the presence of mirror neurons in humans [37]. Furthermore, these neurons constitute a network called the *mirror neuron system* (MNS), which has an extensive brain representation, including the parietal, premotor, and subcortical areas.

As well as being triggered when doing or observing an action, the MNS is also triggered when individuals simulate or imagine that they are performing the actual action. For example, in one experiment of dancers observing and imagining another dancer's body movements, the brain regions that were activated have been related to the MNS. Interestingly, this study also showed that the greatest activation of one of these regions (inferior parietal lobule) was observed when participants imagined dance steps they had practiced for some time, what also means that the MNS is sensitive to previous corporal experience [23].

Similarly, besides being fired when doing, observing or imagining an action, the MNS may also be triggered when a subject listens to actions that can be related to motor skills. For example, one study showed that, when the participants heard sentences describing mouth-, hand- or leg-actions, there was an activation of the same regions that are triggered by observation and execution of those actions [101].

Furthermore, the MNS may be activated by more "indirect stimuli", such as observing the outcomes of a manipulative task, rather than the direct ongoing action. For example, a study by Longcamp et al. [57] found that a brain region connected to the MNS was activated more strongly during observation of handwritten letters as opposed to printed letters. It appears that this brain area can react to handwritten text, because it is an outcome of a hand action. In other words, for this activation it was not necessary to observe or execute the writing action itself [57]. Another indirect stimulus that may trigger the MNS is the observation of instruments that allow a manipulative task. In fact, parts of the brain motor system have been shown to be triggered with the minimal action of observing manipulable objects [69]. To summarise, the MNS can be activated by executing, observing, imagining, or hearing hand tasks, and also by solely perceiving the end results or the instruments of hand actions.

Finally, the MNS is more strongly triggered when the actions are perceived and undertaken by individuals that belong to the same species. In other words, the MNS is *biologically tuned* to be activated preferentially in contexts of social human-human interactions. For example, a neuroimaging study showed that a brain region related to the MNS was activated only when observing manual grasping actions achieved by a human hand but not by a robotic hand [100]. More evidence was found when participants had to make either horizontal or vertical sinusoidal movements with their arm while observing either congruent or incongruent arm movements made by either a human or a robot. No interferences occurred with the participants' execution of arm movements when they observed movements that were: (1) congruent made by robot, (2) congruent made by human, and (3) incongruent made by robot. The only significant interference on arm movement was recorded when the participants observed incongruent arm movements made by another human, suggesting that the MNS is biologically tuned [49].

3.3 Modelling of Motor Skills and Dynamic Versus Static Visualisations

As it was anticipated in the section on evolutionary educational psychology, instructional design should profit from the high efficiency of biologically primary tasks to facilitate the acquisition of secondary knowledge [34, 76]. In the field of dynamic visualisations, when the biologically primary task involves manipulation or gestures, this approach has been termed the *human movement effect*. This effect proposes that, since learning human movements is a primary skill, humans have evolved to learn it without investing as much WM resources as learning other *mechanical* (non-human movement) skills. As a result, when watching a dynamic visualisation of human movements, the learner has more WM resources available that can be allocated to deal with additional cognitive load caused by transient information [76]. It is proposed that underlying physiological reason why this is possible is that humans have the MNS that enables the observation and emulation of human movement skills [106].

Evidence that learning of human movement can overcome the transitory effect has been found in studies that have compared instructional dynamic visualisations with equivalent static instructions. A meta-analysis of 26 studies, by Höffler and Leutner [43], found that dynamic visualisations were better for learning than the static equivalents. The advantage of animations was more evident in procedural-motor skill learning than in other types of training.

This advantage has been reported in studies that have featured whole body movements. For example, in learning to imitate ballet steps [35]. In another study, subjects had to predict the weights of boxes carried by an actor. The dynamic visualisations of these events resulted in better predictions than the static images, even considering that the statics where longer available for observation [105].

Other studies supporting this advantage have used manipulative-procedural tasks requiring only the use of hands. Dynamic visualisations have been shown to be superior to static conditions in assembling a firearm by watching a television video [93], and in hand-puzzle rings and knot tying tasks [5]. Figure 5 shows a static picture from this study about knot tying skills.

The advantage of dynamic visualisations has also been found with paper-folding tasks. Of particular interest in this study was the finding that representing the hands in the depiction was not necessary for the effect to occur. As long as learners could relate the task to a manual skill, in this case, origami, the hands did not need to be observed [110]. Figure 6 shows a simplified static picture from this study.

Summarising the findings of research that has compared animation with statics, it can be concluded that the depiction of human movement in dynamic visualisations generally aids understanding of whole-body or manipulative tasks. CLT researchers have predicted that, since the transient information of animations is less likely to hinder learning if it includes forms of human movement, dynamic visualisations would be equally effective or more effective than the static counterparts [76, 106].

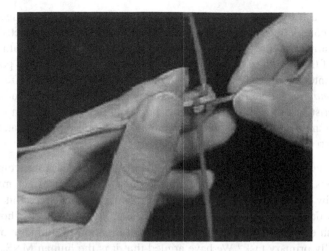

Fig. 5 Frame from a knot tying task (Reprinted from [5] with permission from Elsevier)

Fig. 6 Simplified frame from an origami task (Reprinted from [110] with permission from Elsevier)

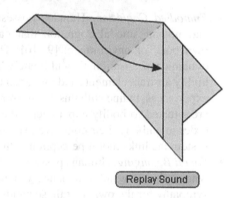

What is more, human movement depictions may help learning beyond whole-body or manipulative tasks, as it has been reviewed recently in animations that incorporate one or more of the following strategies [24]:

- *Gestures.* Learning from dynamic visualisations can be boosted by either: (1) *observing gestures,* or (2) *making gestures.* In the first case, the gestures can be added to the visualisations, for example, by employing an animated character who gesticulates. In the second case, students can follow the motion in the visualisation, for example with their index finger—provided that they do not block the view with their hands [24].
- *Manipulation.* Similarly, learning from animations can be enhanced by either: (1) *observing manipulations,* or (2) *making manipulations to the system depicted.* In the first point, as with gesturing, another person or an animated character can do the manipulations [24]. For example, by showing manipulations to solve ring puzzles, it was observed that the animation condition learned better a related

cognitive task, as compared to the static pictures group [5]. For the second point, actually completing the manipulation can be regarded as a more active strategy, where learners interact with the animation, for example by manipulating a virtual replica of the system to be learnt [24]. However, as observed in the pace-control section, interactivity may act as a moderator of any manipulation effect.

- *Body Metaphors.* Learning from animations can be boosted by adding human characteristics to the moving elements For example, an animation depicting a loader crane could use a picture of a human arm upon the crane, and the hook could be replaced by a finger [24].

To reiterate, the depiction of human movement helps learning from dynamic visualisations. As previously argued, this human movement effect may be underpinned by the simpleness of biologically primary knowledge. But what is it that makes depictions of human movement easy? In other words, how did the human brain evolve to manage the transitory information of human movement as a biologically primary task? We have argued that it is the human MNS. There are two other research areas that support this argument, which we briefly describe next:

- *Embodied Cognition.* Humans possess an *embodied or grounded cognition*, and not an amodal cognition that can operate without references to one's own body or environment [9, 10]. This implies that when learners watch a dynamic visualisation, they ultimately link all the contents—even if they contain highly abstract elements and movements—to their prior bodily or environmental experiences. In line with this, the nearer the animation's components and motions are situated to bodily experiences, the greater the human movement effect. In order to embody their cognitive experiences, humans have evolved the MNS, a system the links motor perception with motor emulation or practice.
- *Social Belonging.* Humans possess a sense of social partnership. In other words, when subjects watch a dynamic visualisation about hands executing an action, sympathy for the owner of those hands may be triggered, which might motivate a greater learning [63, 66]. In line with this hypothesis is social agency theory, which includes empirically supported principles that foster a social belonging, such as the personalisation or the voice principles, previously described [66]. In order to experience social belonging, humans have evolved the MNS, a system the links individual embodied experiences with those of other humans.

We end this chapter by concluding that dynamic visualisations can be very effective media for learning human motor skills. It is likely that this advantage exists because of the MNS, which has evolved to help humans learn quite effortlessly procedural-motor tasks. This conclusion was made possible by considering the limitations and evolution of human cognitive architecture. Through such analysis it will be possible to design more effective dynamic visualisations for instruction.

Acknowledgments This research was supported by an Australian Research Council grant (DP1095685) to the second and third authors.

References

1. Ambadar Z, Schooler JW, Cohn JF (2005) Deciphering the enigmatic face: The importance of facial dynamics in interpreting subtle facial expressions. Psychological Science 16 (5): 403–410
2. Arguel A, Jamet E (2009) Using video and static pictures to improve learning of procedural contents. Computers in Human Behavior 25 (2):354–359. doi:10.1016/j.chb.2008.12.014
3. Ayres P, Paas F (2007) Can the cognitive load approach make instructional animations more effective? Applied Cognitive Psychology 21 (6):811–820. doi:10.1002/acp.1351
4. Ayres P, Paas F (2007) Making instructional animations more effective: A cognitive load approach. Applied Cognitive Psychology 21 (6):695–700. doi:10.1002/acp.1343
5. Ayres P, Marcus N, Chan C, Qian N (2009) Learning hand manipulative tasks: When instructional animations are superior to equivalent static representations. Computers in Human Behavior 25 (2):348–353. doi:10.1016/j.chb.2008.12.013
6. Baddeley A (1992) Working memory. Science 255 (5044):556–559
7. Baddeley A (2000) The episodic buffer: A new component of working memory? Trends in Cognitive Sciences 4 (11):417–423. doi:10.1016/S1364-6613(00)01538-2
8. Baggett P (1987) Learning a procedure from multimedia instructions: The effects of film and practice. Applied Cognitive Psychology 1 (3):183–195
9. Barsalou LW (2008) Grounded cognition. Annual Review of Psychology 59 (1):617–645. doi:10.1146/annurev.psych.59.103006.093639
10. Barsalou LW (2010) Grounded cognition: Past, present, and future. Topics in Cognitive Science 2 (4):716–724. doi:10.1111/j.1756-8765.2010.01115.x
11. Bassili JN (1978) Facial motion in the perception of faces and of emotional expression. Journal of Experimental Psychology: Human Perception and Performance 4 (3):373–379. doi:10.1037/0096-1523.4.3.373
12. Bétrancourt M (2005) The animation and interactivity principles in multimedia learning. In: Mayer RE (ed) The Cambridge handbook of multimedia learning. Cambridge University Press, New York, NY, pp 287–296
13. Bétrancourt M, Chassot A (2008) Making sense of animation: How do children explore multimedia instruction? In: Lowe RK, Schnotz W (eds) Learning with animation: Research implications for design. Cambridge University Press, New York, NY, pp 141–164
14. Blandin Y, Lhuisset L, Proteau L (1999) Cognitive processes underlying observational learning of motor skills. The Quarterly Journal of Experimental Psychology Section A 52 (4):957–979. doi:10.1080/713755856
15. Boucheix J-M (2008) Young learners' control of technical animations. In: Lowe RK, Schnotz W (eds) Learning with animation: Research implications for design. Cambridge University Press, New York, NY, pp 208–234
16. Boucheix J-M, Lowe RK (2010) An eye tracking comparison of external pointing cues and internal continuous cues in learning with complex animations. Learning and Instruction 20 (2):123–135. doi:10.1016/j.learninstruc.2009.02.015
17. Brünken R, Steinbacher S, Plass JL, Leutner D (2002) Assessment of cognitive load in multimedia learning using dual-task methodology. Experimental Psychology 49 (2):109–119. doi:10.1027//1618-3169.49.2.109
18. Carroll WR, Bandura A (1982) The role of visual monitoring in observational learning of action patterns: Making the unobservable observable. Journal of Motor Behavior 14 (2): 153–167
19. Chandler P, Sweller J (1996) Cognitive load while learning to use a computer program. Applied Cognitive Psychology 10 (2):151–170
20. Clark J, Paivio A (1991) Dual coding theory and education. Educational Psychology Review 3 (3):149–210. doi:10.1007/bf01320076
21. Clark RC (2005) Multimedia learning in e-courses. In: Mayer RE (ed) The Cambridge handbook of multimedia learning. Cambridge University Press, New York, NY, pp 589–616

22. Cowan N (2001) The magical number 4 in short-term memory: A reconsideration of mental storage capacity. Behavioral and Brain Sciences 24 (01):87–114. doi:10.1017/S0140525X01003922

23. Cross ES, Hamilton AFdC, Grafton ST (2006) Building a motor simulation de novo: Observation of dance by dancers. NeuroImage 31 (3):1257–1267. doi:10.1016/j.neuroimage.2006.01.033

24. de Koning BB, Tabbers HK (2011) Facilitating understanding of movements in dynamic visualizations: An embodied perspective. Educational Psychology Review 23 (4):501–521. doi:10.1007/s10648-011-9173-8

25. de Koning BB, Tabbers HK, Rikers RMJP, Paas F (2009) Towards a framework for attention cueing in instructional animations: Guidelines for research and design. Educational Psychology Review 21 (2):113–140. doi:10.1007/s10648-009-9098-7

26. de Koning BB, Tabbers HK, Rikers RMJP, Paas F (2010) Learning by generating vs. receiving instructional explanations: Two approaches to enhance attention cueing in animations. Computers & Education 55 (2):681–691. doi:10.1016/j.compedu.2010.02.027

27. di Pellegrino G, Fadiga L, Fogassi L, Gallese V, Rizzolatti G (1992) Understanding motor events: A neurophysiological study. Experimental Brain Research 91 (1):176–180. doi:10.1007/bf00230027

28. Doody SG, Bird AM, Ross D (1985) The effect of auditory and visual models on acquisition of a timing task. Human Movement Science 4 (4):271–281. doi:10.1016/0167-9457(85)90014-4

29. Dowrick PW (1991) Analyzing and documenting. In: Dowrick PW (ed) Practical guide to using video in the behavioral sciences. Wiley Interscience, New York, NY, pp 30–48

30. Dowrick PW (1991) Instructing and informing. In: Dowrick PW (ed) Practical guide to using video in the behavioral sciences. Wiley Interscience, New York, NY, pp 49–63

31. Dowrick PW (1991) Feedback and self-confrontation. In: Dowrick PW (ed) Practical guide to using video in the behavioral sciences. Wiley Interscience, New York, NY, pp 92–108

32. Dowrick PW, Hood M (1981) Comparison of self-modeling and small cash incentives in a sheltered workshop. Journal of Applied Psychology 66 (3):394–397

33. Dowrick PW, Jesdale DC (1991) Modeling. In: Dowrick PW (ed) Practical guide to using video in the behavioral sciences. Wiley Interscience, New York, NY, pp 64–76

34. Geary DC (2002) Principles of evolutionary educational psychology. Learning and Individual Differences 12 (4):317–345. doi:10.1016/s1041-6080(02)00046-8

35. Gray JT, Neisser U, Shapiro BA, Kouns S (1991) Observational learning of ballet sequences: The role of kinematic information. Ecological Psychology 3 (2):121–134. doi:10.1207/s15326969eco0302_4

36. Hall EG, Erffmeyer ES (1983) The effect of visuo-motor behavior rehearsal with videotaped modeling on free throw accuracy of intercollegiate female basketball players. Journal of Sport Psychology 5 (3):343–346

37. Hari R, Forss N, Avikainen S, Kirveskari E, Salenius S, Rizzolatti G (1998) Activation of human primary motor cortex during action observation: A neuromagnetic study. Proceedings of the National Academy of Sciences of the United States of America 95 (25):15061–15065

38. Hasler BS, Kersten B, Sweller J (2007) Learner control, cognitive load and instructional animation. Applied Cognitive Psychology 21 (6):713–729. doi:10.1002/acp.1345

39. Hegarty M (1992) Mental animation: Inferring motion from static displays of mechanical systems. Journal of Experimental Psychology: Learning, Memory, and Cognition 18 (5):1084–1102. doi:10.1037/0278-7393.18.5.1084

40. Hegarty M (2005) Multimedia learning about physical systems. In: Mayer RE (ed) The Cambridge handbook of multimedia learning. Cambridge University Press, New York, NY, pp 447–465

41. Hegarty M, Kriz S (2008) Effects of knowledge and spatial ability on learning from animation. In: Lowe RK, Schnotz W (eds) Learning with animation: Research implications for design. Cambridge University Press, New York, NY, pp 3–29

42. Höffler TN (2010) Spatial ability: Its influence on learning with visualizations—a meta-analytic review. Educational Psychology Review 22 (3):245–269. doi:10.1007/s10648-010-9126-7
43. Höffler TN, Leutner D (2007) Instructional animation versus static pictures: A meta-analysis. Learning and Instruction 17 (6):722–738. doi:10.1016/j.learninstruc.2007.09.013
44. Höffler TN, Schwartz RN (2011) Effects of pacing and cognitive style across dynamic and non-dynamic representations. Computers & Education 57 (2):1716–1726. doi:10.1016/j.compedu.2011.03.012
45. Huk T, Steinke M, Floto C (2010) The educational value of visual cues and 3D-representational format in a computer animation under restricted and realistic conditions. Instructional Science 38 (5):455–469. doi:10.1007/s11251-009-9116-7
46. Jamet E, Gavota M, Quaireau C (2008) Attention guiding in multimedia learning. Learning and Instruction 18 (2):135–145. doi:10.1016/j.learninstruc.2007.01.011
47. Kalyuga S (2009) Managing cognitive load in adaptive multimedia learning. IGI Global, Hershey, PA
48. Kalyuga S, Chandler P, Sweller J (1999) Managing split-attention and redundancy in multimedia instruction. Applied Cognitive Psychology 13 (4):351–371
49. Kilner JM, Paulignan Y, Blakemore SJ (2003) An interference effect of observed biological movement on action. Current Biology 13 (6):522–525. doi:10.1016/s0960-9822(03)00165-9
50. Kirschner PA, Sweller J, Clark RE (2006) Why minimal guidance during instruction does not work: An analysis of the failure of constructivist, discovery, problem-based, experiential, and inquiry-based teaching. Educational Psychologist 41 (2):75–86. doi:10.1207/s15326985ep4102_1
51. Kitsantas A, Zimmerman BJ, Cleary T (2000) The role of observation and emulation in the development of athletic self-regulation. Journal of Educational Psychology 92 (4):811–817. doi:10.1037/0022-0663.92.4.811
52. Knippels M-CPJ, Severiens SE, Klop T (2009) Education through fiction: Acquiring opinion-forming skills in the context of genomics. International Journal of Science Education 31 (15):2057–2083. doi:10.1080/09500690802345888
53. Koroghlanian C, Klein JD (2004) The effect of audio and animation in multimedia instruction. Journal of Educational Multimedia and Hypermedia 13 (1):23–46
54. Kozma R, Russell J (2005) Multimedia learning of chemistry. In: Mayer RE (ed) The Cambridge handbook of multimedia learning. Cambridge University Press, New York, NY, pp 409–428
55. Lin L, Atkinson RK (2011) Using animations and visual cueing to support learning of scientific concepts and processes. Computers & Education 56 (3):650–658. doi:10.1016/j.compedu.2010.10.007
56. Linek SB, Gerjets P, Scheiter K (2010) The speaker/gender effect: Does the speaker's gender matter when presenting auditory text in multimedia messages? Instructional Science 38 (5):503–521. doi:10.1007/s11251-009-9115-8
57. Longcamp M, Tanskanen T, Hari R (2006) The imprint of action: Motor cortex involvement in visual perception of handwritten letters. NeuroImage 33 (2):681–688. doi:10.1016/j.neuroimage.2006.06.042
58. Low R, Sweller J (2005) The modality principle in multimedia learning. In: Mayer RE (ed) The Cambridge handbook of multimedia learning. Cambridge University Press, New York, NY, pp 147–158
59. Lowe RK, Schnotz W, Rasch T (2011) Aligning affordances of graphics with learning task requirements. Applied Cognitive Psychology 25 (3):452–459. doi:10.1002/acp.1712
60. Lusk DL, Evans AD, Jeffrey TR, Palmer KR, Wikstrom CS, Doolittle PE (2009) Multimedia learning and individual differences: Mediating the effects of working memory capacity with segmentation. British Journal of Educational Technology 40 (4):636–651
61. Mayer RE (2004) Should there be a three-strikes rule against pure discovery learning? American Psychologist 59 (1):14–19. doi:10.1037/0003-066x.59.1.14

62. Mayer RE (2005) Principles for reducing extraneous processing in multimedia learning: Coherence, signaling, redundancy, spatial contiguity, and temporal contiguity principles. In: Mayer RE (ed) The Cambridge handbook of multimedia learning. Cambridge University Press, New York, NY, pp 183–200
63. Mayer RE (2005) Principles of multimedia learning based on social cues: Personalization, voice, and image principles. In: Mayer RE (ed) The Cambridge handbook of multimedia learning. Cambridge University Press, New York, NY, pp 201–212
64. Mayer RE (2005) Principles for managing essential processing in multimedia learning: Segmenting, pretraining, and modality principles. In: Mayer RE (ed) The Cambridge handbook of multimedia learning. Cambridge University Press, New York, NY, pp 169–182
65. Mayer RE (2001) Multimedia learning. Cambridge University Press, New York, NY
66. Mayer RE (2008) Research-based principles for learning with animation. In: Lowe RK, Schnotz W (eds) Learning with animation: Research implications for design. Cambridge University Press, New York, NY, pp 30–48
67. Mayer RE, Chandler P (2001) When learning is just a click away: Does simple user interaction foster deeper understanding of multimedia messages? Journal of Educational Psychology 93 (2):390–397. doi:10.1037/0022-0663.93.2.390
68. Mayer RE, Hegarty M, Mayer S, Campbell J (2005) When static media promote active learning: Annotated illustrations versus narrated animations in multimedia instruction. Journal of Experimental Psychology: Applied 11 (4):256–265. doi:10.1037/1076-898x.11.4.256
69. Mecklinger A, Gruenewald C, Besson M, Magnié M-N, Von Cramon DY (2002) Separable neuronal circuitries for manipulable and non-manipulable objects in working memory. Cerebral Cortex 12 (11):1115–1123. doi:10.1093/cercor/12.11.1115
70. Meichenbaum DH (1971) Examination of model characteristics in reducing avoidance behavior. Journal of Personality and Social Psychology 17 (3):298–307
71. Miller GA (1956) The magical number seven, plus or minus two: Some limits on our capacity for processing information. Psychological Review 63 (2):81–97. doi:10.1037/h0043158
72. Moreno R (2007) Optimising learning from animations by minimising cognitive load: Cognitive and affective consequences of signalling and segmentation methods. Applied Cognitive Psychology 21 (6):765–781
73. Moreno R (2008) Animated pedagogical agents: How do they help students construct knowledge from interactive multimedia games? In: Lowe RK, Schnotz W (eds) Learning with animation: Research implications for design. Cambridge University Press, New York, NY, pp 183–207
74. Moreno R, Mayer R (2007) Interactive multimodal learning environments. Educational Psychology Review 19 (3):309–326. doi:10.1007/s10648-007-9047-2
75. Moreno R, Mayer RE, Spires HA, Lester JC (2001) The case for social agency in computer-based teaching: Do students learn more deeply when they interact with animated pedagogical agents? Cognition and Instruction 19 (2):177–213
76. Paas F, Sweller J (2012) An evolutionary upgrade of cognitive load theory: Using the human motor system and collaboration to support the learning of complex cognitive tasks. Educational Psychology Review 24 (1):27–45. doi:10.1007/s10648-011-9179-2
77. Paas F, Tuovinen JE, Tabbers H, Van Gerven PWM (2003) Cognitive load measurement as a means to advance cognitive load theory. Educational Psychologist 38 (1):63–71
78. Park O-C, Hopkins R (1992) Instructional conditions for using dynamic visual displays: A review. Instructional Science 21 (6):427–449. doi:10.1007/BF00118557
79. Peterson LR, Peterson MJ (1959) Short-term retention of individual verbal items. Journal of Experimental Psychology 58 (3):193–198. doi:10.1037/h0049234
80. Pramling N (2009) The role of metaphor in Darwin and the implications for teaching evolution. Science Education 93 (3):535–547. doi:10.1002/sce.20319
81. Rebetez C, Bétrancourt M, Sangin M, Dillenbourg P (2010) Learning from animation enabled by collaboration. Instructional Science 38 (5):471–485. doi:10.1007/s11251-009-9117-6
82. Rieber LP (1990) Using computer animated graphics in science instruction with children. Journal of Educational Psychology 82 (1):135–140

83. Rizzolatti G, Craighero L (2004) The mirror-neuron system. Annual Review of Neuroscience 27:169–192. doi:10.1146/annurev.neuro.27.070203.144230

84. Roncarrelli R (1989) The computer animation dictionary: Including related terms used in computer graphics, film and video, production, and desktop publishing. Springer-Verlag, New York, NY

85. Ross D, Bird AM, Doody SG, Zoeller M (1985) Effects of modeling and videotape feedback with knowledge of results on motor performance. Human Movement Science 4 (2):149–157. doi:10.1016/0167-9457(85)90008-9

86. Rummer R, Schweppe J, Fürstenberg A, Seufert T, Brünken R (2010) Working memory interference during processing texts and pictures: Implications for the explanation of the modality effect. Applied Cognitive Psychology 24 (2):164–176

87. Scheiter K, Gerjets P (2007) Learner control in hypermedia environments. Educational Psychology Review 19 (3):285–307. doi:10.1007/s10648-007-9046-3

88. Schnotz W, Lowe RK (2008) A unified view of learning from animated and static graphics. In: Lowe RK, Schnotz W (eds) Learning with animation: Research implications for design. Cambridge University Press, New York, NY, pp 304–356

89. Schnotz W, Rasch T (2008) Functions of animations in comprehension and learning. In: Lowe RK, Schnotz W (eds) Learning with animation: Research implications for design. Cambridge University Press, New York, NY, pp 92–113

90. Schunk DH, Zimmerman BJ (1997) Social origins of self-regulatory competence. Educational Psychologist 32 (4):195–208. doi:10.1207/s15326985ep3204_1

91. Schwan S, Riempp R (2004) The cognitive benefits of interactive videos: Learning to tie nautical knots. Learning and Instruction 14 (3):293–305. doi:10.1016/j.learninstruc.2004.06.005

92. Sharp G (1981) Acquisition of lecturing skills by university teaching assistants: Some effects of interest, topic relevance, and viewing a model videotape. American Educational Research Journal 18 (4):491–502

93. Spangenberg RW (1973) The motion variable in procedural learning. Educational Technology Research and Development 21 (4):419–436

94. Spanjers IAE, van Gog T, van Merriënboer JJG (2010) A theoretical analysis of how segmentation of dynamic visualizations optimizes students' learning. Educational Psychology Review 22 (4):411–423. doi:10.1007/s10648-010-9135-6

95. Sweller J (2008) Instructional implications of David C. Geary's Evolutionary Educational Psychology. Educational Psychologist 43 (4):214–216. doi:10.1080/00461520802392208

96. Sweller J (2009) Cognitive bases of human creativity. Educational Psychology Review 21 (1):11–19

97. Sweller J (2010) Element interactivity and intrinsic, extraneous, and germane cognitive load. Educational Psychology Review 22 (2):123–138. doi:10.1007/s10648-010-9128-5

98. Sweller J, van Merrienboer JJG, Paas F (1998) Cognitive architecture and instructional design. Educational Psychology Review 10 (3):251–296. doi:10.1023/A:1022193728205

99. Sweller J, Ayres P, Kalyuga S (2011) Cognitive load theory. Explorations in the learning sciences, instructional systems and performance technologies. Springer, New York, NY

100. Tai YF, Scherfler C, Brooks DJ, Sawamoto N, Castiello U (2004) The human premotor cortex is 'mirror' only for biological actions. Current Biology 14 (2):117–120. doi:10.1016/j.cub.2004.01.005

101. Tettamanti M, Buccino G, Saccuman MC, Gallese V, Danna M, Scifo P, Fazio F, Rizzolatti G, Cappa SF, Perani D (2005) Listening to action-related sentences activates fronto-parietal motor circuits. Journal of Cognitive Neuroscience 17 (2):273–281

102. Tosi V (1993) El lenguaje de las imágenes en movimiento (How to make scientific audiovisuals for research) (trans: Broissin M). 2nd edn. Grijalbo, México, México

103. Tversky B, Morrison JB, Betrancourt M (2002) Animation: Can it facilitate? International Journal of Human-Computer Studies 57 (4):247–262. doi:10.1006/ijhc.2002.1017

104. Tversky B, Heiser J, Mackenzie R, Lozano S, Morrison JB (2008) Enriching animations. In: Lowe RK, Schnotz W (eds) Learning with animation: Research implications for design. Cambridge University Press, New York, pp 263–285

105. Valenti SS, Costall A (1997) Visual perception of lifted weight from kinematic and static (photographic) displays. Journal of Experimental Psychology: Human Perception and Performance 23 (1):181–198

106. van Gog T, Paas F, Marcus N, Ayres P, Sweller J (2009) The mirror neuron system and observational learning: Implications for the effectiveness of dynamic visualizations. Educational Psychology Review 21 (1):21–30. doi:10.1007/s10648-008-9094-3

107. Watson G, Butterfield J, Curran R, Craig C (2010) Do dynamic work instructions provide an advantage over static instructions in a small scale assembly task? Learning and Instruction 20 (1):84–93. doi:10.1016/j.learninstruc.2009.05.001

108. Wiley J, Ash IK (2005) Multimedia learning of history. In: Mayer RE (ed) The Cambridge handbook of multimedia learning. Cambridge University Press, New York, NY, pp 375–391

109. Williams R (2001) The animator's survival kit. Faber & Faber, New York, NY

110. Wong A, Marcus N, Ayres P, Smith L, Cooper GA, Paas F, Sweller J (2009) Instructional animations can be superior to statics when learning human motor skills. Computers in Human Behavior 25 (2):339–347. doi:10.1016/j.chb.2008.12.012

111. Wouters P, Tabbers H, Paas F (2007) Interactivity in video-based models. Educational Psychology Review 19 (3):327–342. doi:10.1007/s10648-007-9045-4

112. Yang E-m, Andre T, Greenbowe TJ, Tibell L (2003) Spatial ability and the impact of visualization/animation on learning electrochemistry. International Journal of Science Education 25 (3):329–349. doi:10.1080/09500690210126784

Dynamic Visualizations: A Two-Edged Sword?

Richard K. Lowe

Abstract Advances in computer technology have greatly facilitated the generation of dynamic visualizations fuelling a growing preference for animated graphics over their static counterparts. Animated graphics have thus become the visualization of choice in fields as diverse as science and high finance. An assumption underlying this shift to animations is that their direct portrayal of dynamic information is inherently superior to its indirect depiction in a static graphic. However, any benefits of animations must be offset against the costs that may be incurred due to psychological effects of their dynamics. Such problems are particularly likely in traditionally designed animations that depict the subject matter as an entire functioning system. However, alternative design approaches that are informed by the Animation Processing Model [R. Lowe and J-M. Boucheix. Learning from animated diagrams: How are mental models built? In *Diagrammatic Representation and Inference*, pp. 266–281, Springer, 2008.] offer a more effective way of presenting dynamic information. Viewers can be relieved of the burden of decomposing the whole dynamic system by presenting them instead with ready-made relation sets that are tailored to fit the perceptual and cognitive constraints of the human information processing system. Relation sets are designed to facilitate the internal construction processes that people use in composing mental models of the referent subject matter.

1 Animation's Ascendancy

The rise of high powered graphics-oriented computing towards the end of the twentieth century greatly facilitated the production, display, and distribution of animated visualizations. As a consequence, computer-generated animations have

R.K. Lowe (✉)
School of Education, Curtin University, Kent St. Bentley, WA 6102, Australia
e-mail: r.k.lowe@curtin.edu.au

W. Huang (ed.), *Handbook of Human Centric Visualization*,
DOI 10.1007/978-1-4614-7485-2_23, © Springer Science+Business Media New York 2014

been widely adopted as a means for presenting diverse types of information across industry, education, the professions, and society more broadly. These *informative animations*, the focus of this chapter, have a very different role from that of animations whose main purpose is to entertain. Instead, their primary responsibility is to depict information in a manner that is comprehensible, memorable, and useful.

Due to the rapidity with which the computer animation revolution has occurred, empirical research on the effectiveness of informative animations and theorizing about how people process them lag far behind practice. As a result, today's animations tend to be designed not on the basis of scientific research, but rather according to the intuitions of those who commission and develop them. However, evidence is now gradually accumulating that suggests such intuitive approaches may not produce the best results.

This chapter first examines psychological factors that can influence the effectiveness of informative animations. It then provides a research-based critique of current design orthodoxies about how animations should present the subject matter they depict. Finally, it suggests a new approach to designing animations that is informed by the Animation Processing Model [1] and aims to support the internal mental model building processes that are required for comprehension.

2 Benefits of Animation

It seems intuitively reasonable that animated graphics should be superior to their static counterparts for depicting dynamic subject matter (i.e., content that changes over time). Indeed, the conventional wisdom amongst many who design informational graphics today is that the dynamics of the picture should match the dynamics of the content. In other words, static pictures should be used for subject matter that does not move or otherwise change over time while animations should be used for subject matter that is dynamic [2]. By their very nature, static visualizations cannot portray temporal change directly as animations can. Nevertheless, static visualizations are able to provide *indirect* representations of dynamics via a range of depictive conventions that have developed over their long history. Static graphics are widely employed to convey information about dynamic situations. For example, Fig. 1 illustrates how they are used to present safety messages next to the door of a train in Italy. Each of these depictions represents an undesirable event that passengers are urged to avoid by proper behaviour.

The top left graphic warns about the possibility of falling through an opening door. We are able to interpret this static depiction in terms of the dynamic situation to which it refers because we already possess general background knowledge about the everyday world and how it works. Such knowledge lets us infer what the situation would have been like beforehand and to predict what will happen next. In Fig. 2, two extra static frames have been added to make these previous and following aspects explicit. Nevertheless, this additional static information shows only states, not the intervening movements that would be involved. The viewer still needs to make inferences to fully understand how the events unfold.

Fig. 1 Static graphics used in train safety messages to depict dynamic situations

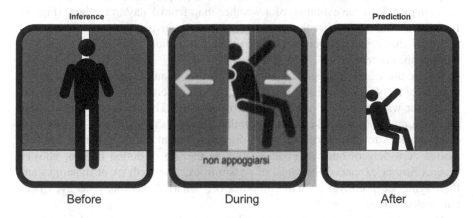

Fig. 2 Background knowledge about the everyday world and how it works allows us to understand the dynamics of familiar subject matter by performing mental animation

One of the earliest techniques for depicting dynamics was to use a series of static graphics ordered across space to represent a set of states that were in reality sequenced in time. The Bayeaux Tapestry exemplifies the use of this practice for expounding a narrative by presenting pictorial 'snapshots' depicting a sequence of key historical events that occurred over a relatively extended time scale. Similarly, more technical types of dynamic content were portrayed using a series of static illustrations capturing successive states of the referent subject matter. In this case,

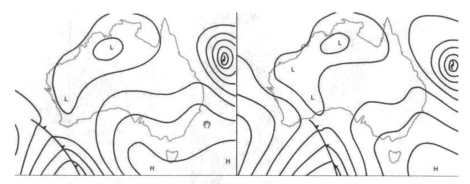

Fig. 3 Weather maps for two successive states. Appreciating the dynamics involved requires the viewer to mentally animate the depicted entities from one state to the next

the time interval represented between successive 'frames' was typically much shorter than for the Bayeaux Tapestry, so providing richer information about the temporal change involved.

This approach continues today with the pictorial assembly instructions used by companies such as IKEA® who supply 'build-it-yourself' products. Multiple static graphics can also be used to portray natural change processes, as exemplified by those depicted for the evolution of a weather map from 1 day to the next (Fig. 3). If viewers wish to visualise the continuous dynamics involved, they must infer the temporal changes that occur between the different depictions in order to mentally animate the content [3, 4].

As the use of multiple static graphics to portray dynamic subject matter developed over the years, additional depictive techniques were introduced to refine and extend the way temporal change could be represented. The incorporation of arrows into pictures [5, 6] was a key innovation that considerably increased the power of static graphics to inform viewers about the changes that occur in dynamic subject matter. A variety of other approaches such as the use of dotted lines or ghosted images further expanded the range of possibilities. The analysis of dynamics and their symbolic portrayal by static graphics (in 'smart depictions') has even been automated in recent years [7].

However, there is a practical limit to what these additions to a static depiction are capable of showing, especially for complex subject matter. At some point, they will become so numerous or intrusive that their net effect is to produce distracting visual clutter rather than to clarify the content. Further, all of these ancillary devices for supplementing the basic content portrayal with dynamic information rely for their success on the viewer being able to interpret them appropriately [8]. A fundamental requirement for such interpretation is that viewers have learned the meaning of these devices and the conventions for their use. This is unlikely when the visualization uses highly specialized symbols and viewers are domain novices. Even if they know what these symbols mean and how to use them, viewers still have to infer dynamics that are represented only indirectly (c.f. [9]).

When simple, familiar subject matter (such as throwing a ball) is depicted via static graphics, inferring the changes that occur over time is relatively straightforward. This is the case for both multiple static graphics and for ancillary devices that indicate dynamics. However, when the subject matter is complex and unfamiliar, inference tends to be far less certain and more prone to error. This is because our general background knowledge about common events in the everyday world does not provide a secure basis for making correct inferences about specialized domains. Further, because the demands of mentally animating the static information under these circumstances can be considerable, they may also prejudice the cognitive processing the viewer needs to undertake in order to understand the referent content. More holistically, static depictions are also unable to offer the perceptual experience of smooth continuous change that is available with animations.

In contrast to static graphics, animations portray dynamic information directly and so do not require the viewer to decode special symbols, apply particular interpretative conventions, or engage in taxing mental animation. Their explicit presentation of dynamics gives viewers the opportunity, in principle at least, to 'read' off this information from the display. They do not rely on the viewer already knowing how things depicted in the display behave and how those behaviours change over time. Further, cognitive processing capacity is freed up so it is available for the central task of understanding the content itself, rather than being bound up in the peripheral task of working out how it changes over time. In effect, an animation fills in information about the myriad micro steps occurring between the key frames that mark major stages of a process [10]. Static graphics simply cannot provide this detailed level of intervening dynamic information. In recent years, researchers have become increasingly interested in testing the relative effectiveness of animated and static graphics [11].

3 Animation Can Have Costs

Despite the widespread popularity of animations, their intuitive appeal, and plausible psychological arguments for their utility, their undoubted potential benefits do not necessarily come without certain costs [10]. With some tasks, animations may even hinder rather than help performance [12]. Animation's downsides come to the fore particularly when the depicted subject matter is complex and unfamiliar to the viewer. Ironically, these are the same attributes that give people trouble with static depictions of dynamic content. Under these circumstances, viewers of an animation can face 'overwhelming' [13] by a deluge of information that is beyond their capacity to process adequately under the prevailing conditions. Conversely, there can also be a danger that animations give viewers a false impression that they comprehend what is shown because their explicit depiction seems self explanatory. Under these circumstances, viewers can be 'underwhelmed' by the depiction and so neglect to process the presented material as deeply as is required.

The key feature of animations that distinguishes them from static graphics is their capacity to present changing information. As already noted, this can be beneficial because it makes the referent dynamics explicit. However, the illusion of continuous change upon which this benefit is based can be sustained only if an animation presents its constituent frames in rapid succession. This severely limits the amount of time that the viewer has to extract information from each individual frame. In contrast, there is no such inbuilt time pressure with static graphics – in principle, viewers have as much time as they need to process the available information.

The issue of limited processing time is particularly crucial when key items of information that need to be attended to and interrelated are presented simultaneously in different regions of the display. With animated graphics, the inbuilt time constraints can make it difficult or impossible for a viewer to process multiple changes that occur simultaneously and are dispersed across the display [14]. There can simply be insufficient time available to attend to all the changes that are occurring in different places, let alone compare and contrast those changes in order to establish any relationships there may be between them. The demands of carrying out the necessary processing therefore exceed the viewer's capacity for execution within the prevailing constraints

The typical psychological response to excessive demands from animation is for the viewer to become highly selective about how the presented information is processed. Rather than trying to deal with all aspects of the animated display, the viewer institutes what is essentially an information triage approach. The result is that over the animation's time course, the viewer attends to only a subset of the available information while neglecting other aspects depicted in the display. In effect, the viewer cuts out a limited attentional region from each of the animation's frames so that only part of its total available information is attended to. Collectively, these individual cut-outs can be conceptualized as an attentional core which 'tunnels' through the entire set of frames [2]. This core information constitutes raw material that is subsequently available for further processing by the viewer. It is ultimately incorporated into the mental model [15] that the viewer constructs from the animation.

If viewers are already well versed in the domain of the depicted subject matter, this selectivity may work to their advantage. It allows them to ignore aspects that are irrelevant to their current interpretative task and concentrate their information processing resources on aspects with high task relevance. This is typically the approach used by experts who recruit their extensive domain specific knowledge to judge what is relevant and what is not [16]. The core information they select provides optimal material for further processing. However, those without this expertise are typically not in a position to make these informed and appropriate judgements. Because they do not possess the requisite specialist background knowledge, their basis for selecting or neglecting information from the display can prejudice their processing of an animation. This can occur even when animations provide user control that allows viewers to tailor the presentation regime to their own needs [17, 18]. In essence, they do not know where to look, when to look, or what to look at.

Lacking domain *specific* knowledge to guide them, non-experts who are faced with animations depicting complex, unfamiliar subject matter fall back on what resources they do have at their command. One way they can respond is to invoke domain *general* knowledge and use that as a basis for attending to, or ignoring, different aspects of the animated display. Consider the case of an animation showing the weather map changes summarised in Fig. 3. A non-meteorologist could characterise the cold front and high pressure cell on the map as objects (rather than air masses) and attribute to them the properties of solid bodies in the everyday world around them [14]. Selecting these features would involve neglecting other markings. By this 'object' characterisation, an animation based on Fig. 3 could be interpreted in terms of the cold front 'pushing' the adjacent high pressure cell from west to east. However, this interpretation is meteorological nonsense because air is fluid (not solid) and because the movements of two features arise from separate causes. Meteorologists would never make such an interpretation because their processing of this information is conducted in the context of domain specific rather than domain general knowledge.

Another way people can deal with the information presented in complex, unfamiliar animations is to respond according to the basic *perceptual* qualities of the display [19]. In this case, the visuospatial and temporal properties of the display's constituent entities and events can play a dominant role in how the viewer's selective attention is allocated. Certain of these various constituents will stand out from their neighbours because of the compelling effect they have on the human visual system [20]. This contrast may be either because their appearance is in some way distinctive or because of their distinctive behaviour. The net effect is that the aspects concerned have a higher *perceptual salience* than other parts of the animated display and so preferentially attract the viewer's attention.

With regard to appearance, an entity that is relatively large, differently shaped, brightly coloured, and in the centre of the viewer's visual field will tent to attract more attention than its less distinctive neighbours. Further, if an object is engaged in events where it behaves very differently from its neighbours, it will be particularly conspicuous because our visual system tends to privilege movement over other perceptual characteristics [21].

The types of perceptual contrast described here (visuospatial contrast and dynamic contrast) can interfere with novices' processing of animation. This is because the aspects of an animated display that attract visual attention most strongly (i.e., have the highest perceptual salience) are not necessarily those that are the most important (i.e., have the highest *thematic relevance*). A consequence of such misalignment between perceptual salience and thematic relevance is that domain novices can be distracted by high salience, low relevance aspects of the animation and so neglect low salience, high relevance aspects. The resulting failure to identify and extract key information from the animation can mean their subsequent processing of the animation will be compromised. This is because they lack essential raw material required to build a high quality mental model of the referent content.

Considering that animations are increasingly used to explain subject matter that may be unfamiliar or otherwise difficult to comprehend, mismatching of perceptual salience and thematic relevance can be a serious cost offsetting animation's potential benefits. It may be possible to ameliorate these costs by interventions such as cueing that are designed to help improve the alignment between salience and relevance. One approach is simply to borrow established techniques that are common with non-animated forms of information [22]. For example, static graphics have long used colour cueing to direct attention to key aspects of the depiction. Application of this technique to animations involves signalling inconspicuous but crucial entities by giving them a bright colour that makes them contrast markedly with their surroundings. The intention here is to draw the viewer's attention to information that may otherwise go un-noticed. However, as discussed below, what works in static graphics does not necessarily work in the dynamic context of an animation. The basis for this difference lies in the distinctive characteristics of animation and how they can influence the psychological processes we use to comprehend them.

4 Comprehension of Animations

Until relatively recently, little research attention was given to the psychological processes by which people internalize and understand the information presented in animated displays. However, the widespread uptake of multimedia by the education community prompted researchers to consider how animations might contribute to learning in text-picture combinations. Researchers into multimedia learning had already developed general theoretical models of what would be required for students to integrate verbal and visual information. Undoubtedly, [23] Cognitive Theory of Multimedia Learning (CTML) has been the most influential of these models.

Although originally devised to address learning from multimedia materials that contained static graphics, the CTML was later recruited for multimedia in which the graphics were animated [24–26]. The main emphasis in this model is upon the processes involved in combining the verbal and visual information supplied by a multimedia resource rather than upon the details of how each component is processed in its own right. In particular, the CTML does not aim to present a comprehensive account of how the graphics themselves are processed. Further, as its name implies, the primary focus of the CTML is upon *cognitive* aspects of processing with little consideration being given to the role of perception.

However, the perceptual properties of animations (and in particular their dynamics) appear to be a powerful influence upon how such depictions are processed by viewers [27]. Early research on multimedia learning was based on the implicit assumption that verbal information (written or spoken text) was the primary source of information being presented, while the accompanying graphics served an ancillary role in the explanation. However, in recent years the traditional dominance of text-based information has been challenged by the increasing reliance now being placed on explanatory visualizations in general and animations in particular. It is therefore appropriate to move beyond existing models of multimedia learning to consider the fine-grained aspects of how people process animated graphics independent of the presence of accompanying text.

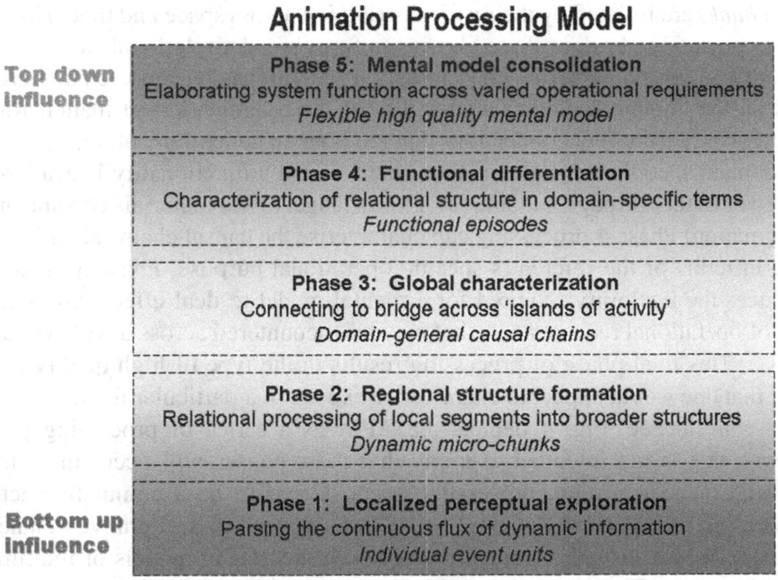

Animation Processing Model

Top down influence

Phase 5: Mental model consolidation
Elaborating system function across varied operational requirements
Flexible high quality mental model

Phase 4: Functional differentiation
Characterization of relational structure in domain-specific terms
Functional episodes

Phase 3: Global characterization
Connecting to bridge across 'islands of activity'
Domain-general causal chains

Phase 2: Regional structure formation
Relational processing of local segments into broader structures
Dynamic micro-chunks

Phase 1: Localized perceptual exploration
Parsing the continuous flux of dynamic information
Individual event units

Bottom up influence

Fig. 4 Summary of the animation processing model showing five phases involved in constructing a mental model from an animated presentation

The Animation Processing Model (APM) proposed by Lowe and Boucheix [1] was developed to provide a more comprehensive account of the perceptual and cognitive processes that occur during learning from animation (Fig. 4). According to the APM, learning from animation involves five main phases which encompass both bottom-up (stimulus driven) and top-down (knowledge driven) contributions to processing [28]. The result of successfully processing an animation with respect to all five phases is assumed to be a high quality mental model of the depicted referent content.

Most conventional educational animations present their referent subject matter in an all-at-once fashion as a fully operational dynamic system containing all the necessary components and actions. With an animation of even moderate complexity, the first task a learner therefore faces is that of parsing the continuous flux of information presented into manageable subsets. The APM characterises Phase 1 processing as decomposition of the animation into *event units,* where each event unit consists of a graphic entity plus its associated behaviour. At a bottom-up level, parsing of an animation into event units is carried out on the basis of the perceptual characteristics of material constituting the display. The event units extracted during Phase 1 processing provide the raw material from which the learner ultimately builds a mental model of the referent subject matter portrayed by the animation.

In Phase 2 processing, event units that have been identified by the viewer are connected on a local basis (via aspects such as gestalt relations of proximity, similarity, etc.) to form small regional clusters of activity. Initially, these *dynamic micro chunks* are isolated by their wide distribution across space and time. However, Phase 3 processing bridges these clusters to form broader relational structures that map out causal chains across the animation. At this stage, these chains are not based on the domain specific functionality of the referent subject matter. Rather, they involve cause-effect connections that are domain general in nature.

If domain specific information about the referent's functionality is available to the viewer (either through suitable prior knowledge, or via materials accompanying the animation) Phase 4 processing can characterise the causal chains already established in terms of the referent's specific operational purpose. Phase 5 processing introduces the flexibility required for a mental model to deal effectively with the range of operational requirements that may be encountered across a wide variety of contexts. This final phase of processing results in the type of high quality mental model that one would expect to characterise experts in a particular field.

For convenience, Fig. 4 depicts the APM as a series of processing phases. However, this is not intended to imply that these phases will occur in a strictly sequential manner. Rather, processing is considered to be a cumulative activity in which the mental model is progressively constructed in an iterative manner. In contrast with established cognitively-dominated theoretical models of multimedia processing, the APM also gives particular emphasis to the role that perceptual influences play in the processing of animations. Specifically, it addresses the contribution that the dynamic attributes of animations have in directing the viewer's attention. However, the APM also takes account of the crucial role that prior knowledge of the depicted content can play in the building of a high quality mental model.

The APM was originally formulated to target conventional animations that provide no more than a faithful depiction of the subject matter and how it behaves. However, the theoretical framework provided by the APM has since proven useful in accounting for how learners process more 'designed' types of educational animations that go beyond mere subject matter depiction. For example, the APM has been used to explain why the traditional types of cues that are so effective for directing learner attention to key information in static graphics tend to be relatively ineffective when used with animations [29, 30].

On the basis of the insights provided by the APM, new types of cues have been devised that are better suited to the demanding processing context learners encounter with complex animated displays [31]. Rather than relying solely on visuospatial contrast (such as highlighting colours) to signal high relevance information, *progressive path* and *local coordinated* cues also provide dynamic contrast to make these key aspects more conspicuous. Because they recruit high salience movement and colour change into their cueing action, these types of cues are able to compete more successfully than normal static cues with the intrinsic dynamics of the animated content for the learner's attention.

Although the APM can be used to suggest alternative forms of guidance for supporting more effective processing of animations, it also raises fundamental questions about current approaches to animation design. For example, cues (whether they be static or dynamic) are added to ameliorate the negative effects of traditionally-designed animations that present entire functioning systems. However, these attention-directing ancillary devices tackle only the symptoms of a far more fundamental problem. They do not directly address the underlying issue of mismatches between human processing capacity and the attributes of such animations. This problem can be traced back to the way animations are conventionally designed.

5   Animation Design

Various design 'orthodoxies' have become established regarding how an animation should present its referent subject matter (c.f. [32]). Many animations produced since computer-based production methods were introduced are essentially animated versions of existing static visualizations. It seems the arguments and common sense canvassed at the beginning of this chapter convinced designers that these depictions would be far more effective if they were converted into animations. The material that has undergone such conversion ranges widely in its visuospatial treatment of the subject matter. In some cases, the original static versions gave the content a realistic appearance, while in others the originals could be highly abstract diagrams. In the latter case, rather than being completely faithful to the superficial appearance of the referent, the depiction was a highly manipulated version of reality.

There is a widespread impression amongst users and graphic designers that photorealism makes displays more effective. This belief, termed 'naïve realism', can be applied to both the visuospatial and temporal properties of a display [33]. However, empirical research shows that performance is often impaired by increasing realism [34]. Hundreds of years of evolution of static visualization design have shown that selective omission, simplification, distortion, and rearrangement of the subject matter's true visuospatial features can greatly facilitate a depiction's effectiveness. Such visuospatial manipulation can make key aspects of the content much clearer and far more accessible. The hydraulic circuit diagram shown in Fig. 5 is an example. Here, instead of depicting the system's hydraulic components in a photorealistic manner, extreme abstractions are used to depict the actuators, valves, pumps and hydraulic lines. Not only are many of the features visible on a real hydraulic system omitted, the portrayal is also highly simplified with lines, circles and other geometric shapes used to represent the components. Further, the components' dimensions have been distorted and their layout re-arranged so that the relationships between them are made clearer.

Overall, the degree of visuospatial realism in this hydraulic diagram is low due to these manipulations. However, despite no longer having the same appearance as an actual hydraulic system, such manipulated depictions have considerable benefits for those who know how to interpret them. For example, the operational connections between the hydraulic components and other relationships that are fundamental to the system's operation are far easier to follow in an hydraulic diagram. This is because the diagram's arrangement is not constrained by practical exigencies of the physical nature of the components (such as their sizes, the available space, etc.).

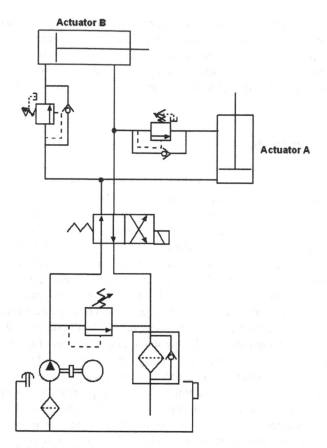

Fig. 5 Many technical visualizations, such as this hydraulic circuit diagram, deliberately manipulate aspects of the original referents to make the subject matter clearer

Rather, the arrangement can be determined by what makes it easiest to perceive all the components and determine the relationship between them.

However, when abstract static depictions such as these are converted into animations, the added dynamics are typically a faithful portrayal of the actual temporal changes that occur in the referent system. For example, an animated visualization of the hydraulic system shown in Fig. 5 would have the actuators and valves behave (relatively speaking) in much the same way as they do in reality. Compared with the extent to which the visuospatial characteristics of the subject matter have been manipulated, there is essentially no manipulation of the temporal characteristics. In this *behaviourally realistic* approach, entities that exhibit high levels of change in the referent system also change markedly in the animated visualization. As we have seen above, this tends to make such aspects especially noticeable because of their high perceptual salience relative to the rest of the display. Conversely, aspects that in reality undergo little visible change are depicted with correspondingly small changes in the animation and so are relatively inconspicuous. When these relatively inconspicuous aspects are crucial to understanding, failure to extract them from an animation is likely to prejudice the quality of the mental model the viewer ultimately builds of the referent subject matter. The problems in this situation are compounded by conventional animation's presentation of events strictly as they occur, with no allowance being made for the negative effects of competition for attention amongst different areas of the display.

One way to address these problems is to question the seemingly inviolable status of temporal information in animated visualizations. As we saw above, when designers develop *static* visualizations, they show no reluctance to drastically alter the *visuospatial* features of the referent subject matter. They perform these manipulations in the name of making depictions more effective for the viewer. However, if the graphic is to be animated, it seems that corresponding manipulations of the *temporal* characteristics are not permissible. It is unclear at this stage why the behaviour of the subject matter (as opposed to its appearance) is sacrosanct. Perhaps it is because those who produce the animations are concerned that the depiction does not contain dynamic inaccuracies that could damage their reputations. Perhaps it is because they are worried that the viewers might misunderstand the presented information if it does not strictly reflect what actually happens in the referent situation. Perhaps it is because they are content to bask in the achievement of being able to depict the movement in such realistic and convincing manner. Whatever the reason, the implicit quarantining of an animation's dynamic aspects from manipulation has potentially profound consequences for viewer processing.

6   Challenges from Current Designs

Current approaches to designing animations typically involve collaboration between a subject matter expert and a graphic designer. The concern of the subject expert is for the animation to provide an accurate, comprehensive depiction that is faithful to the referent content. It is the graphic designer's job to fulfil these requirements while at the same time producing a professional-looking and aesthetically pleasing result. Both contributors are therefore focused on the attributes of the *external* representation that will be presented to the viewer as a completed visualization. They are largely reliant on their intuitions and professional judgements to guide their design activity. Although such intuitively designed animations may be satisfactory when the subject matter portrayed is relatively simple and familiar, this tends not to be the case when more complex and unfamiliar content is involved.

Intuitively designed animations developed as described here usually portray the target subject matter in its entirety, with all entities, events, and relationships depicted much as they would appear in a fully functioning system. When the animation's subject matter is complex and unfamiliar, the results of this holistic approach to depiction present the viewer with a rich and varied array of dynamic information. More specifically, the viewer is faced with a myriad of individual entities distributed across the display area that are engaged in a wide range of different behaviours. These dynamics not only overlap in time but can also change as the animation progresses.

According to the APM, a fundamental task facing viewers of an animation is to parse it into individual event units that will provide suitable raw material for subsequent mental model construction processes. However, if the animation is of the type described above, decomposing it appropriately can be a very demanding task and one that may have uncertain results. If the viewer fails to decompose the animation effectively, subsequent mental model building processes will be compromised. Conventional animations designed intuitively with the intention of providing a comprehensive, faithful portrayal of the subject matter and its behaviour therefore can have the unintended consequence of neglecting the psychological processing needs of the viewer.

The APM characterises the mental model construction process as involving the progressive composition of event units into larger knowledge structures through an iterative and cumulative process. The quality of the mental model composed in this way fundamentally depends on (i) the nature of the raw material that the viewer extracts from the animated display, and (ii) the effective interconnection of this raw material via appropriate relationships to form a coherent hierarchical knowledge structure. What ultimately counts in terms of viewer comprehension is the nature of the *internal* representation (i.e., mental model) that is composed during processing of the animation. If the external representation (i.e., the animation itself) presents information in a manner that makes it difficult for the viewer to extract and internalise, the result is likely to be a poor quality mental model and a consequent lack of understanding.

We have noted that decomposing the rich flux of dynamic information encountered in a conventionally designed animation can impose considerable processing demands on viewers. Further, it is highly likely that factors such as a lack of viewer background knowledge and misalignment between perceptual salience and thematic relevance can be barriers to the appropriate decomposition of the animation. Researchers have explored possibilities such as cueing as a means of helping viewers to decompose animations more appropriately and so ultimately improving the quality of the mental model they construct. However, they have not thus far questioned the necessity of requiring viewers to perform this decomposition process at all. Viewer decomposition of an animation is currently required because conventional design orthodoxy is that animations should present a comprehensive portrayal of their subject matter.

An alternative and somewhat radical approach would be to relieve viewers of the responsibility for this decomposition and instead allow them to devote their processing capacities to activities involved in composing a high quality mental model This alternative would shift responsibility for decomposition from the viewer to the designers of the animation. It would be the task of those who produce the animation to determine how to supply key event units to the viewer in a ready-made fashion that facilitates their composition into hierarchical knowledge structures. Such an approach would demand a considerable departure from current animation design orthodoxies by requiring depiction that extensively manipulates the subject matter. The next section begins by discussing ways in which static graphics have long been manipulated to facilitate comprehension and then identifies possibilities by which helpful manipulations might also be introduced into animations.

7 Exploring Manipulation Possibilities

Graphic representations vary greatly in how closely they resemble their referent subject matter. In some cases, the mapping between the representation and referent can be very close (as in photographs or photo-realistic digital imaging) while in other cases (such as diagrams and graphs), this mapping is far less close. In this section, we consider not only visuospatial mapping (how close the representation and referent are in terms of *appearance*) but also temporal mapping (how close these two are in terms of *behaviour*).

Figure 6 presents a simple conceptual portrayal of these two aspects when the mapping is close. In Fig. 6a, the geometric symbols inside the top rectangle stand for the various *entities* that constitute the referent subject matter while those in the bottom rectangle stand for the graphics representing those entities in a visualization of that subject matter. The vertical lines running between corresponding elements in the two rectangles indicate that there is a close mapping between the referent and

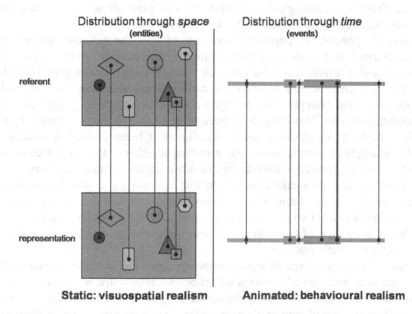

Distribution through *space*
(entities)

Distribution through *time*
(events)

referent

representation

Static: visuospatial realism   Animated: behavioural realism

Fig. 6 Conceptual portrayals of (**a**) visuospatial realism, and (**b**) behavioural realism. Vertical links show close mapping of representation entities and events to those of referent representation – all entities in upper top rectangle map directly onto those in the lower rectangle. The referent entities and their corresponding representations have the same shapes, sizes, colours, orientations, and arrangements. This relationship thus shows *visuospatial realism* (the case for a photograph or photo-realistic image).

Figure 6b shows an analogous situation for temporal properties. In this case the horizontal bars on the referent and representation timelines stand for different *events*, with some short, some long, some early, and some late. Again, the vertical connecting lines indicate there is direct mapping between the upper and lower timelines meaning that events in the representation have the same properties and arrangement as those in the original referent situation. This relationship shows *behavioural realism*. On its own, Fig. 6b could be the case of an animated diagram that realistically portrays the workings of a device. If Fig. 6a and b are considered together, they would exemplify a video recording or a photo-realistic animation exhibiting both visuospatial and behavioural realism because of the close visuospatial and temporal mapping between referent and representation.

Although visualizations that present their subject matter realistically are common, strict realism is abandoned in many others. We noted earlier that visualizations designed to inform or instruct often deliberately manipulate properties of the referent in order to portray key aspects of the subject matter more effectively. These manipulations include changes in size, shape, colour, orientation and arrangement that are introduced to support viewer comprehension. For example, in Fig. 5, the layout of the components and hydraulic lines has been manipulated to help certain aspects of the system stand out and to make the relations between them more evident. From a psychological viewpoint, such manipulations can facilitate the processes required to make sense of the depicted system.

Figure 7a presents a conceptual portrayal of how the referent content can be manipulated in a static educational graphic. Comparison of the upper and lower rectangles reveals that although some visuospatial properties present in the content have been preserved in the representation, others have been manipulated in various ways. One example is the hexagon entity which has been exaggerated by manipulating its size. This could have been done so that its visual salience better matches its thematic relevance and its likelihood of being noticed is increased. Another example of manipulation is the grouping together of the circular elements that are actually separated in the referent. In this case, the entities could have been rearranged so that their spatial proximity better reflects their close relationship on some other more abstract dimension. Other examples include the teasing apart of the square and triangle (which overlap in the referent), the removal of the rectangle, and the re-orientation of the diamond. These various types of manipulations are commonplace in static graphics.

Figure 7b portrays a type of manipulation approach that is far less common. In stark contrast to the case with visuospatial properties, informative and instructive depictions have mostly avoided manipulating the temporal properties of their referent content. This absence of temporal manipulation is especially evident in animations. As noted earlier, the usual approach is for the graphic entities in an animated visualization to behave in much the same way as their referent counterparts behave.

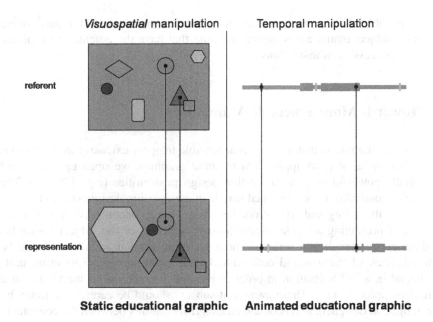

Fig. 7 Conceptual portrayals of (**a**) visuospatial manipulation, and (**b**) temporal manipulation. Few vertical links indicate little direct mapping between representation and referent

In other words, there is close mapping between the dynamics of the referent and the representation. We have seen that this slavish preoccupation with behavioural realism is not exclusive to animations that have a high degree of visuospatial realism. It is also the general rule with highly abstract visualizations such as animated diagrams. Despite their appearance having being highly manipulated, the behaviour of the depicted entities closely parallels that of their referents.

While those who currently design informative animations are perfectly comfortable with manipulating the visuospatial properties of the referent in order to make the representation more tractable for viewers, this is not the case with the referent's dynamics. However, as we have seen, a major challenge that viewers face in dealing animations is the restricted time available for carrying out the necessary processing. A consequence of viewers responding to this challenge by focussing their finite processing resources on a subset of the presented information can be that they allocate those resources inappropriately.

When information relevant to the construction of a high quality mental model is neglected as a result of inappropriate resource allocation, comprehension of the depicted content is likely to be compromised. Further, an inadequate mental model of the referent subject matter will also prejudice the ability to carry out novel tasks such as prediction and problem solving. Persisting with behavioural realism in an animation (even when visuospatial realism has been abandoned) effectively rules out a range of possible temporal manipulations that could make animations more tractable. In the next section, we consider possibilities for partitioning dynamic subject matter across space and time that have the potential to facilitate viewers' processing of animations.

8 Towards More Tractable Animations

If we accept the notion that it is just as acceptable to apply extensive manipulations to animations as it is to apply them to static graphics, we open up a new and potentially powerful range of animation design possibilities (e.g., [35–37]). The particular possibility to be examined here is that animation designers could relieve viewers of the taxing and error prone task of having to decompose an animation. Instead of presenting a single comprehensive depiction of the subject matter (as is done conventionally), a series of more far more limited animations could be offered. Each of these would deal with only a small subset of the event units contained in a full animation in order to present the material to the viewer in a 'pre-decomposed' form. These event unit subsets should be carefully selected by the animation designers to (i) contain meaningful relations between the constituent event units and (ii) be capable of linkage to previous and subsequent event unit subsets. Because within-set and between-set relations are an essential feature of these groupings, they are termed *relation sets* [38].

The assumption here is that if viewers were provided with appropriate relation sets (so as to relieve them of the decomposition burden), their main task would then be to compose them into a high quality mental model. In this approach, viewers are essentially supplied with a 'kit of parts' that are to be assembled, rather than having to extract the needed parts from a full animation. The number of event units included in each relation set should be small enough to be readily accommodated within the limits of human processing capacity. The relations within the set would be chosen by the animation designers to confer coherence on the grouping so that viewers could treat it as an operational chunk. Each relation set would also be deliberately designed to provide opportunities for linkage to other relation sets so as to facilitate composition activities in the viewer's mind. The progressive accumulation of these between-set connections establishes the hierarchical knowledge structures required for a high quality mental model.

The designing of effective relation sets would require not only detailed systematic analysis of the referent content, but also an informed consideration of the psychological processes required to internalise aspects of the content that are essential for composing a high quality mental model. Although current approaches to designing animations give some attention to content analysis, little or no consideration is given to the internal composition processes of the viewer. For example, the main concern of those who design educational animations is with the accuracy and aesthetics of the depiction rather than with the information processing activities learners need to undertake during mental model construction. In a relation set approach to designing an educational animation, the purpose of content analysis would not be solely to ensure there were no errors in the content portrayal. It would also be the basis for determining matters such as which event units should be combined into relation sets, how interconnection between different relation sets could be fostered, and in what order the relation sets should be presented.

The APM provides a robust framework for making the decisions required to implement a relation set approach to designing educational animations. At a broad level, it distinguishes between (i) the initial decomposition processes (Phase 1) that are demanded of learners faced with conventionally-designed animations and (ii) the progressive composition processes (Phases 2 onwards) that are the 'real business' of constructing a high quality mental model. A relation set approach dispenses with much Phase 1 processing by offering learners carefully selected groups of event units that both are high in relevance (irrespective of their intrinsic perceptual salience) and provide important relation-building opportunities. The constitution of each relation set should be such that the psychological processing demands involved are well within the capacities of the learner. This is necessary in order to avoid the problems associated with domain novices not being equipped to select appropriate information from traditional animations which present the subject matter as an entire functioning system.

When faced with a conventional animation, those with expertise in the subject area use their domain-specific background knowledge to determine which aspects (amid the many displayed) deserve their attention. Pre-determined relation sets give domain novices the benefit of such expertise by encapsulating required knowledge in the constituent groups of event units that comprise these sets. In this way, errors associated with domain novices' inappropriate decomposition of an animation's rich flux of information are avoided. Because they are no longer subject to the misleading signals given by aspects of the display that have high perceptual salience but low thematic relevance, the raw material from which they begin mental model composition processes gives them a far better start on their journey of constructing a quality internal representation. Ultimately, it this internal representation of the referent subject matter that really counts (rather than its external representation in a displayed animation).

The concept of relation sets can be illustrated by considering the Newton's Cradle, a device used as an aid for teaching Newton's Laws of Motion (and also as an executive toy). This example is especially illuminating because Newton's Cradle is a system that is visuospatially simple but behaviourally complex. Animations of this device used for elementary science instruction are typically simplified, idealized versions that omit details of how the motions of its components change over time. However, for our purposes, we will consider an animation that accurately depicts the re-distribution of energy as it occurs with a real Newton's Cradle. Our example thus has a high degree of behavioural realism and this has been found to pose considerable interpretative challenges to learners [39].

A typical Newton's Cradle consists of a row of five balls suspended from a frame by threads so that they can swing freely. The balls are identical and their arrangement when at rest is highly regular. Because there are few components and because of the structure's uniformity, the device is very easily characterised

Fig. 8 Selected Newton's Cradle animation frames representing Patterns 1, 2, and 4

in a visuospatial sense. However, once the set of balls is set in motion, things change radically. From the moment a raised terminal ball is released to initiate movement, until the device finally comes to a complete standstill, the five balls exhibit highly individualistic behaviours that are interrelated in a host of different and hierarchically embedded ways. The complexity of these dynamics is further compounded by the fact that the behaviours of the balls and relationships amongst them vary across time. A complete operational cycle of the Newton's Cradle can be broadly divided into an overlapping succession of behaviourally distinctive movement patterns:

Pattern 1 – the balls at the end of the row swing alternately;
Pattern 2 – motion is transmitted via activity of the middle balls;
Pattern 3 – the amplitude of end ball swinging progressive decreases;
Pattern 4 – all balls swing jointly as a unit.
Pattern 5 – the amplitude decreases until all motion stops

A key aspect of this activity is the transition from Pattern 1 to Pattern 4 during which the movement of the balls changes completely (from swinging alternately to swinging jointly). This is due to the redistribution of energy, with the initially static middle three balls of the row gaining energy so that their movement progressively increases as the amplitude of the alternating end balls decreases. An appreciation of this energy redistribution concept is essential for a proper understanding of the way a real Newton's Cradle operates.

When shown a behaviourally realistic animation of Newton's Cradle that portrays this complex activity, students typically fail to extract important aspects of the available information. In particular, they miss the movements involved in the conceptually crucial energy redistribution via the middle balls. This is because these movements are much less readily perceptible than the initial swings of the end balls or the subsequent joint swing of all five balls (see Fig. 8). When embedded in the context of a full behaviourally realistic animation that depicts the Newton's Cradle as an entire functioning system, learners appear quite unable to decompose the presentation appropriately. They are therefore ill equipped to perform the subsequent composition processes required to build a high quality mental model of the referent subject matter.

Fig. 9 Early frames from possible relation set for alternating swings of terminal balls

Rather than presenting a comprehensive portrayal of the Newton's Cradle behaviour, a relation set approach would present the subject matter incrementally, offering learners carefully prepared portions of the information designed to facilitate mental model composition. Coherence within each of these relation sets would be established by grouping together a number of event units that were connected by strong, clear relationships. Individual relation sets would be 'incomplete' external representations of the referent phenomenon. For example, as summarized in Fig. 9, one relation set could include only the two terminal balls of the row and how their behaviour varied across the whole time course of the animation (showing alternating to-and-fro swings that progressively decrease in amplitude and eventually became synchronized). Removal of the middle balls and their associated dynamics makes the presentation far more tractable by improving the match between the processing demands of the depiction and the processing capacities of the viewer. Presented free of the context of how the other three balls behave during this progression, the overall pattern of how these end balls move could be internalized as a dynamic micro chunk to be used for subsequent compositional activities.

Although it may be possible to treat this terminal ball behaviour as a relation set in its own right, other more complex aspects of the total phenomenon may have to be further subdivided to produce relation sets that mesh well with the capacities and attributes of the human information processing system. For example, it is likely Pattern 2 which involves visually subtle changes in movements of the three middle balls would benefit from being split into a series of smaller relation sets in order to keep the processing demands within reasonable limits. Considering that the behaviour of the middle balls during the energy redistribution stage goes through a series of episodes, temporally based segmentation of this stage would seem appropriate. Once again, the combination of specific event units chosen for these relation sets would be determined by the need to produce a coherent grouping bound together by strong, clear linkages.

During the composition process, the various individual relation sets that have been internalized by the learner need to be connected into appropriate super-ordinate mental structures. By this means, higher order relations such as causal chains are established. To facilitate this interconnection (APM Phase 3 processing), relation sets need both to provide the necessary linkage opportunities and to be presented in a sequence that is optimal for efficient, effective composition. For example, the motion of the middle balls in Pattern 2 is embedded both spatially and temporally within Pattern 1 – there needs to be a clear causal connection established between them. In terms of sequencing, it may be better to present the cause (alternations of terminal balls) before the effect (development of middle ball motions).

9 Conclusion

We have seen that with respect to informative purposes, animation can be a two-edged sword. Its undoubted potential benefits with respect to the explicit representation of dynamics can come at the cost of considerable processing challenges. When conventionally-designed animations present complex, unfamiliar subject matter, viewers who lack expertise in the depicted domain typically cope with the high level of processing demands imposed by applying their visual attention in a highly selective manner. However, their lack of domain-specific background knowledge coupled with salience-relevance misalignment in such portrayals can result in them failing to extract information that is necessary to construct a high quality mental model of the depicted content. The use of ancillary support approaches such as cueing is only partially effective in ameliorating this problem.

A more fundamental cause underlies failures in animation processing – the assumptions on which animations are currently designed. Conventional approaches to animation design reflect a misplaced commitment by those who produce them to faithful representation of the subject matter, particular regarding its behavioural aspects. This preoccupation with behavioural realism is inconsistent with designers' preparedness to sacrifice visuospatial realism in the name of facilitating understanding. By eschewing temporal manipulation, designers forego a host of additional and potentially powerful design possibilities that could greatly improve the effectiveness of informative animations. Relation set approaches designed on the basis of the Animation Processing Model offer one way by which the attributes of animated presentation might be better matched to the constraints and affordances of human information processing.

References

1. R. Lowe and J-M. Boucheix. Learning from animated diagrams: How are mental models built? In *Diagrammatic Representation and Inference*, pp. 266–281, Springer, 2008.
2. W. Schnotz and R. Lowe. A unified view of learning from animated and static graphics. In *Learning with Animation: Research Implications for Design,* chapter 14, Cambridge University Press, 2008.
3. M. Hegarty. Mental animation: Inferring motion from static displays of mechanical systems. In *Journal of Experimental Psychology: Learning, Memory, and Cognition,* volume 18, pp. 1084–1102. 1992.

4. M. Hegarty, S. Kriz and C. Cate. The roles of mental animations and external animations in understanding mechanical systems. In *Cognition and Instruction*, volume 21, pp. 209–249. 2003.
5. K. Van der Waarde and P. Westendorp. The functions of arrows in user instructions. *Proceedings of IIID Expert Forum on Manual Design* (Eskilstuna, Sweden, November, 2000), pp. 1–7, 2000.
6. J. Heiser and B. Tversky. Arrows in comprehending and producing mechanical diagrams. In *Cognitive Science*, volume 30, pp. 581–592. 2006.
7. M. Nienhaus and J. Döllner. Depicting dynamics using principles of visual art and narrations. In *IEEE Computer Graphics and Applications,* volume 25, pp. 40–51, 2005.
8. M. Bétrancourt, S. Ainsworth, E. de Vries, J-M. Boucheix and R.K. Lowe. Graphicacy: Do readers of science textbooks need it? Proceedings of the Text and Graphics Comprehension Conference (Grenoble, France, August 29–31, 2012), In Press.
9. N. H. Narayanan and M. Hegarty. Multimedia design for communication of dynamic information. In *International Journal of Human-Computer Studies*, volume 57, pp. 279–315. 2002
10. B. Tversky, J.B. Morrison and M. Bétrancourt, M. Animation: Can it facilitate? In *International Journal of Human-Computer Studies*, volume 57, pp. 247–262. 2002.
11. T. N. Höffler and D. Leutner. Instructional animation versus static pictures: A meta-analysis. In *Learning and Instruction*, volume 17, pp. 722–738. 2007.
12. R. K. Lowe, T. Rasch and W. Schnotz. Aligning affordances of graphics with learning task requirements. In *Applied Cognitive Psychology*, volume 25, pp. 452–459. 2010.
13. R. K. Lowe. Animation and learning: selective processing of information in dynamic graphics. In *Learning and Instruction*, volume 13, pp. 157–176. 2003.
14. R. K. Lowe. Extracting information from an animation during complex visual learning. In *European Journal of Psychology of Education*, volume14, pp. 225–244. 1999.
15. P.N. Johnson-Laird. *Mental Models*, Cambridge University Press, 1983.
16. J-M. Boucheix, R.K. Lowe and A. Soirat, A. (2006, September). One line processing of a complex technical animation: Eye tracking investigation during verbal description. *Proceedings of the Text and Graphics Comprehension Conference*, (Nottingham, UK, September, 2006), pp. 14–17. 2006.
17. R. K. Lowe. Interrogation of a dynamic visualization during learning. In *Learning and Instruction*, volume 14, pp. 257–274. 2004.
18. R. K. Lowe. Learning from animation: Where to look, when to look. In *Learning with Animation: Research Implications for Design*, chapter 3, Cambridge University Press, 2008.
19. R. Lowe. Perceptual and cognitive challenges to learning with dynamic visualizations. Proceedings of International Workshop on Dynamic Visualizations and Learning (Tübingen, Germany, July 17–18, 2002), 2002.
20. W. Winn. Perception principles. In *Instructional Message Design: Principles from the Behavioural and Cognitive sciences*, pp. 55–126, Educational Technology Publications, 1993.
21. J.M. Wolfe and T.S. Horowitz). What attributes guide the deployment of visual attention and how do they do it? In *Nature Reviews Neuroscience*, volume 5, pp. 1–7. 2004.
22. B.B de Koning, H.K. Tabbers, R. M. J. P. Rikers and F. Paas. Towards a framework for attention cueing in instructional animations: Guidelines for research and design. In *Educational Psychology Review,* volume 21, pp. 113–140. 2009.
23. R. E. Mayer. *Multimedia learning*. Cambridge University Press. 2001.
24. R. E. Mayer and R.B. Anderson. The instructive animation: Helping students build connections between words and pictures in multimedia learning. In *Journal of Educational Psychology*, volume 84, pp. 444–452. 1992.
25. M. Bétrancourt. The animation and interactivity principles in multimedia learning. In *The Cambridge Handbook of Multimedia Learning*, chapter 18, Cambridge University Press, 2005.
26. R. E. Mayer. Research-based principles for learning with animation. In *Learning with Animation. Research Implications for Design,* chapter 2, Cambridge University Press, 2008.
27. R.K. Lowe. Multimedia learning of meteorology. In *The Cambridge handbook of multimedia learning*, chapter 27, Cambridge University Press, 2005,

28. S.Kriz and M. Hegarty. Top-down and bottom-up influences on learning from animations. In *International Journal of Human Computer Studies*, volume 65, pp. 911–930. 2007.
29. J-M. Boucheix and R.K. Lowe. An eye tracking comparison of external pointing cues and internal continuous cues in learning with complex animation. In *Learning and Instruction*, volume 20, pp. 123–135. 2010.
30. R. K. Lowe and J-M. Boucheix. Cueing complex animations: Does direction of attention foster learning processes? In *Learning and Instruction*, volume 21, pp. 650–663. 2011.
31. J-M. Boucheix and R.K. Lowe. Cueing animations: Dynamic signaling aids information extraction and comprehension. In press.
32. J.L. Plass, B.D. Horner, and E.O. Hayward. Design factors for educationally effective animations and simulations. *Journal of Computing in Higher Education*, volume 21, 31–61. 2009.
33. H.S. Smallman and M. St. John. Naïve realism: Misplaced faith in realistic displays. *Ergonomics in Design*, volume 13, pp. 6–13. 2005.
34. H.S. Smallman, & M. Cook. Naïve realism: Fallacies in the design and use of visual displays. In *Topics in Cognitive Science*, volume 3, pp. 579–608. 2011.
35. S. Fischer, R.K. Lowe and S. Schwan. Effects of presentation speed of a dynamic visualization on the understanding of a mechanical system. In *Applied Cognitive Psychology*, volume 22, pp. 1126–1141. 2008.
36. B. Tversky, J. Heiser, S. Lozano, R. MacKenzie and J. Morrison. Enriching animations. In *Learning with animation: Research and design implications*, chapter 12, Cambridge University Press. 2008.
37. R. Ploetzner, and R. Lowe. A systematic characterisation of expository animations. In *Computers in Human Behavior*, volume 28, pp. 781–794. 2012.
38. R.K. Lowe and J-M. Boucheix. Dynamic diagrams: A composition alternative. In *Diagrammatic Representation and Inference*, pp. 233–240, Springer. 2012.
39. R.K. Lowe. Changing perceptions of animated diagrams. In *Diagrammatic Representation and Inference*, pp. 168–172, Springer. 2006.

Simultaneous and Sequential Presentation of Realistic and Schematic Instructional Dynamic Visualizations

Michelle L. Nugteren, Huib K. Tabbers, Katharina Scheiter, and Fred Paas

Abstract An experiment was conducted to investigate the effects of combining realistic and schematic dynamic visualizations of mitosis. Ninety-two students from four different biology classes were randomly assigned to one of four conditions. Participants in the simultaneous condition studied both a realistic and a schematic visualization of mitosis that were presented simultaneously; participants in the sequential condition studied these two visualizations sequentially; and participants in the schematic-only condition and the realistic-only condition studied only one of the visualizations. Afterwards, participants made a verbal and visual recognition test, and rated the difficulty and comprehensibility of the visualizations. The results showed that the conditions did not differ on verbal and visual recognition. Only on the schematic questions of the visual recognition test, the realistic-only condition scored significantly lower than the other three conditions. Also, no differences were found on the difficulty and comprehensibility ratings. It is concluded that studying multiple representations of a dynamic process is not necessarily better than studying only one representation.

M.L. Nugteren (✉) • H.K. Tabbers • F. Paas
Department of Psychology, Erasmus University Rotterdam, P.O. Box 1738,
Rotterdam 3000, The Netherlands
e-mail: m.l.nugteren@hotmail.com

K. Scheiter
Multimedia Lab, Knowledge Media Research Center, Schleichstraße 6,
Tübingen 72076, Germany
e-mail: k.scheiter@iwm-kmrc.de

F. Paas
Faculty of Education, University of Wollongong, Wollongong, NSW 2522, Australia
e-mail: paas@fsw.eur.nl

W. Huang (ed.), *Handbook of Human Centric Visualization*,
DOI 10.1007/978-1-4614-7485-2_24, © Springer Science+Business Media New York 2014

1   Introduction

Instructional dynamic visualizations can show either a realistic or schematic representation of a dynamic process. For instance, the process of mitosis (cell division) can be visualized with many details in a real life video of the process, as seen through a microscope. This would be highly realistic. On the other hand, mitosis can also be visualized as an animated line drawing showing only the most basic elements and processes. This would be a more schematic representation than a video [1]. Examples of these two different types of visualizations can be seen in Fig. 1. Both visualization types are frequently used in education, and therefore it is important to understand which type leads to better learning outcomes and under which conditions. The current study will focus specifically on the potential benefits of combining realistic and schematic dynamic visualizations.

According to Goldstone and Son [2], presenting information with either schematic or realistic visualizations has different advantages. Which type of visualization is more effective depends on the learning outcomes that need to be accomplished. An advantage of realistic visualizations is that they can encourage deeper learning, because realistic representations show more details and therefore provide more information to illustrate concepts and processes. Furthermore, learners generally find concrete concepts easier to remember and reason with. It is sometimes also easier to activate prior knowledge when looking at realistic visualizations, because they contain more familiar elements than schematic visualizations. Finally, Goldstone and Son argued that learning with realistic visualizations is more motivating, because learners often enjoy studying with them [2].

Schematic visualizations on the other hand, can also be beneficial to the learning process. A schematic visualization will show less detail, and therefore contains less distracting elements. This will lower cognitive load, because the learner can focus on the important features of the visualization [2]. This is in line with the coherence principle, which states that removing irrelevant words, sounds, or pictures from a multimedia presentation will improve subsequent learning outcomes [3–6]. Accord-

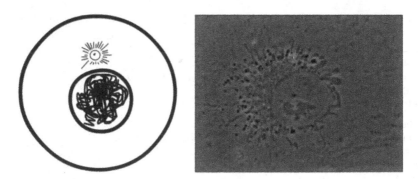

Fig. 1 Screen shots from two different dynamic visualizations of the process of mitosis. On the *left* is the schematic visualization, the *right* image shows the realistic visualization

ing to Butcher [7], this principle also applies to diagrams. When irrelevant details from diagrams are removed, learning outcomes will improve. Another advantage of schematic visualizations is that they can be especially useful for promoting transfer between domains. The abstract elements in schematic visualizations can be more easily applied in different domains than realistic elements [2].

Thus, whether information should be presented schematically or realistically depends on the desired learning outcomes. For complex information, it is better to start learning with schematic representations, because they contain few to no irrelevant details. When the goal is to reach a deeper understanding of a concept, realistic representations can be more useful, because they present learners with a more complete view of the concept.

In educational practice, dynamic visualizations are usually presented either schematically or realistically. Höffler and Leutner [8] investigated with a meta-analysis whether it is better to have more or less realism in a dynamic visualization. At first sight, more realistic videos seemed to benefit learning more than schematic animations, but if controlled for type of visualization (representational versus decorational), this relative advantage disappeared. In another review of the literature, Tversky, Morrison, and Betrancourt [9] were more decisive in this matter, stating that animations should contain as little realism as possible, because realistic details (although appealing) interfere with the clarity of the visualization. By only showing the most essential elements, it is easier for students to perceive the relevant movements in a dynamic visualization. Therefore, it would be more effective to study with schematic dynamic visualizations, than with realistic dynamic visualizations [9].

However, it could also be argued that combining a realistic and a schematic visualization is more effective than presenting only one type of visualization. According to Ainsworth [10], multiple representations in learning environments can serve three functions: to complement, to constrain, and to construct. Multiple representations can complement each other, because each representation provides different information. For example, when looking at the different visualizations of mitosis in Fig. 1, the schematic representation gives a better view on the movements during the different phases of mitoses, whereas the realistic representation shows what the different parts actually look like in real life. Complementary processes occur when the learner combines these different pieces of information and integrates them mentally. It is also possible to combine all these different pieces of information for the learner into one single representation, but such a complete representation can be very difficult to interpret [10]. Multiple representations can also constrain each other. When learners are confronted with different representations of the same process, their interpretation of one representation can be constrained by the other representation [10]. In the example in Fig. 1, the realistic representation is harder to understand at first. The schematic representation can be used as a constraint, because it presents the information more simplistically. Thus, students can use it as a starting point for understanding the movements in the realistic visualization. Finally, the use of multiple representations leads to the construction of deeper understanding. By studying several representations of the same concept, abstraction is promoted as well as generalization, leading to a higher order mental representation of this concept [10]. For example, by studying both a schematic visualization and a realistic visualization of mitosis, learners will construct a more integrated mental representation of the process of mitosis that can be applied in many new situations.

These functions of multiple representations illustrate the potential benefits of learning with both a schematic and realistic visualization of the same concept. However, to obtain the largest benefits, it is important to consider whether it is best to present these representations simultaneously or sequentially. According to Ainsworth [10], sequential presentation has the advantage that learners can focus on the separate aspects of the different representations. This method is best when the relation between the representations is not a primary learning goal. If students would try to find relations under these circumstances, time and effort will be taken away from the actual learning goals [10]. However, when the relations between the representations are important, it is better to present them simultaneously. That way, it is easier for learners to process and comprehend these relations.

Nevertheless, using multiple representations can also cause problems. Each representation is a separate information element that is related to the other representation. But due to their separation, split-attention effects can occur [e.g., 11–14]. This means that students will have to divide their attention across the separate elements to be able to construct an integrated cognitive schema of the elements. Several studies have shown that the need to focus attention on too many elements at the same time may overload the learner's working memory [e.g., 15–17]. Due to this overload, no working memory resources are available to connect the different elements together [18]. Related to the split-attention effect are the spatial contiguity and temporal contiguity principles [e.g., 4, 19]. According to the spatial contiguity principle, higher learning outcomes are achieved when the informational elements are presented in close proximity, than when they are placed further apart. The temporal contiguity principle states that learning outcomes are higher when the different elements are not separated through time [e.g., 4, 19]. In general, sequential presentation of multiple representations will result in a temporal split-attention effect, and possibly a spatial split-attention effect if the different presentations are spatially separated. Simultaneous presentation of multiple representations will result in a spatial split-attention effect, but not in a temporal split-attention effect because the different representations can be observed simultaneously. When these representations are also placed in close proximity of each other, the spatial split-attention effect will be cancelled out as well. Based on this account, it can be expected that sequential presentations of multiple representations will lead to lower learning outcomes than simultaneous presentations.

Thus, having multiple representations can have both advantages and disadvantages. They can support knowledge construction, but can also cause split-attention effects. In the current study, it is investigated how effective multiple representations are, when combining multiple dynamic visualizations. There have been a limited number of studies that have investigated the effects of presenting both schematic and realistic representations to learners. For example, Moreno, Ozogul, and Reisslein [20] presented participants with worked-out examples and practice problems including realistic and schematic diagrams of electrical circuits. For some participants, the diagrams were either realistic or schematic. For the other participants, the realistic and the schematic diagrams were presented simultaneously. Results showed that participants who had studied the schematic diagrams performed better on transfer than participants who had studied the realistic diagrams. Furthermore, participants who had received both type of diagrams simultaneously, performed significantly better on the problem solving practice task than participants who had only studied schematic or realistic diagrams, and they also performed better on transfer than participants who had only studied realistic diagrams. According to Moreno et al., participants receiving both types of diagrams performed best, because the multiple representations had a constraining function [10]. In this case, the realistic diagram was the starting point, because it matched better with the participants' prior knowledge than the schematic diagram. With their understanding of the realistic diagrams, participants were able to better comprehend the schematic diagrams.

A study that investigated the effect of studying both realistic and schematic *dynamic* visualizations was done by Scheiter et al. [1]. In their first experiment, participants studied either a realistic or a schematic animation of mitosis (see Fig. 1). Subsequently, participants were tested on the functions of cell structures, definitions, and their understanding of the different phases of mitosis. Results showed that learning performance for the schematic animation condition was superior to the realistic animation condition. In their second experiment, Scheiter et al. investigated whether presenting participants with both visualizations (schematic and realistic) in succession would make them more successful on subsequent tests. The visualizations were presented twice under different conditions. Participants either watched the schematic animation twice, the realistic animation twice, the realistic animation first and then the schematic animation, or the schematic animation first and then the realistic animation. Results showed that participants from all conditions performed equally well, except that participants who had watched the realistic animation twice performed significantly worse than the other participants.

Thus, although Moreno et al. [20] found that studying with both a realistic and schematic *static* visualization is beneficial, the results from Scheiter et al. [1] seem to suggest that having both a realistic and schematic *dynamic* visualization does not benefit learning. However, an important difference between these studies is that Moreno et al. found the largest benefits when participants were presented with both types of visualizations simultaneously. This condition was not used in the study by Scheiter et al., who presented both types of visualizations only sequentially. It could be argued that simultaneous presentation would have better allowed the constraining and construction functions of multiple representations [10], leading to a more integrated mental representation of mitosis. The focus of the current study is to present participants with the same realistic and schematic dynamic visualizations as used by Scheiter et al., but unlike their study, the current goal is to present them simultaneously to participants to investigate whether these lead to the same benefits as simultaneous presentations did in the study by Moreno et al.

The realistic animation used by Scheiter et al. [1] was very complex: it was somewhat blurry and contained many details. Hence, for some concepts mentioned in the accompanying narration, it was hard to recognize these parts in the animation.

The schematic animation on the other hand was much easier to comprehend. Different concepts were clearly drawn and easy to distinguish. However, if the goal is to transfer the knowledge obtained from the animation into real life practice, it is unlikely that this will occur when students have only studied the schematic representation. The schematic representation is so simplistic, that it is unlikely that students will recognize the concepts from the schematic visualization when seeing, for instance, a microscopic view of the actual process of mitosis for the first time. When students study with both representations, they can relate the schematic representation to the realistic representation, and vice versa. This integration of both representations will help students recognize this information in new situations.

Thus, the goal of the current study was to further investigate whether the positive effects of having multiple representations, as predicted by the theory of Ainsworth [10], would also occur when schematic and realistic dynamic visualizations are presented simultaneously. Our main research question was whether presenting schematic and realistic dynamic visualizations simultaneously or sequentially, indeed has different effects on how difficult and comprehensible the visualizations are and on how much students learn from them. The first hypothesis was that participants who had studied both types of visualizations simultaneously would score higher on visual and verbal recognition and comprehension than participants who had studied the visualizations sequentially. This was based on the idea that it is easier for learners to integrate different representations when presented simultaneously than when presented sequentially, because temporal split-attention effects are less likely to occur during a simultaneous presentation [e.g., 21]. Moreover, learners can fully exploit the benefits of learning from multiple representations, leading to better recognition and better comprehension.

The second hypothesis was that participants in the simultaneous and sequential groups would perform better on visual recognition and comprehension than participants who had studied only one type of visualization (either schematic or realistic). Participants who had studied multiple representations were expected to be able to answer questions about both these representations. Participants who had only studied one representation would have more difficulty answering questions about the representation they did not study, which should result in lower scores on those questions.

The third hypothesis was that participants who had studied both visualizations simultaneously would judge the visualizations they studied as less difficult and easier to understand, than participants who had studied the visualizations sequentially and than participants who had studied only one type of visualization.

To investigate these hypotheses, an experiment was set-up that was an adaptation of the study done by Scheiter et al. [1]. The main difference was that Scheiter et al. had presented the realistic and schematic visualizations sequentially, whereas in our study, the visualizations were also presented simultaneously. Also, participants in the current sample were second-year secondary school students, whereas participants in the study by Scheiter et al. were university students. We opted for secondary school students, as they were still unfamiliar with the process of mitosis and were therefore expected to have very little prior knowledge. Another difference between this study and the study by Scheiter et al. was that in the current study, visual recognition of the realistic visualization was measured more extensively on the final tests. It was expected that a more extensive measure would be more suitable to measure the expected higher recognition when students had studied with both the realistic and schematic visualization.

In the current study, participants studied the process of mitosis through different dynamic visualizations. These visualizations were either schematic or realistic. Each participant was assigned to one of four conditions. In the first condition, participants studied the schematic and realistic visualization simultaneously. For the second condition, participants also studied both the schematic and realistic visualization, but sequentially instead of simultaneously. The third en fourth conditions were control conditions. They were used to check whether having multiple representations would have any benefit at all. Participants in the third condition only studied the schematic visualization, and in the fourth condition only the realistic visualization. After studying the visualizations, all participants filled in a questionnaire, used to check how difficult and comprehensible they judged the different visualizations, and a final test, which measured visual and verbal recognition.

2   Method

2.1   Participants

Second-year students from four different biology classes from a secondary school in Oud-Beijerland, the Netherlands, were approached to participate in this experiment. These four classes were taught by two different teachers. A total of 113 students and their parents or guardians were asked for informed consent. Ninety-three students returned correctly signed forms and were hence allowed to participate. Of these 93 students, one was absent on the day the tests were taken, resulting in a final sample of 92 students (51 women, 41 men). Their mean age was 13.84 years ($SD = 0.54$). Per teacher, participants were randomly allocated to one of the four conditions using a blocked design, resulting in 23 participants in each condition.

2.2   Design

A between-subjects design was used, in which participants were allocated to either the sequential, simultaneous, schematic-only, or the realistic-only condition. Participants in all conditions watched a dynamic visualization representing the process of mitosis. In the sequential condition, participants first watched a schematic visualization and then a realistic visualization. In the simultaneous condition, participants watched both visualizations simultaneously, as they were presented next to each other. For the sequential condition, the presentation order was based on the constraining function of multiple representations [10]. The schematic visualization served as a starting point for learners unfamiliar with the content, because the relevant elements in the process of mitosis were easier to recognize in the schematic visualization than in the realistic visualization. After studying the schematic visualization, the realistic visualization would be easier to understand than the other way around. For this same reason, in the simultaneous condition, the schematic visualization was presented on the left and the realistic visualization on the right. The schematic-only and realistic-only conditions were the control conditions, in which participants studied only either the schematic or realistic visualization. To keep presentation times equal, participants in the simultaneous, schematic-only, and realistic-only conditions watched their visualization twice.

Prior knowledge was measured as a control variable, to check whether participants from the four conditions differed significantly on prior knowledge. The dependent variables were experienced difficulty and comprehensibility of the dynamic visualizations, as well as verbal and visual recognition and comprehension of the information from the visualizations and narration.

2.3 Materials

All participants received a prior knowledge assessment, an introductory text that explained some simple facts about mitosis, a visualization showing the process of mitosis, an evaluative questionnaire, and a final test. All materials were translated (German to Dutch) versions of the materials used by Scheiter et al. [1]. Adaptations were made to better fit the current research question, sample, and settings.

The 13 multiple-choice questions from the original German prior knowledge assessment were replaced with one single question: "Write down everything you know about mitosis". This replacement was made, because the original questions were thought to be too complicated for the current sample. We expected that with the open question, participants would have better opportunity to show their current knowledge on this topic than with restricted multiple-choice questions. The question was presented as a paper-and-pencil test. The participants were allowed to draw pictures if they wanted. One point was awarded for every correctly named item from a scoring model that described essential facts about the process of mitosis. Points were given for each correct fact mentioned and for correct drawings. These points were added to make a total score, ranging between 0 and 40 points.

The introductory text was also presented on paper and gave a short explanation about deoxyribonucleic acid (DNA), the physical appearance of animal cells, and the cell cycle. This information made it easier to understand the subsequent information presented in the visualizations. The first three paragraphs of the original German introductory text were removed in the current experiment due to expected time constraints during the test sessions. A picture with text was added that showed snapshots from the first few seconds of the realistic and schematic visualizations.

Items in these pictures were labeled to help students understand what they would be looking at in the visualizations. This information was added to the original text to better match the prior knowledge of the students in the current sample.

The visualizations used in this experiment were the same ones as used in the experiments by Scheiter et al. [1]. The realistic and schematic visualization both showed a dynamic representation of the process of mitosis, accompanied by a narrated explanation. Both visualizations lasted 5 min. The realistic visualization consisted of video recordings through a microscope. The schematic visualization showed the process of mitosis in a simplistic black-and-white animation. In all conditions, the visualizations were accompanied by the same verbal explanation, which was translated as literally as possible from the original German narration from Scheiter et al. [1]. The narration exemplified what happened in each phase. The visualizations were presented to the participants on laptops, and headphones were used to present the narration.

The evaluative questionnaire was presented and filled in on paper. The questions from this questionnaire concerned the experienced difficulty and comprehensibility of the studied materials. In the original German materials, these questions were very specified, with each question focusing on a different aspect of the learning process. Due to expected time constraints during the test sessions, the questions in the current experiment were compressed into several general questions about the learning materials. The first four questions from this questionnaire were the same for all participants. Participants from the simultaneous and sequential conditions received five extra questions concerning their experience of watching both the realistic and schematic visualizations. Participants from the schematic-only and realistic-only conditions received two extra questions concerning their experience watching their visualizations twice. All questions were answered on a five-point Likert scale, except one question for the simultaneous and sequential conditions. This extra question, posed in a multiple-choice format, asked which images were most helpful for learning mitosis: The schematic images, the realistic images, or the combination of both images.

Scores were calculated by dividing the answers over two constructs: Comprehensibility and difficulty. A higher score on comprehensibility indicated that participants judged the material they studied as more comprehensible than participants with lower scores. A higher score on difficulty indicated that participants had experienced more difficulty when studying the material than participants with lower scores. The mean of two questions was used to calculate the score on comprehensibility, whereas the mean of the other questions was used to form the difficulty score. In the simultaneous and sequential conditions, the multiple-choice question on helpfulness was analyzed separately from the other constructs.

The final knowledge test was also a paper-and-pencil test and consisted of two parts. The first part was a pictorial test, which measured visual recognition and comprehension. The second part consisted only of verbal questions and tested verbal recognition and comprehension.

For all eight questions in the pictorial test, one or more pictures of mitosis were used. The five questions using schematic pictures were all taken from the original

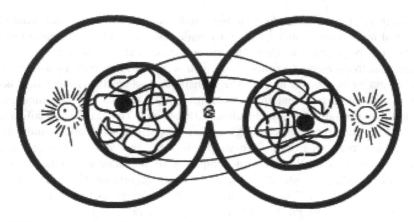

Fig. 2 Example question using a schematic picture from the pictorial part of the final test. Participants were instructed to mark the mistake made in this drawing

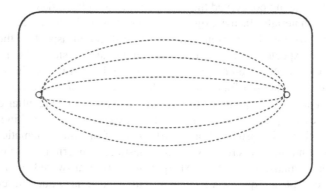

Fig. 3 Example question using a schematic picture from the pictorial part of the final test. Participants were instructed to draw the equatorial plate

materials. For two questions, participants had to mark mistakes made in existing pictures. An example of such a question is shown in Fig. 2. Three other questions asked participants to draw new features in existing pictures. An example question can be seen in Fig. 3. The original test only contained one question with realistic pictures, and this question did not seem very challenging. Thus, three new questions were constructed using photographs from cells. Participants were either asked to name the phase of mitosis, and/or to name a labeled part of the cell. These three questions were posed as multiple-choice questions. An example question is shown in Fig. 4. Scoring was done by awarding points for correct answers. Points from the questions with realistic pictures were added up to form a realistic score. Points from the questions with schematic pictures were added up to form a schematic score.

The second part of the test consisted of eight purely verbal multiple-choice questions, without pictures. These questions concerned general facts about mitosis, based on the notion that the participants would need both the narration and

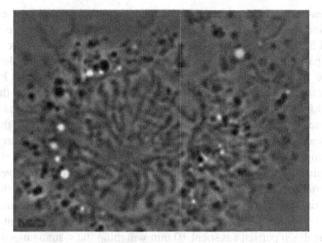

Fig. 4 Example question using a photograph from the pictorial part of the final test. Participants were instructed to name which phase of mitosis this cell is in

visualizations for answering these questions. Also, again due to time constraints, a selection of questions from the original test was made. More difficult questions were left out because of the lower level of prior knowledge of the current sample. An example question is: "In which phase are the sister chromatids unconnected?". Scoring was done by awarding one point for each correct answer. These points were added up to form a final score on the verbal test.

Because subjectiveness during scoring might be an issue for the prior knowledge test and the schematic questions on the visual test, a random sample of 20 participants was taken and their scores were checked by a second rater. The intraclass correlation coefficients for the prior knowledge assessment and for the schematic questions were both 0.81, which was considered sufficiently high.

2.4 Procedure

Participants were tested with their entire biology class during one biology lesson at their own school. In short, the course of the experiment was as follows. After verbal instructions, participants started with the prior knowledge assessment. Next, they read the introductory text and watched the visualizations. Finally, they filled in the evaluative questionnaire and the final test.

Each lesson lasted 50 min. When participants entered the classroom at the beginning of the lesson and all were seated, they received verbal instructions from the experimenter and their teacher explaining to them the course of the experiment. Next, the prior knowledge assessment was handed out and participants received a maximum of 5 min to complete it. All participants finished within these 5 min.

When all participants had finished the prior knowledge assessment, the tests were taken in, and they received instructions about the introductory text. Participants were told that it was very important to study the text carefully, since they needed the information to understand the visualization they were going to see. They were also told to study the text for 5 min; if they were done reading within these 5 min, they had to keep on rereading until the 5 min were up. After these instructions, the texts were handed out and the students read the text for 5 min.

After 5 min, the texts were taken in. Then, each participant received a laptop with the shortcut to the visualization they needed to view placed on the desktop, according to the condition they had been assigned to. After some final instructions, participants were asked to put on their headphones and click the shortcut to start the visualizations. Each participant spent 10 min watching the visualizations.

When participants had finished watching the visualizations, they were instructed to shut down the laptops and raise their hands to show they had finished. When participants raised their hands, they were provided with the evaluative questionnaire and the final test. They had until the end of the lesson to complete them. All participants finished on time.

3 Results

Means, standard deviations, and ranges of the different test scores and answers to the evaluative questionnaire are given in Table 1.

To check whether all conditions had comparable prior knowledge at the beginning of the experiment, the scores on the prior knowledge assessment were analyzed with a one-way Analysis of Variance (ANOVA). The independent variable was the condition participants were assigned to (simultaneous, sequential, schematic-only, or realistic-only). The dependent variable was the score obtained on the prior knowledge assessment. Results showed that the conditions did not differ significantly on prior knowledge, $F(3,88) = 0.30$, ns.

The first hypothesis was that participants from the simultaneous group would score significantly higher on visual and verbal recognition and comprehension than participants from the sequential group. The second hypothesis was that participants in the simultaneous and sequential groups would perform better on the verbal and visual recognition and comprehension test than participants from the schematic-only and realistic-only groups.

To test these hypotheses, first a one-way Multivariate Analysis of Covariance (MANCOVA) was performed, with condition (simultaneous, sequential, schematic-only, and realistic-only) as the independent variable, and realistic score and schematic score as dependent variables. Prior knowledge was used as a covariate. Results suggest that, using Pillai's Trace, participants from the four condi-

Table 1 Means and standard deviations of the scores on the prior knowledge assessment, the pictorial test, the multiple-choice test, and the evaluative questionnaire for each condition

	Simultaneous M (SD)	Sequential M (SD)	Schematic-only M (SD)	Realistic-only M (SD)
Prior knowledge				
Total score	1.30	1.43	1.57	1.43
(0–40)[a]	(1.22)	(0.84)	(0.79)	(0.84)
Pictorial test				
Realistic score	2.91	2.61	2.83	2.52
(0–10)[a]	(1.88)	(1.37)	(1.75)	(1.31)
Schematic score	4.83	4.43	4.48	3.17
(0–8)[a]	(2.17)	(1.81)	(1.50)	(1.27)
Verbal test				
Total score	2.09	1.57	1.83	1.78
(0–8)[a]	(1.13)	(1.12)	(1.27)	(1.04)
Evaluative questionnaire				
Comprehensibility	2.80	2.20	2.80	2.65
(1–5)[a]	(0.94)	(0.70)	(0.91)	(0.87)
Difficulty	2.70	2.82	2.72	2.50
(1–5)[a]	(0.56)	(0.61)	(0.75)	(0.52)

Note. $n = 23$ for all tests across all conditions, except for the scores on the evaluative questionnaire in the sequential condition, where $n = 22$
[a]Potential range

tions scored marginally different on the pictorial tests, $V = 0.13$, $F(6,174) = 1.96$, $p = 0.07$, partial $\eta2 = 0.06$. Univariate ANOVA's were used as a follow-up. The realistic score did not differ significantly for the different groups, $F(3,87) = 0.32$, ns. However, the schematic score differed significantly for the different groups, $F(3,87) = 4.05$, $p = 0.01$, partial $\eta2 = 0.12$. Planned contrasts revealed that for the schematic score, participants from the simultaneous and sequential conditions did not differ significantly, $t(87) = 0.78$, ns. Also, participants from the schematic-only condition did not differ significantly on the schematic score from participants in the simultaneous and sequential groups, $t(87) = 0.37$, ns. However, participants from the realistic-only group did score significantly lower on the schematic questions than the other three groups, $t(87) = 3.39$, $p = 0.001$.

The scores on the verbal recognition and comprehension test were analyzed with a one-way Analysis of Covariance (ANCOVA). The independent variable was the condition (simultaneous, sequential, schematic-only, and realistic-only), and the dependent variable was the total score on the verbal test. Prior knowledge was again used as a covariate. Results showed that prior knowledge was significantly related to scores on the verbal test, $F(1,87) = 4.60$, $p < 0.05$, partial $\eta2 = 0.05$. However, the four conditions did not differ significantly on the verbal test, $F(3,87) = 0.97$, ns.

Thus, according to these results, the first hypothesis was not confirmed. Participants from the simultaneous group did not differ significantly from the sequential group on a test of visual and verbal recognition and comprehension. The second

Table 2 Frequencies and percentages of the answers given to the question which visualization was most helpful in the simultaneous and sequential conditions

	Simultaneous		Sequential	
	Frequencies	Percentages	Frequencies	Percentages
Schematic	16	69.6	13	56.5
Realistic	1	4.3	0	0.0
Combination	6	26.1	10	43.5
n	23		23	

hypothesis was only partially confirmed. There was no difference between all four conditions on the verbal recognition and comprehension test and on the realistic questions of the visual recognition and comprehension test. There was only a significant difference on the schematic questions of the visual recognition test. The realistic-only group scored significantly lower than the other three groups.

The third hypothesis was that participants who had studied both visualizations simultaneously would judge the visualizations they studied as less difficult and easier to understand, than participants who had studied the visualizations sequentially, or who had studied only one type of visualization. To analyze the scores on the evaluative questionnaire, a Multivariate Analysis of Variance (MANOVA) was performed with condition (simultaneous, sequential, schematic-only, and realistic-only) as independent variable. The first dependent variable was difficulty, the second was comprehensibility. Using Pillai's Trace, results suggested that participants from the four different groups did not evaluate the test materials significantly different, $V = 0.11$, $F(6,174) = 1.69$, ns.

Thus, the third hypothesis was not confirmed. Participants from the simultaneous group did not judge their visualizations to be less difficult or more comprehensible than participants from the other groups.

In Table 2, answers on the multiple-choice question from the evaluative questionnaire are given. Participants from the simultaneous and sequential conditions had to answer the question which images they found most helpful for understanding the process of mitosis. The alternatives were the schematic images, the realistic images, or the combination of both the schematic and realistic images. As can be seen, most participants from the simultaneous condition judge the schematic images to be most helpful, followed by the combination of both images. In the sequential condition, about half of the participants rated the schematic images as more useful and the other half the combination. Most notably, across all participants from both conditions, only one participant rated the realistic images as more helpful than the schematic images or the combination.

4   Discussion and Conclusion

The research question was whether presenting schematic and realistic dynamic visualizations simultaneously or sequentially has different effects on how much students learn from the visualizations and how difficult and comprehensible students judge these visualizations. The results suggest that presenting schematic and realistic dynamic visualizations simultaneously or sequentially does not have different effects on students learning outcomes. Participants from the sequential group did not score significantly different from the sequential group on tests of verbal and visual recognition and comprehension. The only significant difference in scores was found for the realistic-only group, who scored significantly lower than the other three groups on the schematic questions from the pictorial test. This difference seems logical, because participants from the realistic only group did not study the schematic visualization. This gave them a disadvantage when answering the schematic questions. However, it is interesting to note that participants from the schematic group did not suffer from this same disadvantage on the realistic questions from the pictorial test, while they had not studied the realistic visualization. These results are in line with the results from Scheiter et al. [1], who also suggested that studying only the realistic visualization is less beneficial than studying the schematic visualization. However, in the study by Scheiter et al., the pictorial test contained only one question with realistic pictures, so the test was biased towards the schematic visualization. Therefore, the results from the current study give stronger support to the conclusion from Scheiter et al., because extra questions were added with realistic pictures, and realistic and schematic scores were analyzed separately.

But the main question remains: why didn't we find a benefit of simultaneous presentation of a schematic and a realistic dynamic visualization. Scheiter et al. [1] already found that participants who were presented with realistic and schematic visualizations sequentially did not seem to have obtained extra learning benefit over a schematic-only visualization. However, based on the results from the study of Moreno et al. [20], it was expected that presenting visualizations simultaneously would cause extra benefits over presenting visualizations sequentially. Our current results however do not confirm this idea. Possibly, this was caused by the difficulty of the visualizations. The visualizations used by Moreno et al. were static diagrams. The visualizations used for this study and Scheiter et al.'s study were dynamic, and because of this more complex. It takes more effort to study a dynamic visualization than a static visualization. Furthermore, the realistic visualization on mitosis was particularly difficult to study, because it proved to be very hard to identify the separate elements from the cell. For the simultaneous visualizations to be successful, integration between the two visualizations must occur. It is possible that studying the separate visualizations already cost so much working memory load that there was not enough capacity left for integrating the separate visualizations.

Taken together, the current study and the study by Scheiter et al. both fail to show a benefit for combining a schematic and realistic representation of a dynamic process like mitosis. So apparently, the multiple representations have not served the functions of complementing, constraining, and constructing, as suggested by Ainsworth [10]. This lack of benefits is possibly caused by the occurrence of split-attention effects [e.g., 11–14]. Participants from the sequential group may have been hindered by a temporal split-attention effect, because the two representations were separated in time. Participants were required to keep the schematic visualization in memory and to be able to make connections between this visualization and realistic visualization. This may have raised the cognitive load of the participants, which might explain why their learning performance did not exceed the performances of participants from single-representation groups. Participants in these groups did not experience split-attention effects. And although participants from the simultaneous group did not have to deal with a temporal split-attention effect, they still may have experienced a spatial split-attention effect. They were required to attend to two different items on the computer screen: the schematic visualization and the realistic visualization. This may also have raised cognitive load and prevented them to outperform the students who only saw a single visualization.

Finally, most participants from the simultaneous and sequential conditions rated the schematic visualization as the most helpful one, and only one participant thought the realistic visualization most helpful. As mentioned above, the realistic visualization was difficult to study with, so it seems logical that hardly any participant rated this visualization as helpful. However, that does make it interesting to see that the combination of the two visualizations was still rated by many participants as most helpful, despite the difficulty of the realistic visualization. Especially in the sequential group, more than 40 % of the participants preferred the combination of visualizations, indicating that the order of first schematic and then realistic made some sense to them.

Of course, the current study has some limitations. First of all, the materials used for this experiment may have been too difficult for our sample. Throughout the experiment, many participants complained about the difficulty of the visualizations and the questions on the final test. Especially the scores on the verbal test are so low, that a floor effect may have occurred. In the experiment by Scheiter et al. [1], the sample consisted of university students. Although the materials were made easier to better match the level of the current sample, maybe the materials were still too difficult and more adaptations would have been necessary to better fit the knowledge level of these participants. A second limitation is that the realistic visualization was rather difficult. It seemed to be hard to identify the different parts of the cell in the visualization. Perhaps adding visual cues would have helped students to learn more from this visualization. For instance, arrows could point to certain parts of the cell when they were mentioned in the narration.

In conclusion, our results suggest that multiple representations of dynamic processes are not 'just' superior over a single representation. Participants in this experiment who have seen both a schematic and a realistic visualization of the process of mitosis do not outperform participants who have studied only one type of dynamic visualization. It seems that the benefits of multiple representations do not yet outweigh the complexity of having to process two dynamic visualizations either sequentially or simultaneously. On the other hand, the design of the specific visualizations used in the current study may be further optimized. The resulting reduction in processing load may lead to a better integration of both visualizations, and thus better learning. Nevertheless, this promise of multiple dynamic representations has yet to be fulfilled, so for the moment, the designer of dynamic visualizations should be cautious in expecting too great a benefit from the combination of realistic and schematic visualizations.

Acknowledgements Many thanks to Eveline Osseweijer, Vincent Hoogerheide, and Sandra Visser for their help in the design and performance of this study. Also thanks to the teachers and students who participated in this study.

References

1. Scheiter, K., Gerjets, P., Huk, T., Imhof, B., & Kammerer, Y. (2009). The effects of realism in learning with dynamic visualizations. *Learning and Instruction, 19,* 481–494.
2. Goldstone, R. L., & Son, J. Y. (2005). The transfer of scientific principles using concrete and idealized simulations. *The Journal of the Learning Sciences, 14,* 69–110.
3. Mayer, R. E., Heiser, J., & Lonn, S. (2001). Cognitive constraints on multimedia learning: When presenting more material results in less understanding. *Journal of Educational Psychology, 93,* 187–198.
4. Mayer, R. E., & Moreno, R. (2002). Animation as an aid to multimedia learning. *Educational Psychology Review, 14,* 87–99.
5. Mayer, R. E., & Moreno, R. (2003). Nine ways to reduce cognitive load in multimedia learning. *Educational Psychologist, 38,* 43–52.
6. Moreno, R., & Mayer, R. E. (2000). A coherence effect in multimedia learning: The case for minimizing irrelevant sounds in the design of multimedia instructional messages. *Journal of Educational Psychology, 92,* 117–125.
7. Butcher, K. R. (2006). Learning from text with diagrams: Promoting mental model development and inference generation. *Journal of Educational Psychology, 98,* 182–197.
8. Höffler, T. N., & Leutner, D. (2007). Instructional animation versus static pictures: A meta-analysis. *Learning and Instruction, 17,* 722–738.
9. Tversky, B., Morrison, J. B., & Betrancourt, M. (2002). Animation: Can it facilitate? *International Journal of Human-Computer Studies, 57,* 247–262.
10. Ainsworth, S. (1999). The functions of multiple representations. *Computers & Education, 33,* 131–152.
11. Chandler, P., & Sweller, J. (1991). Cognitive load theory and the format of instruction. *Cognition and Instruction, 8,* 293–332.
12. Ginns, P. (2006). Integrating information: A meta-analysis of the spatial contiguity and temporal contiguity effects. *Learning and Instruction, 16,* 511–525.
13. Mayer, R. E., & Moreno, R. (1998). A split-attention effect in multimedia learning: Evidence for dual processing systems in working memory. *Journal of Educational Psychology, 90,* 312–320.
14. Tarmizi, R. A., & Sweller, J. (1988). Guidance during mathematical problem solving. *Journal of Educational Psychology, 80,* 424–436.
15. Chandler, P., & Sweller, J. (1996). Cognitive load while learning to use a computer program. *Applied Cognitive Psychology, 10,* 151–170.
16. Kalyuga, S., Chandler, P., & Sweller, J. (1998). Levels of expertise and instructional design. *Human Factors, 40,* 1–17.

17. Yeung, A. S., Jin, P., & Sweller, J. (1998). Cognitive load and learner expertise: Split-attention and redundancy effects in reading with explanatory notes. *Contemporary Educational Psychology, 23,* 1–21.
18. Sweller, J., Van Merriënboer, J. J. G., & Paas, F. G. W. C. (1998). Cognitive architecture and instructional design. *Educational Psychology Review, 10,* 251–296.
19. Moreno, R., & Mayer, R. E. (1999). Cognitive principles of multimedia learning: the role of modality and contiguity. *Journal of Educational Psychology, 91,* 358–368.
20. Moreno, R., Ozogul, G., & Reisslein, M. (2011). Teaching with concrete and abstract visual representations: Effects on students' problem solving, problem representations, and learning perceptions. *Journal of Educational Psychology, 103,* 32–47.
21. Mayer, R. E., & Anderson, R. B. (1991). Animations need narrations: An experimental test of a dual-coding hypothesis. *Journal of Educational Psychology, 83,* 484–490.

How Do You Connect Moving Dots? Insights from User Studies on Dynamic Network Visualizations

Michael Smuc, Paolo Federico, Florian Windhager, Wolfgang Aigner,
Lukas Zenk, and Silvia Miksch

Abstract In recent years, the analysis of dynamic network data has become an increasingly prominent research issue. While several visual analytics techniques with the focus on the examination of temporal evolving networks have been proposed in recent years, their effectiveness and utility for end users need to be further analyzed. When dealing with techniques for dynamic network analysis, which integrate visual, computational, and interactive components, users become easily overwhelmed by the amount of information displayed—even in case of small sized networks. Therefore we evaluated visual analytics techniques for dynamic networks during their development, performing intermediate evaluations by means of mock-up and eye-tracking studies and a final evaluation of the running interactive prototype, traceing three pathways of development in detail: The first one focused on the maintenance of the user's mental map throughout changes of network structure over time, changes caused by user interactions, and changes of analytical perspectives. The second one addresses the avoidance of visual clutter, or at least its moderation. The third pathway of development follows the implications of unexpected user behaviour and multiple problem solving processes. Aside from presenting solutions based on the outcomes of our evaluation, we discuss open and upcoming problems and set out new research questions.

M. Smuc (✉) • F. Windhager • L. Zenk
Department for Knowledge and Communication Management, Danube University Krems,
Dr.-Karl-Dorrek-Straße 30, 3500 Krems, Austria
e-mail: michael.smuc@donau-uni.ac.at; florian.windhager@donau-uni.ac.at;
lukas.zenk@donau-uni.ac.at

P. Federico • W. Aigner • S. Miksch
Institute of Software Technology and Interactive Systems, Vienna University of Technology,
Favoritenstrasse 9-11/188, 1040 Vienna, Austria
e-mail: federico@ifs.tuwien.ac.at; aigner@ifs.tuwien.ac.at; silvia@ifs.tuwien.ac.at

W. Huang (ed.), *Handbook of Human Centric Visualization*,
DOI 10.1007/978-1-4614-7485-2_25, © Springer Science+Business Media New York 2014

1   Introduction

The analysis of dynamic network data has become an increasingly important
research field with promising application areas in different real-world domains,
including the analysis of organizational knowledge and collaboration networks [25].
As the temporal dimension is adding a new level of complexity, the demand on
computational methods—and the cognitive efforts for their users—are even higher
than they are in static network analysis anyway [7, 33].

While several visual and computational methods addressing the examination
of temporally evolving networks have been proposed in recent years, their effectiveness and utility for end users needs to be further analyzed. Considering the
increasing complexity and the novelty of all these methods, adopting participatory
design strategies can be beneficial. These strategies can help to improve the methods
and particularly their application to real world scenarios [4, 35] by bringing users'
needs and experiences into the development process. Moreover, by analyzing
users' preferences and performances when dealing with such methods in specific
scenarios, it is possible to gain insights that might be applicable in a more general
context.

Following this approach, we evaluated a visual analytics method for dynamic
networks along its development process. First, we performed an intermediate
evaluation by the means of mock-up studies and second, we conducted a qualitative
evaluation of the final interactive prototype.

In the following sections, we want to discuss related work, summarize insights
gained by a mock-up study, give an overview of the main results of the prototype
evaluation, highlight examples for pathways of (participatory) development and
design and bundle the outlined issues into conclusions and future research questions.

1.1   Related Work

While several methods for the visualization of static networks have been proposed
in Graph Drawing [11], Information Visualization [20], and Data Mining [10]
communities, the interactive visual analysis of networks evolving over time is
an emerging research field. Besides the choice of a visual representation for
the relational data (e.g. node-link diagrams or matrix-based representation), an
important issue for time-varying networks is the appropriate visual encoding of
the temporal dimension [2]. At least four different approaches exist: animation
[16], superimposition [5], juxtaposition [3]; and two-and-a-half-dimensional view
[6, 12]. But finding an adequate visual encoding for the time dimension is not
sufficient to solve the issue of visualizing dynamic networks. Another important
aspect is to obtain a sequence of diagrams that facilitates the perception of changes,
by preserving the user's mental map [13]: it must minimize unnecessary changes
while emphasizing temporal trends or patterns. An early formulation of the problem

is sketched by [33], while [7] discuss it systematically from a graph drawing perspective. Several computational methods, which descend from Social Network Analysis (SNA) [45], can be integrated into visualizations. A common approach is to compute some static SNA metrics associated to nodes and edges and then encode them to a chosen visual variable or exploit them to perform dynamic filtering [34].

To test prototypes in the field of visual analytics, various methods for empirical user studies were discussed in recent years. Especially in the visual analytics community, the usage of highly standardized quantitative methods (see [4]) was criticized of being too rigid resulting in artificial results [14]. Therefore more qualitative approaches were favored [22, 39, 40]. Methods which allow to gain insights into which problems occur and why they occur [27] should also engage users to "search to learn" and show real behavior instead of using simple "lookup tasks" [28]. Another necessary step to avoid artificial results when covering users' exploration process [41] is to use real world data with context [47]. Therefore the selection of expert groups who have to deal with (often ill-defined) real-data is favored by some authors [21, 23]. A rather novel trend to analyze exploration behavior when using visual methods is to analyze exploration focusing on the multiple ways of problem solving processes of the users [30].

1.2 Visual Analytics Methods

The prototype at hand is aimed at the examination of dynamic social networks and has been designed and implemented on the basis of a visual analytics approach [15]. It features the integration of visual, analytical and interactive techniques, led by some basic perceptual principles, and it is tailored for small longitudinal network datasets (up to 50 nodes), manually collected by the means of questionnaires (discrete time domain). Even though it is definitely far from being of general applicability and covering all recent developments in the field, its integration of some different techniques provides us the opportunity to observe how users exploit, alternate between, or combine them for the means of visual network data exploration.

The visualization is based on node-link diagrams, and three ways to map the temporal dimension into it leads to three different views: juxtaposition (JX), superimposition (SI) and two-and-a-half-dimensional (2.5D) view.

The JX view (see Fig. 1) is obtained by mapping time to space (the horizontal temporal axis), i.e. by placing the diagrams of different time-slices side by side. It applies the principle of small multiples [43] and allows the reader to directly compare the time-slices. Coordinated zooming and panning and coordinated highlighting further facilitate comparison.

The data which is shown by the dynamic network visualizations in this report— and which was shown in all empirical studies we refer to—is a real world data set, which covered eight different relations of knowledge communication structures at a university department [48] with four time steps and 33–34 nodes per time point

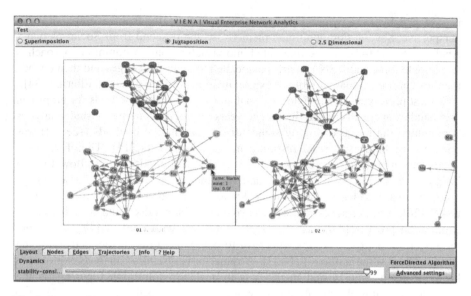

Fig. 1 Juxtaposition view (*JX*)

(38 in total). Relational questions included content related and technical advice, intensive collaboration, awareness of individual knowledge, knowledge substitution, discussion of new ideas and suggested communication that should be intensified.

The SI view is obtained by superimposing the node-link diagrams (see Fig. 2). It can be described as mapping time to a visual variable, namely the transparency, which is employed to differentiate between time-slices, so that more recent elements are more opaque. It requires less screen space than the previous view, but is affected by more visual clutter and occlusion. To reduce these problems, at first only nodes are shown to reduce occlusion and visual clutter, but edges can be displayed on demand.

In the 2.5D view (see Fig. 3), diagrams for each time-slice are drawn on separate transparent planes, stacked along the horizontal time axis, orthogonally. It can be seen as the mapping of time to an additional spatial dimension, along which more information can be displayed, as described in the following. 3D zooming, rotating and panning controls allow the user to set the best viewpoint.

In order to preserve the user's mental map and provide a common context for the interactive exploration of the three views, they are built upon a consistent spatial metaphor, which also drives smooth animated transitions between them (see Fig. 3): the sheets, on which the diagrams are drawn, are stacked upon each other in the SI view, then translated alongside the time axis in the JX view, and finally rotated by 90° around their vertical axes in the 2.5D view.

As for the layout of the node-link diagrams (i.e. the way nodes are arranged), the prototype adopts a continuously running force-directed layout that also ensures the preservation of the mental map over different time slices. The user can interactively

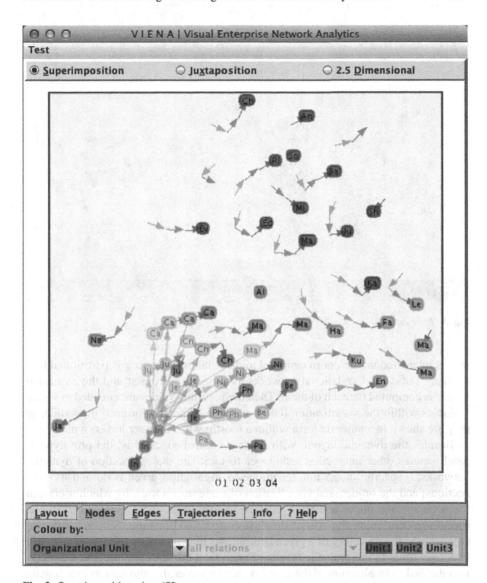

Fig. 2 Superimposition view (*SI*)

control the amount of preservation: a simple slider in the Graphical User Interface (GUI) allows users to select stability (maximum mental map preservation) or consistency (independent layouts) and to pass from one to the another through stepwise transitions (see Fig. 1 at the bottom).

An integrated SNA computational component provides the calculation of SNA metrics on demand (e.g. different types of centralities). In this way the user can interactively select a certain SNA metric to be computed for a certain type of relation

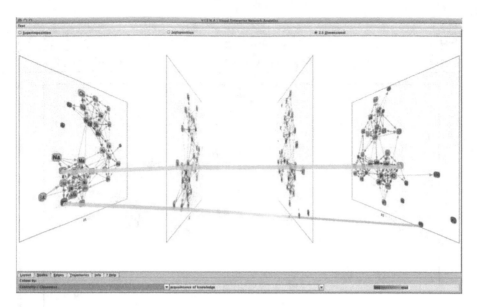

Fig. 3 2.5D view

s/he is interested in; the entire temporal multi-relational network is partitioned into as many static single-relational networks as time-slices consist and the requested metric is computed for each of them. Then the resulting values are encoded to visual variables within the visualization (color and size of nodes) for each time-slice, or they are shown in a numeric form within a tooltip when the user hovers a node.

Besides the dynamic layout, with its user-controlled stability, the prototype at hand features other interaction techniques to facilitate the exploration of dynamic networks: a specific interaction technique to highlight a given node and its connections; and the on-demand visualization of node trajectories, by which users can focus on specific nodes and track their evolution. In the 2.5D view, for example, trajectories run along the spatial dimension dedicated to time. Shading different colors along the trajectory of a given node shows how its values for a certain metric vary over time. In this way, the results of analytical methods are integrated directly into the main visualization of the network, aiming to enable the user to examine its relational and temporal aspects simultaneously without any additional diagram.

1.3 Overview

To match the described features with the needs of the intended user group, the development process of the prototype at hand crossed six main stages, with three of them bringing user expectations, evaluations and participatory elements into play (marked with a star below):

1. Assessment of the State of the Art,
2. User and Task Analysis*,
3. Design,
4. **Mock-up Study***,
5. Implementation, and
6. **Prototype Evaluation**.*

While a initially executed user and task analysis had the function to bundle state of the art options on the targeted group of users and align possible features with their real world tasks and needs (see [15, 48]), the next section focuses on the participatory part of a mock-up study on dynamic network layouts, while the following sections will turn towards the empirical results of the implemented feature evaluation.

2   Mock-Up Study

The aim of the mock-up study was to test three early sketches of non-interactive dynamic network visualizations (one JX view and two versions of SI views) on their comprehensibility, visual design and utility.

2.1   Study Design

Therefore we conducted an experiment with a sample size of 38 participants, including 10 experts (with at least 2 years SNA experience) and 28 non-experts in the field of social network analysis. Each participant was tested individually for about an hour and had to solve four open tasks as well as four pre-defined tasks. These tasks were similar to the tasks 2–7 used in the evaluation study (see Table 1) and their thinking aloud and viewing behavior were observed, recorded and analyzed.

Our real world data on two time points of knowledge communication at a university department was firstly visualized in a JX view. The network structure of the two layers differed, since it was computed for each network with a medium stability-consistency balance. This also applied to the first variant of a SI view, in which the two layers were displayed as stacked overlay and the nodes were additionally connected by trajectories over time (see Fig. 4). We will refer to this view as comet plot. In the second SI view (to be referred to as SPOCC plot—Stable POsitions, Color Coded), nodes kept a fixed position over time, but the relations and nodes were color coded on their temporal attribute: red for relations or nodes only existent in timepoint 1 (t1), green for relations or nodes appearing in timepoint 2 (t2) and blue for relations or nodes that were constant over t1 and t2 (see Fig. 4, on the right hand side).

Table 1 Mock-up study plan

#	Task	Material	Minutes
	Introduction and calibration	Slides	5′
1	Open task	1 network (static)	5′
2	Open task	2 networks (JX)	5′
3	Open task	2 networks (comet plot)	5′
4	Open task	2 networks (SPOCC plot)	5′
5	Pre-defined task	Individual vs. structure	2 × 3′
6	Pre-defined task	Dynamics (JX)	2 × 3′
7	Pre-defined task	Dynamics (comet plot)	2 × 3′
8	Pre-defined task	Dynamics (SPOCC plot)	2 × 3′
9	Interpretation task	All networks so far	3′
10	Derived countermeasures	All networks so far	3′
11	(Only non-experts) pre-defined task	Individual vs. structure	2 × 3′
12	Pre-defined task	Dynamics (JX)	2 × 3′
11	Post questionnaire	All networks	10′
	Planned Total Time	—	~66′

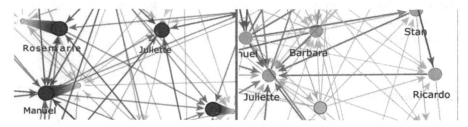

Fig. 4 Closeups of two variants of a superimposition (SI) view (see also http://www.smuc.at/cometandspocc/): comet plot (*left hand side*) and SPOCC plot (*right hand side*). While the comet plot shows relations of time point t1 in orange and t2 in blue, it allows certain shifting of the nodes due to the consistency of the temporarily changing network structure and its force-directed visualization by the chosen spring embedder layout. In contrast, the SPOCC plot holds all node positions stable, but codes temporal changes with the colors *green* (nodes or ties which emerged at t2), *red* (nodes or ties which vanished after t1) and *blue* (nodes or ties present at both time points)

The analysis of visual information processing is necessary to examine how users gain insights into network visualizations. In this study, we used an eye tracking technology to analyze on which parts users focused to understand the network. To examine visual information processing, eye tracking technology provides a means to observe a viewer's point-of-gaze (e.g., [36]). In the past, eye tracking focused mainly on scene perception and reading under laboratory conditions [18, 36]; only in the last years, applications in more everyday settings [30] became possible with the emergence of more usable technology.

Central eye-movement measures are fixations and saccades. Saccades are shifts from one point of gaze to another; fixations indicate visual attention to that information [36]. In scene perception, top-down and bottom-up influences control where one looks [18, 46]. Bottom-up influences are stimulus-driven, whereas top-down influences are viewer-driven. Bottom-up influences are mainly based on the visual salience of the stimulus, i.e., color, saturation, and [28]. Top-down influences on the other hand are a viewer's knowledge about the stimulus, his or her domain knowledge, and his or her goals [18]. Another top-down influence is the viewer's domain knowledge [9] showed that due to their higher knowledge on possible configurations experts in chess can easier create chunks of information.

Eye movements were recorded using an SMI iView X™ RED eye tracker at a temporal resolution of 60 Hz. It tracks the corneal reflection of the pupils and allows relatively free movement of the head when seated approximately 60 cm from the tracking device. As it allows eye tracking with glasses and contact lenses, a wide range of participants could be included. Each participant was tested individually. After an explanation on the purpose of the study, the functionality of the eye tracking device was explained to the participants. The device was calibrated using a nine-point-calibration. Participants viewed the scenes on the 17″ computer screen, integrated in the eye tracking device. The experimenter was seated next to the participant with a control screen of the participant's gazes to intervene, if the gaze was lost by the eye tracking system.

Think aloud notes were used to study the participants' problem solving strategies and to gain deeper insights about their exploration behavior. Using this method, we logged participants' interaction, tracked their eye movements, observed their behavior, and asked them to think aloud during the experiment. We integrated these data sources, segmented them according to the tasks, and documented the users' success levels.

Eye tracking data were analyzed with BeGaze™ analysis software from SMI. We segmented the recordings based on single tasks and extracted the fixations (number and duration) and saccades (number and amplitude). To analyze the visual attention given to highly informative regions, the scenes were coded in accordance to predefined Areas of Interest (AOIs) similar to [19], dependent on the tasks [24].

The Mock-up study plan (see Table 1) consisted of four large parts: An introductory/calibration phase, four open tasks session with a static network and the three different mock-ups where participants were allowed to explore the networks freely while thinking aloud and get familiar with the visualizations and the GUI (task 1–4). Then the users had to solve and interpret a pre-defined set of tasks, from rather basic tasks up to more complex structural analysis (task 5–12). Finally, in the post questionnaire, user could provide additional feedback, discuss problems and suggest improvements (if they had not done it before). This test plan consisted of two slightly varying variants of network data the mock-ups were based upon to avoid rigor in the visualizations. The results for the two variants were merged for the following analysis and for the eyetracking results.

2.2   Study Results

Overall, the feedback from participants was promising that dynamic network visualizations can be made comprehensible with such graphs and allowed for further fine tuning and interactive enrichment of at least two out of three variants. All participants were able to comprehend the JX view fast and easily, even the non-experts. The comet plot was the most difficult to comprehend, only some experts caught the concept behind this visualization at a first glance, without an explanation how to read the graphs. Many of the non-experts asked for an explanation and some of them could not utilize the structural visualization in the intended way. The SPOCC plot was easier to understand, but suffered from visual clutter to a high extent.

We will come back to these user problems, eyetracking results and resulting design decisions in a more detailed manner further down (see section about visual clutter).

3   Prototype Evaluation

The qualitative prototype evaluation was conducted to evaluate the prototype's usability and the comprehensibility of the different views and interaction techniques, and to cover users' exploration process [41].

In contrast to the mock-up study described above, the sample consisted of nine experts who work in the field of social network research as pre- and postdocs, mainly as computer scientists or as graph theorists plus another computer scientist from the visual analytics field. None of the participants had prior knowledge about the prototype or has been tested in the mock-up study.

In the first phase of this study, the prototype was presented to the participants in an interactive session together with an instructor. Participants were encouraged to explore the functions of the prototype, ask questions, give feedback about the prototype's usability and express their ideas and suggestions for improvement. In the second phase, they had to solve seven tasks, which were derived from [1] and were selected on the basis of our experiences with the mock-up study (see Table 1).

These tasks (see Table 2) included lower-level activities like the identification and comparison of the relations of a single node at two time points as well as higher-level activities [29] like the description of structural group changes over time. In the first task participants were allowed to use all prototype functions and views freely. In all other tasks they were compelled to work with a preselected initial view. In the third and last phase, participants were asked to summarize their impressions and give additional feedback.

The material consisted of the same real world data set we used in the mock-up study (as proposed by [21, 23]), except that we used four instead of two time steps. The verbal comments and a screen cast were recorded during all phases—

Table 2 List of tasks for the prototype evaluation. The tasks were named according to the scheme proposed by Ahn et al. [1]

Task name	Task description	Predefined settings
T1: Network—Growth	Has the total density of the network increased or decreased from t1 to t2?	Open
T2: Group—Stability	Which groups/clusters do you detect? How do they change?	JX
T3: Node/Link—Growth	Had Leonard (Le) more relations at t1 or at t2?	JX + SNA
T4: Node/Link—Single Occurency	Please identify the outdegree of actor Hans at t3	JX + SNA + tooltip
T5: Node/Link—Growth	Please identify the change (increase/decrease) of Ines' eigenvector centrality (from t1 to t4)	2.5D + SNA + single trajectory
T6: Node/Link—Birth Death	Who has joined / who has left the network (causing relational consequences)?	2.5D + SNA + all trajectories
T7: Node/Link—Peak/Valley	Are there significant shifts of single actors from cores to peripheries or vice versa?	SI

which lasted about 1.5 up to 2 h in total. Notes were taken by an observer during these sessions. These notes were jointly analyzed by a team of three usability experts who were also part of the testing-team. The notes were segmented in single observations, which were categorized and counted as presented in the following section. First we want to present an overview about users' feedback and observed problems during the introduction phase, later we will describe our main insights that derived from task analysis.

3.1 Evaluation Results

The evaluation results are structured as a matrix, with the main visual, computational and interactive features of the prototype as rows and columns (see Fig. 5).

The feedback was segmented into 255 distinct observations, which were categorized as problems (118), positive feedback (45) and ideas for improvement (109). It has to be noted that similar observations were counted multiple times, so that we could identify 155 unique observations in total.

To give an overview we will focus mainly on areas in which many observations have been made, leaving bugs and too specific implementation issues aside. In all views, many participants stated that the transitions are too slow, although the idea to maintain the mental map by transitions yielded consistently positive feedback. In the

	all			SI			JX			2.5d			GUI		
transitions	6	7	3	4	0	0	0	0	0	0	1	1	7	1	2
highlighting	8	2	15	3	1	9	0	2	1	4	0	0	0	0	1
SNA measures	5	0	12	1	0	0	0	0	0	4	0	0	0	0	2
(s-c) layout	8	3	18	4	0	0	0	0	0	0	0	0	16	0	1
relations	0	0	10	0	0	0	0	0	0	0	0	0	0	0	0
info on demand	5	4	15	0	0	0	0	1	0	8	0	0	4	0	2
navigation	3	0	3	1	1	0	2	0	2	10	2	5	0	0	0
graph comprehension	2	0	0	1	0	1	1	0	2	3	0	1	1	0	1
trajectories	0	0	0	0	0	1	1	0	2	6	0	1	0	0	1

Fig. 5 Frequency of observations which feature problems (*red, left hand side of each column*), positive feedback (*green, center of each column*) and ideas for improvement (*blue, right hand side of each column*)—which could be identified for all views, for single views (i.e. SI-, JX-, and 2.5D view) and for the GUI itself

case of the highlighting feature, participants recommended additional interactions to make comparisons easier by highlighting more than one node at a time.

SNA measures of nodes like centralities were always double coded by size and by color in the prototype. Many users expressed the wish to have more freedom in selecting how these measures are displayed, and they preferred to use their favorite color palette. This applies for the relations too, where different types of relations should be visualized by different visual features like color or line style.

Concerning the main views, the juxtaposition view (JX) was rated as the most comprehensible by users comments and we detected the least problems in this view.

In the superimposition view (SI), participants mainly struggled following transitions and dealing with visual information overload. We will describe these problems in detail in a later section.

For the 2.5D view, users reported navigational problems as being too slow, not responsive enough and they missed an immediate feedback of the prototype when they zoom, pan or rotate. Most users suffered from perspective distortion when comparing node sizes and they mentioned legibility issues since the node labels and tool tips were distorted to a high extent in the two middle layers. Many users also mentioned a visual information overload as soon as many of the (too boldly styled) trajectories were displayed in 2.5D view.

Regarding the GUI, most users reported serious problems in understanding some of the labels, especially those of the dynamic views. Seven of eleven users reported (all of them no native English speakers) comprehension problems for the naming of the views (mainly "Superimposition" and "Juxtaposition"), two proposed to use icons instead of names. Only one person made sense of all the chosen view names. When dealing with user feedback seriously, this could be also seen as a hint that the untested transfer of technical terms (here: from the InfoVis community) via a prototype to an audience without that specific domain knowledge could have a negative impact on usability.

Aside this summary of problem oriented feedback, all users focused on implementation and in general, nearly all participants expressed a remarkably positive assessment of the prototype in their overall summary.

Table 3 List of tasks for the prototype evaluation. The tasks were named according to the scheme proposed by Ahn et al. [1]

Task mean (stddev)	T1	T2	T3	T4	T5	T6	T7	Overall
Correctness %	62 (51.75)	100 (—)	100 (—)	88 (35.36)	100 (—)	88 (35.36)	88 (35.36)	89
Confidence %	40 (53.45)	75 (46.29)	100 (—)	100 (—)	100 (—)	71(48.79)	71 (48.79)	80

3.2 Task Completion Analysis

We used two indicators to assess the effectiveness of our prototype's features in supporting users to solve assigned tasks: correctness and confidence. The correctness is defined as the conformity of user's answer to the answer we obtained by numerical methods and, for certain tasks, also by our previous knowledge of the real-world network at hand. The confidence differentiates between affirmative certain answers, and uncertain answers expressed in vague forms (e.g. "I would say", "I guess", "I am not sure"). We disregarded the task completion time, because we were more interested in the reasoning process, and asked users to think aloud and explain how they conceived the answer rather than to give the fastest answer.

The overall correctness of the answers was 89% (see Table 3). Half of the incorrect answers were given to task 1, but they might be ascribed to the task openness (without any default settings of the view and other parameters) and to the fact that it was intrinsically hard to solve, demanding the detection of a very slight variation of the network density. As for the confidence, 82% of the correct answers overall were also certain answers, with the highest value for task 3 and task 5, and the lowest also in this case for task 1.

As a general conclusion, we observed that most of the users were able to provide correct, complete and confident answers for task 2 to task 7 (see Table 3), mostly by using the combination of visual, analytical and interactive options we had set, with noticeable exceptions and unexpected behavior that we discussed (see section about multiple problem solving strategies). For some users, their performances on given tasks also affected their initial preference about a given view, for example some users initially were skeptical about the 2.5D and the SI views, but changed their mind after they realized they had been able to solve task 6 and task 7 by using them.

4 Pathways of Development

To illustrate how the process of prototype development was related to participatory aspects and the results of the final evaluation outlined above, we want to use this section to trace some pathways of development in detail. The first one will be referred to as maintenance of the mental map, the second one as avoidance of visual clutter, and the third one is following the implications of multiple problem solving and unexpected user behaviors.

4.1 Maintenance of the Mental Map

Within the context of dynamic network visualization the general visualization principle of "preserving the mental map" [32] predominantly refers to the challenge that the layout randomness, which is introduced by random steps of spring embedder algorithms, has to be brought under control. Starting with network data of a given time point, force directed layout algorithms usually generate node-link arrangements, that are driven by the overall aim of stress minimization (or majorization). This procedure reliably reproduces global patterns like clusters or local configurations like node neighbourhoods, but still could be realized by infinite specific detail arrangements, all solving the overall equation of stress minimization. This means, that even two instances of a barely evolving network tend to look quite different—if no further methods of layout preservation take care for visual comparability.

To still allow for the visual analysis of network dynamics, the spring embedder layout of a second instance has to be coordinated with the first layout solution, so that the mental map, which a user generates when viewing the first instance, could be preserved and leveraged to also analyze (stabilities or changes) within the second or third instance. Hence the sequence of layouts of the different instances of an evolving network has to provide a minimum amount of graph stability, whereas structural changes and the shifting of single nodes (due to a consistent layout solution at a certain time point) should not be overly suppressed. This means that an appropriate trade-off between inter-time stability and layout consistency has to be found [38]. The solution which was implemented in the prototype at hand allows the user to control this balance by herself—depending on the data and tasks which are at hand [15].

Beyond that solution, the basic requirement of maintaining mental network maps was generalized and pursued as an overall aim for all cases of interactions, which re-arrange the structure of a node-link diagram. This led to the implementation of three kinds of methods which maintain the mental map within the linked view architecture of the (superimposed, juxtaposed or 2.5 dimensionally stacked) time panels of each dynamic view:

- Maintain the mental map over time: aside the dynamic layout control mentioned above, a continuously running real-time layout provides smooth structural transformations after all kinds of user triggered structural changes.
- Maintain the mental map throughout user interactions: implemented methods include the coordinated highlighting of single nodes or neighbouring nodes after hovering a node on any panel (i.e. the visual linking of the same nodes at different time points), as well as coordinated positional shifts after dragging & dropping nodes on a single panel.
- Maintain the mental map amongst the three different views: a feature of smooth transitions was implemented, which allows for animated transits from one view to the other.

The basic idea of this general line of development was evaluated considerably positive. Suggested improvements were mainly addressed to detail or implementation issues like the speed of the continuous layout, the duration of its re-stabilization or of the transitions between views. Still the continuously provided visual integration and visual feedback, which arises from the combination of (A) and (B) was consistently rated positively. The highlighting function was appreciated for connecting different instances of evolving nodes or patterns across the time layers—hence helping to strongly reduce visual work. When the mock-up-study showed that finding the same node on other layers was quite time consuming (even in spite of the given layout stability), the prototype feature of linked highlighting (red for the focal node, hovered on any layer, yellow for all neighbouring nodes) solved this problem entirely.

Similarly, the method of smooth transitions between different views was consistently rated as supportive for the understanding of the operational principles of the different dynamic views. On the one hand, the way how the display architecture of a view is working, could be inferred just from observing the smooth transitions and how layers are visibly re-arranged. As one participant of the prototype evaluation put it, in the case of being new to the tool, the transitions could save hours to be spent with reading a manual otherwise. On the other hand, several subjects pointed out, that this feature should be made optional for the purpose of daily use, where the mental maps of all views already would have been successfully deployed. Aside these functional evaluations, the transitions of the tested prototype version were rated as being too slow for efficient use.

By analysing strategies adopted by users to solve the assigned tasks, we conclude that the techniques implemented to maintain the mental map were in general working effectively. For example, considering the mental map preservation amongst views (C), we looked at task 3, 4 and 5. These three tasks have a sub-task in common, namely finding a certain node by visually inspecting the network. Predefined settings provided JX view for task 3 and task 4, and 2.5D view for task 5. Even if we have not explicitly measured the task completion time, the 'finding' sub-task resulted to be much harder in 2.5D, because of perspective distortion of the node-link diagrams, according to users' oral feedback. We observed that some users reminded the position of the user to be found in task 5 from previous explorations in a different view, and this supports the idea that not only the mental map, but also the learning curve is somehow preserved amongst views. Moreover, one user switched to the JX view to find the requested node, and then back to the 2.5D to track its temporal evolution; this observation suggests that the mental map is also preserved when switching views for solving subtasks of a more complex task.

4.2 Avoidance of Visual Clutter

For static node-link diagrams, there is no a priori criterion for determining topological or geometric properties, but several "good" layout approaches have been proposed based both on computation and comprehension aspects [36]. Much research is based on optimizing the graph layout to enhance perception and comprehension [17] like minimizing edge-crossing, preserving symmetry, minimizing edge bends, minimizing edge length. There is also a research trend focused to optimize consumability of huge networks [44].

Techniques to avoid clutter for static graphs are a pressing issue even for small dynamic networks with 30–50 nodes since dynamics could multiply the information to be displayed by time steps and relations over time. In the following section we want to describe our efforts and insights at some decisive points during the development process.

With the help of the mock-up study, we wanted to gain first insights into how the perception of visual clutter (for a definition see [37]) can be influenced by different layouts and where comprehension or interpretation problems arise if information was hidden or compressed to reduce clutter. As described earlier, we used only two time steps for the construction of the mock-ups and about 35 nodes—but even in this case participants frequently reported clutter problems ("There are so many lines, I can't see anything").

Our approach consisted of two analysis stages: At stage one, we collected some basic behavioral indicators that have a relation to visual clutter, answering questions how easy a single node can be found, how easy its number of relations can be compared and how cognitively demanding this comparison process was. This behavioral data was analysed by using data of users viewing behaviour with eye-tracking methods. On the second stage, we looked for more subjective measures and analysed users feedback and their reported problems. In contrast to the first stage, also more complex tasks like structural analysis could be taken into account and presumably more top-down processes in users cognition were involved.

Based on some selected results of the eye-tracking data, we want to provide a first overview about the viewing behaviour of our participants with emphasis on visual clutter. We selected a task where the relational dynamics of the actor Leonard (Le) should be analyzed (see Table 1, tasks 6–8) for the three mock-ups with three different network questions. We choose one of the simplest, least demanding tasks to analyse effects of clutter on a near to perception level and to be able to include non-experts who had no problem to solve these basic tasks. Another advantage to select simple tasks for this analysis was that all participants used the same strategy to solve this task (see next section).

The first sub-task was to find the actor Leonard within the node-link diagram (see Table 4). For this and further analysis, we used a subset of the test sample consisting of experts and non-experts (n = 18), leaving out the group of involved participants to prevent biases caused by the usage of previous knowledge.

Interestingly, the median duration to find Leonard with the help of the JX and SPOCC plot were similar and higher than in the comet plot. Possible explanations could be that the JX plot is small sized, since two separated networks require more space than two merged networks. The SPOCC plot has less lines, but the coloring might be responsible to have caused some distraction, since there is evidence from other graph based eyetrackinb studies that lines are mainly ignored during the search

Table 4 Amount of time to find Leonard (n = 18); medians for experts non-experts

	Median duration to 1st fixation on Leonard in milliseconds	Median number of fixations needed to find Leonard
JX	3,043.0	12.5
SPOCC	3,457.5	12.0
Comet plot	1,693.0	7.0

Table 5 Glance durations on the Area of Interest (AOI) around Leonard in milliseconds (Sum of fixation duration on node Leonard and Leonard's direct neighbourhood), n = 18

	JX_left	JX_right	JX summed	SPOCC	Comet
Median	4,535.2	6,326.7		8,702.5	7,577.7
Mean	6,762.4	8,059.8		11,189.3	11,427.2
stddev	5,125.1	5,536.2		7,624.8	10,318.9
Sum			237,155.2	179,029.5	182,835.8

process [26]. The comet plot might be the one where the actors are most salient, since the comet tails emphasize the nodes visually. A problem with this analysis could have been learning effects, but it has to be noted that the participants had the chance to become acquainted with the layouts and have seen and analyzed similar structures in previous tasks.

After the users had found Leonard, the next sub-task was to compare Leonard's connections for two time points. To gather insights into the effort needed for this sub-tasks, we summed all fixation durations on the node Leonard or Leonard's direct neighbourhood.

The SPOCC and the comet plot showed only small differences in the descriptive statistics compared to the JX plot. At the JX plot, mean durations at one AOI were clearly shorter which is a positive result for the readability of the network. This finding is also in line with users feedback. To make a fair comparison, we have to sum the left AOI and the right AOI in JX plot to compare the sub-tasks with the other plots (see Table 5, row at the bottom). In summary, the amount of time needed to solve the sub-task is clearly higher for the JX plot, possibly due to the additional effort to close the lateral gap (see also Fig. 6, upper plot).

In Fig. 6 a comparison of the scan paths for each plot is shown—on the left the original plots, on the right the same plots with paths overlaid. The thick red lines denote the saccades or jumps of the eye, red blobs denote fixations and the size of the blobs their duration. In the upper plot (JX plot), there are three main visual attractors: actor Leonard on the left side and on the right hand side and the legend on the lower right corner. Concerning the fact that the legend is frequently used—which is a typical viewing behaviour when viewing graphs [42]—interestingly, only the legend on the right side was used. Most of the smaller red blobs are short fixations during the initial scanning for the label "Leonard". During this scanning process, both networks where scanned to find Leonard, but all of the 18 participants have first fixated the AOI of Leonard on the left network.

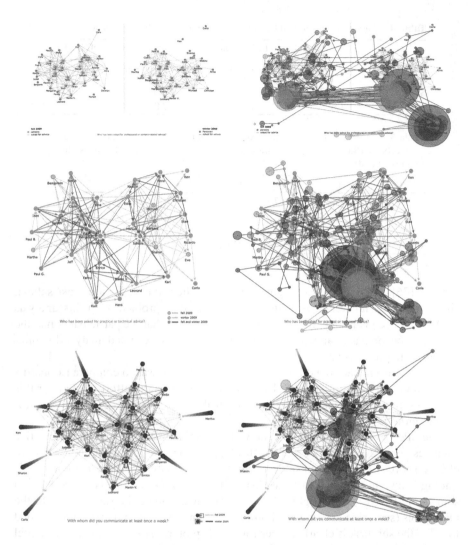

Fig. 6 A comparison of JX plot (*upper figures*), SPOCC plot (*middle figures*) and comet plot (*lower figures*). On the left hand side, the original plots are displayed, on the right hand side, scanpaths are shown as overlays. *Red lines* denote scan paths, blobs the fixations and blob's size the fixation duration. The paths of the first half of the group of participants is shown (n = 9), since the plots for the other half differed slightly to vary the experimental condition

As noted earlier, the size of the red blobs indicates the fixation duration, which is also a marker of cognitive effort [24]. We analyzed also the fixation duration statistically, since "a longer fixation duration indicates difficulty in extracting information, or it means that the object is more engaging in some way." [36]. We have found significant differences between the three mock-ups, for experts as well

Table 6 Median fixation
duration in milliseconds

	Experts	Non-experts
JX plot	318.0	378.0
SPOCC plot	330.0	318.0
Comet plot	252.0	278.0

as non-experts: statistics showed shortest durations for comet plot and with a clear difference to the two other plots (experts: c2 (2, N = 720) = 12.54, p = 0.002; non-experts: c2 (2, N = 530) = 9.19, p = .01), longer durations for both SPOCC and JX plot (no significant difference for experts c2 (1, N = 488) = .23, p = .63, but significantly shorter durations when non-experts used the SPOCC plot: c2 (1, N = 388) = 4.12, p = .042). In Table 6, the median fixation durations are presented, with indicators for the demand of more cognitive effort for SPOCC and JX plot.

To sum up the results of eye-tracking data, the comet plot seems to have some advantages over the two other plots when used for simple, very elementary tasks. Especially for the SPOCC plot we found hints for its high visual density (corresponding to high cognitive requirements) during the scanning process (sub-task 1) and sub-task 2.

In the second analysis stage, we also assessed think-aloud notes and feedback from the participants about their impressions when working with the three mock-ups: For the comet plot (see Fig. 4 for a zoomed view, left hand side) overlay problems were reported frequently: One problem arises when a node doesn't change its position over time. In this case the node at t1 (orange circle, see Juliette) was hidden by the node at t2. This kind of projection problem was difficult to grasp since it was unclear where the orange relations belonged to and was mentioned by many users. Users also reported that in case the node changed its position over time, some relations at t1 got masked by the trajectories. It has to be noted that, as stated earlier, users also frequently reported to have general comprehensibility problems and needed more time to figure out how to interpret the dynamics than in the other layouts.

Regarding the mock-up of the SPOCC plot (see Fig. 4, right hand side), users frequently reported problems to follow the coloring scheme. The aim of color-coding was to avoid visual clutter by using colors to compress information. Exemplarily we presented only one blue relation instead of two relations if the relation was stable. But color-coding gets easily difficult when used in network dynamics. For example, if A relates to B in t1 (orange) and B relates to A in t2 (green), we have to display two relations (or define a new color for it, and this is not the only combination). For this mock-up we decided to overlay such relations with some transparency (alpha = .7), but many users mentioned that it is stressful to deal with these "brownish lines". An example where an orange and a green line are mixed can be found in Fig. 4 (right hand side) with one of Juliette's relations coming nearly vertically at 1 o'clock from top. Users had to find the arrows to decompose the direction. These findings, the knowledge that more time points will make the whole topic even more complex, and the observation that many non-expert users had difficulties to find structural changes by color based macro-reading led to the decision to close this branch of development.

From a design point of view, we gained the impression that users easily get into problems with clutter in SPOCC plot but also the comet plot, where many users reported to have troubles to understand the visualization at first glance and solve more than basic tasks. So we identified color coding of ties to visualize dynamics as a dead end road concerning the design decisions of the mock-up study, since it worked neither for simple nor for complex tasks. Regarding the indicators of movements with tails or trajectories, we decided to rework the visualization. For prototype implementation, where interaction components can help to reduce clutter, we therefore recommended the strict motto to hide as much information as possible and make it available on demand only. But how much and which information can and should be hided to be beneficial and where are the drawbacks?

In the final prototype evaluation, the superimposition view—initially displayed with temporal trajectories like the comet plot, but without relations—got reasonable positive feedback regarding clutter by many experts at first sight. However, some of them stated that this reduced view does not provide enough information to be interpreted safely since no relational information is available. Hence they are forced to rely on movement information alone, which was seen as a general drawback of this visualization. The interaction feature to highlight and show previous temporal relations on demand, was therefore highly welcome for most participants, but seen as too limited, since there was no opportunity for multiple selection.

To sum up the participative design of the SI view, both strategies—to show too much or too little relations—have been criticized due to specific advantages and disadvantages. As color coding of ties turned out to be very difficult to be used for differentiating temporal dynamics only additional interaction methods will be able to deliver the basis for a user-controlled solution. Such methods would help to control the current amount of complexity to be shown (from no relations up to all relations), as well as continuously adjustable node highlighting (from single nodes and their relations up to multiple node selections), as well as various graph lenses [5]. In our view, these methods will have to be combined and further fine tuned to meet users full acceptance.

By providing this description how to deal with visual clutter, we wanted to provide some insights about the difficulties, possibilities, and limitations to find appropriate pathways through the methods and design space in the realm of dynamic network analysis by the means of empirical user feedback during the development process.

4.3 Multiple Problem Solving Strategies

Concerning the collected data of the final prototype tests, we analyzed the think-aloud audio recordings and the prototype interaction screencasts, using a categorization scheme during observation to extract information how their interaction is related to their insights and categorizing the different solution strategies for every task. This is comparable to the work of [31], who used a similar procedure with insights analyzing open tasks. We found interesting empirical results besides correctness and confidence of users' answers: we noticed multiple problem solving strategies spanning both tasks and users, pointing out relevant differences in either the alternative or the combined use of several prototype features.

The first empirical result refers to the integration of visual and analytical methods and their balance. When addressing task 1, that was the toughest to accomplish and registered the lowest correctness, most users looked for or asked for an analytical method directly providing the numeric answer (that was actually missing, since the given SNA component computes only node-level measures so far, and does not provide any network-level measure such as density). Also for task 6, one user said that a binary table would have helped her in tracking nodes' presence more than any visualization. Conversely, we noticed an opposite and unexpected behavior for task 3 and task 4. Task 3 required to compare the degree of a given node over time, and these values are mapped to the color and size of nodes, by default settings; task 4 required to find the out-degree of a given node at a given time, and this value pops out as a numerical tooltip on mouse over, by default settings. By analyzing these tasks, we observed that some users disregarded the analytical hints and preferred to find the (out-)degree just by counting the adjacent nodes, with the help of the highlighting interaction. This observation would lead us to infer that users prefer to visually solve those tasks they think they can manage, and to have recourse to analytical methods for harder tasks. It is worth noting that for task 4 users who counted were as confident as users who looked at the numeric tooltip, but the former were less often correct than the latter. The analysis of task 5 showed us another interesting user behavior: after finding the sign of the variation of the eigenvector centrality by looking at the provided visual mapping of the analytical value (color and size of the trajectory), some users rotated the 2.5D view or switched to the JX view in order to verify whether the network topology was compliant and confirmed the answer they gave.

Another interesting result, which we found from the analysis of the task completion procedures, concerns the recourse to different views to solve the same task. For task 6, for example, we provided a predefined setting with the 2.5D view and all trajectories activated. Most users solved the task looking at the interruptions of trajectories (an interruption on the left side indicates a node who has joined the network, an interruption on the right side indicates a node that has left the network, and an interruption in the middle of two trajectory segments indicates a short leave). In answering the second part of the task, about the relational consequences of these changes, some users switched forth and back to other views, while others kept on using the predefined view and 3D navigation. For task 7, some users switched from the predefined SI view to the 2.5D view, where they looked at the slopes of trajectories to investigate movements of nodes. In Fig. 7 we can compare the exploration strategies of two subjects dealing with task 1. This figure not only shows a temporal view of the interaction logs (i.e. which views and which interactions were used or performed and when), but also task lengths and occurrences of insights. In this context, an insight is meant as a guess, a partial answer (in the sense of knowledge-building insights; see [8]), or any additional

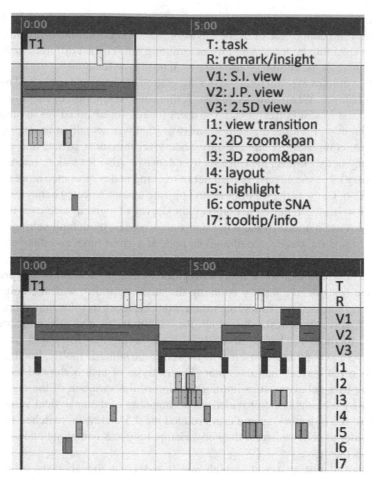

Fig. 7 Interaction graph for task 1 of the prototype test. While the first two rows show task duration (*gray, row T*) and occurred insights (*light yellow, row R*), rows 3–5 show the usage of the main views (SI = *blue* = row V1, JX = *green* = row V2, 2.5D = *red* = row V3), whereas the remaining rows show the usage of different interaction and exploration techniques

remark. In particular, the first two rows show the task duration (gray) and the sequence of insights (light yellow). Then, the following three rows correspond to the three views: superimposition (blue), juxtaposition (green), and 2.5D (red). Finally, remaining rows show the sequence of different types of interactions: transition between views (purple), 2D navigation (light green), 3D navigation (light blue), change of layout stability (pink), highlighting nodes and/or their trajectories (light purple), computing an SNA metric (orange), and showing additional details in the info area or in the tooltip (light orange). Comparing the exploration behavior of these two subjects, namely Subject A (top) and Subject B (bottom), we see that A was faster, used as few interactions as possible and solved the task straightforwardly.

Conversely, B seemed to exploit task 1 to play with the tool, exploring most of its views and interactions. It is worth noting that both were possible expected behaviors, given the openness of the first task.

In Fig. 8 we compare the exploration strategies of the same pair of subjects dealing with tasks 2–7. While the correctness of answers is the same (100%) and completion times are similar, we can identify very different patterns of interaction and few similarities. Overall, we note that subject A switched views quite often, while subject B never changed the pre-defined view that was offered by the experiment setup. Moreover, A never performed 3D panning or rotation, which were used quite often by B; conversely, A made an intensive use of the tooltip while B used it very seldom. Nevertheless, there is a similarity in the layout adjustment, which was performed by both users only for task 2 (clusters and their stability) and task 7 (shifts from core to periphery).

Looking in detail at specific tasks, we also find more differences than similarities. When dealing with task 2, for example, subject A started the analysis at a local level with a lot of 2D zooming and panning and then some layout adjustments, while subject B set the layout at first and then analyzed the network at a global level, during a long visual reasoning phase without any interaction, besides some highlighting in the end. For task 4 (node outdegree), both subjects had the same interaction pattern, but with a difference: A gave an answer only after reading the SNA value in the tooltip, while B counted highlighted nodes and gave the correct answer, then used the tooltip to confirm it. Tasks 5 and 6, whose predefined view was the 2.5D, also showed differences: subject A explored the view by highlighting nodes and trajectories, while subject B navigated it in the 3D space. For task 5, in particular, we can see as both subjects launched an SNA computation, but A looked at the numeric value in the tooltip, while B looked at its visual mapping (as explained above).

In general, considering all subjects of our study, we found that the different views complement each other, and the 'best' view does not depend only on the data and task, but also on users, who might have different strategies even if they belong to a homogeneous group with a common background. Furthermore, even a single user dealing with a single task, might find beneficial switching from one view to another to ensure the correctness and/or the completeness of her insights. Similarly, for many tasks there is no perfect choice between only visualization (node-link diagrams) and only computation (numbers and tables), but the best choice is to integrate both of them to support the visual analytics reasoning process. This approach enables the user to exploit his/her preferred problem solving strategy at best and to gain complex insights from a multifaceted methods approach to network analysis. A possible disadvantage is that beginners might choose a wrong way, and in more complex cases training might be needed to help them switching to the fastest and most accurate strategy, but in general flexibility and multiple options seem to be beneficial.

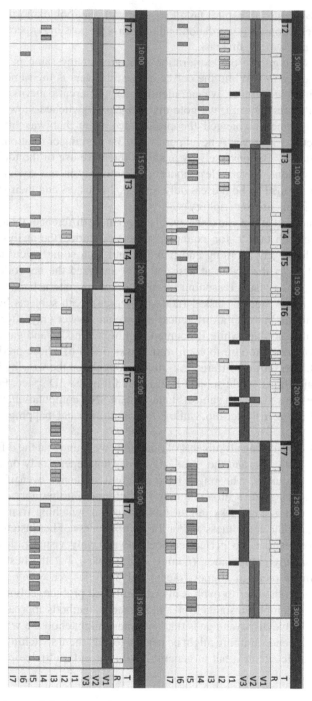

Fig. 8 Interaction graph for the tasks 2–7 of the prototype test

5 Conclusions and Future Research Questions

We have presented the main results of the evaluation of a visual analytics approach to dynamic network analysis. After the description of the features of the existing prototype, we provided insights into its participatory development process and focused on the results of a final prototype evaluation. The main contribution of this approach is to be seen in its consistent focus on users of visual analytics methods as one of the crucial factors for a method's utility and success in real world tasks and data scenarios.

In case of the prototype we examined, this evaluation approach provided a fine grained matrix, showing specific strengths, weaknesses and further development suggestions. Aside from technical implementation issues, the continuation of the outlined issues is obvious following three main strands we have pointed out:

Maintenance of the mental map: as a general aim, the extended preservation of the analysts mental map is to be seen as one of the major challenges for future developments of complex visual analytics methods and technologies. The aim is to free the analysts cognitive capacities from visual efforts of searching, matching or navigating, but focusing it earlier on intended tasks of pattern analysis and exploration. Still, as our results show, the control of the amount of mental map maintenance should be made optional for the purpose of daily use, where it could have also been already successfully deployed as cognitive scripts and schemes.

Avoidance of visual clutter: from our empirical results, visual clutter turned out to be a decisive aspect when developing visualizations for dynamic network analysis, even for smaller networks and a small number of time points. For color-coding of temporal aspects we could not derive a satisfying and sustainable solution. Interactions could be of help to avoid cluttering, but they have to be carefully aligned with users interactions patterns and exploration behavior. Extended interaction techniques might include lens techniques or other smart interactions to open up more space for all that ink that temporal visualization requires.

Multiple Problem Solving Strategies: we observed many different problem solving strategies in a small group of subjects, sharing the same skills and coping with the same task on the same data. In particular, subjects made different recourse at either visual or computational methods. From this observation, we can preliminary infer that a seamless integration of several views and computations in a consistent framework seems to give better results than optimizing the design of a single technique. As a result, we want to advocate the implementation of multiple problem solving options and methods into complex information visualization tools and technologies, for only this strategy seems to be able to cope with the diversity of future users and their (evolving) tasks.

Acknowledgements This research was supported by the Austrian FFG research program FIT-IT Visual Computing (No. 820928, Research project ViENA, Visual Enterprise Network Analytics, http://fitit-viena.org) and the Centre for Visual Analytics Science and Technology (No. 822746, CVAST, http://www.cvast.tuwien.ac.at) funded by the Austrian Federal Ministry of Economy, Family and Youth in the exceptional Laura Bassi Centres of Excellence initiative. For their support of the prototype evaluation we want to particularly thank the algorithmics research group at Konstanz University, Department of Computer and Information Science, Prof. Ulrik Brandes.

References

1. Ahn, J., Plaisant, C., Shneiderman, B.: A task taxonomy of network evolution analysis. Tech Report HCIL-2011-09 (2011)
2. Aigner, W., Miksch, S., Schumann, H., Tominski, C.: Visualization of Time-Oriented Data, 1st edn. Human-Computer Interaction. Springer Verlag (2011)
3. Andrews, K., Wohlfahrt, M., Wurzinger, G.: Visual graph comparison. In: Information Visualisation, 2009 13th International Conference, pp. 62–67 (2009)
4. Bertini, E., Perer, A., Plaisant, C., Santucci, G.: Beliv'08: Beyond time and errors: novel evaluation methods for information visualization. In: CHI'08 extended abstracts on Human factors in computing systems, pp. 3913–3916. ACM (2008)
5. Bier, E. A., Stone, M. C., Pier, K., Buxton, W., DeRose, T. D.: Toolglass and magic lenses: the see-through interface. Proceedings of the 20th annual conference on Computer graphics and interactive techniques (S. 73–80). ACM (1993)
6. Brandes, U., Corman, S.: Visual unrolling of network evolution and the analysis of dynamic discourse. Information Visualization 2(1), 40–50 (2003)
7. Brandes, U., Indlekofer, N., Mader, M.: Visualization methods for longitudinal social networks and stochastic actor oriented modeling. Social Networks (2011)
8. Chang, R., Ziemkiewicz, C., Green, T. M., & Ribarsky, W.: Defining insight for visual analytics. Computer Graphics and Applications, IEEE, 29(2), 14–17 (2009)
9. Chase, W. G., Simon, H. A. (1973). Perception in chess. Cognitive Psychology, 4, 55–81 (1973)
10. Correa, C.D., Ma, K.L.: Visualizing social networks. In: C.C. Aggarwal (ed.) Social Network Data Analytics, pp. 307–326. Springer US (2011)
11. Di Battista, G., Eades, P., Tamassia, R., Tollis, I.G.: Algorithms for drawing graphs: an annotated bibliography. Computational Geometry 4(5), 235–282 (1994)
12. Dwyer, T., Gallagher, D.R.: Visualising changes in fund manager holdings in two and a halfdimensions. Information Visualization 3(4), 227–244 (2004)
13. Eades, P., Lai, W., Misue, K., Sugiyama, K.: Preserving the mental map of a diagram. Technical Report IIAS-RR-91-16E, International Institute for Advanced Study of Social Information Science, Fujitsu Laboratories Ltd. (1991)
14. Ellis, G., Dix, A.: An explorative analysis of user evaluation studies in information visualisation. In: Proceedings of the 2006 AVI workshop on BEyond time and errors: novel evaluation methods for information visualization, pp. 1–7. ACM (2006)
15. Federico, P., Aigner, W., Miksch, S., Windhager, F., Zenk, L.: A visual analytics approach to dynamic social networks. In: Proceedings of the 11th International Conference on Knowledge Management and Knowledge Technologies, i-KNOW 11, pp. 47:1–47:8. ACM, New York, NY, USA (2011)
16. Friedrich, C., Houle, M.: Graph drawing in motion ii. In: P. Mutzel, M. Jnger, S. Leipert (eds.) Graph Drawing, Lecture Notes in Computer Science, vol. 2265, pp. 122–125. Springer Berlin / Heidelberg (2002)
17. Freeman, L. C.: Visualizing social networks. Journal of social structure, 1(1), 4 (2000)
18. Henderson, J. M.: Regarding scenes. Current Directions in Psychological Science 16, pp. 219–222 (2007)
19. Helsen, W. F., Starkes, J. L.: A multidimensional approach to skilled perception and performance in sport. Applied Cognitive Psychology, 13, pp. 1–27 (1999)
20. Herman, I., Melancon, G., Marshall, M.: Graph visualization and navigation in information visualization: A survey. Visualization and Computer Graphics, IEEE Transactions on 6(1), 24–43 (2000)

21. Isenberg, P., Tang, A., Carpendale, S.: An exploratory study of visual information analysis. In: Proceeding of the twenty-sixth annual SIGCHI conference on Human factors in computing systems, pp. 1217–1226. ACM (2008)

22. Isenberg, P., Zuk, T., Collins, C., Carpendale, S.: Grounded evaluation of information visualizations. In: Proceedings of the 2008 conference on BEyond time and errors: novel evaLuation methods for Information Visualization, p. 6. ACM (2008)

23. Jonassen, D.: Instructional design models for well-structured and ill-structured problem-solving learning outcomes. Educational Technology Research and Development 45(1), 65–94 (1997).

24. Just, M. A., Carpenter, P. A.: Eye fixations and cognitive processes. Cognitive Psychology 8(1), 441–480, (1976)

25. Kilduff, M., Brass, D.: Organizational social network research: Core ideas and key debates. The Academy of Management Annals 4(1), 317–357 (2010)

26. Körner, C. (2004). Sequential processing in comprehension of hierarchical graphs. Applied Cognitive Psychology, 18(4), 467–480. doi:10.1002/acp.997

27. Landauer, T., Prabhu, P., Helander, M.: Handbook of Human-Computer Interaction. Elsevier Science Inc. (1998)

28. Mahapatra, D., Winkler, S., Yen, S. C.: Motion saliency outweighs other low-level features while watching videos. In Society of Photo-Optical Instrumentation Engineers (SPIE) Conference Series, 6806, pp. 68060P1-68060P10 (2008)

29. Marchionini, G.: Exploratory search: from fnding to understanding. Communications of the ACM 49(4), 41–46 (2006)

30. Mayr, E., Knipfer, K., Wessel, D.: In-sights into mobile learning. An exploration of mobile eye tracking methodology for learning in museums. In G. Vavoula, N. Pachter, & A. Kukulska-Hulme (Eds.), Researching mobile learning: Frameworks, methods, and research designs, pp. 189–204. Oxford, UK: Peter Lang (2009)

31. Mayr, E., Smuc, M., Risku, H.: Many roads lead to rome: Mapping users problem-solving strategies. Information Visualization 10(3), 232–247 (2011)

32. Misue, K., Eades, P., Lai, W., Sugiyama, K.: Layout adjustment and the mental map. Journal of Visual Languages and Computing 6(2), 183–210 (1995)

33. Moody, J., McFarland, D., Bender-deMoll, S.: Dynamic network visualization. American Journal of Sociology 110(4), 1206–1241 (2005)

34. Perer, A., Shneiderman, B.: Integrating statistics and visualization: case studies of gaining clarity during exploratory data analysis. In: Proceedings of the twenty-sixth annual SIGCHI conference on Human factors in computing systems, CHI 08, pp. 265–274. ACM, New York, NY, USA (2008)

35. Plaisant, C.: The challenge of information visualization evaluation. In: Proceedings of the working conference on Advanced visual interfaces, pp. 109–116. ACM (2004)

36. Poole, A., Ball, L. J.:. Eye tracking in human-computer interaction and usability research: current status and future prospects. Encyclopedia of human computer interaction, pp. 211–219 (2005)

37. Rosenholtz, R., Li, Y., Mansfeld, J., Jin, Z.: Feature congestion: a measure of display clutter. In: Proceedings of the SIGCHI conference on Human factors in computing systems, CHI '05, pp. 761–770. ACM, New York, NY, USA (2005)

38. Saffrey, P., Purchase, H.: The "mental map" versus "static aesthetic" compromise in dynamic graphs: a user study. In: Proceedings of the ninth conference on Australasian user interface - Volume 76, AUIC 08, pp. 85–93. Australian Computer Society, Inc., Darlinghurst, Australia (2008)

39. Saraiya, P., North, C., Duca, K.: An insight-based methodology for evaluating bioinformatics visualizations. Visualization and Computer Graphics, IEEE Transactions on 11(4), 443–456 (2005)
40. Shneiderman, B., Plaisant, C.: Strategies for evaluating information visualization tools: multidimensional in-depth long-term case studies. In: Proceedings of the 2006 AVI workshop on BEyond time and errors: novel evaluation methods for information visualization, pp. 1–7. ACM (2006)
41. Springmeyer, R., Blattner, M., Max, N.: A characterization of the scienti?c data analysis process. In: Proceedings of the 3rd conference on Visualization'92, pp. 235–242. IEEE Computer Society Press (1992)
42. Trafton, J. G., Marshall, S., Mintz, F., & Trickett, S. B.: Extracting and implicit information from complex visualizations (pp. 206–220). In M. Hegarty, B. Meyer, & H. Narayanan (Eds.) Diagrammatic Representation and Inference. Diagrammatic Representation and Inference (2002)
43. Tufte, E. R.: The Visual Display of Quantitative Information. Graphics Press, Cheshire, CT (1983)
44. von Landesberger, T., Kuijper, A., Schreck, T., Kohlhammer, J., Van Wijk, J., Fekete, J., & Fellner, D.: Visual analysis of large graphs. Proceedings of Euro-Graphics: State of the Art Report, 2 (2010)
45. Wasserman, S., Faust, K.: Social Network Analysis: Methods and Applications, 1 edn. No. 8 in Structural analysis in the social sciences. Cambridge University Press (1994)
46. Ware, C.: Information Visualization - Perception for Design, Morgan Kaufmann (2004)
47. Whiting, M., Haack, J., Varley, C.: Creating realistic, scenario-based synthetic data for test and evaluation of information analytics software. In: Proceedings of the 2008 conference on BEyond time and errors: novel evaLuation methods for Information Visualization, p. 8. ACM (2008)
48. Windhager, F., Zenk, L., Federico, P.: Visual enterprise network analytics-visualizing organizational change. Procedia-Social and Behavioral Sciences 22, 59–68 (2011).

Part VII
Interaction

Interaction Taxonomy for Tracking of User Actions in Visual Analytics Applications

Tatiana von Landesberger, Sebastian Fiebig, Sebastian Bremm, Arjan Kuijper, and Dieter W. Fellner

Abstract In various application areas (social science, transportation, or medicine) analysts need to gain knowledge from large amounts of data. This analysis is often supported by interactive Visual Analytics tools that combine automatic analysis with interactive visualization. Such a data analysis process is not streamlined, but consists of several steps and feedback loops. In order to be able to optimize the process, identify problems, or common problem solving strategies, recording and reproducibility of this process is needed. This is facilitated by tracking of user actions categorized according to a taxonomy of interactions.

Visual Analytics includes several means of interaction that are differentiated according to three fields: information visualization, reasoning, and data processing. At present, however, only separate taxonomies for interaction techniques exist in these three fields. Each taxonomy covers only a part of the actions undertaken in

T. von Landesberger (✉)
Head of Junior Research Group Within the Graphics Interactive Systems Group at Technische Universität Darmstadt, Darmstadt, Germany
e-mail: tatiana.von-landesberger@gris.tu-darmstadt.de

S. Fiebig
Student at Technische Universität Darmstadt, Darmstadt, Germany
e-mail: sebastian.fiebig@gris.tu-darmstadt.de

S. Bremm
PhD Student at Technische Universität Darmstadt, Darmstadt, Germany
e-mail: sebastian.bremm@gris.tu-darmstadt.de

A. Kuijper
Private Lecturer at Technische Universität Darmstadt, Darmstadt, Germany

Research Coach at Fraunhofer IGD, Darmstadt, Germany
e-mail: arjan.kuijper@igd.fraunhofer.de

D.W. Fellner
Full Professor at Technische Universität Darmstadt, Darmstadt, Germany

Leading Graphics-Interactive Systems Group, Head of Fraunhofer IGD, Darmstadt, Germany
e-mail: d.fellner@igd.fraunhofer.de

W. Huang (ed.), *Handbook of Human Centric Visualization*,   653
DOI 10.1007/978-1-4614-7485-2_26, © Springer Science+Business Media New York 2014

Visual Analytics. Moreover, as they use different foundations (user intentions vs. user actions) and employ different terminology, it is not clear to what extent they overlap and cover the whole Visual Analytics interaction space. We therefore first compare them and then elaborate a new integrated taxonomy in the context of Visual Analytics.

In order to show the usability of the new taxonomy, we specify it on visual graph analysis and apply it to the tracking of user interactions in this area.

1 Introduction

Large amounts of data are analyzed in various disciplines such as economics, earth sciences, or social sciences. This analysis is often supported by interactive Visual Analytics tools [7, 11]. Visual Analytics combines techniques from data processing, information visualization, human-computer interaction, and human reasoning fields for efficiently gaining insights from the data.

Analysis of large data sets consists of several steps and feedback loops, which are interactively steered by the user (see Fig. 1). In this process, various tools are used to support gaining knowledge from the data. In order to be able to optimize the process, identify problems, or common problem solving strategies, process recording and reproducibility is needed.

The **reproducibility of the analytical process** is supported by tracking of user activities. The tracking records all actions taken by the user while interacting with the Visual Analytics application. The recorded actions can then be analyzed

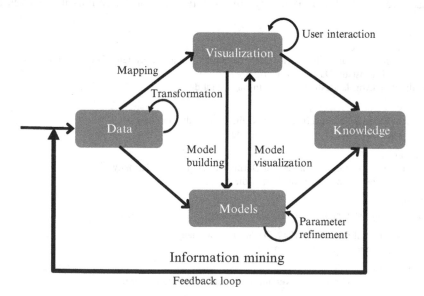

Fig. 1 Visual Analytics process by Keim et al. [8]

Fig. 2 Interaction in Visual
Analytics as integrated
interaction with visualization,
data processing, and
reasoning and among them

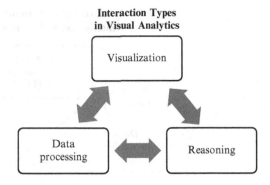

algorithmically or shown to the user for their visual inspection. It enables better understanding of analytical process. Moreover, the recorded actions can also be reused later for repeating the analytical process.

As Visual Analytics interactively integrates three main fields (data processing, information visualization, and reasoning), capturing user actions should cover all three areas (see Fig. 2). An important aspect is the unifying element of these areas by allowing interaction to take place between all three areas. The integration of the three interaction areas enhances the flexibility and ease when changing the analytical subtask or focus, without losing sight of the overarching analytical process. It also entails the seamless change from one part into another, such as from algorithmic data processing to visual exploration of the data and analytic annotation of the found insights. For example, after running a clustering algorithm, the results may directly be presented in a visual way and then can be further interactively explored in the view. Vice versa, changes of parameters in the view may directly invoke new algorithmic calculations. Thereby, the analytical loop is closed.

Tracking of an analytical process is based on the underlying **interaction taxonomy**. At present, however, only separate taxonomies for interaction techniques exist in the three Visual Analytics fields [1, 4, 13, 14] (see Fig. 3). None of them covers the whole Visual Analytics area, so the interwoven character of the Visual Analytics process cannot be captured. Their application in complex Visual Analytics systems is therefore limited. Moreover, they use different terminology so the analysis of their scope and overlap is very difficult.

We therefore present a new taxonomy for user interaction in Visual Analytics applications. For this purpose, we first compare the existing interaction taxonomies. The new taxonomy provides an integrated view on the analytical process. In order to show the usability of the new taxonomy, we specify it to visual graph analysis and apply it to tracking of user interaction in visual analysis of large graphs.

This chapter is structured as follows: Sect. 2 reviews and compares relevant categorizations of interaction techniques according to the three fields of interest to Visual Analytics. Section 3 presents our new taxonomy which alters and combines the described methodologies. Section 4 specifies the taxonomy to the visual analysis of graphs. Section 5 presents the application of the new taxonomy on tracking of the Visual Analytics process. Finally, Sect. 6 concludes the chapter.

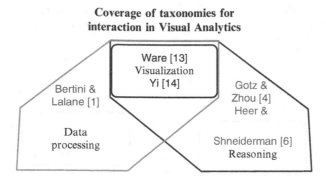

Fig. 3 Coverage of related taxonomies of interaction in Visual Analytics. Note that the *arrows* between the parts are missing. It symbolizes the missing integration between the three fields

2   Comparison of Related Taxonomies

In this section, we present relevant categorizations of interaction techniques according to the three fields information visualization, reasoning, and data processing, which are of interest to Visual Analytics.

2.1   Overview

The seminal study by Ware et al. [13] extensively described interaction techniques and levels of interaction in information visualization. Recently, four papers dealing with the theory of interaction in information visualization [14], reasoning and information visualization in sense of Visual Analytics [4, 6] and visual data mining [1] have been published. In the following, we present the taxonomies of all the above-mentioned studies and compare them (see Fig. 3).

2.2   Information Visualization

In information visualization, interaction plays a major role for gaining insight into the data via exploration and navigation. It provides also means for overcoming the scalability and complexity problems – e.g., occlusion, screen limitations and fundamental human perception limitations.

According to Ware [13], interactive visualization includes feedback loops of three classes from lowest to highest level:

1. *Data manipulation loop*: objects are selected and moved using eye-hand coordination.

2. *Exploration and navigation loop*: finding a way in the data space and thereby building a mental model of the data.
3. *Problem solving loop*: analyst forms hypothesis about the data and refines them through an augmented visualization process.

Exploration and navigation in information visualization include

- *View navigation*: changing view control, e.g., panning, walk-through,
- *Focus, context and scale*: moving between views on different scale including distortion techniques, rapid and semantic zooming, multiple windows, elision,
- *Rapid interaction with the data*: being in direct contact with the data, such as

  - *Change of data mapping*: e.g., mapping of data attributes to x and y axis, to shape, color, etc.,
  - *Change of data transformation for presentation*: e.g., changing of parameters of transfer function,
  - *Dynamic queries*, i.e., filter: limit range of data that is visible, and
  - *Brushing (and linking)*: highlighting subsets of data interactively (in multiple windows).

This simple but expressive taxonomy on an abstract level includes the main classic interaction techniques for data exploration and navigation. It does not distinguish clearly between these two types of actions. It is more centered on user actions, not on user intentions (i.e., it is a low-level taxonomy). This categorization is however in line with the information visualization reference model of Card et al. [2], where interaction actions follow the information visualization pipeline.

Yi et al. [14] present an extensive overview of interaction taxonomies, tasks and dimensions for information visualization. They propose a new taxonomy based on user intentions. It includes seven types of intentions and the associated techniques. In the following, we match this taxonomy with the taxonomy of Ware [13].

- *Select*: marks something as interesting. Corresponds to brushing in [13].
- *Explore*: enables users to examine a different subset of data, includes, for example, panning and direct walk. Corresponds to view navigation in [13].
- *Reconfigure*: provides users with different perspectives of the data by changing the spatial arrangement of representation. This includes sorting and rearranging columns, changing attributes assigned to x and y-axes, reducing occlusion (e.g., rotation, jitter). This intention corresponds to rapid interaction with the data in [13]. However, it includes techniques for solving occlusion by view navigation, which does not seem intuitive.
- *Encode*: alters the fundamental visual representation of the data including visual appearance. Although there is no direct correspondence to one of the items, it entails attributes of changes in data transformations for representation and data mapping.
- *Abstract/Elaborate*: adjusts the level of abstraction of a data representation, e.g., details on demand, focus on data, tooltips, and zooming. This intention corresponds to focus, context and scale in [13].

- *Filter*: changes the set of data items presented, e.g., dynamic queries. This intention corresponds to dynamic queries in [13].
- *Connect*: includes techniques used to (a) highlight associations and (b) show hidden data items that are relevant to a specific item e.g., multiple views that reveal items that are not directly shown (e.g., show children of a node in a graph). It does not have a direct correspondence in the taxonomy of [13], but it is similar to explore (also according to Yi et al. [14]).

In their paper, Yi et al. [14] mention further techniques that were not categorized such as undo and redo, change system configuration, threshold highlighting (could be part of select), semantic zooming (fulfilling multiple intents). Many of these techniques may not have been classifiable owing to their proximity with elements of interaction belonging to reasoning and to data processing.

In general, from the strong correspondence of the two types of taxonomies (by Ware [13] and Yi et al. [14]), it can be seen that the user intentions and low level interaction techniques correspond to a very large extent when talking about interaction in information visualization.

2.3 Reasoning

The taxonomies in information visualization focus on the manipulation of visualizations and therefore do not include further analytical (insight) elements such as annotation and change in view history. These features are deemed relevant for user interaction from an analytic (reasoning) point of view. They are mentioned in the taxonomies for Visual Analytics concentrating on reasoning and visualization [4] and [6].

Gotz and Zhou [4] create a taxonomy of user activities in the analytic process with three main categories: exploration actions, insight actions, and meta actions (see Fig. 4). Exploration actions are divided into data exploration (Filter, Inspect Query, Restore) and visual exploration (Brush, Change-Metaphor, Change-Range, Zoom, Pan, Merge, Sort, Split); insight actions into visual (annotate, bookmark) or knowledge-based (remove, modify, create) and meta actions include the following types: redo, undo, revisit, delete. Taken together, they define four distinct intents: (1) data change, (2) visual change, (3) notes change, and (4) history change.

We see that the exploration actions roughly correspond to the interaction actions discussed above [13, 14] with respect to information visualization. Moreover, with respect to analytic activity Gotz and Zhou enhance the types of interactions with insight and meta actions, which were mostly neglected before as they refer to the reasoning part of Visual Analytics. Both Yi [14] and Gotz and Zhou [4] use user intentions to characterize interactions.

Heer and Shneiderman [6] propose a taxonomy of interactive dynamics for visual analysis. It is composed of three main categories: Data & View Specification, View Manipulation and Process & Provenance. The first two categories deal with changes

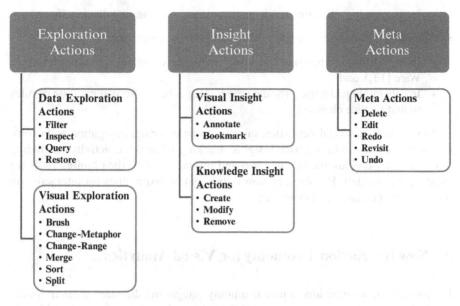

Fig. 4 The action taxonomy according to Gotz and Zhou [4]

in visual representation of the data (visualize data, filter, sort, derive data, navigate, select, link and organize views). The third category deals with reasoning actions (record analysis histories, annotate found patterns, share views and guide users through the analytical process). Interesting aspect is the inclusion of collaborative aspects of analytical process (sharing views and guiding users). In the reasoning part, it is similar to Gotz and Zhou's taxonomy.

2.4 Data Processing

All the above mentioned taxonomies do not consider interaction with data processing (data mining in particular) tools as presented in [1]. With regard to visual data mining tools, Bertini and Lalanne [1] discriminate between pre- and post model interventions to change the scheme or manipulating the scheme for both visualization and data mining. They thereby take a different approach to the categorization of interaction techniques. They look at whether the user action only changes the current state (tuning, change of parameters) or changes it completely (change of the scheme). Both means are presented as follows:

- *Manipulating and tuning*: change parameters within the context of a given scheme

    – In visualization: changing representation parameters (e.g., zooming, etc.), and

- In data mining: changing model parameters (e.g., changing distance function).

* *Changing the scheme*: changing the data model or representation

  - In visualization: changing the visual mapping or visual representation (see Ware [13]), and
  - In data mining: changing the data model (e.g., change from generation of rules to finding data clusters).

In summary, in Visual Analytics, interaction incorporates navigation and exploration as well as capturing user insights, tracking of analytic activity (including navigation in previous analytic actions) and interaction with data mining tools for creating data models. However, by now only separate taxonomies for interaction in individual fields have been presented.

3   New Interaction Taxonomy for Visual Analytics

In this section, we introduce a new taxonomy integrating the three areas in Visual Analytics – visualization, reasoning and data processing (data mining) (see Fig. 2).

We categorize user interaction according to the *action* that is taken by the user. This categorization is more suitable for dividing interaction techniques into categories than division according to user intention, as user actions can be directly measured by the computer (by mouse clicks). Additionally, as shown in Sect. 2, user actions correspond to user intentions in most of the cases, so they are a very good way of finding out about the analytical process.

Our unified approach covers all areas (Visualization (V), Reasoning (R), and Data processing (DP)). In each area it consists of two main subcategories: changes in the data and changes in the respective representation (see Fig. 5). The *changes in the data* affect the presented or underlying data set and *changes in representation* refer to other forms of interaction. Changes in the data are divided into two subcategories: changes affecting the selection of the data set (in particular filtering) and changes to the data set introduced by the user (e.g., by editing the data or by annotating). The representation changes are divided (in line with Bertini et al. [1]) into changes of representation parameters and changes of scheme.

This simple but powerful classification allows us to highlight the corresponding elements in user actions across the three areas. Moreover, user actions can be grouped into a hierarchy with two consistent main categories across areas. Note that this categorization is in line with the information visualization reference model of Card et al. [2].

Please note that these types of interaction are often closely connected. For example, data manipulation may automatically lead to changes of visual parameters (e.g., data filtering can influence the graph layout, or zooming can be combined with data filtering forming a type of semantic zooming).

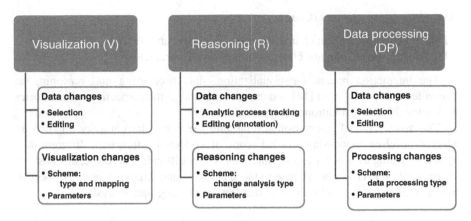

Fig. 5 A new unified interaction model showing two analogous interaction means with visualization (V), reasoning (R), and data processing (DP) as well as their interplay

V Visualization

V1 *Data changes*

V1.1 Selection: includes filtering (dynamic queries), details on demand
V1.2 Editing: data editing via direct manipulation

V2 *Visualization changes* (Changes in representation)

V2.1 Scheme changes: include inter alia change of visualization type or visual mapping.
V2.2 Parameter changes: include view changes (e.g., zooming, panning) or change of the used color scheme

R Reasoning

R1 *Data changes*

R1.1 Analytic process tracking
R1.2 Editing: annotation

R2 *Reasoning changes* (Changes in representation)

R2.1 Scheme changes: change type of analysis
R2.2 Parameter changes: undo/redo, etc.

DP Data Processing

DP1 *Data changes*

DP1.1 Selection: filtering of data used for data processing
DP1.2 Editing: creation of new data through data processing (e.g., data transformation)

DP2 *Processing changes* (Changes in representation)

DP2.1 Scheme changes: change of processing algorithm type
DP2.2 Parameter changes: change of algorithmic parameters

The interaction means for visualization, data processing and reasoning as presented in [1, 4, 13], and [14] and can be assigned to the respective subcategories following the above mentioned categorization.

The assignment of interaction actions into the particular category on such an abstract level encompassing a broad scope of systems dealing with different data types, including a broad variety of functions is very difficult. Therefore, this abstract categorization needs to be adapted to the specific Visual Analytics areas. In Sect. 4, we provide an example of such specification for visual graph analysis.

4 Specification of the Taxonomy for Visual Graph Analysis

Visual Analytics covers a broad range of areas where large data sets need to be examined. Many use cases deal with graph data. For example, social scientists analyze social networks, economists examine shareholder networks, or researchers explore relationships between publications for finding trends in literature.

Graph analysis includes specific tasks. There is a need to understand global and local structure of the graph, to examine connections between entities, or to find clusters of highly connected entities. Such analysis requires a combination of interactive visual presentations with algorithmic graph analysis methods. The analysis process often consists of several steps where user reasons about the data at hand.

Owing to the specific needs of visual graph analysis, the general taxonomy of user interaction needs to be specified. We present such specification based on the recent review of visual graph analysis approaches [9].

4.1 Visualization

Visualization is one of the main means of exploratory graph analysis, where the users can interactively navigate and change the data space as well as adapt the visual representation to their needs.

V1 *Data changes*

V1.1 *Data selection:*
The users can choose which data to show on the screen. Specifically for graphs, the filtering mechanisms are applied to either nodes or edges. Moreover, the users can apply graph algorithms for expanding/shrinking the current data selection (e.g., showing/hiding children of a selected node).

V1.2 *Data editing:*
    Data editing changes the data set used for the analysis. In visual graph
    analysis, there are two main ways of data editing:
    V1.2a Graph editing: adding/removing nodes or edges from the graph. Note
    that using graph algorithms with node deletion also implies deleting or
    redirecting the adjacent edges.
    V1.2b Graph aggregation: In graph aggregation, the user can select a group
    of nodes, or run a graph clustering algorithm making a group of nodes, which
    are then merged into one "supernode" as well as merging the corresponding
    edges into a "superedge". In this way, a new graph for analysis is created.

V2 *Visualization changes*

V2.1 *Visualization scheme*

V2.1.1 *Visual representation:*
    The users can change the currently employed visual representation. They
    can switch between matrix or a node-link diagram.
V2.1.2 *Visual mapping:*
    The users may, according to their analytical needs, map various data
    attributes to node and edge representations such as size, shape, or color.

V2.2 *Visual parameters*

V2.2.1 *View change:*
    The users can adjust the view on the data by navigating in the view (e.g.,
    zooming, panning).
V2.2.2 *Visual parameters:*
    In graph visualization, the user can adapt the data presentation to her needs
    by changing visual parameters. There are two main types:
    V2.2.2a Visual mapping parameters: The users can adapt visual mapping,
    e.g., change the employed color scheme, change the parameters of edge
    bundling algorithm, etc.
    V2.2.2b Graph layout: One characteristic of graph visualization is the
    need for algorithmic determination of the position of nodes and edges
    on the screen. There are various algorithms available, that the user can
    choose from and thereby change the presentation of the data for improved
    readability of the drawing.

4.2   Reasoning

The focus in the reasoning part is on supporting the "human internal" analytical
process – support changes in the analytical reasoning and reproduce the progress of
the analysis.

R1 *Data changes*

Data changes refer to changes in the obtained data about the analytical process.

R1.1 *Analytic process tracking:* Each step of the user's visual graph analysis process is tracked and saved for future review. The data changes with new analytical steps.

R1.2 *Editing:* The users can make annotations to the analytical steps. Advanced Visual Analytics tools may also allow to interactively change the user history on demand. A whole user's action chain could be replayed or deleted from history.

R2 *Reasoning changes*

R2.1 *Change of scheme* The users can change the focus of their analysis, e.g., from analyzing relationships between nodes to analyzing graph topology or clustering of graph nodes. Definition of reasoning schemes is important for merging of user actions into meaningful groups.

R2.2 *Change of parameters* Within the current analytical process, the users can change the type of analytical tools employed.

4.3 Data Processing

Many graph analysis steps are supported by data processing – graph analysis algorithms. These can be either induced by a user action in visualization (e.g., graph layout, select new data to be visualized based on neighborhood relationships) or reasoning (e.g., analyze graph clusters). The results of these algorithms are then visualized according to the algorithm type. Often graph algorithms induce further data changes or changes in visual data representation or even changes in reasoning scheme. Data processing interaction refers to the changes of data for the algorithmic processing via e.g., filtering or editing and changes to the algorithm parameters, changes of the type of processing algorithm or method.

DP1 *Data changes*

Data changes refer to changes in the data used for graph algorithmic analysis.

DP1.1 *Selection*: The user can choose which data to use for calculation. For example, which node should be used as a start for breadth search, and whether to use only the visible part of the graph or the whole graph.

DP1.2 *Editing:* Results of graph algorithms (e.g., clustering) may result in new data that needs to be visualized.

DP2 *Processing changes*

DP2.1 *Change of scheme* The user can choose which graph algorithm to use for her current analytical tasks. Specific graph algorithms used in visual graph analysis include: finding paths between nodes, calculate graph topology measures, identification of important nodes (e.g., hubs and authorities), finding graph motifs, graph clustering, calculation of graph similarity, matching of graph substructures.

DP2.2 *Change of parameters* Many algorithms require setting of parameters (e.g., number of clusters). The users can interactively change these parameters and rerun the algorithms.

5   Application to Tracking of an Analytical Process in Visual Graph Analysis

We now present the application of our new taxonomy on the tracking of an user analytical process in analysis of graphs. We record user activities and show their history in a graphical representation (see Sect. 5.1). Graphical display of activities is commonly used in visual analytical reasoning [3,5,10]. It displays the user's past actions for exploring the process. It also offers the possibility to go back to a certain process stage and to annotate process steps.

The example shows analysis of a large set of documents, where the network represents relationships between the documents. Users explore the network in order to identify which documents are important to read and which documents form clusters within a specific area. Our example uses documents from PubMed database [12]. Each document is a node in the network and an edge denotes neighborhood or similarity of the documents. Document similarity is shown using edge width and document's main topic is denoted by node greyscale/color (print/digital version) (see Fig. 6).

5.1   View on the History of the Analytical Process

In our system, we show the activity history as a graph, where each node represents an action and directed edges denote sequences of actions. The nodes are ordered in flow of analysis from top to bottom. The start node is colored in blue (the top node), intermediate nodes are shown in red. The node of the current state of the process is colored in green. The users can annotate the steps using free text editing (see Fig. 7 left). All steps are labeled according to the type of action taken by the user.

The users have the possibility to review their analytical process, go back and follow a different analytical path. In this case, a branch in the analytical history is created (see Fig. 7 right). The old part of the process (inactive part) is then colored in gray – showing the user' current analytical state.

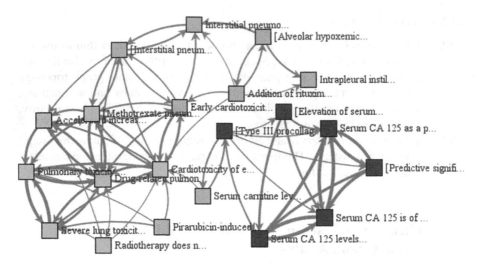

Fig. 6 Visualization of document networks. Each node is a document. Node greyscale/color (print/digital version)represents represents document's topic. Edges connect related documents. Edge width denotes document similarity

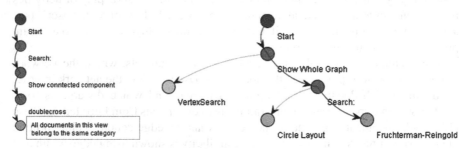

Fig. 7 Graphical representation of user history using graphs. The users can annotate each analytical step (*left*). Users can go back in the process and follow a different analytical path (*right*)

As the analytical process can be long and may consist of many subtasks, we simplify the view by automatic aggregation of actions. The aggregation is based on the classification presented in Sect. 4. The aggregated nodes are colored in yellow and can be disaggregated on demand. For automatic aggregation, we analyze if a sequence of actions has the same type (e.g., layout change). When an action of a different type is performed (e.g., motif analysis), this sequence is merged into one node (see Fig. 8).

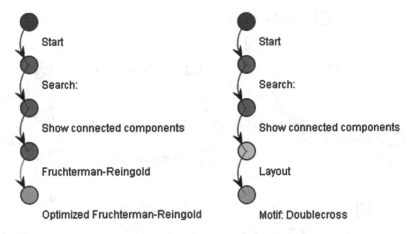

Fig. 8 Aggregation of history steps. *Left*: before aggregation. *Right*: after merging

5.2   Example Analytical Process

As an application example, we describe a prototypical user process for analyzing the document collection. The process is documented and the corresponding parts of the taxonomy are denoted as [.].

A user starts the process by searching for top 10 documents in her favorite topic (antibodies for liver cancer). After these documents are shown [V1.1], the view is adjusted by several layout changes [V1.2.2b] and zooming [V2.2.1]. These steps are tracked [R1.1] and actions within the same category are automatically merged (see Fig. 9 right). The user then changes her analysis focus from gaining overview of the data to analysis of the topological relationships between the documents [R2.1] by motif analysis using various motifs [DP2.1 and DP2.2] and finds out that the central document within this topic is the paper Nr. 3266164 entitled "Monoclonal antibodies of the ICO series" (see Fig. 10 left). She than wants to explore related documents [R2.1] and therefore uses neighborhood search [DP2.1] to make the neighbors of the document "Mouse monoclonal antibodies" visible and lays them out [V1.2.2b] so that the view is clear (see Fig. 11 left). Then she decides to explore a different part of the document space [R2.2], therefore she goes back to one of the previous steps [R2.2] and then shows the neighbors of a different node ("Production and characterization of monoclonal antibodies") [DP2.1]. This makes a second path in the process (see Fig. 12 right).

The tracking of this process, which includes all types of user actions (visualization changes, reasoning changes, and data processing changes) would not be possible with any of the previous taxonomies. They would cover only parts of the process (e.g., visualization and reasoning but not data processing).

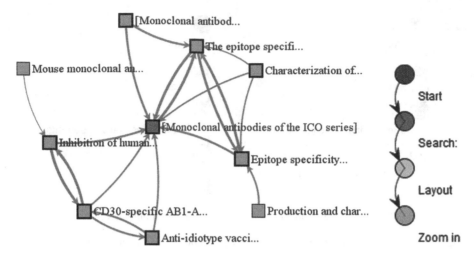

Fig. 9 Analytical history after loading the graph and adjusting its presentation

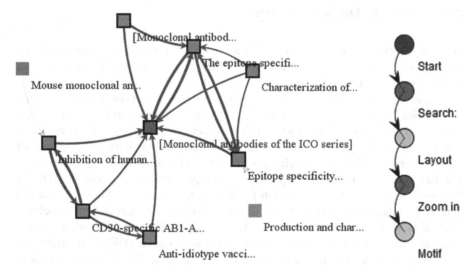

Fig. 10 Analytical history after motif-based analysis. Note that actions of the same type are automatically merged

6 Conclusions

In this chapter, we have presented a new taxonomy of user actions in Visual Analytics processes. It merges and extends previous taxonomies that existed only in separate areas (information visualization, data mining, and analytical reasoning). We reviewed and compared them showing the advantages of our integrated approach. As taxonomies are often defined on an abstract level so that

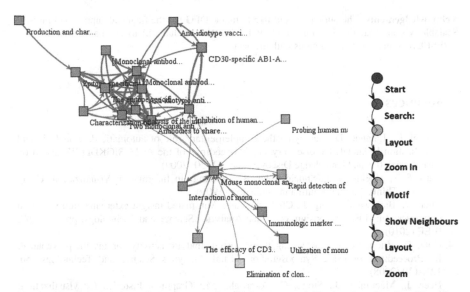

Fig. 11 Analytical history after a change of analysis type to neighborhood exploration. Note that actions of the same type are automatically merged

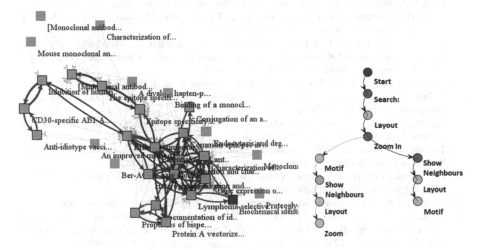

Fig. 12 Analytical history after going back in the process and following a new analytical path

they cover a broad variety of use cases, we specialized the taxonomy for visual analysis of large graphs and presented its application on a use case from digital libraries domain.

In this chapter, we did not explicitly consider interaction on large displays or collaborative interaction, although many of the proposed categories can be applied also to these interaction means.

Acknowledgements The authors would like to thank DFG for the financial support within SPP Scalable Visual Analytics Programme (SPP 1335). We are thankful to our partners within the THESEUS program for providing us with the data.

References

1. Bertini, E., Lalanne, D.: Surveying the complementary role of automatic data analysis and visualization in knowledge discovery. In: Proceedings of the ACM SIGKDD Workshop on Visual Analytics and Knowledge Discovery, pp. 12–20 (2009)
2. Card, S.C., Mackinlay, J., Shneiderman, B.: Readings in Information Visualization: Using Vision to Think. Morgan Kaufmann Publishers (1999)
3. Chen, Y., Barlowe, S., Yang, J.: Click2annotate: Automated insight externalization with rich semantics. In: IEEE Symposium on Visual Analytics Science and Technology, pp. 155–162. IEEE (2010)
4. Gotz, D., Zhou, M.: Characterizing users' visual analytic activity for insight provenance. In: Proceedings of IEEE Symposium on Visual Analytics Science and Technology, pp. 123–130 (2008)
5. Heer, J., Mackinlay, J., Stolte, C., Agrawala, M.: Graphical histories for visualization: Supporting analysis, communication, and evaluation. Visualization and Computer Graphics, IEEE Transactions on **14**(6), 1189–1196 (2008)
6. Heer, J., Shneiderman, B.: Interactive dynamics for visual analysis. Queue **10**(2), 30 (2012)
7. Keim, D., Kohlhammer, J., Ellis, G., Mansmann, F.: Mastering The Information Age-Solving Problems with Visual Analytics. EuroGraphics (2010)
8. Keim, D.A., Mansmann, F., Schneidewind, J., Thomas, J., Ziegler, H.: Visual analytics: Scope and challenges, visual data mining: Theory, techniques and tools for visual analytics. Lecture Notes In Computer Science (lncs) **4404**, 76–90 (2008)
9. von Landesberger, T., Kuijper, A., Schreck, T., Kohlhammer, J., van Wijk, J., Fekete, J.D., Fellner, D.: Visual analysis of large graphs: State-of-the-art and future research challenges. Computer Graphics Forum **30**(6), 1719–1749 (2011)
10. Lipford, H., Stukes, F., Dou, W., Hawkins, M., Chang, R.: Helping users recall their reasoning process. Proc. VAST'10 pp. 187–194 (2010)
11. Thomas, J.J., Cook, K.A.: Illuminating the Path: The Research and Development Agenda for Visual Analytics. National Visualization and Analytics Center (2005)
12. US National Library of Medicine National Institutes of Health : PubMed database. online. URL http://www.ncbi.nlm.nih.gov/pubmed/
13. Ware, C.: Information visualization: Perception for Design. Morgan Kaufmann (2000)
14. Yi, J.S., Kang, Y.a., Stasko, J., Jacko, J.: Toward a deeper understanding of the role of interaction in information visualization. IEEE Transactions on Visualization and Computer Graphics **13**(6), 1224–1231 (2007). DOI http://dx.doi.org/10.1109/TVCG.2007.70515

Common Visualizations: Their Cognitive Utility

Paul Parsons and Kamran Sedig

Abstract Visualizations have numerous benefits for problem solving, sense making, decision making, learning, analytical reasoning, and other high-level cognitive activities. Research in cognitive science has demonstrated that visualizations fundamentally influence cognitive processing and the overall performance of such aforementioned activities. However, although researchers often suggest that visualizations support, enhance, and/or amplify cognition, little research has examined the cognitive utility of different visualizations in a systematic and comprehensive manner. Rather, visualization research is often focused only on low-level cognitive and perceptual issues. To design visualizations that effectively support high-level cognitive activities, a strong understanding of the cognitive effects of different visual forms is required. To examine this issue, this chapter draws on research from a number of relevant domains, including information and data visualization, visual analytics, cognitive and perceptual psychology, and diagrammatic reasoning. This chapter identifies and clarifies some important terms and discusses the current state of research and practice. In addition, a number of common visualizations are identified, their cognitive and perceptual influences are examined, and some implications for the performance of high-level cognitive activities are discussed. Readers from various fields in which a human-centered approach to visualization is necessary, such as health informatics, data and information visualization, visual analytics, journalism, education, and human-information interaction, will likely find this chapter a useful reference for research, design, and/or evaluation purposes.

P. Parsons (✉) • K. Sedig
Western University, London, Canada
e-mail: pparsons@uwo.ca; sedig@uwo.ca

W. Huang (ed.), *Handbook of Human Centric Visualization*,
DOI 10.1007/978-1-4614-7485-2_27, © Springer Science+Business Media New York 2014

1 Introduction

It is well known that visualizations have numerous benefits for supporting problem solving, sense making, decision making, learning, analytical reasoning, and other high-level cognitive activities (see, e.g., [40, 45, 59, 62, 63, 77]). Because of their numerous benefits, visualizations are used in nearly all information-based domains including, but not limited to, science, engineering, journalism, education, public health, finance, medicine, and insurance [34, 37, 66, 67, 71, 77]. The term visualization can have different meanings depending on the context in which it is used—sometimes it refers to a computational tool, sometimes to the process of encoding and representing information, and sometimes to the visual representation that is displayed to users at the interface of a tool. To avoid ambiguity, the terms visualization tool (VT) and visual representation (VR) are used throughout this chapter to refer to an entire tool, and to the visual form of information that a user perceives and acts upon, respectively (see Sect. 2.1 for more detail). Unless stated otherwise, the use of the term visualization in this chapter is synonymous with VR. Examples of VRs are radial diagrams, network graphs, tables, scatterplots, parallel coordinates, geographic maps, and any other visual form that encodes and organizes information.

Research in cognitive science has demonstrated that cognitive processes are distributed across internal, mental representations of information (e.g., mental models) and external representations of information (e.g., VRs) [45, 86]. In addition, psychologists have discovered that VRs represent information (e.g., concepts, events, objects, relationships) in such a way that there is a semantic connection between the VR and the underlying information (see [11]). In other words, from the perspective of the user, there is a unity of meaning between a VR and what it represents. Consequently, VRs influence cognitive processes and the construction of mental models in a very fundamental manner. A strong understanding of human cognition is therefore necessary for proper research and design of VRs. Although such is the case, visualization researchers often place undue emphasis on VRs alone, and do not focus enough on how VRs affect internal representations (e.g., mental models) of users [45]. What often results is an assumption that VRs automatically amplify cognition, without any critical analysis of how and why this may be the case [4]. Moreover, when research does examine cognitive effects of VRs, the focus is often on perceptual and low-level cognitive processes, rather than on higher-level cognitive processes and activities [23].

Although some relevant research does exist, the scattering of pertinent research findings across numerous disciplines makes it difficult for researchers and designers to think systematically about how VRs influence the overall performance of high-level cognitive activities. In this chapter, we draw on research from information and data visualization, visual analytics, cognitive and perceptual psychology, learning sciences, and diagrammatic reasoning to identify some perceptual and cognitive effects of common VRs, and to discuss their utility for performing high-level cognitive activities. As there are countless instances of VRs, we will discuss some of their common features and cognitive effects. Where applicable, we will discuss studies that have been conducted in various contexts on the cognitive effects of VRs. We will also identify some common visualization techniques, provide some examples of existing visualizations, and discuss some of their effects on the performance of cognitive activities. In doing so, this chapter will identify and integrate research that is not often acknowledged in the visualization sciences, yet is applicable and useful. This chapter can thus serve as a point of reference for researchers and practitioners interested in visualizations that support high-level cognitive activities.

2 Background

This section will provide an overview of background concepts and terms that are necessary for understanding the rest of the chapter. Section 2.1 identifies some commonly used terms, clarifies their meanings, and sets their usage for the remainder of the chapter. Section 2.2 distinguishes between perception and cognition, suggests the necessity of viewing them as connected but distinct, and identifies some shortcomings in visualization research regarding claims about cognitive support and amplification of visualizations. Section 2.3 introduces the notion of complex cognition, contrasts it with simple cognition, and briefly characterizes some complex cognitive activities. Finally, Sect. 2.4 discusses the current state of research in light of the preceding three subsections.

2.1 Visualization Tools, Visualizations, and Visual Representations

There is no commonly agreed upon set of terms and meanings that are used in visualization research. For example, when the term 'visualization' is used, it is often not clear whether the meaning is the whole tool, the process of visually encoding (i.e., visualizing) information and data, or the visual representation that sits at the interface of a tool. To add more clarity and precision to the discussion, we make a distinction between visualization tools (VTs), visual representations (VRs), and the process of visual encoding. VTs are electronic computational tools that encode data and/or information in visually perceptible forms—i.e., VRs. Unlike static media, VTs are interactive, allowing users to perform actions upon VRs and receive reactions. Furthermore, VTs can store information, perform analytic operations on the stored information (e.g., as in visual analytics tools), and manipulate the information in numerous ways. In this manner, VTs can take on an active information-processing role to facilitate the performance of complex cognitive activities (see [57]). Figure 1 depicts the components of a VT. VTs receive or retrieve information (e.g., from

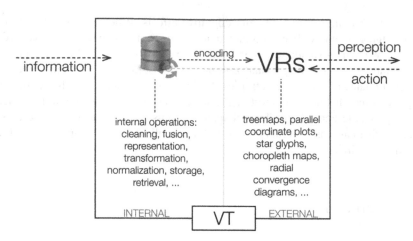

Fig. 1 Internal and external components of a visualization tool

databases, sensors, images), perform some information processing, encode the information in VRs, and allow input from users in the form of actions. The depiction in Fig. 1 is for conceptualization purposes, and does not suggest that a VT has to be an isolated artifact. A webpage, for example, could be considered a VT if it displays interactive VRs and engages in some internal information processing. In such a case, the information processing may take place on a remote server, but could still be conceptualized as the internal component of the VT. As this chapter takes a human-centric approach, the primary concern is the relationship between VRs and users—what aspects of information VRs emphasize and communicate to users, how perceptual and cognitive processes are affected, and how high-level cognitive activities are supported. Indeed, it is the focus on human mental-state changes that characterizes all research in human-centric informatics [49].

2.2 Perception and Cognition

The underlying motivation for doing visualization research is that VRs amplify human cognition. The classic definition of information visualization from Card, Mackinlay, and Shneiderman states that information visualization is "the use of computer-supported, interactive, visual representations of data to amplify cognition" ([8], p. 6). Similarly, Ware has stated that information visualization is "the use of interactive visual representations of abstract data to amplify cognition" ([81], p. xvii). More recently, Mazza has suggested that "visualization is a cognitive activity, facilitated by external visual representations from which people build an internal mental representation of the world" ([50], p. 7). Although most definitions suggest that the purpose of VRs is to amplify cognition, much of the research that is done in the visualization sciences (e.g., data visualization, information visualization, visual analytics), especially in the context of design guidelines and frameworks, focuses more on perception than it does on cognition. It is often the case that the relationship between perception and cognition is not discussed, or when discussed, it is done so in a nebulous manner. This is a likely consequence of the fact that the human perceptual system is a less complex phenomenon than the cognitive system, and its features are more easily observable and amenable to measurement and experimentation. Indeed, while research conducted throughout the past century has characterized many of the features of the human perceptual system, research findings are not as clear about cognition in general, and especially about higher-order cognition [59]. It should be noted that perception and cognition are indeed components of the overall human cognitive system, and that they cannot be entirely separated. For scientific purposes, however, it is necessary to make a clear distinction among them. As Pylyshyn notes, "visual perception does indeed merge seamlessly with reasoning and other aspects of cognition", but cautions that this view is "too broad to be of scientific value" ([59], p. 50).

This is not to say that an extensive understanding of the perceptual system, and how features of VRs affect perception, is not necessary. The human perceptual system does have its own independent characteristics that can be exploited if VRs are designed properly, and many such characteristics are important to have an awareness of in order to create effective visualizations. For example, our visual systems pre-attentively process many features within our visual field in less than 250 milliseconds—quickly, effortlessly, and in parallel—without requiring any conscious mental effort [29, 78, 79]. Such features include, among others, length, orientation, width, hue, curvature, and intersection (see [29]). Being aware of these features and their perceptual effects is therefore necessary for proper design of VRs. In addition, Harnad [27] suggests that the basic properties of our perceptual system may in fact give rise to our higher-order cognition. For instance, although colors differ only in their wavelengths, and exist along a continuum, our visual system detects qualitative changes, from red to yellow to orange, and so on. These bounded categories that are created by our pre-attentive processing may be what provides the groundwork for higher-order cognition [27]. Thus, what is important for visualization research is having the requisite knowledge to take both an analytic and a synthetic approach to perception and cognition to have a more comprehensive understanding of how VRs affect the performance of high-level cognitive activities.

2.3 Cognitive Activities and Complex Cognition

To engage in human-centric visualization research, researchers must understand the difference between simple and complex cognition. Simple cognition to low-level cognitive and perceptual processes such as recognizing objects in a visual field and identifying and comparing them. Complex cognition is emergent—that is, it results from the combination and interaction of simple processes such as perception and memory [21]—and is concerned with how humans mentally represent and think with and about information [41, 75]. When complex cognitive processes involve an active component, as in decision making and problem solving, they can be referred to as complex cognitive activities (see [57]). In this chapter, we have used the terms complex cognitive activities and high-level cognitive activities interchangeably. Although they are not technically synonymous, for the purposes of this chapter they have the same meaning.

Many of the phenomena that VTs are nowadays being applied to, such as climate change, insurance fraud, and disease outbreaks, necessarily involve the performance of complex cognitive activities. Consequently, to develop a more adequate understanding of the cognitive utility of visualizations, one of the necessary lines of research is to explicate this level of cognition in the context of the activities in which users engage. Researchers and practitioners require a sense of which cognitive activities a VT is intended to support, and what the characteristics of such an activity are, in order to effectively design and evaluate VRs. Another chapter in this volume [65] has identified a number of complex cognitive activities, characterized them, and has discussed implications for research and design of VTs. Readers are referred to this other chapter for more information (see also [63]). Also required is knowledge of what tasks can be carried out while performing an activity, and what low-level interactions are performed with visualizations to achieve the goals of such tasks. Such interaction-related concerns are beyond the scope of this paper, however, and readers are directed to other publications [57, 64, 65] that address these concerns more fully.

2.4   Current State of Research and Practice

To date, research has identified and characterized many of the pre-attentive and elementary influences that VRs have on perceptual and cognitive processing. For instance, Bertin's seminal work identifying "visual variables" [6], which have been further studied by subsequent researchers (e.g., [46, 47, 53]); Tukey's work on exploratory data analysis [80]; Cleveland and McGill's experimentation with elementary perceptual tasks [10]; and other similar work that is oft-cited in visualization literature has provided us with a good idea of how features such as color, shape, orientation, length, and texture affect simple cognitive processes. Not as well understood, however, are the effects of VRs on higher-order cognitive processes and their influence on the performance of complex cognitive activities.

Although the need for a deeper understanding of cognitive effects of VRs has been identified in the literature (see [63]), many recent research contributions still place most or all of their focus on low-level considerations, providing guidance for choosing appropriate layouts and visual encodings, focusing on position, layout, axes, color, size, proximity, gestalt laws, depth cues, and so on (e.g., [38, 50, 51, 82]). Survey-style articles (e.g., [7, 30]) also do not discuss effects on higher-order cognition. In their research and development agenda for visual analytics, while acknowledging that this aforementioned type of work is valuable, Thomas and Cook suggested that cognitive principles relevant to analytical reasoning with VRs must be "better understood", and that "we are far from having a complete, formally developed theory of visual representations" ([77], p. 71). While this was stated several years ago, other researchers have recently made similar claims, suggesting that this problem still exists. For instance, in the context of geovisual analytics, Fabrikant states that "we still know little about the effectiveness of graphic displays for space-time problem solving and behavior, exploratory data analysis, knowledge exploration, learning, and decision-making" ([17], p. 139). Green and Fisher have also recently suggested that "there is still a lack of precedent on how to conduct research into visually enabled reasoning. It is not at all clear how one might evaluate interfaces with respect to their ability to scaffold higher-order cognitive tasks." ([22], p. 45). In other words, we still know little about researching and designing VRs that effectively support complex cognitive activities.

In industry, the problem may be even worse, as there seems to be a disconnect between the research that does exist and how visualizations are typically designed. On this topic, Few [18] suggests that "products... promote data visualization in ways that feature superficially appealing aesthetics above useful and effective data exploration, sense-making, and communication." Even relatively recent books that are geared towards practitioners (e.g., [32, 36]) mention only perceptual concerns and do not provide any guidance for facilitating cognitive activities. The next section attempts to address some of these problems by discussing how some common visualizations influence the performance of high-level cognitive activities.

3 Common Visualizations and Their Utility

This section identifies some common visualizations—i.e., common categories in which instances of VRs can be placed—and their perceptual and cognitive influences, with a particular attempt to identify utility for complex cognitive activities. Perceptual effects are described to help give a more complete picture of the overall utility of VRs. This is especially true for perceptual effects that are not typically referenced in the visualization literature. Because research on high-level cognitive utility is not comprehensive or universal, some VRs have received more attention than others in the existing literature. Since we cannot report research that has not been conducted, some sections have more support than others. However, where appropriate, we will make inferences about the possible utility of some VRs and techniques. In addition, where applicable, we will provide empirical evidence based on studies that have been conducted in areas such as cognitive and perceptual psychology and learning sciences. Each category will be briefly characterized and then its utility discussed.

3.1 Visual Encodings and Marks

Visual encodings, also known as visual marks, are atomic visual entities such as lines, dots, and other simple shapes. Visual encodings are the building blocks of more complex VRs. Thus, their cognitive utility does not typically arise from their isolated existence and is best discussed in the context in which they are employed in more complex VRs.

3.2 Glyphs and Multidimensional Icons

Glyphs, also known as multidimensional icons, combine and integrate a number of encodings into one visual entity. Glyphs often make use of multiple visual variables such as color, shape, size, length, and orientation, to encode multiple properties of information items. In doing so, glyphs exploit the perceptual system's ability to discern finely resolved spatial relationships and differences [60]. Wickens and Carswell [83] have investigated how the integration of multiple encodings into one visual entity (e.g., a glyph) can result in emergent features that cannot be communicated with the encodings alone. While such emergent features may be detected pre-attentively by the perceptual system, Wickens and Carswell note that glyphs may engender conscious cognitive processing of information by having attention drawn to such features. That is, glyphs may demand conscious cognitive effort and facilitate higher-order cognitive processes due to the salience of their emergent features. Therefore, although some of the utility of glyphs comes from their perceptual effects that exploit the features of the human visual system, their utility for performing cognitive activities can also be considered. This may be especially true when multiple glyphs are combined within a VR. For example, consider the ClockMap technique [19] shown in Fig. 2. The empty (white) chunk of the glyph in the middle of the VR draws attention and engenders higher-level cognitive activities such as knowledge discovery and sense making. Users may pose questions, engage in exploratory analysis, drill further into the information space to look for patterns, and so on. Since glyphs allow multiple dimensions to be parsed and compared quickly by the perceptual system, they are useful for facilitating the identification of trends and patterns that can assist in high-level cognitive activities such as decision making [48]. In a decision making activity, the attention that is drawn to emergent features may facilitate the choice of one among a number of alternatives within an information space. In one study, Spence and Parr [72] found that subjects who used glyphs for a decision making activity required half the time as subjects who used other VRs.

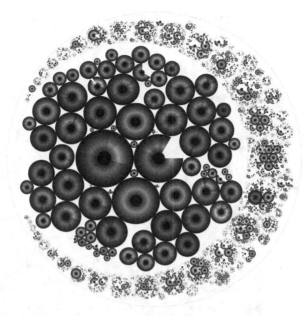

Fig. 2 A VR that uses the ClockMap technique (CC BY-SA 3.0, Fabian Fischer, http://ff.cx/clockmap/)

3.2.1 Techniques

Some common techniques are Chernoff faces [9], Clockeye glyphs [19], multidimensional icons, stick figures, star glyphs, timewheels, multicombs, spikeglyphs, stardinates, kite diagrams, and whisker glyphs. Figure 2 shows a VR that uses the ClockMap technique, which uses a number of small Clockeye glyhps.

3.3 Plots and Charts

Plots, also known as charts, map information onto coordinate systems. Plots help users to think about the distribution of information by depicting the location of information items relative to an axis. For example, plots can facilitate the perception of anomalies, deviations, and outliers, and thus facilitate the performance of cognitive activities that involve reasoning about trends and patterns within an information space [28, 33, 58]. Because of such perceptual suggestions, the solution to a problem involving linear functions, for instance, can be much more apparent when the information is encoded with a plot than with mathematical symbols [3]. For example, an equation such as $y = x2 + 5x + 3$, fails to make explicit the variation which is perceptually evident in a conceptually-equivalent plot [1]. Different types of plots serve different perceptual and cognitive functions.

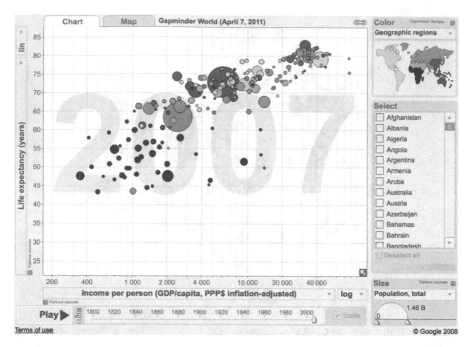

Fig. 3 A VR that uses the scatterplot technique (Used with permission from www.gapminder.org. Data from www.gapminder.org/data)

For instance, bar graphs can be used to facilitate comparisons among discrete or categorical information, whereas line graphs can be used to facilitate reasoning about trends and linear relationships [28, 68, 85].

3.3.1   Techniques

Some common techniques are scatterplots, column, line, circular, bar, area, point, trilinear, vector, radar, nomograph, contour, wireframe, and surface graphs, and parallel coordinates. Figures 3 and 4 show VRs that use scatterplot and parallel coordinate techniques respectively. By distributing the information relative to axes, users can quickly perceive outliers and anomalies within an information space. This can assist with knowledge discovery, for example, by helping users explore an information space to identify the distribution and dispersion of information items, formulate hypotheses based on observed patterns, trends, anomalies, and outliers, and discover new and unsuspected correlations and potential causal relationships.

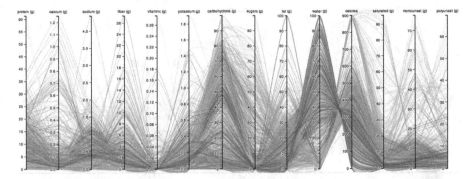

Fig. 4 A VR that uses a Parallel Coordinates technique (Created by Kai Chang, Mary Becica and Vaibhav Bhawsar, http://exposedata.com/intake/. Used with permission under GNU General Public License V3.0. Data from USDA Nutrition Database)

3.4 Maps

Maps spatially distribute information in such a manner that geometric properties of the VR correspond to geometric properties in the information space. Maps can represent both concrete and abstract information spaces. An obvious example of a concrete information space is a VR of a geographical area. In such cases geometric properties of the information space map naturally to geometric properties of the VR. In the case of abstract information, entities within an information space, such as concepts and ideas, may be encoded such that locations, relative distances, and other geometric properties depict semantic distance, categorical similarity, and so on. Representing abstract information spaces in such a way may facilitate the development of mental models of an information space, considering that research in cognitive science has demonstrated that spatial metaphors form a foundation upon which all conceptual structures are built (see [42]). Maps are effective at facilitating high-level cognitive activities requiring spatial reasoning, route planning, and spatial navigation [5, 15, 31, 44]. Consequently, maps also facilitate better decision making and problem solving regarding geographic information—while making decisions about land use, for example [54]. Likewise, maps can facilitate reasoning and higher-order thinking about spatial patterns such as the spread of disease, distribution of mortality rates, and weather patterns [2, 25]. John Snow's map of cholera, for instance, reportedly helped to identify the cause of the cholera outbreak in London in 1854. In a study, Smelcer and Carmel [70] found that for tasks that require geographic, spatial analysis, problems were more effectively solved when the information was encoded with a map than other VRs.

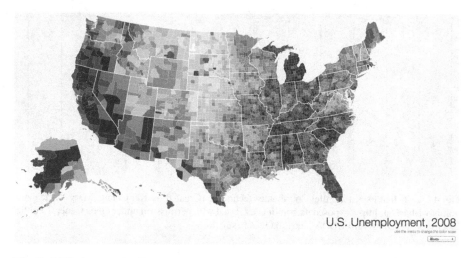

U.S. Unemployment, 2008

Fig. 5 A VR that uses the Choropleth technique (Created with D3.js, http://mbostock.github.com/d3/talk/20111018/choropleth.html)

Fig. 6 A VR that uses the heatmap technique (Used with permission under CC BY 3.0 License, http://www.bluemoon.ee/~ahti/touristiness-map/)

3.4.1 Techniques

Some common techniques are thematic, nautical, weather, geologic, topographic, choropleth, and isarithmic maps, cartograms, Themescapes [84], Self Organizing Maps (e.g., [69]), and heatmaps. Figures 5 and 6 show VRs that use choropleth and heatmap techniques respectively. By representing an information space in such a way, users can quickly perceive the distribution and density of information such that cognitive activities are supported. The VR in Fig. 5, for instance, by using color to encode density, facilitates the quick processing of the distribution of information across different geographic regions. Such a VR could then help policymakers to make decisions about job growth strategies or to solve problems regarding the optimal geographic distribution of employment centers.

3.5 Graphs, Trees, and Networks

Graphs, also known as trees and networks, connect information items with lines, arrows, and other shapes. The connections that are used in graphs are readily detected by our perceptual systems at the level of Gestalt organization [56], and thus powerfully express relationships [81]. This perceptual utility can facilitate high-level cognitive activities that involve reasoning about relationships within an information space. For example, Spence [71] recounts a case of mortgage fraud investigation where eight person-years were spent identifying a perpetrator using typical linguistic representations on static, paper-based media. The same information space was later investigated with a visualization that used a radial technique to represent the relationship network among purchases, lenders, and solicitors. Using this VR, the perpetrator was found within four weeks by a single investigator—a time improvement of a factor of about 100. Spence submits that the explicit relationships depicted by the VR played a valuable role in understanding the relationships within the information space. Cognitive psychologists Novick and Hurley [52] conducted a study investigating structural features of VRs, and concluded that the structural properties of graphs facilitate inference-making about movement, transition, or relation more so than other types of VRs. Furthermore, their study indicated that the structure of graphs and networks facilitates reasoning and problem solving when subjects already mentally represent the information space in graph and network forms.

3.5.1 Techniques

Some common techniques are tree, arc, radial, network, node-link, flow, decision, layout, circuit, concept, and decision diagrams, cone trees [61], and hyperbolic trees [43]. Figures 7 and 8 show two VRs that use a typical node-link technique and a radial convergence technique respectively. The VR in Fig. 7 depicts the relationships among different classes in a java program. The connections between items, as well as encoding techniques such as color and size, exploit the user's perceptual system and thus facilitate higher-order cognition. In a sense making activity, for example, the object and string classes immediately stand out and draw attention, allowing the user to identify questions that further the activity, such as why certain classes have more connections, leading to the development of mental models that incorporate the structure and texture of the information space.

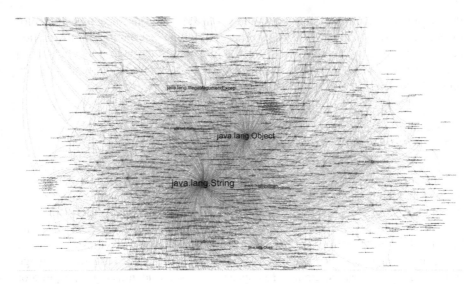

Fig. 7 A VR that uses a typical node-link technique (Created with Gephi, www.gephi.org)

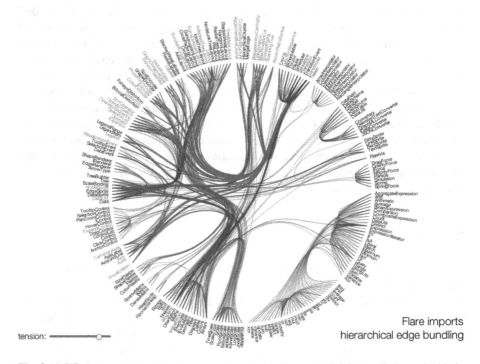

Fig. 8 A VR that uses the radial convergence technique (Created with D3, http://mbostock.github.com/d3/talk/20111116/bundle.html)

3.6 Enclosure Diagrams

Enclosure diagrams place information items in different regions of space. There is a strong perceptual tendency to see information as being inside or outside an enclosed region [81], and thus such regions perceptually suggest commonality [55]. Although VRs such as Venn diagrams exploit properties of the perceptual system, they do so in such a way as to facilitate higher-order cognition [59]. For instance, research in psychology and diagrammatic communication and reasoning suggests that techniques such as Venn and Euler diagrams, which contain or segment information, can help with reasoning about class or set membership and inclusion and exclusion of information [20, 24, 26, 76]. When their perceptual and cognitive effects are combined, enclosure diagrams can act to constrain the set of possible inferences that can be made, and thus facilitate and canalize higher-order thinking processes [14, 74]. Accordingly, such VRs can facilitate high-level cognitive activities such as decision making and problem solving by constraining the set of alternatives that one must consider during a decision making activity, and specifying paths and commonalities among different problem states within an information space.

Enclosure diagrams such as tables and various types of matrices are useful for representing precise and indexical information, both in a quantitative and qualitative manner. Their utility is due to the fact that they facilitate the extraction of exact numerical values or single bits of information, even within large sets of information [13, 25, 70]. In doing so, such VRs facilitate problem solving, by making missing information explicit (e.g., with empty cells) and directing attention to unsolved parts of a problem [14]. In Mendeleev's periodic table of elements, for example, the empty cells that the table displayed helped to predict the discovery of yet unknown elements, and the properties of known and unknown elements could be predicted from their positions within the table [39]. Such VRs also tend to support quicker and more accurate read-off, and highlight patterns and regularities across cases or sets of values [1].

3.6.1 Techniques

Some common techniques are Venn, Voronoi, and Euler diagrams, adjacency matrices, analytical, frequency, percent, contingency, time, and bidirectional tables, TreeMaps [35], SunBursts [73], and DocuBursts [12]. Figure 9 shows a VR that uses the TreeMap technique. By enclosing the different food groups, commonality is naturally suggested to the user's perceptual system. Moreover, visual variables such as size and color also make perceptual suggestions, and can facilitate high-order cognition. For example, in a sense making activity, the user must identify and reason about the structure and texture of the information space, identify the major divisions and groups within the information space, and ask important questions (e.g., why is there only one green box in the 'fats' category?), all of which are made easier due to the manner in which the VR exploits features of the user's perceptual system.

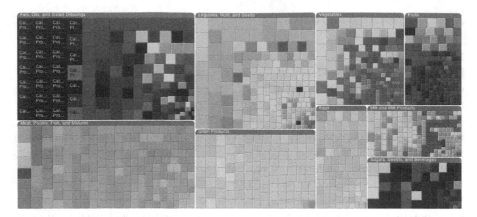

Fig. 9 A VR that uses the Treemap technique (www.hivegroup.com. Used with permission)

4 Summary and Future Research

Much of the visualization research to date has focused on low-level perceptual and cognitive effects of VRs. While important, this addresses only part of the picture. What is needed is further research into how VRs affect higher-order cognition and the performance of complex cognitive activities such as sense making, knowledge discovery, and decision making. The development of an adequate body of research in this regard requires moving beyond the familiar territory of visual variables and pre-attentive perceptual processes. What is needed is a systematic investigation of how visualizations influence cognition to, for instance, canalize and scaffold higher-order thinking processes, assist in discovering the structure and texture of an information space, and facilitate the investigation and selection of alternatives in a decision making process. Much future work is still required, and this chapter has attempted to make a contribution to this research challenge by discussing the utility of some common visualizations for performing complex cognitive activities.

As was mentioned earlier, research into the effects of VRs on the performance of complex cognitive activities is still in the early stages. As a consequence, there are many different lines of future research that can be identified. One is concerned with systematic research being done in visualization disciplines on how visualization impact higher-order cognition. Much of the support that has been provided in this chapter is from disciplines not concerned specifically with interactive VTs and/or VRs. Although in itself this is not a major problem, it would likely be beneficial for more visualization researchers to study these issues in the context of modern visualizations for at least two reasons. First, research in cognitive science and other areas does not examine the techniques particular to data visualization, visual analytics, and similar areas, and thus the application of research from such disciplines has limitations. Second, and of particular importance, is that visualizations nowadays are very often interactive and/or dynamic, which adds many new research questions and challenges that are not addressed by older research on static VRs.

This chapter has grouped a number of common VRs loosely based on some common features to discuss their utility for performing cognitive activities. Future research should develop more comprehensive and elaborate taxonomies and catalogs of VRs that bring more order and structure to the visualization landscape. Additionally, as has been previously suggested by researchers in information visualization (e.g., [16]) and visual analytics (e.g., [77]), one useful but challenging line of research is in developing a pattern language to guide and bring systematicity to the design process.

Still another area of needed research is the performance of more empirical studies that examine the effects of different VRs on high-level cognitive activities. Although some studies have been mentioned in this chapter, many are outdated and were not carried out in the context of modern visualizations. Also needed is a closer investigation of the relationship between perception and cognition and how perceptual effects of certain visualizations may naturally facilitate certain cognitive activities.

Finally, a grand challenge of such research must be to develop comprehensive descriptive and prescriptive frameworks that integrate the aforementioned lines of research. Descriptive frameworks can capture a broad range of considerations to help thinking about the utility of all kinds of visualizations for many different cognitive activities. Prescriptive frameworks can provide visualization design guidance by identifying principles for supporting cognitive activities in general, as well as for supporting particular activities, users, contexts, and domains. Carrying out these aforementioned lines of research can make a positive contribution to the visualization literature, and can help develop a more comprehensive understanding of how visualizations can and should be used to support high-level cognitive activities.

Acknowledgements The authors would like to thank the Natural Sciences and Engineering Research Council of Canada for their financial support.

References

1. Ainsworth, S.: The functions of multiple representations. Computers and Education **33**(2/3), 131–152 (1999)
2. Anselin, L., Syabri, I., Kho, Y.: GeoDa: An introduction to spatial data analysis. Geographic Analysis **38**(1), 5–22 (2006)
3. Arcavi, A.: The role of visual representations in the learning of mathematics. Educational Studies in Mathematics **52**(3), 215–241 (2003)
4. Arias-Hernandez, R., Green, T.M., Fisher, B.: From cognitive amplifiers to cognitive prostheses: Understandings of the material basis of cognition in visual analytics. Interdisciplinary science reviews **37**(1), 4–18 (2012)

5. Berendt, B., Barkowsky, T., Freksa, C., Kelter, S.: Spatial representation with aspect maps. In: C. Freksa, C. Habel, K.F. Wender (eds.) Spatial Cognition—An Interdisciplinary Approach to Representing and Processing Spatial Knowledge. Berlin: Springer.
6. Bertin, J.: Sémiologie Graphique. Les diagrammes, les réseaux, les cartes. With Marc Barbut [et al.]. Paris: Gauthier-Villars. (Translation 1983. Semiology of Graphics by William J. Berg.) (1967)
7. Blackwell, A.: Visual Representation. In M. Soegaard & R.F Dam (eds.) Encyclopedia of Human-Computer Interaction. Aarhus, Denmark: The Interaction Design Foundation. Available online at http://www.interaction-design.org/encyclopedia/visual_representation.html (2011)
8. Card, S.K., Mackinlay, J., Shneiderman, B.: Readings in Information Visualization: Using Vision to Think. San Francisco, CA: Morgan Kauffman (1999)
9. Chernoff, H.: The Use of faces to represent points in k-dimensional space graphically. Journal of the American Statistical Association **68**, 361–368 (1973)
10. Cleveland, W.S., McGill, R.: Graphical perception: Theory, experimentation, and application to the development of graphical methods. Journal of the American Statistical Association **79**(387), 531–554 (1984)
11. Cole, M., Derry, J.: We have met technology and it is us. In: R.J. Sternberg, D.D Preiss (eds.) Intelligence and Technology: The impact of tools on the nature and development of human abilities, 210–227. Mahwah, NJ: Lawrence Erlbaum (2005)
12. Collins, C., Carpendale, S., Penn, G.: DocuBurst: Visualizing document content using language structure. In: Proceedings of Eurographics/IEEE-VGTC Symposium on Visualization (EuroVis '09) **28**(3), 1039–1046 (2009)
13. Cox, R.: Analytical reasoning with multiple external representations. Doctoral Dissertation, University of Edinburgh (1996)
14. Cox, R., Brna, P.: Supporting the use of external representations in problem solving: The need for flexible learning environments. Journal of Artificial Intelligence in Education **6**(2/3), 239–302 (1995)
15. Egenhofer, M.J., Mark, D.M.: Naive geography. In A.U. Frank, W. Kuhn (eds.) Spatial information theory. A theoretical basis for GIS. LNCS 988, 1–15. Berlin: Springer (1995)
16. Elmqvist, N., Moere, A.N., Jetter, H-C., Cernea, D., Reiterer, H., Jankun-Kelly, T.J.: Fluid interaction for information visualization. Information Visualization **10**(4), 327–340 (2011)
17. Fabrikant, S.I.: Persistent problem in geographic visualization: Evidence of geovis(ual analytics) utility and usefulness. ICA Geovis Commission (ICC2011) Workshop, Paris, France (2011)
18. Few, S.: Data visualization for human perception. In: M. Soegaard, R.F. Dam (eds.) Encyclopedia of Human-Computer Interaction. Aarhus, Denmark: The Interaction Design Foundation. Available online at http://www.interaction-design.org/encyclopedia/data_visualization_for_human_perception.html (2010)
19. Fischer, F., Fuchs, J., Mansmann, F.: ClockMap: Enhancing Circular Treemaps with Temporal Glyphs for Time-Series Data. In: Proceedings of the Eurographics Conference on Visualization (EuroVis 2012), 5 pp. (2012)
20. Fitter, M., Green, T.R.: When do diagrams make good computer languages? International Journal of Man-Machine Studies **11**, 235–261 (1979)
21. Funke, J.: Complex problem solving: A case for complex cognition? Cognitive Processing **11**(2), 133–42 (2010)
22. Green, T.M., Fisher, B.: The personal equation of complex individual cognition during visual interface interaction. In: A. Ebert, A. Dix, N. Gershon, M. Pohl (eds.) Human Aspects of Visualization, LNCS 6431, 38–57 (2011)
23. Green, T.M., Ribarsky, W., Fisher, B.: Building and applying a human cognition model for visual analytics. Information Visualization **8**(1), 1–13 (2009)
24. Gurr, C.A.: Effective diagrammatic communication: Syntactic, semantic and pragmatic issues. Journal of Visual Languages and Computing **10**, 317–342 (1999)

25. Guthrie, J.T., Weber, S., Kimmerly, N.: Searching documents: Cognitive processes and deficits in understanding graphs, tables, and illustrations. Contemporary Educational Psychology **18**, 186–221 (1993)
26. Harel, D.: On visual formalisms. In: J. Glasgow, N.H. Naryanan, B. Chandrasekaran (eds.) Diagrammatic Reasoning: Cognitive and Computational Perspectives, 235–271. Cambridge, MA: MIT Press (1995)
27. Harnad, S.: Categorical Perception: The Groundwork of Cognition. NY: Cambridge University Press (1990)
28. Harris, R.: Information Graphics: A Comprehensive Illustrated Reference. Atlanta, GA: Management Graphics (1999)
29. Healey, C.G., Enns, J.T.: Attention and visual memory in visualization and computer graphics. IEEE Transactions on Visualization and Computer Graphics **18**(7), 1170–1188 (2012)
30. Heer, J., Bostock, M., Ogievetsky, V.: A tour through the visualization zoo: A survey of powerful visualization techniques, from the obvious to the obscure. ACM Queue **8**(5), 1–22 (2010)
31. Hutchins, E.: Cognition in the Wild. MIT Press (1995)
32. Illinsky, N., Steele, J.: Designing Data Visualizations. Sebastopol, CA: O'Reilly (2011)
33. Jarvenpaa, S., Dickson, S.: Graphics and managerial decision making: research-based guidelines. Communications of the ACM **31**(6), 764–774 (1988)
34. Johnson, C., Moorhead, R., Munzner, T., Pfister, H., Rheingans, P., Yoo, T.: NIH-NSF Visualization Research Challenges. Tech. rep., Los Alamitos, CA (2006)
35. Johnson, B., Shneiderman, B.: Tree-maps: A space-filling approach to the visualization of hierarchical information structures. In: Proceedings of Information Visualization, 284–291 (1991)
36. Johnson, J.: Designing with the Mind in Mind: Simple Guide to Understanding User Interface Design Rules. Burlington, MA: Morgan Kaufmann (2010)
37. Keim, D., Kohlhammer, J., Ellis, G.: Mastering The Information Age-Solving Problems with Visual Analytics (2010)
38. Kelleher, C., Wagener, T.: Ten guidelines for effective data visualization in scientific publications. Environmental Modelling & Software **26**(6), 822–827 (2011)
39. Kemp, M.: Visualizations. The Nature Book of Arts and Science. Berkeley, CA: University of California Press (2000)
40. Kirsh, D.: Thinking with external representations. AI & Society **25**, 441–454 (2010)
41. Knauff, M., Wolf, A.G.: Complex cognition: The science of human reasoning, problem-solving, and decision-making. Cognitive Processing **11**(2), 99–102 (2010)
42. Lakoff, G., Johnson, M.: Metaphors We Live By. University of Chicago Press (1996)
43. Lamping, J., Rao, R.: Laying out and visualising large trees using a hyperbolic space. In: Proceedings of ACM UIST '94, 13–14 (1994)
44. Liben, L., Kastens, K., Stevenson, L.: Real-world knowledge through real-world maps: A developmental guide for navigating the educational terrain. Developmental Review **22**, 267–322 (2002)
45. Liu, Z., Stasko, J.T.: Mental models, visual reasoning and interaction in information visualization: A top-down perspective. IEEE transactions on visualization and computer graphics **16**(6), 999–1008 (2010)
46. MacEachren, A.M.: How Maps Work: Representation, Visualization, and Design. NY: Guilford Press (1995)
47. Mackinlay, J.: Automating the design of graphical presentations of relational information. ACM Transactions on Graphics **5**(2), 110–141 (1986)
48. Mahan, R.P., Wang, J., Yanchus, N., Elliot, L.R. Redden, E.S., Shattuck, R.: Iconic representation and dynamic information fidelity: Implications for decision support. U.S. Army Research Laboratory (Report ARL-CR-0580). Aberdeen Proving Ground, MD (2006)
49. Marchionini, G.: Information Concepts: From Books to Cyberspace Identities. Bonita Springs, FL: Morgan & Claypool (2010)
50. Mazza, R.: Introduction to Information Visualization. Springer, London (2009)

51. Nazemi, K., Breyer, M., Kuijper, A.: User-oriented graph visualization taxonomy: A data-oriented examination of visual features. In: M. Kurosu (ed.) Human Centered Design, LNCS 6776, 576–585. Berlin: Springer-Verlag (2011)
52. Novick, L.R., Hurley, S.M.: To matrix, network, or hierarchy: That is the question. Cognitive Psychology 42(2), 158–216 (2001)
53. Nowell, L.T.: Graphical encoding for information visualization: Using icon color, shape, and size to convey nominal and quantitative data. Doctoral Dissertation. Virginia Polytechnic Institute and State University (1997)
54. O'Looney, J.A.: Beyond Maps: GIS and Decision Making in Local Government. Redlands, CA: ESRI (2000)
55. Palmer, S.E.: Common region: A new principle of perceptual grouping. Cognitive Psychology 24, 436–447 (1992)
56. Palmer, S.E., Rock, I.: Rethinking perceptual organization: The role of uniform connectedness. Psychonomic Bulletin and Review 1(1), 29–55 (1994)
57. Parsons, P., Sedig, K.: Distribution of information processing while performing complex cognitive activities with visualization tools (this volume)
58. Peebles, D., Cheng, P.-H.: Extending task analytic models of graph-based reasoning: A cognitive model of problem solving with cartesian graphs in ACT-R/PM. In: E. Altmann, A. Cleeremans, C. Schunn, W. Gray (eds.) Proceedings of the 2001 Fourth International Conference on Cognitive Modeling, 169–174 (2001)
59. Pylyshyn, Z.: Seeing and Visualizing: It's Not What You Think. MIT Press (2003)
60. Ribarsky, W., Foley, J.: Next-generation data visualization tools. GVU Technical Report GIT-GVU-94-27, Georgia Institute of Technology. Retrieved from http://smartech.gatech.edu/bitstream/handle/1853/3594/94-27.pdf (1994)
61. Robertson, G. Card, S., Mackinlay, J.: Cone trees: Animated 3D visualizations of hierarchical information. In: Proceedings of CHI '91, 189–194 (1993)
62. Scaife, M., Rogers, Y.: External cognition: how do graphical representations work? International Journal of Human-Computer Studies 45(2), 185–213 (1996)
63. Sedig, K., Parsons, P.: Interaction design for complex cognitive activities with visual representations: A pattern-based approach. AIS Transactions on Human-Computer Interaction (in press)
64. Sedig, K., Parsons, P., Babanski, A.: Towards a characterization of interactivity in visual analytics. Journal of Multimedia Processing and Technologies, Special Issue on Theory and Application of Visual Analytics 3(1), 12–28 (2012)
65. Sedig, K., Parsons, P., Dittmer, M., Haworth, R.: Human-centered interactivity of visualization tools: Micro- and macro-level considerations (this volume)
66. Sedig, K., Parsons, P., Dittmer, M. Ola, O.: Beyond information access: Support for complex cognitive activities in public health informatics tools. Online Journal of Public Health Informatics 4(3) (2012)
67. Segel, E., Heer, J.: Narrative visualization: Telling stories with data. IEEE Transactions on Visualization and Computer Graphics 16(6), 1139–1148 (2010)
68. Shah, P., Mayer, R. E., Hegarty, M.: Graphs as aids to knowledge construction: Signaling techniques for guiding the process of graph comprehension. Journal of Educational Psychology, 91(4), 690–702 (1999)
69. Skupin, A.: A cartographic approach to visualizing conference abstracts. IEEE Computer Graphics and Applications 22(1), 50–58 (2002)
70. Smelcer, J., Carmel, E.: The effectiveness of different representations for managerial problem solving: Comparing tables and maps. Decision Sciences 28(2), 391–420 (1997)
71. Spence, R.: Information Visualization: Design for Interaction. (2nd ed.). Pearson Education Limited, Essex, England (2007)
72. Spence, R., Parr, M.: Cognitive assessment of alternatives. Interacting with Computers 3(3), 270–282 (1991)

73. Stasko, J., Catrambone, R., Guzdial, M., McDonald, K.: An evaluation of space-filling information visualizations for depicting hierarchical structures. International Journal of Human-Computer Studies **53**, 663–694 (2000)
74. Stenning, K., Lemon, O.: Aligning Logical and Psychological Perspectives on Diagrammatic Reasoning. Artificial Intelligence Review **15**, 29–62 (2001)
75. Sternberg, R.J., Ben-Zeev, T.: Complex Cognition: The Psychology of Human Thought. NY: Oxford University Press (2001)
76. Suwa, M., Tversky, B.: External representations contribute to the dynamic construction of ideas. In: M. Hegarty, B. Meyer, & N. Narayanan (eds.) Diagrammatic representation and inference, LNCS 2317, 149–160. Berlin: Springer (2002)
77. Thomas, J., Cook, K.: Illuminating the path: The research and development agenda for visual analytics. IEEE Press (2005)
78. Treisman, A.: Preattentive processing in vision. Computer Vision, Graphics, and Image Processing **31**(2), 156–177 (1985)
79. Treisman, A.: Features and objects in visual processing. Scientific American **255**(5), 114–125 (1986)
80. Tukey, J.W.: Exploratory Data Analysis. Addison-Wesley (1977)
81. Ware, C.: Information Visualization: Perception for Design (2nd ed.). Waltham, MA: Morgan Kaufmann (2004)
82. Ware, C.: Information Visualization: Perception for Design (3rd ed.). Waltham, MA: Morgan Kaufmann (2012)
83. Wickens, C.D., Carswell, C.M.: The proximity compatibility principle: Its psychological foundation and relevance to display design. Human Factors **37**(3), 473–494 (1995)
84. Wise, J.A., Thomas, J.J., Pennock, K., Lantrip, D., Pottier, M., Schur, A., Crow, V.: Visualizing the non-visual: Spatial analysis and interaction with information from text documents. In: N. Gershon, S. Eick (eds.) Proceedings IEEE Visualization 95, 51–58 (1995)
85. Zacks, J., Tversky, B.: Bars and lines: A study of graphic communication. Memory and Cognition **27**(6), 1073–1079 (1997)
86. Zhang, J.: External representations in complex information processing tasks. In: A. Kent (ed.) Encyclopedia of Library and Information Science Vol. 68, 164–180. NY: Marcel Dekker (2000)

Distribution of Information Processing While Performing Complex Cognitive Activities with Visualization Tools

Paul Parsons and Kamran Sedig

Abstract When using visualization tools to perform complex cognitive activities, such as sense-making, analytical reasoning, and learning, human users and visualization tools form a joint cognitive system. Through processing and transfer of information within and among the components of this system, complex problems are solved, complex decisions are made, and complex cognitive processes emerge—all in a manner that would not be easily performable by the human or the visualization tool alone. Although researchers have recognized this, no systematic treatment of how to best distribute the information-processing load during the performance of complex cognitive activities is available in the existing literature. While previous research has identified some relevant principles that shed light on this issue, the pertinent research findings are not integrated into coherent models and frameworks, and are scattered across many disciplines, such as cognitive psychology, educational psychology, information visualization, data analytics, and computer science. This chapter provides an initial examination of this issue by identifying and discussing some key concerns, integrating some fundamental concepts, and highlighting some current research gaps that require future study. The issues examined in this chapter are of importance to many domains, including visual analytics, data and information visualization, human-information interaction, educational and cognitive technologies, and human-computer interaction design. The approach taken in this chapter is human-centered, focusing on the distribution of information processing with the ultimate purpose of supporting the complex cognitive activities of human users of visualization tools.

P. Parsons (✉) • K. Sedig
Western University, London, Canada
e-mail: pparsons@uwo.ca; sedig@uwo.ca

W. Huang (ed.), *Handbook of Human Centric Visualization*, 693
DOI 10.1007/978-1-4614-7485-2_28, © Springer Science+Business Media New York 2014

1 Introduction

The use of visualization tools (VTs) is on the rise in many spheres of human activity. Such tools are increasingly being used in the areas of insurance, education, science, finance, public health, emergency management, journalism, business, and others (see, e.g., [23, 32, 56, 57, 64]). In this chapter, a VT is an electronic computational tool that visually represents (i.e., encodes and visualizes) data and/or information at its visually perceptible interface to help human users[1] analyze data, solve problems, make decisions, and perform other such complex cognitive activities. Therefore, visualizations—whether simple or complex—that do not have an electronic, computational component that allows them to be interactive, and to potentially perform computational and analytic operations, are not considered VTs. Examples of visualizations that are not considered VTs are info-graphics and other static information representations. Furthermore, this chapter is concerned with VTs that support the performance of complex cognitive activities—activities that involve higher-order cognitive processes and occur under complex conditions. Examples of complex cognitive activities are problem solving, sense-making, learning, decision making, and analytical reasoning. VTs have been referred to in the literature by different names, including, but not limited to, cognitive activity support tools, decision support tools, knowledge discovery tools, visual analytics tools, educational tools, and cognitive tools and technologies. This chapter is concerned with all such tools, and does not confine itself to discussion of tools in only one domain. Therefore, any tool—whether used in the context of science, business, insurance, education, libraries, or journalism—that is electronic, computational, and encodes and displays data and/or information in visual forms at its visually perceptible interface is considered a VT.

When using VTs to support the performance of complex cognitive activities, users and VTs are coupled together to form a joint cognitive system [53, 55]. Because of this, information processing that is required to perform complex cognitive activities is distributed across the components of the human-VT system. Moreover, unlike static information-based tools, VTs can take on an active role in the processing of information. For example, VTs can perform data analysis and engage in data mining and knowledge discovery, and can store, manipulate, and encode data in numerous forms. However, as this is a relatively young research area, we still have very little understanding of how to best distribute the information-processing load during the performance of different complex cognitive activities. For example, consider an emergency manager performing time-critical risk and impact analysis of a pending natural disaster. Consider also a university student learning about subatomic particle interactions as part of an undergraduate course. In both cases, a VT could greatly benefit the performance of the activity; however, the ideal distribution of information-processing load in each case would almost

---

[1] In this chapter, the terms human and user are used interchangeably.

certainly be much different. In the latter case, the user would benefit from being required to take on much of the information processing—that is, being required to engage in deep and effortful mental processing of the information to develop sophisticated mental models of the phenomena. In the former case, however, the user would likely benefit most from having the VT shoulder much of the information-processing load, through data analysis and other computational processes, to simply communicate the result for quick decision making. This chapter will more fully explicate the underlying issue, describe some of the features that affect the ideal distribution of information processing, and provide some high-level suggestions as to how information-processing may be best distributed in different contexts.

As this handbook is concerned with human-centric visualization, this chapter assumes a human-centric perspective on information processing in complex cognitive activities. While computational agents may communicate and work together to analyze data, make decisions, and so on, we are interested in joint cognitive systems that have a human core. In such systems, although some information processing may not be taken on by the human, the human is an essential component of the system, and often has the majority of control in the performance of the activity. Moreover, the goal of using VTs is usually to ultimately alter the mental state of the human user. In other words, the performance of an activity results in some change in the user's knowledge, understanding, worldview, schemas, mental models, or other forms of internal, mental representation.

The structure of this chapter is as follows: Sect. 2 will discuss some necessary background concepts and terminology, including distributed cognition and interactive coupling, complex cognitive activities, information processing and human-information interaction, and types and functions of VTs. Section 3 will examine previous work that has categorized the human-VT system into five spaces—information, computing, representation, interaction, and mental space—and will discuss how information is processed in each of the spaces. Section 4 will identify some of the factors that contribute to the ideal distribution of information processing, including activities, information spaces, users, and VTs, and will include a discussion of how researchers and designers can think about the distribution of information processing according to each of these factors. Finally, Sect. 5 will summarize the ideas discussed in the chapter.

2 Background

This section will examine some necessary background concepts and terminology. Four main issues will be briefly examined: (1) distributed cognition and interactive coupling, (2) complex cognitive activities, (3) information processing and human-information interaction, and (4) types and functions of VTs.

2.1 Distributed Cognition and Interactive Coupling

Until recently, the unit of analysis for human cognition—i.e., that which was considered the necessary scope of study to understand human cognition—was typically limited to internal mental structures and processes. Recent research in the cognitive sciences, however, has led scholars to posit that the unit of analysis should be extended to include the body, the external environment, and interactions with objects (e.g., people, artifacts) within the external environment. Numerous subdomains and areas of interest have emerged from this consequential and far-reaching revision in our understanding of human cognition, each emphasizing different aspects of cognition and using slightly different terminology—e.g., suggesting that cognition is embodied, extended, embedded, distributed, and/or situated (see [4, 8, 21, 31, 48]). Whatever the preferred terminology, these recent perspectives all characterize human cognition as an emergent feature of interactions among the internal mental space, the rest of the body, and the external environment and its objects and processes.

One of the better-known theories to emerge from the shifting landscape in cognitive science research is known as the theory of distributed cognition. This theory proposes that cognitive processes are distributed across the internal mental space and the external environment. Hollan et al. [19] posit that this distribution occurs in three main ways: socially, temporally, and across internal (mental) representations and external representations of information. In order to perform complex cognitive activities, such as those performed with the support of VTs, one often combines and processes information from both internal and external representations, in an integrative and dynamic manner [68].

These new perspectives on cognition are not simply changes in terminology; rather, they effect changes in the methodology of cognitive science research and in the explanatory methods of human cognition [8]. Thus, there are vast implications for how we conceptualize the use of VTs for performing complex cognitive activities. Not only do VTs help with memory offloading and computation; they are also integral components of a joint cognitive system, and fundamentally shape and alter cognitive processes [33, 52, 53]. Kirsh [25] suggests that a key tenet of distributed cognition is that of coordination—that the user-VT relation is dynamic, involving reciprocal action and harmonious interaction. That is, the key interactive relation between the user and VT has to do with coordination rather than control. In other words, there is a partnership between a user and VT that results in reciprocal causal influence. This notion of computational technologies acting as partners in cognition has been around for some time now. For instance, based on the theory of distributed cognition, Salomon and colleagues [48, 50] emphasized the idea of thinking *with* computers two decades ago (see [49] for a more recent discussion of the issue). While such ideas have been discussed in the field of education and psychology for some time, visualization researchers have been slow to catch up.

When two entities reciprocally interact—i.e., changes in one cause changes in the other—and the state trajectory of one is dependent on the state trajectory of the other, the two entities are considered to be closely coupled [26]. In the context of a user-VT system, there is reciprocal interaction and mutual causal influence between the user and the VT. Furthermore, since cognitive processes are intrinsically temporal and dynamic, interactive VTs can create a harmony and a close temporal coupling with cognitive processes [4, 24, 25, 53]. This close and interactive coupling is significant, as the user and the VT each have a causal influence on one another (see [7, 52]). Due to their emergent nature, *the performance of any complex cognitive activity is fundamentally affected by the characteristics of both the user and the VT, as well as the strength of the coupling between them* (see [55] for a discussion of what contributes to the strength of this coupling). Therefore, to study the performance of complex cognitive activities with VTs, the unit of analysis must be the user-VT system.

While computers may communicate and work together to analyze data, make decisions, and so on, we are interested in joint cognitive systems that have a human core. In user-VT systems, although the user does not do all of the information processing, he or she is an essential component of the system. Moreover, the goal of using VTs is usually to alter the mental state of the user and to help carry out complex cognitive activities. This primary focus on mental state changes is a fundamental aspect of all human-centric informatics research (see Sect. 2.3).

2.2 Complex Cognitive Activities

Cognitive scientists make a distinction between complex cognition and simple cognition. Whereas simple cognition refers to elementary cognitive and perceptual processes, complex cognition refers to high-level, emergent cognitive processes, such as decision making and problem solving, that take place in complex environments and/or under complex conditions [30, 51, 63]. To emphasize the active aspect of such cognitive processes, and to emphasize their complex nature, they can be referred to as complex cognitive activities (see, e.g., [12]). Although there are numerous complex cognitive activities that can be performed, some of the more common ones are decision making, problem solving, planning, analyzing, forecasting, reasoning, learning, and sense-making [53]. Section 4.1 will elaborate on some of some of these.

2.3 Information Processing and Human-Information Interaction

Complex cognitive activities involve the engagement of human beings in goal-directed information processing [15, 30, 63]. What is meant by the term 'information processing', however, varies according to the context and domain in which it is used. Moreover, such a definition also depends on the definition of information itself, which also varies according to the domain and context. We adopt Bates' [1] definition of information as the pattern of organization of matter and energy. Information processing, then, has to do with changes in the organization of matter and energy. In the context of this chapter, however, the concern is more specific— information processing refers to any change in a mental or physical state (i.e., organization) of the user-VT system. Just as physical state changes provide the basis for classical information theory, it is the focus on mental state changes that characterizes human-centric informatics [35]. Thus, although information processing done by VTs can by analyzed through the lens of classical information theory, because the approach here is human-centric, and because VTs necessarily involve human users, information-processing that results in at least some mental-state changes in a human user is the focus of this chapter.

It should be noted that there is debate that crosses multiple disciplines including cognitive science, artificial intelligence, and philosophy of mind, regarding the information-processing theory of cognition. Much of the disagreement that exists seems to be with the way that 'information' and 'information processing' are defined (see [5]). In this chapter, however, we are not endorsing any particular take on this issue, nor are we concerned with technical definitions in these different fields. Regardless of the adopted theory and terminology, in accordance with the definitions given above, users and VTs share the load of the requisite information processing during the performance of complex cognitive activities.

2.4 Types and Functions of VTs

As mentioned previously, VTs have the following necessary characteristics: they are electronic, computational, and encode and display data and/or information in the form of visual representations (VRs) at their visually perceptible interface. As this characterization is general, there is a broad range of VTs, with varying functions and levels of sophistication, to which the ideas in this chapter are applicable. Some VTs have tremendous computing power (e.g., those that are connected to distributed computer networks), while others have comparatively little power (e.g., some tablet computers). This dictates their ability to perform complex computational analysis, and thus, also determines their potential functions. For instance, a tablet-based VT cannot (as of now) be used to sequence the human genome and display VRs to offer a visual data-mining component for further genome analysis. However, such a VT can support doctors in their decision making by performing simple analysis on patient data and displaying and inviting actions upon VRs. Depending on the context of use, some VTs do not need to perform much computational analysis, and simply need to display VRs and respond to actions from the user (see Sect. 4.4). In addition, VTs offer all kinds of different possibilities for interaction with the underlying information. Some invite the performance of only one or two interactions (e.g., selecting, filtering), whereas others offer many (see [53] for more discussion of this issue). Another consideration still is the types of interactions that are performed. These range from allowing the user to only access existing information, to annotate it (i.e., add a layer of personal meta-information), to modify properties of the existing information, to insert completely new information into the VT, or any combination thereof. Still another dimension in which VTs differ is the degree to which a VT takes a proactive role in information processing. VTs can simply wait for user input before responding, can engage in computational processing in the background, or can actively prompt the user with some information that the VT deems to be appropriate.

Aside from their processing power, storage, and other such characteristics, VTs are also used in a wide variety of domains. For instance, VTs can be used in educational, financial, scientific, journalistic, insurance, emergency response, healthcare, national defense, and many other settings. Thus, the necessary demands of each domain require VTs with different characteristics; however, this chapter is relevant across all domains in which VTs are used to support the performance of complex cognitive activities.

3   Structure of the Human-VT Cognitive System

In order to discuss the distribution of information processing in a human-VT system, there must be a clear division of the different components of the system. Although there is necessary overlap, each component must have a clear function. In previous work, the authors have categorized the human-VT system into five spaces: (1) information space, (2) computing space, (3) representation space, (4) interaction space, and (5) mental space. Information space refers to the body of information with which users interact while performing complex cognitive activities. Computing space is concerned with encoding and storing internal representations of items from an information space and performing operations upon them. Representation space is concerned with encoding and displaying VRs of information so as to be visually perceptible to the user. Interaction space is where actions are performed and consequent reactions occur. Finally, mental space refers to the space in which internal mental events and operations take place (e.g., apprehension, induction, deduction, memory encoding, memory storage, memory retrieval, judgment, classification, categorization). Readers are referred to [54] for a more detailed discussion of these different spaces. Figure 1 depicts this categorization of the human-VT cognitive system.

According to this categorization, the information processing that occurs in each space can be examined in relative isolation. The following three subsections will examine information processing in (1) mental space, (2) computing space, and (3) representation and interaction space. Because of the necessary dependence that exists between interaction and representation space (i.e., actions are performed on VRs, reactions are perceived from changes in VRs), we have chosen to examine them together. However, for other purposes, such as interaction design and represen-

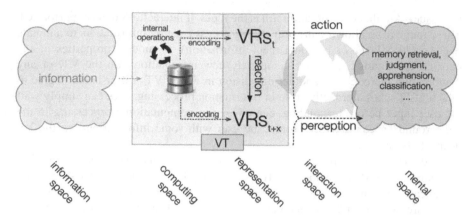

Fig. 1 Categorization of the human-VT cognitive system into five spaces

tation design, it can be important to examine these two spaces independently [53]. Although information processing occurs in all dynamic information spaces (e.g., genetic mutation, social interaction, and financial trading all involve information processing), it is not necessary to examine in light of the goals of this chapter— namely, to examine the distribution of information processing in a user-VT system. It should be noted that decomposing the user-VT system to examine information processing in each of these spaces is useful as an analytical, conceptual tool to facilitate systematic thinking about this issue. In practice, information processing often occurs simultaneously, and in an interdependent manner, in each of these different spaces. Therefore, analyzing the distribution of information processing in this manner serves primarily to assist researchers and designers with conceptualization, rather than to offer prescriptive design guidelines.

3.1 Information Processing in Mental Space

Information processing in mental space consists of changes in the mental state of a user. Changes in one's mental state can take place by working with only internal, mental representations, or by acting upon external representations—i.e., VRs—to co-ordinate and adjust internal mental representations (see [11, 14, 27, 53]). When information processing in mental space involves working with external representations, such as VRs, perception—i.e., awareness of and interpretation of external stimuli—becomes part of the information processing. This perceptual processing of information is a bridge between the internal mental space of the user and the external world. Therefore, the manner in which information is processed by the human perceptual system is an important consideration for research, design, and evaluation of any VT. Figure 2 depicts the different stages of information processing in mental space. The earliest stage of the process can be referred to

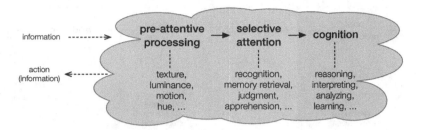

Fig. 2 Information processing in mental space

as pre-attentive processing (also called early vision by some scientists). Although there is considerable debate regarding the boundary of vision and cognition, and the degree to which pre-attentive processing is influenced by cognition (see [20, 43]), pre-attentive processing is typically considered to operate largely independent of conscious cognitive processing and prior knowledge. Therefore, some universal principles of pre-attentive processing can be identified. These are very important for the effective design of VTs—especially for the design of VRs—as some of the features of pre-attentive processing can be exploited with proper visualization design. For example, our visual systems pre-attentively process many features within our visual field in less than 250 milliseconds without requiring any conscious cognitive effort [17]. Such features include length, orientation, width, hue, curvature, and intersection, among others (see [17]). The second stage of information processing in mental space is the stage of selective attention, where attention is concentrated to a specific area in the visual field. This stage can generally be considered as the bridge between perception and cognition [44]. In the context of most visualization research, the first two stages are typically considered as being part of perception. Researchers have identified many principles and guidelines for VR design that are in accordance with the first two stages of processing (e.g., see [9, 34, 38, 65]).

The third stage involves conscious and deliberate information processing to adjust, add to, create, or remove mental representations, models, and/or schemas. In this stage, users consciously perform tasks such as generating hypotheses, comparing them to existing mental structures, constructing analogies, chaining items of information together through inference, categorizing information items, and many others. This is also where metacognitive awareness and regulation take place. That is, one plans cognitive tasks, monitors the performance of such tasks, and evaluates the outcomes of such tasks. Although many visualization researchers have suggested a need for deeper understanding of cognitive—rather than just perceptual—issues in VT use (see [13, 16, 33, 37, 64]), many visualization researchers still consider only the first two stages in their research (see [40] for more discussion). As a result, there is a lack of research in the visualization literature that deals with information processing in mental space in a comprehensive manner.

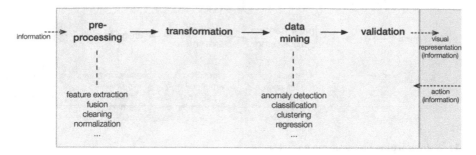

Fig. 3 Information processing in computing space

3.2 Information Processing in Computing Space

Information processing in computing space is concerned primarily with encoding data and other information items into data representations, performing operations upon such representations, and organizing and storing such representations.[2] The ability to computationally process information makes VTs much more powerful mediators of human thinking than static media. Indeed, the speed and precision of information processing in computing space allows VTs to perform all kinds of information processing tasks that would be difficult or impossible for a human user to perform. Figure 3 depicts the stages of information processing in computing space. Although there is no commonly agreed upon set or number of steps, and the labeling changes according to the domain and application of use, a number of potentially present stages can be identified. Depending on the VT, certain stages may not be present or may be skipped during some portion of an activity.

First, information comes from an information space and is input into the VT—this information can come from many sources: textual sources such as e-mails, web pages, and other documents; databases; images and videos; and sensors, such as gyroscopes, altimeters, particle detectors, barometers, and others. Incoming information must often first be pre-processed. Depending on the context, pre-processing can include sub-processes such as cleaning, filtering, fusion, integration, normalization, and others. In other words, this stage processes the information so that it is consistent; free from errors, missing values, and duplicates; and so that both the VT and the user can further process it in a meaningful way. Statistical and mathematical procedures, such as data transformations, also take place within computing space (the transformation stage is sometimes considered as part of the pre-processing stage and not as its own stage). In the most basic sense, data transformations are computational procedures that convert between data representations. These data transformations serve multiple functions. First,

---

[2]Representations in computing space (i.e., data representations) are not visually perceptible to users and should not be confused with visual representations in representation space.

data transformations can create new, derived information from the existing data. In this sense, data transformations create a new subspace of information that is derived from the information space. Second, data transformations can convert data into representational forms that are best suited for encoding in VRs. Third, data transformations can result in representations that are better suited to a user's tasks (for more discussion of data representations and transformations in the context of visualization see [22]). The third step of information processing in computing space involves processing information to discover meaningful patterns—also known as data mining. This stage often involves the performance of computational tasks such as classification, clustering, regression, and anomaly detection. Finally, if the data-mining stage does occur, it is often necessary to check whether the patterns discovered by the data mining algorithms are valid. If there are training samples to facilitate the data-mining step, for example, there may be over-fitting of the model.

Aside from the typical challenges of information processing in computing space, due to the influx of information in all domains, the increase in computing power, and the increasing demand for analytics, new challenges are emerging. For example, incoming information is often heterogeneous, presenting many challenges for existing relational database systems, computational algorithms, and other well-established architectures and techniques (see [23, 66, 67] for more discussion of these issues).

As computing space is only one component of the user-VT system, it receives from and transmits information to other spaces—namely, representation space and interaction space. An additional step of information processing, which bridges computing space and representation space, is the encoding of information in visually perceptible forms (i.e., VRs) for the user to perceive and act upon. This space also receives information from the user in the form of actions. Actions performed by the user can influence and/or be components of any of the stages of information processing in this space (e.g., as in interactive visual data mining).

3.3 Information Processing in Representation and Interaction Space

Because information stored in computing space is not directly accessible to users, and because the form in which information is represented in computing space is not meaningful to humans, VTs encode information from computing space into meaningful visual forms in representation space. These visual forms (i.e., VRs) are the primary means through which users access, work with, and interpret the underlying information. Examples of common VRs are geo-spatial maps, network diagrams, natural and formal languages, treemaps, glyphs, and Venn diagrams. Cognitive scientists have studied VRs for many years, and have discovered numerous benefits that VRs provide for our thinking and reasoning processes (see, e.g., [27]). Furthermore, because VTs inherently have interactive potential, VRs can be made malleable, providing numerous benefits to the user for performing complex cognitive activities [53]. The back-and-forth flow of information that occurs in these two spaces is critical to human-information discourse. Information processing occurs at the boundary of computing space and representation space, where information is encoded into VRs. This information is then perceptually detected by the user and further processed in mental space. Information processing in these spaces also occurs when users input information by performing actions, and VRs are removed, created, or modified in the representation space.

As was mentioned in Sect. 2.4, VTs vary in terms of how pro-active they are in their information processing. One underexplored area of research is to what degree VTs should shoulder the information-processing load in the context of a single interaction that is taking place. For instance, if the user wishes to perform an annotating action, the VT can be completely passive, requiring the user to perform all of the work, or can be active, making suggestions or performing automatic annotations based on the user's action history.

One of the challenges for researchers and designers of VTs is knowing what interactive possibilities can and should be made available to users, and how such interactions impact cognitive and perceptual processes during the performance of complex cognitive activities [64]. Sedig and Parsons [53] have recently developed a framework to address this challenge. The framework includes a comprehensive catalog of fundamental action patterns that users perform when engaged in complex cognitive activities. While each of these actions necessarily impacts information processing in different ways, another important factor to consider is the manner in which interactions are operationalized. Different ways of operationalizing the action and reaction component of an interaction have different cognitive effects, and ultimately influence information processing throughout the human-VT system in different ways. Another chapter of this book (see [55]) presents a framework dealing with macro- and micro-level interactivity considerations for visualization tools— where interactivity refers to the quality of interaction between a user and a VT. As part of the framework, 12 micro-level interactivity elements, which collectively give structure to an individual interaction, are identified and characterized. Each element has different operational forms, and each is identified and briefly discussed. As any individual interaction is composed of an action and a reaction component, some of the elements pertain to the action component and some pertain to the reaction component. The operationalization of each of these elements constitutes part of the information processing that occurs in these two spaces. For instance, one element that is present in both action and reaction is *flow*. Flow is concerned with how an action or reaction is parsed in time. Flow has two operational forms: discrete and continuous. If flow is discrete, the action or reaction occurs instantaneously and/or is punctuated. If flow is continuous, the action or reaction occurs over a span of time in a fluid manner. The manner in which flow is operationalized affects information processing in representation space and interaction space (see [55] for further detail), and has been shown to have a significant impact on the performance of complex cognitive activities (see Sect. 4.5 for some discussion of this issue). Figure 4 depicts some of these aforementioned aspects of information processing in interaction and representation space.

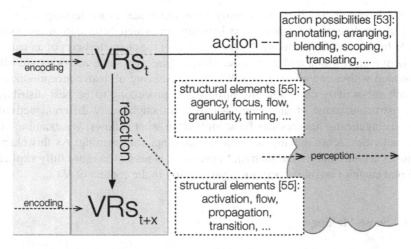

Fig. 4 Information processing in interaction and representation space

4 Factors Affecting the Ideal Distribution of Information Processing

No systematic treatment of how to best distribute the information-processing load during the performance of complex cognitive activities is available in the existing literature. Although some previous research has identified some relevant principles and guidelines, the relevant information is not integrated and is scattered across many disciplines, such as cognitive psychology, educational psychology, information visualization, data analytics, and computer science. Because of the highly variable nature of VTs and complex cognitive activities, any ideal prescription on this matter is very much activity-, user-, and VT-dependent. For example, an intelligence analyst making time-critical decisions, a scientist making sense of research collections over a long period of time, and a student learning about biological processes, would each benefit from different distributions of the requisite information-processing load. To get a sense of the factors that contribute to this ideal distribution, we examine four interrelated considerations on which the ideal distribution is dependent: (1) activities, (2) users, (3) information spaces, and (4) visualization tools. Following the examination of each of these considerations, we will discuss some of the implications for the distribution of information processing.

4.1 Activity-Dependent

Although there are overlaps among different complex cognitive activities, and activities are often embedded within one another during the performance of an overall activity, it is still beneficial to identify their individual characteristics (see also [53]). Depending on the activity that is being performed, information processing should be distributed differently. Having an idea of the characteristics of an activity can help with research and design decisions. Moreover, since activities are often embedded within one another as sub-activities, having a clear conceptualization of each sub-activity can allow the information processing to be best distributed for a particular stage of an overall activity. Although many different activities with distinguishing features can be identified, in what follows, we examine only three activities: sense-making, analytical reasoning, and learning. As this chapter is mostly a preliminary examination, future work is needed to more fully explicate different complex cognitive activities, particularly in the context of VTs.

4.1.1 Sense-Making

Sense-making is a term that has often been used in a somewhat nebulous manner. In an attempt to clarify its meaning, Klein et al. [28] briefly examine five common concepts through which sense-making can be primarily understood: creativity, curiosity, comprehension, mental modeling, and situation awareness. Although each of these can be considered as facets of sense-making, they suggest that modern researchers typically mean something more than just these. They then posit that the additional characteristics that are implicit in the modern use of the term sense-making are "*motivated, continuous effort* to *understand connections* ... in order to *anticipate their trajectories* and *act effectively*" (p. 71, italics added). In a companion paper [29], they attempt to explicate the process of sense-making, suggesting that it begins with some mental model—how-ever minimal—that involves tasks such as elaborating on the model by adding to it, questioning the model and its assumptions, rejecting the model and replacing it, and comparing different models. They posit that the sense-making process involves a closed-loop transition sequence between mental model formation and mental simulation. In response to Klein et al.'s approach to sense-making, Blandford et al. [2] suggest that not enough attention is given to the conceptual structures that people work with when making sense of an information space. They describe sense-making as an 'information journey' that typically starts with either identifying a need (a gap in knowledge) or encountering some information that addresses a latent need or interest. Whatever the particular theory or perspective on sense-making, there are a number of typically present characteristics that can be identified: motivated, continuous, active, inquisitive, open-ended, anticipatory, connective, constructive, and exploratory. Tasks that are typically performed during a sense-making activity include scanning the information space, assessing the relevance of items within the space, selecting items for further attention, examining them in more detail and integrating them into mental models, establishing questions to be asked, determining how to organize the answers, searching for pieces of information, filtering aspects of information, and categorizing items of information [45–47].

4.1.2   Analytical Reasoning

Analytical reasoning is a special type of reasoning, and is based on rational, logical analysis and evaluation of information [53]. Because of its intrinsic analytic nature, an analytical reasoning activity typically involves decomposing or deconstructing an information space to clearly identify its components and their relationships. Analytical reasoning is a core concern of visual analytics, and is an activity performed by analysts from numerous domains, including finance, insurance, national defense, intelligence, climate science, and others [18, 64]. Although some aspects of an analytical reasoning activity can be open-ended and exploratory, one feature that distinguishes it from other complex cognitive activities is its more focused, closed-ended nature that requires the performance of tasks that have a limited set of viable, definite solutions. For instance, an analyst is often presented with a claim and must examine the evidence to either confirm or contradict the claim [58]. Other typical tasks involved in an analytical reasoning activity include examining an information space to find alternative or conflicting evidence in order to challenge an assumption or claim, asserting and testing key assumptions, testing biases, assessing alternatives, comparing and contrasting different hypotheses with the goal of identifying the most plausible one, detecting causal relationships and determining the nature of the supporting evidence, determining which available resources to use, tracing and identifying cause-effect relationships, predicting future states of an information space, identifying the variables within an information space, supporting conclusions and statements with adequate data or evidence, and elaborating an argument and developing its implications [18, 42, 58, 64].

4.1.3   Learning

Learning refers to an activity in which one gains knowledge of an information space and develops skills and capabilities to function in the space [36]. During the process of learning, information is converted into knowledge and assimilated into pre-existing mental models, thereby creating new or revised knowledge structures [3]. Although the exact mechanisms by which one learns are not well understood, a number of mental-state changes that typically occur during a learning activity can be identified. Chi and Ohlsson [6] characterize such changes as potentially occurring along seven different dimensions: (1) growth in mental representation (e.g., mental model or schema) of the information space about which one is learning; (2) denser connectedness among knowledge elements that exist in mental space; (3) increased consistency between mental representations and the information space; (4) finer grain of mental representation of an information space along with appreciation of emergent features of the information space; (5) greater complexity of mental representations; (6) mentally re-representing information items at higher levels of abstraction; and (7) developing multiple perspectives of an information space in order to shift vantage points. In other words, during a learning activity, mental models or schemas change along multiple dimensions: size, connectedness, consistency, grain, complexity, abstraction, and perspective. Although these dimensions are separated for analytical purposes, Chi and Ohlsson suggest that when learning is a complex—rather than simple—cognitive activity, there are typically mental-state changes along multiple dimensions simultaneously.

Learning shares characteristics of both sense-making and analytical reasoning. For instance, some aspects of learning are exploratory, inquisitive, and open-ended, whereas others are focused, analytical, and closed-ended. Unlike other complex cognitive activities that may be time-sensitive (e.g., decision making in a disaster scenario), or may not require a deep understanding of the information space (e.g., browsing a dataset to identify outliers), complex learning requires effortful and reflective information processing with the ultimate goal of understanding. In other words, a distinguishing feature of learning as a complex cognitive activity is that the ultimate focus is on creating lasting and meaningful changes in mental space that give one new knowledge and/or skills.

4.2 User-Dependent

While there are obvious differences among users that must be considered for design and evaluation of any VT, such as age and physical or mental ability, there can be significant differences among users even within these typical categories. For example, users differ based on cognitive, thinking, and learning styles. Sternberg [62] identifies 13 prevalent thinking styles—that is, preferred ways of thinking and of using the abilities that one has, and posits that they fundamentally influence one's cognitive performance. Green and Fisher [16] note that individual user differences can have significant effects on problem solving behaviors, tasks such as categorization and information search, and rationality and reasoning (see also [39, 60, 61]). Not only do users have different personalities and cognitive and thinking styles, they also have different levels of knowledge and expertise that fundamentally influence courses of action in any given situation. That is, the extent of one's knowledge, and the sophistication of one's conceptual structures, *determine what action choices one has to draw from when performing a complex cognitive activity* [10]. Petre and Green [41] also emphasize the roles of training and experience in the interpretation of VRs. People see VRs differently—some people see abstract structure, while others see more concrete configurations and detect different features [59]. Users should thus be given different VRs to work with, and should be given multiple action possibilities to work with the given VRs [53]. Treatment of this issue in the visualization literature is sparse, making it difficult to suggest any concrete design guidelines for VTs at this point.

4.3 Information-Space-Dependent

Certain characteristics of an information space fundamentally influence the appropriate distribution of information processing. Such characteristics include the size of the information space, whether information is static or dynamic, homogeneous or heterogeneous, and structured or unstructured. If the information space consists solely of a relatively small dataset that does not require any pre-processing or data mining, for example, most of the information processing occurs in mental space, interaction space, and representation space. In such a case, the function of computing space is simply to encode information in VRs and respond to input from the user's actions (as is the case in many data and information visualization tools). Many of these concerns have been discussed previously in Sect. 3.2. Aside from such concerns, however, are the density, complexity, and other characteristics of an information space that can place a burden on a user's mental space while performing an activity. In other words, some information spaces are more difficult to understand, conceptualize, mentally navigate, and make sense of. For instance, an information space containing complex mathematical concepts and an information space containing simple sports statistics do not require the same amount of mental effort to understand. In the latter case, it may be desirable to place much of the information-processing load on mental space, since a typical activity would require only relatively simple operations (e.g., inferences, categorizations) to be made. In the former case, however, the required information-processing operations may be much more sophisticated, necessitating the transfer of information processing onto other spaces. Although in both cases the user and VT form a joint cognitive system, the former case requires more sophisticated coordination between the user and the VT. In such situations, the manner in which the transactions between the user and VT take place (i.e., via information processing in interaction and representation space) are of critical importance. The manner in which such transactions are operationalized can determine the strength of the coupling of the user-VT system and ultimately affect the quality of the activity being performed (see [55] for more on this issue). Although some recent work has been done in this area, we still do not have a principled understanding of how to best distribute the load of information processing according to the features of the information space.

4.4 VT-Dependent

The ideal distribution of information processing is naturally dependent on the VT that is mediating the human-information discourse. Numerous characteristics of the VT and its underlying technology must be considered. These include processing power, storage capacity, battery power, display resolution, and display size. Furthermore, the activity must be appropriate for the technology so that the necessary tasks can be carried out. If the underlying technology is a handheld device, for example, the possibilities for intense information processing in computing space are limited compared to those of a desktop computer. The underlying technology on which a VT is built also affects the interaction possibilities that can be offered, and the manner in which information can be processed in interaction and representation space. For instance, certain technologies may limit the possible operational forms that interaction elements can take (see Sect. 3.3).

4.5 Discussion

While more research in this area is required before precise descriptions or pre-scriptions can be given, some previously conducted studies can shed some light on this matter and provide an empirical basis for discussion. Two studies conducted by Sedig and colleagues (see [32, 52]), have not only confirmed that cognitive processes are distributed across users and VTs, that tools shape thinking and reasoning processes and canalize them in certain directions, and that the design of VTs fundamentally influences the performance of complex cognitive activities, but have also made some findings that go against conventional wisdom in the visual-ization literature. Such wisdom has often promoted ease of use and intuitiveness as hallmarks of well-designed VTs. In other words, the suggestion is often that the load of information processing in mental space should be minimized. However, these two studies have shown that subscribing to such wisdom while designing VTs can actually result in negative effects on the performance of complex cognitive activities. VTs designed according to such advice may unintentionally communicate to the user that he or she need not invest much mental effort to consciously process information and plan his or her actions with care.

Based on the results of the study reported in [52], the authors suggested that VTs should reduce the mental information-processing load while users perform tasks that are not directly focused on information that needs to be integrated into mental space. These include working with menus, buttons, and other interface elements that are not encodings of items within the information space. In contrast, it can be beneficial to increase the mental information-processing load for some tasks that are directly focused on the information space—e.g., tasks that require forming hypotheses about items within the space, comparing and assessing alternatives within the space, and drawing inferences about causal relationships. The second study [32] examined the cognitive effects of different operational forms of the element of *flow* (see Sect. 3.3) while using a VT to support a complex learning activity. The results of the study suggest that ease and intuitiveness of use are not necessarily conducive to deep thinking, and can cause the processing of information in mental space to be more automatic and shallow. Results of the study showed that *participants who used the most intuitive and easy to use version of the VT had significantly lower scores* on post-tests that assessed cognitive performance. Moreover, an extra finding of the study was that the manner in which flow was operationalized significantly affected the amount of time required to complete tasks. The groups using discrete actions—

the more difficult and less intuitive ones—were actually more efficient in completing the tasks. In contrast, *the group that had the most intuitive and easy to use version of the tool took significantly more time* than the other groups. The researchers concluded that this is likely due to the cost associated with performing interactions. When users could undo actions with ease, they were not forced to develop premeditated strategies and reflect carefully before performing an action. Therefore, although counterintuitive according to conventional wisdom in the visualization literature, not only did placing more information-processing load on mental space result in better cognitive performance, but it also resulted in faster completion of tasks.

While such issues are rarely discussed in the visualization literature, they have vast implications for all VTs that support the performance of complex cognitive activities. By understanding the distinction among different types of tasks as highlighted in [52], for instance, designers of VTs can deliberately alter the distribution of information processing according to the tasks that are being performed, and can free up mental resources for the most important information-processing tasks. For example, consider the design of a visual analytics tool for intelligence analysis. If designers are aware of the characteristics of analytical reasoning, as described in Sect. 4.1.2, they can then adjust the load of information processing according to the tasks an analyst is likely to perform. For instance, the information processing required to identify and categorize potential threats can be placed mostly on the computing space. The tasks of assessing a hypothesis to determine its validity and then comparing it to other hypotheses, however, requires very careful and effortful processing of information. As a human analyst has more expertise and better judgment skills than any VT, more of the information-processing load for such a task should be placed on the mental space. Moreover, by providing specific action possibilities (see [53]), and by constraining the operationalization of these action possibilities in particular ways (e.g., as in the study reported in [32]), the thinking processes of the analyst can be canalized in certain directions to result in more effective analysis of the information space. While visualization researchers are often focused on "building impressive tools" [13], discovering and studying the types of issues mentioned above are necessary if we are to develop highly coordinated, strongly coupled user-VT systems.

5   Summary

This chapter has examined the distribution of information processing that occurs when VTs are used to support the performance of complex cognitive activities. When engaged in the performance of such activities, information processing occurs simultaneously in multiple spaces in the joint user-VT cognitive system. Furthermore, the processing that occurs in these different spaces is often interdependent. In order to design and/or evaluate VTs in an effective manner, the issues identified in this chapter must be well understood. This chapter has drawn on research from multiple disciplines, including cognitive and perceptual psychology, computer science, information visualization, visual analytics, and educational technologies, to provide an initial examination of the aforementioned issue of information-processing distribution for complex cognitive activities. By identifying and discussing some key concerns, integrating some fundamental concepts, and highlighting some current research gaps that require future study, this chapter lays some groundwork for future research in this area. The issues examined in this chapter are of importance to many domains, including visual analytics, data and information visualization, human-information interaction, and educational and cognitive technologies.

As society's production of information increases, and the desire to analyze and make sense of this information also increases, the issues discussed in this chapter are becoming more pertinent to all areas of endeavor. Whether in insurance, finance, education, medicine, public health, journalism, science, or other information- and knowledge-based enterprises, humans need to work with VTs to perform their information-based activities. Having a principled understanding of how to best distribute the load of information-processing, according to the considerations identified in this chapter, will allow for the development of VTs that more effectively support the complex cognitive activities of their users.

Acknowledgements The authors would like to thank the Natural Sciences and Engineering Research Council of Canada for their financial support.

References

1. Bates, M.J.: Information and knowledge: An evolutionary framework for information science. Information Research **10**(4), (2005)
2. Blandford, A., Faisal, S., Attfield, S.: Conceptual design for sense-making (this volume)
3. Bransford, J.D., Brown, A.L., Cocking, R.R.: How People Learn: Brain, Mind, Experience, and School. Washington, DC: National Academies Press (2000)
4. Brown, J.S., Collins, A., Duguid, P.: Situated cognition and the culture of learning. Educational Researcher **18**(1), 32–42 (1989)
5. Chemero, A.: Information for perception and information processing. Minds and Machines **13**, 577–588 (2003)
6. Chi, M.T.H., Ohlsson, S.: Complex declarative learning. In: K.J. Holyoak, R.G. Morrison (eds.) The Cambridge Handbook of Thinking and Reasoning, 371–399 (2005)
7. Clark, A.: Time and mind. The Journal of Philosophy **95**(7), 354 (1998)
8. Clark, A., Chalmers, D.: The extended mind. Analysis **58**(1), 7–19 (1998)
9. Cleveland, W.S., McGill, R.: Graphical perception: Theory, experimentation, and application to the development of graphical methods. Journal of the American Statistical Association **79**(387), 531–554 (1984)
10. Cox, R.: Analytical reasoning with multiple external representations. Doctoral Dissertation, University of Edinburgh (1996)
11. Crowley, S., Vallée-Tourangeau, F.: Thinking in action. AI & Society **25**, 469–475 (2010)
12. Ericsson, K.A., Hastie, R.: Contemporary approaches to the study of thinking and problem solving. In: R. Sternberg (ed.) Thinking and Problem Solving, 37–80. San Diego, CA: Academic Press (1994)

13. Fabrikant, S.I.: Persistent problem in geographic visualization: Evidence of geovis(ual analytics) utility and usefulness. ICA Geovis Commission (ICC2011) Workshop, Paris, France (2011)
14. Fischer, A., Greiff, A., Funke, J.: The process of solving complex problems. The Journal of Problem Solving 4(1), 19–42 (2012)
15. Funke, J.: Complex problem solving: A case for complex cognition? Cognitive Processing 11(2), 133–42 (2010)
16. Green, T.M., Fisher, B.: The personal equation of complex individual cognition during visual interface interaction. In: A. Ebert, A. Dix, N. Gershon, M. Pohl (eds.) Human Aspects of Visualization, LNCS 6431, 38–57 (2011)
17. Healey, C.G., Enns, J.T.: Attention and visual memory in visualization and computer graphics. IEEE Transactions on Visualization and Computer Graphics 18(7), 1170–1188 (2012)
18. Heuer, R.: Psychology of Intelligence Analysis. Washington, DC: Center for the Study of Intelligence (1999)
19. Hollan, J., Hutchins, E., Kirsh, D.: Distributed cognition: Toward a newfoundation for human-computer interaction research. ACM Transactions on Computer-Human Interaction (TOCHI) 7(2), 174–196 (2000)
20. Hollingworth, A., Henderson, J.M.: Vision and cognition: Drawing the line. Behavioral and Brain Sciences 22(3), 380–381 (1999)
21. Hutchins, E.: Cognition in the Wild. MIT Press (1995)
22. Jasik, D.J., Ebert, D., Lebanon, G., Park H., Pottenger, W.M.: Data transformations and representations for computation and visualization. Information Visualization 8(4), 275–285 (2009)
23. Keim, D., Kohlhammer, J., Ellis, G.: Mastering The Information Age-Solving Problems with Visual Analytics (2010)
24. Kirsh, D.: Interactivity and multimedia interfaces. Instructional Science 25(2), 79–96 (1997)
25. Kirsh, D.: Metacognition, distributed cognition and visual design. Cognition, Education, and Communication Technology, 1–22 (2005)
26. Kirsh, D.: Distributed cognition: A methodological note. Pragmatics & Cognition 14(2), 249–262 (2006)
27. Kirsh, D.: Thinking with external representations. AI & Society 25, 441–454 (2010)
28. Klein, G., Moon, B., Hoffman, R.: Making sense of sense-making 1: Alternative perspectives. IEEE Intelligent Systems 21(4), 70–73 (2006)
29. Klein, G., Moon, B., Hoffman, R.: Making sense of sense-making 2: A macrocognitive model. IEEE Intelligent Systems 21(5), 88–92 (2006)
30. Knauff, M., Wolf, A.G.: Complex cognition: The science of human reasoning, problem-solving, and decision-making. Cognitive Processing 11(2), 99–102 (2010)
31. Lakoff, G., Johnson, M.: Philosophy In The Flesh: the Embodied Mind and its Challenge to Western Thought. NY: Basic Books (1999)
32. Liang, H.N., Parsons, P., Wu, H.C., Sedig, K.: An exploratory study of interactivity in visualization tools: Flow of interaction. Journal of Interactive Learning Research 21(1), 5–45 (2010)
33. Liu, Z., Stasko, J.T.: Mental models, visual reasoning and interaction in information visualization: A top-down perspective. IEEE transactions on visualization and computer graphics 16(6), 999–1008 (2010)
34. Mackinlay, J.: Automating the design of graphical presentations of relational information. ACM Transactions on Graphics 5(2), 110–141 (1986)
35. Marchionini, G.: Information Concepts: From Books to Cyberspace Identities. Bonita Springs, FL: Morgan & Claypool (2010)
36. Mason, R.: How do People Learn? London: CIPD (2002)
37. Meyer, J., Thomas, J., Diehl, S., Fisher, B.: From visualization to visually enabled reasoning. In: H. Hagen (ed.) Scientific Visualization: Advanced Concepts, 227–245 (2010)

38. Nowell, L.T.: Graphical encoding for information visualization: Using icon color, shape, and size to convey nominal and quantitative data. Doctoral Dissertation. Virginia Polytechnic Institute and State University (1997)
39. Palmer, J.: Scientists and information: II. Personal factors in information behaviour. Journal of Documentation 3, 254–275 (1991)
40. Parsons, P., Sedig, K.: Common visualizations: Their cognitive utility (this volume)
41. Petre, M., Green, T.R.G.: Learning to read graphics: Some evidence that 'seeing' an information display is an acquired skill. Journal of Visual Languages and Computing 4, 55–70 (1993)
42. Powers, D.E., Enright, M.K.: Analytical reasoning skills in graduate study: Perceptions of faculty in six fields. The Journal of Higher Education 58(6), 658–682 (1987)
43. Pylyshyn, Z.: Is vision continuous with cognition? The case for cognitive impenetrability of visual perception. Behavioral and Brain Sciences 22, 341–423 (1999)
44. Pylyshyn, Z.: Seeing and Visualizing: It's Not What You Think. MIT Press (2003)
45. Pirolli, P., Card, S.K.: The sensemaking process and leverage points for analyst technology as identified through cognitive task analysis. In: 2005 International Conference on Intelligence Analysis, 6 pp. (2005)
46. Qu, Y., Furnas, G.: Sources of structure in sense making. In: CHI Extended Abstracts, 1989–1992 (2005)
47. Russell, D., Stefik, M. Pirolli, P., Card, S.K.: The cost structure of sense making. In: Proceedings of ACM Conference on Human Factors in Computing Systems, 269–276 (1993)
48. Salomon, G. (ed) Distributed Cognitions: Psychological and Educational Considerations. Cambridge, UK: Cambridge University Press (1993)
49. Salomon, G., Perkins, D.: Do technologies make us smarter? Intellectual amplification with, of, and through technology. In: R. Sternberg, D.D. Preiss (eds.) Intelligence and Technology: The Impact of Tools on the Nature and Development of Human Abilities, 71–86. Mahwah, NJ: Lawrence Erlbaum (2005)
50. Salomon, G., Perkins, D., Globerson, T.: Partners in cognition: Extending human intelligence with intelligent technologies. Educational researcher 20(3), 2–9 (1991)
51. Schmid, U., Ragni, M., Gonzalez, C., Funke, J.: The challenge of complexity for cognitive systems. Cognitive Systems Research 12, 211–218
52. Sedig, K., Klawe, M., Westrom, M.: Role of interface manipulation style and scaffolding on cognition and concept learning in learnware. ACM Transactions on Computer-Human Interaction (TOCHI) 8(1), 34–59 (2001)
53. Sedig, K., Parsons, P.: Interaction design for complex cognitive activities with visual representations: A pattern-based approach. AIS Transactions on Human-Computer Interaction (in press)
54. Sedig, K., Parsons, P., Babanski, A.: Towards a characterization of interactivity in visual analytics. Journal of Multimedia Processing and Technologies, Special Issue on Theory and Application of Visual Analytics 3(1), 12–28 (2012)
55. Sedig, K., Parsons, P., Dittmer, M., Haworth, R.: Human-centered interactivity of visualization tools: Micro- and macro-level considerations (this volume)
56. Sedig, K., Parsons, P., Dittmer, M. Ola, O.: Beyond information access: Support for complex cognitive activities in public health informatics tools. Online Journal of Public Health Informatics 4(3) (2012)
57. Segel, E., Heer, J.: Narrative visualization: Telling stories with data. IEEE Transactions on Visualization and Computer Graphics 16(6), 1139–1148 (2010)
58. Shrinivasan, Y.B., van Wijk, J.J.: Supporting the analytical reasoning process in information visualization. In: Proceedings of the SIGCHI Conference on Human Factors in Computing Systems (CHI '08), 1237–1246 (2008)
59. Sloman, A.: Diagrams in the mind? In: M. Anderson, B. Meyer, P. Olivier (eds.) Diagrammatic Representation and Reasoning, 7–28. London: Springer (2002)
60. Stanovich, K.E.: Who is rational? Studies of individual differences in reasoning. Mahwah, NJ: Lawrence Erlbaum (1999)

61. Stanovich, K.E., West, R.F.: Individual differences in reasoning: Implications for the rationality debate? Behavioral and Brain Sciences **23**, 645–726 (2000)
62. Sternberg, R.J.: Thinking Styles. NY: Cambridge University Press (1997)
63. Sternberg, R.J., Ben-Zeev, T.: Complex Cognition: The Psychology of Human Thought. NY: Oxford University Press (2001)
64. Thomas, J., Cook, K.: Illuminating the path: The research and development agenda for visual analytics. IEEE Press (2005)
65. Ware, C.: Visual Thinking for Design. Morgan Kaufmann Series in Interactive Technologies. Morgan Kaufmann (2008)
66. Westphal, C.: Data Mining for Intelligence, Fraud, & Criminal Detection: Advanced Analytics & Information Sharing Technologies. CRC Press (2009)
67. Witten, I.H., Frank, E., Hall, M.A.: Data Mining: Practical Machine Learning Tools and Techniques. The Morgan Kaufmann Series in Data Management Systems. Elsevier (2011)
68. Zhang, J.: External representations in complex information processing tasks. In: A. Kent (ed.) Encyclopedia of Library and Information Science Vol. 68, 164–180. NY: Marcel Dekker (2000)

Human-Centered Interactivity of Visualization Tools: Micro- and Macro-level Considerations

Kamran Sedig, Paul Parsons, Mark Dittmer, and Robert Haworth

Abstract Visualization tools can support and enhance the performance of complex cognitive activities such as sense making, problem solving, and analytical reasoning. To do so effectively, however, a human-centered approach to their design and evaluation is required. One way to make visualization tools human-centered is to make them interactive. Although interaction allows a user to adjust the features of the tool to suit his or her cognitive and contextual needs, it is the quality of interaction that largely determines how well complex cognitive activities are supported. In this chapter, interactivity is conceptualized as the quality of interaction. As interactivity is a broad and complex construct, we categorize it into two levels: micro and macro. Interactivity at the micro level emerges from the structural elements of individual interactions. Interactivity at the macro level emerges from the combination, sequencing, and aggregate properties and relationships of interactions as a user performs an activity. Twelve micro-level interactivity elements and five macro-level interactivity factors are identified and characterized. The framework presented in this chapter can provide some structure and facilitate a systematic approach to design and evaluation of interactivity in human-centered visualization tools.

K. Sedig (✉)
Associate Professor, Computer Science & Information and Media Studies,
Western University, London, ON, Canada
e-mail: sedig@uwo.ca

P. Parsons • R. Haworth
Ph.D. Student, Computer Science, Western University, London, ON, Canada
e-mail: pparsons@uwo.ca; rhaworth@uwo.ca

M. Dittmer
M.Sc. Student, Computer Science, Western University, London, ON, Canada
e-mail: mdittmer@uwo.ca

W. Huang (ed.), *Handbook of Human Centric Visualization*,   717
DOI 10.1007/978-1-4614-7485-2_29, © Springer Science+Business Media New York 2014

1 Introduction

A human-centered approach to visualization requires the consideration of a number of issues, including the perceptual and cognitive characteristics of users, their goals and needs, and the nature of human tasks and activities. One way to make visualization tools (VTs) human-centered is to design them with interactive features, so that users can engage in a dialogue with a VT through a back-and-forth flow of information. In this manner, users can adjust visualizations to suit their needs and preferences. Although it is widely acknowledged that making VTs interactive enhances their utility, the degree of utility depends upon the quality of the interaction with a VT.

Numerous contextual and ideological factors have influenced the manner in which the use of interactive technologies has been studied. For instance, Crystal and Ellington [13] discuss how task analysis in human-computer interaction research has its historical roots in early studies of physical activity, organizational management, and human factors. These influences led to a system-centric approach when characterizing interaction with technologies. Paul Dourish [14] describes the influence of computational models of the mind on early HCI research, and how the result was wide adoption of procedural accounts of human activity. Consequently, such models largely informed the conceptualization and design of interactive technologies. These views, however, have been challenged by recent research in cognitive science, which posits that cognitive activity is distributed, embodied, and generally far richer and more complex than previous models suggest. In addition, much research in the past involving human activity has overlooked complexity in order to make objects of study 'researchable' [17]. Such approaches have only limited utility, and although potentially useful as analytic frameworks, cannot adequately characterize human cognitive activity.

In addition to these shifting views, interactive technologies are nowadays being used to engage in more complex activities than the simple structured tasks of the early days of HCI [2, 3, 22, 28, 47, 53, 56]. More recently, researchers in various domains related to human-centered visualization and informatics (e.g., [3, 17, 28, 45]) have begun to focus more on the needs, characteristics, and activities of users. Technological advances in recent years have led to the development of highly interactive visualization tools that are used to engage in complex and unstructured activities. For instance, visualization tools are being used to support sense making of large and complex social networks, solving open-ended problems in science, and making decisions regarding global distribution of resources. While researchers now understand what leads to effective visualizations for simple and well-defined tasks, we still know very little about the dynamics of effective VTs for complex activities [22, 53].

Part of the challenge of designing effective VTs is the lack of comprehensive frameworks to inform the conceptualization and discussion of interactivity. Researchers investigating different aspects of human-centered informatics have recently been emphasizing the need for theoretical frameworks. For instance,

Kaptelinin and Nardi [28] have proposed that there is currently "marked interest in frameworks that can provide an explanation of why and how certain subjective phenomena are taking place in situations surrounding the use of interactive technologies". A number of other researchers have identified the lack of theoretical frameworks as a major research problem in information visualization, human-computer interaction, visual analytics, and other related areas over the past decade (see [16, 27, 29, 42, 54, 57, 68]).

Frameworks that thoroughly and methodically characterize interactivity in VTs can greatly assist designers and evaluators. Presently, there is no common vocabulary for discussing the interactivity of VTs, and frameworks can provide such a vocabulary. This chapter makes a contribution to address this need, and is part of a larger research plan to develop a comprehensive framework for design, analysis, and evaluation of interactive tools that mediate and facilitate the performance of complex cognitive activities. This large framework is called EDIFICE (Epistemology and Design of human InFormation Interaction in complex Cognitive activitiEs). This chapter characterizes some aspects of interactivity in visualization tools, and is therefore referred to as EDIFICE–IVT—where IVT stands for interactivity in visualization tools. Interactivity is not exhaustively characterized here, as such an endeavor is beyond the scope of a single chapter. However, this chapter does approach interactivity in a methodical manner and can therefore provide some systematicity to interactivity research and design. Section 2 provides some necessary background information regarding interaction, interactivity, and some cognitive considerations. Section 3 examines some of the challenges encountered by researchers when discussing interaction and interactivity, and proposes a categorization of interaction and interactivity into multiple levels to deal with some of these challenges. Section 4 identifies and characterizes elements and factors of interactivity at a micro and at a macro level, and provides a design scenario. Finally, Sect. 5 provides a summary of the chapter.

2 Background

Modern visualization tools are used to support the performance of activities such as analyzing terrorist threats [68], making sense of climate change patterns [29], and learning about complex mathematical concepts and structures [39, 40]. Such activities involve mental processes that derive new information from given information in order to reason, solve problems, make decisions, and plan actions [36]. As such activities emerge from the combination and interaction of elementary processes (e.g., perception, memory encoding and retrieval), and take place under complex conditions, they are referred to in the cognitive science literature as complex [18]. In this chapter, we are concerned with how VTs best support complex cognitive activities rather than simpler and lower-level cognitive and perceptual processes. Of particular concern is the manner in which such activities emerge from interactions with VTs. While using VTs, users interact with representations of information displayed at the visually perceptible interface of the tool. Henceforth these are referred to as visual representations (VRs). Examples of VRs include radial diagrams, network graphs, tables, scatterplots, parallel coordinates, maps, and any other visual form that encodes and organizes information. What constitutes a VR within an interface can vary depending on the level of granularity at which the interface is viewed. For instance, the totality of an interface can be considered a VR. However, the interface may also be said to contain a number of distinct VRs (e.g., a map, a table, a scatterplot, and so on). Furthermore, each of these could be considered to be made up of different VRs (e.g., the map may contain any number of glyphs). In this chapter we are concerned specifically with interactive VRs and not with static representations. Henceforth, the term 'VR' implies 'interactive VR'.

2.1 Interaction and Interactivity

Broadly speaking, interaction refers to a reciprocal active relationship—that is, action and reaction. The suffix 'ity' is used to form nouns that denote a quality or condition. In this chapter, interactivity refers to the quality of interaction between a user and a VT. By defining interaction and interactivity in this manner, a clear distinction is made between them and each can be analyzed and developed in relative independence. The distinction is important since a VT may be highly interactive, but if the quality of the interaction is not good, the system will not support the cognitive activities of its users effectively.

One way to conceptualize this difference is in the context of a user performing an individual interaction. An interaction may be thought of as having both an ontological and an operational aspect. The ontological aspect is concerned with *what* the interaction is and what its goal is. For instance, filtering refers to a user acting upon a VR to have only a subset of it displayed according to some criteria. The operational aspect is concerned with *how* an action is performed. For instance, a user may issue a textual command through a keyboard to operationalize the filtering interaction. On the other hand, the user may click and drag on a slider to achieve the same result. The manner in which an interaction is operationalized has been shown to have a significant effect on the quality of a user's interaction with a VT (see [38]). The ontological aspect—what an interaction is and what its characteristics are—is concerned with the interaction itself, whereas the operational aspect—how the interaction is put into use—is the concern of interactivity.

The concept of interactivity has been discussed previously in the literature of various domains; its use and characterization, however, has often been vague and haphazard. Within the past decade researchers have referred to the characterizations of interactivity in the literature as "exceedingly scattered and incoherent" [31], "vague and all-encompassing" [21], "blurry" [1], "lacking in an underlying model" [44], "lack[ing] a common language" [60], and "undertheorized" [8]. Although some researchers have attempted to characterize interactivity, much of the research has been done in the context of media and communication studies (e.g., [8,15,26,31,37]) and advertising and marketing (e.g., [20, 41, 72]). The focus of such research is often on human-human communication, brand perception, communication medium, and social information exchange. As a result, the research in these areas does not necessarily transfer well to the domain of interactive visualizations.

Although visualization researchers have been focusing on different elements of *interaction* in recent years (e.g., [5, 19, 43, 52, 66, 71]), very little attention has been paid to *interactivity*. Some effort has been made to characterize interactivity in the context of educational technologies (e.g., [60]). However, as the function of educational technologies is often very specific (e.g., engaging users in deep and effortful processing of information), such research is not necessarily generalizable to all VTs. In addition, these previous characterizations have not been exhaustive, and the research community would likely benefit from a more systematic and thorough characterization of interactivity that is applicable to all VTs.

As research from a human-centered perspective is fundamentally concerned with how VTs best support human cognition, it is helpful to briefly examine developments within cognitive science research and their implications for the design, use, and evaluation of VTs.

2.2 Cognitive Considerations

Research in various branches of cognitive science over the past few decades has demonstrated that human cognition is fundamentally influenced by the environment in which one is situated [7, 11, 12, 24, 35, 55]. Recent characterizations of human cognition as a phenomenon that emerges from interactions among the brain, body, and external environment have supplanted older models depicting human cognition as an internal phenomenon consisting of symbolic computation—a type of 'software' running on neural 'hardware'. Indeed, research that has been conducted on the use of external resources for cognitive purposes has demonstrated that human cognition is deeply intertwined with phenomena that are external to the brain and body (e.g., see [35]).

Although a deep understanding of human cognition is necessary for research in human-centered visualization, the development of VTs is often uninformed by research in cognitive science [22]. This condition is being noticed by researchers in the visualization community. For instance, recently Arias-Hernandez et al. [6] have stated that "these understandings [in visualization research] still rely on traditional cognitive models that focus on universalisms and assumptions of humans as passive cognitive agents while downplaying recent models that emphasize the situated-ness and active role of humans in tight couplings with external representations-processes." A more systematic incorporation of cognitive science research would certainly benefit visualization researchers and practitioners.

One recent development in the study of human cognition that is particularly relevant for VTs is the theory of distributed cognition. This theory posits that the unit of analysis for cognition should include elements external to one's brain and body that contribute to cognitive processes. Cognition may be socially distributed, temporally distributed, and/or distributed across internal and external structures and processes [23]. Consequently, the unit of analysis of cognition is not restricted to the brain or even the body alone—it includes socio-technical systems such as the bridge of a ship [24] or an airline cockpit [25], and human–artifact systems such as a person using a pencil and paper [9]. The theory of distributed cognition is being used more and more in recent years in the visualization community to conceptualize various aspects of design and evaluation of VTs (e.g., see [29, 42, 52, 53, 59, 60, 62, 69]). In this chapter we are concerned with the distribution of cognitive processes across an individual and a VT. As a result, the unit of analysis is the human–VT system, and of particular concern is the strength of the coupling among these two components. The quality of interaction—the interactivity—of a VT is a direct result of the strength of the coupling of this human–VT system. In another chapter of this book, the distribution of information processing that occurs within the human–VT system during the performance of complex cognitive activities is analyzed in detail (see [50]).

When users interact with VTs, cognitive processes emerge from a coupling that is formed between the internal representations and processes of the user and external representations and processes at the interface [11, 34, 55]. Although research has determined that the quality of this coupling is vital to the performance of complex cognitive activities, visualization researchers have tended to overemphasize the importance of external representations (i.e., VRs) and underplay the importance of internal representations and how they are coupled through interaction [6]. This is not to say that proper design and analysis of VRs is trivial; rather, the point is that the user and the VT must be considered as a dynamic system, and the effects of each on the other must be given appropriate consideration (see [49] for a discussion of the cognitive utilities of different VRs).

Although working with a static representation to support cognitive activities engages external cognition and creates a coupling, the coupling is not very strong. During the performance of complex cognitive activities, users are forced to adapt to the characteristics of static representations and to make extrapolations regarding information that is not encoded. When representations are made interactive, however, there is potential for strong coupling, and users can adjust VRs to meet their contextual and cognitive needs. In addition, as cognitive processes are intrinsically temporal and dynamic, interactive VRs potentially create a harmony and a tight temporal coupling with cognitive processes [32, 33]. As part of this dynamically coupled cognitive system, the user and the VT each have a causal influence—in other words, the user and the VT are continuously affecting and simultaneously being affected by each another (see Clark's discussion of continuous reciprocal causation in [10]).

The ultimate implication here is that complex cognitive activities are circumscribed by the features of the environment and, in particular, by the strength of the coupling between internal mental processes and external representations of information. As interaction with VTs forms a coupled system in which there is reciprocal causal influence, we cannot understand or discuss complex cognitive

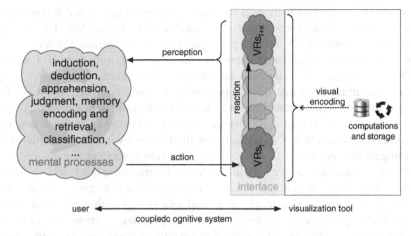

Fig. 1 The coupling that is formed between a user and a visualization tool

activities without examining the ways in which such activities are constrained, canalized, or enhanced by the tools that are supporting the activities. It is important to realize that the process of performing activities is as much driven by the characteristics of tools as it is by the characteristics of users [33]. It is necessary, therefore, to examine the elements and factors that affect the quality of interaction with a VT. Some elements and factors will be identified and developed in Sect. 4 (see Fig. 1).

3 Levels of Interaction

The interaction that takes place between a user and a VT can be characterized at multiple levels of granularity. Here we propose that interaction be categorized into four main levels: activities, tasks, actions, and events. Activities occur at a high level and are often complex and open-ended (e.g., problem solving, decision making, and forecasting); tasks are goal-oriented behaviors that occur at a lower level during the performance of activities (e.g., categorizing, identifying, and ranking); interactions occur at an even lower level and involve actions that are performed upon an interface (e.g., annotating, drilling, and filtering) and their consequent reactions; events occur at the lowest-level of physical interaction with a VT and are the building blocks of interactions (e.g., mouse clicks, keyboard presses, and finger swipes). There are also minor levels among these major ones. Activities often involve sub-activities; tasks often involve sub-tasks and visual tasks; and interactions involve lower-level conceptual steps and implementation techniques.

One of the main challenges of characterizing interaction and interactivity is that there are many levels at which they may be viewed. This presents an especial challenge for using language that accurately conveys the level of granularity that is being discussed. For instance, consider comparing as an interaction. A user can compare at the level of an individual interaction, by acting upon a VR and receiving a reaction that communicates its degree of similarity to another VR. The user can also compare at the higher level of performing a task, by combining and linking multiple interactions together to determine the degree of similarity of a number of VRs. Additionally, although not an interaction, the user can compare at the level of perceptual tasks that involve pre-attentive visual comparisons. It is often the case in the visualization literature that no distinction is made among these levels. This simple example highlights the necessity of having an accurate language for discussing interaction. Conceptualizing interaction as having multiple levels can mitigate this issue and facilitate more consistent discourse using a common vocabulary. In this chapter, we have attempted to discuss these levels in a consistent manner. The catalog discussed below, for instance, attempts to give some structure to interaction at a particular level—at a level that is higher than physical events, low-level conceptual steps, and interaction techniques, but is lower than tasks and complex activities. This type of consistency can help to clarify and give structure to the landscape of interaction and interactivity design.

In previous years, researchers were concerned with designing and evaluating interactive technologies to effectively support relatively simple and highly structured tasks, such as entering data into spreadsheets, composing letters and other documents, sending emails, locating particular files on a hard drive, and organizing files and folders. Numerous models were constructed and/or used to characterize user activity. Hierarchical task analysis, GOMS, and cognitive task analysis are examples of models that were used in the HCI community to characterize user interaction with technology. The utility of such models is their rigorous and highly structured characterizations of user activity. Their descriptive and prescriptive abilities, however, seem to fall short in the context of open-ended, unstructured, and complex activities. Visualization researchers have also devised descriptive models of user activity. Examples include the Information Seeking Mantra [67]: *overview, zoom, filter, and details on demand* and the Visual Analytics Mantra [30]: *analyze first, show the important, zoom, filter and analyze further, details on demand*. In a similar manner, these models are not sufficient for capturing the richness of deep and complex activities.

Characterizing interaction at multiple levels in order to discuss interactivity can help deal with some of the challenges mentioned above. For example, the dynamics of complex cognitive activities in a user-VT system can be conceptualized in terms of both embeddedness and emergence. That is, lower levels are embedded within one another. A ranking task, for instance, may involve the actions of filtering, selecting, and arranging. Each action may involve any number of low-level conceptual steps and physical events. Conceptualizing interactivity in terms of embeddedness allows for clear decomposition of phenomena into constituent component parts at lower levels. This analytical approach is typical of much of the early research in interaction as mentioned in the previous paragraph. Such an approach is highly useful for analysis, and allows for precise characterizations of interaction, especially at lower levels of granularity. Unlike simple tasks, however, complex cognitive activities are nonlinear and emergent phenomena [46].

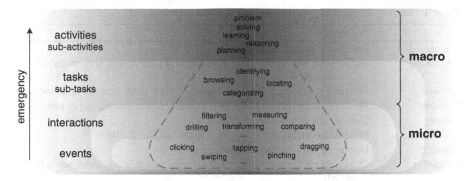

Fig. 2 Interaction categorized into four main levels

Accordingly, the manner in which users perform interactions with a VT to engage in complex activities are often nonlinear and do not follow a pre-determined path [32, 63, 68]. During the course of such activities, although user goals typically have some stability over time, they often undergo changes—in other words, the constituents of an activity are not fixed but can dynamically change as conditions of the activity change [48]. If interaction with tools is treated as a rigid and formulaic process, then we risk missing the dynamics that emerge from sustained interaction with a tool [32]. Therefore, at higher levels, emergent properties occur as a result of the combination of phenomena at lower levels. As complex cognitive activities often do not follow a pre-determined plan, precisely describing the path of an activity is not possible. What can be done, however, is to understand the elements and factors that contribute to the interactivity of a VT and create an environment that best supports the emergence of complex cognitive activities. Such an approach combines the strength of an analytic strategy of decomposing lower levels of activity into component states and processes as well as a synthetic strategy that supports emergence at higher levels (Fig. 2).

In this chapter, interactions are viewed at the level of general patterns of action and reaction that have an epistemic benefit, rather than as more concrete techniques or instantiations of patterns. Sedig and Parsons [63] have recently devised a framework that contains a catalog of over 30 epistemic action patterns, and have discussed the utility of each for performing complex cognitive activities. Table 1 provides a list of some of these patterns. In this catalog, each action is characterized in a conceptual, pattern-based fashion in terms of its epistemic benefit. As a result, there can be many variations of a pattern and many techniques for implementing an instance of an action pattern in a VT. For example, consider the action pattern of *arranging* that is identified and characterized in the catalog. This action pattern refers to a user acting upon VRs to change their ordering and spatial organization within the interface. Variations of this pattern include moving, ranking, ordering, and sorting. In other words, each of these variations consists of a user acting upon VRs to change their ordering and spatial organization within the interface. In addition, each of these may consist of any number of conceptual steps and/or events at the physical level of interaction with the VT.

Table 1 Some action patterns from the catalog of Sedig and Parsons [63]

Action	Description
Animating	Generating movement within VRs
Annotating	Augmenting VRs with additional visual marks and coding schemes, as personal meta-information
Arranging	Changing the ordering and organization of VRs, either spatially or temporally
Blending	Fusing VRs together such that they become one indivisible, single, new VR
Cloning	Creating one or more copies of VRs
Drilling	Bringing latent, interior information to the surface of VRs
Filtering	Displaying only a subset of information in VRs
Measuring	Quantifying VRs in some way (e.g., by area, length, mass, temperature, or speed)
Searching	Seeking out the existence of or locating information in VRs
Scoping	Dynamically working forwards and backwards to view the compositional development and growth of VRs
Transforming	Changing the geometric form of VRs
Translating	Converting VRs into alternative informationally- or conceptually-equivalent forms

4 Characterizing Interactivity

Just as interaction can be conceptualized as having multiple levels, so can interactivity be conceptualized in this manner. In this chapter we categorize interactivity into two main levels: a micro level and a macro level. Interactivity at the micro level emerges from the structural elements of individual interactions. Interactivity at the macro level emerges from the combination of individual interactions to perform tasks and activities. Sections 4.1 and 4.3 will characterize and discuss some considerations of interactivity at the micro level and at the macro level respectively.

4.1 Micro-level Interactivity

As we are concerned with individual interactions at a general, pattern-based level (see Sect. 3), any interaction has a number of elements that collectively give it structure. Interactivity at the micro-level emerges from these elements. As discussed earlier, each individual interaction has two components: action and reaction. The manner in which the action and reaction components of an interaction are operationalized affects the strength of the coupling between a user and a VT. Each element has different operational forms, and varying the operationalization of these structural elements determines the quality of the interaction. Currently, we have identified 12 elements—6 for action and 6 for reaction. The elements of action are: presence, agency, granularity, focus, flow, and timing. The elements of reaction are: activation, flow, transition, spread, state, and context. In what follows, we will characterize each element and discuss some possible ways in which each can be operationalized (Fig. 3).

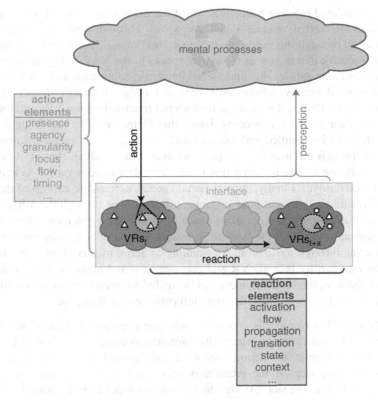

Fig. 3 Some structural elements of an individual interaction

4.1.1 Action Component

There are at least six elements that make up the action component of an interaction. These are discussed next.

Presence. This element is concerned with the existence and advertisement of an action. In other words, this element is about the cue or signal from the visualization used to prompt the user or advertise the existence of the interaction. Two of the main forms of this element are: explicit and implicit. If presence of an action is explicit, the availability, existence, or provision of the interaction is explicitly advertised by the tool, such as when a label or tool tip is used to let the user know that the interaction exists. When presence is implicit, the interaction exists, but its availability is either not easily perceptible by the user, or it is not visible at the interface level. In this case, the user must know of the existence of the interaction, or it would seem to the user that the interaction were non-existent.

Agency. This element is concerned with the metaphoric agency through which an action is expressed. Once the user knows of the existence of an action, the action must be articulated in some manner. Some of the main forms that this element can assume include: verbal, manual, pedal, and aerial. Through verbal agency, actions are expressed through the user's 'mouth' (e.g., text menus, commands, or natural-language speech); that is, it is as if the user speaks to the VRs. Using this form, the VRs are viewed as entities that understand linguistic commands and react to them. Through manual agency, actions are expressed through the user's 'hands' (e.g., a pointing cursor); that is, it is as if the user's hand reaches into the VRs and touches and grasps their visual components. Using this form, the VRs are thought of as objects that can be handled and manipulated. Through pedal agency, actions are expressed through the user's 'feet' (e.g., an avatar that walks); that is, it is as if the user walks on a terrain. Using this form, the VRs can be regarded as maps on which the user moves. Finally, through aerial agency, actions are expressed through the user's 'wings'; that is, it is as if the user flies over or through the VRs. Using this form, the VRs are thought of as space through which the user can navigate. It is important to note that the last two forms are very similar as they both express an interaction through navigation. An example of aerial means of an interaction is when the user can fly in a 3D VR and gets near a visual element. Upon reaching a certain distance, the visual element can be drilled to provide extra information to the user. As the user flies away, the extra information can disappear.

Granularity. This element is concerned with the constituent steps of an action. There are two main forms of granularity: atomic and composite. If the granularity of an action is atomic, the action cannot be decomposed into further steps—i.e., there is only one step. If the granularity of an action is composite, the action requires more than one step. As the interaction construct is at a higher level than low-level physical events, an action may be operationalized in different ways such that there are different granularities in different contexts. In other words, since interactions are not characterized at the lowest possible level, the constituent parts of an action pattern are variable. To clarify and illustrate this element, let us examine a VT, Super Tangrams (see [58]). This is an interactive game in which children use transformation geometry operations to rearrange visual shapes and fit them into an outline without the shapes overlapping. Solving each puzzle requires a set of interactions. Consider the user moving a shape—a variation of the *arranging* interaction pattern discussed above. In order to move the shape (i.e., perform one interaction), the user must go through the following steps: choose the shape, choose an operation (e.g., rotation), adjust the parameters of the operation (e.g., angle of rotation and center of rotation), and finally press a 'Go' button. In this case, the action has composite granularity. This same interaction can be designed to have atomic granularity. For instance, the user can choose a shape that has pre-determined parameters simply by clicking on the shape, and then the reaction ensues automatically without the need to press a 'Go' button. In other words, the action in this case cannot be decomposed into multiple steps.

Focus. This element is concerned with the focal point of an action. Two of the main forms of focus are: direct and indirect. If focus of an action is direct, the action is expressed by the user directly acting upon the VR. If the focus of action is indirect, the action is expressed by the user operating on other intermediary interface representations in order to communicate with and cause a change in the VR of interest. As an example, consider a VR of a human heart that a user wishes to slice open to make sense of its internal components. If the focus of action is direct, the user could click on the VR to open the heart. If the focus of action is indirect, the user may select an anatomical feature from a list to have the VR of the heart open to expose that feature.

Flow. This element is concerned with how an action is parsed in time. Two main forms of flow are: discrete and continuous. If the flow of an action is discrete, the action occurs instantaneously and/or is punctuated. If flow of an action is continuous, the action occurs over a span of time in a fluid manner. For example, a user may be viewing a VR of a scientific co-citation network for the year 2005 and want the VR to display the network for 2010. The user may click on a button that says '2010'—that is, the action flow is discrete. The user may also click on a slider at its current position and drag it until it is at 2010—an example of continuous action flow. One study found that the manner in which action flow is operationalized has a significant impact on the cognitive processes of the user (see [38]).

Timing. This element is concerned with the amount of time the user is given to compose and/or commit an action. There are two main forms of action timing: user-paced and system-paced. User-paced timing allows the user to compose and commit an action at his or her own pace. Using this form of timing, the user has as much time as needed to think about and examine a situation before committing an action. Even when the flow of action is discrete, the user may choose to take any amount of time before the discrete submission of the action. If action timing is system-paced, however, the user has a limited time to compose and perform an action.

4.1.2 Reaction Component

There are at least six elements that make up the reaction component of an interaction. Collectively these six elements can also be referred to as feedback. Even though feedback is discussed by many researchers, it is often presented as an all-encompassing construct that does not distinguish between different levels of interaction and interactivity. Visualization researchers and practitioners would benefit from having a clearer characterization of the elements that make up the structure of feedback at the level of each interaction. As this chapter is concerned with human-centered interactivity, reaction refers to the effects of an action that are visually perceptible at the interface, and not those that may take place internally in the VT and are hidden from view of the user. In addition, as users and VTs are coupled into one cognitive system, using language from systems theory can facilitate conceptualization of the reaction component and its elements. Interfaces are subsystems of a broader user-VT system. Interfaces are also open systems—they receive some input from the user (i.e., an action) and provides some output to the user (i.e., a reaction). During the reaction process, the interface goes through fluctuations before reaching equilibrium. As a result, some of the reaction elements deal with the reaction during fluctuation while others deal with the reaction as the interface reaches equilibrium.

Activation. This element is concerned with the point at which the reaction begins. There are at least three main forms of activation: immediate, delayed, and on-demand. If activation is immediate, the interface reacts to the user's action instantaneously. If activation is delayed, there is a temporal gap between the user's action and the reaction. Finally, if activation is on-demand, the reaction does not take place until requested by the user. Immediate activation of reaction is often discussed in the literature—and is often referred to as 'immediate feedback'—as the only desirable form of activation. While this may be true for most productivity VTs, there are applications in which delayed and on-demand activation are useful (see [4]).

Flow. This element is concerned with how a reaction is parsed in time. There are two main forms of flow: discrete and continuous. In discrete flow, the reaction occurs instantaneously and/or is punctuated. In continuous flow, the reaction occurs over a span of time in a fluid manner. For example, consider a user making sense of climate change patterns with a 3D VR of the earth. If the user is viewing temperature patterns at a point in time (e.g., 1950), during the activity she can perform an action to request that the VR display the temperature at a different point in time (e.g., 2010). The flow of the reaction may be discrete—that is, the temperature patterns for 2010 appear instantaneously or the change is punctuated and has discrete intervals. On the other hand, the flow of the reaction may be continuous—that is, the change from 1950 to 2010 takes place over a span of time in a fluid manner. One study found that the manner in which reaction flow is operationalized has a significant impact on the cognitive processes of the user (see [38]).

Transition. This element is concerned with how change is presented. As an interactive VR is a spatio-temporal entity, its changes can be presented either by distorting its temporal dimension or its spatial dimension. Hence, there are two general types of transition: stacked and distributed. With stacked transition, changes are sequentially stacked on top of one another so that only the current frame of the changing VR is visible. In distributed transition, a number of visualizations capture and preserve instances of the changing VR and present them spatially—in other words, the temporal dimension of the changing VR is distorted and is presented as parallel visualizations distributed in space. To examine the difference between the two forms, consider an educational VT that supports learning about molecular biology. Such a VT may display a VR of a cell with which the user can learn about mitosis. A user can act upon the VR so that there is a transition from the current state of the cell to the end of a mitosis process. If the transition of the reaction is stacked, the subsequent states of the mitosis process will be displayed on top of one another. If the transition is distributed, subsequent phases of the transition will be displayed spatially in different locations. One study found that the different forms of transition had significantly different effects on cognitive processes of the user (see [64]).

Spread. This element is concerned with the spread of effect that an action causes. When an action is performed, it can cause change not only in the VR of interest, but also in other VRs. There are two main forms of spread: self-contained and propagated. In the self-contained form, the VR of interest is the only VR that is affected by the action. In the propagated form, the effect of the action propagates to other VRs in the interface. Consider a VT that supports forecasting of financial outcomes for a company. The interface may contain five separate VRs—one for each of accounting records, projected revenue, sales data, market indicators, and the period of time that is being considered. If an action is performed, the spread may affect only the VR of interest (e.g., acting upon the VR of market indicators to show or hide a subset of the possible indicators). The effect may also be propagated to other VRs (e.g., acting upon the VR of accounting records and having the change spread to the VR of projected revenue so that it is updated).

State. This element is concerned with the conditions of the interface (i.e., the interface's VRs) once the reaction process is complete and the interface reaches equilibrium. There are three main states that VRs affected by an action can take: created, deleted, and altered. VRs that have been affected by an action may be in a created state—that is, they did not exist before the activation of the reaction, but were created during the reaction process and are now visually perceptible at the interface. VRs that have been affected by an action may also be in a deleted state— that is, they did exist before the activation of reaction, but were deleted during the reaction process and are no longer visible. Finally, VRs that have been affected by an action may be in an altered state—that is, they did exist before the activation of reaction, and still exist as the interface reaches equilibrium, but some of their properties have been altered.

Context. This element is concerned with the general context in which VRs exist as the interface reaches equilibrium. Before the activation of a reaction, there is some context in which a VR exists. A reaction either maintains this general context or effects a change in context. Hence, there are two main forms of this element: changed and unchanged. There is an important difference between context and state. A VR may be created or deleted, for instance, but the general context in which the VR exists can remain unchanged. As an example, consider a VT for public health informatics. A user can perform an annotating action on a VR by highlighting or attaching a note to it. As the interface reaches equilibrium (i.e., the reaction has occurred and the annotation is displayed), the context in which the VR exists is unchanged and is the same as it was prior to the reaction. The user may perform a drilling action on the same VR that results in new information about a particular disease appearing and temporarily replacing the previous information, thus changing the context.

As was mentioned above, these 12 elements collectively contribute to the structure of any individual interaction. In addition, as discussed in Sect. 2, the interface of a VT can contain any number of individual VRs. Consequently, the different forms of these structural elements are not necessarily mutually exclusive. For instance, in the study conducted by Sedig et al. [64], in one of the test versions of the VT, an action resulted in both forms of transition in two different VRs.

Table 2 Micro-level interactivity considerations

Component	Element	Concern	Forms
Action	Presence	Existence and advertisement of action	Explicit, implicit
	Agency	Metaphoric agency through which action is expression	Verbal, manual, pedal, aerial
	Granularity	Constituent steps of action	Atomic, composite
	Focus	Focal point of action	Direct, indirect
	Flow	Parsing of action in time	Discrete, continuous
	Timing	Time available for user to compose and/or commit action	User-paced, system-paced
Reaction	Activation	Point at which reaction begins	Immediate, delayed, on-demand
	Flow	Parsing of reaction in time	Discrete, continuous
	Transition	Presentation of change	Stacked, distributed
	Spread	Spread of effect that action causes	Self-contained, propagated
	State	Condition of VRs as interface reaches equilibrium	Created, altered, deleted
	Context	Context in which VRs exist as interface reaches equilibrium	Changed, unchanged

The study showed that operationalizing these different forms simultaneously in different VRs had a significant effect on the cognitive processes of the users. Moreover, different operational forms can be combined in a single interaction. In another tool, Super Tangrams [58], an action that has composite granularity can exhibit both continuous and discrete flow as the steps are put together to perform the action. These combinations and their effects on the strength of coupling between users and VTs requires further explication in future research. When these elements and their forms are brought together and used in the design and evaluation of VTs that support cognitive activities many combinations are possible. For instance, an interaction may be operationalized with direct focus of action, discrete flow of action, and continuous flow of reaction. The same interaction may be operationalized with indirect focus of action and the same forms of action and reaction flow. Alternatively, the focus of action may be direct, the flow of action discrete, and the flow of reaction also discrete. Section 4.2 gives a design scenario to facilitate thinking about the combination of different operational forms of the structural elements that are listed in Table 2.

4.2   EDIFICE–IVT: Design Scenario for Micro-level Interactivity

The following scenario illustrates the potential for micro-level interactivity considerations to inform the design of visualization tools. An awareness of the different interaction elements and some of their possible forms enables designers

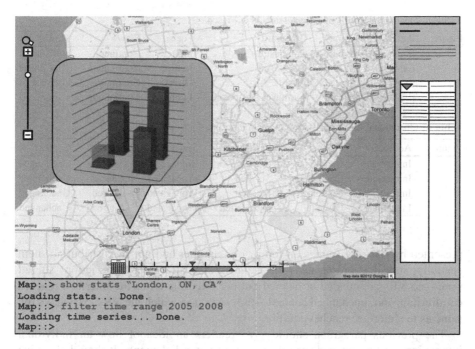

Fig. 4 A user filtering a VR

to operationalize them in a deliberate manner. Consider the design of a map-based visualization tool that supports analysis and sense making of sales data for different cities in a region. At some point the user will need to drill into the map VR to bring to the surface information about sales data for particular cities. Figure 4, for example, depicts an interface after the user has drilled into a city, causing two VRs to appear: a three-dimensional bar graph and a table. The rest of the scenario will describe how further interactions with such a VT may be operationalized using different forms. Figures 4–7 depict different ways of operationalizing the same interaction: filtering out all of the sales data except for a particular time period.

Figure 4 depicts the user filtering the VR to show sales data during the period of 2005–2008 only. The operational forms of the action component are: implicit presence—the user performs an unadvertised action and must know the proper input command; verbal agency—the user types a linguistic command to the VR; composite granularity—there are two constituent steps of the filtering interaction: typing the command and then pressing 'enter'; indirect focus—the VR of interest is the bar graph and the focal point of action is the command line; continuous flow—the action takes place over a period of time while the user is typing; and, user-paced timing—the user has no time limit on committing the action. The operational forms of the reaction component are: immediate activation—once the user commits the action the reaction begins without any delay; continuous flow—the reaction occurs fluidly (shown as transparent sections of the bars); stacked transition—change is

Table 3 Operational forms of action elements in Figs. 4 –7

Figure	Presence	Agency	Granularity	Focus	Flow	Timing
4	Implicit	Verbal	Composite	Indirect	Continuous	User-paced
5	Explicit	Manual	Composite	Indirect	Continuous	User-paced
6	Explicit	Manual	Atomic	Direct	Continuous	User-paced
7	Explicit	Manual	Atomic or composite	Indirect	Continuous	User-paced

Table 4 Operational forms of reaction elements in Figs. 4 –7

Figure	Activation	Flow	Transition	Spread	State	Context
4	Immediate	Continuous	Stacked	Propagated	Altered	Unchanged
5	Immediate	Continuous	Stacked	Propagated	Altered	Unchanged
6	Immediate	Continuous	Stacked	Propagated	Altered	Unchanged
7	Immediate	Discrete	Distributed	Propagated	Altered	Unchanged

presented by stacking frames on top of one another and not by distributing them spatially; spread is propagated—the table is also affected by the action; altered state—no VRs are created or deleted but are only altered as the interface reaches equilibrium; and, unchanged context—the context in which the VRs exist stays the same as the interface reaches equilibrium.

The previous paragraph should give readers an idea of how an individual interaction can be analyzed according to its structural elements. It should also give the reader a sense of how the operationalization of these elements can affect the interactivity of a VT at a micro level. For the sake of brevity, the next few examples will not identify the form of every element, but will discuss only a subset. All of the elements and their operational forms are listed, however, in Tables 3 and 4.

Figure 5 depicts the user performing the same interaction—filtering—but with a different operationalization. In this case, the user chooses the filtering period by dragging a slider from one spot to another (2005–2008) and then presses a 'go' button. Unlike the interaction described above and shown in Fig. 4, the action presence is explicit and the agency is manual; the granularity, focus, flow, and timing, however, are all the same. In this case, the operational forms of reaction elements are exactly the same as in Fig. 4. In the case of Fig. 6, the focus of action is direct—the user acts directly upon the bar graph VR to filter it—and the granularity is atomic. The other elements are operationalized the same as in Fig. 5. The reaction elements are also operationalized in the same manner. In the case of Fig. 7, the action is the same, but the reaction is operationalized differently. In all previous examples the transition of reaction was stacked; in this case, however, the transition is distributed. Multiple visualizations capture and preserve instances of the changing VR and present them spatially—that is, the temporal dimension of the changing VR is distorted and is presented as parallel visualizations distributed in space.

The examples discussed above do not constitute an exhaustive design scenario. This section only briefly explores a small number of design options to demonstrate how EDIFICE–IVT can facilitate analysis and design of micro-level interactivity in visualization tools. Using such a framework, designers can methodically analyze

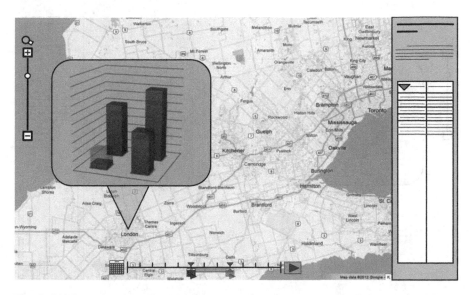

Fig. 5 A different operationalization of the same interaction shown in Fig. 4

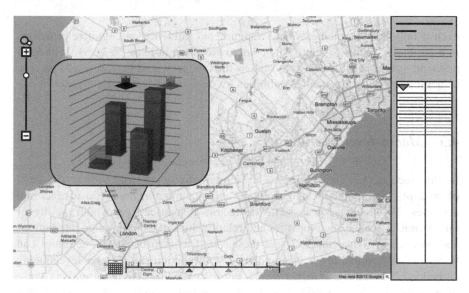

Fig. 6 An example of direct focus of action

the combinatorial possibilities that the operational forms of interaction elements create in terms of design variations for VTs. For example, if each interaction has 12 elements, each of which has at least 2 forms, the number of possible ways to operationalize an interaction is at least 2^{12}, or 4,096. It should be noted that not all elements are applicable or have significant cognitive effects in every VT. However,

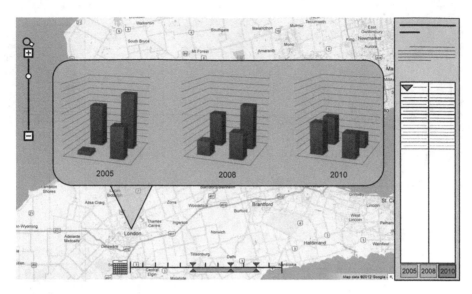

Fig. 7 An example of distributed transition of change

even if only half the elements have a significant influence on cognitive processes in a particular context, the possible combinations are at least 2^6, or 64. Without a descriptive, analytical framework, such as the one presented here, it would be very difficult to consider the many possibilities for design in a systematic manner.

4.3 Macro-level Interactivity

In this section, we analyze macro-level interactivity—interactivity at the level of multiple interactions being combined and put together to perform tasks and activities. This analysis deals with the factors that affect the overall quality of interaction, as well as the properties and relationships of its aggregated interactions and how interactivity emerges from these. As discussed in Sect. 3, activities can be viewed at multiple levels of granularity, with phenomena at lower levels being embedded within those at higher levels. For instance, the activity of making sense of a complex 3D geometric structure may include a task, such as identifying different objects and sub-structures. This task may in turn have several interactions embedded in it, such as filtering, scoping, and annotating [63]. Therefore, macro-level interactivity emerges from the whole interface of a VT—that is, the properties of all its interactions and the relationships of these interactions with each other. In what follows, we will characterize and discuss five of the factors that we believe affect the quality of interaction at a macro level. These five factors are: diversity, complementarity, fitness, flexibility, and genre.

Diversity. This factor is concerned with the number and diversity of interactions that are available to the user. A multiplicity of interactions allows the user to perform different types of cognitive tasks. Some studies show that providing a diverse set of interactions in a VT can have a positive effect on reasoning and other cognitive activities [40, 65]. Such diversity can encourage more autonomous and self-regulated cognitive processes. However, van Wijk [70] points out that although interaction is generally good it should be used carefully, as there are costs associated with the number of interactions. Both too few as well as too many interactions can be costly. In an empirical study involving mathematical visualizations, Liang and Sedig [40] demonstrate that lack of interactions can make exploration of the visualizations ineffective and inefficient. The same study also shows that having many interactions may result in some costs due to high time consumption and cognitive demand. When there are too many interactions, the user may need to spend time trying out all available interactions, figuring out their functions and benefits, and remembering how and when to use them. As such, even though some degree of diversity can generally result in positive benefits, it should be balanced.

Complementarity. This factor is concerned with harmonious and reciprocal relationships among interactions, and how well they work with and supplement each other. This factor affects the quality of interaction of a tool by allowing the user to conduct more coordinated and integrated cognitive activities. That is, although each individual interaction independently supports one particular action, collectively the interactions can work together and assist the user to perform more complicated tasks and activities. For instance, in a study [40] of an interactive VT for exploring and reasoning about 3D lattices, it was observed that two interactions, filtering and annotating, were used to complement each other in performing certain tasks. Annotating was used to reason about paths by providing mechanisms for labeling and tracing nodes and edges of lattices, while filtering was used to isolate and focus on certain node types and patterns within 3D lattices.

Fitness. This factor is concerned with the appropriateness of interactions for the given VRs, the tasks and the activity, and the user's needs and characteristics. This is a complex and multi-faceted factor, each of its facets may need analysis. Some of these facets include: semantic-fitness, task-fitness, user-fitness, and context-fitness [61]. The first facet, semantic-fitness, deals with whether an interaction can enhance the communicative and semantic utility of a VR. For instance, a VR may be designed to display a 3D geometric shape to communicate its structure—that is, its constituent polygonal faces and their relationships. In such a case, providing rotation as an interaction may not be good for better communicating the semantic features of the 3D shape. This is because the 3D shape is symmetric; rotation may only allow the user to observe partial structures of the VR, while some parts may remain occluded from view. This can make it difficult for the user to fully perceive the structural semantics of the 3D shape. Providing decomposing as an interaction, however, can be more semantically appropriate for communicating the structural semantics of this 3D shape, as it allows the user to break the shape apart and display it as a flat 2D representation, thereby allowing the user to observe and examine its structural semantics with more ease—e.g., observing that the shape has 20 faces, and these are all triangles. In a similar manner, the other facets allow an analysis of the fitness of interactions: task-fitness deals with the suitability of interactions to support a task that involves given VRs; user-fitness deals with whether interactions match the cognitive needs and characteristics of the user (e.g., a child versus an adult); and context-fitness is about whether interactions support the psychological, cognitive, and structural requirements of an environment (e.g., a visual game versus a visual analytics tool). These facets provide a more organized way of thinking about the relevance, conceptual correspondence, cognitive cost, and appropriateness of interactions—and hence the quality and utility of interactions.

Flexibility. This factor is concerned with the range and availability of adjustability options. A highly flexible tool provides options for the user to be able to adjust the properties of the interface to suit his/her needs, characteristics, and goals. For instance, a tool that allows the user to adjust the dimensions of VRs, such as their appearance or density (see [51]), is more flexible than one that does not. Another facet of flexibility is with regard to the order of interactions when performing a task or activity. Some tools can have a very rigid sequencing and path of interactions. However, the final goal of many complex activities can be reached via different trajectories through the representation space of a tool. This is called the principle of equifinality. The interactive features of a flexible tool support this principle. Yet another facet of flexibility is the degree of control that the user has over the micro-level forms of some of the elements of interaction, such as agency, flow, activation, and transition. The flexibility factor can play an important role in the overall quality of interaction.

Genre. This factor is concerned with the types of transactions that are available to the user—that is, interactions through which the user makes exchanges with the VRs. The types of interactions that are provided can be placed on a continuum: allowing the user to only access VRs to allowing the user to only create VRs. As such, a VT's interactions can be classified into different genres: access-based, annotation-based, modification-based, construction-based, and combination-based. Using access-based interactions, the user accesses the stored, available, existent VRs already contained in the tool. Using annotation-based interactions, the user adds further notations or codes to the existing VRs. Using modification-based interactions, the user alters the properties of existing VRs such as by adding to or removing from them. Using construction-based interactions, the user constructs new VRs— VRs that are not necessarily provided in the tool, but rather created, synthesized, and composed from scratch. Finally, using combination-based interactions, the user operates upon VRs with two or more of the previous types of transactions. Consider the interactions listed in Table 1. Arranging, drilling, and filtering are all examples of access-based interactions. With these interactions a user typically does not create, destroy, add to, or modify VRs in any way. Annotating is an example of an annotation-based interaction. The user is not inserting new information in the tools,

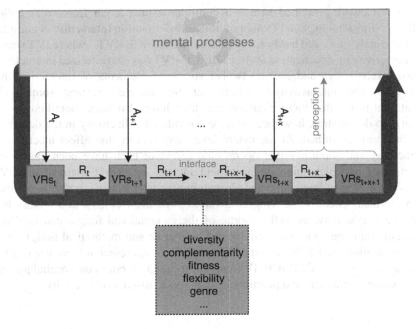

Fig. 8 Some macro-level interactivity factors

but rather is adding meta-information—i.e., a layer of information that highlights and describes—to the existing VRs. Assigning, transforming, and inserting are all examples of modification-based interactions. With these interactions a user adds properties to VRs, removes properties from VRs, adjusts the value of the properties of VRs, and so on. Composing is an example of a construction-based interaction. Once again, as can be seen, the genre of interactions has an overall effect on the macro-level interactivity of a VT (Fig. 8).

5 Summary

For visualization tools to be human-centered, they must be designed with a well-informed understanding of human cognition. However, visualization research is often based on traditional models of cognition that do not emphasize its situated nature and the role that interaction with the external world plays in performing complex cognitive activities. When users interact with visualization tools, cognitive processes emerge from a coupling that is formed between the internal representations and processes of the user and the external representations and processes that exist at the tool's interface. In this chapter, interactivity has been conceptualized as the strength of the coupling—in other words, the quality of the interaction—between a user and a visualization tool.

The framework presented here is a component of a larger framework called EDIFICE (Epistemology and Design of human InFormation Interaction in complex Cognitive activitiEs), and has been referred to as EDIFICE–IVT—where IVT stands for interactivity in visualization tools. EDIFICE–IVT has characterized interactivity at two levels: micro and macro. Twelve structural elements of interaction that affect micro-level interactivity have been identified and characterized. Some of the operational forms that these elements can take have also been identified, and a scenario to demonstrate how these may be considered collectively in the design of VTs has been examined. At the macro level, five factors that affect macro-level interactivity and some possible operational forms of each have been examined. The manner in which these elements and factors are operationalized in a VT affects the quality of interaction and ultimately affects how well cognitive activities are performed. Therefore, having an awareness of the elements and factors that influence interactivity, as well as some of their operational forms, can facilitate systematic thinking about interactivity and deliberate and methodical design practices. In addition, as the discussion of interactivity in the research literature is often vague and inaccurate, EDIFICE–IVT can contribute to a common vocabulary that visualization researchers and practitioners can use to discuss interactivity.

References

1. Aigner, W.: Understanding the role and value of interaction: First steps. In: S. Miksch, G. Santucci (eds.) International Workshop on Visual Analytics (2011)
2. Albers, M.: Design for effective support of user intentions in information-rich interactions. Journal of Technical Writing and Communication 39(2), 177–194 (2009)
3. Albers, M.: Human-information interactions with complex software. Design, User Experience, and Usability. pp. 245–254 (2011)
4. Alessi, S.M., Trollip, S.R.: Multimedia for Learning: Methods and Development. Allyn and Bacon (2001)
5. Amar, R., Eagan, J., Stasko, J.: Low-level components of analytic activity in information visualization. IEEE Symposium on Information Visualization, 2005. INFOVIS 2005. pp. 111–117 (2004)
6. Arias-Hernandez, R., Green, T., Fisher, B.: From cognitive amplifiers to cognitive prostheses: Understandings of the material basis of cognition in visual analytics. Interdisciplinary Science Reviews 37(1), 4–18 (2012)
7. Brown, J., Collins, A., Duguid, P.: Situated cognition and the culture of learning. Educational researcher 18(1), 32 (1989)
8. Bucy, E.: Interactivity in society: Locating an elusive concept. The Information Society 20(5), 373–383 (2004)
9. Clark, A.: Microcognition: Philosophy, Cognitive Science, and Parallel Distributed Processing. Explorations in cognitive science. MIT Press (1991)
10. Clark, A.: Time and mind. The Journal of Philosophy 95(7), 354 (1998)
11. Clark, A.: Supersizing the Mind: Embodiment, Action, and Cognitive Extension. Philosophy of Mind Series. Oxford University Press (2008)
12. Clark, A., Chalmers, D.: The extended mind. Analysis 58(1), 7–19 (1998)
13. Crystal, A., Ellington, B.: Task analysis and human-computer interaction: Approaches, techniques, and levels of analysis. In: Tenth Americas Conference on Information Systems, pp. 1–9. New York, New York, USA (2004)

14. Dourish, P.: Where The Action Is: The Foundations Of Embodied Interaction. Bradford Books. MIT Press (2004)
15. Downes, E.J., McMillan, S.J.: Defining interactivity: A qualitative identification of key dimensions. New Media & Society 2(2), 157–179 (2000)
16. Fabrikant, S.I.: Persistent problem in geographic visualization: Evidence of geovis(ual analytics) utility and usefulness. In: ICA Geovis Commission ICC2011, vol. 44, pp. 2009–2011. Paris, France (2011)
17. Fidel, R.: Human Information Interaction: An Ecological Approach to Information Behavior. MIT Press (2012)
18. Funke, J.: Complex problem solving: A case for complex cognition? Cognitive Processing 11(2), 133–42 (2010)
19. Gotz, D., Zhou, M.: Characterizing users visual analytic activity for insight provenance. IEEE Symposium on VAST pp. 123–130 (2008)
20. Hang, H., Auty, S.: Children playing branded video games: The impact of interactivity on product placement effectiveness. Journal of Consumer Psychology 21(1), 65–72 (2011)
21. Hannon, J., Atkins, P.: All about interactivity. Tech. rep., Victoria, Australia (2002)
22. Hegarty, M.: The cognitive science of visual-spatial displays: Implications for design. Topics in Cognitive Science 3(3), 446–474 (2011)
23. Hollan, J., Hutchins, E., Kirsh, D.: Distributed cognition: Toward a new foundation for human-computer interaction research. ACM Transactions on Computer-Human Interaction (TOCHI) 7(2), 174–196 (2000)
24. Hutchins, E.: Cognition in the Wild. Bradford Books. MIT Press (1995)
25. Hutchins, E., Klausen, T.: Distributed cognition in an airline cockpit. Cognition and Communication at Work pp. 15–34 (1996)
26. Jensen, J.: Interactivity: Tracking a new concept in media and communication studies. Nordicom Review 19(1), 185–204 (1998)
27. Johnson, C., Moorhead, R., Munzner, T., Pfister, H., Rheingans, P., Yoo, T.: NIH-NSF Visualization Research Challenges. Tech. rep., Los Alamitos, CA (2006)
28. Kaptelinin, V., Nardi, B.: Activity Theory in HCI. Morgan & Claypool Publishers (2012)
29. Keim, D., Kohlhammer, J., Ellis, G.: Mastering The Information Age-Solving Problems with Visual Analytics (2010)
30. Keim, D., Mansmann, F., Schneidewind, J., Thomas, J., Ziegler, H.: Visual analytics: Scope and challenges. In: Visual Data Mining: Theory, Techniques and Tools for Visual Analytics, lncs edn. Springer (2008)
31. Kiousis, S.: Interactivity: a concept explication. New Media & Society 4(3), 355–383 (2002)
32. Kirsh, D.: Interactivity and multimedia interfaces. Instructional Science 25(2), 79–96 (1997)
33. Kirsh, D.: Metacognition, distributed cognition and visual design. Cognition, Education, and Communication Technology pp. 1–22 (2005)
34. Kirsh, D.: Problem solving and situated cognition. The Cambridge handbook of situated cognition pp. 264–306 (2009)
35. Kirsh, D., Maglio, P.: On distinguishing epistemic from pragmatic action. Cognitive Science: A Multidisciplinary Journal 18(4), 513–549 (1994)
36. Knauff, M., Wolf, A.G.: Complex cognition: The science of human reasoning, problem-solving, and decision-making. Cognitive Processing 11(2), 99–102 (2010)
37. Laine, P., Phil, L.: Explicitness and interactivity. Proceedings of the 1st international symposium on Information and communication technologies p. 426 (2003)
38. Liang, H.N., Parsons, P., Wu, H.C., Sedig, K.: An exploratory study of interactivity in visualization tools: Flow of interaction. Journal of Interactive Learning Research 21(1), 5–45 (2010)
39. Liang, H.N., Sedig, K.: Can interactive visualization tools engage and support pre-university students in exploring non-trivial mathematical concepts? Computers & Education 54(4), 972–991 (2010)
40. Liang, H.N., Sedig, K.: Role of interaction in enhancing the epistemic utility of 3D mathematical visualizations. International Journal of Computers for Mathematical Learning 15(3), 191–224 (2010)

41. Liu, Y., Shrum, L.J.: A dual-process model of interactivity effects. Journal of Advertising **38**(2), 53–68 (2009)
42. Liu, Z., Nersessian, N., Stasko, J.: Distributed cognition as a theoretical framework for information visualization. IEEE transactions on visualization and computer graphics **14**(6), 1173–80 (2008)
43. Liu, Z., Stasko, J.: Mental models, visual reasoning and interaction in information visualization: a top-down perspective. IEEE transactions on visualization and computer graphics **16**(6), 999–1008 (2010)
44. Mann, S.: Conversation as a basis for interactivity. In: Proceedings of the 15th Annual Conference of the National Advisory Committee on Computing Qualifications, pp. 281–288. Hamilton, NZ (2002)
45. Marchionini, G.: Information Concepts: From Books to Cyberspace Identities (2010)
46. McClelland, J.L.: Emergence in cognitive science. Topics in Cognitive Science **2**(4), 751–770 (2010)
47. Mirel, B.: Interaction Design for Complex Problem Solving: Developing Useful and Usable Software. The Morgan Kaufmann Series in Interactive Technologies. Morgan Kaufmann (2004)
48. Nardi, B.: Studying context: A comparison of activity theory, situated action models, and distributed cognition. Context and consciousness: Activity theory and human-computer interaction pp. 69–102 (1996)
49. Parsons, P., Sedig, K.: Common visualizations: Their cognitive utility (this volume)
50. Parsons, P., Sedig, K.: Distribution of information processing while performing complex cognitive activities with visualization tools (this volume)
51. Parsons, P., Sedig, K.: Properties of visual representations: Improving the quality of human-information interaction in complex cognitive activities. Journal of the American Society for Information Science and Technology (under review)
52. Pike, W.a., Stasko, J., Chang, R., OConnell, T.A.: The science of interaction. Information Visualization **8**(4), 263–274 (2009)
53. Pohl, M., Wiltner, S., Miksch, S., Aigner, W., Rind, A.: Analysing interactivity in information visualisation. KI - Künstliche Intelligenz **26**(2), 151–159 (2012)
54. Purchase, H., Andrienko, N., Jankun-Kelly, T., Ward, M.: Theoretical foundations of information visualization. In: A. Kerren, J.T. Stasko, J.D. Fekete, C. North (eds.) Information Visualization: Human-Centered Issues and Perspectives, *Lecture Notes in Computer Science*, vol. 4950, pp. 46–64. Springer Berlin Heidelberg, Berlin, Heidelberg (2008)
55. Scaife, M., Rogers, Y.: External cognition: how do graphical representations work? International Journal of Human-Computer Studies **45**(2), 185–213 (1996)
56. Scholtz, J.: Beyond usability: Evaluation aspects of visual analytic environments. In: IEEE Symposium on Visual Analytics Science and Technology, pp. 145–150 (2006)
57. Sedig, K.: Need for a prescriptive taxonomy of interaction for mathematical cognitive tools. Lecture Notes in Computer Science pp. 1030–1037 (2004)
58. Sedig, K.: From play to thoughtful learning: A design strategy to engage children with mathematical representations. Journal of Computers in Mathematics and Science Teaching **27**(1), 65–101 (2008)
59. Sedig, K., Klawe, M., Westrom, M.: Role of interface manipulation style and scaffolding on cognition and concept learning in learnware. ACM Transactions on Computer-Human Interaction (TOCHI) **8**(1), 34–59 (2001)
60. Sedig, K., Liang, H.N.: Interactivity of visual mathematical representations: Factors affecting learning and cognitive processes. Journal of Interactive Learning Research **17**(2), 179 (2006)
61. Sedig, K., Liang, H.N.: On the design of interactive visual representations: Fitness of interaction. In: C. Seale, J. Montgomerie (eds.) World Conference on Educational Multimedia, Hypermedia and Telecommunications, pp. 999–1006. AACE (2007)
62. Sedig, K., Liang, H.N.: Learner-information interaction: A macro-level framework characterizing visual cognitive tools. Journal of Interactive Learning Research **19**(1), 147–173 (2008)

63. Sedig, K., Parsons, P.: Interaction design for complex cognitive activities with visual representations: A pattern-based approach. AIS Transactions on Human-Computer Interaction (2013, to appear)
64. Sedig, K., Rowhani, S., Liang, H.N.: Designing interfaces that support formation of cognitive maps of transitional processes: an empirical study. Interacting with Computers 17(4), 419–452 (2005)
65. Sedig, K., Rowhani, S., Morey, J., Liang, H.N.: Application of information visualization techniques to the design of a mathematical mindtool: A usability study. Information Visualization 2(3), 142–159 (2003)
66. Sedig, K., Sumner, M.: Characterizing interaction with visual mathematical representations. International Journal of Computers for Mathematical Learning 11(1), 1–55 (2006)
67. Shneiderman, B.: The eyes have it: A task by data type taxonomy. Tech. rep., University of Maryland, College Park (1996)
68. Thomas, J., Cook, K.: Illuminating the path: The research and development agenda for visual analytics. IEEE Press (2005)
69. Ware, C.: Visual Thinking for Design. Morgan Kaufmann Series in Interactive Technologies. Morgan Kaufmann (2008)
70. van Wijk, J.J.: Views on visualization. IEEE transactions on visualization and computer graphics 12(4), 421–32 (2006)
71. Yi, J., Kang, Y., Stasko, J., Jacko, J.: Toward a deeper understanding of the role of interaction in information visualization. IEEE Transactions on Visualization and Computer Graphics 13(6), 1224–1231 (2007)
72. Yoo, W.S., Lee, Y., Park, J.: The role of interactivity in e-tailing: Creating value and increasing satisfaction. Journal of Retailing and Consumer Services 17(2), 89–96 (2010)

Printed in the United States
By Bookmasters